Blueprint Reading and Technical Sketching for Industry

2nd Edition

See
5,6,7

Blueprint Reading and Technical Sketching for Industry

THOMAS P. OLIVO
with contributions from
C. Thomas Olivo

2nd Edition

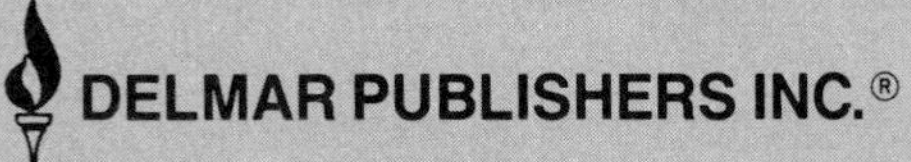

DELMAR PUBLISHERS INC.®

NOTICE TO THE READER

Publisher does not warrant or guarantee any of the products described herein or perform any independent analysis in connection with any of the product information contained herein. Publisher does not assume, and expressly disclaims, any obligation to obtain and include information other than that provided to it by the manufacturer.

The reader is expressly warned to consider and adopt all safety precautions that might be indicated by the activities described herein and to avoid all potential hazards. By following the instructions contained herein, the reader willingly assumes all risks in connection with such instructions.

The publisher makes no representations or warranties of any kind, including but not limited to, the warranties of fitness for particular purpose or merchantability, nor are any such representations implied with respect to the material set forth herein, and the publisher takes no responsibility with respect to such material. The publisher shall not be liable for any special, consequential or exemplary damages resulting, in whole or in part, from the readers' use of, or reliance upon, this material.

Delmar Staff:
Executive Editor: Michael A. McDermott
Associate Editor: Kevin Johnson
Project Editor: Andrea Edwards Myers
Production Coordinator: Wendy Troeger
Design Supervisor: Susan C. Mathews

For information, address Delmar Publishers Inc.
3 Columbia Circle, PO Box 15015
Albany, New York 12212-5015

Delmar Publishers' Online Services
To access Delmar on the World Wide Web, point your browser to:
http://www.delmar.com/delmar.html
To access through Gopher: gopher://gopher.delmar.com
(Delmar Online is part of "thomson.com", an Internet site with information on more than 30 publishers of the International Thomson Publishing organization.)
For information on our products and services:
email: info@delmar.com
or call 800-347-7707

Printed in the United States of America
Published simultaneously in Canada
by Nelson Canada,
a division of The Thomson Corporation

10 9 8 7

Library of Congress Cataloging-in-Publication Data

Olivo, Thomas P.
Blueprint reading and technical sketching for industry / Thomas P. Olivo ; with contributions from C. Thomas Olivo. — 2nd ed.
p. cm.

Includes index.
ISBN 0-8273-5077-5
1. Blueprints. 2. Freehand technical sketching.
I. Olivo, C. Thomas. II. Title.
T379.044 1992 604.2—dc20 91-38722 CIP

PREFACE

Blueprints and sketches provide a universally accepted language for communicating information about simple or complex parts, mechanisms, systems, and processes. **Blueprint Reading and Technical Sketching for Industry** is refreshingly new and is the most comprehensive textbook available. Its contents are based on extensive occupational studies and careful analyses of current occupational practices, trends, and needs. Equally important, attention is directed to the personal needs of students/trainees in school and institutional courses in preparation for and advancement in employment at many different levels ranging from operators to craftspersons, programming technicians, designers, supervisors, and engineers and including specialized marketing and sales personnel.

CONTENTS AND ORGANIZATION

The text contents incorporate functional teaching and learning practices that have proven to be effective in school/institutional courses and in industrial, military specialty, and other types of occupational training programs. The text provides for self-paced instruction.

There is a deliberate progression in the instructional units from the simple to the relatively complex. The Units are grouped into Sections that contain common elements. The starting point in Section 1 deals with foundations of needs within many occupational areas. The succeeding Sections and Units develop broad understandings of principles that are applied in the Assignment for each Unit to a representative occupational drawing.

The first sections and units relate to: lines; standard orthographic and SI metric projection and views; dimensioning practices; section views; basic and geometric tolerancing; and machine processes. As experience is developed in interpreting and applying drawing standards, principles, and drafting room techniques, the remaining sections deal with machine elements (actuators); production working drawings; machine and tool drawings; and pictorial drawings.

Special consideration is given to numerical control (NC), computer numerical control (CNC), computer-aided design and drawing (CADD) and computer-aided manufacturing (CAM) systems and products. Particular attention is focused in Section 14 on the development of print reading and sketching skills for a variety of manufacturing and occupational drafting systems. The occupational areas include: welding, casting, and forging; plastic materials and fabrication; piping, electrical installation, architectural, and structural design; aeronautical and aerospace industries; fluid power and instrumentation industries; precision sheet metal forming occupations; and electronics industries applications.

A new series of 18 frames that provide enrichment materials relate to the interlocking of state-of-the-art *High Technology Applications* with the latest advances in computer-aided drawing and design (CADD) and automated occupational systems, processes, materials, and products.

TECHNICAL SKETCHING SKILLS

Part 2 deals entirely with principles, techniques, and practices of preparing technical illustrations. The units are organized according to the knowledge and skills that are required on-the-job in sketching simple lines and other geometric forms, parts, and components; shading; and using appropriate lettering for dimensioning, notes, and other specifications. Section 17 on *industrial pictorial sketching* includes four units on orthographic, oblique, isometric, and perspective sketching.

COMPLEMENTARY RESOURCE MATERIALS

The contents of this book conform to standards of the American National Standards Institute (ANSI), Canadian Standards Association (CSA), and other applicable occupational area standards. SI metric (ISO) standards are applied throughout the text to simulate the acceptance and use of these standards in industry. Selected drawings also include dual dimensioning in Customary inch and SI metric units of measurement. A note is made that deviations from ANSI, CSA, ISO, and other association standards, are found on some of the representative industrial drawings used in the text.

The *Appendix* includes a *Glossary of Technical Terms*, *Standard Drawing Abbreviations*, a new list of *Standards, Codes, Symbols: Reference Sources*, and selected content-referenced *Handbook Tables*. The detailed *Index* permits easy reference to major content items.

TESTING FOR SKILL DEVELOPMENT IN PRINT READING AND SKETCHING

Each unit is written in clear, direct, technical language as experienced at the workplace. Also, provision is made in the *Assignment* for each unit for the continuous measurement of skills developed by each student under conditions that simulate those found in occupational settings. The orderly flow of content in the units and sections permits flexibility in programming to meet diverse training requirements.

A companion **Instructor's Guide** provides solutions to all test items and problems and permits quick and accurate checking of assignments. *Progress charts* are included to suggest simple procedures for recording group and individual achievement in interpreting blueprints and preparing technical sketches.

ABOUT THE AUTHOR

Thomas P. Olivo has served successfully as a teacher in secondary schools and post-secondary colleges and as assistant professor in vocational-industrial teacher education at Long Beach State College. Mr. Olivo worked in manufacturing and construction occupations and in drafting and design. He made significant contributions in instructional materials development in relation to area technical schools, institutes, and colleges on state and regional levels through his work in the Curriculum Planning Laboratory at Clemson University and, as a former state supervisor of technical education, he contributed to occupational competency testing and instructional supervision.

Mr. Olivo currently teaches courses in engineering drafting, architectural drawing and design, CAD, and blueprint reading and technical sketching. He also serves as Executive Director, Industrial/Vocational Training Consultants.

His extensive writings include: *Basic Technical Mathematics*, *Fundamentals of Applied Physics*, *Basic Blueprint Reading and Sketching*, and other texts.

ABOUT THE COLLABORATOR

Dr. C. Thomas Olivo is recognized as one of the nation's experienced authors and a foremost leader in vocational-technical education and human resource development training in industry. Dr. Olivo entered teaching following service in industry as a highly skilled craftsperson. Dr. Olivo served at successively higher levels of responsibility as teacher, instructional supervisor, curriculum development specialist and C & I materials laboratory director, state technical institute director, branch chief of staff at the US Armed Forces Technical Institute (England), state director of vocational-technical education, and R & D specialist and executive director of the National Occupational Competency Testing Institute (NOCTI).

Dr. Olivo's authorship includes extensive writings in machine tool technology and manufacturing, related technical mathematics, applied science, blueprint readings and sketching, and other professional publications. Dr. Olivo collaborated with the author of this text in the areas of research, organization, technical content, and skills assessment.

Contents

Section 4: Dimensioning Using Customary Units of Measure

Section 5: SI Metric System of Projection and Dimensioning

Section 6: Machine Elements and Processes

Section 7: Sectional Views

Section 8: Tolerancing

Section 9: Machine Elements (Actuators)

Section 10: Pictorial Drawings

Section 11: Industrial Production Working Drawings

Section 12: Machine and Tool Drawings

Section 13: Computerized Systems: NC, CNC, CAD, and CAM

Section 14: Manufacturing and Occupational Drafting Systems

PART 2 TECHNICAL ILLUSTRATION: SKETCHING

Section 15: Basic Lettering Techniques

Section 16: Fundamentals of Industrial Sketching

Section 17: Industrial Pictorial Sketching

HIGH TECHNOLOGY APPLICATIONS FRAMES

APPENDIX: REFERENCE TERMS, SYMBOLS, SOURCES, AND TABLES

GUIDELINES FOR STUDY

This revised, updated, and expanded second edition of **Blueprint Reading and Technical Sketching for Industry** provides in a single volume the depth of technical content and the range of occupational drawings that must be successfully mastered in order to develop essential skill competencies.

CONTINUING OBJECTIVES

The *prime* overall *objective* remains constant. The content is designed to develop speed, skill, and accuracy in reading and interpreting shop and laboratory drawings, technical illustrations, and other graphics, and in preparing occupational sketches, as required. The *secondary objective* is to develop parallel skills so as to be able to transfer technical information conveyed by drawings and sketches to functional on-the-job applications. These applications relate to design features, sizes and measurements, manufacturing and construction processes and systems, inspection practices, and other activities dealing with the making, assembling, and testing of parts and components. *Specific objectives for each unit* are readily identified from the descriptive title of the unit.

OUTSTANDING FEATURES OF THIS EDITION

- Updating is consistent with the latest ANSI, ISO Metric, AWS, ASME, CNA, United Building Codes (UBC), Code Administration and Building Official (CABO), and other occupational standards and practices.
- Basic tolerancing and geometric tolerancing standards, principles, and applications reflect common concepts and uses for dimensioning, controlling tolerances, and statistical process (quality) control.
- The architectural and structural drawing unit is expanded into two units. One unit covers a wide range of architectural design and construction features. The second unit deals with structural technology and applications, including prestructures, reinforced structural brick, steel, aluminum, and steel building materials, and structural design features.
- The electrical unit is reorganized. Electrical circuitry and house wiring is covered in one unit. Another unit deals with electronic technology. This expansion reflects a more realistic separation between high voltage equipment and installations and micro electronic devices, chips, logic systems, symbols, and diagrams.
- There is an additional drawing assignment to the Plastics: Materials, Fabrication, and Design unit.
- Another unit is added to deal with *Fluid Power Technology and Diagrams.*
- Significantly, 18 *High Technology Applications* frames of resource materials are now included. Each frame deals with state-of-the-art products, processes, equipment, and systems that interface with CADD, computer-aided manufacturing (CAM), numerical control (NC), computer numerical control (CNC), robotics, systems within flexible integrated manufacturing (FIM), and other occupational applications.
- *New Reference Tables* are included in the Appendix that relate to drafting room standards, symbols, and occupational codes; fasteners; classes of fits; and measurement conversion.
- The print reading and sketching problems and test items represent the most *comprehensive battery* of resource materials and testing programs that is available.

ORGANIZATION OF THE TEXTBOOK

The *Table of Contents* is especially arranged to show how the text is organized and the progression from one learning experience to the next. The text consists of two parts: Part 1 covers blueprint reading; Part 2, technical illustration and sketching.

Part 1 has 14 Sections. Part 2 contains Sections 15, 16, and 17. A section represents all of the technology, drafting room techniques, skills of interpretation, and other experiences which must be learned within a major category.

Each section contains a series of units. These subdivisions provide new learning elements which are the building blocks for mastering each new principle. Each unit has three elements: (1) a *basic principle*, (2) a specially selected industrial *blueprint or technical illustration* to show how each principle is applied, and (3) an *assignment*. The problems and test items in each assignment follow a particular pattern that provides for the continuous interlocking of new experiences into successive units.

OVERVIEW OF THE CONTENTS

Part 1, Section 1 provides *Foundations for Interpreting Blueprints and Making Technical Drawings*. Section 2 covers *Lines for Technical Drawings and Sketching*; Section 3, *Basic Principles of Projection: Views*; Section 4, *Dimensioning Using Customary Units of Measure*; Section 5, *SI Metric System of Projection and Dimensioning*; Section 6, *Machine Elements and Processes*; Section 7, *Sectional Views*; Section 8, *Tolerancing*; Section 9, *Machine Elements (Actuators)*; Section 10, *Pictorial Drawings*; Section 11, *Industrial Production Working Drawings*; Section 12, *Machine and Tool Drawings*; Section 13, *Computerized Systems: NC, CNC, CAD, and CAM*; Section 14, *Manufacturing and Occupational Drafting Systems*.

The three sections in Part 2 include Section 15, *Basic Lettering Techniques*; Section 16, *Fundamentals of Industrial Sketching*; and Section 17, *Industrial Pictorial Sketching*. The units in this final section deal with principles and practices of orthographic, oblique, isometric, and perspective sketching.

The 18 *High Technology Applications* frames that were previously mentioned are distributed throughout the textbook.

SKETCHING TECHNOLOGY, PROCESSES, AND APPLICATIONS

Part 2 is devoted to basic principles and techniques for preparing technical illustrations. Fundamentals are presented for drawing straight and curved lines and combining these to produce freehand sketches of any object. The importance of properly formed numerals and letters in different styles is described with illustrations. Finally, underlying principles and practices in making industrial sketches according to orthographic, oblique, isometric, and perspective techniques are presented. This technical background and the hands-on assignments provide additional experience in interpreting drawings and making freehand sketches.

APPENDIX RESOURCE MATERIALS

The resource materials contained in the *Appendix* include sample standard abbreviations, representative drawing symbols, measurement conversion and tooling information tables, and a detailed *Index*. The *Glossary* provides practical, short, technical descriptions of terms commonly used. The abbreviations and symbols relate to materials, processes, and products for many industries.

The tables are valuable for soft and hard metric conversions, and for establishing dimensional values and tolerances. The selected tables (including formulas) are used to complement the unit assignments. The Index provides an additional resource for quickly locating techniques of representation and dimensioning, processes, products, and other major items relating to blueprint reading and technical sketching.

INSTRUCTOR'S GUIDE

Solutions to all test items and sketching assignments are contained in an **Instructor's Guide**. Sample charts are provided for recording group or class progress, as well as individual achievement.

ACKNOWLEDGMENTS

Blueprint Reading and Technical Sketching for Industry is the end product resulting from a cooperative working relationship between the author and a cadre of devoted experts. The following individuals, with special teaching and assessment expertise, reviewed all or part of the manuscript and provided important feedback on scope, depth, and technical accuracy.

Warren H. Anderson, South Shore Regional Vocational High School, Hanover, MA
Robert N. Brown, Chabot College, Hayward, CA
Calvin L. Christen, Spokane Community College, Spokane, WA
Raymond W. Cross, Ferris State College, Big Rapids, MI
Peter Fricano, Triton College, River Grove, IL
Jack Johnson, Texas State Technical Institute, Waco, TX
Dietrich R. Kanzler, Santa Ana Community College, Santa Ana, CA
Mark A. Knott, Texas State Technical Institute, Harlingen, TX
Adam J. Machuga, Butte College, Oroville, CA
Charles E. Mattingley, Pikes Peak Community College, Colorado Springs, CO
Raymond J. Noga, Westfield Vocational High School, Westfield, MA
James W. Phillips, Penta County Joint Vocational School, Perrysburg, OH
Vard A. Roper, Utah Technical College, Provo, UT
David Steinhauer, Tidewater Community College, Portsmouth, VA
Richard Sunsdahl, Faribault Area Vocational Technical Institute, Faribault, MN
Andrew T. Surratt, Harper College, Palatine, IL

In addition to the important contributions of the product users and reviewers who participated, recognition is made of the host of companies and professional associations who provided technical materials, drawings, and other supportive services. Grateful appreciation is expressed to the plant training personnel, supervisors, managers, design/programming technicians, research and development engineers, marketing specialists, architects, and other lead persons who assisted from the following participating companies.

Allen-Bradley Company
American National Standards Institute (ANSI)
(Sperry Rail Service Division)
Automation Industries, Inc.
Balzers Tool Coating, Inc.
(The) Barden Corporation
Beech Aircraft Corporation
Brown and Sharpe Manufacturing Company
Charles Stark Draper Laboratories, Inc.
Cincinnati Milacron, Inc.
Clausing Machine Tools
Cornerstone Architectural Designers, Inc.
Danly Machine Corporation
(Sterling Instruments Division)
Designatronics, Inc.
DoALL Company
Eimeldingen Corporation
Farrel Company
Federal Products Corporation
General Motors Corporation, Terrex Division
Gleason Machine Division, The Gleason Works
Hardinge Brothers, Inc.
Incom International, Inc., Boston Gear Division
JAKA, Inc.
Jarvis Products Corporation
Langer, Dion, and Morse Associates
Lodge and Shipley Company
McDonnell Douglas Corporation
Mite Corporation, Heli-Coil Products Division
Moore Special Tool Company
National Joint Steamfitter-Pipefitter Apprenticeship Committee
National Tooling and Machining Association
J.P. Owens, Kevin Roche, John Dinkeloo, and Associates
Parker Hannifin Corporation
Perkin-Elmer Corporation
Schlumberger CAD/CAM Division

Shore Instrument and Manufacturing Company
Sodick, Inc. (EDM)
(The) L. S. Starrett Company
Summagraphics Corporation
Thermolyne Corporation
Trumpf Industrial Lasers, Inc.
(Greenfield Tap and Die Division) TRW, Inc.
Union Carbide Corporation
Universal Vise and Tool Company, Swartz Fixture Division

A personal word of "thanks" is expressed by the author to the following persons who went that *extra mile* to produce or secure new product designs, systems information, and typical occupational drawings and technical illustrations.

- John Paul and Ed Lagrange, Brown and Sharpe Manufacturing Company, for technical data and art work on digital electronics measuring instruments and data input systems in statistical process (quality) control.
- Douglas Carter, Cincinnati Milacron, Inc., for the newest robotic design illustrations, line configurations in CADD/CNC actuated machining centers, FIM systems, and other high technology applications.
- Judi Matz, DoALL Company, for technical information on machine tools, machining centers, and tooling.
- Roy A. Schlunz, The Charles Stark Draper Laboratories, for current data on CADD/CAM developments and representative drawings and art copy.
- Arthur W. Mannette, Federal Products Corporation, for high magnification comparator materials and hydraulic diagram, logicial schematic, and power distribution schematic drawings used in instrumentation applications.
- Mike Majlak, Moore Special Tool Company, for special numerical control drawings and technical materials on high-precision machine tools.
- John J. Murphy, The Barden Corporation, for reference data on geometric tolerancing and additional information on high technology developments.
- Joseph D. Hornak, Sperry Rail Service Division, Automation Industries, for an unusual series of detail drawings and technical illustrations.
- William D. Downing, Heli-Coil Products Division of Mite Corporation, for basic representative industrial working and detail drawings.
- J. P. Owens, Kevin Roche, John Dinkeloo, and Associates, for blackline prints of a specially selected structural drawing.
- Gary Roosa, Cornerstone Architectural Designers, Inc. for residential plan views and elevations.
- Frank Glaser and Robert Taylor, Swartz Fixture Division, Universal Vise and Tool Company, for distinctive orthographic and pictorial drawings representing fixtures and other tooling.

The author is grateful to the Delmar Publishers staff: Michael McDermott, executive editor; Kevin Johnson, associate editor; Marlene McHugh Pratt, editorial supervisor; Frederick J. Sharer, Director of Manufacturing; and others, for personal interest, enthusiastic support, and expert services in advancing the final manuscript through all stages of assessment, editing, and production, leading to this finished publication.

A special "thank you" is expressed to my wife, Stephanie B. Olivo, and to our children for sharing precious time throughout all stages of research and manuscript preparation. Recognition is made to my mother, Hilda G. Olivo, for applying her editorial expertise in reviewing all final copy.

A special citation is reserved for my father, Dr. C. Thomas Olivo. A distinguished author, always generous in sharing his talents and leadership experiences, he guided the writing at each successive stage of development. This book is a tribute to him for lifelong dedication to excellence in teaching and learning in his quest to develop each individual to his or her highest potential, consistent with the constantly changing technology and skill requirements of industry and of society.

Thomas P. Olivo

PART 1

Industrial Blueprint Reading

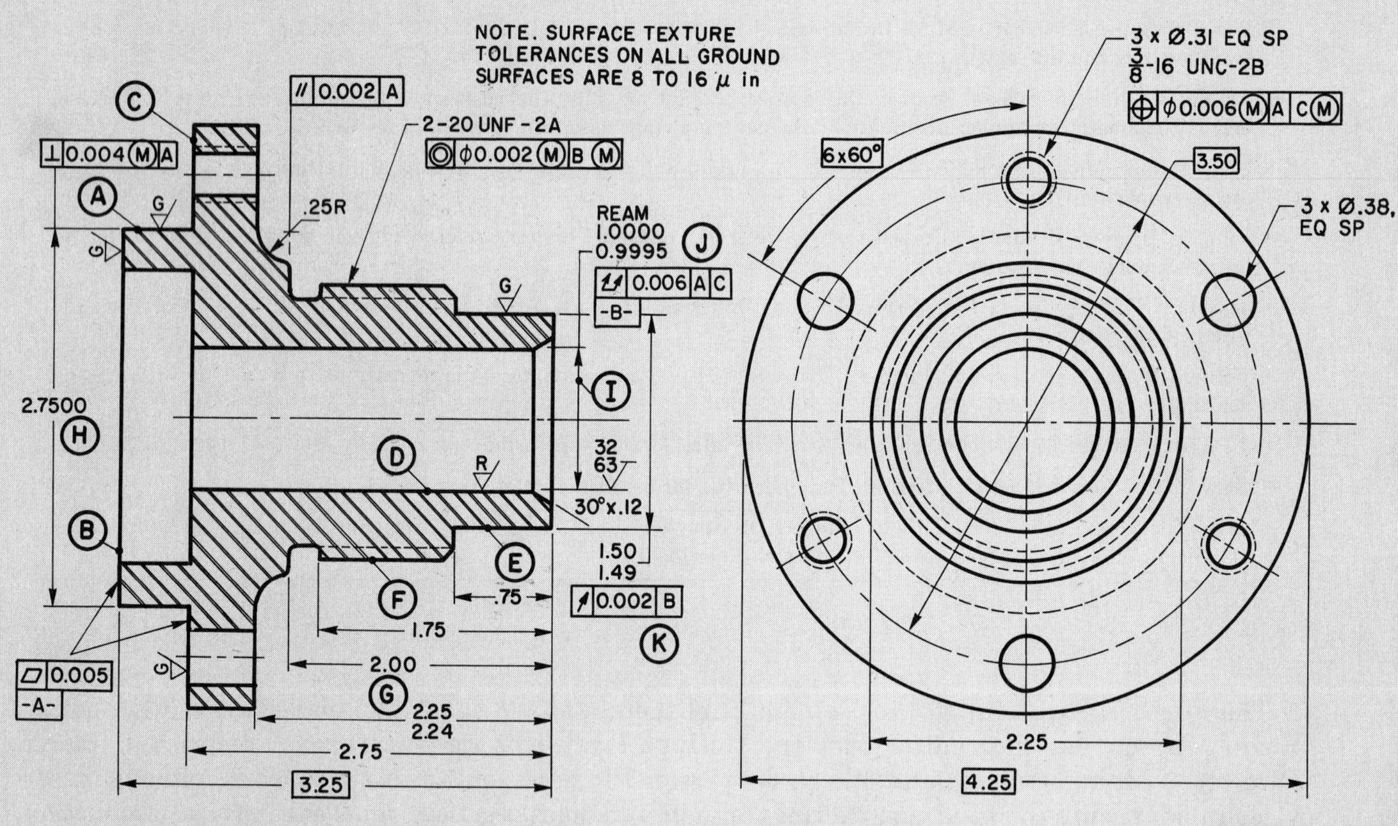

SECTION 1
Foundations for Interpreting Blueprints and Making Technical Sketches

UNIT 1
Industry Dependence on Technical Drawings

The American system of mass production with interchangeable parts was successfully demonstrated between 1813 and 1815. The first United States government contract *ever to specify* interchangeable parts in manufacturing a product was issued in 1813 to Simeon North. The contract required "...that component parts of pistols are to correspond so exactly that any limb or part of one pistol may be fitted to any other pistol of the Twenty Thousand...." During, and at the completion of this contract in 1815, government inspectors from two national armories inspected and assessed the North plant and manufacturing system. North then introduced the system of mass production with interchangeable parts at the Harper's Ferry Arsenal.

AMERICAN SYSTEM OF MANUFACTURING

With the spread of such manufacturing, the system came to be known abroad as the "American System of Manufacturing." Thus, Simeon North fulfilled the first contract of mass producing interchangeable parts. In the process, North made significant changes in design features of machine tools, accessories, cutting tools, and work-holding and work-positioning devices. These changes resulted in additional mechanization of existing equipment and processes. The evolving American system encouraged the further development of machine tools, permitting the manufacture of parts to ever-higher limits of geometric and dimensional accuracy and quality of surface finish, thereby eliminating most of the early hand labor.

In summary, the first completed contract recorded in the United States to mass-produce a product with interchangeable parts in 1815 resulted from modifications of equipment, cutting tools, and accessories; changed manufacturing methods; and ever higher precision capability.

There are three basic elements to the American system that continue through today:

- Parts production is broken down according to the nature of the work.
- Machines are designed for one or a number of operations within a particular category.
- The required number of machine units is proportional to the number of machine processes to be performed and quantities to be manufactured.

COMMUNICATION THROUGH TECHNICAL DRAWINGS

Technical drawings played a key role prior to, during, and following the worldwide movement toward industrialization. Technical drawings were

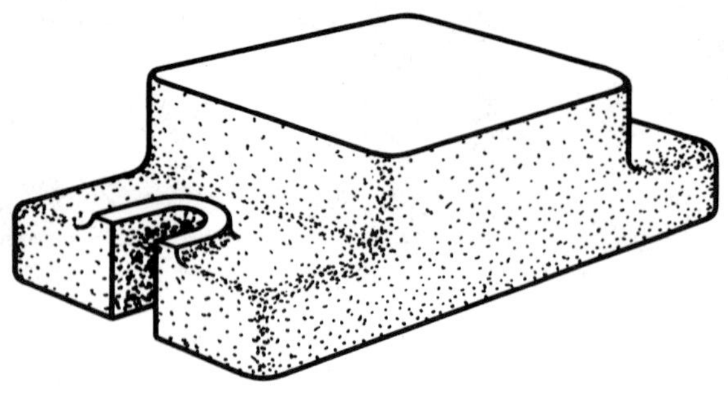

FIGURE 1-1 SKETCH USED TO DESCRIBE THE FORM OF A DIE BLOCK

and are the *universal language.* Parts information and specifications are communicated by technical drawings in such detail that a single part, a complete mechanism, or quantities may be mass-produced to uniform standards anywhere in the world.

While a picture provides visual information about a part, it is the technical drawing which gives design details; dimensions for producing, machining, and fitting; and other specifications. The sketch of the Die Block in figure 1-1 quickly shows the shape. If foundry dimensions were added, a pattern could be made for casting purposes. Other machining dimensions might be included.

An example of a technical drawing which is more widely used to convey precise information from the designer or engineer to the draftsperson, skilled mechanic, and technician, is shown in figure 1-2.

Original (master) technical drawings may be produced "mechanically" with the aid of drafting instruments or automated drafting machines. Design information and statistical data provide computer input. The computer then controls and programs automated drafting equipment. The original drawings are usually filed to record subsequent changes and for record purposes.

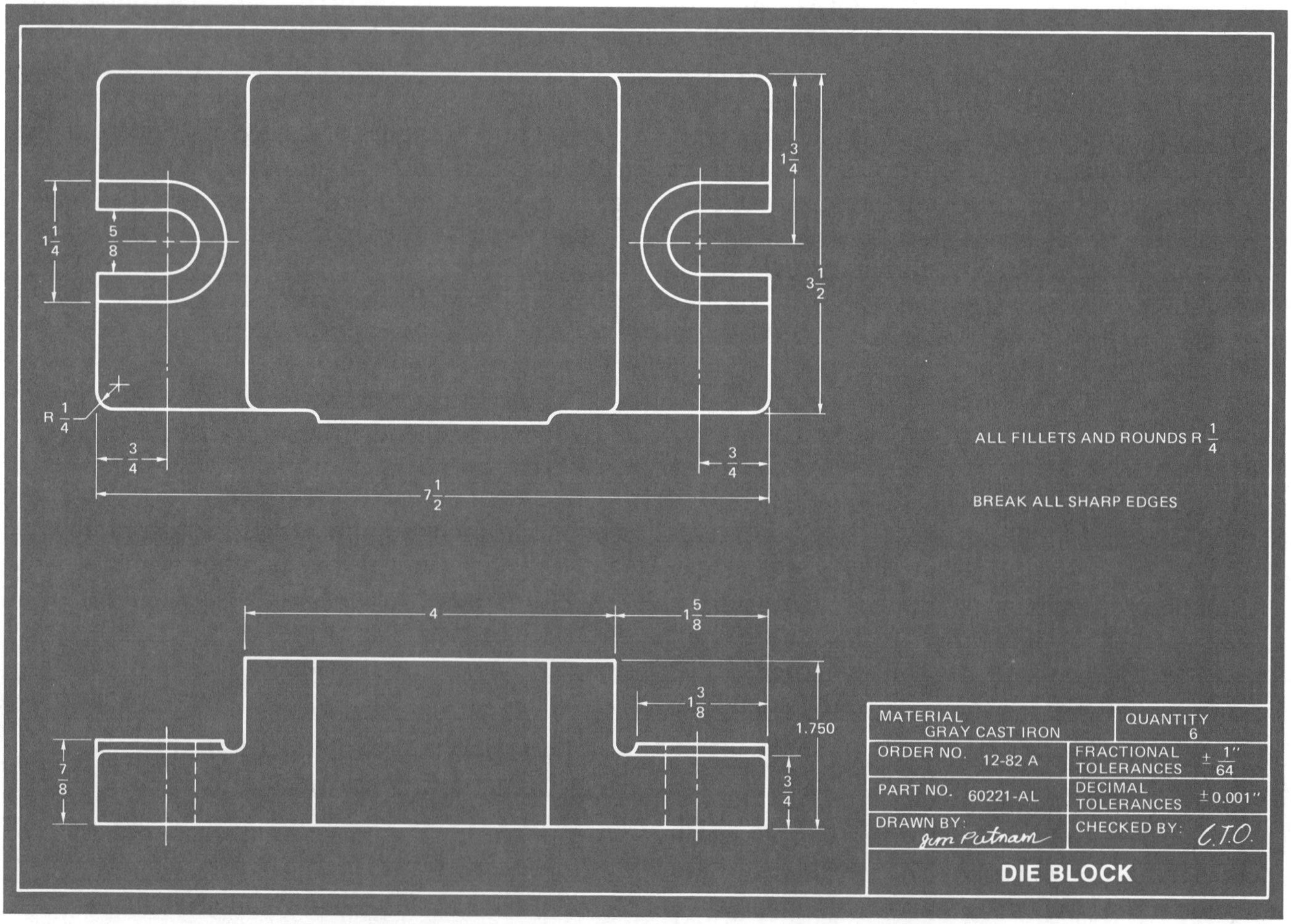

FIGURE 1-2 MECHANICAL DRAWING OF DIE BLOCK PROVIDES COMPLETE MACHINING INFORMATION

Prints of the original are made for daily use in development and manufacturing. A print is generally called a *blueprint*, regardless of the color of the print or the reproduction process. In reading a blueprint a person is able to:

- *Visualize* what a part looks like.
- *Relate each feature* to a production, inspection, or assembly process.
- *Make other* complementary *drawings or sketches* in order to produce the part or mechanism.
- *Understand* the *limits of accuracy* for dimensional measurements.

SI METRIC AND AMERICAN STANDARD DRAWINGS (BLUEPRINTS)

Parts are represented and dimensioned on technical drawings primarily according to American National Standards Institute (ANSI), international SI metric (ISO), Canadian Standards Association (CSA), and British standards. In some instances, the SI system of representation and dimensioning is used for the entire drawing, as shown in figure 1-3A.

Similarly, most drawings in the United States, Canada, and British Commonwealth nations use ANSI and customary inch units of measurement and the American/British system of representation. The blueprint of the Guide Bracket (fig-

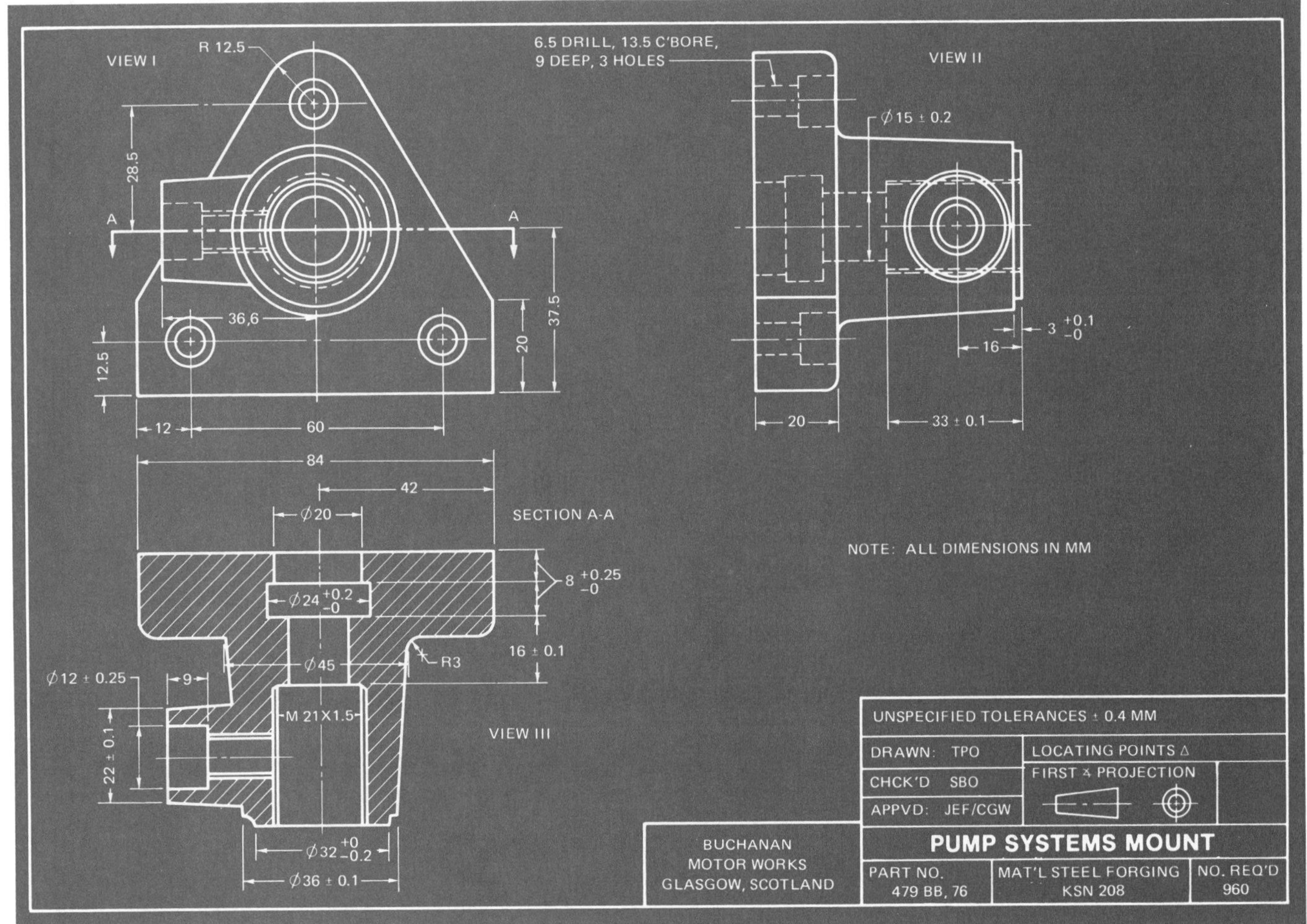

FIGURE 1-3A BLUEPRINT OF A TECHNICAL WORKING DRAWING FOLLOWING SI METRIC STANDARDS OF REPRESENTATION AND DIMENSIONING (REDUCED SIZE)

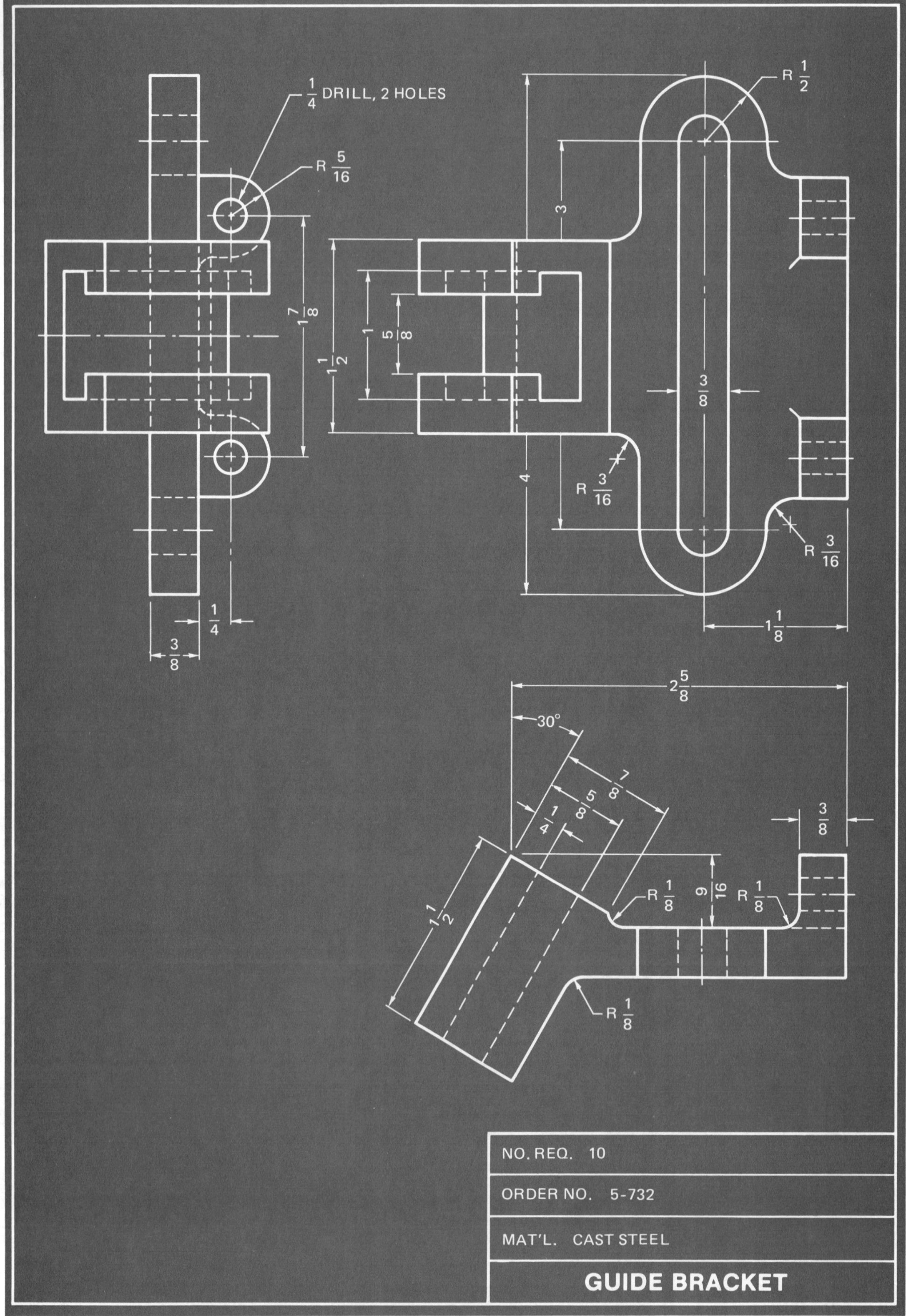

FIGURE 1-3B TYPICAL BLUEPRINT WHICH MEETS ANSI STANDARDS, USING CUSTOMARY INCH UNITS OF MEASUREMENT

ure 1-3B) provides an example of three-view drawing representation and application of customary inch units of measurement.

ELEMENTS COMMON TO ALL DRAWINGS

The interpretation of what is represented and specified on a technical drawing is based on a set of common principles. Examination of a blueprint, or other reproduction of a master drawing or sketch, reveals that drawings contain many of the following major elements:

- *Lines* of varying shapes and thicknesses.
- *Views* which alone, or in combination, provide full descriptions of external and internal features of a part.
- *Dimensions* for position, size, and surface finish measurements.
- *Sections* which show internal and often complicated details.
- *Processes* represented according to particular standards.
- *Techniques* of making projection drawings.
- *Characteristics* of newer numerical control and computer numerical control (CNC) drawings.
- *Practices* and *procedures* for producing technical illustrations and simple shop/laboratory sketches.

This textbook on *Blueprint Reading and Technical Sketching for Industry* covers basic principles related to each of the common elements. Practical applications are provided in each unit to develop skill in interpreting industrial blueprints and to make everyday sketches.

REPRODUCTION OF TECHNICAL DRAWINGS

As stated earlier, technical drawings are reproduced in order to have duplicates available for use in planning, manufacturing, and for other purposes, and to preserve the original drawing. Reproductions may be made to the same size, reduced, or enlarged. Exact *positives* or reverse *negatives* may be produced. Generally, reproduced copies of drawings are called *blueprints* or *whiteprints*.

Blueprints

A *blueprint* is an exact duplicate of a technical drawing or an illustration, except that the lines are white against a blue background and there is a slight variation in size due to shrinkage. Blueprints are made by placing the original drawing on a chemically coated blueprint paper. The drawing and paper are fed and exposed under ultraviolet light. The chemical action produced by the light rays causes the areas of the blueprint paper which are not shielded by lines on the drawing to turn blue.

After exposure to light rays, the blueprint paper is washed in water to bring out the blue background to clearly show against the white areas. The print is then passed through a *fixative* to set the colors and, finally, is dried.

Whiteprints (Xerox and CADD Generated)

Drawings that are reproduced with a white background are classified as *whiteprints*. Xerographic copies of drawings (generally produced by xeroxing©) are a common form of whiteprint. Standard single- and double-size sheet drawings may be conveniently enlarged, reduced, or be produced to the same size as an original drawing by xeroxing.

In the xeroxing process, the drawing is subjected to light rays. All lines, letters, and other markings on a drawing are instantaneously reproduced on a specially-coated, heat sensitive, xerox paper.

Lines and details may also be produced in blue, black, maroon, and other colors by using papers that are coated with different color dyes. The technical drawing and whiteprint background paper are exposed to ultraviolet light and a gas or liquid developer. This older process is identified as the *diazo dye process*.

CADD Generated Prints

Prints of drawings that are generated by computer-aided design and development systems (CADD) are available either as black and white prints or in multiple colors. A computer-generated drawing is reproduced electronically by feeding design and other information about a part or component to a CADD printer. Some printers are designed with pens for producing only black and white drawings. Other multiple-pen printers have capacity to produce multiple-color drawings and other information. Single or multiple copies may be produced.

Photographic Silver Reproduction Processes

The *silver process* of reproduction requires the use of photographic techniques and materials. Common products of the photographic process (the oldest of the reproduction processes) are known as *photocopy*, *microfilm*, and *photostat*®. The process provides flexibility to produce a positive print from a positive original, a negative print from a positive, or any other combination.

Prints are processed in a variety of ways: by directly exposing sensitized materials which are in direct contact with the original, by passing light rays through the master copy, or by reflecting light rays to the sensitized coating from the original.

The photographic process is adapted to enlarging or reducing the size of reproduced copy. In addition, a drawing may be reproduced on any surface that can take a coating of a sensitized solution and can be photographed, developed, and fixed.

TRANSMITTING INFORMATION (DRAWING AND DATA)

CADD systems permit the instantaneous transmission of drawings and other data to any reproducing electronic unit that is interlocked into the system. Drawings may be transmitted and reproduced at one or more work stations within a plant or spread across many remote plants.

Another popular system of transferring drawing and manufacturing information is referred to as *FAX copying*. A *facsimile* of any form of typed, plotted, drafted, or sketched drawing may be electronically transmitted and received over *long distance xerography* (*LDX*). This system of communication uses an *LDX scanner* to convert images from a drawing to video transmission signals. The signals are electronically restored at the destination to which the information is being sent by an *LDX printer*. The printer produces a black-on-white copy of the original drawing. The drawing is fed into a FAX machine at the sending end and a facsimile is reproduced at the receiving end.

STORING DRAWING INFORMATION AND DATA

Visual Microfilm Systems

Although CADD and other design and development systems have memory and storage capability, a back-up security system is often required.

Many drawings and records are inexpensively placed on *microfilm* for safe storage and as a practical means of communicating information throughout the country and world. A microfilm is a photographic negative of a technical drawing which may be reduced up to sixty times from the original size.

The photographic negative may be placed on a roll or individually on an *aperture card*. The aperture card may also have additional information punched into it. A microfilm negative, which looks like a 16 mm, 35 mm, or 105 mm film negative, or a negative mounted in an aperture card, is generally read by using a special *reader-printer*. The reader-printer projects an enlarged image of the original drawing on a screen for rapid viewing and easy reading.

Microfilm copy is adaptable to electrostatic, photocopy, and other photographic duplicating processes. Microfilm systems are being locked into electronic computer banks or machine control systems for transmission of technical information to plants and shops in different geographic locations. Limited storage space requirements, easy access to technical data and drawing specifications, and the preservation of original drawings are a few of the advantages of using microfilm copy.

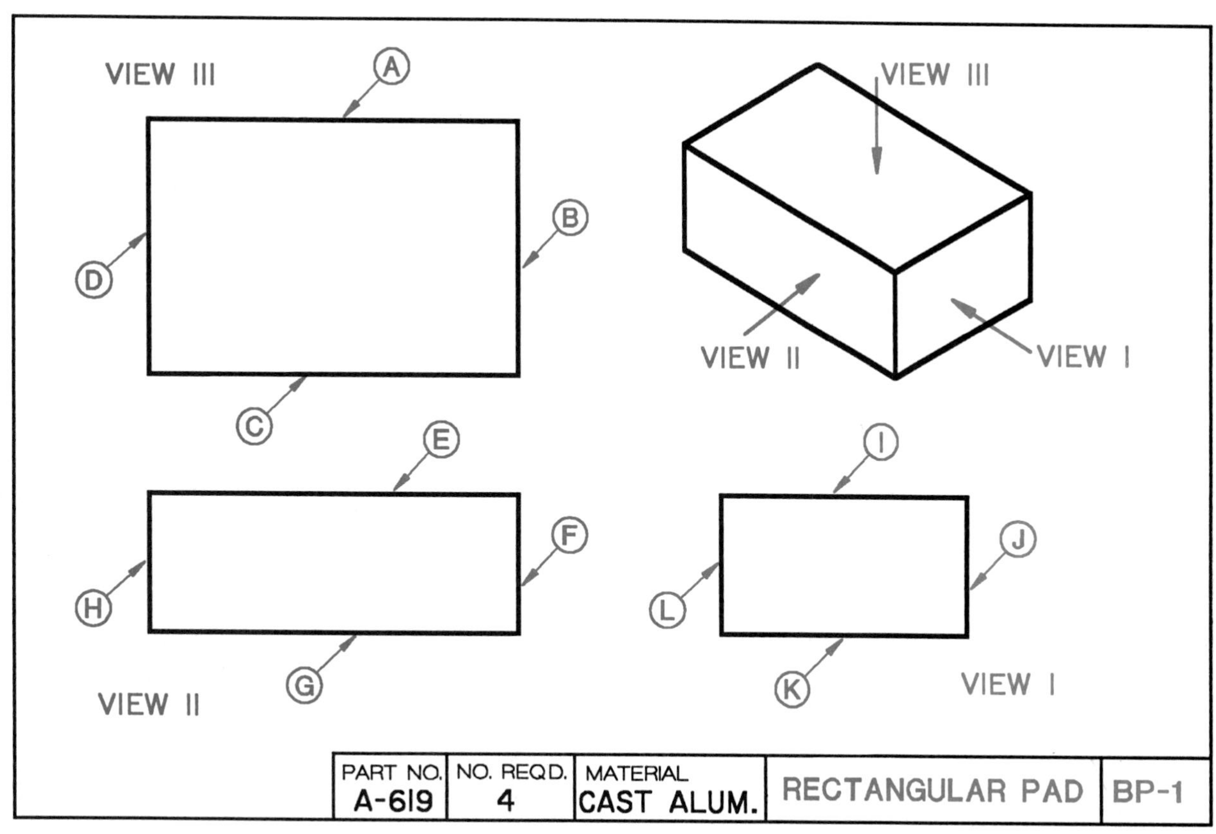

ASSIGNMENT—UNIT 1

RECTANGULAR PAD (BP-1)

1. Name the two key factors on which the "American System of Manufacturing" was founded.
2. State two skills to be developed through the mastery of blueprint reading and sketching.
3. List three different types of blueprints and/ or drawing reproduction processes.
4. Cite two advantages of microfilms over regular blueprints.
5. Name two standards systems for uniformly representing and dimensioning a part on a technical drawing.
6. Give (a) the part name on the drawing above and (b) the part number.
7. Indicate (a) the kind of material to be used for the part and (b) the required number.
8. List the lettered lines which show (describe) the shape of the part in (a) **VIEW I**, (b) **VIEW II**, and (c) **VIEW III**.

Student's Name ____________________

1. a. ____________________
   b. ____________________
2. a. ____________________
   ____
   b. ____________________
   ____
3. a. ____________________
   b. ____________________
   c. ____________________
   ____
4. a. ____________________
   ____
   b. ____________________
   ____
5. a. ____________________
   b. ____________________
6. a. ____________________
   b. ____________________
7. a. ____________________
   b. ____________________
8. a. **VIEW I** ____________________
   b. **VIEW II** ____________________
   c. **VIEW III** ____________________

UNIT 2
Universal Systems of Measurement

Together with standardized techniques of representation, dimensions on technical drawings constitute another major part of the universal language of industry. Dimensions are used to define distance, size, form, and other specifications of an object. *Dimensional measurements* relate to plane surfaces, multiple-surface objects, straight and curved lines, areas, angles, and volumes.

Dimensions are essential in design, fabrication, machining, assembly, inspection, and maintenance of parts, mechanisms, and total structures. Dimensions are generally taken directly by measurement of the part and by comparison with the design dimension given on a technical drawing. In some instances, dimensions are computed.

PURPOSES OF DIMENSIONAL MEASUREMENT

The three main purposes of dimensional measurement relate to:

- *Design* and further description of a part or mechanism (in addition to the representation of surface features on a drawing).
- *Construction* of a part, component, or complete unit.
- *Controls* for producing a single part or for mass-manufacturing interchangeable parts.

DIMENSIONAL ACCURACY AND THE PRECISION OF MEASURING INSTRUMENTS

The accurate control of dimensional measurements and surface finishes is necessary to ensure uniformity of size and shape. The engineer, designer, skilled mechanic, and technician must be able to express, interpret, and to uniformly measure. Manufacturing is controlled by being able to reproduce identical measurements regardless of where a part is fabricated.

The required accuracy of a stated dimension provides a key to the kind of measuring tool or instrument to use. For example, measurement accuracies to within 1/64″, or 1/100″, or 1/2 mm are within the range of and may be taken directly with a steel rule. Dimensions within 0.001″ or 0.02 mm are generally measured with a standard micrometer. Accuracies of 0.0001″ or 0.002 mm require a more precise vernier micrometer. Still higher precision measurements with accuracies in the one-millionth-of-an-inch (0.00002 mm) range are made under controlled temperature conditions. Optical flats, highly precise pneumatic, electronic, and other instruments and measuring machines are used.

CHARACTERISTICS OF MEASUREMENTS

Linear (length) measurements are the most widely used dimensional measurements. A linear

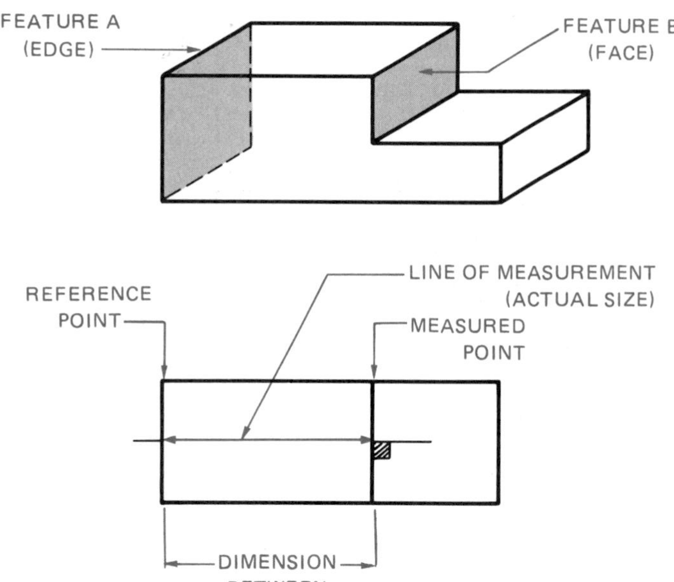

FIGURE 2-1 CHARACTERISTICS OF LINEAR MEASUREMENTS

measurement represents a straight line distance between two points or surfaces. This distance represents the *direction* of the line of measurement. Each measurement of a dimension begins at a specific point called a *reference point* (figure 2-1). Each measurement ends at a *measured point*. The straight line distance between these two points is the *line of measurement*.

The reference point and measured point are known as *references*. Each reference point or surface identifies a *feature*. A part consists of a number of features which, collectively, make up the part. Measurements are usually taken from an outside surface to an inside surface of the feature to be measured. The dimension on a drawing states the designer's requirements of an *exact size*. A dimensional measurement refers to the *actual size* between features.

The terms *accuracy* and *precision* are widely used with dimensions and measurements. *Accuracy* deals with the number of measurements which conform to a specified standard. *Precision* relates to the degree of exactness. The two terms are generally used interchangeably in shop and laboratory practice. *Tolerances*, which indicate maximum and minimum dimensions for machining, fitting, and operating, are treated in detail in later units.

BASIC SYSTEMS OF MEASUREMENT

The three basic systems of linear measurement include:

- The customary inch standard unit of measurement in the British/United States system.
- The decimal inch system.
- The SI metric system.

The Customary Inch Standard Unit of Measurement

The *inch* is the standard unit of linear measure in the British/United States system. The inch is divided into *common fractions* and *decimal fractions*. Measurements, accurate to 1/8″, 1/16″, 1/32″ and 1/64″, and 1/50″ and 1/100″, are taken directly with steel rules. Figure 2-2 shows fractional parts of an inch which are widely used on technical drawings and in shops.

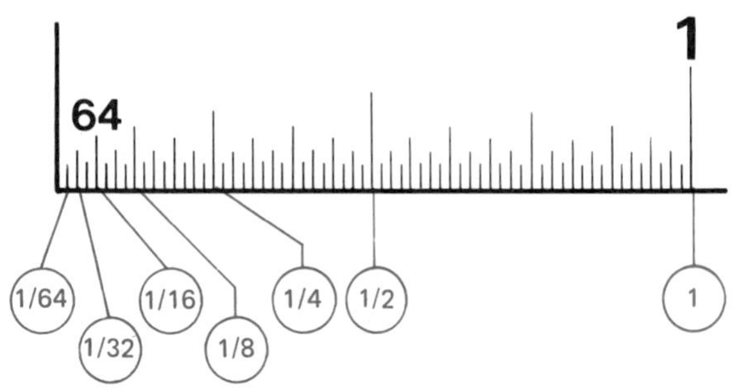

FIGURE 2-2 COMMONLY USED FRACTIONAL PARTS OF AN INCH (ENLARGED)

The Decimal Inch System of Dimensioning

This system was promoted by the Ford Motor Company in the late 1920s. It was accepted by such industrial societies as the Society of Automotive Engineers. Dimensions are given with subdivisions of the inch in multiples of ten. The thousandth part of an inch is a *mil.* The 1/100″ graduations of a standard steel rule are shown enlarged in figure 2-3.

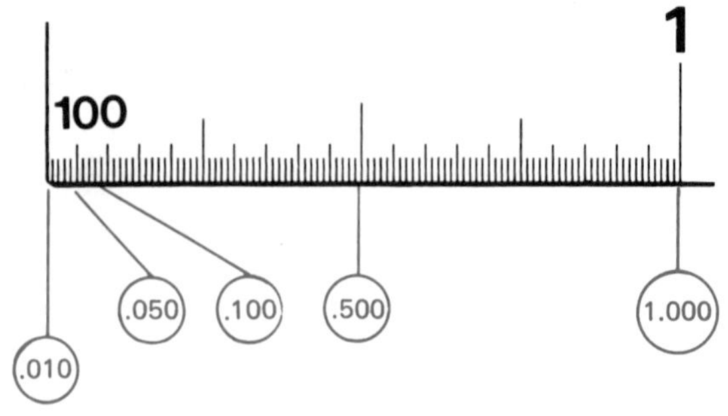

FIGURE 2-3 DECIMAL PARTS OF AN INCH (ENLARGED)

SI Metric Standard Units of Linear Measurement

The *meter* is the basic unit of linear measurement in SI metrics. For practical measurement purposes, the meter is broken down into smaller units of linear measure. These units are a multiple of 10, 100, and 1,000 for most shop and laboratory applications.

Prefixes like *deci* (10), *centi* (100), and *milli* (1,000) express the multiple of a meter. For example, one meter = 10 *deci*meters (dm) or 100 *centi*meters (cm) or 1,000 *milli*meters (mm). While the centimeter and millimeter are primarily European metric system prefixes, and they are not recommended in SI metrics, they are actually functional measurements which are applied in industry.

The meter is for practical shop purposes the equivalent of 39.37″. One centimeter (cm) = 0.3937″; one millimeter, 0.0394″. Values of 2.54 cm = 1″ and 25.4 mm = 1″ are generally used.

Examples of the graduations on each of two customary inch measuring rules are illustrated in figure 2-4 and SI metric, figure 2-5.

Courtesy of the L.S. STARRETT COMPANY

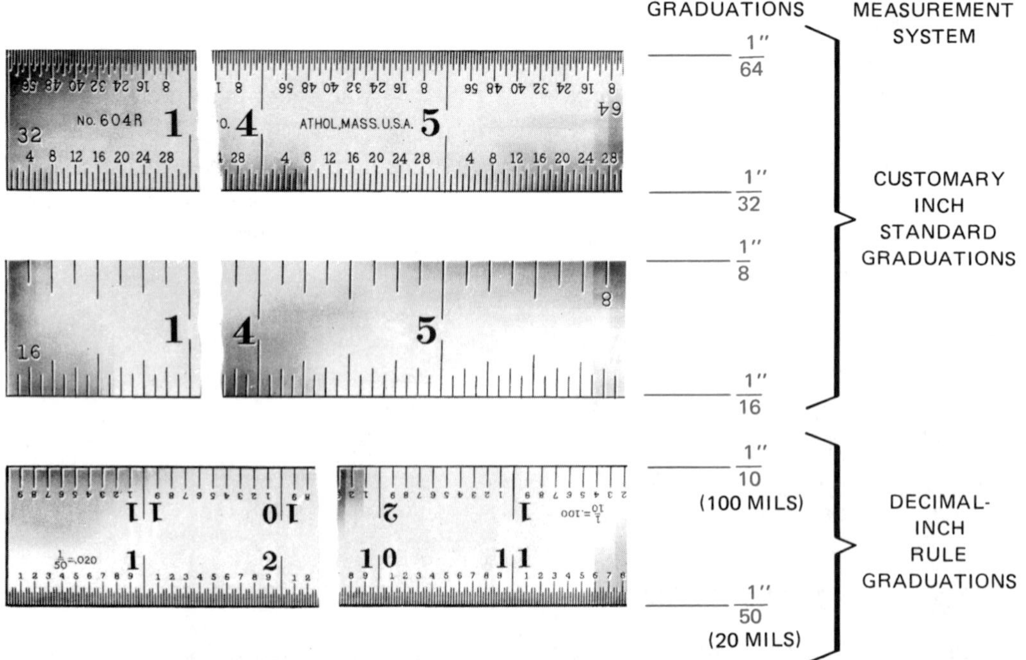

FIGURE 2-4 STANDARD GRADUATIONS ON CUSTOMARY INCH AND DECIMAL INCH STEEL RULES

Courtesy of the L.S. STARRETT COMPANY

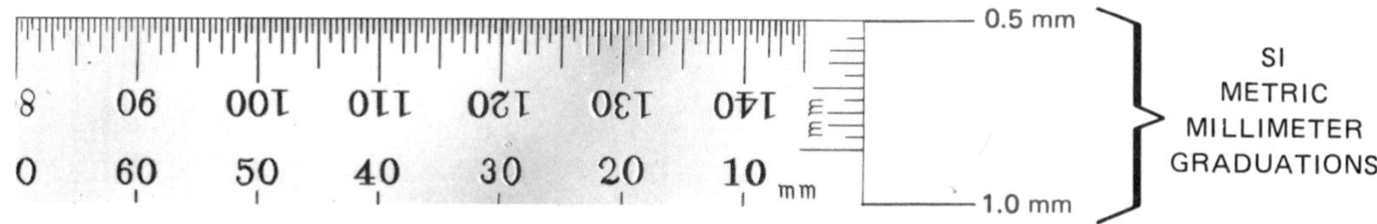

FIGURE 2-5 COMMON SI METRIC GRADUATIONS ON STANDARD STEEL RULES

LIMITATIONS OF CUSTOMARY INCH AND METRIC MEASUREMENT SYSTEMS

One main advantage of metric system measurements over customary inch measurements is the ease with which decimals may be used for computations and for expressing quantities. Drafting room and manufacturing standards are based on both metric and customary inch units of measurement.

Steel rules that are graduated in half-millimeters (0.5 mm) and millimeters, as shown in figure 2-5, are commonly used in shops and laboratories. The 0.5 mm graduation (0.020″) is not as precise as the 1/64″ graduation (0.016″) on a customary inch rule or a 1/100″ (0.010″) graduation on a decimal inch rule.

Dimensions that are given on drawings in terms of one-thousandth and one ten-thousandth part of an inch are realistic, functional for quality control, and cost efficient. The 0.001″ and 0.0001″ are more precise than metric dimensions expressed in 0.1 mm and 0.001 mm (0.004″ and 0.0004″).

CONVERSION TABLES OF LINEAR UNITS OF MEASUREMENT

A linear unit of measurement in one system may be easily converted to an equivalent unit in another required system. Charts with information similar to the partial table 2-1 are found in trade handbooks and manufacturers' technical data sheets. Complete tables are also contained in the appendix. Equivalent values are read directly.

Conversion factors are also contained in these publications and the appendix. Equivalent measurement values may be easily computed by using the appropriate conversion factor and carrying out a simple mathematical process. For example, to convert one inch to equivalent millimeters, multiply by 25.4. In like manner, to change a millimeter value to an equivalent inch measurement, simply divide by 25.4.

TABLE 2-1 PARTIAL CONVERSION TABLE OF STANDARD MEASUREMENT UNITS

MILLIMETER	DECIMAL INCH	FRACTIONAL INCH	EQUIVALENT DECIMAL INCH
0.1			0.00394
0.2			0.00787
	0.01		0.01000
0.3			0.01181
0.397		1/64	0.01563
0.4			0.01575
0.5			0.01968
	0.02		0.02000
0.6			0.02362
0.7			0.02756
	0.03		0.03000
0.794		1/32	0.03125
0.8			0.03149
0.9			0.03543
1.0			0.03937
	0.04		0.04000

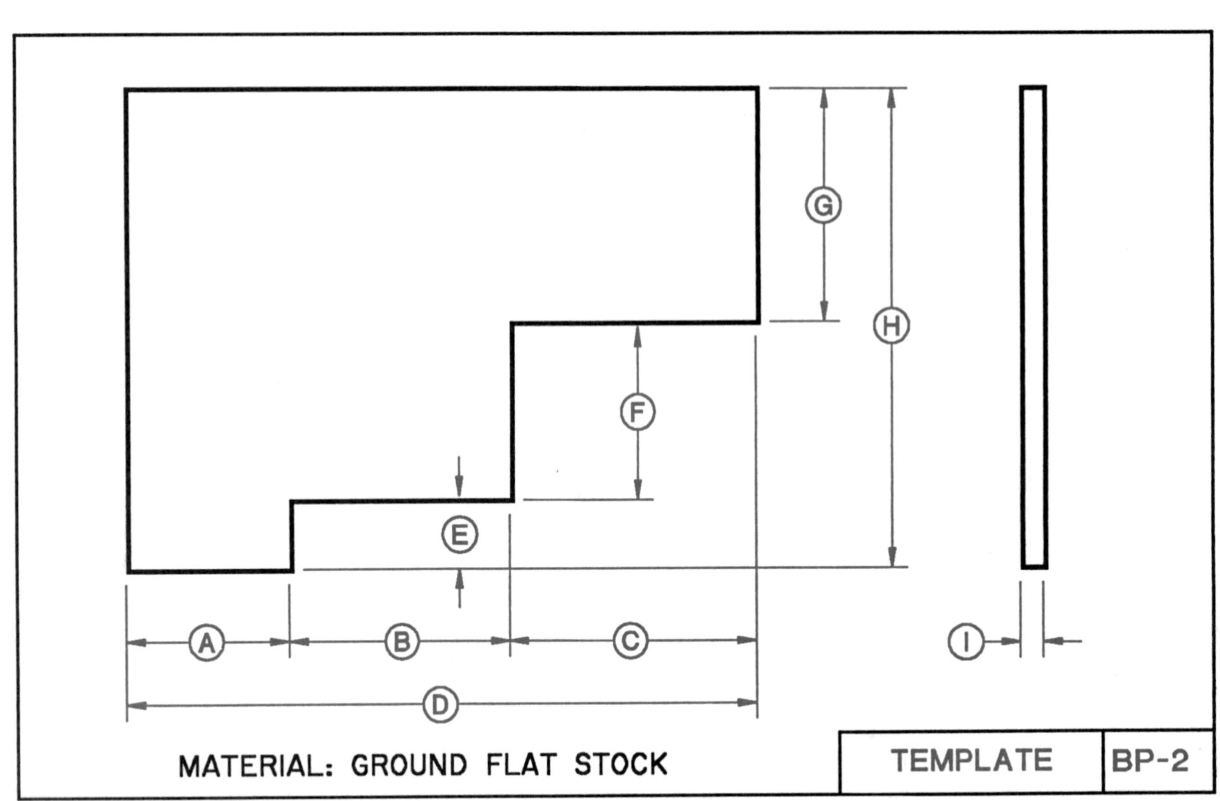

ASSIGNMENT—UNIT 2

TEMPLATE (BP-2)

1. State two main functions of dimensional measurements.
2. Cite three characteristics of a dimension that must be considered in order to accurately take a linear measurement.
3. Describe briefly the term "feature" as applied to a technical drawing.
4. Indicate two methods of changing measurements from customary inch to SI metric values.

   Note. In industrial practice, measurements are not taken directly from a technical drawing. However, for purposes of providing a measurement exercise only, assume for problems 5 and 6 that the Template is machined and the size of each feature is to be checked.
5. Take measurements Ⓐ through Ⓘ from the drawing of the machined Template. Use a steel rule graduated in customary inch units of 1/64″. Measure to the nearest 1/64″ and record.
6. Retake measurements Ⓐ through Ⓘ. Use a steel rule graduated in metric units of 1/2 mm. Record each measurement correct to 1/2 mm.

Student's Name ______________________

1. a. ______________________

   b. ______________________

2. a. __________ b. __________

   c. __________

3. ______________________

4. a. Method 1 ______________________

   b. Method 2 ______________________

5. Customary Inch Measurements		6. SI Metric Measurements	
Ⓐ	______	Ⓐ	______
Ⓑ	______	Ⓑ	______
Ⓒ	______	Ⓒ	______
Ⓓ	______	Ⓓ	______
Ⓔ	______	Ⓔ	______
Ⓕ	______	Ⓕ	______
Ⓖ	______	Ⓖ	______
Ⓗ	______	Ⓗ	______
Ⓘ	______	Ⓘ	______

SECTION 2
Lines for Technical Drawings and Sketching

UNIT 3
The Alphabet of Lines: Visible (Object) Lines

Technical drawings depend upon a standard set of *lines (symbols)*. Lines are used alone or in combination with one another. Lines describe and/or provide other essential information about design features and dimensions for constructing or assembling a part or a complete mechanism of many parts.

Lines are of different shapes, thicknesses, and lengths. Each type of line has its own particular line symbol and meaning. In order to design, manufacture, inspect, or service an object, the lines which appear on a drawing must be uniformly interpreted. Lines must convey the same meaning to the designer, engineer, craftsperson, or reader of the drawing or print.

ANSI ALPHABET OF LINES

Fortunately, a universally accepted *alphabet of lines* is used by industry and throughout the world. The alphabet, which is often referred to as *line conventions*, has been standardized by the American National Standards Institute (ANSI). In Canada, the Canadian Standards Association (CSA) conducts similar standards-setting functions.

The most widely used lines are given in figure 3-1. A description with applications of *visible (object)*, *hidden*, *center*, *extension*, and *dimension lines*, and combinations of these lines, is included in this Section.

Additional lines, like those illustrated in figure 3-2, are used to show complex internal details and

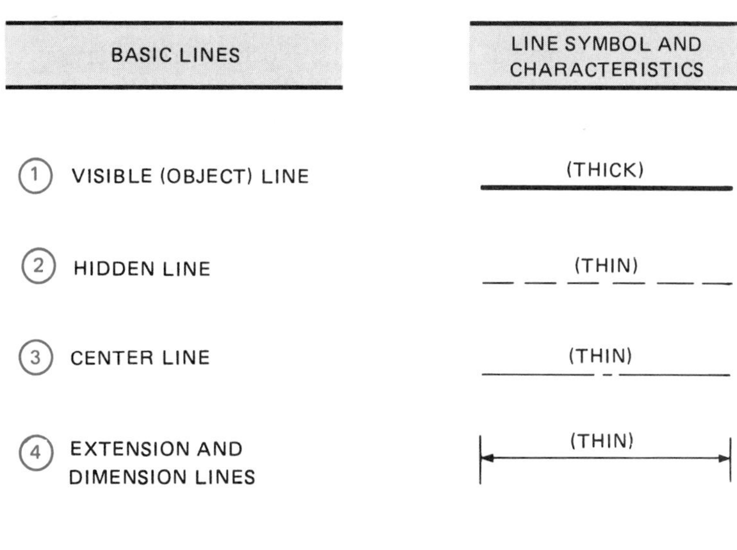

BASIC LINES
LINE SYMBOL AND CHARACTERISTICS
1 VISIBLE (OBJECT) LINE
(THICK)
2 HIDDEN LINE
(THIN)
3 CENTER LINE
(THIN)
4 EXTENSION AND DIMENSION LINES
(THIN)

FIGURE 3-1 ANSI BASIC LINES, LINE SYMBOLS, AND CHARACTERISTICS

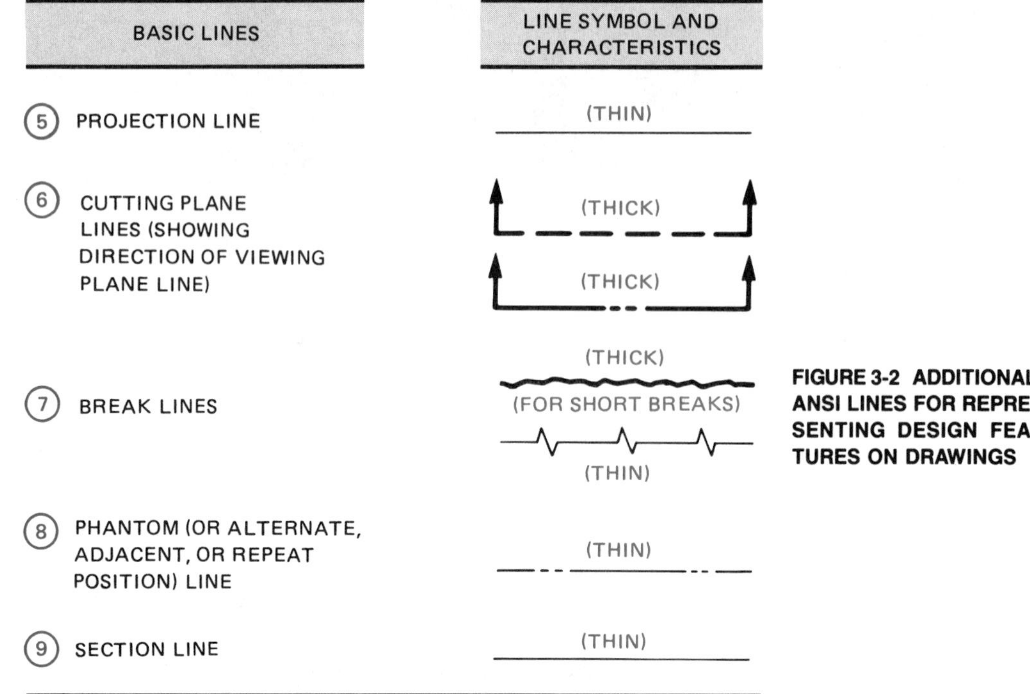

BASIC LINES
LINE SYMBOL AND CHARACTERISTICS
5 PROJECTION LINE
(THIN)
6 CUTTING PLANE LINES (SHOWING DIRECTION OF VIEWING PLANE LINE)
(THICK)
(THICK)
7 BREAK LINES
(THICK)
(FOR SHORT BREAKS)
(THIN)
8 PHANTOM (OR ALTERNATE, ADJACENT, OR REPEAT POSITION) LINE
(THIN)
9 SECTION LINE
(THIN)

FIGURE 3-2 ADDITIONAL ANSI LINES FOR REPRESENTING DESIGN FEATURES ON DRAWINGS

other design features and to provide information to produce a part. Each different type of line (⑤ through ⑨) is covered in detail in later units.

Standards for Line Widths (Weights)

Two weights of lines, *thick* and *thin*, are recommended by ANSI. The width of a thick line for either mechanical drawings or sketches may vary from 0.030″ to 0.038″ (0.8 mm to 1.0 mm); a fine line, from 0.015″ to 0.022″ (0.2 mm to 0.6 mm). The line widths depend on the size and complexity of the part to be represented. Consideration is also given to whether the drawing is to be reduced or blown up in size. However, once a line width is determined, it must be constant throughout the drawing.

VISIBLE (OBJECT) LINES

The outline or shape of an object is represented on a drawing by a thick, dark, continuous line (figure 3-3A). These lines are known as *visible* or *object lines*. Figure 3-3B shows the use of visible lines to represent the external shape of a rectangular block with one corner cut at an angle. Similarly, thick, dark, continuous visible lines are used in the pictorial sketch (figure 3C) of the same object.

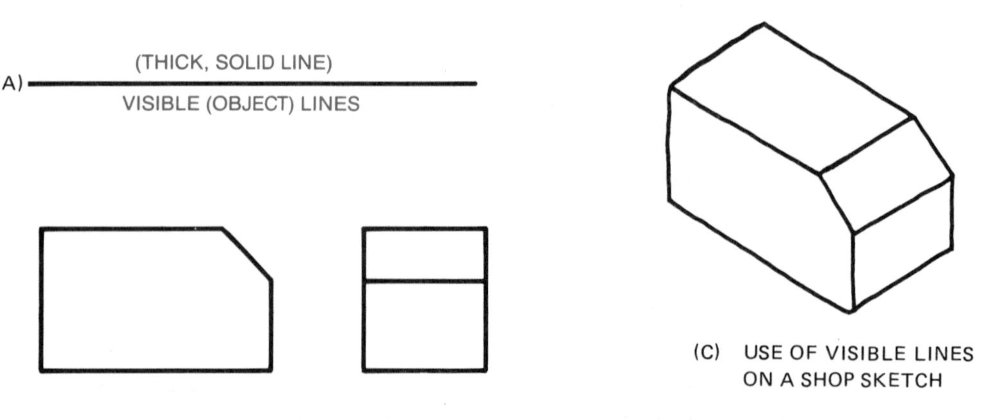

FIGURE 3-3 VISIBLE LINES APPLIED IN DRAFTING AND SKETCHING

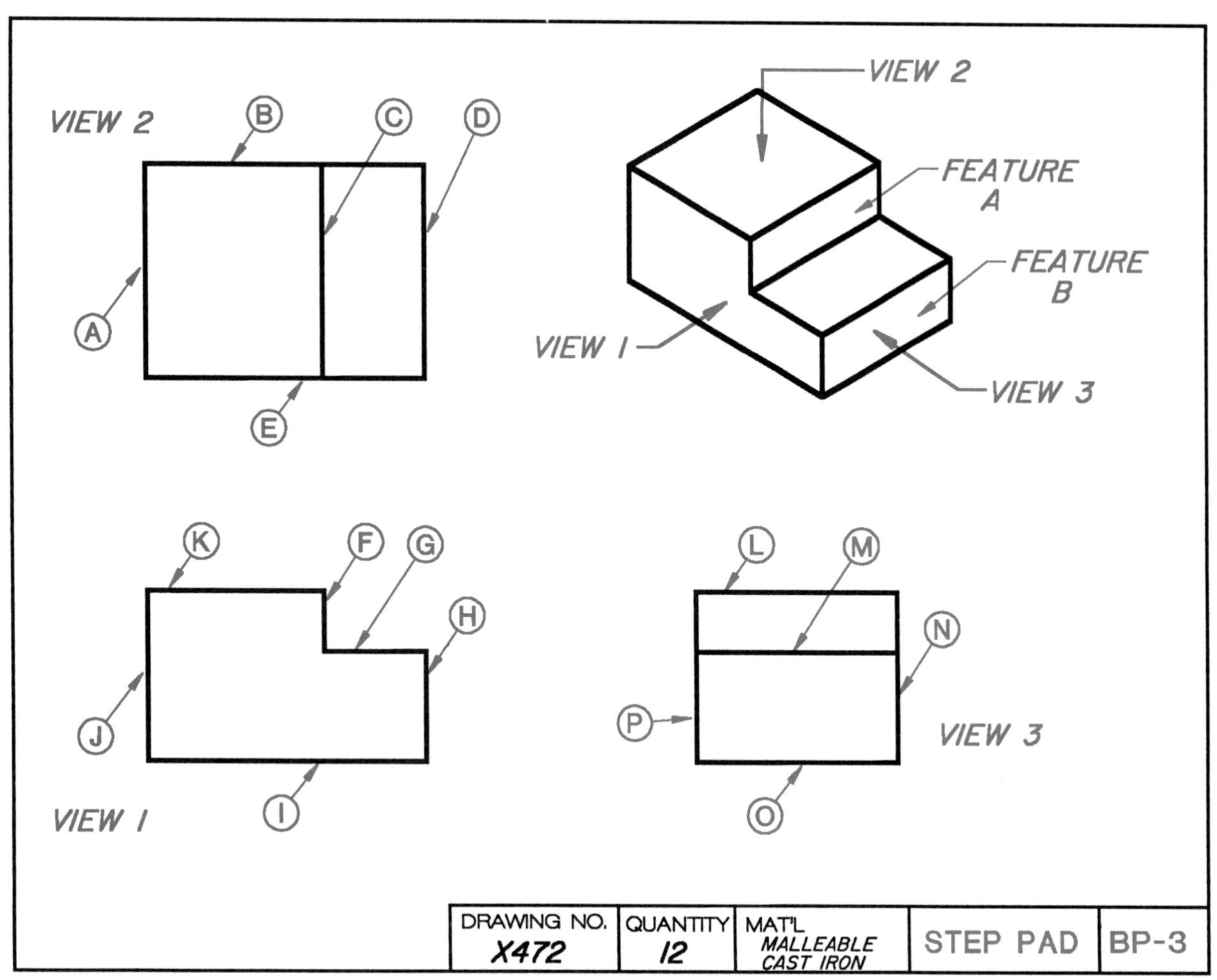

ASSIGNMENT—UNIT 3

STEP PAD (BP-3)

1. Give the name of the part.
2. Identify the drawing number.
3. Tell how many parts are required.
4. Indicate the kind of material to be used for the part.
5. Name the kind of line which is used on the drawing to represent the shape of the part.
6. List the lettered lines which represent the shape of the Step Pad in:
   (a) **VIEW 1**
   (b) **VIEW 2**
   (c) **VIEW 3**
7. Identify the letters which describe Features A and B in **VIEW 3**.

Student's Name ____________________

1. ____________________
2. ____________________
3. ____________________
4. ____________________
5. ____________________
6. a. **VIEW 1** ____________________
   b. **VIEW 2** ____________________
   c. **VIEW 3** ____________________
7. **VIEW 3** ____________________

UNIT 4
Hidden Lines and Center Lines

HIDDEN LINES

Object lines describe the shape and size of an object as viewed from the outside. Internal features and details which are covered by some portion of the object are represented on technical drawings and illustrations by *hidden lines*. A hidden line consists of a series of thin, evenly spaced, short dashes (figure 4-1).

The hidden line is used in combination with object and other lines. In short, hidden lines represent:

- Edges and surfaces which are hidden from the worker in a particular view, or
- Internal details which are not visible from the outside.

Two general applications of hidden lines are illustrated in figure 4-2.

FINE, ALTERNATELY SPACED, SHORT DASHES

FIGURE 4-1 CHARACTERISTICS OF HIDDEN LINES

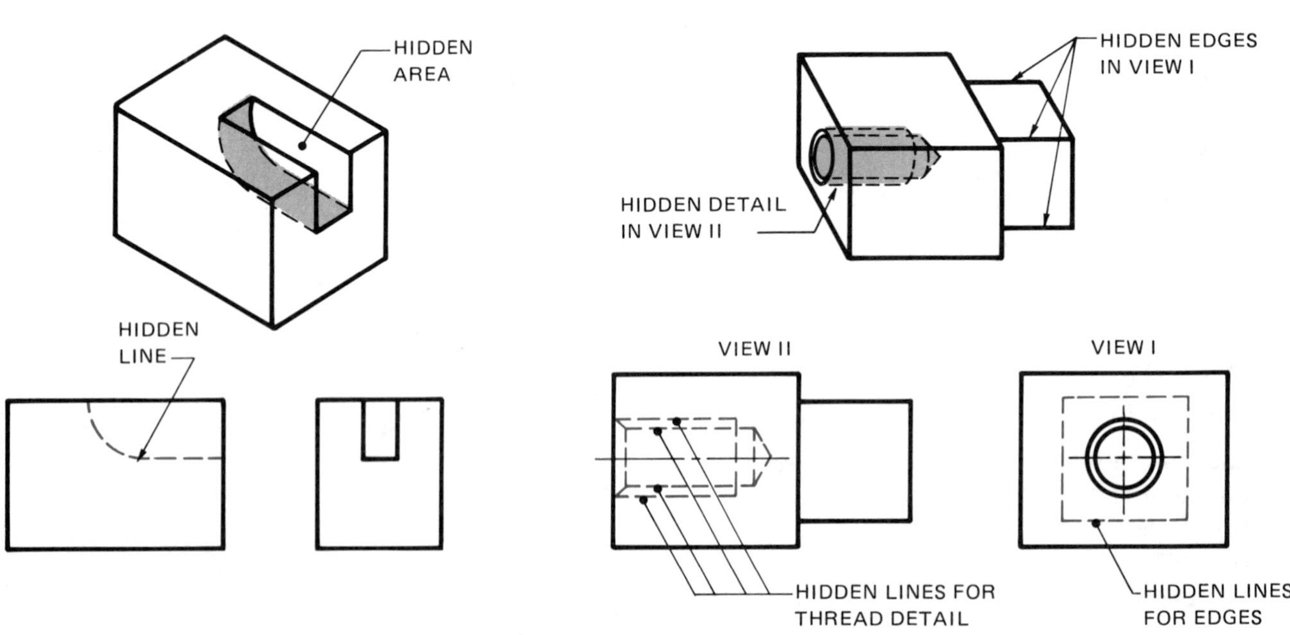

FIGURE 4-2 EXAMPLES OF HIDDEN LINES TO REPRESENT HIDDEN AREAS, EDGES, AND OTHER INTERNAL DETAILS

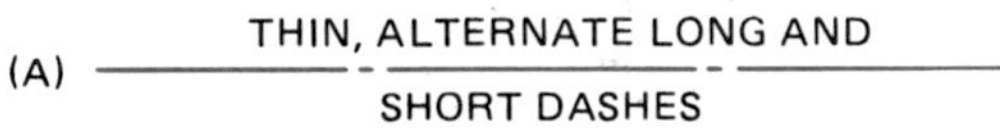

REPRESENTS CENTERS OF HOLES, SYMMETRICAL OBJECTS, AND PATH OF MOTION

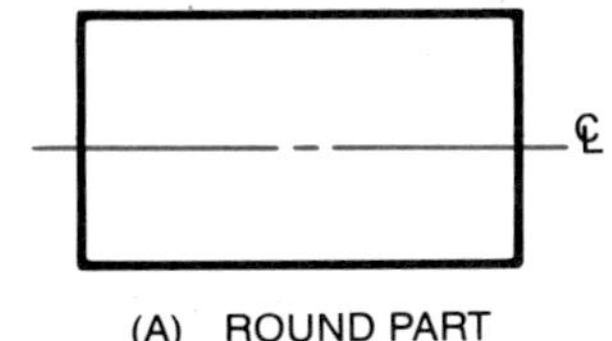

(A) ROUND PART

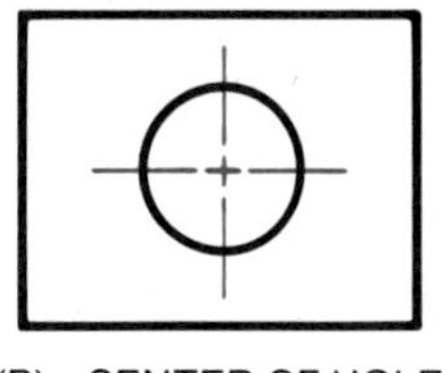

(B) CENTER OF HOLE

FIGURE 4-3 COMMON APPLICATIONS OF CENTER LINES

CENTER LINES

Center lines are used on drawings and layouts to show the center of round and symmetrical parts. A part is said to be *symmetrical* (SYMM) when the shape and all other features of an object on both sides of the center line are identical. Drawings of such parts are often simplified by showing just one-half of the part.

Center lines appear on drawings as a series of thin-width, alternate long and short dashes, as shown in figure 4-3A. The center line is occasionally accompanied by the symbol ℄.

Common applications of the center line are shown in figure 4-3 at (A) and (B). Center lines identified with a round part are shown at (A). The use of center lines for positioning a hole are given at (B).

Two other applications of center lines are illustrated in figure 4-4. The use of a center line and the addition of the symmetrical symbol (SYMM) at (A), indicates that the part is symmetrical. In other words, all features, dimensions, finish, etc. for the part are the same even though only half of the part is drawn. The center line application at (B) is used to show the path of motion.

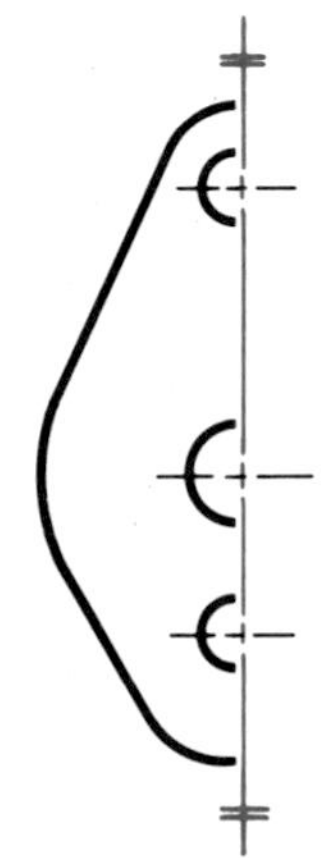

(A) CENTER LINE INDICATES PART IS SYMMETRICAL

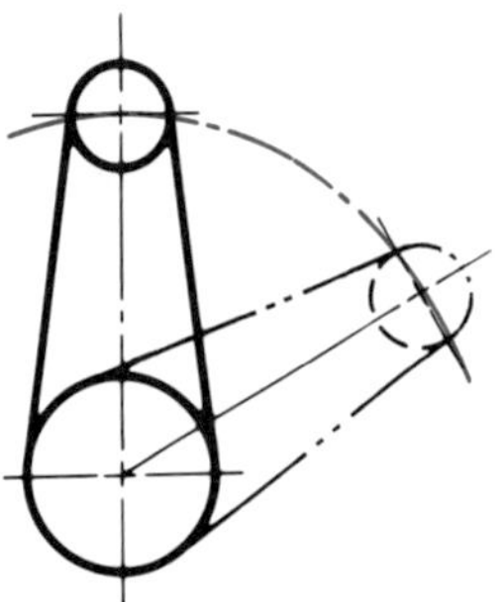

(B) CENTER LINE SHOWS PATH OF MOTION AND ALTERNATE POSITION

FIGURE 4-4 SYMMETRICAL AND PATH OF MOTION APPLICATIONS OF CENTER LINES

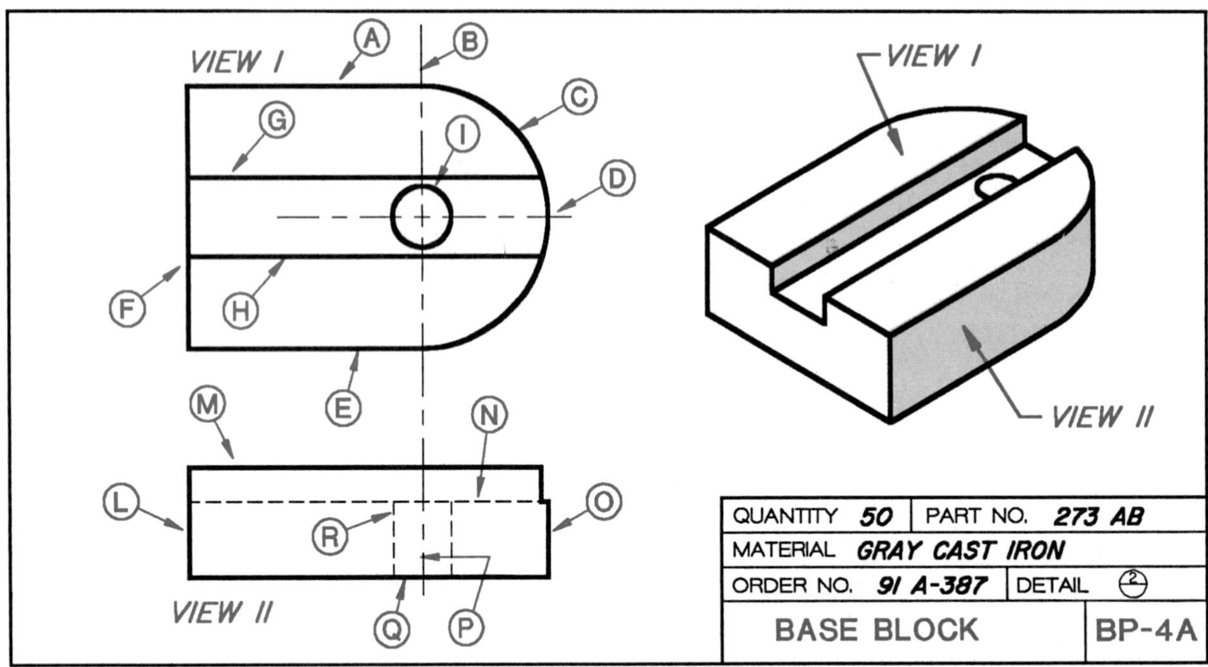

ASSIGNMENT A—UNIT 4

BASE BLOCK (BP-4A)

Student's Name ____________________

1. Give the letters (in order) of the object lines in **VIEWS I** and **II** that represent the shape of the **BASE BLOCK**.
2. Provide the following title block information:
   a. The kind of material used for the part.
   b. The required quantity.
   c. The drawing detail indentification.
   d. The part number.
3. Name the following two lines that are used on mechanical drawings:
   a. Lines that show the location of the center of a hole or symmetrical surfaces.
   b. Internal details of a part that are not visible externally.
4. Identify the kinds of lines that are represented by the following callout letters on the drawing of the **BASE BLOCK**.
   a. (A), (G), and (I).
   b. (B), (D), and (P).
   c. (N) and (R).

1. ____________________

2. a. ____________________
   b. ____________________
   c. ____________________
   d. ____________________

3. a. ____________________
   b. ____________________

4. a. ____________________
   b. ____________________
   c. ____________________

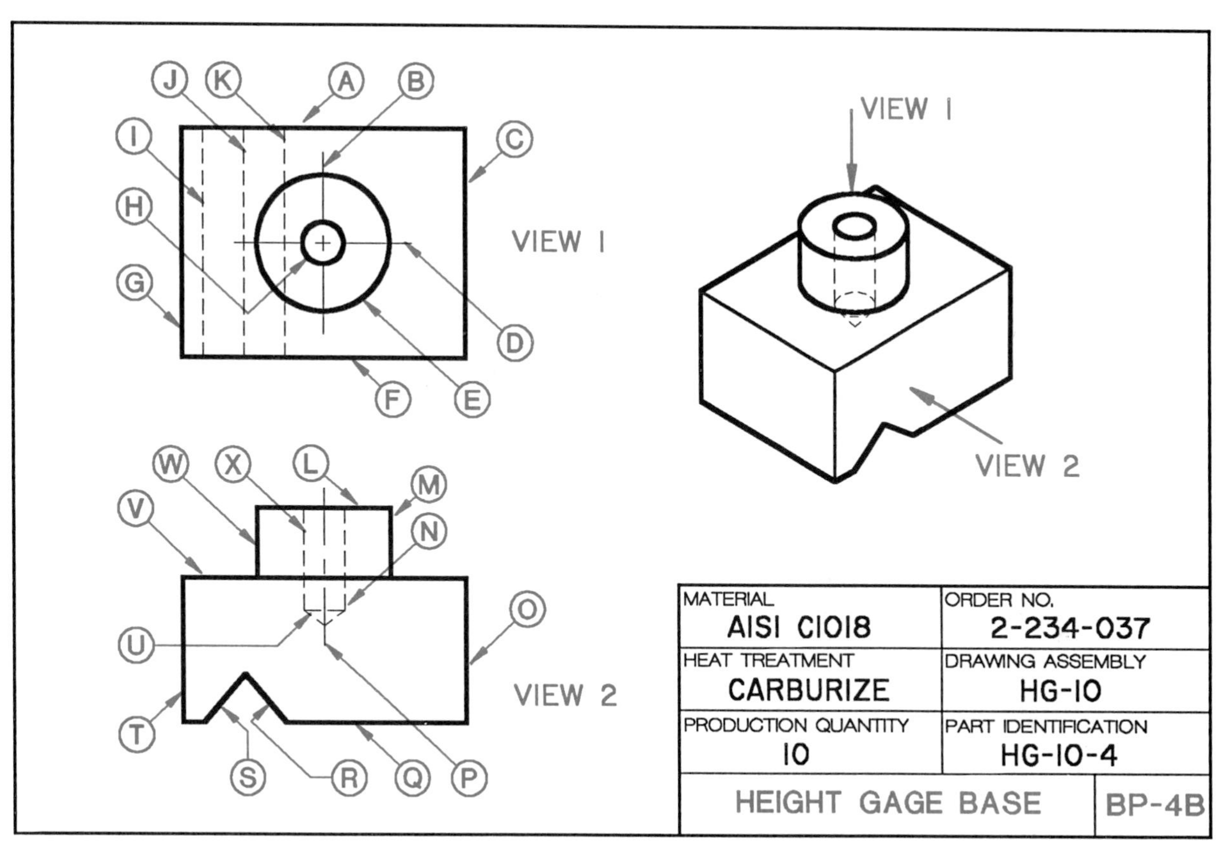

ASSIGNMENT B—UNIT 4

HEIGHT GAGE BASE (BP-4B)

1. Name the kind of line used to represent the shape of the part.
2. Specify the kind of material from which the Height Gage Bases are to be machined.
3. Provide the following information:
   a. Order number.
   b. Drawing assembly number.
   c. Part identification number.
4. Tell how many of the part are to be produced.
5. Name the kind of line used to represent invisible surfaces or inside details.
6. Give the letters of the center lines in **VIEW I** which locate the center of the hole.
7. List the letters of the object lines which represent the Height Gage Base in
   a. **VIEW I** b. **VIEW II**
8. Give the letters of the hidden lines in (a) **VIEW I** and (b) **VIEW II** which represent invisible surfaces or edges.
9. Identify the hidden lines in **VIEW II** which represent hole (H) in **VIEW I**.

Student's Name ______________________

1. ______________________
2. ______________________
3. a. Order number ______________________
   b. Drawing assembly ______________________
   c. Part identification ______________________
4. ______________________
5. ______________________
6. ______________________
7. a. **VIEW I** ______________________
   b. **VIEW II** ______________________
   ____
8. a. **VIEW I** ______________________
   b. **VIEW II** ______________________
9. ______________________

UNIT 5
Extension Lines, Dimension Lines, and Leaders

Dimensions are represented on technical drawings by combining the following elements: *extension lines*, *dimension lines*, *arrowheads* or *dots*, and *leaders*. *Notes* and *symbols* are used with leaders to provide additional specifications.

EXTENSION LINES

Extension lines locate the *reference point* and the *measured point* at which a dimension line begins and ends. An extension line is a thin, dark, solid line as shown in figure 5-1. Extension lines appear on drawings in pairs and are at a right angle to the dimension line. Extension lines start a short distance [usually about 1/16″ (1.5 mm)] away from an object line (figure 5-2) and extend this same distance beyond the dimension line (figure 5-3). Center lines, when extended as fine, solid lines, are often used as extension lines in dimensioning holes and other details. Figure 5-3 shows such an example.

DIMENSION LINES

Dimension lines show the exact size and location of a dimension. The dimension line is a thin, dark line which terminates at the two extension lines in arrowheads. Dimension lines are drawn parallel to the line of measurement and are broken near the middle to provide a space for the dimension. In structural, architectural, and some technical drawings, the dimension line is an unbroken line with the dimension placed above the line. Dimensioning practices on machine and architectural drawings are illustrated in figure 5-4.

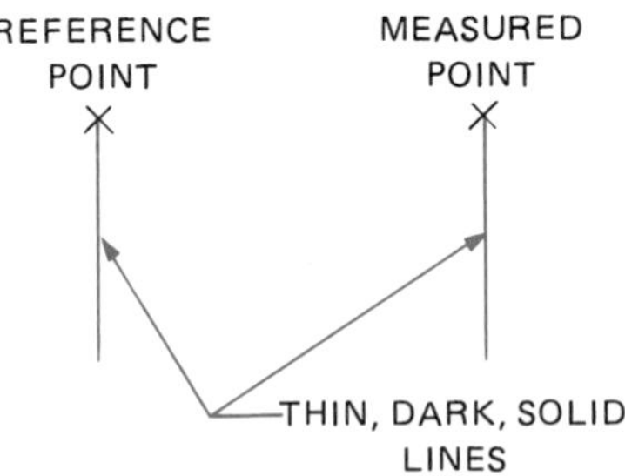

FIGURE 5-1 EXTENSION LINES

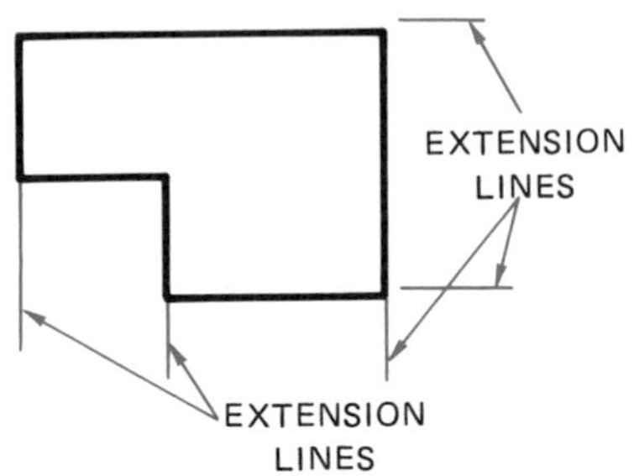

FIGURE 5-2 PRACTICES IN THE PLACEMENT OF EXTENSION LINES

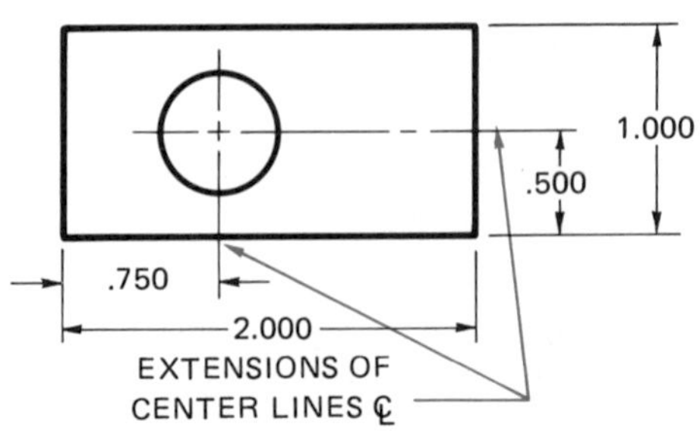

FIGURE 5-3 CENTER LINES USED AS EXTENSION LINES

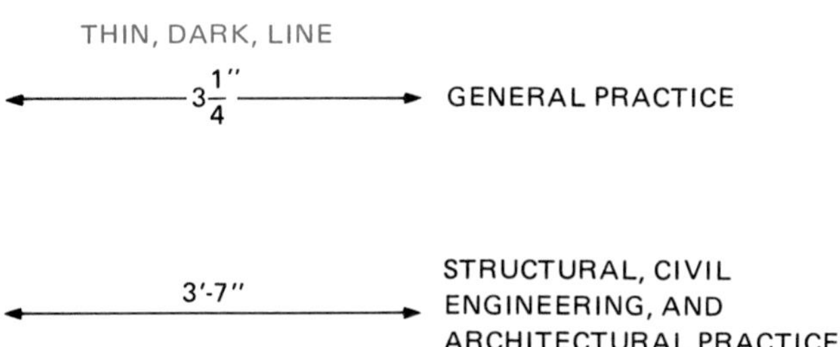

FIGURE 5-4 DIMENSION LINES

LEADERS

A *leader* is used to direct attention to dimensions, notes, or symbols, or to a specific feature on a part. A *leader line* has two parts. The leader starts with a short line which leads to the required information. The body of the leader changes to an angle of 30°, 45°, or 60°. The thin, dark, solid line used for a leader ends with an arrow or dot at the indicated feature. Leaders are drawn *axially* for circles and arcs. This means that if the leader were extended, it would cut through the center of the circle. The leader arrow is placed on the inside of large arcs and outside small arcs. General applications of leaders are illustrated in figure 5-5.

ARROWHEADS

Arrowheads indicate the beginning and end of dimension lines. The point of the arrowhead touches the extension line. In the case of a leader, either an arrowhead or a dot terminates at a specific surface or feature. Arrowheads are of uniform size throughout a drawing and may be drawn open (⟶▷) or closed (⟶▶). Arrowhead sizes vary with the size of a drawing.

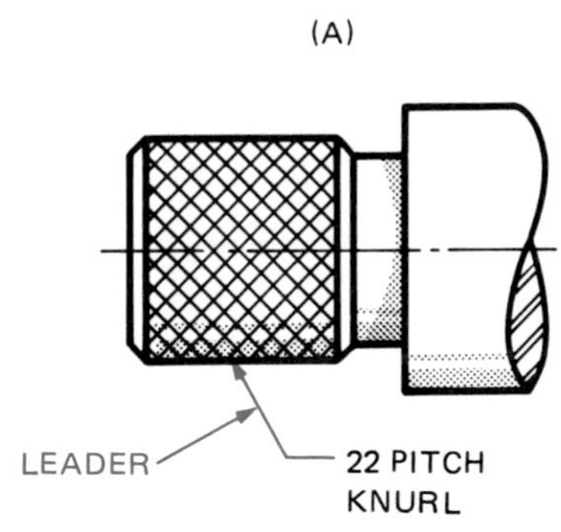

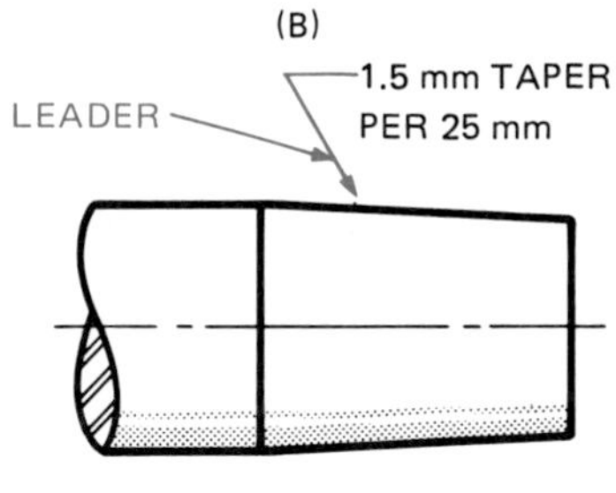

LEADERS PROVIDE PROCESS AND DIMENSIONAL INFORMATION

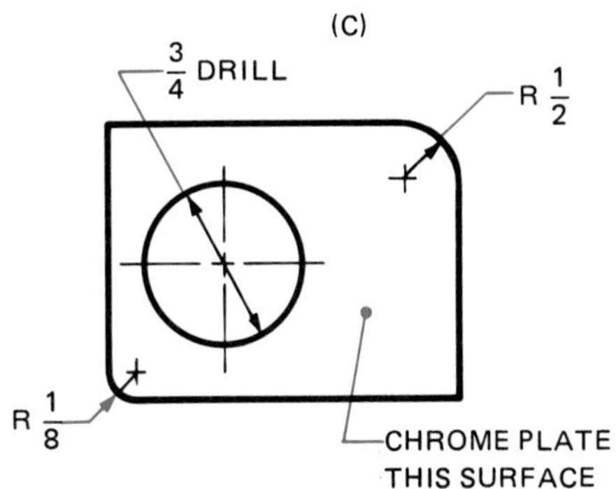

LEADERS USED TO DIMENSION CIRCLE AND ARCS AND TO INDICATE SURFACE TREATMENT

FIGURE 5-5 GENERAL APPLICATIONS OF LEADERS

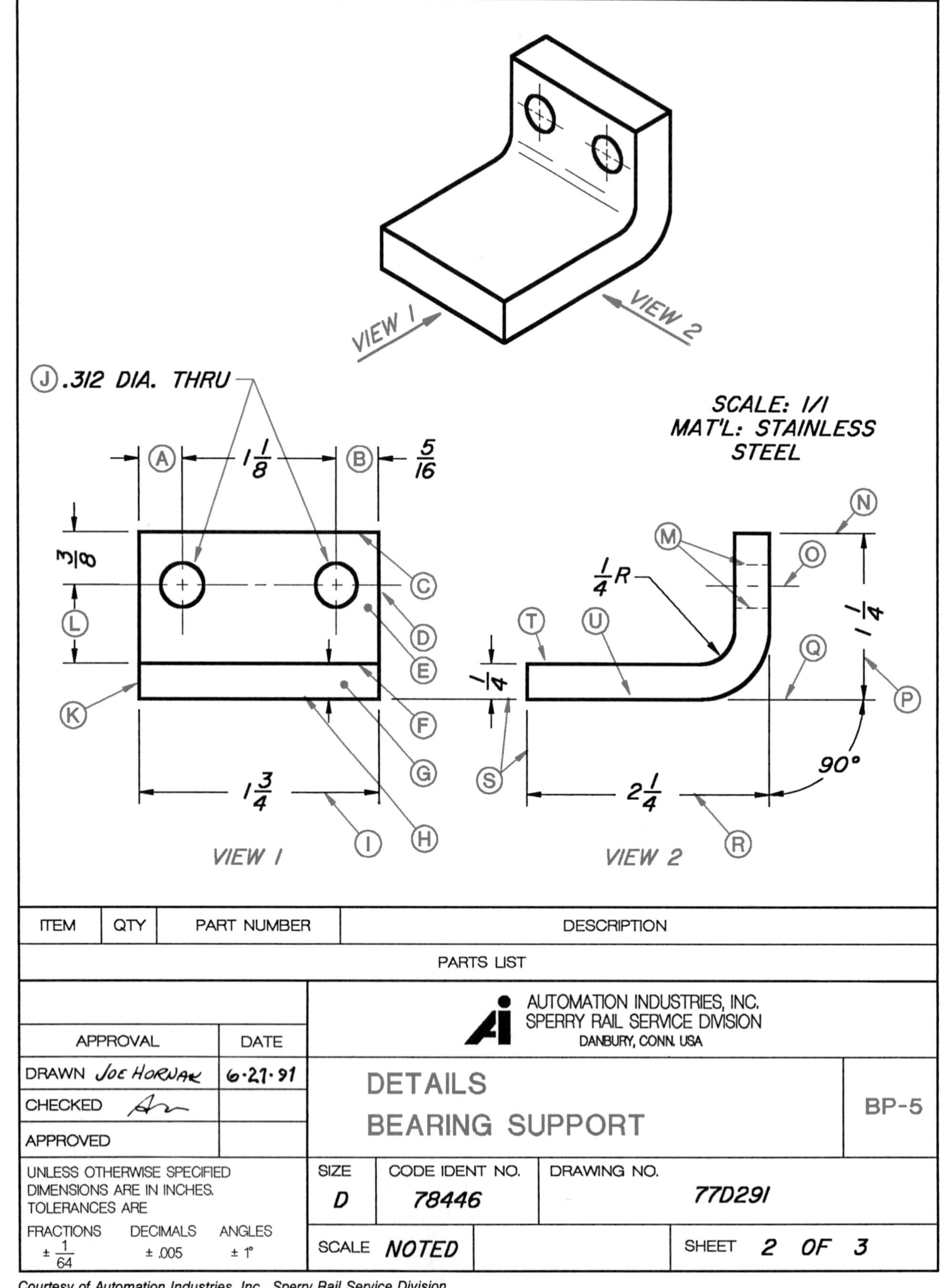

Courtesy of Automation Industries, Inc., Sperry Rail Service Division

ASSIGNMENT—UNIT 5

BEARING SUPPORT (BP-5)

1. State the name of the part.
2. Identify (a) the drawing number and (b) the code identification number.
3. Indicate (a) the kind of material to be used to make the part, and (b) the scale to which the part is drawn.
4. Give the letter which identifies the hidden lines.
5. Identify by letter each visible (object) line which describes the Bearing Support in **VIEW 1**.
6. Give the letter of each lettered extension line in **VIEW 2**.
7. List the letter of each lettered dimension line used in (a) **VIEW 1** and (b) **VIEW 2**.
8. State the function of line Ⓞ
9. Read and record the bend angle of the Bearing Support.
10. Give the inside radius of the bend.
11. Compute the following dimensions:
    a. Ⓐ
    b. Ⓛ
    c. Thickness (height) Ⓖ
12. Give the center-to-center distance of the holes.
13. State what information is provided by the leader in relation to the two holes.
14. a. Refer to the drawing and name each line or reference symbol identified by the letters Ⓒ, Ⓜ, Ⓢ, Ⓘ, Ⓔ, and Ⓞ.
    b. State the characteristics of each line.

Student's Name ______________________

1. ______________________
2. a. ______________________
   b. ______________________
3. a. ______________________
   b. Scale ______________________
4. ______________________
5. Object Lines ______________________
6. Extension Lines ______________________
7. a. **VIEW 1** ______________________
   b. **VIEW 2** ______________________
8. ______________________
   ____
9. ______________________
10. ______________________
11. a. ______________________
    b. ______________________
    c. ______________________
12. ______________________
13. ______________________
14.

Line	(a) Name	(b) Characteristics
Ⓒ		
Ⓜ		
Ⓢ		
Ⓘ		
Ⓔ		
Ⓞ		

SECTION 3 Basic Principles of Projection: Views

UNIT 6 Orthographic Projection: Multiview Drawings

Technical drawings provide a universal system for accurately visualizing and describing features, shapes, and sizes of objects. While such information may be communicated by a photograph or pictorial drawing, it is the technical drawing which gives the most accurate and clearest description of the object.

Technical drawings are prepared according to the standards of *orthographic projection. Ortho-* means at *right angles* and *straight. Graphic* relates to representation by *writing* or *drawing.* Combined, orthographic projection refers to a standardized method of representing an object with details by drawing straight lines perpendicular from the object to two or more planes. Since more than one view is projected by orthographic projection, the drawings produced are called *multiview drawings.*

FUNCTION OF EACH SEPARATE VIEW

In orthographic projection, a series of separate views of an object is arranged so that each view is related to each other view in a specific way. Simply stated, a *view* is a drawing of what a person sees when looking perpendicularly at a face of an object. A view (drawing of one face) produced by orthographic projection shows the exact shape of that face of an object.

PRINCIPLES OF PROJECTION: MULTIVIEW DRAWINGS

Learning to read a technical drawing is then a process of recognizing and interpreting accepted principles of orthographic projection. These principles are used by the designers, drafters, and engineers who select, project, and arrange views, and other skilled workers who must construct, test, and maintain a part or unit.

Basic Elements of Orthographic Projection

There are four basic elements of orthographic projection, as follows:

- The line of sight of the observer.
- The object.
- The plane of projection.
- The visual rays of sight or the projection lines.

These four elements are shown graphically in figure 6-1.

- The object (I) is imagined to be at an infinite distance from the observer (II).

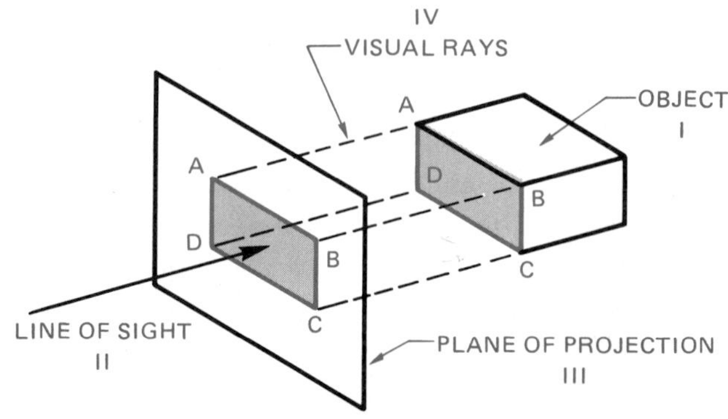

FIGURE 6-1 BASIC ELEMENTS OF ORTHOGRAPHIC PROJECTION

- The *plane of projection* (III) is perpendicular to the line of sight (II).
- The *visual rays* (IV) are parallel.

When the visual rays are also perpendicular to the plane of projection, a view is produced by orthographic projection.

THE FRONT VIEW

The most important view of an object is called the *front view.* Several criteria are considered in selecting the front view, including the following:

- The normal position of the object.
- The view which most clearly shows the unique shape of the object.
- The view which furnishes the best information for constructing the part.

To make a technical drawing, the draftsperson goes through the imaginary process of raising the object to eye level and viewing the front view at a right angle to the line of sight.

An imaginary plane (referred to in figure 6-2 as the *frontal plane*) is placed between the observer and the object. The visual rays are projected at right angles from the object to the frontal plane. The intersecting points are then connected to form the shape of the object. The observer is assumed to be at an *infinite distance* from the object and the visual rays which strike the frontal plane are parallel.

Figure 6-2 illustrates the process of generating the front view. The front view, however, does not show the shape or distance from the front to the rear of the object. Therefore, at least one more view is needed to describe the object.

THE TOP VIEW

The draftsperson goes through a similar procedure to produce a *top view.* Another imaginary transparent plane is placed horizontally above the object, as in figure 6-3. Visual rays are projected by extending perpendicular lines vertically from the object to the *horizontal plane.* The top view of the object is produced by connecting the intersections of these rays on the horizontal plane.

The top view gives the exact shape of the object as viewed by looking squarely down upon it. This view gives distances from the front to the back of the object. Note, also, that the frontal and horizontal planes are perpendicular to each other. On an actual drawing, the horizontal and frontal planes are rotated into the same plane, as shown

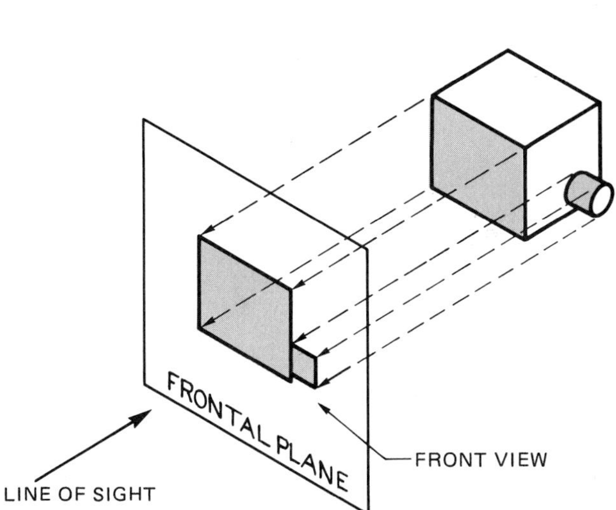

FIGURE 6-2 FRONT VIEW PROJECTION OF OBJECT

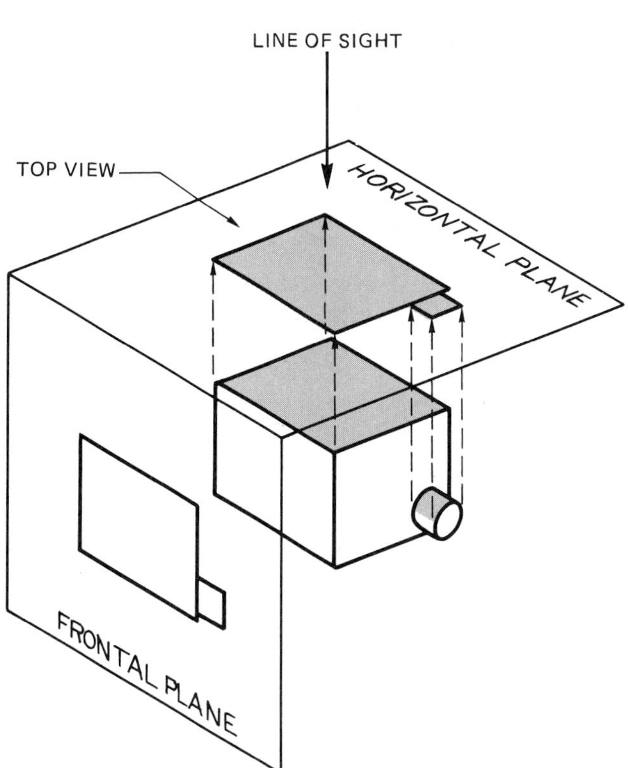

FIGURE 6-3 TOP VIEW PROJECTION OF OBJECT

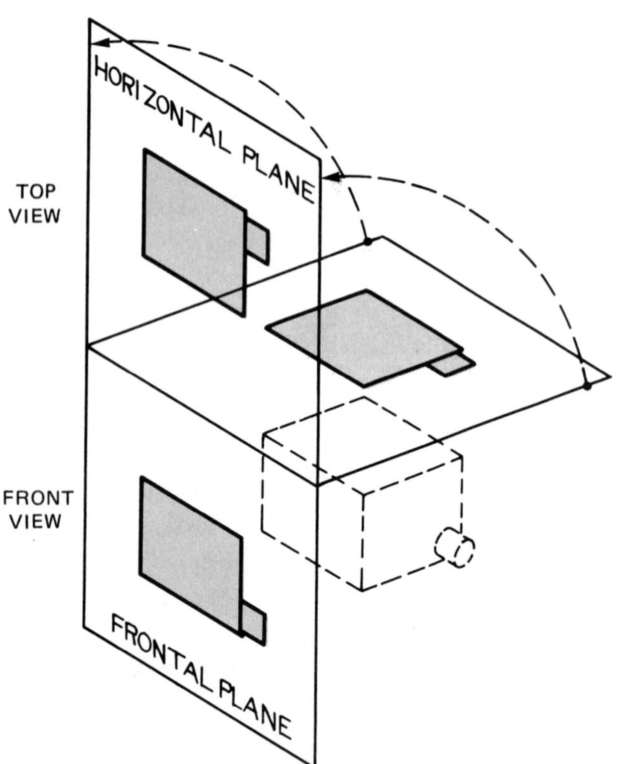

FIGURE 6-4 RELATIONSHIP OF HORIZONTAL AND FRONTAL PLANES

in figure 6-4. The top view is placed directly above the front view.

THE SIDE VIEW

A third imaginary parallel plane of projection is frequently drawn. This plane, which is perpendicular to the frontal and horizontal planes, is called the *profile plane* (figure 6-5). The *side view* of the object is produced on this plane in the same manner as the front and top views. The *profile plane view* (figure 6-5) is needed to show that one section of the object is circular in shape.

FRONT, TOP, AND RIGHT-SIDE VIEWS

By rotating the profile plane (figure 6-6) in the same manner as was done to the horizontal plane, the draftsperson now is able to draw all three views (front, top, and right side) of the object on a single-plane sheet. Figure 6-7 shows the relationship of the front, top, and right-side views as found on a blueprint.

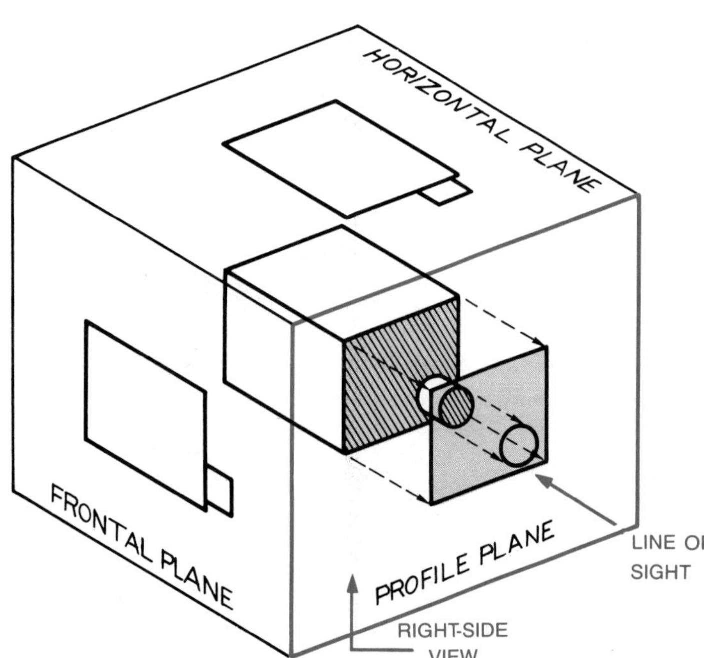

FIGURE 6-5 SIDE VIEW PROJECTION OF OBJECT

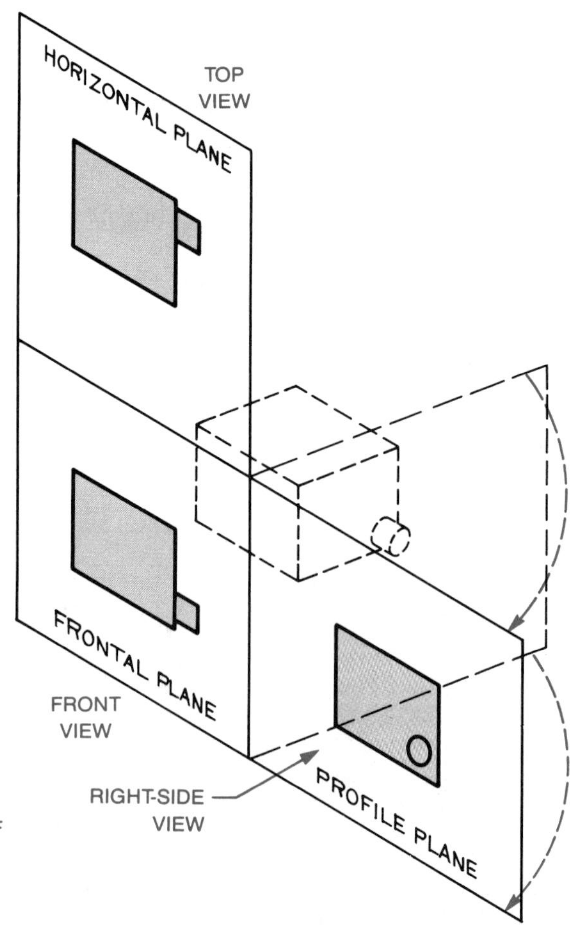

FIGURE 6-6 RELATIONSHIP OF VIEWS ON FRONTAL, HORIZONTAL, AND PROFILE PLANES

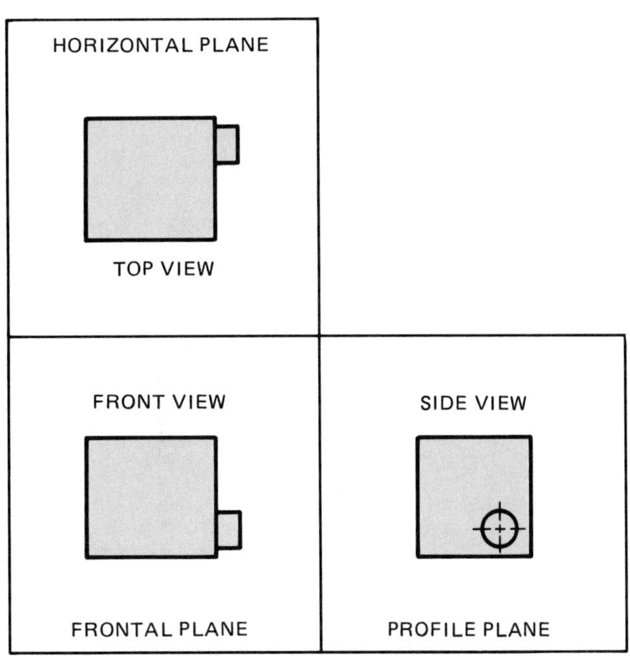

FIGURE 6-7 POSITION OF FRONT VIEW, TOP VIEW, AND SIDE VIEW IN THE SAME PLANE

WORKING DRAWINGS

In actuality, a drawing of a part is constructed without imaginary transparent planes (figure 6-8). The views are drawn as they would be generated if all projection planes were rotated as shown. The exact shape and size of the object are defined then by a series of views which have an established relationship to one another.

A multiview orthographic drawing required for the design, manufacture, construction, or assembly of a machine or structure is called a *working drawing* when it is completely dimensioned and has necessary notes. A *three-view working drawing*, like the one illustrated in figure 6-9, showing the front, top, and right-side views is the most commonly used working drawing for shop and laboratory work.

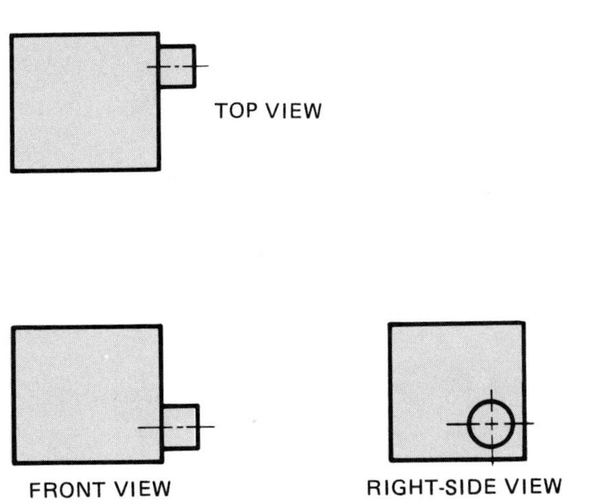

FIGURE 6-8 FRONT, TOP, AND RIGHT-SIDE VIEWS WITH IMAGINARY PLANES REMOVED

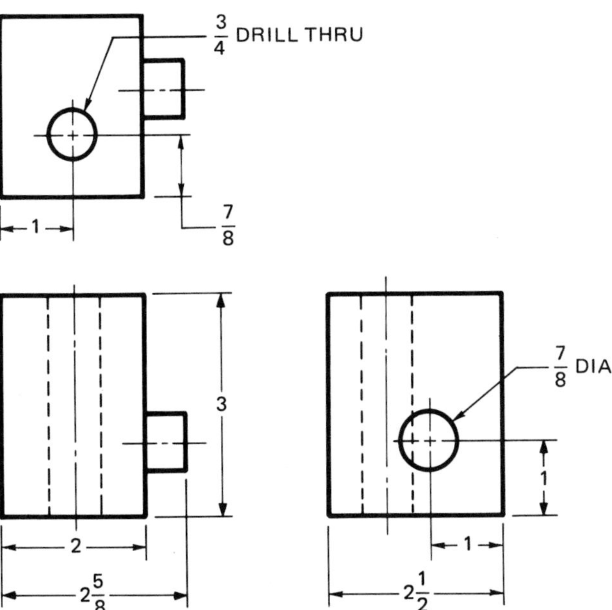

FIGURE 6-9 EXAMPLE OF A WORKING DRAWING

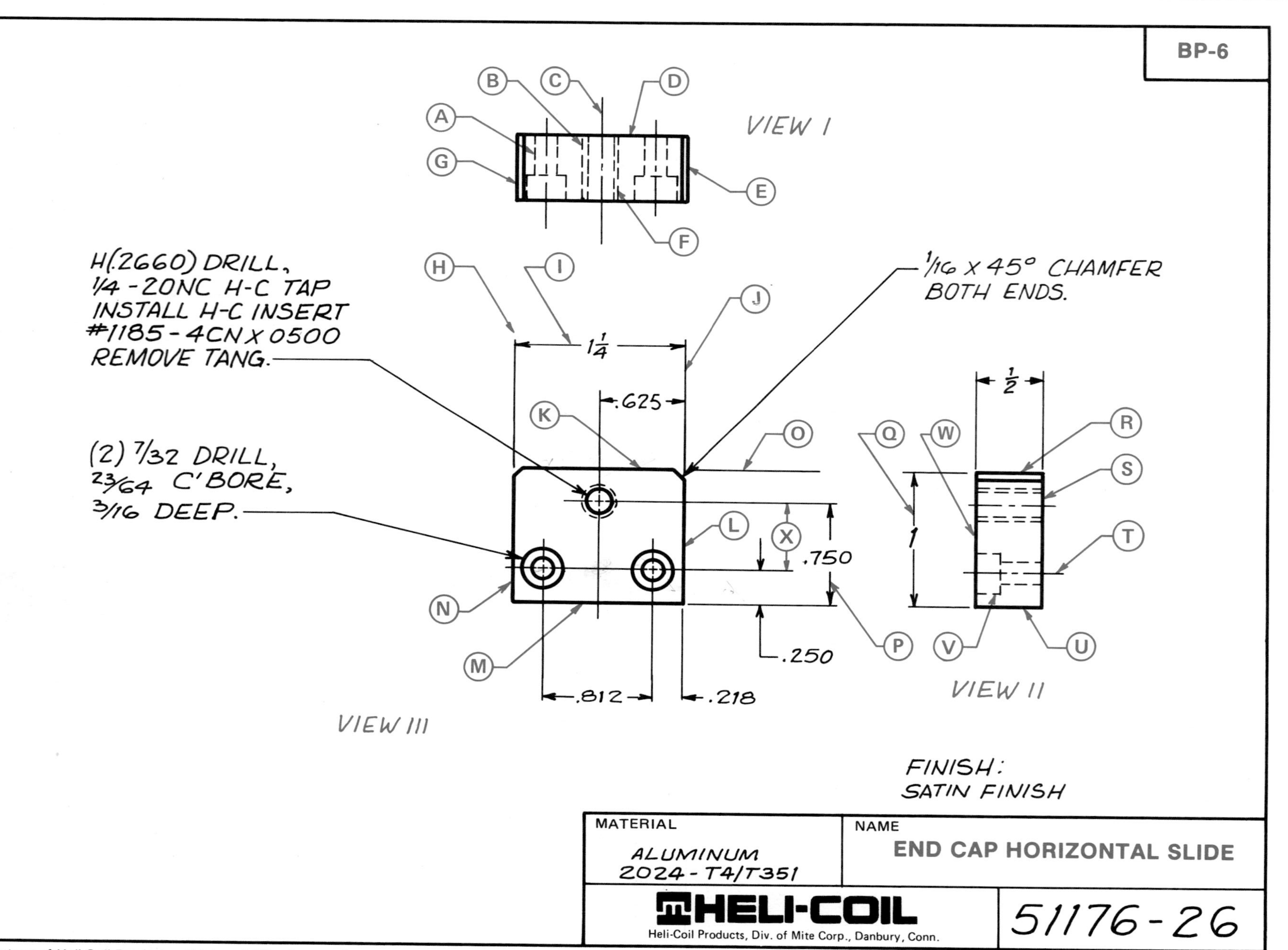

Courtesy of Heli-Coil Products, division of Mite Corporation
(Modified Industrial Blueprint)

ASSIGNMENT—UNIT 6

END CAP HORIZONTAL SLIDE (BP-6)

Student's Name ________________

1. State the name of the part.
2. Indicate the kind of material of which the part is made.
3. Specify the finish of the part.
4. Indicate how many leaders are used for notes on the print.
5. State three imaginary elements which are considered in visualizing an orthographic projection drawing.
6. Explain briefly the difference between a working drawing and an orthographic drawing.
7. Describe four steps the draftsperson uses in making a three-view orthographic drawing.
8. List the letters which identify visible lines in (a) the front view and (b) the side view.
9. Identify the letter(s) of the dimension line(s) in (a) the front view and (b) side view.
10. Give the letters of the hidden lines in View I which represent the 1/4–20 NC tapped hole.
11. Indicate the depth of the counterbored (**C'BORED**) holes in the top view.
12. Compute distance (X).
13. Calculate the center line distance between the centers of the counterbored holes to surface (K).

1. ________________
2. ________________
3. ________________
4. Leaders ________________
5. a. ________________
   b. ________________
   c. ________________
6. ________________
   ____
   ____
   ____
   ____

7. Step 1 ________________
   ____
   ____
   Step 2 ________________
   ____
   ____
   Step 3 ________________
   ____
   ____
   Step 4 ________________
   ____
   ____

8. a. Front View ________________
   b. Side View ________________
9. a. Front View ________________
   b. Side View ________________
10. Hidden Lines ________________
11. Depth ________________
12. (X) = ________________
13. Center to (K) Surface = ________________

UNIT 7
Arrangement of Views: Placement of Dimensions

The selection of which views to include on a technical drawing is largely determined by the complexity of the object and how clearly the drawing may be interpreted. While most drawings include one, two, or three views, complicated parts may require a fourth, fifth, or sixth view. In principle, only those views are drawn which present information that may not be easily shown in any other view.

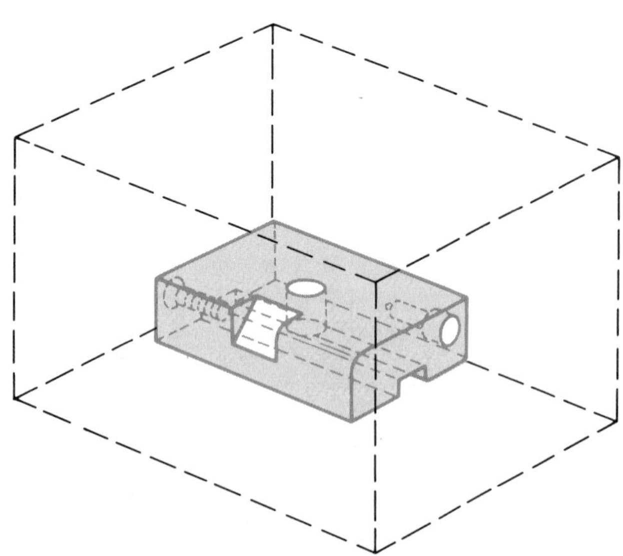

FIGURE 7-1 PICTORIAL DRAWING OF A PART WITHIN AN IMAGINARY PROJECTION BOX

ARRANGEMENT OF SIX PRINCIPAL VIEWS

Six principal views are covered in this unit. Auxiliary and other types of views follow in later units. To illustrate the six principal views, the Slide Block (figure 7-1) is placed in an imaginary, transparent projection box. The projection box consists of six sides, each of which may be considered as a *plane of projection*. Using orthographic projection, six different views may be projected, one on each of the six planes: frontal (2), horizontal (2), and profile (2).

By swinging each projection plane onto a single plane as shown in figure 7-2, six possible views are displayed. Each view is identified by name and position. The six common views, with the abbreviations sometimes used, follow:

- *Front View (F. V.)*
- *Right-Side View (R. V.)*
- *Left-Side View (L. V.)*
- *Top View (T. V.)*
- *Bottom View (Bot. V.)*
- *Back* or *Rear View (B. V.)*

This naming and placing of views provides a standard for uniformly interpreting technical drawings. The back or rear view (B.V.) may be placed to the left of the left-side view or to the right of the right-side view. Occasionally, the back view is placed above the top view.

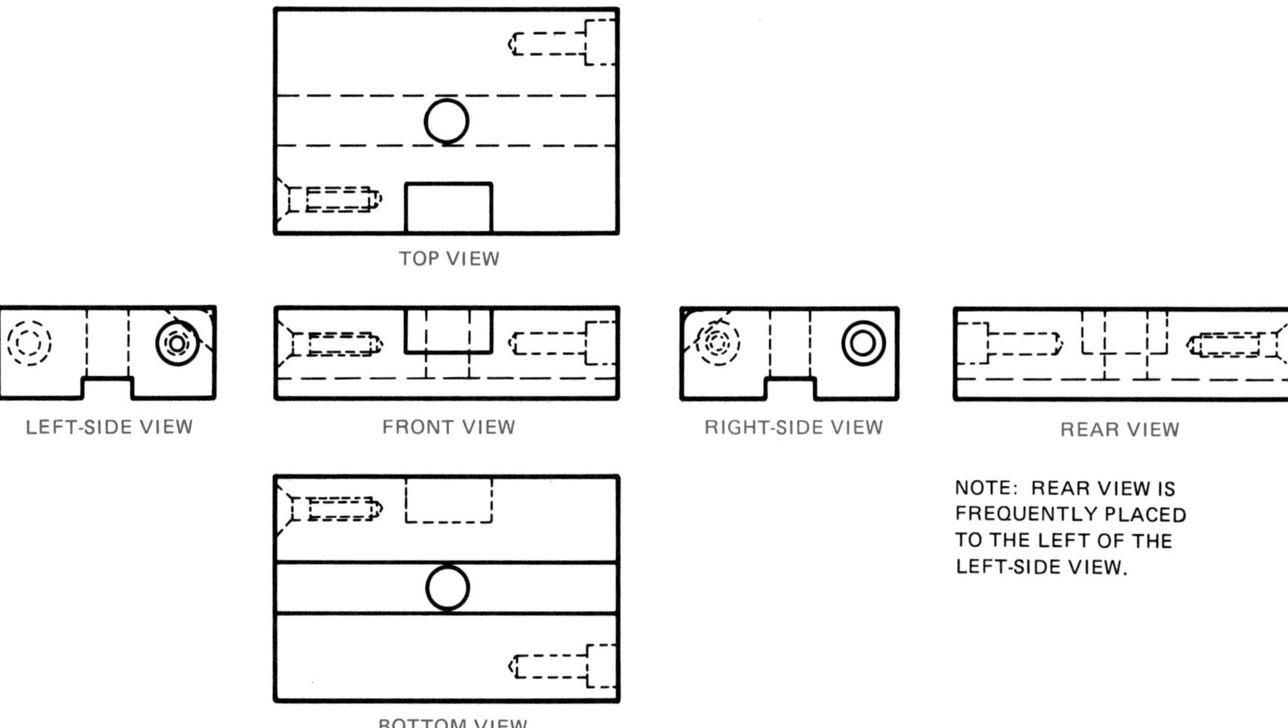

FIGURE 7-2 ARRANGEMENT OF SIX PRINCIPAL VIEWS

PLACEMENT OF DIMENSIONS

The three principal dimensions for linear measurements are *width (W)*, *height (H)*, and *depth (D)*. Only two of these dimensions are included on any one view. The third dimension is found on an adjacent view.

The American National Standards Institute (ANSI) designation of width, height, and depth is illustrated in figure 7-3. *Width* (figure 7-3A) is defined as the perpendicular distance of a line or surface measured between a pair of profile planes. In the shop, the term *length* is often used for this dimension. *Height* (figure 7-3B) and *depth* (figure 7-3C) represent the perpendicular distance of a line or surface as measured between a pair of horizontal planes and frontal planes, respectively. Generally, the terms *thickness* and *width* are used interchangeably.

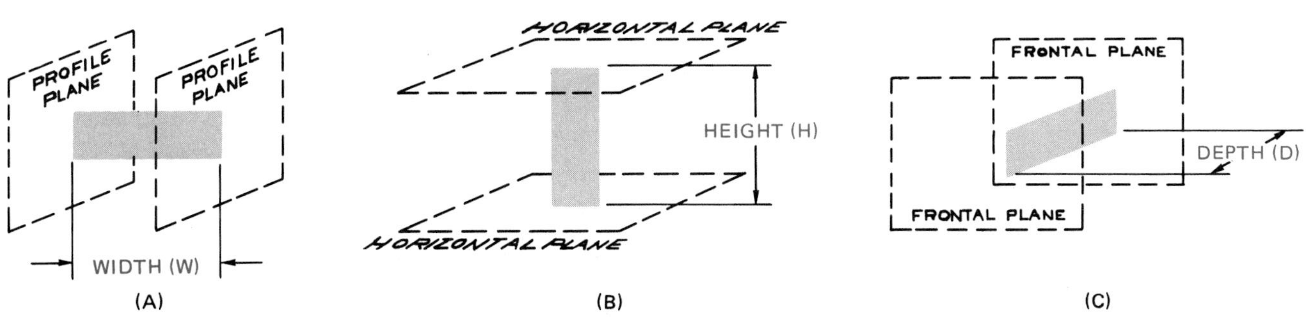

FIGURE 7-3 ANSI DESIGNATION OF WIDTH, HEIGHT, AND DEPTH DIMENSIONS

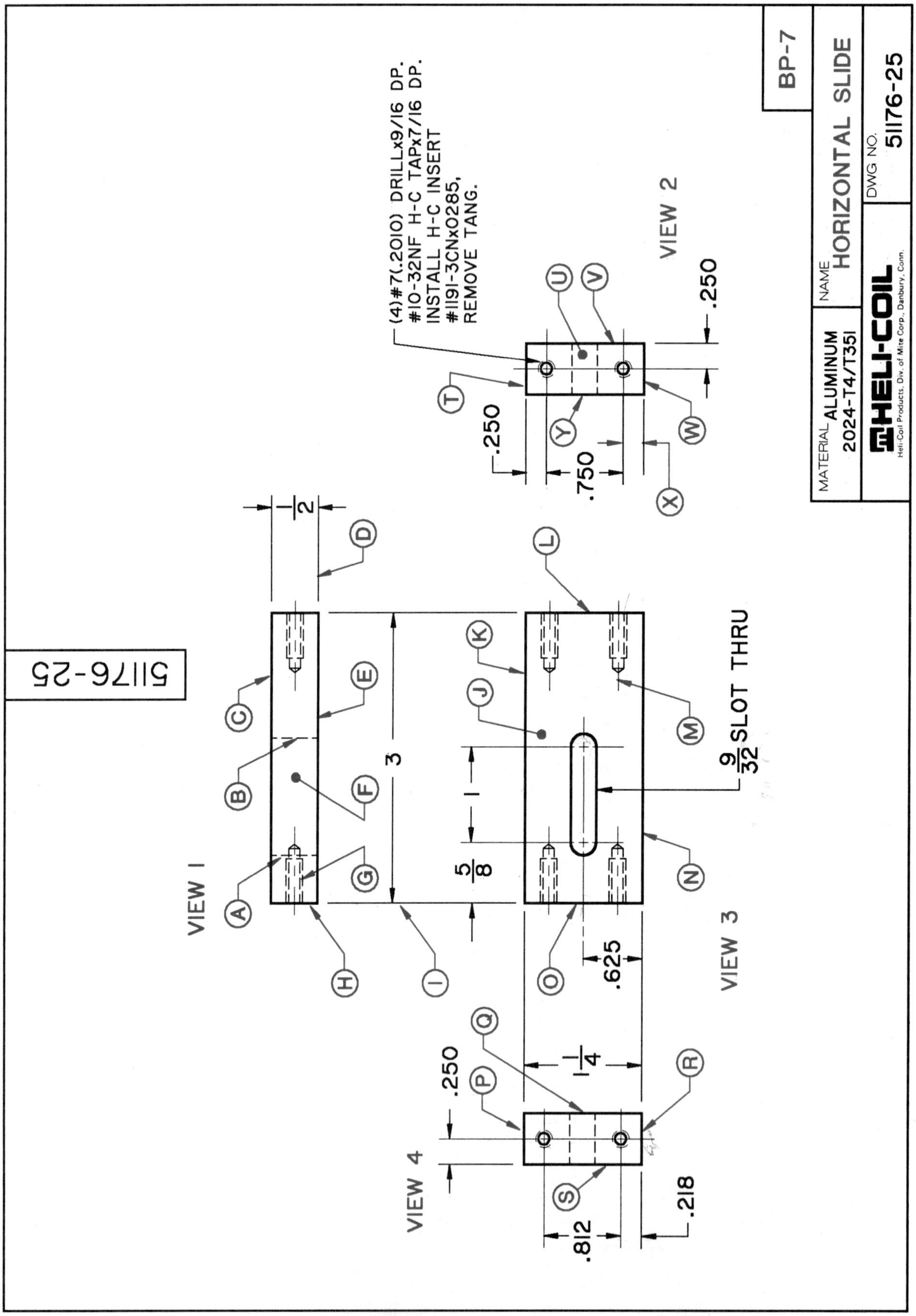

Courtesy of Heli-Coil Products, division of Mite Corporation
(Modified Industrial Blueprint)

ASSIGNMENT—UNIT 7

HORIZONTAL SLIDE (BP-7)

Student's Name ______________________

1. Identify the standard method of projection used to make the Horizontal Slide drawing.
2. Give the name of this type of four-view drawing.
3. Indicate the kind of material of which the Horizontal Slide is made.
4. Specify the part drawing number.
5. State the ANSI designation (name) for
   a. **VIEW 1** c. **VIEW 3**
   b. **VIEW 2** d. **VIEW 4**
6. Tell briefly why four views are used on this drawing instead of two or three views.
7. Identify face (J) in **VIEW 3** in each of the following views:
   a. **VIEW 1** b. **VIEW 2** c. **VIEW 4**
8. Identify end (H) in **VIEW 1** in the Front View.
9. Locate surface (N) in **VIEW 3** in each of the following views:
   a. Top View c. Left-Side View
   b. Right-Side View
10. Give the letters of the following surfaces:
    a. (P) (**VIEW 4**) in the R.V.
    b. (S) (**VIEW 4**) in the T.V.
11. Determine the letters of the hidden lines in the Top View which represent the ends of the elongated slot.
12. Identify the letter in the Top View which relates to the four threaded holes.
13. State what information is supplied by the leader for the following machining or assembling processes:
    a. Drilling c. Insert Assembly
    b. Tapping
14. Compute the distance from the end of the elongated slot to surface (L).
15. Determine the distance from the top of the elongated slot to surface (K).
16. Give dimension (X).
17. Determine (a) dimensions 1, 2, 3, and 4 and (b) name one view in which each dimension appears on the drawing.
    1. Width of elongated slot.
    2. Height between center lines of the #7 drilled holes in **VIEW 2**.
    3. Depth (thickness) of the part.
    4. Height of the Horizontal Slide.

1. ______________________
2. ______________________
3. ______________________
4. ______________________
5. a. **VIEW 1** ______________________
   b. **VIEW 2** ______________________
   c. **VIEW 3** ______________________
   d. **VIEW 4** ______________________
6. ______________________
   ____
   ____
7. a. **VIEW 1** ______________________
   b. **VIEW 2** ______________________
   c. **VIEW 4** ______________________

8. Front View ______________________
9. a. Top View ______________________
   b. Right-Side View ______________________
   c. Left-Side View ______________________
10. a. R.V. ______________________
    b. T.V. ______________________
11. ______________________
12. Top View ______________________
13. a. Drilling ______________________
    ____
    b. Tapping ______________________
    ____
    c. Assembly ______________________
    ____

____
14. ______________________
15. ______________________
16. (X) = ______________________

17.

	Dimensions (a)	Example of View (b)
1. (W) =	________	________
2. (H) =	________	________
3. (D) =	________	________
4. (H) =	________	________

UNIT 8
Two-view Drawings

Although it has been stated that an object may be represented on a technical drawing by as many as six possible views, rarely are six views required. Often, several views duplicate one another. The guiding principle used by the part designer or detailer is to select the combination of views that will show the shape and details of the object clearly and economically. Generally, objects are represented by drawing three views: front, top, and right- or left-side views.

SYMMETRICAL PARTS AND TWO-VIEW DRAWINGS

In a number of cases, essential information about the contour, features, and dimensions of a part are contained within two views. Such drawings are referred to as *two-view drawings.*

Square, rectangular, circular, and curved symmetrical forms may be adequately described in two views. Symmetrical (**SYMM**) means that features on both sides of a center line (℄) are of equal form, dimensional size, and surface quality. Two-view drawings are particularly adapted to such symmetrical forms as shafts, gears, splines, arbors, linkages, handwheels, and handles. Figure 8-1 provides examples of two-view drawings of symmetrical parts. In addition, features such as drilled counterbored, reamed, or tapped holes may be represented by two-view drawings.

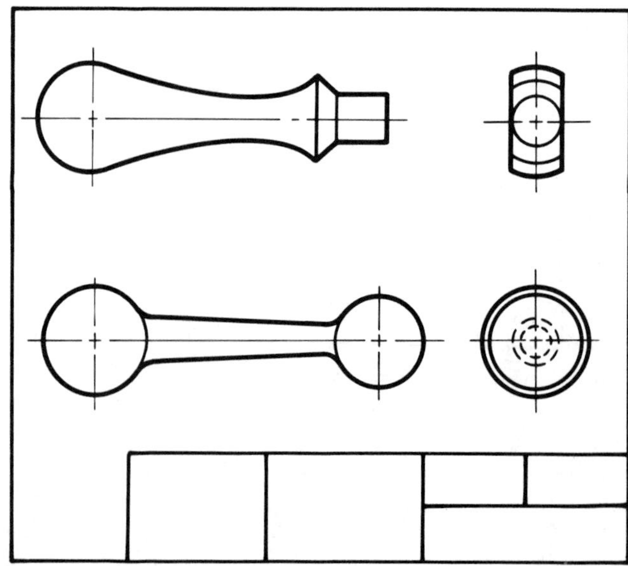

FIGURE 8-1 REPRESENTATION OF SYMMETRICAL PARTS BY TWO-VIEW DRAWINGS

SELECTION OF VIEWS ON TWO-VIEW DRAWINGS

Following the principles of projection, two-view drawings are prepared when the addition of any other view merely duplicates details which are already represented on a drawing (figure 8-2). Note that the top view is identical to the front view. In practice, this top view would be omitted because it provides no additional information.

Views which are commonly displayed in two-view technical drawings are shown in figure 8-3. The views at (A) represent the object as seen from the front and top sides. The center line in the front view and the two center lines in the top view indicate that all surfaces are symmetrical. The front and top views are the *preferred views* when the object is positioned vertically.

The part may also be drawn as shown at (B). Center lines are used in the front and right-side views. Generally, a front view and a right-side view are the *preferred views* for an object which is viewed in the horizontal position as at (B).

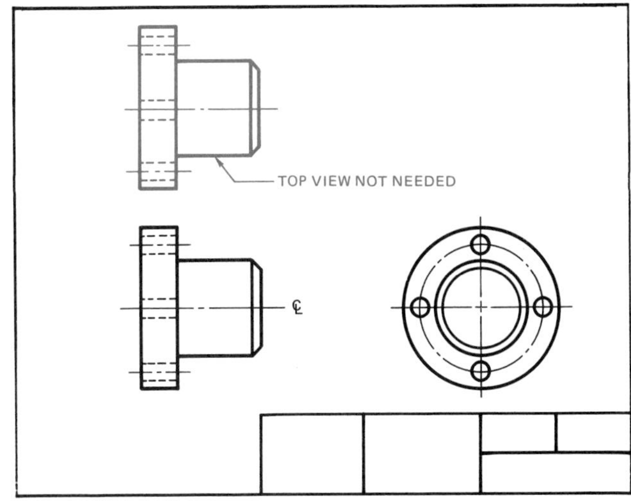

FIGURE 8-2 ESSENTIAL INFORMATION ABOUT A SYMMETRICAL PART PROVIDED ON A TWO-VIEW DRAWING

HIDDEN FEATURES

Hidden lines are used to represent invisible, cut-away, or internal features, as described in an earlier unit. The four holes in the drawing of the Flange (figure 8-2) are represented by three sets of hidden (invisible-edge) lines in the front view. The two views (front and right-side view) in this figure clearly identify the shape of the object and the details of the holes in the most economical manner. Since the top view duplicates the front view, only two views are needed.

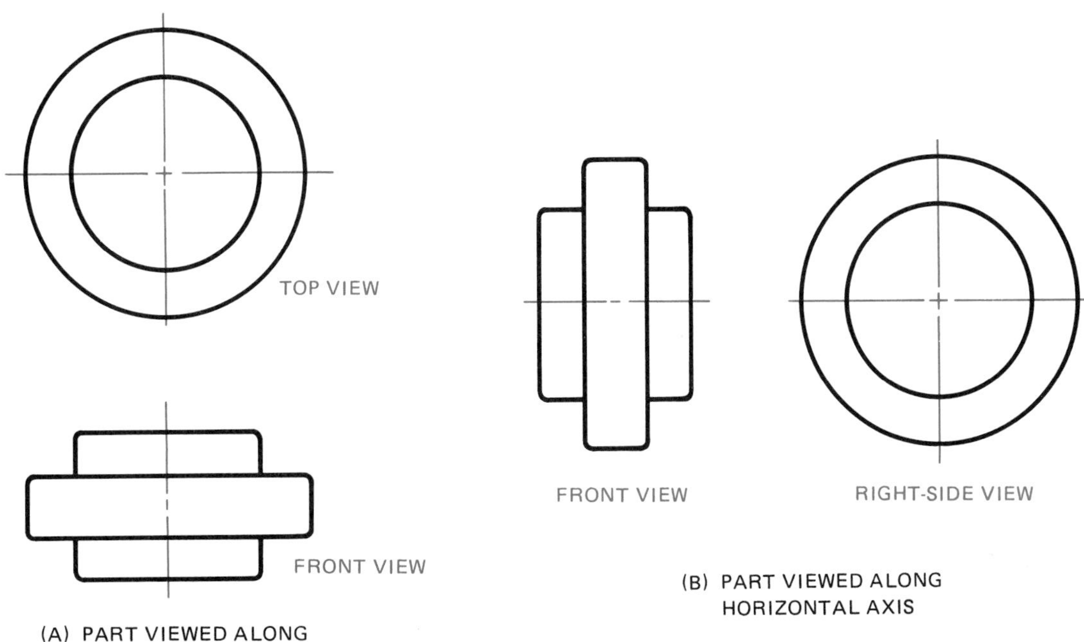

FIGURE 8-3 PREFERRED VIEWS FOR TWO-VIEW DRAWINGS OF A SYMMETRICAL OBJECT

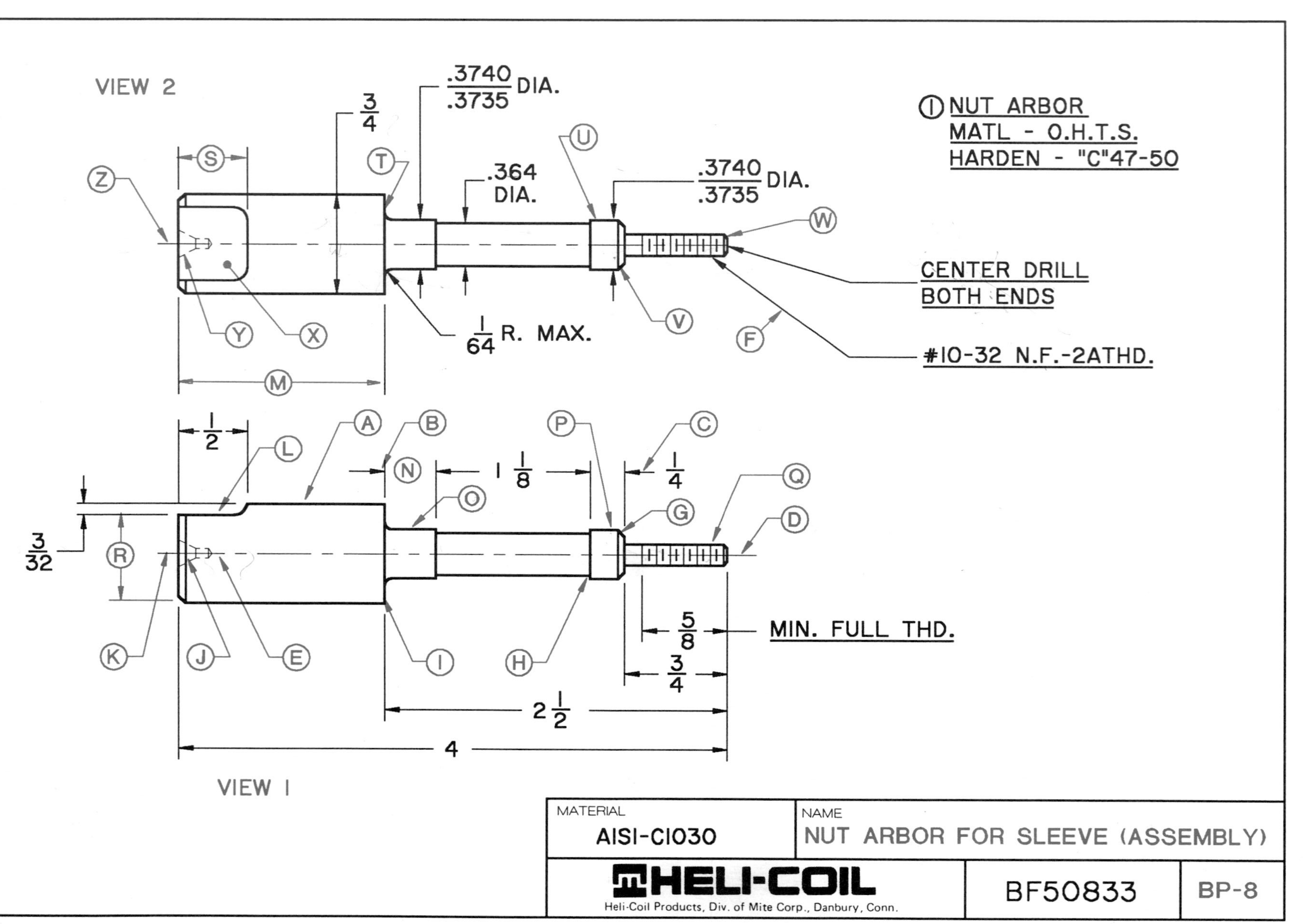

Courtesy of Heli-Coil Products, division of Mite Corporation
(Adaptation of Industrial Drawing)

ASSIGNMENT—UNIT 8

NUT ARBOR (BP-8)

Student's Name ____________________

1. Tell why a two-view drawing is used to represent the Nut Arbor.
2. Explain what the symbol (**SYMM**) means when it is used with a center line.
3. Provide the following information:
   a. Part name.
   b. Drawing number.
   c. Material used for the part.
   d. Hardness requirement.
4. Name
   a. **VIEW 1**
   b. **VIEW 2**
5. Explain briefly the purpose served by the center lines.
6. Name the type of line used on the drawing to represent each of the following circled letter features:
   Ⓐ Ⓒ Ⓔ
   Ⓑ Ⓓ Ⓕ
7. Identify the following part features in **VIEW 1** with the features in the top view which correspond.
   Ⓖ Ⓘ Ⓚ
   Ⓗ Ⓙ Ⓛ
8. Determine the following dimensions or specifications:
   Ⓜ Ⓞ Ⓠ
   Ⓝ Ⓟ Ⓡ
9. Give the full thread length.
10. State the shoulder radius.
11. Give the overall length of the Nut Arbor.
12. Determine (a) the length and (b) the diameter of the relieved area between surfaces Ⓞ and Ⓟ.

1. ____________________
   ____
2. ____________________
   ____
   ____
   ____
3. a. ____________________
   b. ____________________
   c. ____________________
   d. ____________________

4. a. **VIEW 1** ____________________
   b. **VIEW 2** ____________________
5. ____________________
   ____
6. Ⓐ ____________________
   Ⓑ ____________________
   Ⓒ ____________________
   Ⓓ ____________________
   Ⓔ ____________________
   Ⓕ ____________________

7. Ⓖ = __________ Ⓙ = __________
   Ⓗ = __________ Ⓚ = __________
   Ⓘ = __________ Ⓛ = __________
8. Ⓜ = __________ Ⓟ = __________
   Ⓝ = __________ Ⓠ = __________
   Ⓞ = __________ Ⓡ = __________
9. ____________________
10. ____________________
11. ____________________
12. a. __________ b. __________

UNIT 9
One-view Drawings

One-view drawings are commonly used in industry. One-view drawings are supplemented by *notes, symbols, abbreviations*, and *other written information*. Such material is lettered on a drawing as briefly as possible. Where needed, attention is directed by the use of leaders to relate a note to a particular feature of a part.

One-view drawings are adequate to describe the shape of cylindrical, cone-shaped, rectangular, and other symmetrical parts. Thin, flat objects of uniform thickness are represented by one-view drawings. The one-view drawing at (A) in figure 9-1 represents a cylinder of a given length and diameter. The drawing at (B) represents a symmetrical flat part.

SYMBOLS, ABBREVIATIONS, AND NOTES ON ONE-VIEW DRAWINGS

In the two examples in figure 9-1, the use of the abbreviation **DIA** at (A), and the **Ø** and **R** symbols with the thickness note at (B), take the place of second views.

Figure 9-2 shows the use of the abbreviations **HEX** and **SQ** to indicate the hexagon-shaped head and the square form for the opposite end. The combination of the one view and the supplemental information indicates the millimeter sizes for the hexagon head, the body diameters and

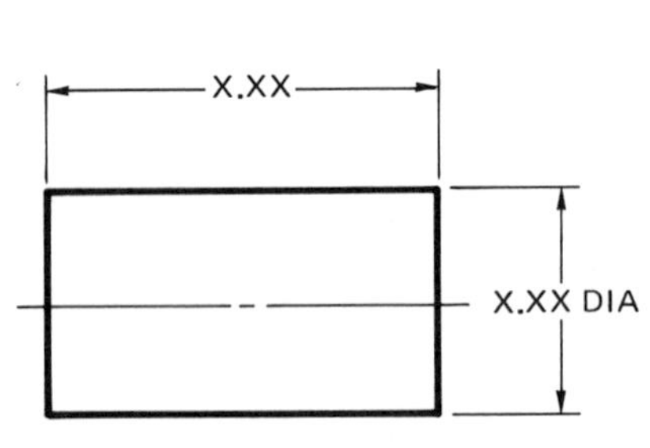

(A) REPRESENTATION OF A CYLINDRICAL PART OF A GIVEN DIAMETER AND LENGTH

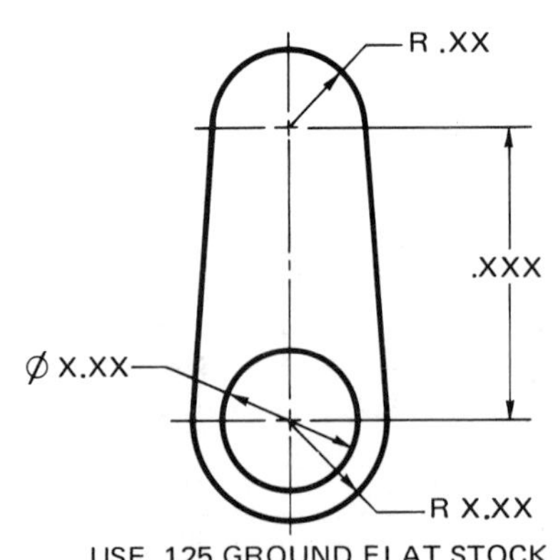

(B) REPRESENTATION OF A FLAT, SYMMETRICAL PART

FIGURE 9-1 REPRESENTATION OF SYMMETRICAL PARTS BY ONE-VIEW DRAWINGS

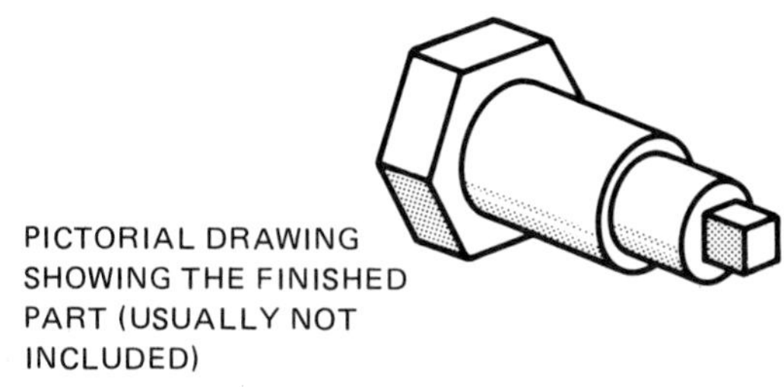

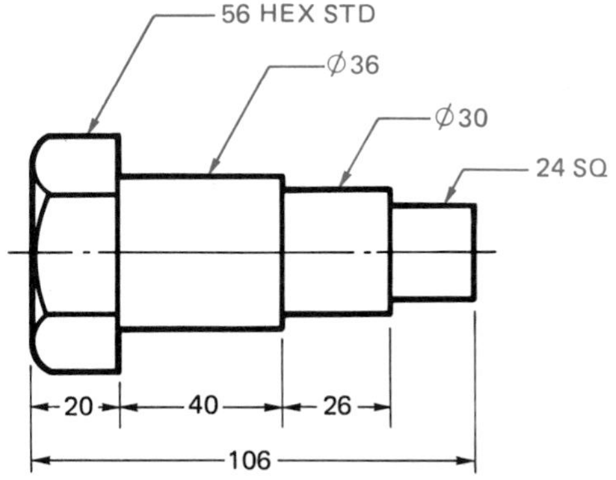

FIGURE 9-2 USE OF ABBREVIATIONS AND SYMBOLS ON A ONE-VIEW DRAWING

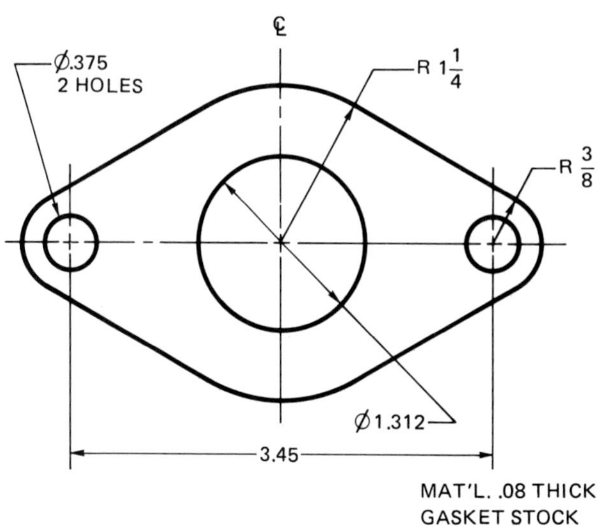

FIGURE 9-3 ONE-VIEW DRAWING OF A GASKET

lengths, the dimensions of the square end, and the overall length. A pictorial drawing, which normally would not be included, shows what the part looks like.

This technique of providing what would be third dimensions (on a second or third view) in the form of supplemental notes on a one-view drawing, is widely used for drawings of gaskets, shims, and other thin, flat plates. Figure 9-3 is a typical one-view drawing of a gasket. Since the third dimension such as thickness (height) can be given as a note, gaskets, shims, and the like, may be represented on a drawing by one view.

Another example of a one-view drawing (where symbols and machining notes are used to supplement the dimensions) is shown in figure 9-4.

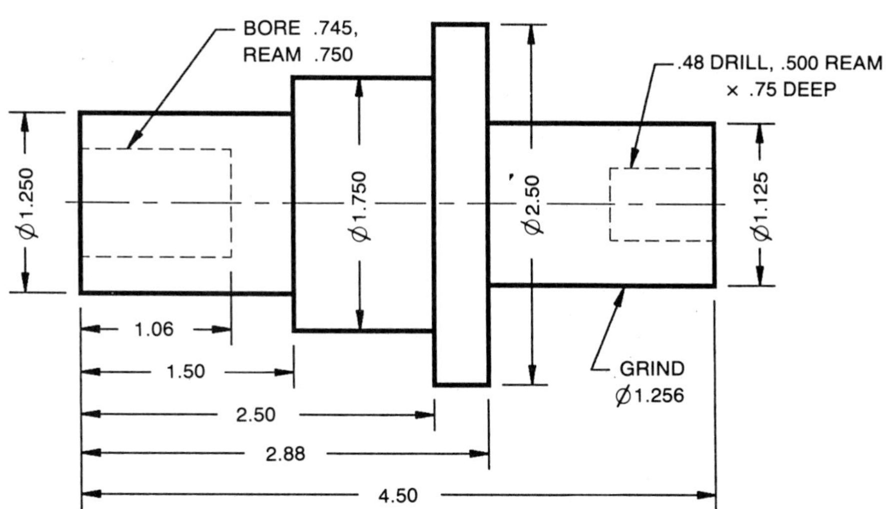

FIGURE 9-4 SYMBOLS AND MACHINING NOTES ON A ONE-VIEW DRAWING OF A CYLINDRICAL PART

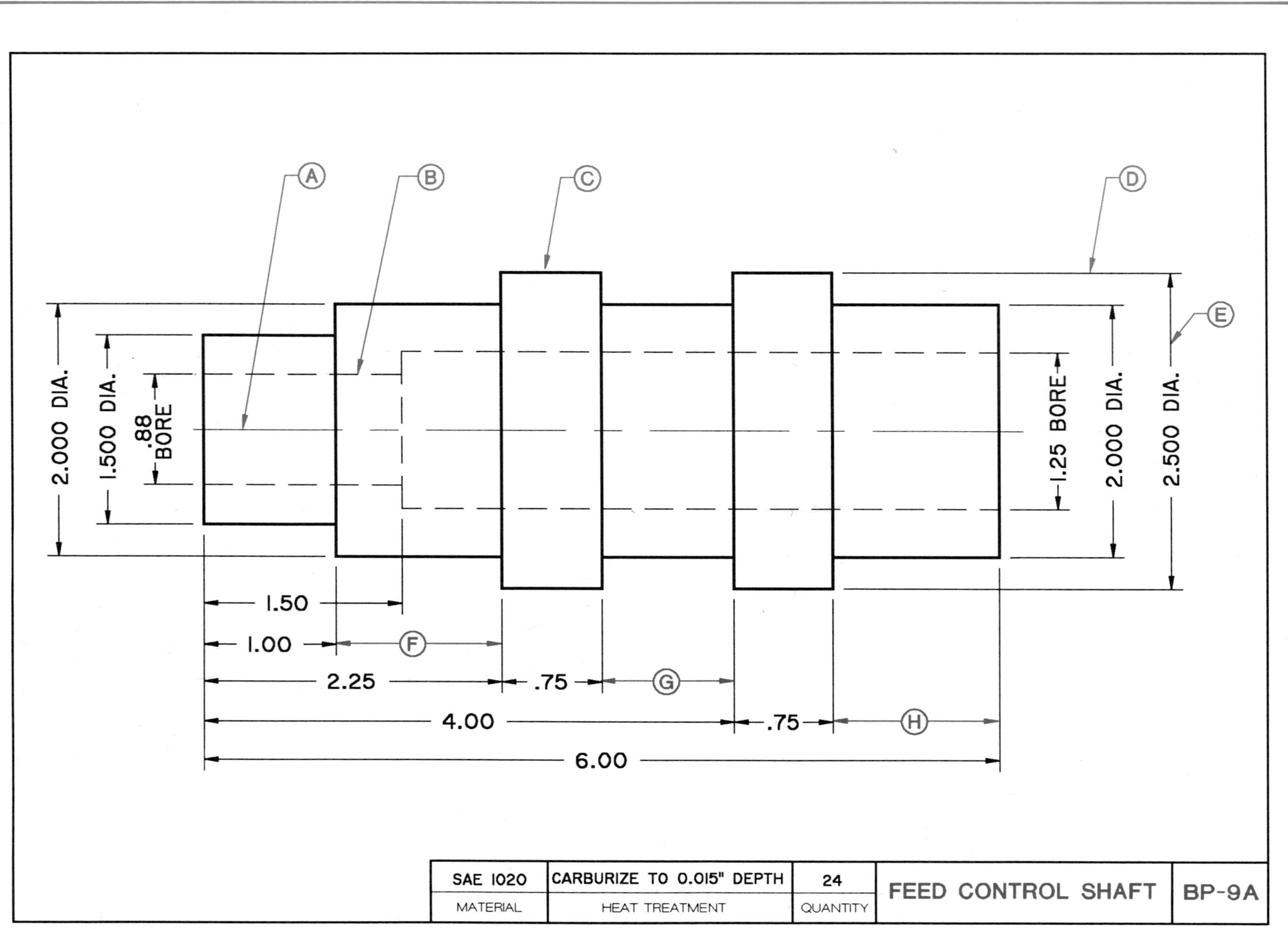
A
B
C
D
E
2.000 DIA.
1.500 DIA.
.88 BORE
1.25 BORE
2.000 DIA.
2.500 DIA.
1.50
1.00
F
2.25
.75
G
4.00
.75
H
6.00
SAE 1020
MATERIAL
CARBURIZE TO 0.015" DEPTH
HEAT TREATMENT
24
QUANTITY
FEED CONTROL SHAFT
BP-9A

ASSIGNMENT A—UNIT 9

FEED CONTROL SHAFT (BP-9A)

Student's Name ____________________

1. Name the type of drawing used to represent the Feed Control Shaft.

2. Give one advantage of using one view on this drawing.

3. a. Indicate the shape of the Feed Control Shaft.
   b. Tell how the shape is determined from reading the print.

4. Explain why all the features are symmetrical.

5. Name each of the following lines:
   (A)
   (B)
   (C)
   (D)
   (E)

6. Identify the following items:
   a. Material used for the part.
   b. Required heat treatment.
   c. Quantity to be machined.

7. Give (a) the largest outside diameter and (b) the smallest outside diameter.

8. State the diameters of the two bored holes.

9. Determine the thickness of the two **2.500″ DIA** sections.

10. Give the lengths of (a) the **.88″ DIA** and (b) the **1.25″ DIA** bored holes.

11. Compute distances (F) (G) (H)

1. ____________________

2. ____________________
____
____

3. a. ____________________
   b. ____________________
____
____

4. ____________________
____
____

5. (A) ____________________
(B) ____________________
(C) ____________________
(D) ____________________
(E) ____________________

6. a. Material ____________________
   b. Heat Treatment ____________________
   c. Quantity ____________________

7. a. Largest Outside Diameter = ______
   b Smallest Outside Diameter = ______

8. a. ____________ b. ____________

9. ____________________

10. a. Length of **.88″ DIA** Hole = ______
    b. Length of **1.25″ DIA** Hole = ______

11. (F) = ____________________
(G) = ____________________
(H) = ____________________

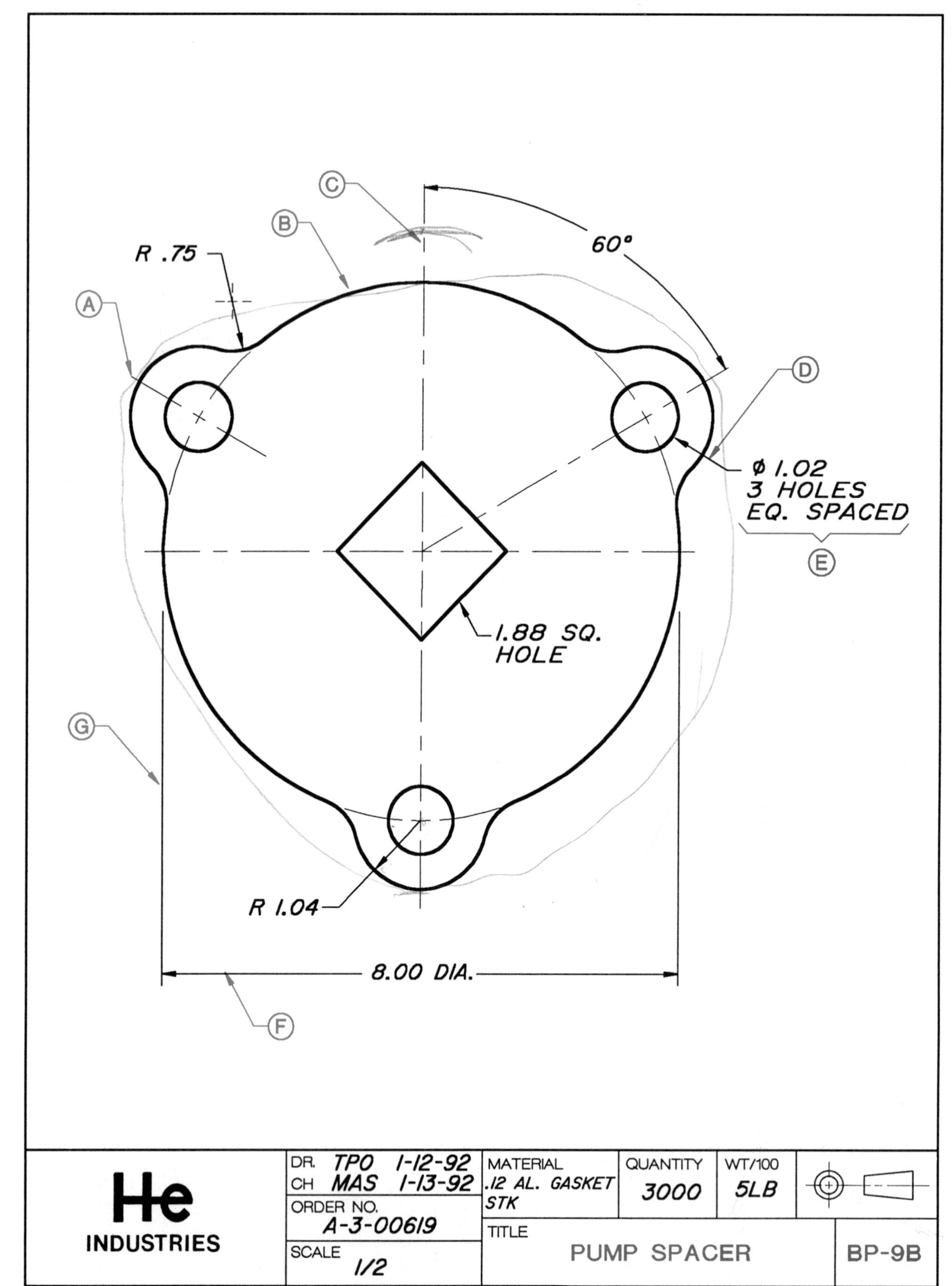

R .75
A
B
C
60°
D
Ø 1.02
3 HOLES
EQ. SPACED
E
1.88 SQ.
HOLE
G
R 1.04
8.00 DIA.
F
He
INDUSTRIES
DR. TPO 1-12-92
CH MAS 1-13-92
ORDER NO.
A-3-00619
SCALE
1/2
MATERIAL
.12 AL. GASKET STK
QUANTITY
3000
WT/100
5LB
TITLE
PUMP SPACER
BP-9B

ASSIGNMENT B—UNIT 9

PUMP SPACER (BP-9B)

1. Name the type of drawing used to represent the PUMP SPACER.
2. Explain briefly why all information required to produce the part may be provided in one view.
3. Tell why the part is symmetrical with the vertical axis.
4. Identify the:
   a. Order number.
   b. Print number.
5. Indicate the following:
   a. The thickness and kind of material in the Pump Spacer.
   b. The quantity to be produced.
6. Calculate the weight of the stamped parts in the order.
7. Identify the letter which relates to each of the following types of lines or function:
   a. Object lines
   b. Center lines
   c. Extension lines
   d. Dimension lines
   e. Leaders
8. State the purpose of a leader.
9. Specify the angular dimension between each of the three center lines for the three equally spaced holes on the Pump Spacer.
10. Give the size of the square hole.
11. Calculate the number of degrees center line Ⓐ is from center line Ⓒ.
12. Interpret the meaning of the note Ⓔ.
13. Compute the radius to the center of the 1.02 holes.
14. Determine the following dimensions:
    a. The body diameter of the Pump Spacer.
    b. The radius of the lugs.
    c. The overall outside diameter.

Student's Name ____________________

1. ____________________
2. ____________________
   ____
   ____
   ____
3. ____________________
   ____
   ____
4. a. Order Number ____________________
   b. Print Number ____________________
5. a. ____________________
   b. Number of Parts ____________________
6. Weight ____________________
7. a. Object Lines ____________________
   b. Center Lines ____________________
   c. Extension Lines ____________________
   d. Dimension Lines ____________________
   e. Leaders ____________________
8. ____________________
   ____
   ____
9. ____________________
10. ____________________
11. ____________________
12. ____________________
    ____
    ____
    ____
    ____
13. Ⓡ = ____________________
14. a. Body Diameter ____________________
    b. Lug Radius ____________________
    c. Overall Outside Diameter ____________________

4 + 1.04

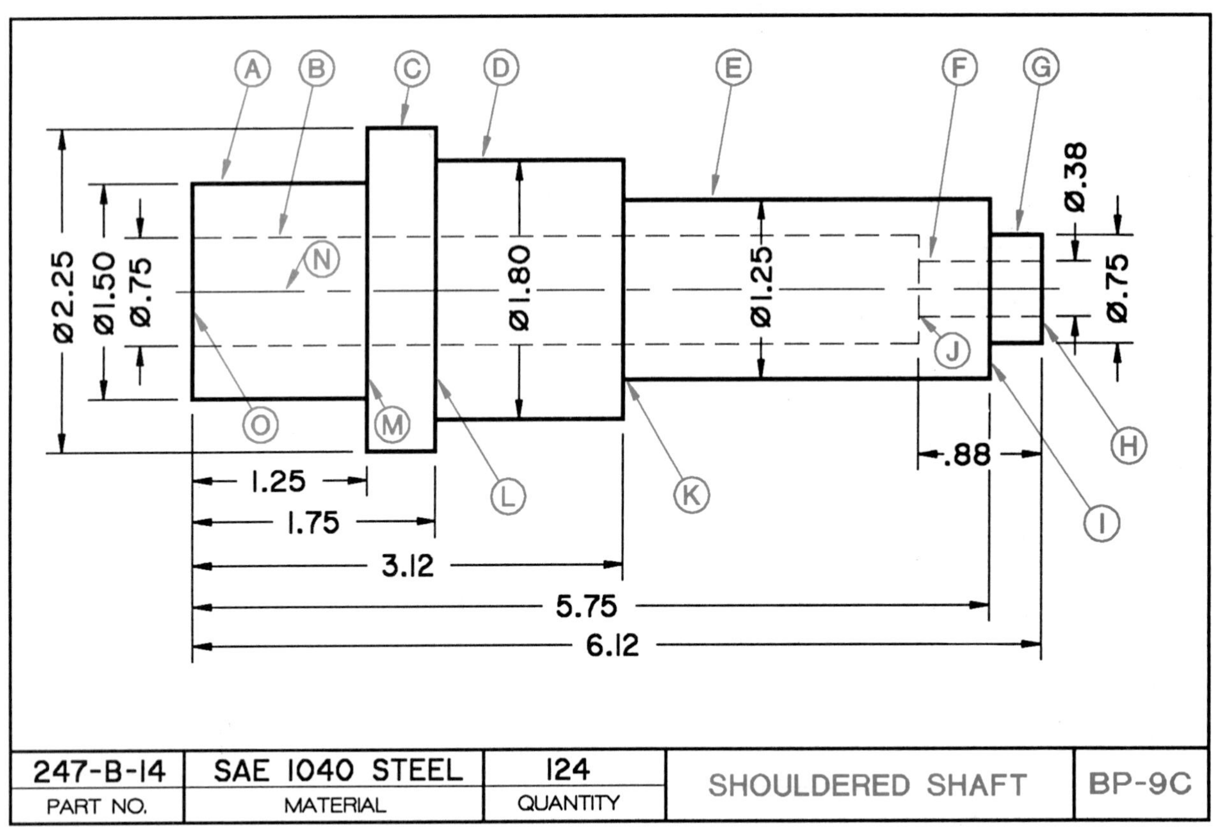

ASSIGNMENT C—UNIT 9

SHOULDERED SHAFT (BP-9C)

1. Name the type of drawing that is used to represent the SHOULDERED SHAFT.
2. Provide the following information about the part:
   a. The specifications of the material from which the parts are to be machined.
   b. The part number.
   c. The required quantity.
3. Determine the shape of the part.
4. Give the dimension of each outside diameter.
5. State the overall length of the SHOULDERED SHAFT.
6. Determine the depth of the .38″ hole.
7. Calculate the lengths of the following features of the part:
   a. The .75″ end.
   b. The 1.25″ diameter section.
   c. The 1.80″ diameter section.
8. Give the letters of the different lines that are used to represent:
   a. The external shape of the part.
   b. The internal details.
   c. The center line.

Student's Name ______________________

1. ______________________
2. a. ______________________
   b. ______________________
   c. ______________________
3. ______________________
4. ______________________
5. ______________________
6. ______________________
7. a. ______________________
   b. ______________________
   c. ______________________
8. a. ______________________
   b. ______________________
   c. ______________________

UNIT 10
Auxiliary Views

Up to this point, the surfaces and features of all objects have been perpendicular to one another. Using any combination of the six principal views, lines, circles, and arcs were projected on horizontal and vertical planes in their true size and shape.

However, there are many objects which have inclined surfaces, holes, or cut-out sections. Such surfaces slant away from the conventional horizontal or vertical planes of projection. If details of an inclined surface are projected onto normal views, the true size and shape are distorted (foreshortened).

Figure 10-1 provides an example of a part with an inclined surface. The features are not accurately represented by any combination of principal views. Note that the front and right-side views show the inclined face foreshortened. The hole is distorted and appears as an ellipse. This means that another type of distortion-free view is needed for a true representation of the surface and features.

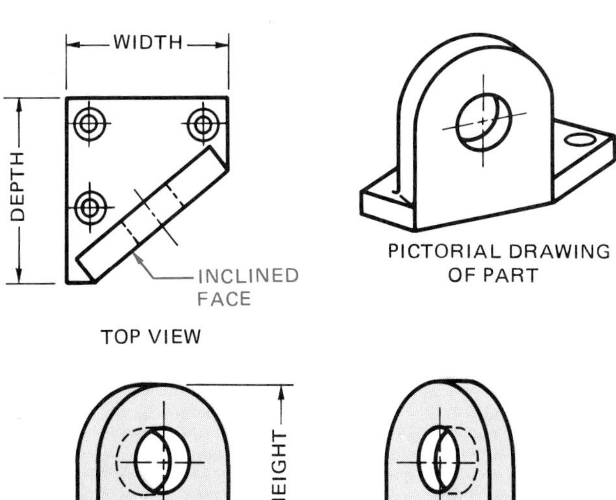

FIGURE 10-1 REGULAR FRONT AND RIGHT VIEWS SHOW INCLINED FACE AND HOLE DISTORTED

AUXILIARY VIEWS

An additional view, known as an *auxiliary view*, permits the true projection of angular surfaces and features. Auxiliary views may be full or partial outlines of an object. Internal details are often represented by full or partial *sectional auxiliary views*.

PROJECTION PRINCIPLES APPLIED TO AUXILIARY VIEWS

Auxiliary views are developed on *auxiliary planes of projection*. The auxiliary plane is perpendicular to one of the normal planes of projection and parallel with the inclined surface (figure 10-2).

The draftsperson visualizes the inclined face (auxiliary view) of an object in much the same manner as the imaginary process used to project details onto any combination of the six principal views. The difference with auxiliary views is that the imaginary plane is the auxiliary plane of projection. To repeat, *projection lines* (which are fine, solid lines) are used to locate and/or outline the features on the inclined surface. When the imaginary auxiliary plane is revolved 90°, the projected auxiliary view becomes an additional view on a technical drawing.

Figure 10-3 shows a revolved auxiliary plane

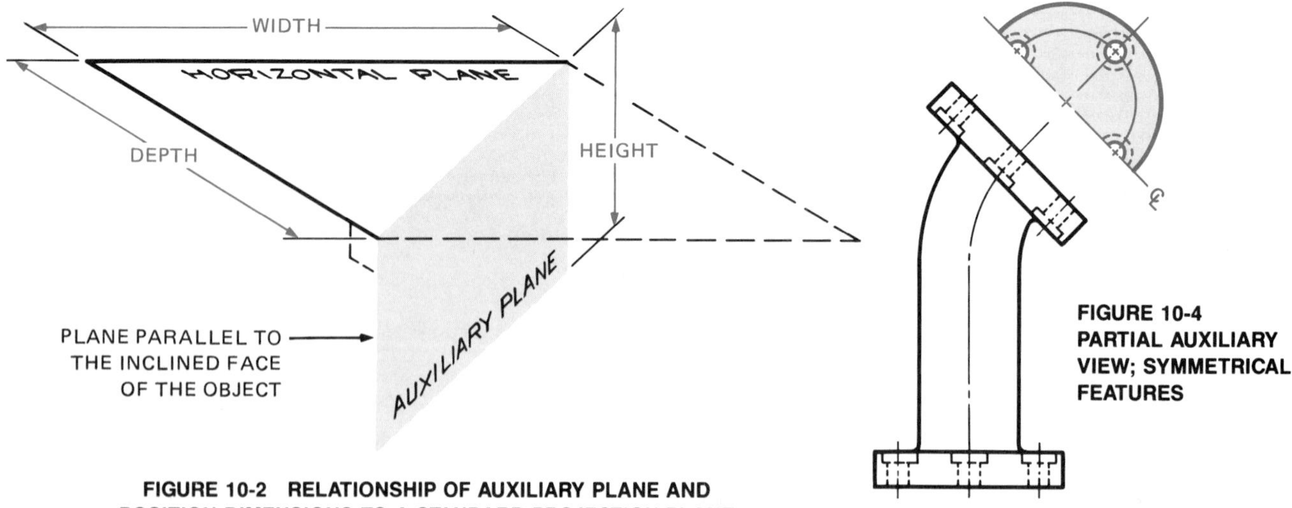

FIGURE 10-4 PARTIAL AUXILIARY VIEW; SYMMETRICAL FEATURES

FIGURE 10-2 RELATIONSHIP OF AUXILIARY PLANE AND POSITION DIMENSIONS TO A STANDARD PROJECTION PLANE

which is parallel to the angular face of the object. Points (a) through (g) are projected at a right angle from the angular face onto the auxiliary plane. Since the height dimensions are accurate in the front view, they may be projected or transferred from the front view to the auxiliary view. The auxiliary view shows all features on the inclined surface in their true size and shape.

Selection of Auxiliary Views

As stated before, the number, type, and position of regular, sectional, and auxiliary views depends on the complexity and/or dimensions of the part. Auxiliary views are named according to the view from which features are projected in their true size and shape. The auxiliary view may be an auxiliary front, top, bottom, right-side, or left-side view. In figure 10-3 the view is a top auxiliary view. Occasionally, auxiliary views are referred to according to the principal dimensions as *height auxiliary*, *depth auxiliary*, or *width auxiliary views*.

When a view of an angular surface is projected directly from a regular view onto an auxiliary plane, this projection produces a *primary auxiliary view*. Complex parts having surfaces and features at compound angles often require the projection of certain details from a primary auxiliary view to a secondary auxiliary plane. The view developed on the secondary plane is referred to as a *secondary auxiliary view*.

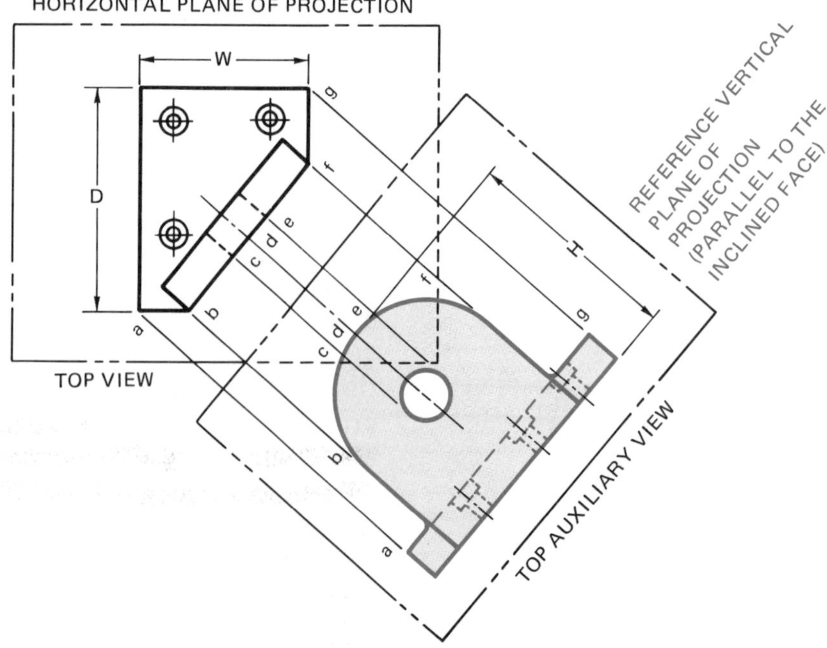

FIGURE 10-3 VERTICAL REFERENCE PLANE REVOLVED TO FORM AUXILIARY VIEW. FEATURES ON INCLINED FACE APPEAR IN TRUE SIZE AND SHAPE.

APPLICATIONS OF AUXILIARY VIEWS

In actual practice, only the projected inclined surface and features are represented on auxiliary views. Sometimes, portions of adjacent features are included to simplify dimensioning and reading the print. Generally, a complete auxiliary view is too confusing. Hidden lines are omitted unless they are needed to clarify certain details. Inclined surfaces with symmetrical features are usually represented by a half-section auxiliary view as illustrated in figure 10-4.

The working drawing of the Angle Bracket (figure 10-5) shows how an auxiliary view is generally used in combination with other regular views. The top view shows the rectangular shape and provides location and other width and depth dimensions. The auxiliary view shows the true shape of the inclined face and the reamed hole and the required height dimensions. Note that only a small cut-away portion of the base is represented. This permits dimensioning the .75 height (or thickness) of the base.

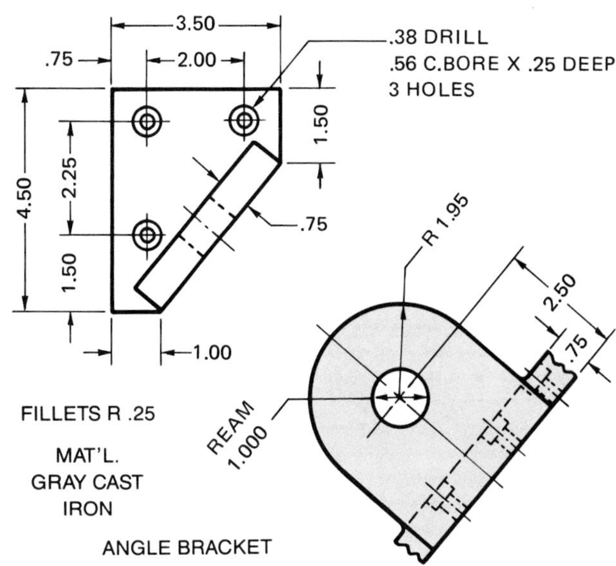

FIGURE 10-5 REGULAR TOP VIEW AND A PARTIAL AUXILIARY VIEW

SECONDARY AUXILIARY VIEWS

Secondary auxiliary views are used on drawings when surfaces and design features cannot be adequately represented by regular views or primary auxiliary views. For example, the part represented by the regular front and top views in figure 10-6 does not show the true shape and sizes for surfaces Ⓐ, Ⓑ, Ⓒ, or Ⓓ and all holes.

Primary auxiliary view (1) shows the true size and shape of inclined surface Ⓐ and the through hole. Primary auxiliary view (2), also projected from the front view, shows the true size of the hole and the shape and size of surface Ⓑ. However, neither primary auxiliary view (1) or (2) shows the true shape and size of surface Ⓒ and Ⓓ.

A *secondary auxiliary view* is needed to show the true shape and size of the surface Ⓒ or Ⓓ and the holes. Auxiliary view (3) is called a secondary auxiliary view since it is projected from a primary auxiliary view. Auxiliary views (1) and (2) are primary auxiliary views that are projected from one of the regular views. As stated earlier, hidden lines are omitted unless they are needed to clarify drawing details.

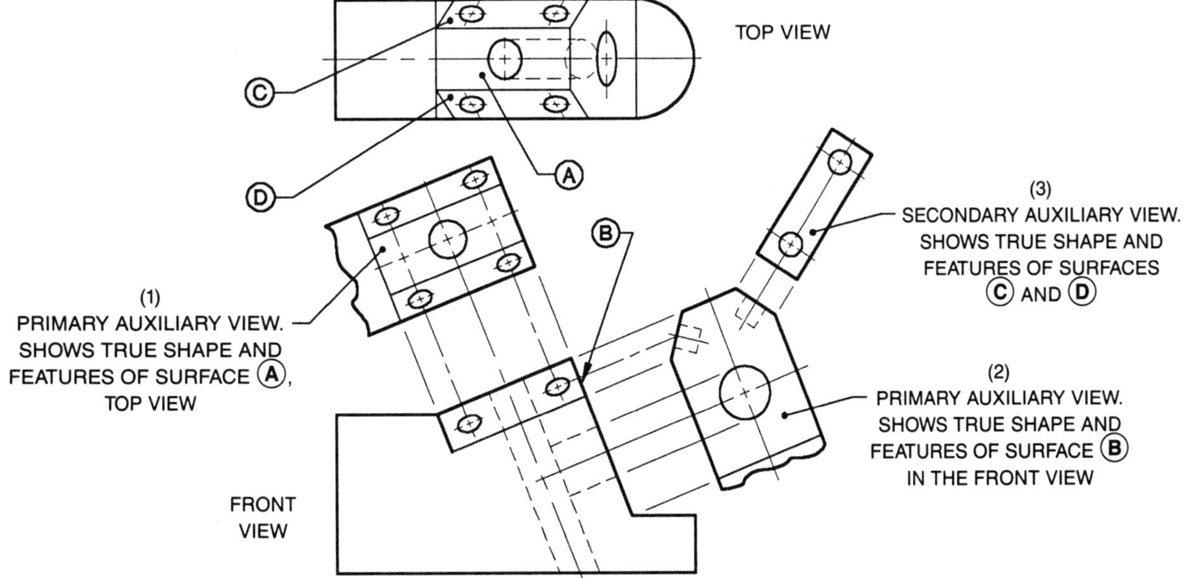

FIGURE 10-6 APPLICATION OF PRIMARY AND SECONDARY AUXILIARY VIEWS

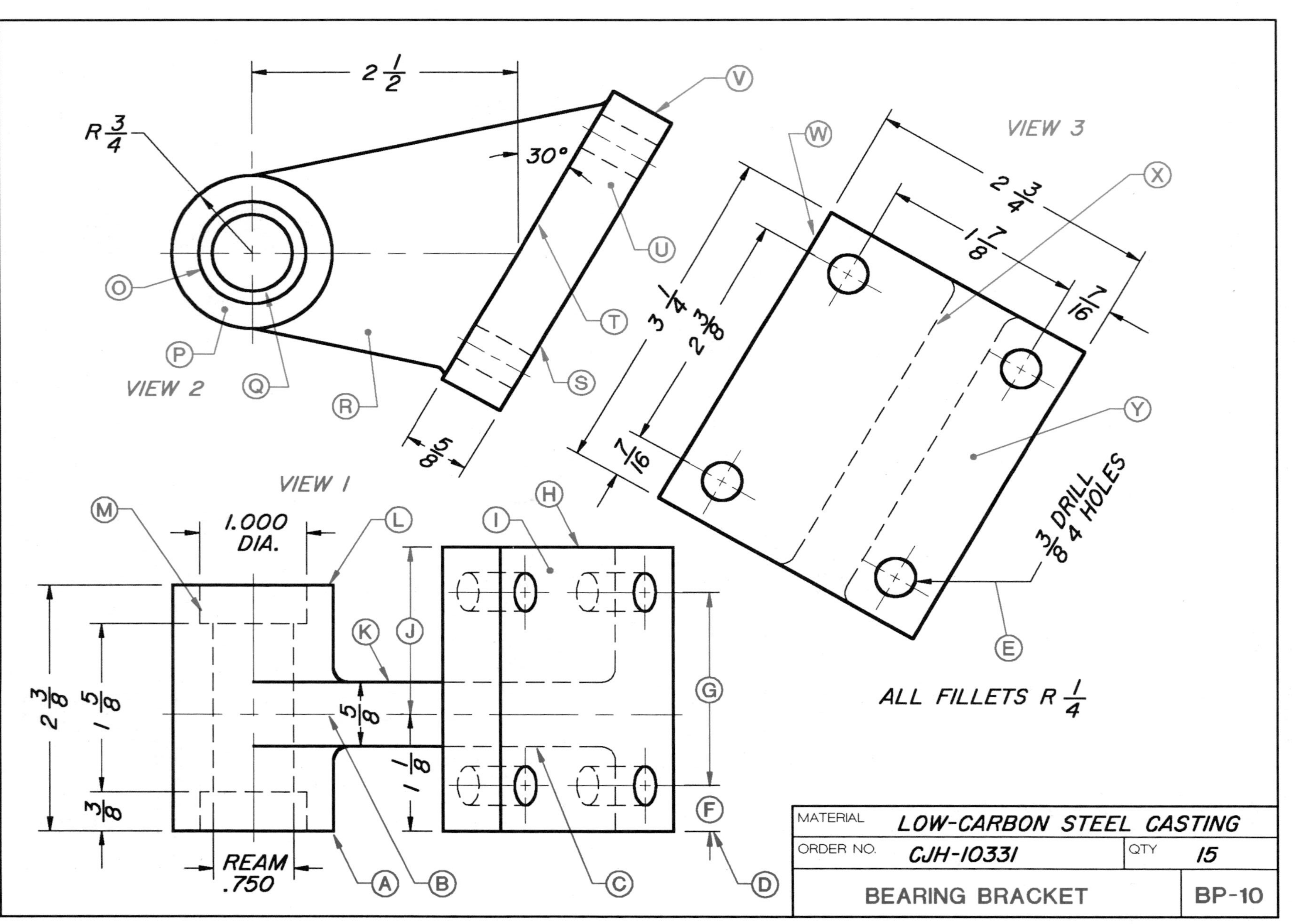

VIEW 1
VIEW 2
VIEW 3
R 3/4
2 1/2
30°
5/8
2 3/4
1 7/8
7/16
3 1/4
2 3/8
3/8 DRILL 4 HOLES
ALL FILLETS R 1/4
1.000 DIA.
REAM .750
1 5/8
1/8
MATERIAL LOW-CARBON STEEL CASTING
ORDER NO. CJH-10331
QTY 15
BEARING BRACKET
BP-10

ASSIGNMENT—UNIT 10

BEARING BRACKET (BP-10) **Student's Name** ______________________

1. Specify the kind of material to be used for the Bearing Bracket castings.
2. Identify (a) the order number and (b) the required quantity.
3. Name each of the three views.
4. Identify the following types of lines or features:
   a. (A) b. (B) c. (C) d. (D) e. (E)
5. Locate the following features in **VIEW 2**:
   a. (H) b. (I) c. (K) d. (L) e. (M)
6. State the purpose of using auxiliary **VIEW 3**.
7. Locate the following features from the auxiliary view in (a) **VIEW 1** and (b) **VIEW 2**:
   a. (W) b. (X) c. (Y)
8. a. Determine the horizontal and vertical center line dimensions of the drilled holes.
   b. Give the specifications of these holes.
9. Determine the following dimensions of the hub section of the Bracket:
   a. Outside diameter
   b. Diameter (O)
   c. Diameter (Q)
10. Give the overall height and width of the rectangular pad section of the Bracket.
11. Determine the web thickness.
12. Compute the following dimensions:
   a. (F) b. (G) c. (J)
13. Interpret the note:

   ALL FILLETS $\frac{1}{4}$ R

1. ______________________

2. a. Order # ______________________
   b. Qty ______________________

3. **VIEW 1** ______________________
   VIEW 2 ______________________
   VIEW 3 ______________________

4. a. (A) ______________________
   b. (B) ______________________
   c. (C) ______________________
   d. (D) ______________________
   e. (E) ______________________

5. | | **VIEW 1** | **VIEW 2** | | **VIEW 1** | **VIEW 2** |
   |---|---|---|---|---|---|
   | a. (H) = | ______ | | d. (L) = | ______ | |
   | b. (I) = | ______ | | e. (M) = | ______ | |
   | c. (K) = | ______ | | | | |

6. ______________________
   ____
   ____
   ____

7. | Auxiliary View | **VIEW 1** (a) | **VIEW 2** (b) |
   |---|---|---|
   | a. (W) | ______ | ______ |
   | b. (X) | ______ | ______ |
   | c. (Y) | ______ | ______ |

8. a. Horizontal Center Line Dimension
   ____
   Vertical Center Line Dimension
   ____
   b. ______________________

9. a. Outside Diameter = ______________________
   b. Diameter (O) = ______________________
   c. Diameter (Q) = ______________________

10. Height ______ Width ______

11. Web Thickness ______________________

12. a. (F) = ______
    b. (G) = ______
    c. (J) = ______

13. ______________________
    ____
    ____
    ____
    ____

HIGH TECHNOLOGY applications

HT-1 HIGH AMPLIFICATION MEASUREMENT COMPARATORS

ELECTRONIC DIGITAL MEASUREMENT/INSPECTION INSTRUMENTS

Digital electronic instruments provide a functional and practical link in precision measurement. These instruments are also used in the inspection of finished parts to establish dimensional accuracy, quality of surface finish, and other geometric tolerancing requirements.

Digital electronic instruments are applied either directly to physical measurement or as part of a system of integrated statistical quality and process control (SPC). In such a system, measurement data from on-the-spot analysis is transmitted as statistical data.

Digital electronic instruments are designed for use with other statistical process control instruments that relate to measurement/inspection data collection, transmission, gathering, processing, storing, and finally, as printed copy readouts.

Electronic Digital Calipers

The electronic digital caliper illustrates a direct reading caliper. Readings may be taken in customary inch and instantly converted to SI metric standard units of linear measurement to accuracies of four decimal places.

The four callouts as illustrated identify tolerance and classifications (modes) of this measuring instrument. For example, the first reading (**0.7500**) at (A) shows the *preset tolerance limit* mode. The **0.7455** (B) indicates an *acceptable tolerance limit*; **0.7480** (C) a *reject*; and **0.7525** (D), a part that requires *reworking*.

The electronic digital caliper may also be used for comparative measurements. That is to say, the instrument may be set for measuring a particular part feature at a zero point in measurement. Deviations below the zero point for a standard measurement are displayed as a minus (−) value; above zero, by a positive (+) value.

Some models of electronic digital calipers (such as the one displayed) are equipped with a light controlled display (LCD). Generally, three colors of lights are used to show dimensional classifications. A green light (B) indicates a *within tolerance* classification. A red light signal (C) shows an *out-of-tolerance* condition; a yellow light signal (D), a *reworkable* machined part.

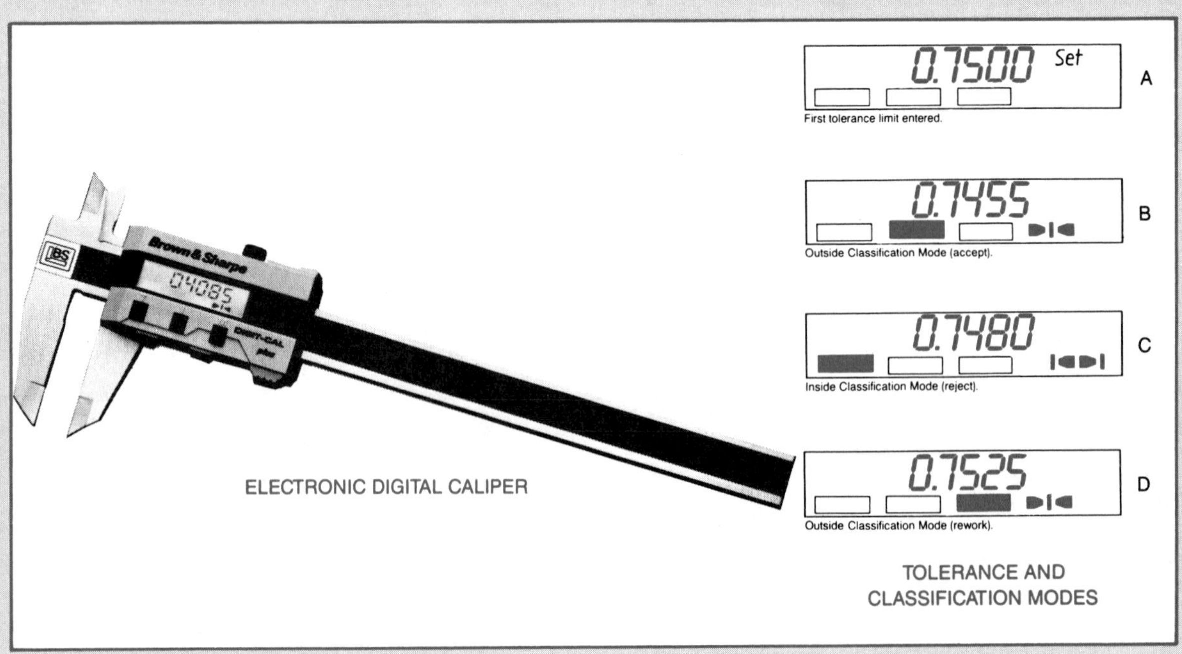

ELECTRONIC DIGITAL CALIPER

TOLERANCE AND CLASSIFICATION MODES

Drawings courtesy of BROWN & SHARPE MANUFACTURING COMPANY

SECTION 4
Dimensioning Using Customary Units of Measure

UNIT 11
Dimensioning Principles and Standards

There are two major parts to a working drawing. The first requires the use of lines and views to describe the outer form and internal details of the object. The second relates to dimensions which govern sizes and distance. This Section begins with the application of selected lines, basic dimensioning practices, and measurements according to customary inch units. Later, size and location dimensions are considered as they are used for dimensioning straight, angle, curved, and combinations of lines and surfaces. The final unit in the Section covers systems of dimensioning and dimensional data which is provided in table form.

The dimensioning of drawings is controlled by standards. Standards established by the American National Standards Institute (ANSI) and Canadian Standards Association (CSA) are used with customary units of measure. Dimensioning practices are extended in Section 5 to cover SI metric (ISO) practices.

SIZES OF PART FEATURES

In simplest form, a part may be a solid prism, cylinder, pyramid, cone, or sphere (figure 11-1). Each form is represented on a drawing by a series of lines which describe the shape. In order to fabricate, machine, inspect, or measure a part or mechanism, dimensions are added to establish distances and sizes.

Dimensions which are under 72″ are usually given in customary units of inches and fractional and/or decimal parts of an inch. In metric, similar

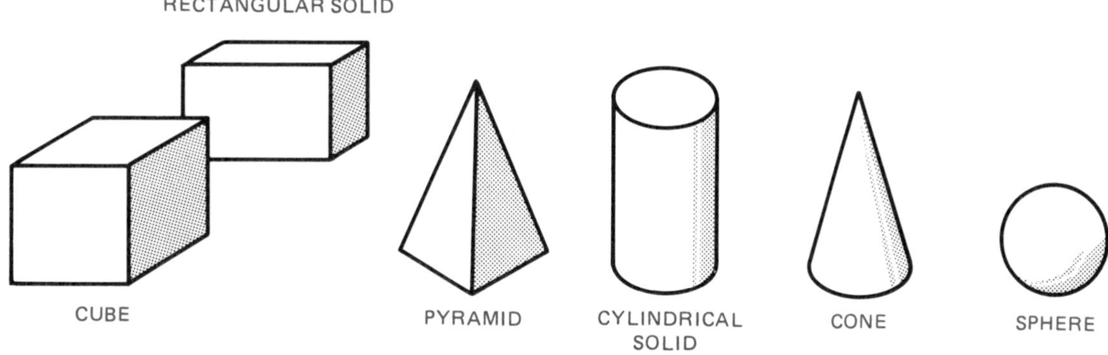

FIGURE 11-1 BASIC GEOMETRIC FORMS

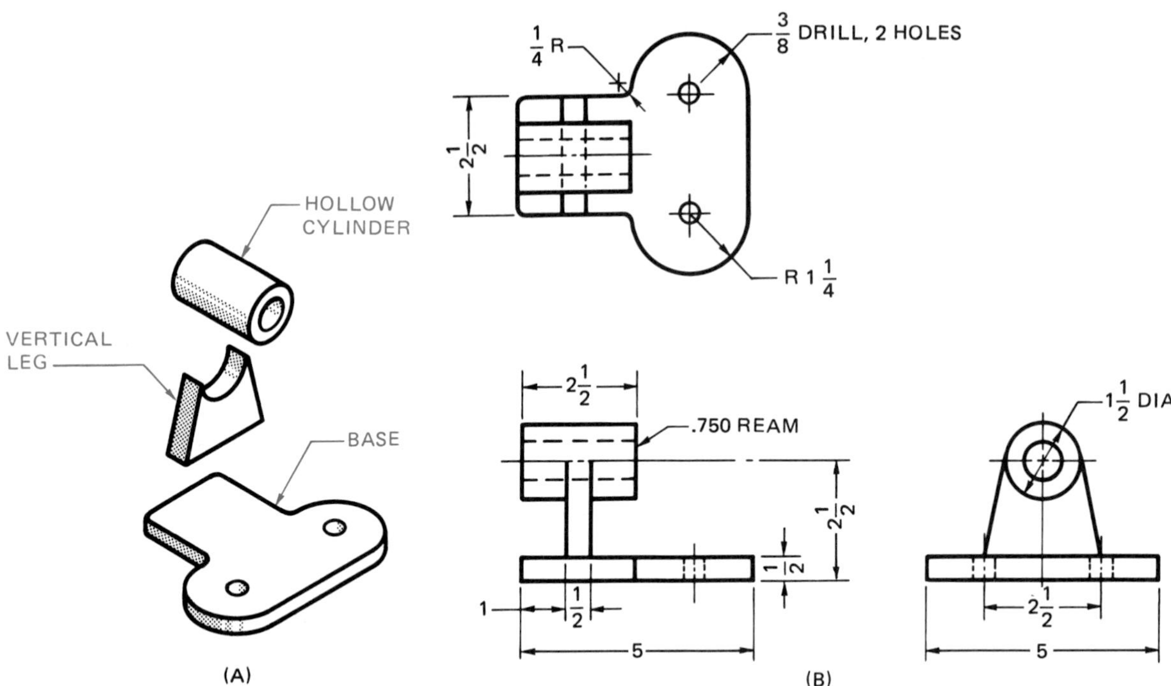

FIGURE 11-2 VISUALIZING AND DIMENSIONING THE GEOMETRIC FORMS OF A PART

dimensions appear in centimeter (cm) and millimeter (mm) units. Dimensions up to 72″ are widely found on drawings used for parts and mechanisms in aerospace, automotive, electrical-electronic, machine and metal products manufacture, and other industries. Dimensions larger than 72″ are expressed in feet and inches. Such dimensions (or metric equivalents) are used in structural, architectural, building construction, and civil engineering applications.

VISUALIZING GEOMETRIC SHAPES WHICH FORM A PART

As the shape of a part starts to vary from a solid form, the interpretation of a drawing may be simplified by considering the part as being constructed of several basic geometric shapes. Dimensions are then related to each shape. Figure 11-2A shows a pictorial drawing. The Bracket may be visualized as including a *base, vertical leg,* and *hollow cylinder.* The three-view working drawing at (B) provides information about size and location dimensions.

The dimensions for the *hollow cylinder* give the outside diameter (1 1/2″), the hole diameter (ream 0.750″), and the length (2 1/2″). The *vertical leg* which extends from the base to the sides of the cylinder is 2″ high and curves around the cylinder. The leg tapers from 1 1/2″ to 2 1/2″ and is 1/2″ thick. The *base* is a combination of two rectangular solids (2 1/2″ × 2 1/2″) and a hollow cylinder consisting of two semicircular sections (figure 11-3). The outside diameter of the cylinder is 2 1/2″; the hole size, 3/8″ diameter. The base is 1/2″ thick. The radius of the corner fillets is 1/4″.

After each geometric shape is visualized for form and size, the part is then considered as a whole. Note that there are no *superfluous* (unnecessary) *dimensions.* For example, even the note

$\frac{3}{8}$ DRILL, 2 HOLES

eliminates the need for giving the drill diameter of each hole.

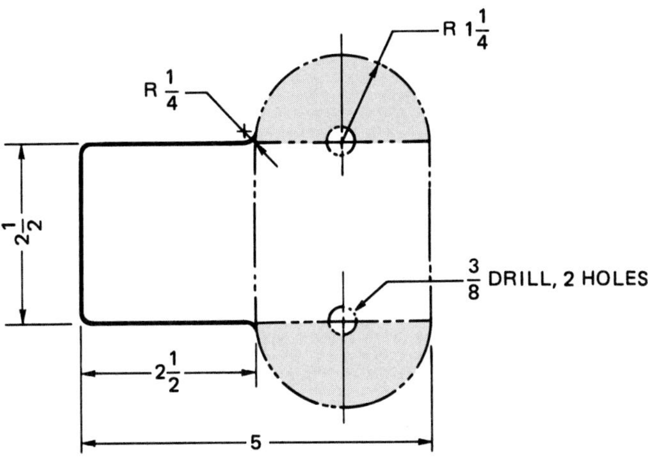

FIGURE 11-3 BASE CONSISTS OF REGULAR FORMS

DIMENSIONS AND DIMENSION LINES

In review, lines used in connection with dimensioning are fine, solid, dark lines. Dimensions are found on drawings between extension lines and/or center lines, or are used with leaders. For dimensions given in decimal parts of an inch, the number of digits to the right of the decimal point indicates the degree of accuracy. Fractional parts like 1/2″ may be dimensioned as 0.50″ or 0.500″. While the use of decimal dimensioning is the preferred practice in today's industry, there still are a number of drawings on which common fractions and decimal dimensions are used. Examples of this practice are included in this text. Common fractions are generally used where accuracies must be no closer than ±1/64″.

When all dimensions on a drawing are given in inches, the inch symbol (″) is omitted. The same practice is followed with millimeter (mm) dimensions. Fasteners, like bolts, nuts, studs, and rivets are generally given as standard fractional designations. Drawings of parts dimensioned in feet and inches or meters and millimeters require the use of feet (′), inch (″), meter (m), and millimeter (mm) symbols.

BASIC DIMENSIONING PRACTICES

Dimensions for the shortest width, height, and depth are usually found closest to the object. Parallel dimensions are arranged so that the largest dimension is farthest from the object. In general practice, dimensions are placed near the view where the features to be dimensioned are shown in true representation and the contour and shape are clearly displayed. Dimensions on large parts are commonly placed within the view when this practice shows the dimension with greatest clarity.

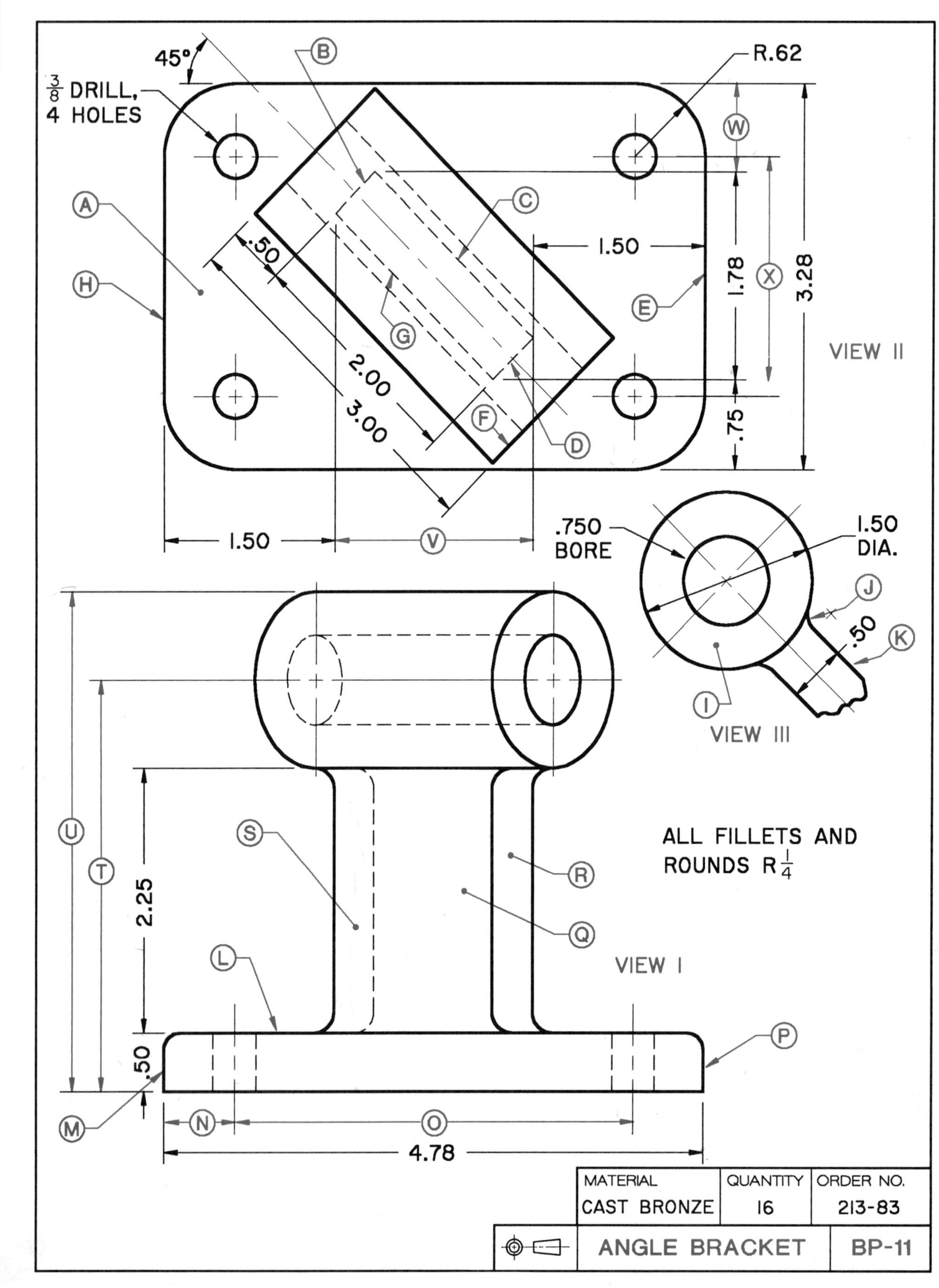
45°
3/8 DRILL,
4 HOLES
B
R.62
W
A
C
H
.50
1.50
E
1.78
X
3.28
G
VIEW II
2.00
3.00
F
D
.75
1.50
V
.750
BORE
1.50
DIA.
J
.50
K
I
VIEW III
U
S
T
2.25
R
ALL FILLETS AND
ROUNDS R 1/4
Q
L
VIEW I
P
.50
M
N
O
4.78
MATERIAL
CAST BRONZE
QUANTITY
16
ORDER NO.
213-83
ANGLE BRACKET
BP-11

ASSIGNMENT—UNIT 11

ANGLE BRACKET (BP-11)

1. a. Name the part.
   b. Give the required quantity.
   c. Identify the material to be used for each part.
   d. State the dimension of all fillets and rounds.
   e. Indicate the system of projection.
2. Name the three views.
3. Identify surfaces Ⓐ, Ⓓ, Ⓔ, Ⓖ, and Ⓗ in **VIEW II** with the letter in **VIEW I** of the corresponding surface.
4. Give the letters of the surfaces in **VIEW III** which correspond with surfaces Ⓒ and Ⓕ in **VIEW II**.
5. Determine which line in **VIEW II** represents surface Ⓢ.
6. Name the three geometric forms which combined produce the shape of the Angle Bracket.
7. Give the overall (a) width, (b) height, and (c) depth of the part.
8. Determine dimension Ⓝ.
9. Compute dimensions Ⓞ, Ⓣ, Ⓤ, Ⓥ, and Ⓦ
10. Calculate the center-to-center distance Ⓧ.
11. Give the length of the bored hole.
12. Determine the specifications for the holes.

Student's Name ______________________

1. a. ______________________
   b. Quantity ______________________
   c. Material ______________________
   d. Fillets/Rounds ______________________
   e. ______________________
2. a. **VIEW I** ______________________
   b. **VIEW II** ______________________
   c. **VIEW III** ______________________
3. Ⓐ = ____________ Ⓖ = ____________
   Ⓓ = ____________ Ⓗ = ____________
   Ⓔ = ____________
4. Ⓒ = ____________ Ⓕ = ____________
5. Ⓢ = ______________________
6. a. ______________________
   b. ______________________
   c. ______________________
7. a. Width (or Length) ______________________
   b. Height ______________________
   c. Depth ______________________
8. Ⓝ = ______________________
9. Ⓞ = ______________________
   Ⓣ = ______________________
   Ⓤ = ______________________
   Ⓥ = ______________________
   Ⓦ = ______________________
10. Ⓧ = ______________________
11. Bored Hole Length ______________________
12. ______________________

UNIT 12
Size and Location Dimensions and Notes

Most industrial drawings are produced to *scale.* This means the geometric forms which are part of the design of the object are drawn accurately. If the features are represented full size, the scale is said to be 1:1, full size, or 1″ = 1″ (or 25 mm = 25 mm). When a drawing is made half size, it is drawn *half scale* or 1/2″ = 1″ (or 12.5 mm = 25 mm).

Scale is the *ratio of the drawing size to the actual size.* Scale drawings show the true relationship among features and ensure that it is feasible to cast, forge, machine, assemble, and produce the part. Other common scales are *one-quarter* (1/4″ = 1″) and *one-eighth* (1/8″ = 1″). Small, fine parts and features are often drawn to a *double-size scale* of 2″ = 1″. Regardless of scale, *actual size and location dimensions* are included on a drawing.

CONSTRUCTION DIMENSIONS

The two basic *dimensions of size and location* are referred to as *construction dimensions.* Each serves a different function. A *size dimension,* as the term indicates, provides the size. A *location dimension* identifies the position of surfaces, lines, or features in relation to one another. Figure 12-1 illustrates the use of size and location dimensions and how they are placed. In this example, size dimensions (S) are given for the rectangular plate and the hole. Location dimensions (L) indicate the position of the hole with respect to the outer surfaces of the part.

Drawings of prisms require three dimensions: width (or length), height (or thickness), and depth. The three dimensions are generally found on two views.

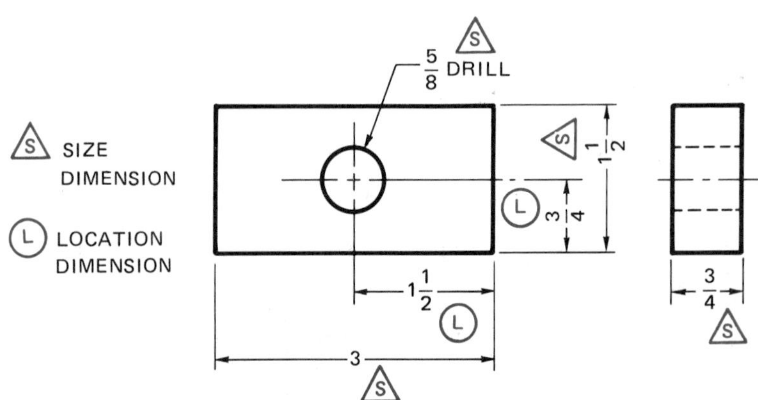

FIGURE 12-1 APPLICATION OF SIZE AND LOCATION DIMENSIONS

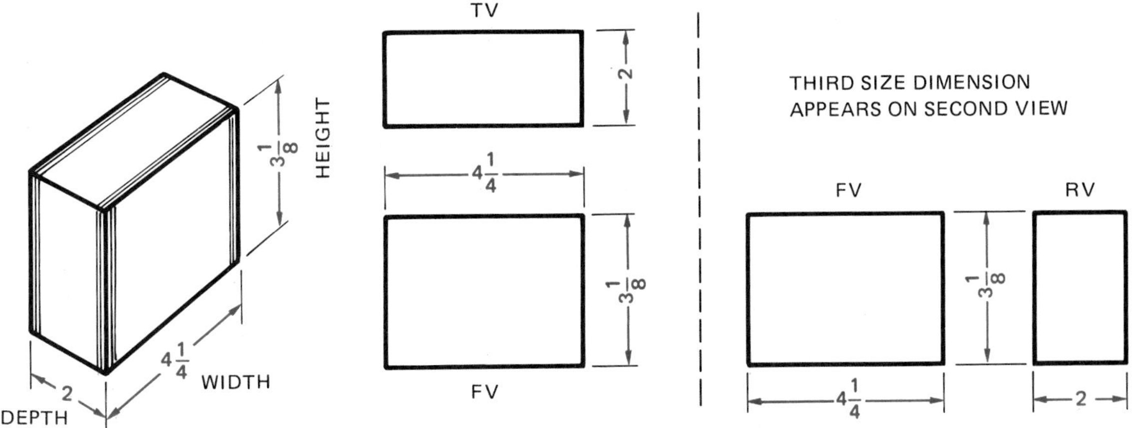

FIGURE 12-2 PLACEMENT OF SIZE DIMENSIONS ON A PRISM

SIZE DIMENSIONS OF PRISMS AND TRUNCATED PYRAMID FORMS

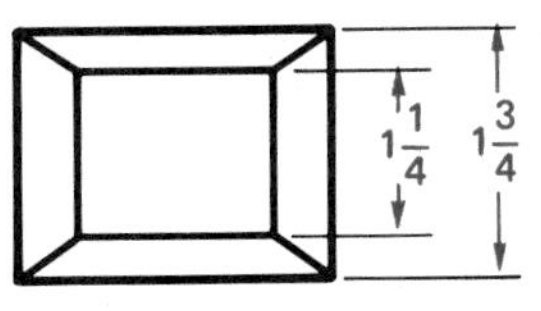

The placement of size dimensions as they are commonly found on drawings of a regular prism are shown in figure 12-2. The dimensioning of an object in the form of a *truncated pyramid* is illustrated in figure 12-3. This form is produced when a plane cuts through the top area of a pyramid and *truncates* (cuts through) it. Size dimensions for width and height are given in the *principal view* which, in this case, is the front view. The size dimensions of depth appear in the top view.

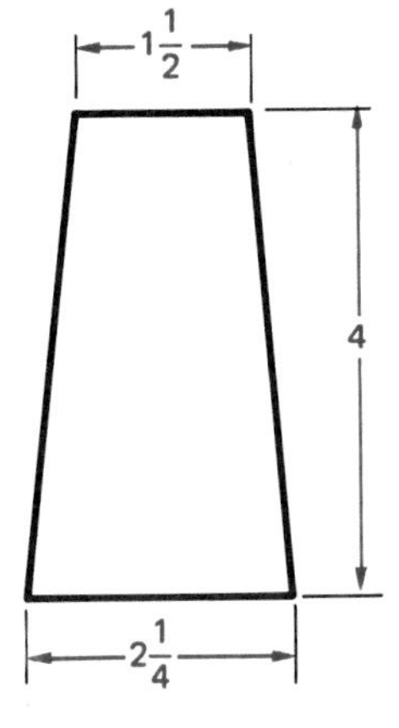

FIGURE 12-3 SIZE DIMENSIONS OF A TRUNCATED PYRAMID FORM

LOCATION DIMENSIONS

Location dimensions establish the relative positions of surfaces, points, and features. Location dimensions are generally given from a finished surface, an axis, or a center line, as shown in figure 12-4.

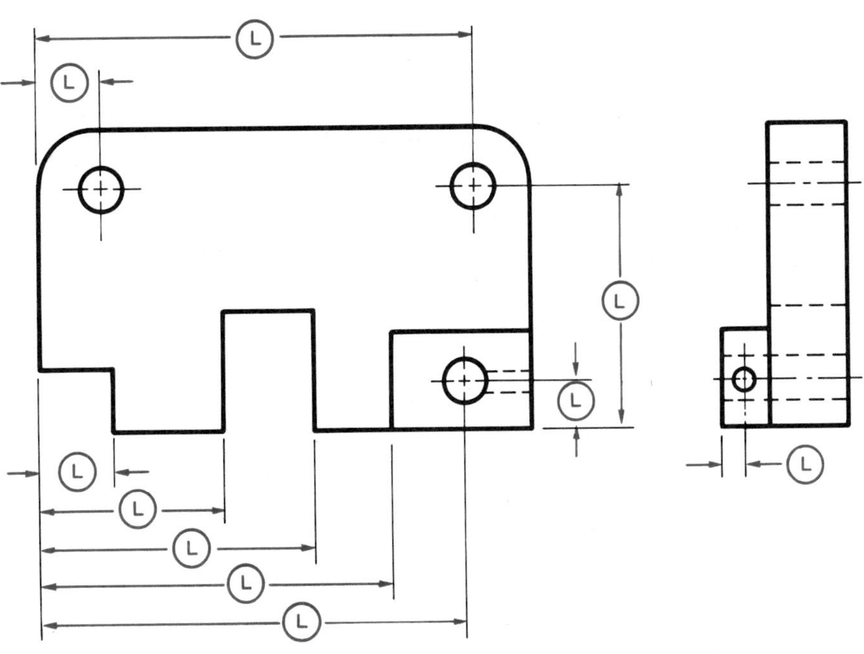

FIGURE 12-4 LOCATION DIMENSIONS ESTABLISH THE POSITION OF FEATURES

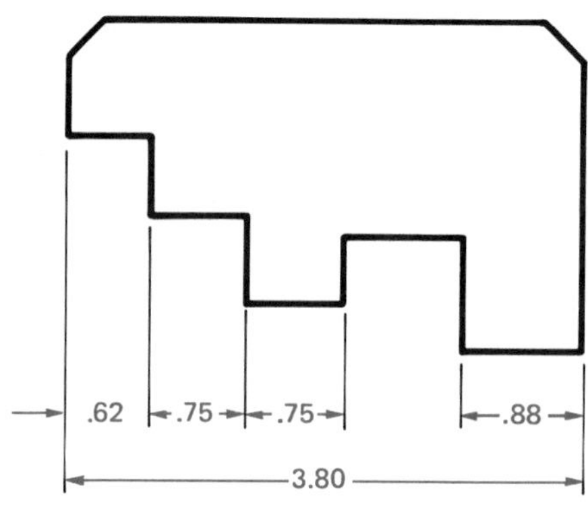

(A) PREFERRED CONTINUOUS DIMENSIONING

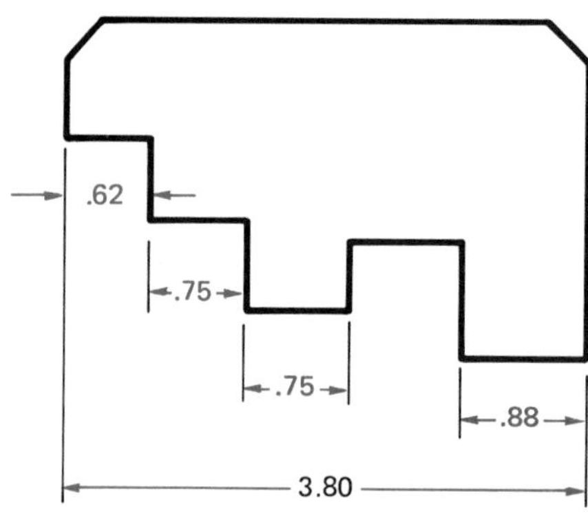

(B) STAGGERED DIMENSIONING (NOT RECOMMENDED)

FIGURE 12-5 RECOMMENDED CONTINUOUS DIMENSIONING

CONTINUOUS DIMENSIONS

In common practice, dimensions are placed between two views. Needed dimensional information is, thus, readily available. Dimensions are spaced for ease and accuracy of reading.

Dimensions which follow one another in a series are placed in a line as *continuous dimensions.* Figure 12-5A illustrates the *preferred method* in which continuous dimensions are provided on a drawing. *Staggered dimensions* (figure 12-5B) are not recommended for continuous dimensioning.

DIMENSIONS IN LIMITED SPACES

Conventional dimensioning of narrow grooves, slots, and details in limited spaces is often confusing and difficult to read. The dimensions may extend beyond extension lines and tend to run together. Under these conditions, dimensions are placed (1) on either side of the extension lines, (2) with a leader, or (3) dots are used. Three examples are included in figure 12-6 to demonstrate dimensioning practices in limited spaces.

DIMENSIONING WITH NOTES

Notes provide a practical technique of including essential technical information which supplements dimensions and other data. Notes are easily and quickly read. The abbreviations used with notes and the form in which they appear on drawings are standardized.

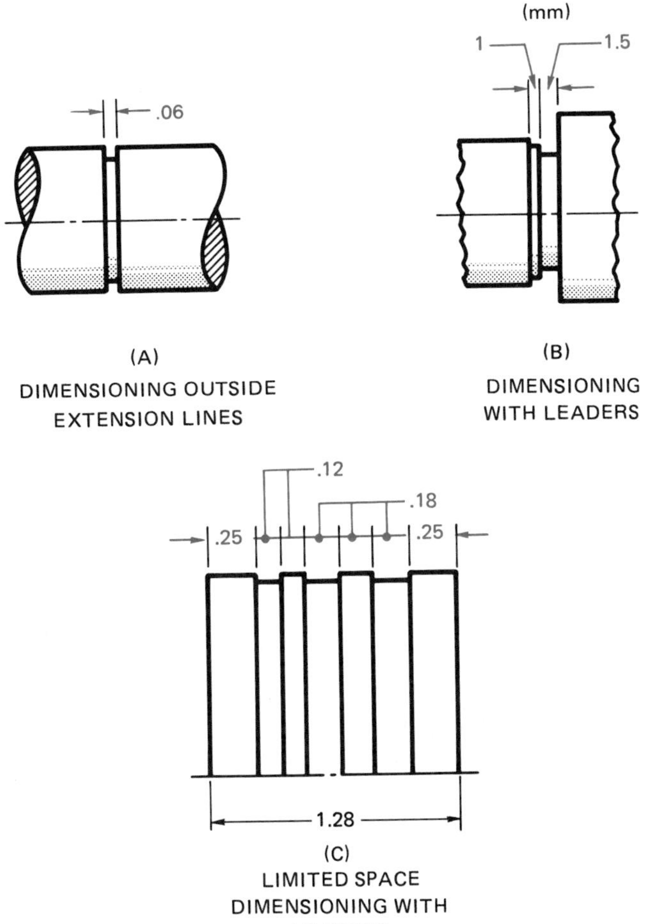

(A)
DIMENSIONING OUTSIDE EXTENSION LINES

(B)
DIMENSIONING WITH LEADERS

(C)
LIMITED SPACE DIMENSIONING WITH LEADERS

FIGURE 12-6 DIMENSIONING IN LIMITED SPACES

General notes relate to an entire part or parts on a drawing. A note like **BREAK ALL EDGES** means that all edges may be rounded with an abrasive stone or file. Such a note may appear near the title block or anywhere convenient on a drawing.

Local notes are connected by a leader with the design feature to which the note applies. Local notes relate to such processes as drilling, boring, reaming, counterboring, and others. Local notes are illustrated, described, and applied throughout this text. For example, notes applying to hole-forming processes are covered in the next unit.

MAXIMUM AND MINIMUM DIMENSIONING NOTES

Maximum and minimum dimensional sizes within which different features of a part must be produced are often established by applying information which is contained in a note. It is common practice for a drawing to include a note, such as:

UNLESS OTHERWISE SPECIFIED, TOLERANCES ON DIMENSIONS ARE:

FRACTIONAL	$\pm \frac{1}{64}''$
DECIMAL	**(XX) ± 0.01″**
	(XXX) ± 0.001″

This general information is used to determine each dimension to which the note applies. For example, using the tolerance limits specified in the note for fractional dimensions, a feature dimensioned $\frac{15}{16}$ is acceptable if machined $\frac{15}{16}'' + \frac{1}{64}''$ $(\frac{61}{64}'')$ or $\frac{15}{16}'' - \frac{1}{64}''$ $(\frac{59}{64}'')$. The $\frac{61}{64}$ is identified as the *maximum* or *upper limit dimension*. The $\frac{59}{64}$ dimension is the *minimum* or *lower limit dimension*. In this instance, the feature may be machined to any dimension between the maximum and minimum.

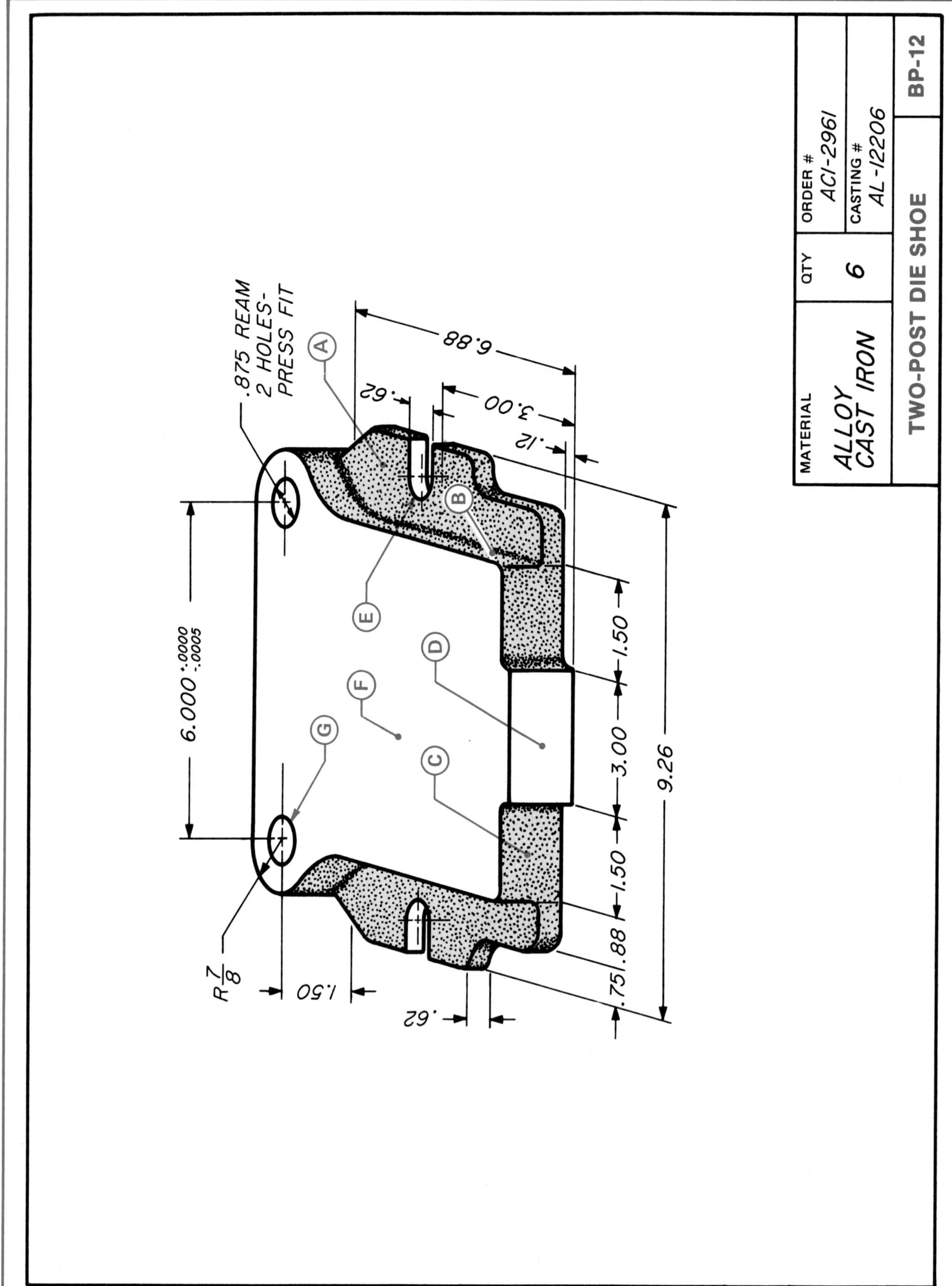
.875 REAM
2 HOLES-
PRESS FIT
6.88
.62
3.00
.12
A
B
C
D
E
F
G
6.000 +.0000 -.0005
1.50
3.00
1.50
.88
.75
9.26
R 7/8
1.50
.62
MATERIAL
ALLOY CAST IRON
QTY
6
ORDER #
ACI-2961
CASTING #
AL-12206
TWO-POST DIE SHOE
BP-12

TWO-POST DIE SHOE (BP-12)

ASSIGNMENT—UNIT 12

Student's Name ____________________

NOTE. STUDY THE SKETCH OF THE DIE SHOE AND **VIEWS I** AND **II**.

1. a. Specify the material used for the Die Shoe.
   b. Identify (1) the order number and (2) the casting number.
   c. Indicate the quantity required.
2. Identify lines (P), (S), (U), and (W), in **VIEW I** according to type.
3. Check dimensions (L), (M), (N), (O), (Q), (R), (T), and (V) in the answer block to classify as size or location dimensions.
4. Give the overall width, height, and depth dimensions.
5. Place the letter representing each surface given on the sketch in the circles in **VIEWS I** and **II** which correspond.
6. Write in the appropriate dimension in each circle provided in **VIEWS I** and **II**.
7. Identify the information which is provided as a note in the sketch.
8. Give the letter of the dimension or feature in **VIEW I** which corresponds with (H), (I), (J), and (K) in **VIEW II**.

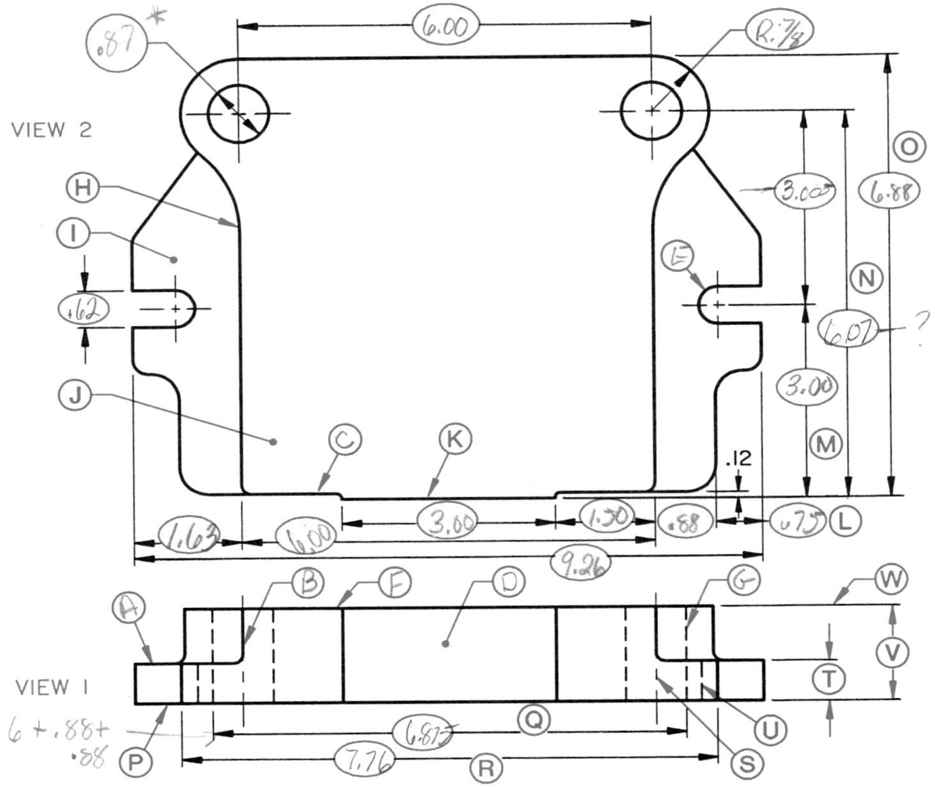

VIEW 2

VIEW I

1. a. Material ____________________
   b. Order # ____________________
   Casting # ____________________
   c. Quantity ____________________

2. Types of Lines
   (P) = ____________________
   (S) = ____________________
   (U) = ____________________
   (W) = ____________________

3.

Drawing Dimension	Type of Dimension	
	Size	Location
(L)		✓
(M)		✓
(N)		✓
(O)	✓	
(Q)		✓
(R)	✓	
(T)		✓
(V)	✓	

4. Width ____________________
   Height ____________________
   Depth ____________________

7. 2 holes, press fit, .875 ream

8. (H) = B
   (I) = A
   (J) = F
   (K) = D

UNIT 13
Dimensioning Cylinders, Holes, Arcs, and Angles

Applications of size and location dimensions are related in this unit to cylindrical parts, circular features, arcs, and holes. From among the many hole-forming processes, drilled, reamed, bored, countersunk, counterbored, and spotfaced holes have been selected for dimensioning. Typical abbreviations for ANSI are used. The dimensioning of angles is treated in relation to angular surfaces and equally and unequally spaced holes. Dimensions are placed and read according to aligned and unidirectional methods.

ALIGNED AND UNIDIRECTIONAL METHODS OF DIMENSIONING

There are two common methods of placing dimensions on drawings: (1) *aligned* and (2) *unidirectional. Aligned dimensions* are generally placed within horizontal or vertical dimension lines to which they refer. Aligned dimensions are read from the bottom or right side (figure 13-1A).

Unidirectional dimensions are read from one direction. All measurements, notes, and other values are placed horizontally regardless of the direction of the dimension line (figure 13-1B). Unidirectional dimensioning is especially practical in aerospace, automotive, and heavy equipment industries where large drawings make it cumbersome to read dimensions unless they are placed horizontally.

DIMENSIONING CYLINDERS AND CIRCLES

The length and diameter dimensions of a cylinder appear on the principal view in which the cylinder is rectangular in form. Figure 13-2A is a two-view drawing of a cylinder. The dimensions on the principal view indicate the part is 2.25″ in diameter and 3.50″ long. The hole size and machining process are identified by a leader. The drill diameter indicates the size and machining process.

The same dimensions and process may be represented on a one-view drawing as illustrated in figure 13-2B. According to ANSI practice, the **∅** diameter symbol may be used in place of the **DIA** abbreviation.

DIMENSIONING ARCS, FILLETS, AND CORNERS

Castings, forgings, and similarly produced parts are designed with corner areas which are strengthened by allowing for additional material to form a *fillet.* Corners are rounded to remove sharp edges for appearance, for safety, and for the removal of a pattern from the mold. A simple note like **.25R ALL FILLETS AND ROUNDS** is placed on a drawing to cover all applications of fillets and rounds on the workpiece.

Arcs are dimensioned by extending a dimension line from the radius center to the arc. A

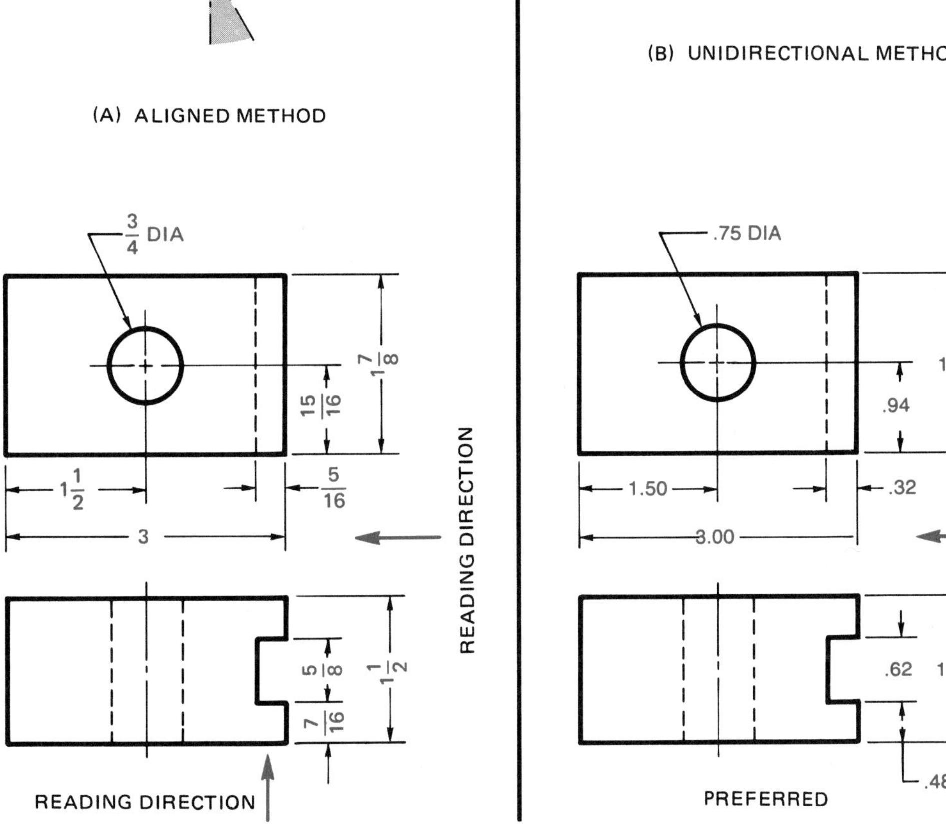

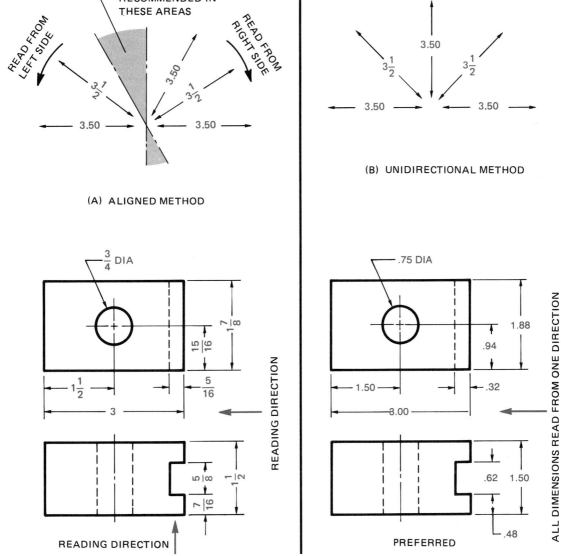

FIGURE 13-1 DIMENSIONING: ALIGNED AND UNIDIRECTIONAL METHODS

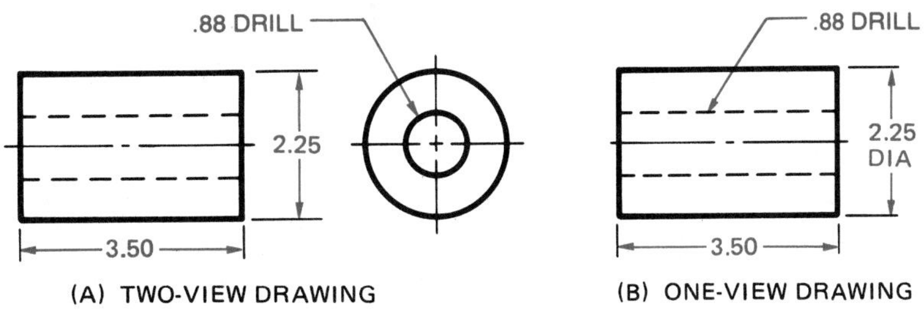

FIGURE 13-2 DIMENSIONING A CYLINDER

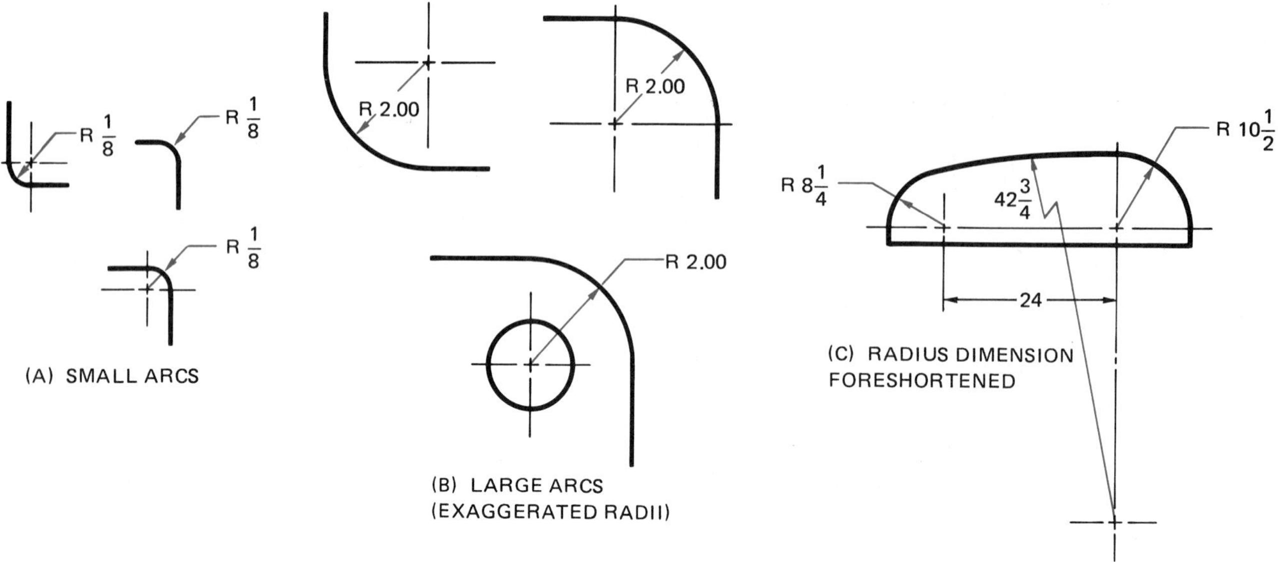

FIGURE 13-3 DIMENSIONING SMALL AND LARGE ARCS

dimension and the abbreviation **R** indicate the size of the radius. Common methods of dimensioning small and large arcs are represented in figure 13-3. When space is limited and it is not possible to draw the radius to full scale, the radius dimension line is *foreshortened* (broken) as illustrated at (C).

DIMENSIONING EQUALLY SPACED HOLES ON A CIRCLE

The position and size of holes which are equally spaced on a circle are represented on a drawing by following four steps:

- Dimension the bolt circle (BC) diameter.
- Locate the position of the first hole.
- Show the position of the remaining holes.
- Provide size, process, and other information as a note.

Figure 13-4 provides an example of how equally spaced holes on a circle are represented and dimensioned.

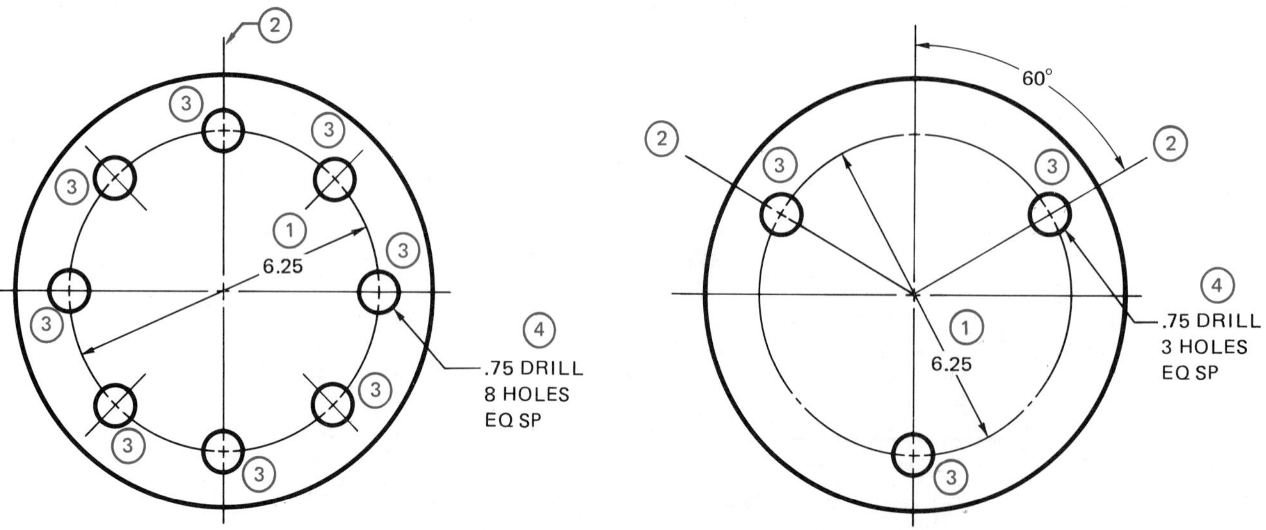

FIGURE 13-4 DIMENSIONING EQUALLY SPACED HOLES ON A CIRCLE

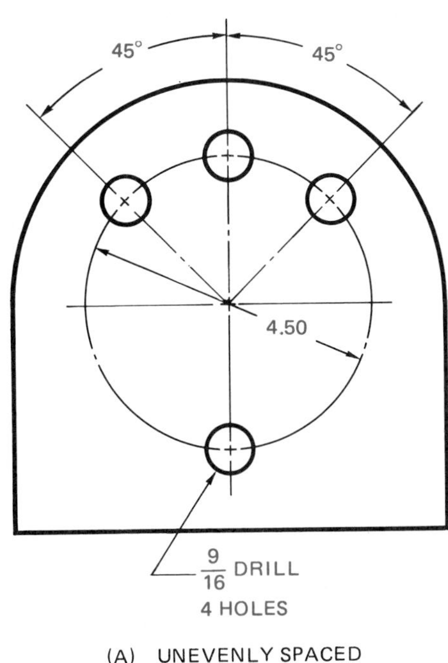

(A) UNEVENLY SPACED HOLES ON A CIRCLE

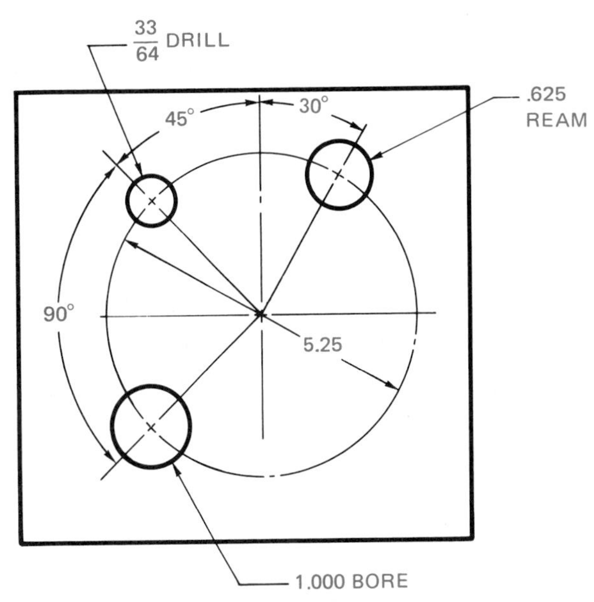

(B) UNEVENLY SPACED HOLES ON A CIRCLE: VARYING DIAMETERS AND MACHINING PROCESSES

FIGURE 13-5 REPRESENTING AND DIMENSIONING UNEQUALLY SPACED HOLES AND MACHINING PROCESSES

DIMENSIONING UNEQUALLY SPACED HOLES ON A CIRCLE

Holes which are unequally spaced on a circle are usually positioned by one or more angle dimensions (figure 13-5A). When all holes are of the same diameter, a leader and note provide information about the size and number of holes. However, if there is a difference in hole size or machining process, each of the unequally spaced holes is dimensioned for position, size, and process as shown in figure 13-5B.

REPRESENTING AND DIMENSIONING MACHINED HOLES

Holes are represented and dimensioned on drawings in many different ways. ANSI, CSA, and SI metric standards are used to identify the shape

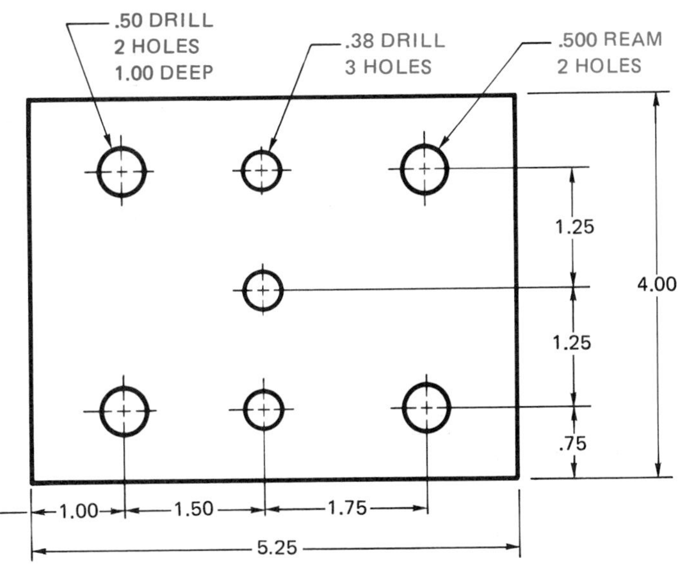

FIGURE 13-6 WORKING DRAWING WITH NOTES FOR HOLE SPECIFICATIONS

of a hole, its size, position, and the machining process. Holes may be straight, tapered, beveled, recessed, or threaded. Holes may be formed by manufacturing processes such as casting, forging, stamping, or pressing. Holes may be machined by drilling, boring, reaming, threading, countersinking, counterboring, and other processes.

Drilled, Reamed, and Bored Holes

Single or multiple holes may be located and accurately machined by reading the size and location dimensions, the diameter, and the number of required holes. These specifications are identified by a leader. Figure 13-6 is a typical dimensioned drawing. Note that after the hole centers are located, the leaders provide information about the hole size, machine process, and number of holes.

Unless a hole depth is indicated by a dimension or note, it is assumed the hole goes through the part. In figure 13-6, the 0.50″ drill holes are 1.00″ deep. The depth dimension refers to the length of the full diameter in the hole.

DIMENSIONING COUNTERBORED, COUNTERSUNK, AND SPOTFACED HOLES

When the form of a hole is altered, it is necessary to provide additional information on a drawing. Sometimes, a second and larger diameter is required in order to *recess* (enlarge) a hole so a bolt, nut, or shoulder may fit against a flat machined surface. Such a hole is *counterbored* and is specified in a note by the abbreviation **C'BORE**. A hole which is machined with a cone-shaped opening is identified as a *countersunk* (**CSK**) hole. When a limited diameter area is machined to provide a flat surface for a washer or nut or other shouldered part, the surface is said to be *spotfaced* (**SF**).

These are three common hole-forming processes. In each case, a leader is used with additional dimension and process information. Specified are the diameter, depth, process, and number of holes. Figure 13-7 shows how counterbored, countersunk, and spotfaced holes are represented and dimensioned.

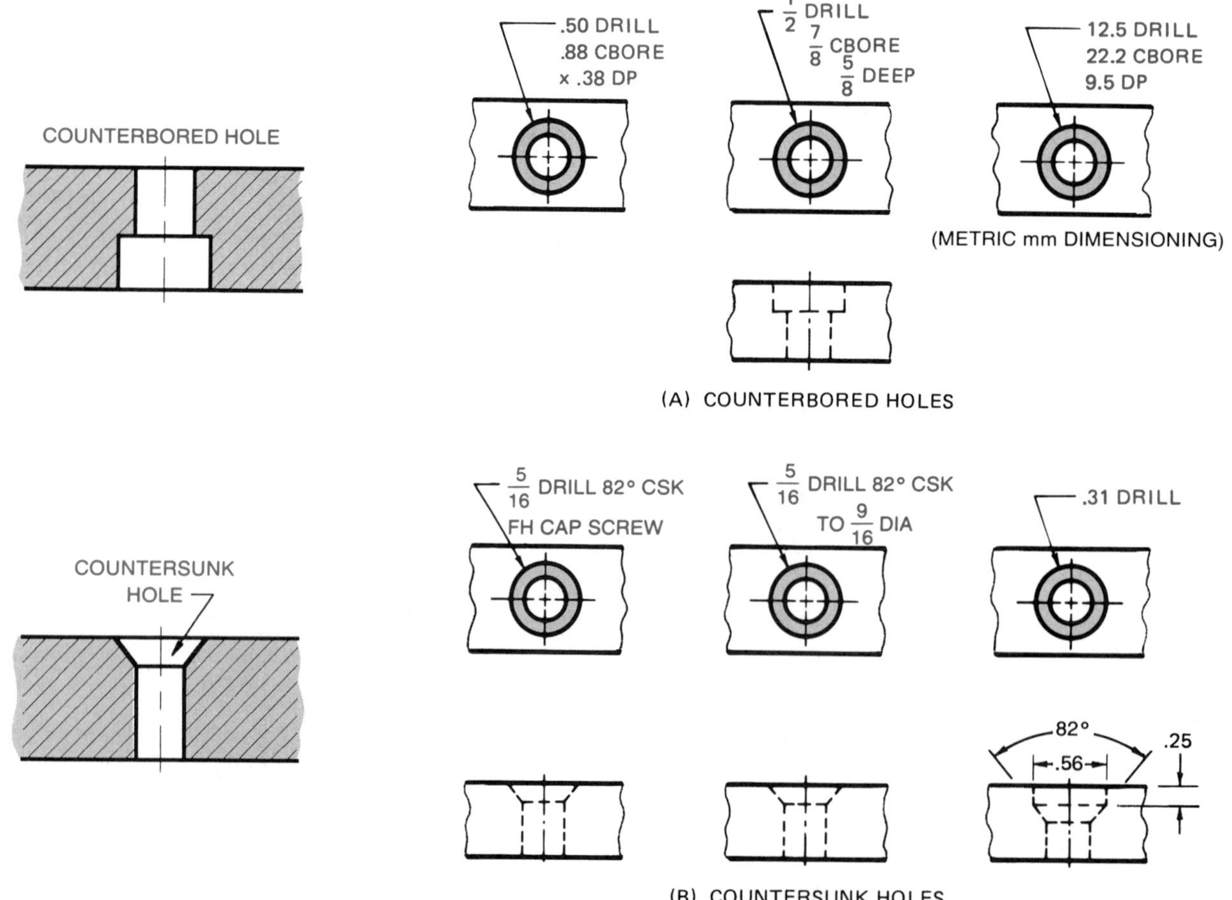

FIGURE 13-7 REPRESENTATION AND DIMENSIONING OF MACHINED HOLES

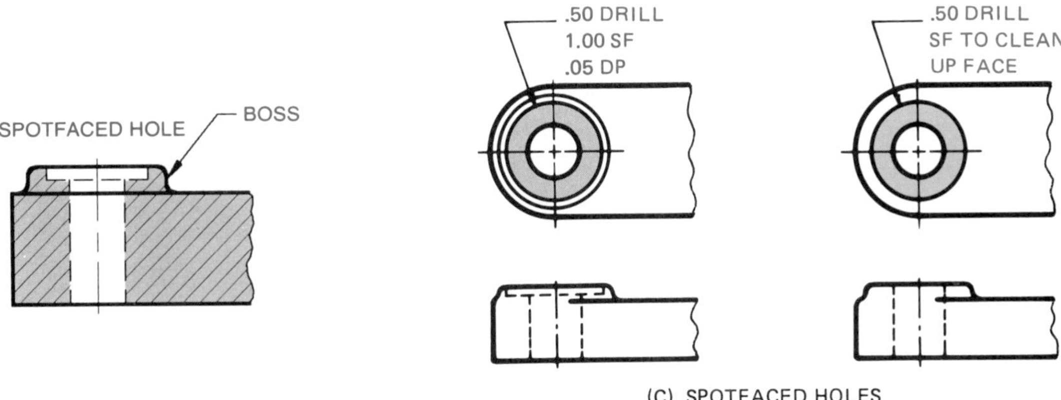

FIGURE 13-7 REPRESENTATION AND DIMENSIONING OF MACHINED HOLES (CONTINUED)

DIMENSIONING ANGLES

Two methods are generally used to dimension angles. The first method involves the use of linear dimensions, as illustrated in figure 13-8A. The second method requires the use of a numerical angle dimension to indicate the divergence (the amount one line diverges [moves away] from another) and a dimension to locate where the angle begins. A simple example is shown at figure 13-8B.

Angle dimensions are expressed in degrees (°) or fractional parts of a degree, like minutes (′) or seconds (″) of arc or decimal parts of a degree. The symbols for degrees or parts of a degree are placed after the numerical value of the angle. Note that a dimension line for an angle terminates in arrowheads.

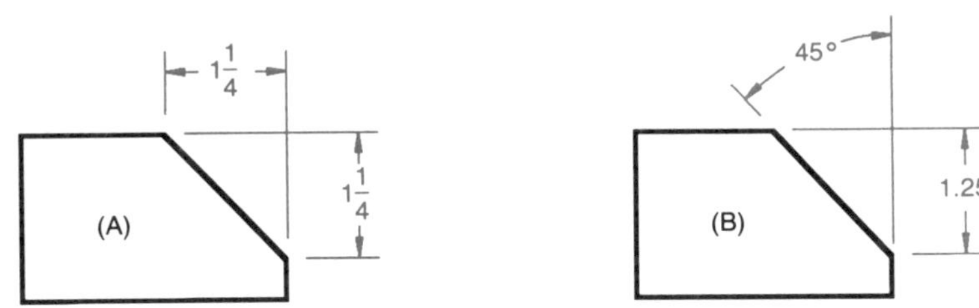

FIGURE 13-8 METHODS OF DIMENSIONING ANGLES

Figure 13-9A provides examples of how angle dimensions appear and are read on drawings. Where there is adequate space for dimensioning, the angle dimension may be placed along the arc as illustrated in figure 13-9B.

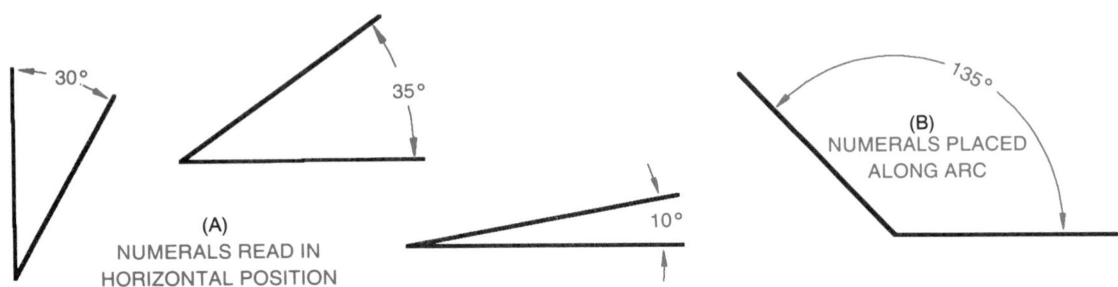

FIGURE 13-9 PLACING ANGULAR DIMENSIONS

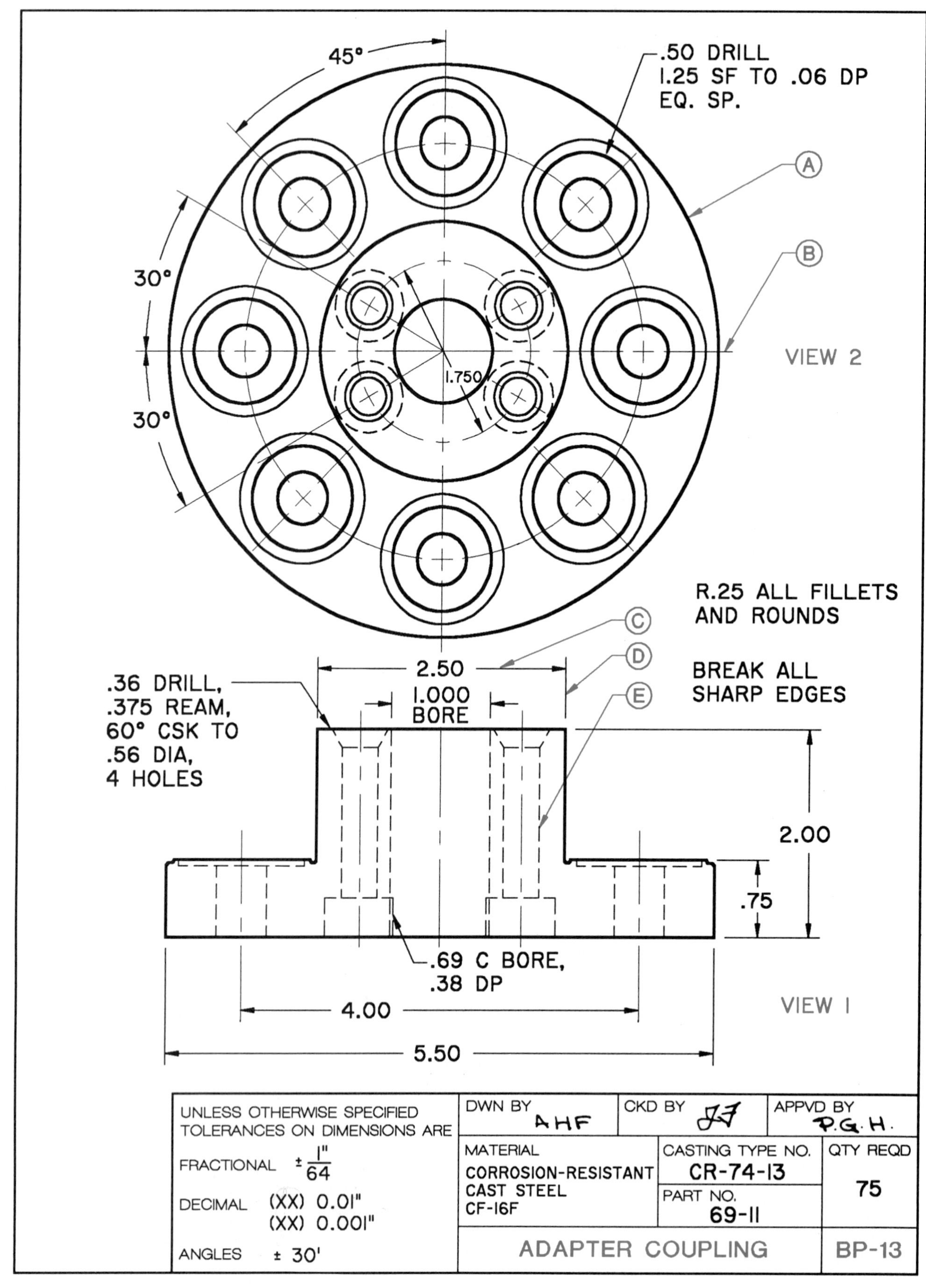
45°
30°
30°
.50 DRILL
1.25 SF TO .06 DP
EQ. SP.
A
B
1.750
VIEW 2
R.25 ALL FILLETS
AND ROUNDS
C
D
E
BREAK ALL
SHARP EDGES
2.50
1.000
BORE
.36 DRILL,
.375 REAM,
60° CSK TO
.56 DIA,
4 HOLES
2.00
.75
.69 C BORE,
.38 DP
4.00
5.50
VIEW 1
UNLESS OTHERWISE SPECIFIED
TOLERANCES ON DIMENSIONS ARE
FRACTIONAL ± 1"/64
DECIMAL (XX) 0.01"
(XX) 0.001"
ANGLES ± 30'
DWN BY AHF
CKD BY
APPVD BY P.G.H.
MATERIAL
CORROSION-RESISTANT
CAST STEEL
CF-16F
CASTING TYPE NO.
CR-74-13
PART NO.
69-11
QTY REQD
75
ADAPTER COUPLING
BP-13

ASSIGNMENT—UNIT 13

ADAPTER COUPLING (BP-13)

1. a. Identify the kind of material used in the production of the Adapter Couplings.
   b. Give the type number of the casting, part number, and required quantity.
2. Name **VIEW 1** and **VIEW 2**.
3. Identify the type of line represented at Ⓐ, Ⓑ, Ⓒ, Ⓓ, and Ⓔ.
4. a. State the method of dimensioning used on the drawing.
   b. Describe briefly one advantage of this method of dimensioning.
5. Provide the following nominal dimensions:
   a. Outside diameter of flange section.
   b. Outside diameter of hub section.
   c. Flange height.
   d. Overall Coupling height.
6. Identify the machining process required to produce all of the holes in the Coupling.
7. Determine the maximum and minimum dimensions for each of the following features:
   a. Bored hole.
   b. Circle diameter for the drilled holes in the flange.
   c. Circle diameter for the reamed holes in the hub.
   d. Diameter of the reamed holes.
8. Indicate the upper and lower limit angle dimensions for locating the first hole in the flange.
9. Give the dimensions and specifications for machining the four center holes.
10. Interpret the meaning of the note:

    .50 DRILL
    1.25 SF to .06 DP,
    8 HOLES
    EQ SP

11. Give the radius for all corners and fillets.
12. Indicate how all edges are to be finished.

Student's Name ______________________

1. a. ______________________
   b. Type No. ______________________
   Part No. ______________________
   Quantity Required ______________________
2. **VIEW 1** __________ **VIEW 2** __________
3. Type of Line
   Ⓐ __________ Ⓓ __________
   Ⓑ __________ Ⓔ __________
   Ⓒ __________
4. a. ______________________
   b. ______________________
5. a. Flange Diameter ______________________
   b. Hub Diameter ______________________
   c. Flange Height ______________________
   d. Coupling Height ______________________
6. Machining Processes
   ____
   ____
   ____
   ____

7.

	Maximum Diameter	Minimum Diameter
a.		
b.		
c.		
d.		

8.

Upper Angle Limit	Lower Angle Limit

9. ______________________
   ____
   ____
10. ______________________
    ____
    ____
11. ______________________
12. ______________________

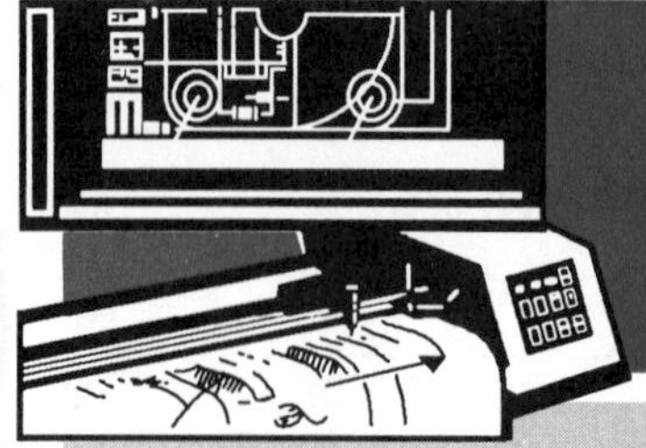

HIGH TECHNOLOGY applications

HT-2 HIGH AMPLIFICATION MEASUREMENT COMPARATORS

ELECTRONIC COMPARATORS IN MANUFACTURING

Electronic comparators have great versatility in precision measurement/inspection applications. The absence of moving parts (as in the case of mechanical comparators) permits the use of electronic comparators in taking measurements of moving features. Other advantages of electronic comparators include the following.

- High sensitivity and speed in taking measurements.
- Adaptability for measuring at multiple amplification ranges.
- Ease with which adjustments may be made.
- Comparative accuracy and repeatability of measurements.
- Minimum amount of time for self-checking.
- The use of at least two electronic measurement units and two gaging heads makes it possible to establish measurement variations in part sizes, thickness, diameter, flatness, and other dimensional and geometric tolerancing requirements.

Common magnification ranges are from ±0.010″ (±0.200 mm) to ±0.0001″ (±0.0020 mm). The two basic readout models include a highly sensitive dial face and a direct digital readout display.

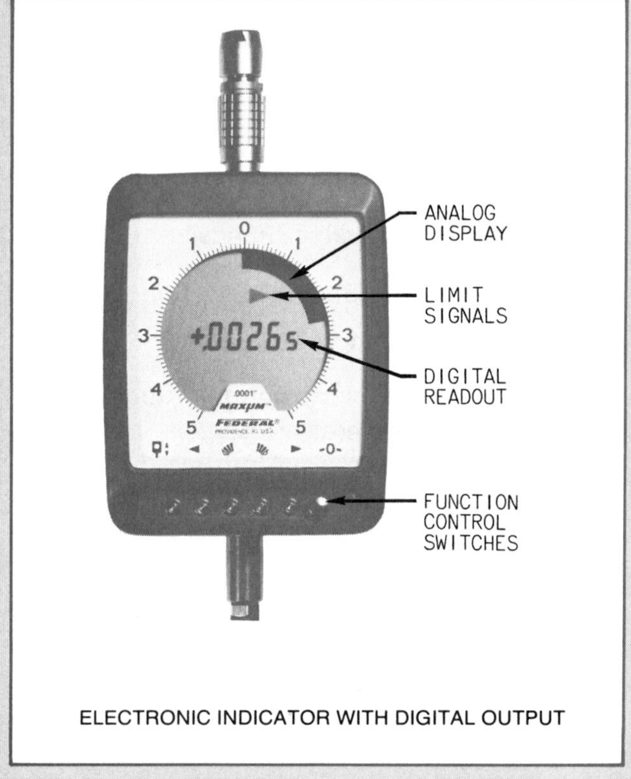

ELECTRONIC INDICATOR WITH DIGITAL OUTPUT

Electronic Indicator with Digital Output

The features of this type of instrument may be interfaced with other computer-aided hardware and software components. The model illustrated provides a clear digital display and an accurate analog reading of a measured feature of a part. In addition, a light limit signal flasher indicates when an analog reading is *under* or *over* the tolerance limit.

This instrument may be connected with a hand-held or mainframe computer. Measurement data may be interfaced with other CADD-CAM functions. Data may be fed into a system for analysis, decoding, transmission, storage, statistical process control in manufacturing, and to produce hardcopy printouts.

Some of the major features of the electronic indicator with digit output are shown by the illustration callouts. Fast, accurate readings appear on the *digital readout display.* Measurement deviations from zero (*over* and *under* size) are shown on the analog display graduation by *flashing limit signals.*

The combination of light-emitting diode control elements relate to measurement conditions. Six different measurement settings are provided by the *function control switches.* The six functions relate to automatic zero, true spindle position, upper limit, lower limit, count up, and count down.

The general *measurement range* of digital electronic indicators is from ±0.10000″ to ±0.0400″. The minimum *measurement value range* is from 0.00005″ to 0.0001″. Analog graduations range from 0.0001″ to 0.0010″.

The *analog display range* is from ±0.0050″ to ±0.040″. The comparable metric indicator range is from ±0.199 mm to ±1.000 mm (with minimum digit values of 0.001 mm). Minimum analog graduations are 0.001 mm and 0.010 mm, respectively.

Photo courtesy of FEDERAL PRODUCTS CORPORATION

UNIT 14
Base-Line/Datum, Ordinate, and Tabular Dimensioning

In general dimensioning practice, dimensions relate one feature with another. This kind of dimensioning is referred to as *point-to-point*. In other words, each dimension locates a surface or feature in relation to another surface, point, or feature.

As parts become more complicated in design, or high degrees of precision in measurements are required, or there are a number of similar features to dimension, it may be necessary to use other dimensioning techniques. This unit covers the following three different dimensioning practices:

- Base-line or datum dimensioning
- Ordinate dimensioning
- Tabular dimensioning

Using these techniques, drawings may be dimensioned without producing *cumulative errors* (errors that add). Each dimension is taken in relation to an exact reference point, line, plane, feature, or source known as a *datum* or a *datum line*. Essential information is derived from a datum, sometimes called a *base line*. By using any one of the three techniques, a dimension may be held to a closer tolerance than is possible by conventional dimensioning practices.

DATUM OR BASE-LINE DIMENSIONING

All measurements in *base-line* or *datum dimensioning* originate at a finished (reference) surface or a center line. The point of origin of a datum is identified on a drawing as *point zero* (0). Sometimes, the reference point is identified as a base line or datum line; the "0" as a datum.

Three general applications of base-line dimensioning are illustrated in figure 14-1. Note at (A) that each position dimension originates at a base (datum) line. The datum line at (B) is the center line. At (C) the vertical (height) dimensions for the contour start at the machined datum line.

Base-line dimensioning is used when a number of features are to be accurately laid out and measured within precise limits.

ORDINATE DIMENSIONING

Ordinate dimensioning is an *arrowless, rectangular datum dimensioning system*. All dimensions originate from two or three datum planes which are mutually related (figure 14-2). The datum planes in ordinate dimensioning are indicated as *zero coordinates*. Dimensions from a zero coordinate to a particular feature are given at the end of a thin, extension line after it passes through a datum line. Thus, dimension lines and arrowheads are not required for dimensioning.

Ordinate dimensioning is especially useful for programming numerically controlled machine tools and processes. In other machining and tooling operations, once the initial point of origin is established, graduated collars on the machine lead screws may be adjusted to zero. Positions of subsequent operations may then be read directly along the *X* (longitudinal), *Y* (transverse), or *Z* (vertical) axes. The settings correspond to the ordinate dimensions given on the drawing.

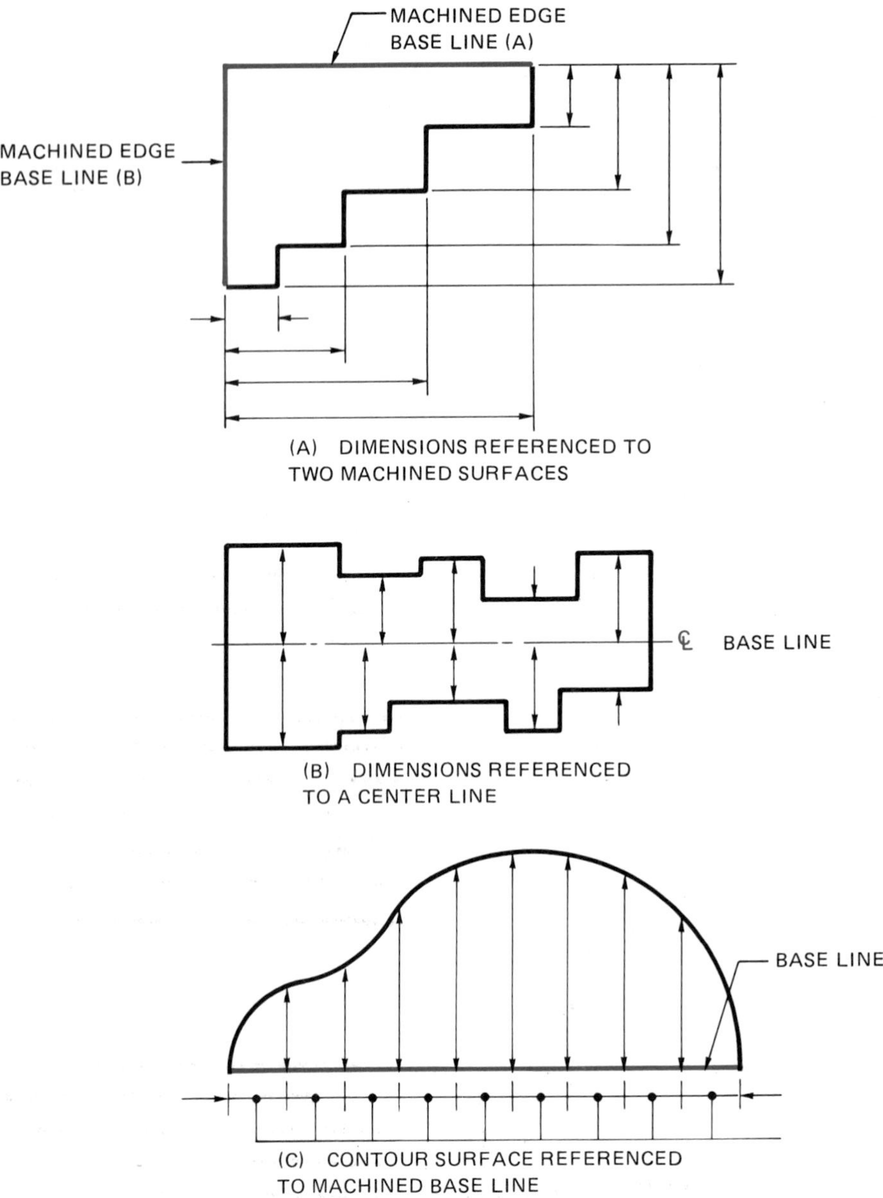

FIGURE 14-1 GENERAL APPLICATIONS OF BASE-LINE DIMENSIONING

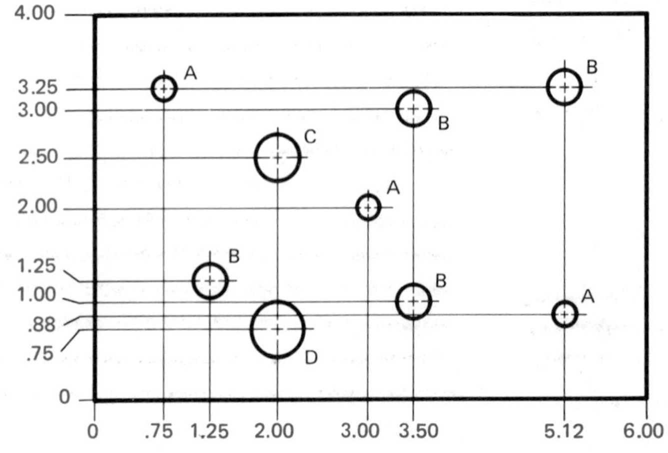

HOLE SYMBOL	DRILL SIZE
A	#6
B	3/8
C	1/2
D	0.56

FIGURE 14-2 ORDINATE DIMENSIONING

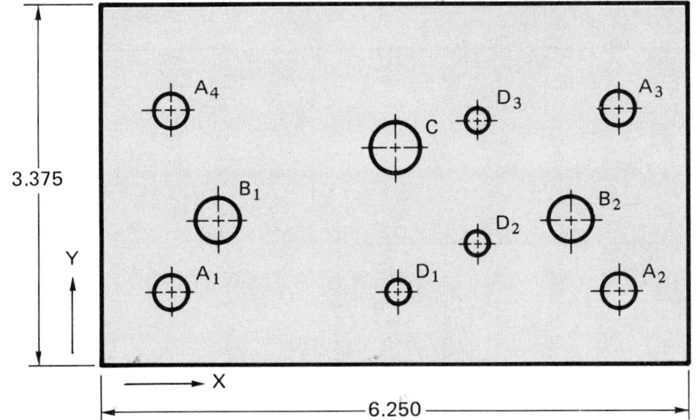

(A) DIMENSIONING TECHNIQUE

HOLE		LOCATION (AXES)	
SYMBOL	SIZE	X→	Y↑
A_1	3/8	0.75	0.75
A_2	3/8	5.50	0.75
A_3	3/8	5.50	2.63
A_4	3/8	0.75	2.63
B_1	7/16	1.25	1.50
B_2	7/16	5.00	1.50
C	1/2	3.12	2.25
D_1	15/64	3.12	0.75
D_2	15/64	4.00	1.25
D_3	15/64	4.00	2.50

(B) COORDINATE CHART OF TABULAR DIMENSIONS

FIGURE 14-3 TABULAR DIMENSIONING

Further simplification is possible by using a complementary table when there are a number of features represented on a drawing. Figure 14-2 provides an example of a drawing with ordinate dimensioning. The table lists the diameter of each of the four holes. These holes are identified by letters which correspond with those on the drawing.

TABULAR DIMENSIONING AND COORDINATE CHARTS

Tabular dimensioning is recommended as another form of datum dimensioning for drawings of parts which have many hole sizes and repetitive details. If conventional drafting practices were followed, it would be difficult to draw each center line and/or surface position and still be able to interpret a drawing clearly and accurately.

Datum lines are usually identified as *location axes*. All holes or repeat features are identified by a symbol, mostly by a letter. Each hole of the same size is identified by a letter and a subscript numeral.

Figure 14-3A is a *tabular dimensioned drawing*. There are four different hole sizes shown: A (3/8″), B (7/16″), C (1/2″), and D (15/64″). There are, however, four hole locations for the "A" holes. These are shown on the drawings as A_1, A_2, A_3, and A_4.

Similarly, there are two hole positions for B (B_1 and B_2), one position for C, and three positions for D (D_1, D_2, and D_3). The only dimensions which appear on the drawing are the overall width of 6.250″ and length of 3.375″.

The position dimensions for each hole in relation to the datum axes (X and Y) are given in chart form (figure 14-3B). The chart presents an identification of each hole, its size, and positional dimensioning along the X→ and Y↑ axes. Operations requiring dimensions along a Z ↗ axis are represented in a second view. In such cases, the dimensional information is contained in an additional *"Z" axis column* in the chart.

Dimensions in base-line, ordinate, and tabular dimensioning may be expressed in customary inch or SI metric units of measurement, or in both systems. A drawing may also include another chart to provide additional specifications for each feature. For example, if a number of hole-forming operations were required to produce a part, the added information would deal with drilling, reaming, tapping, and so forth.

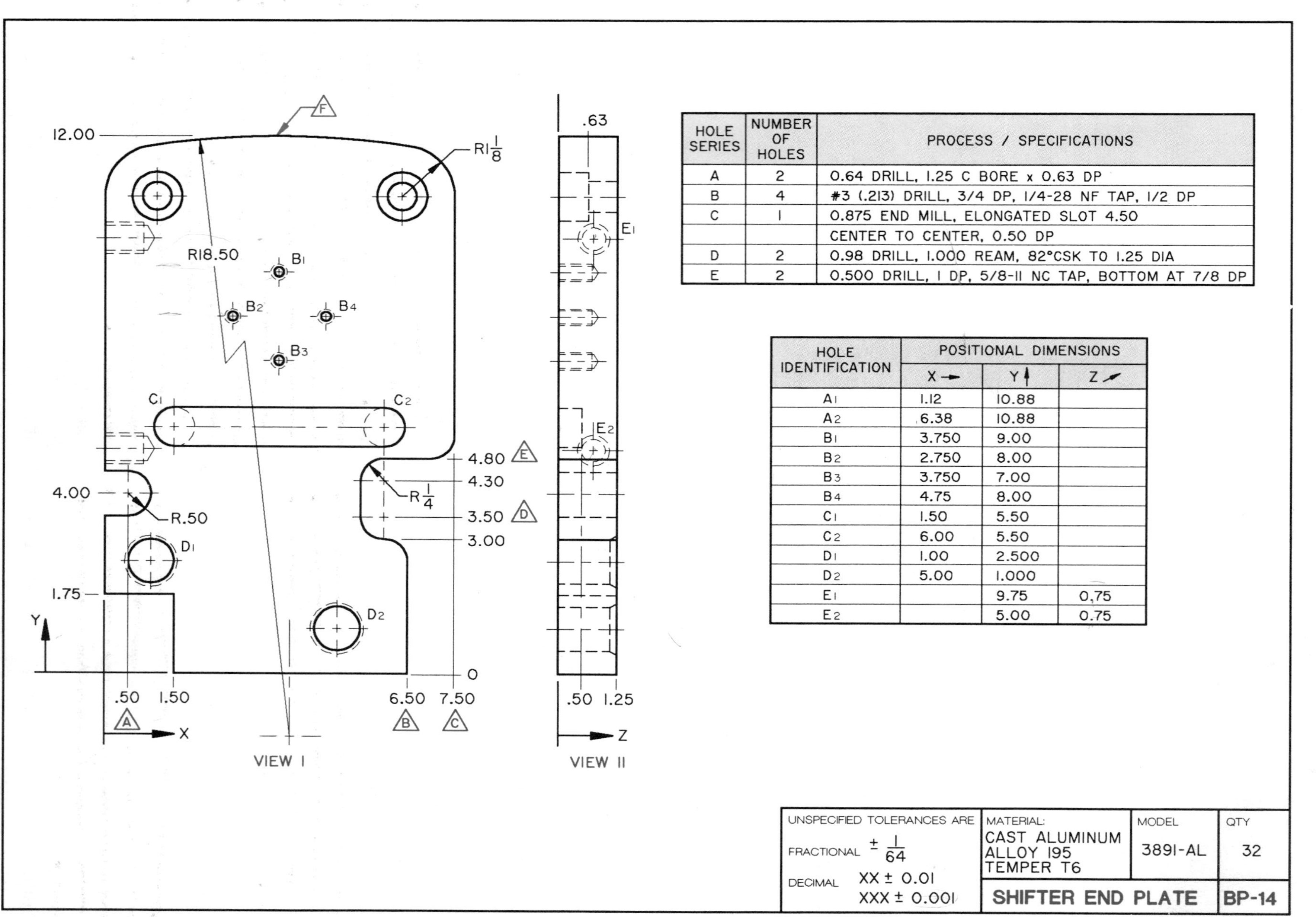

HOLE SERIES	NUMBER OF HOLES	PROCESS / SPECIFICATIONS
A	2	0.64 DRILL, 1.25 C BORE x 0.63 DP
B	4	#3 (.213) DRILL, 3/4 DP, 1/4-28 NF TAP, 1/2 DP
C	1	0.875 END MILL, ELONGATED SLOT 4.50
		CENTER TO CENTER, 0.50 DP
D	2	0.98 DRILL, 1.000 REAM, 82°CSK TO 1.25 DIA
E	2	0.500 DRILL, 1 DP, 5/8-11 NC TAP, BOTTOM AT 7/8 DP

HOLE IDENTIFICATION	POSITIONAL DIMENSIONS		
	X →	Y ↑	Z ↗
A_1	1.12	10.88	
A_2	6.38	10.88	
B_1	3.750	9.00	
B_2	2.750	8.00	
B_3	3.750	7.00	
B_4	4.75	8.00	
C_1	1.50	5.50	
C_2	6.00	5.50	
D_1	1.00	2.500	
D_2	5.00	1.000	
E_1		9.75	0.75
E_2		5.00	0.75

UNSPECIFIED TOLERANCES ARE	MATERIAL:	MODEL	QTY
FRACTIONAL ± 1/64 DECIMAL XX ± 0.01 XXX ± 0.001	CAST ALUMINUM ALLOY 195 TEMPER T6	3891-AL	32
	SHIFTER END PLATE		BP-14

ASSIGNMENT—UNIT 14

SHIFTER END PLATE (BP-14)

Student's Name ______________________

1. a. Give the material specifications for the Shifter End Plate.
   b. Indicate the model number and required quantity.
2. a. Name **VIEW 1** and **VIEW 2**.
   b. State briefly why **VIEW 2** is used.
3. Give the overall sizes for the width, height, and depth of the Plate.
4. Classify dimensions (A), (B), (C), (D), and (E) according to type.
5. a. Identify the two dimensioning methods used on the Shifter End Plate drawing.
   b. State two distinguishing characteristics for arrowless dimensioning.
6. a. Give the number of datums used.
   b. Indicate how each datum is specified.
7. a. State the number of holes which are to be machined for each series of holes.
   b. Name the hole-forming process(es) required to machine each hole series.
8. a. Give the letter of each series of tapped holes.
   b. Provide complete specifications for each series of tapped holes.
9. a. Determine the center (position) dimensions for holes **D_1** and **D_2**.
   b. State the machining specifications for the **D** series holes.
10. a. Give the locational dimensions for the center of **C_1** to start the end-milling cut for the elongated slot.
    b. Determine (1) the nominal height and (2) nominal overall width of the slot.
    c. State the minimum and the maximum dimensions for the height and width of the slot.
11. Give the maximum radius for (a) arc (F) and (b) the rounded corners on the arc portion.
12. Compute the minimum center-to-center **Y** axis dimensions between the following holes:
    a. **B_1** and **B_2** c. **D_1** and **D_2**
    b. **B_1** and **B_3** d. **E_1** and **E_2**
13. Compute the maximum center-to-center **X** axis dimensions between the following holes:
    a. **A_1** and **A_2** b. **B_2** and **B_4** c. **D_1** and **D_2**
14. Determine the following maximum and minimum depths along the **Z** axis:
    a. The elongated slot b. The counterbored hole

---

1. a. ______________________
   ____
   b. Model ________ Quantity ________
2. a. **VIEW 1** ______________________
   VIEW 2 ______________________
   b. ______________________
   ____
   ____
   ____
3. Maximum: width ________; height ________; depth ________
4.

Dimension	Type	
	Size	Position
(A)		
(B)		
(C)		
(D)		
(E)		

Center-position

5. a. (1) ______________________
   (2) ______________________
   b. (1) ______________________
   ____
   ____
   (2) ______________________
   ____
6. a. ______________________
   b. ______________________
   ____
   ____

7.

Series	# of Holes	Machining Process(es)

8. Series (a) Specifications (b)
   ____
   ____
   ____
   ____
9. a. **X** Axis **Y** Axis
   D_1 ________ ________
   D_2 ________ ________
   b. ______________________
   ____
10. a. **X** Axis ________ **Y** Axis ________
    b. (1) Nominal Height = ________
    (2) Nominal Width = ________
    c. Height Width
    Minimum ________ ________
    Maximum ________ ________
11. a. Maximum ________
    b. Maximum Corners ________
12. **Y** Axis Minimum Dimensions
    a. ________ c. ________
    b. ________ d. ________
13. **X** Axis Maximum Dimensions
    a. ________ c. ________
    b. ________
14. Max. Depth Min. Depth
    a. Slot ________ ________
    b. C'bored Hole ________ ________

SECTION 5
SI Metric System of Projection and Dimensioning

UNIT 15
SI Metric Measurements and Dimensioning

The *Conference Générale des Poids et Mesures* is an international body which is responsible for all matters concerning the metric system. In 1960, this General Conference on Weights and Measures named the evolving metric system, which was approved by leading industrial nations, the *Systeme International d'Unites*. It is this system which is abbreviated in all languages as *SI* (metric).

The SI system is a consolidation and refinement of other metric systems. SI is based on decimals and avoids fractions. By multiplying or dividing a base unit it is possible to form units of different sizes.

There are seven *base units of measure* in the SI system. The three which are widely used in physical mechanics computations include the base quantities of *length (meter, m)*, *mass (kilogram, kg)*, and *time (second, s)*. The other four base units are the *ampere (A)* for *electric current*, the *Kelvin (K)* for *temperature*, the *candela (cd)* for *light*, and the *mole (mol)* for the amount of substance. All other units of measurement in SI metric, which include *supplementary* and *derived units*, are founded on the seven base units.

ADVANTAGES OF SI METRICS

- Each base unit is accurately defined in terms of physical measurement. Each base unit can be accurately produced throughout the world.
- SI provides well-defined symbols, abbreviations, precise definitions, and techniques of representation. These make it possible to read metric drawings internationally with uniform interpretation.
- SI retains the decimal-arithmetic relationship which exists among the older metric systems. Units of different sizes for each physical quantity are formed by multiplying or dividing a single base value by adding zeros or shifting decimal points.

THE SI SYSTEM OF MEASUREMENT

To be able to interpret drawings prepared according to SI standards, it is necessary to understand differences in representation techniques, units of measure, dimensioning practices, converting measurement values, and dual dimensioning. These key items are covered in this Section.

Quantities are easily computed in SI metrics by using *prefixes*. Each different prefix indicates a specific measurement value. For example, the prefix *milli* in *milli*meter means the measurement is one-thousandth part of one meter (milli = 1/1000). The prefix *centi* in *centi*meter means one-hundredth part of a meter. A *deci*meter = 1/10 meter; *deci*liter = 1/10 liter.

Larger dimensions are designated by other prefixes. The most common is the *kilo*. *Kilo* means one thousand times greater than the base unit. One *kilo*meter = 1000 meters. One *kilo*gram = 1000 grams.

(A) CONVERSION OF FRACTIONAL INCH VALUES TO METRIC UNITS

Fractional Inch	*mm Equivalent*	*Fractional Inch*	*mm Equivalent*	*Fractional Inch*	*mm Equivalent*	*Fractional Inch*	*mm Equivalent*
1/64	0.397	17/64	6.747	33/64	13.097	49/64	19.447
1/32	0.794	9/32	7.144	17/32	13.494	25/32	19.844
3/64	1.191	19/64	7.541	35/64	13.890	51/64	20.240
7/32	5.556	15/32	11.906	23/32	18.256	31/32	24.606
15/64	5.953	31/64	12.303	47/64	18.653	63/64	25.003
1/4	6.350	1/2	12.700	3/4	19.050	1	25.400

(B) CONVERSION OF METRIC TO INCH-STANDARD UNITS OF MEASURE

mm Value	*Inch (decimal) Equivalent*	*mm Value*	*Inch (decimal) Equivalent*	*mm Value*	*Inch (decimal) Equivalent*	*mm Value*	*Inch (decimal) Equivalent*
.01	.00039	.34	.01339	.67	.02638	1	.03937
.02	.00079	.35	.01378	.68	.02677	2	.07874
.03	.00118	.36	.01417	.69	.02717	3	.11811
.23	.00906	.56	.02205	.89	.03504	23	.90551
.24	.00945	.57	.02244	.90	.03543	24	.94488
.25	.00984	.58	.02283	.91	.03583	25	.98425

(C) DECIMAL EQUIVALENTS OF FRACTIONAL, WIRE GAGE, LETTER AND METRIC SIZE DRILLS
(0.0059" (0.15 mm) TO 1.000" (25 mm)

Decimal	*Inch*	*Wire*	*mm*	*Decimal*	*Inch*	*Wire*	*mm*	*Decimal*	*Inch*	*Wire*	*mm*	*Decimal*	*Inch*	*Wire*	*mm*	*Decimal*	*Inch*	*Wire*	*mm*
.0059		97	.15					.0532			1.35					.1495		25	
.0063		96	.16	.0217			.55	.0550		54		.0938	3/32			.1496			3.80
.0067		95	.17	.0225		74		.0551			1.40	.0945			2.40	.1520		24	
.0071		94	.18	.0236			.60	.0571			1.45	.0960		41		.1535			3.90
.0075		93	.19					.0591			1.50					.1540		23	
.0079		92	.20	.0240		73		.0595		53		.0965			2.45	.1562	5/32		
.0083		91	.21	.0250		72		.0610			1.55	.0980		40		.1570		22	

FIGURE 15-1 PARTIAL CONVERSION TABLES (CUSTOMARY INCH AND SI METRIC EQUIVALENTS)

Symbols are also used to identify measurement values and to express them quickly in shortened form. Measurements in meters are identified by the symbol (m); millimeters, (mm); micrometers, (μm); and kilometers (km). Similarly, each base, supplementary, and derived unit of measurement is identified by a symbol.

TABLES OF CONVERSION VALUES

It is common practice for a manufacturer's literature to include measurement information in both customary inch and SI metric units. However, standards within the two basic systems differ for design features such as thread forms, sizes and series, tapers, gearing, drill sizes, and so forth. It is necessary for the technician and craftsperson to be able either to locate information in a trade or engineering handbook or to compute needed values.

Figures 15-1A and 15-1B have been extracted from two of the most widely used simple conversion tables. Decimals and fractions in customary inch units are cross-referenced with SI metric millimeter values. Figure 15-1C is a portion of another representative conversion table. Standard fractional, decimal, wire gage number, or letter drills in the customary inch system are listed with an equivalent millimeter size. It must be emphasized that the millimeter equivalents are not representative of the metric sizes which are commercially available.

A few selected complete tables relating to linear units of measurement and conversion factors are included in the appendix as tables A-1 through A-4.

BASE UNIT OF LINEAR MEASURE

In SI, the base unit of linear measure is the meter (m). One meter is the equivalent of 39.37 inches. The most widely used, functional SI linear unit is the millimeter. One millimeter (mm) = 0.03937″ or 1/25.4 part of an inch. 0.001″ = 0.0254 mm; 1.000″ = 25.4 mm. This means that a millimeter value equivalent of a customary inch measurement may be computed by multiplying a decimal inch value by 25.4. The reverse process of conversion requires dividing an SI measurement in millimeters by 25.4 to obtain the equivalent customary inch value. Unless otherwise stated, SI metric values indicated on a drawing are assumed to be in millimeters.

ROUNDING OFF DECIMALS

A typical note on a drawing reads

UNLESS OTHERWISE SPECIFIED, TOLERANCES ON DIMENSIONS ARE:

XX ±0.02 mm
XXX ±0.002 mm

The number of digits in the note indicates the extent to which a computed dimension needs to be *rounded off*. The degree of accuracy to which a part is machined affects production time and costs. Therefore, it is important (where practical) to round off the limits of each dimension. Each design feature should be held to the least number of digits which will provide the greatest tolerance and still ensure interchangeability.

One common practice in rounding off values is to increase the last required digit by one (1) when the digit following on the right is five (5) or greater.

Examples

7.6245″	rounded to three decimal places = 7.625″
16.7374 mm	rounded to three decimal places = 16.737 mm
1.8349″	rounded to two decimal places = 1.83″
1.8351 mm	rounded to two decimal places = 1.84 mm.

When the last decimal digit number is five (5) followed by zeros, the last required digit is increased by one (1) *if it is an odd number*. The last digit remains unchanged *if it is an even number*. For example, 1.835 is rounded to 1.84. However, 1.825 is rounded to 1.82.

As previously stated, factors are available in handbook tables to convert one physical quantity in one system to a quantity in the other system. The kilogram, as a unit of mass, may be converted to an equivalent value in pounds. Conversion factors are included in handbooks for square, cubic, dry, and liquid units and all other physical quantities.

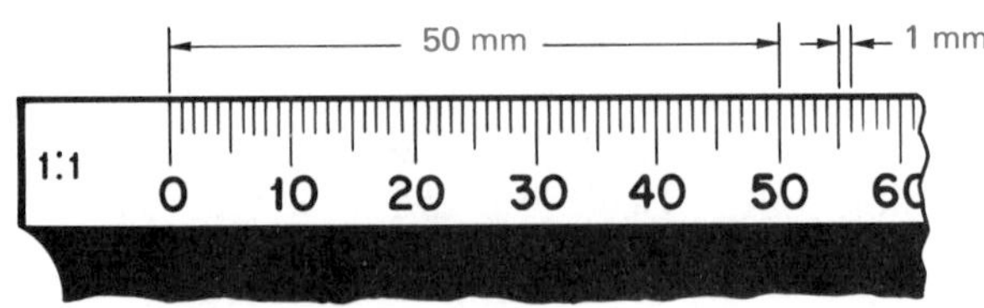

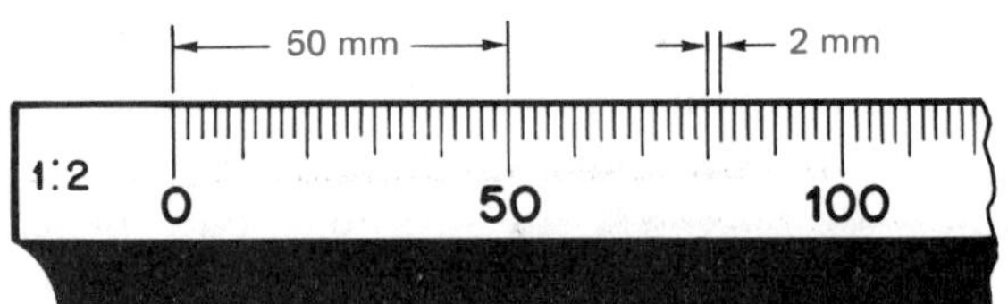

COMPARISON OF A 50-mm MEASUREMENT ON FOUR DIFFERENT METRIC SCALES

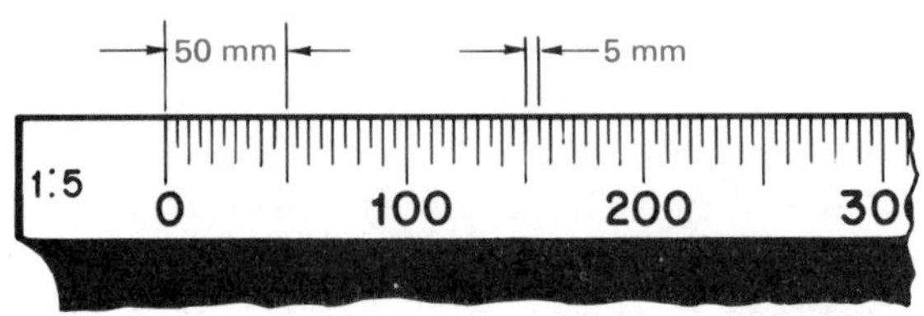

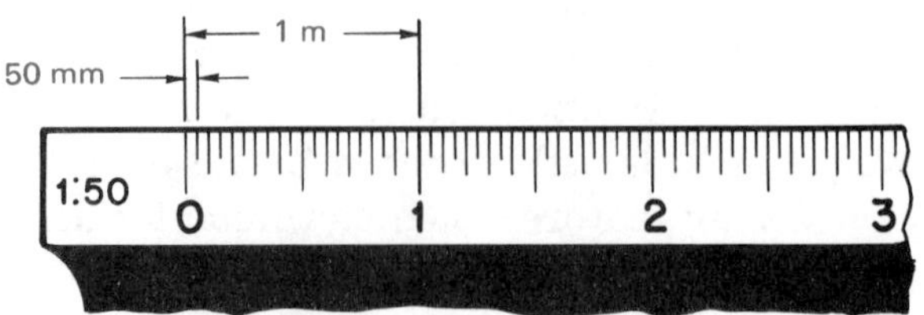

FIGURE 15-2 SAMPLE METRIC SCALES ON DRAFTING RULES

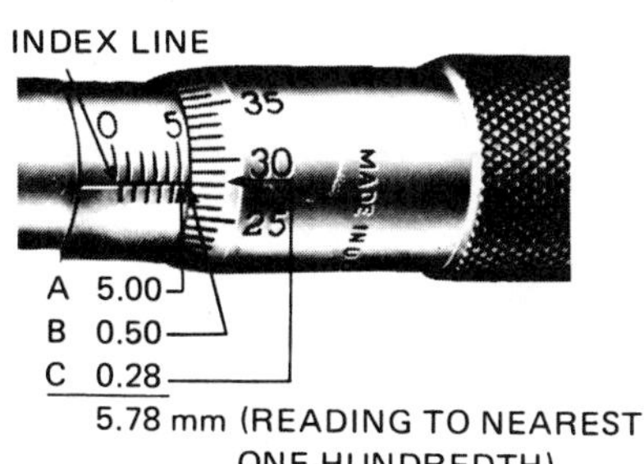

FIGURE 15-3 DETERMINING A METRIC MICROMETER MEASUREMENT (COURTESY OF THE L.S. STARRETT COMPANY)

SCALES OF METRIC DRAWINGS

Metric drawings which cannot be drawn full scale may be represented by reducing the scale to 1/2, 1/5, 1/10, 1/20, 1/50, 1/100, and up to 1/1000 size. Fine, small, precision objects are drawn to scales which are 2, 5, or 10 times the actual size. Samples of 1:1, 1:2, 1:5, and 1:50 metric scales are shown in figure 15-2.

TAKING LINEAR MEASUREMENTS

The most functional tools used for taking linear metric measurements are the steel rule, engineer's rule, and the metric micrometer. Precision measurements may be taken within 1/100th of a millimeter using a standard micrometer. Direct readings are made from the position of the graduations on the *sleeve* and *thimble of the micrometer.* These two features are shown in figure 15-3. The reading in this illustration includes the number of whole millimeters (5), plus the number of half millimeters (0.50), plus the hundredths reading on the thimble (0.37).

5.00	(whole millimeters)
0.50	(half millimeters)
0.28	(hundredths of a millimeter)
5.78 mm	(reading to the nearest one-hundredth of a mm)

Finer measurements may be taken to accuracies of 0.002 mm with a vernier graduated metric micrometer.

METRIC DIMENSIONING PRACTICES

Many drawings with metric dimensions usually contain a note indicating the units are **METRIC**. In European practice, a comma is used instead of a decimal point to indicate decimals. Drafting practice in the United States and other nations using customary inch units is to use the decimal point instead of a comma.

Features like Unified and American standard screw threads, pipe sizes, diameters of number drills, and metal sheets are identified by such established sizes as:

$\frac{1}{2}$ – 13 NC TAP, $\frac{3}{8}$" PIPE THREAD, #7 DRILL SIZE

Another practice on European metric drawings is to use spaces instead of commas when a value has more than four digits on either side of the decimal point. A value like 4,928 is written on a metric drawing as 4 928; 0.12853 appears as 0.128 53. Also, a zero is used to precede a decimal point for values smaller than one (1). A measurement of .437 mm is written as 0.437 mm.

Fractional parts of metric units are expressed as a decimal value and not as common fractions. For instance, the correct form is 0.5 mm, not 1/2 mm; 0.75 mm, not 3/4 mm.

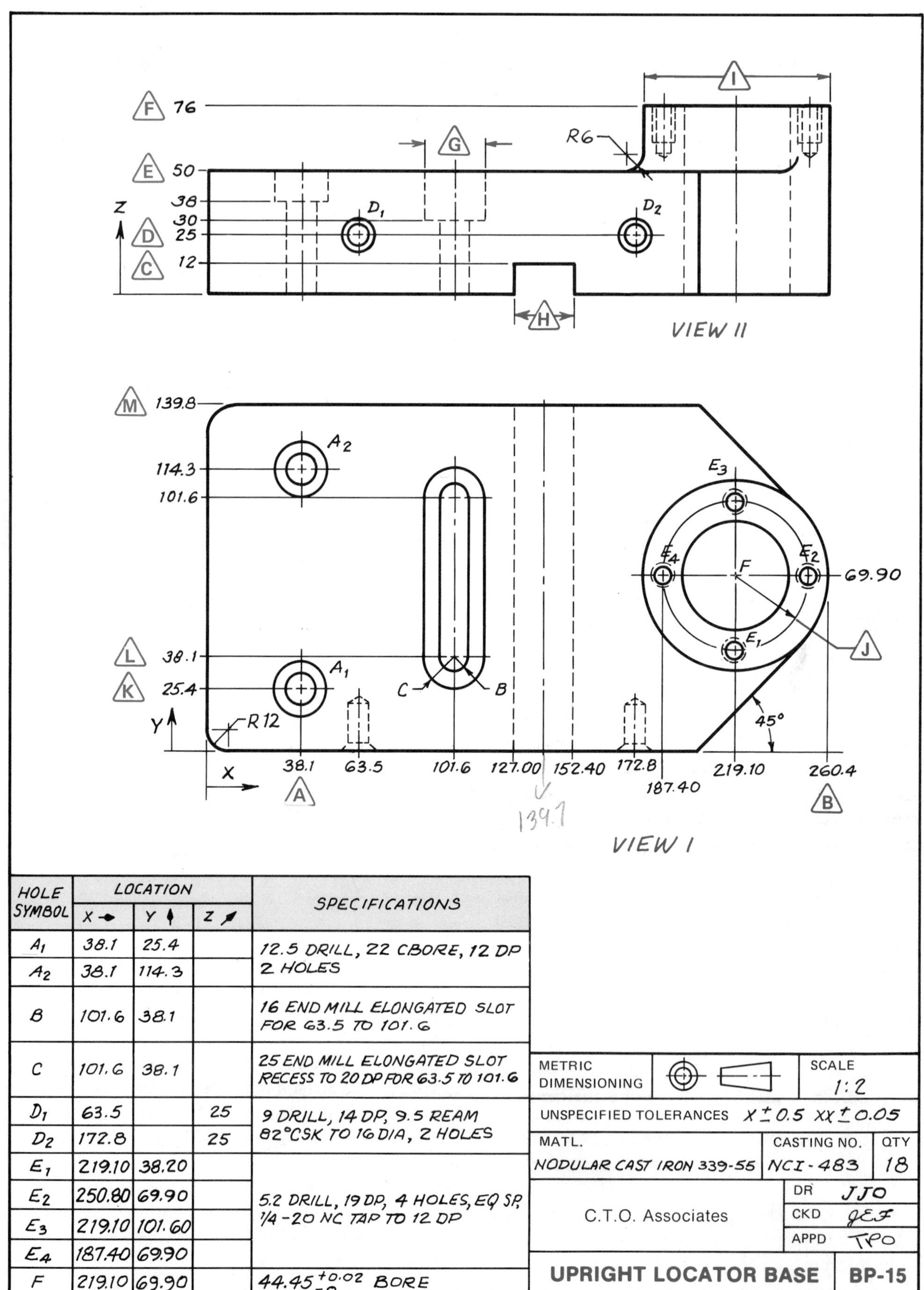

HOLE SYMBOL	LOCATION			SPECIFICATIONS
	X	Y	Z	
A_1	38.1	25.4		12.5 DRILL, 22 CBORE, 12 DP 2 HOLES
A_2	38.1	114.3		
B	101.6	38.1		16 END MILL ELONGATED SLOT FOR 63.5 TO 101.6
C	101.6	38.1		25 END MILL ELONGATED SLOT RECESS TO 20 DP FOR 63.5 TO 101.6
D_1	63.5		25	9 DRILL, 14 DP, 9.5 REAM 82°CSK TO 16 DIA, 2 HOLES
D_2	172.8		25	
E_1	219.10	38.20		5.2 DRILL, 19 DP, 4 HOLES, EQ SP, 1/4-20 NC TAP TO 12 DP
E_2	250.80	69.90		
E_3	219.10	101.60		
E_4	187.40	69.90		
F	219.10	69.90		$44.45^{+0.02}_{-0}$ BORE

UPRIGHT LOCATOR BASE (BP-15)

1. Identify the base unit of measure used on the Upright Locator Base drawing.
2. Give (a) the material specification for the part and (b) the casting number.
3. Determine the following nominal casting sizes:
   a. Length
   b. Height (thickness) of boss
   c. Height (thickness) of base
   d. Width of base
4. Classify dimensions A through F (position or size).
5. a. Name the two methods used in dimensioning the drawing.
   b. How many reference surfaces are used in dimensioning?
   c. Give the symbol used for each datum.
6. a. List each different series of holes.
   b. Determine the number of holes in each hole series which are to be machined.
   c. Indicate the machining process for forming each hole.
7. Furnish the specifications for the **A** and **D** series of holes.
8. Compute the nominal center-to-center dimension between the following sets of holes:
   a. A_1 and A_2 (**Y** axis)
   b. D_1 and D_2 (**X** axis)
   c. **A** and **C** (**X** axis)
   d. **A** and **F** (**X** axis)
9. a. Determine the maximum overall widths and lengths of the stepped elongated slot.
   b. Give the maximum depth to which the recessed shoulder portion of the slot is machined.
   c. Compute the maximum distance from the center of the elongated slot to the center of the grooved feature.
10. Compute the minimum and maximum widths for features G and H.
11. Determine the maximum outside diameter of feature I and the minimum diameter of circle J.
12. Convert linear dimensions K, L, and M to customary inch equivalents.
    Note. Round off each decimal value to two places.
13. Refer to appendix table A-4 for drill sizes.
    a. Give the metric drill diameter for holes **A**, **D**, and **E**.
    b. Use the table to establish the closest inch standard drill number, letter, or fractional equivalent for drills **A**, **D**, and **E**.
14. Give the maximum and minimum diameter for boring hole F.

ASSIGNMENT—UNIT 15

Student's Name ______________________

1. Measurement Unit ______________________
2. a. ______________________
   b. Casting # ______________________
3. a. __________ c. __________
   b. __________ d. __________
4. A __________ D __________
   B __________ E __________
   C __________ F __________
5. a. ______________________
   b. ______________________
   c. ______________________
6.

(a) Hole Series	(b) No. of Holes	(c) Machining Process(es)

7. **A** ______________________
   D ______________________
8. Holes — Center-to-center Distance
   a. A_1 and A_2 __________
   b. D_1 and D_2 __________
   c. **A** and **C** __________
   d. **A** and **F** __________
9. Max. Width — Overall Length
   a. Bottom Area __________ __________
   Top Recessed Area __________ __________
   b. Max. Depth __________
   c. Center Distance __________
10. Dimensions Min. Max.
    G __________ __________
    H __________ __________
11. I = __________ J = __________
12. K = __________ L = __________
    M = __________
13.

Hole	Metric Drill Size	Customary Inch Drill Size
A		
D		
E		

14. Max. Dia. = __________ Min. Dia. = __________

UNIT 16
SI Metric: First-Angle Projection

The SI metric system of orthographic projection also provides for an object to be represented graphically on a flat surface by a line drawing. There are *three planes* on which views are projected: *horizontal* (X axis), *frontal* (Y axis), and *profile* (Z axis). The number and direction of each plane is the same as for third-angle projection.

INTERNATIONAL USE OF FIRST-ANGLE AND THIRD-ANGLE PROJECTION

Most drawings in the United States and Canada use third-angle projection. By contrast, European and other nations generally use first-angle projection for SI metric drawings. There is also a great deal of interchange of drawings among industries within a nation and with foreign companies. Thus, regardless of the projection system used, it is necessary for each worker to be able to interpret drawings accurately in both first-angle and third-angle projection.

REVIEW OF THIRD-ANGLE PROJECTION

The three planes of projection in third-angle projection are reviewed in figure 16-1. The planes form four quadrants: I, II, III, and IV. All drawings up to this unit have been considered as being projected from quadrant III. Third-angle projection was adopted by the United States and Canada because each projected view and all features are easily identified and interpreted. Figure 16-2 reviews the three basic planes of projection and the position of the three most widely used views.

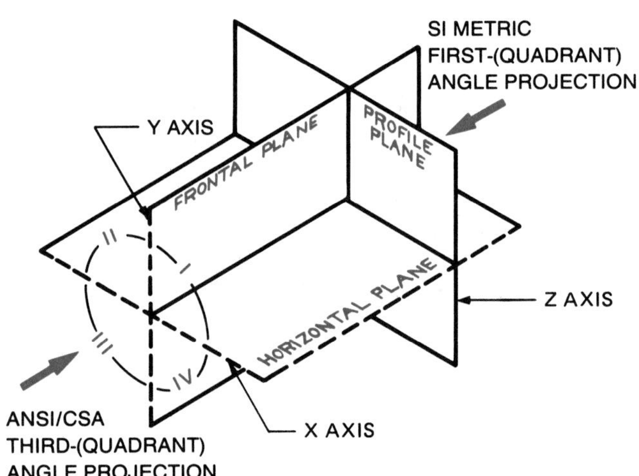

FIGURE 16-1 QUADRANTS AND PLANES USED IN ORTHOGRAPHIC PROJECTION

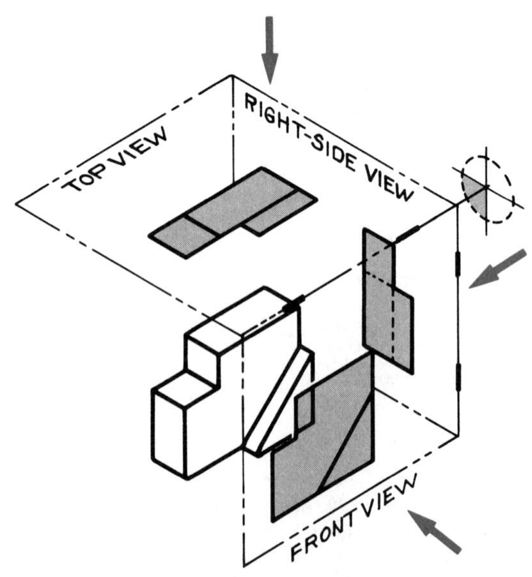

FIGURE 16-2 ANSI THIRD-ANGLE PROJECTION AND POSITION OF COMMON VIEWS

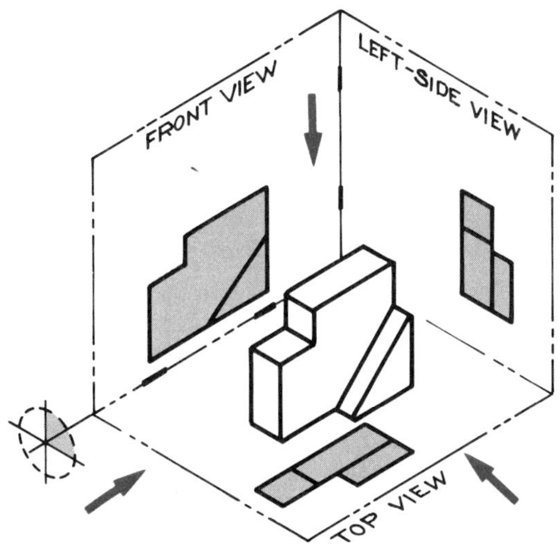

FIGURE 16-3 SI FIRST-ANGLE PROJECTION AND POSITION OF COMMON VIEWS

FIRST-ANGLE SI METRIC PROJECTION SYSTEM

Drawings which conform to SI metric standards are composed of views projected from *quadrant I*. The system is known as *first-angle projection*. The three planes of projection are still the *horizontal*, *vertical*, and *frontal planes*. In this case, the planes are positioned in quadrant I as seen in figure 16-3.

In first-angle projection, each view is projected *through the object*. Each surface, point, and feature is projected away from the surface of the object being viewed to a corresponding plane which is in back of or below the object.

First-angle projection is not natural and makes the preparation, visualization of details, and the interpretation of a drawing more difficult than third-angle projection.

The three most common views in first-angle projection are shown in figure 16-3. It is important to note the *difference in the position of each view* in comparison with third-angle projection. If the planes of projection in quadrant I are revolved onto a flat plane, the relative position of each view is indicated as a *front view*, *left-side view*, and *top view*.

By contrast, the view that normally is the left-side view in third-angle projection is placed to the right of the front view. The top view in first-angle projection occupies the position of the bottom view in third-angle projection. The six principal views are used in both systems. Stated again, one-view, two-view, and three-view drawings are common. The number and selection of views depend on the complexity of the part. Also, auxiliary views are used to show the true size and shape of features on angular surfaces. Figure 16-4 shows the six principal views in SI metric projection.

FIRST-ANGLE PROJECTION SYMBOL

A symbol is used on SI metric drawings to avoid confusion and simplify the reading of the drawing. The symbol indicates that the views are projected according to first-angle projection. The enlarged line drawings in figure 16-5 show the ANSI symbol for third-angle projection drawings and the SI symbol for metric drawings. The symbols are usually placed adjacent to or within the title block. Sometimes, the word **METRIC** appears with the first-angle projection symbol.

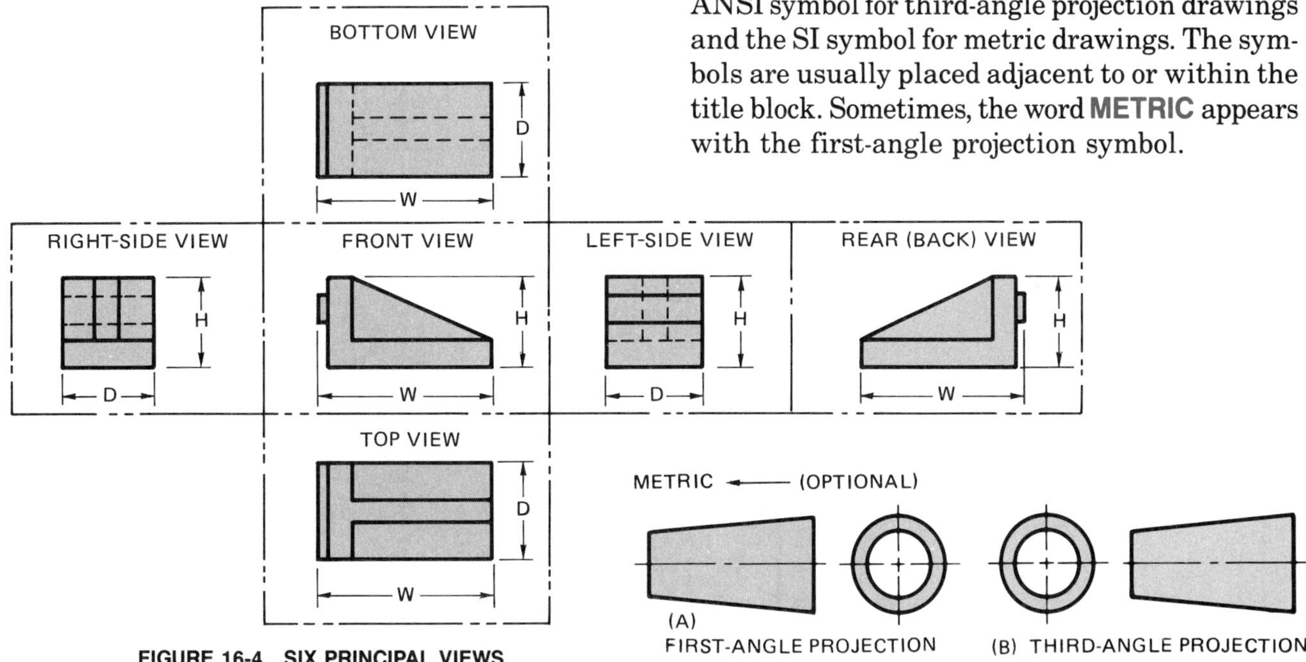

FIGURE 16-4 SIX PRINCIPAL VIEWS IN FIRST-ANGLE PROJECTION

FIGURE 16-5 SI/ANSI PROJECTION SYMBOLS (ENLARGED)

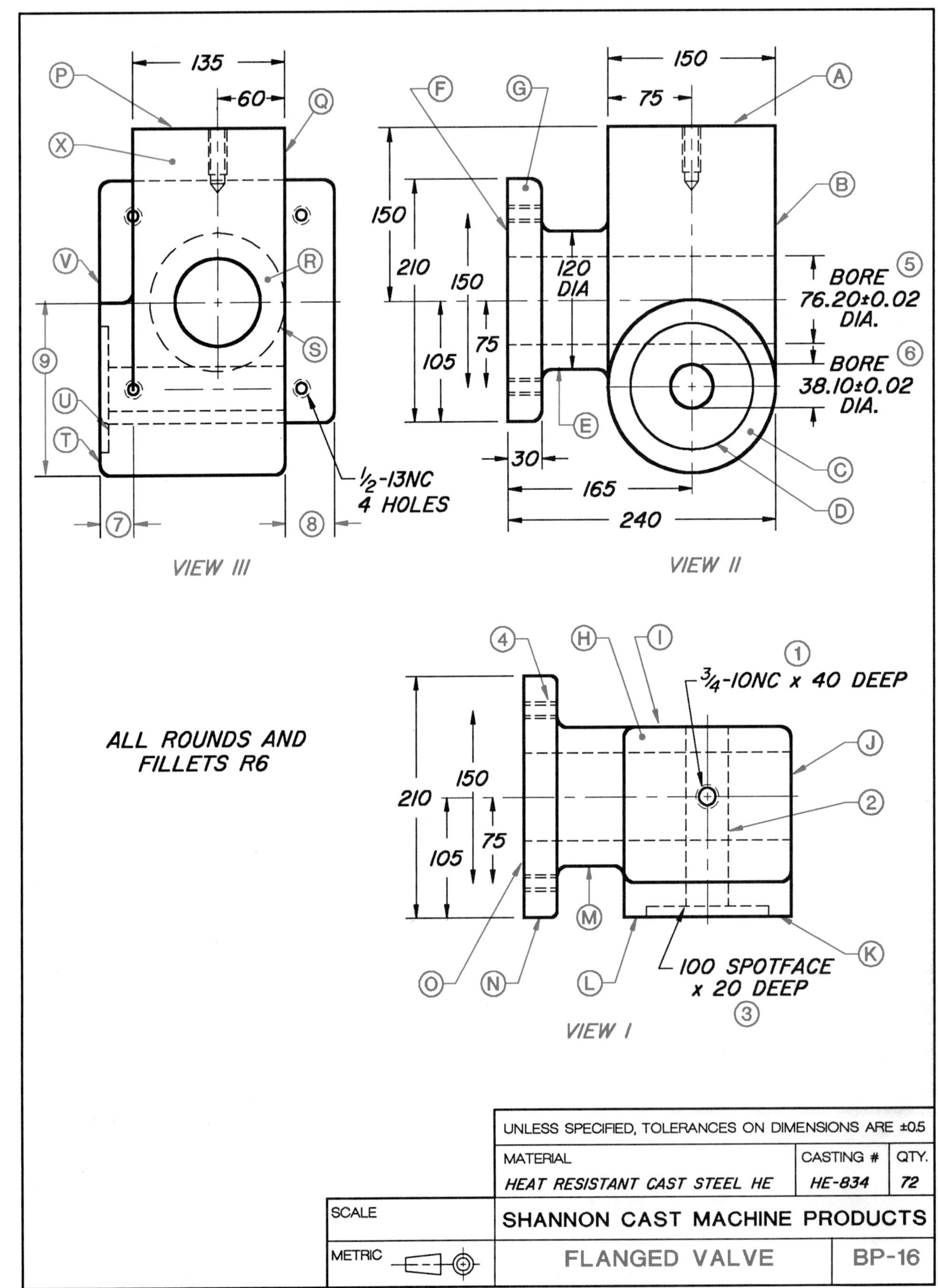

UNLESS SPECIFIED, TOLERANCES ON DIMENSIONS ARE ±0.5		
MATERIAL HEAT RESISTANT CAST STEEL HE	CASTING # HE-834	QTY. 72

SCALE	SHANNON CAST MACHINE PRODUCTS	
METRIC	FLANGED VALVE	BP-16

ASSIGNMENT—UNIT 16

FLANGED VALVE (BP-16)

Student's Name ____________________

1. a. Indicate the material used to manufacture the Flanged Valve.
   b. Identify the casting number and the required quantity.

2. a. Name the system of projection.
   b. State the name of each view.
   c. Identify the system and the unit of measurement used on the drawing.

3. Locate surfaces (A), (B), and (C) in **VIEW I** and **VIEW III**.

4. Identify features (K), (M), and (N) in **VIEW II** and **VIEW III**.

5. Give the letter(s) and number(s) in **VIEW I** which relate to the following features:
   a. Flange
   b. Flange hub
   c. Flange screw threads
   d. Threads in block
   e. 38.10 bored hole

6. Determine the overall machined (a) width, (b) height, and (c) depth of the part.

7. Write the specification of the tapped hole in the block.

8. a. Determine the maximum width, height, and thickness of the flange.
   b. Give the size of the corner rounds and fillets.

9. Establish the minimum and maximum center-to-center distances (along X and Y axis) between the tapped holes in the flange plate.

10. a. Calculate the minimum and maximum diameter of bored holes (5) and (6).
    b. Give the length of bored holes (5) and (6).

11. Calculate the following nominal dimensions:
    a. (7)
    b. (8)
    c. (9)

---

1. a. ____________________
   b. Casting # ________ Quantity ________

2. a. ____________________
   b. **VIEW I** ____________________
   VIEW II ____________________
   VIEW III ____________________
   c. ____________________

3.

	VIEW I	**VIEW III**
(A) =	________	________
(B) =	________	________
(C) =	________	________

4.

	VIEW II	**VIEW III**
(K) =	________	________
(M) =	________	________
(N) =	________	________

5.

	Identification (**VIEW I**)
a. Flange	________
b. Flange Hub	________
c. Flange Screw Threads	________
d. Threads in Block	________
e. 38.10-Bored Hole	________

6. a. Width ________ b. Height ________
   c. Depth ________

7. ____________________

8. a. Maximum:
   Width ________
   Height ________
   Thickness ________
   b. Rounds and Fillets ________

9. Minimum ____________________
   Maximum ____________________

10. a. (5) Min. ________ Max. ________
    (6) Min. ________ Max. ________
    b. (5) ________
    (6) ________

11. (7) = ________ (9) = ________
    (8) = ________

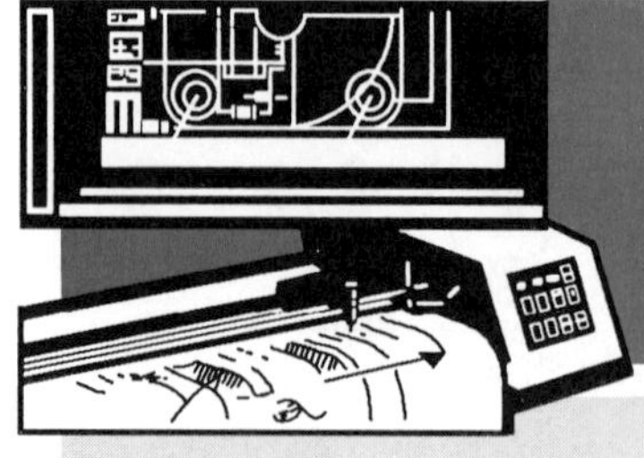

HIGH TECHNOLOGY applications

HT-3 STATISTICAL PROCESS (QUALITY) CONTROL

FOUNDATIONS FOR SPC

Statistical Process (Quality) Control (SPC) is widely used in production to ensure that each part, each unit, each assembly of parts, and each complete machine or mechanism meets specifications and specific standards. SPC deals with quality of product: dimensional control, surface texture characteristics, class of fit, conformance to geometric tolerancing; and other design, material, and operational requirements of engineering. Also, data from SPC provides useful input for controlling quality in production, cost effectiveness, and the need to redesign the product.

Normal Frequency Distribution Curve

When precisely measured, recorded, and plotted on a graph, the size variations from a nominal measurement are almost equally divided on both sides of a centerline. The greatest number of parts meet a nominal (basic) size and are concentrated around the centerline. The lesser number of parts that are above or below size are distributed according to a normal frequency curve. An example of these conditions is illustrated at (A) according to micrometer measurements of 50 machined pieces.

Analysis of the curve shows that the variations in the measured parts falls within six groups; three on each side of the centerline. The range of measurement groups is called *zones*. Each zone is identified by the Greek word *sigma* or letter σ as shown at (B).

Statistically, in a normal frequency curve of distribution, 99 3/4% of the machined parts fall within a natural tolerance limit of ± 3 sigma (σ). Sixty-eight percent of the parts, as shown graphically at (A), measure between 0.9374″ (-1σ) and 0.9376 ($+1\sigma$), or ± 0.001″. At $\pm 2\sigma$, 95 1/2% of the parts range between ± 0.002″ (0.9373″ and 0.9377″. At 3σ, 99 3/4% are ± 0.003″ (between 0.9372″ and 0.9278″.

Sampling Plans

The acceptable tolerance range for class of fit, dimensional accuracy, geometric and other part characteristics is specified in the title block or as notes on drawings. Since 100% inspection of all parts is costly and not practical, a *product sampling plan* is used. This type of plan is based on establishing an *acceptable quality level* (AQL). This means that a specific number of marginally defective parts may be accepted.

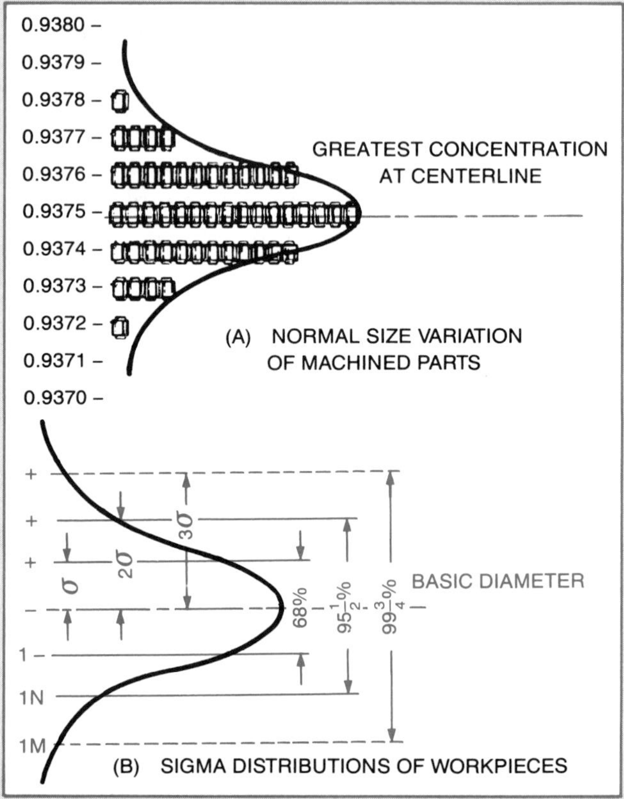

Single-, double-, and sequential-sampling plans for parts inspection and lot sizes are established to ensure that the quantity of approved parts meets the AQL. Lots that fail to meet sampling requirements are reinspected, and defective parts are replaced with acceptable parts.

Quality Control Charts

Four basic types of quality control charts used in SPC are identified by the letters: **c**, **p**, $\overline{X}$, and $\overline{R}$.

- The number of defects in a part are plotted on **c** charts.
- The percent defective parts in a sample are shown on **p** charts.
- A graph of variations in the averages of the samples is provided on $\overline{X}$ charts.
- Variations in the range of samples is shown on $\overline{R}$ charts.

Adapted from Machine Tool Technology and Manufacturing Processes

Courtesy of C. THOMAS OLIVO ASSOCIATES

UNIT 17
Measurement Conversions and Dimensioning Practices

It is mathematically possible to convert a customary inch value to a precisely equivalent SI metric unit of measurement (and vice versa). However, it may not be practical to produce a part to the dimensions in both systems. For this reason, it is important to understand soft and hard conversion, direct reading conversion scales, the use of diametral symbols and nonsignificant zeros, and reference dimensions.

SOFT CONVERSION OF CUSTOMARY AND SI METRIC UNITS OF MEASUREMENT

Soft conversion means that a unit of measurement in one system may be directly converted to another system. Soft conversion provides equivalent measurements in two systems without regard to standardized sizes or to the availability of materials, cutting tools, products, etc. For instance, a feature on a part which is 0.625″ may also be dimensioned 15.875 mm. This is a soft conversion. However, a three-place decimal measurement of 0.875 mm may be an incorrect reading of the required degree of accuracy. In fact, a three-place decimal reading is outside the range of a standard metric micrometer or measuring tool.

Equal precision (between 0.625″ and 15.875 mm) may be approached by requiring one less decimal place in an SI metric dimension than in the same customary inch dimension. The 0.625″ measurement would then be stated as 15.88 mm.

HARD CONVERSION

As discussed before, in some instances, there are no comparable sizes of cutting tools in both measurement systems. A part designed for drilling reaming, threading, and other processes in customary inch measurements may not be machined with SI metric designed cutting tools. A one-inch hole requires a 25.4 mm drill. A 5/8-11 NC tap may not be interchanged with a standard metric size tap.

While features can be dimensioned by soft conversion between the two measurement systems, the actual sizes and forms are slightly different. *Hard conversion* means design changes. Features of an object are *designed to conform to standard sizes in the other system of measurement.* Under these conditions, parts designed by hard conversion are not interchangeable between the two measurement systems.

This is apparent if the one-inch drill is considered again. The diameter in hard conversion is changed to 25 mm. This size metric drill is available. However, the equivalent diameter drill in customary units of measurement is 0.984″. This is a special-size drill. Hard conversion requires size and/or other design changes, including *preferred metric sizes* which may introduce still other variations in dimensions. Hard conversion is one reason why many ANSI standards are not identical with published ISO standards.

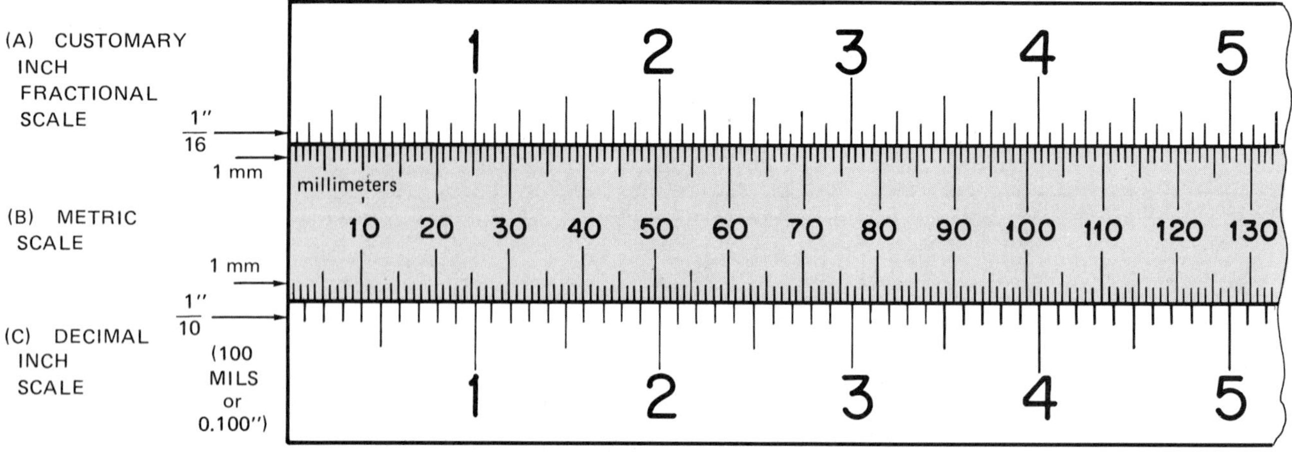

FIGURE 17-1 RELATIONSHIP OF METRIC (MM) AND CUSTOMARY INCH FRACTIONAL AND DECIMAL MEASUREMENTS ON A DIRECT-READING CONVERSION TABLE

DIRECT READING CONVERSION SCALE

A *direct reading conversion scale* provides a quick, easy method of changing measurements between different systems. The scales permit close, but not precise conversions. Figure 17-1 shows three different scales. The scale at (A) is graduated in customary units of 1/16″. The metric scale at (B) shows graduations in millimeters. The decimal inch scale at (C) is graduated in tenths of an inch or 100 mils (0.100″).

A measurement in any one of the three units is converted to its equivalent by a corresponding reading on the required scale. As an example, the 5″ graduation on the top scale corresponds with the 127 mm line graduation. A 40 mm reading corresponds with the 1.6″ graduation on the decimal inch scale. Mathematically, 40 mm = 1.5748″. Such a measurement would have to be taken and read with a vernier micrometer or other precise measuring tool. The 1.6″ direct conversion scale reading is generally adequate.

DIAMETRAL MEASUREMENTS

Cylindrical surfaces, particularly those represented on one-view drawings, are identified by the use of the abbreviation **DIA** following the dimension. The symbol **Ø** also indicates round, cylindrical features. While the **Ø** symbol precedes a dimension on SI metric drawings, it is common practice to use the symbol following a *diametral dimension.*

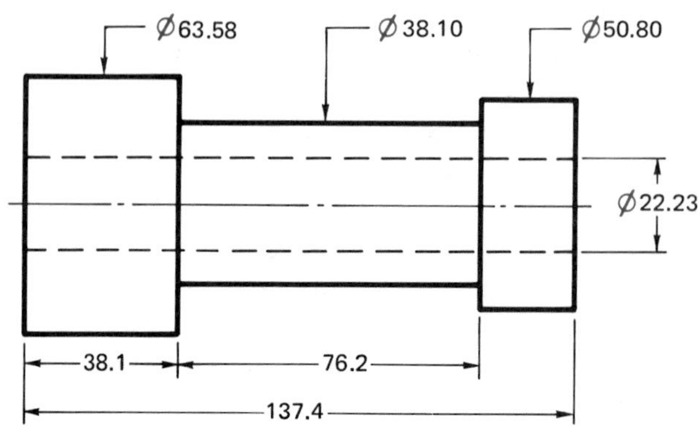

FIGURE 17-2 USE OF DIAMETRAL DIMENSION SYMBOL TO IDENTIFY CYLINDRICAL SURFACES

The four symbols on the one-view drawing (figure 17-2) show that there are three outside diameters and one hole diameter. The diametral dimension follows each symbol on this drawing.

NONSIGNIFICANT ZEROS

Nonsignificant zeros provide information which simplifies the interpretation of the required dimensional accuracy. A nonsignificant zero has no value in a number. Nonsignificant zeros are added to the right of a decimal dimension in customary units so that the decimal digits in the dimension and the tolerance are the same. The following shows the use of nonsignificant zeros.

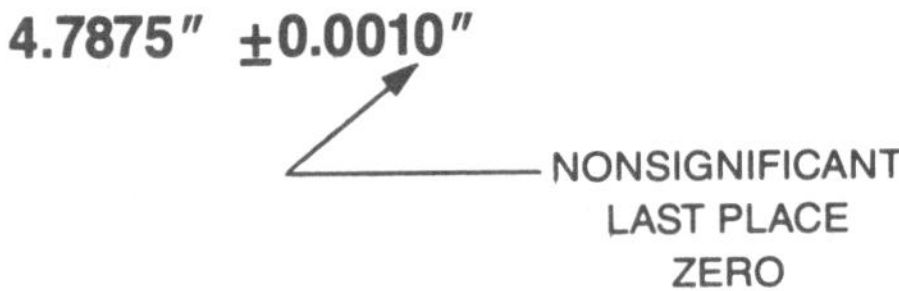

Nonsignificant zeros *are not added* to the decimal value in metric dimensioning. Drawings frequently show a different number of digits between a millimeter dimension and the tolerance. A part may be dimensioned **112.6 mm ±0.025 mm**. In this case, it is understood that the part may be machined to a maximum size of 112.625 mm and a minimum size of 112.575 mm.

REFERENCE DIMENSIONS

Reference dimensions are given on drawings to provide general information only. No tolerances are indicated, as reference dimensions are not used in the manufacture, layout, or inspection of the part.

Reference dimensions were earlier indicated by the abbreviation REF following a dimensional value (figure 17-3A). Present standards eliminate the use of REF, replacing it with parentheses or brackets to enclose the reference dimension. The preferred method of using parentheses or brackets is illustrated in figure 13-3 at (B) and (C).

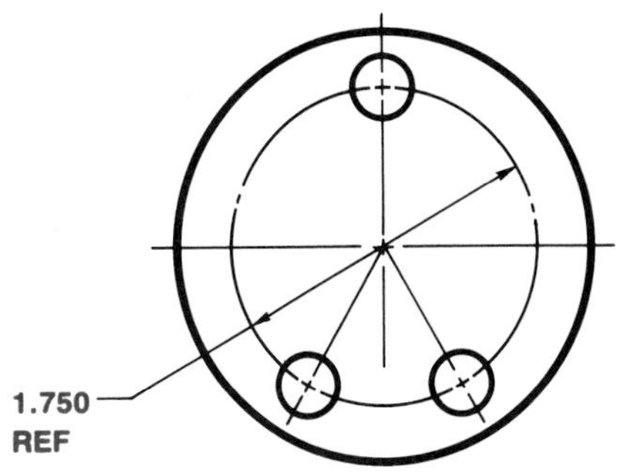

(A) USE OF ABBREVIATION (EARLIER METHOD)

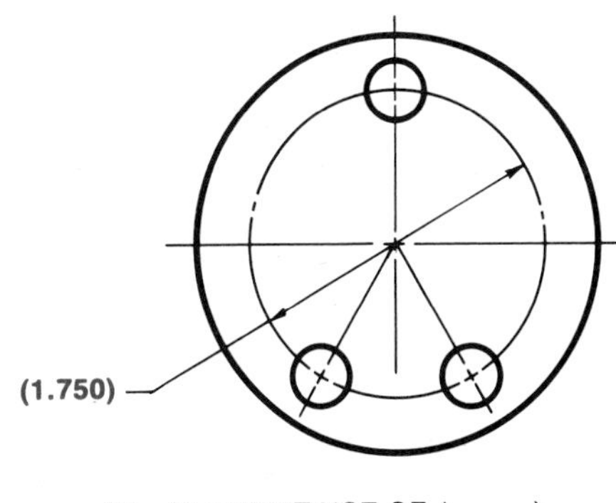

(B) CURRENT USE OF ( )

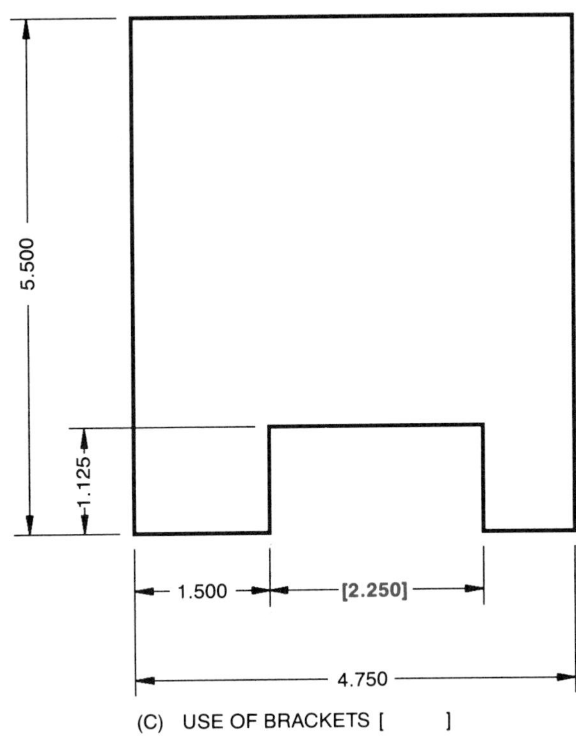

(C) USE OF BRACKETS [ ]

FIGURE 17-3 METHODS OF IDENTIFYING A REFERENCE DIMENSION

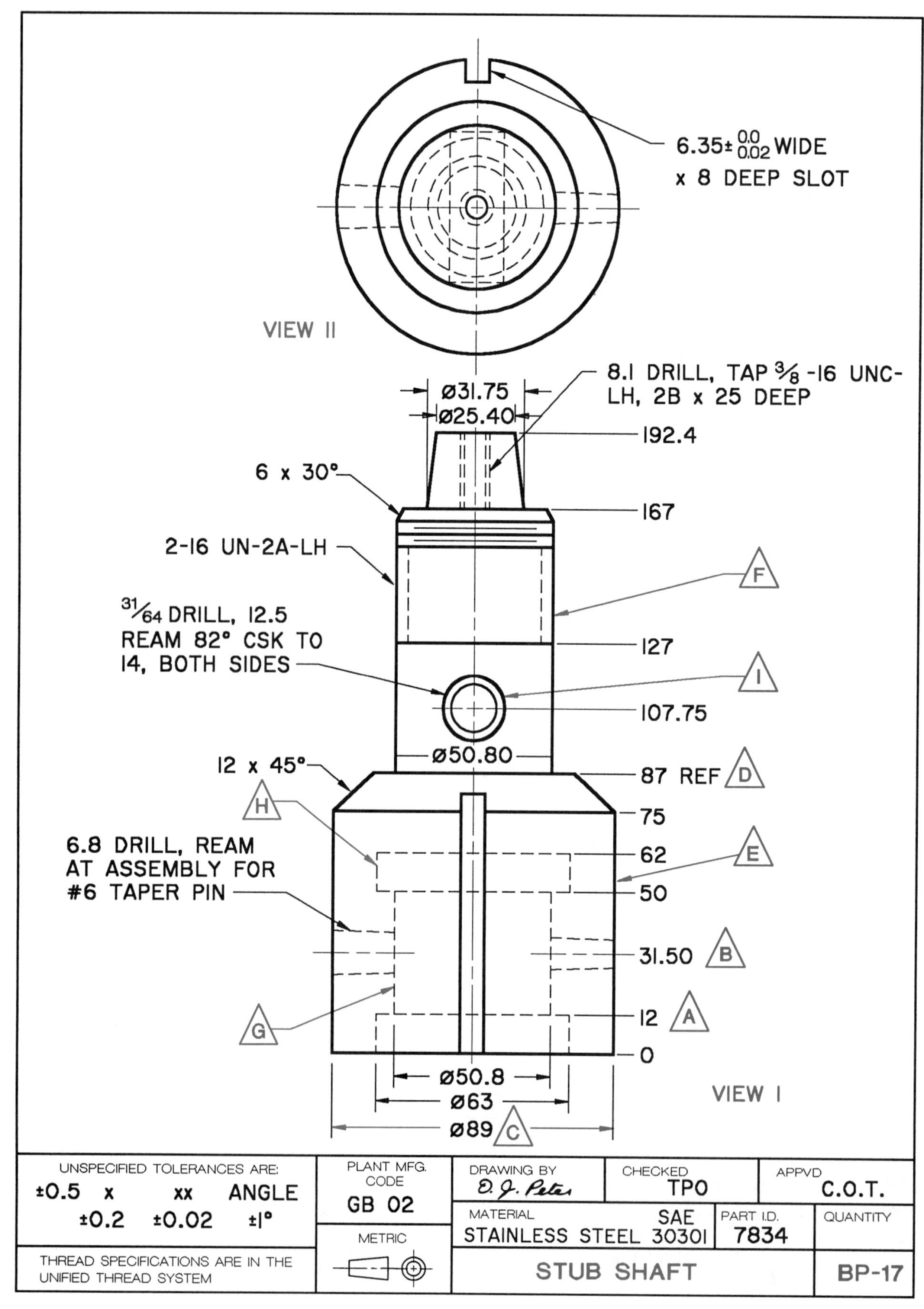
6.35± 0.0 0.02 WIDE
x 8 DEEP SLOT
VIEW II
8.1 DRILL, TAP 3/8 -16 UNC-
LH, 2B x 25 DEEP
Ø31.75
Ø25.40
192.4
6 x 30°
167
2-16 UN-2A-LH
F
31/64 DRILL, 12.5
REAM 82° CSK TO
14, BOTH SIDES
127
I
107.75
Ø50.80
12 x 45°
87 REF
D
H
75
6.8 DRILL, REAM
AT ASSEMBLY FOR
#6 TAPER PIN
62
E
50
31.50
B
G
12
A
0
Ø50.8
Ø63
Ø89
C
VIEW I
UNSPECIFIED TOLERANCES ARE:
±0.5 X XX ANGLE
±0.2 ±0.02 ±1°
THREAD SPECIFICATIONS ARE IN THE
UNIFIED THREAD SYSTEM
PLANT MFG. CODE
GB 02
METRIC
DRAWING BY
D. J. Peter
CHECKED
TPO
APPVD
C.O.T.
MATERIAL
SAE
STAINLESS STEEL 30301
PART I.D.
7834
QUANTITY
STUB SHAFT
BP-17

STUB SHAFT (BP-17)

1. a. State what dimensioning system is used.
   b. Give two advantages of this dimensioning system over conventional dimensioning.
2. Identify the function of dimensions A, B, and C as to size or location.
3. a. Describe briefly the use of dimension D.
   b. Tell what function the symbol ∅ serves.
4. a. Identify the number of threaded features.
   b. Give the specifications of the tapped hole.
   c. Translate what each symbol or dimension in the tapped hole specifications means.
5. a. Give the unspecified tolerances.
   b. Determine the minimum and maximum outside diameters of features E and F and the inside diameters of G and H.
6. a. Give the maximum diameter at (1) the small end, (2) the large end, and (3) the length of the taper.
   b. Compute the taper per foot in mm.
   c. Convert the mm taper per foot to the equivalent customary inch measurement.
7. a. Give the specifications for hole I.
   b. Explain the meaning of each dimension and symbol in the specifications.
   c. Give a reason why a fractional inch drill is specified.
8. Compute the maximum and minimum center-to-center distance between holes B and I.
9. a. Give the required taper pin size.
   b. Explain the meaning of the taper pin note.
10. Give (a) the nominal width (height) and (b) the maximum outside diameter of feature H.
11. Determine the maximum and minimum width and depth of the slot.
12. Indicate the effect of hard conversion and preferred dimensioning on the interchangeability of a part between the two major dimensioning systems.

ASSIGNMENT—UNIT 17

Student's Name ______________

1. a. ______________
   b. (1) ______________
   (2) ______________
2. A ______________
   B ______________
   C ______________
3. a. D ______________
   b. ∅ ______________
4. a. ______________
   b. ______________
   c. ______________
5. a. ______________
   b.

	Minimum	Maximum
E		
F		
G		
H		

6. a. (1) Small Diameter = ______________
   (2) Large Diameter = ______________
   (3) Taper Length = ______________
   b. ______________
   c. ______________
7. a. ______________
   b. ______________
   c. ______________
8. Maximum ______________ Minimum ______________
9. a. ______________
   b. ______________
10. a. Nominal Width ______________
    b. Maximum Outside Diameter ______________
11. Max Width = ______________ Min. = ______________
    Max Depth = ______________ Min. = ______________
12. ______________

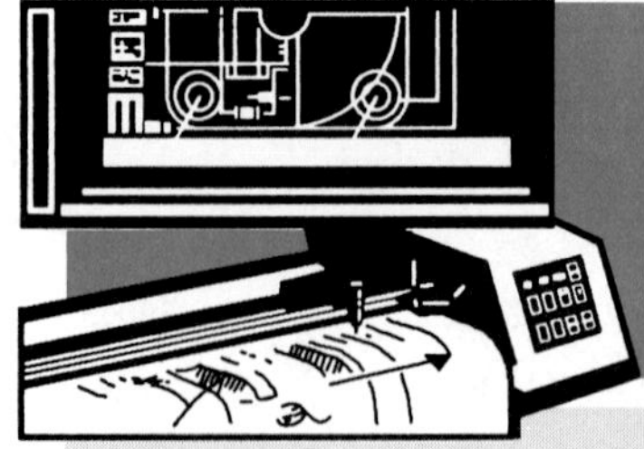

HIGH TECHNOLOGY applications

HT-4 STATISTICAL PROCESS (QUALITY) CONTROL

DIGITAL ELECTRONIC INSTRUMENTS FOR INTEGRATED SPC SYSTEMS

Digital electronic instruments and equipment may be interlinked so that dimensional and other measurement/inspection data may be taken, gathered, transmitted, or used on-spot for analysis or to provide a digital readout or a printout.

The four basic units that are needed to perform the above functions include the following pieces of equipment/instruments.

- Electronic digital gages, measuring and inspection instruments, and other special devices for data gathering on-the-job.
- Electronic transmitting equipment.
- Electronic data processing equipment with capability to store measurement/inspection data.
- Electronic software for data analysis and documentation.

Some of the commonly used precision measuring electronic equipment, with a brief statement of advantages, follows.

- Digital electronic indicators which eliminate counting revolutions and graduations, as is the case with other models of indicators.
- Digital electronic micrometers for direct display of + or − measurements or GO-No GO comparative measurements in relation to a preset nominal dimension.
- Digital electronic column gages with switching controls that permit the following processes.
  - Setting upper and lower limits.
  - Count up and count down.
  - Adjusting for true spindle position.

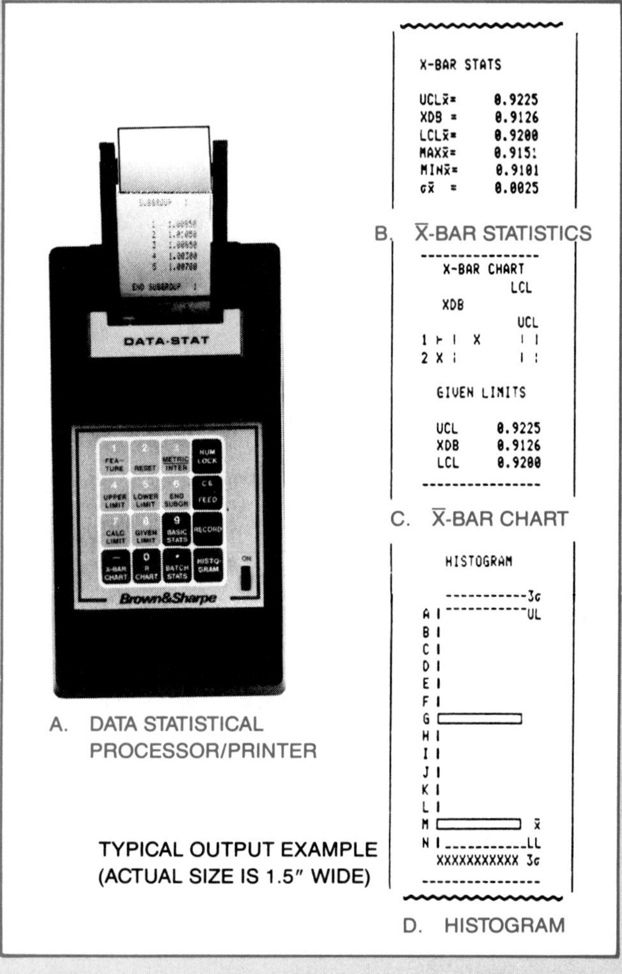

A. DATA STATISTICAL PROCESSOR/PRINTER

B. $\overline{X}$-BAR STATISTICS

C. $\overline{X}$-BAR CHART

D. HISTOGRAM

TYPICAL OUTPUT EXAMPLE (ACTUAL SIZE IS 1.5″ WIDE)

Features for SPC Data Processing Instruments

Other complementary electronic units permit the instantaneous processing of measurement data to produce tape printouts. *Data processor instruments*, such as the model illustrated, have the capability to perform the following SPC tasks.

- Collect, store, and analyze a number of measurements as generated by compatible indicators, micrometers, height gages, calipers, and other measurement instruments and fixturing devices.
- Generate $\overline{X}$ and $\overline{R}$ charts and histograms (as shown at (B), (C), and (D)) based on SPC measurements gathered on a part lot.
- Calculate maximum and minimum part sizes and store control limits in memory.
- Make comparisons of upper and lower limit part measurements with specified part size limits.
- Print statistics, charts, graphs, and other required SPC reports.
- Permit interfacing with other mainframe computers for purposes of programming essential machining adjustments to correct size and/or product quality variations.
- Provide linkages for the storage in memory of SPC measurement and inspection data.

Courtesy of BROWN & SHARPE MANUFACTURING COMPANY

UNIT 18
Dual Dimensioning Principles and Practices

Dual dimensioned drawings provide for the manufacture of parts and mechanisms according to the dimensioning system which is in operation in the producing country and industrial plant. *Dual dimensioning* involves equivalent dimensions in customary inch and SI metric units of measurement. *Equivalent dimensions are based on soft conversion.* Thus, parts designed, produced, and measured according to dimensions in either system are *compatible.* In other words, they are interchangeable.

CONTROLLING DIMENSIONS

When the dimensions on a drawing are given in both customary inch and SI metric units, one system is singled out for the *controlling dimensions.* These dimensions indicate the measurement system in which the product is designed. The controlling dimension appears first, followed by the equivalent dimension in the second measurement system. Figure 18-1 illustrates this fact.

DUAL DIMENSIONING

Dual dimensioning permits workers to read and interpret a drawing in either measurement system, regardless of whether one or both systems are used in the industry. On older drawings, controlling dimensions in the primary system appear above a dimension line. Equivalent dimensions in the second system follow or are placed under the primary dimension. Figure 18-2 shows simple examples of dual dimensioning. At (A) the controlling dimension of 78 mm is placed above the line. The equivalent inch dimension of 3.070″ is placed under the metric dimension.

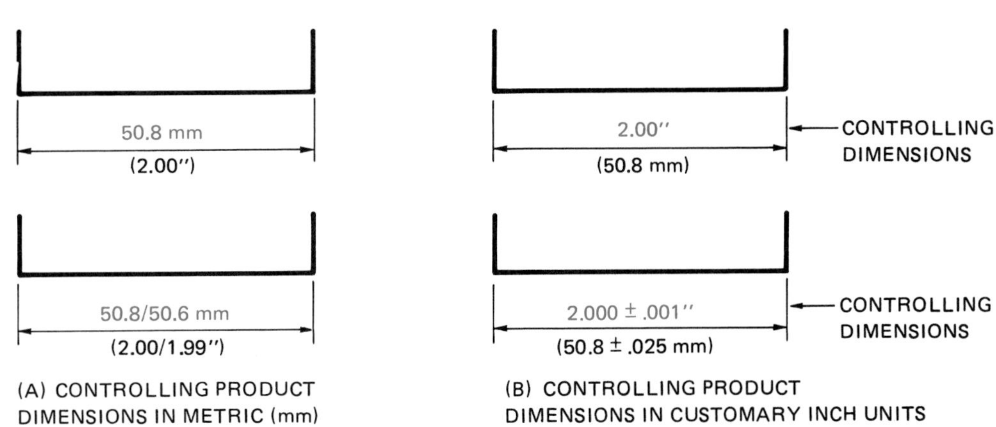

FIGURE 18-1 EXAMPLES OF CONTROLLING DIMENSIONS

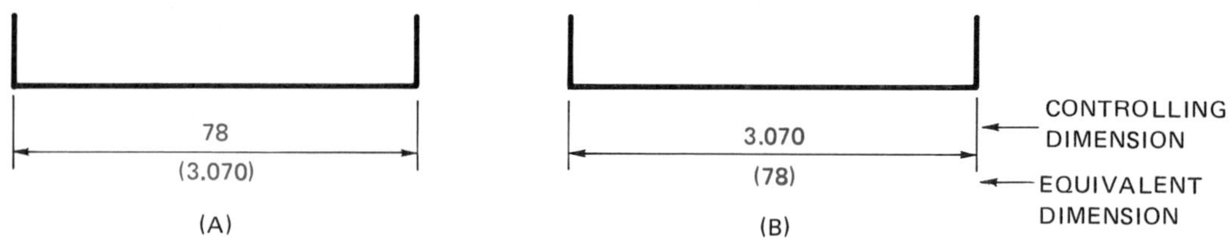

FIGURE 18-2 SIMPLE DUAL DIMENSIONING

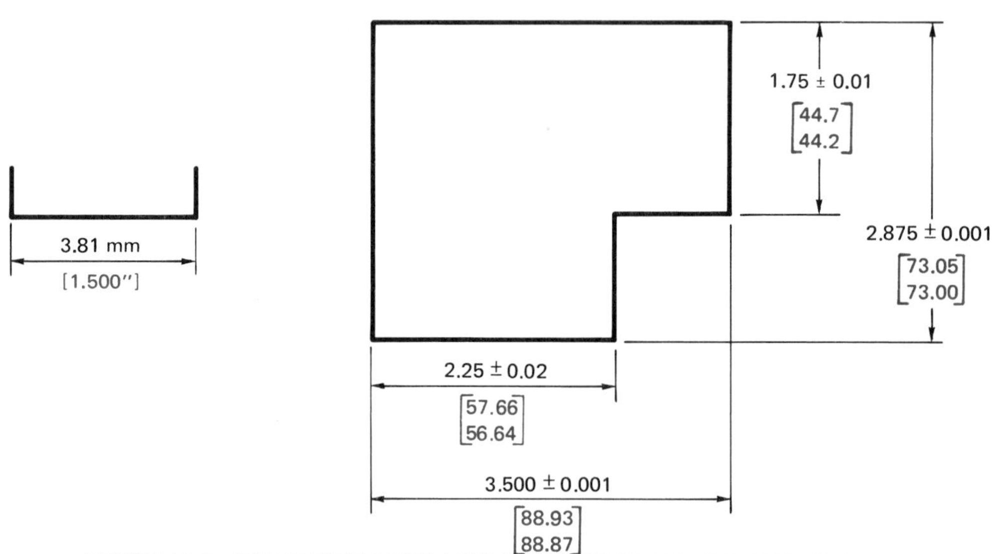

FIGURE 18-3 APPLICATION OF THE BRACKET METHOD OF DUAL DIMENSIONING

The measurement units must be clearly specified so that the person reading the drawing has no difficulty establishing which measurement system is used.

Bracket Method

In the *bracket method*, the design dimension representing the primary measurement system is given first. The equivalent dimension in the second measurement system is included in brackets. The brackets and dimension may be placed either after or below the controlling dimension (figure 18-3).

Position Method

Some industries use the *position method* to distinguish between a controlling dimension and the second system measurement. The primary design or control dimension is given first. It is placed before or above the secondary equivalent measurement as shown in figure 18-4. The dimension may appear as shown at (A), by using a slash symbol (B), or in brackets (C). A general note

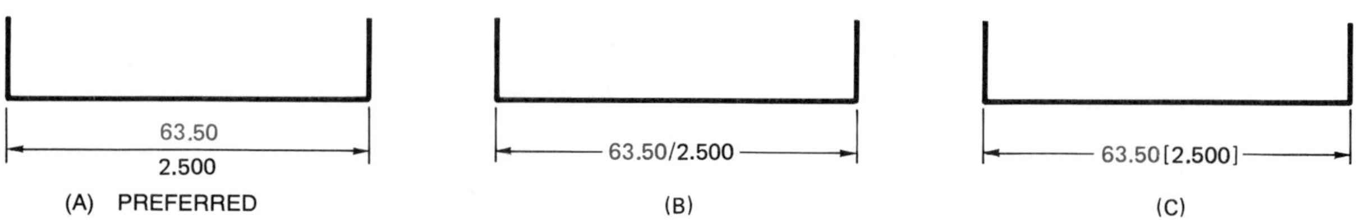

FIGURE 18-4 POSITION METHOD OF DIMENSIONING

such as the following is generally provided on a drawing.

$$\frac{\text{MILLIMETER}}{\text{INCH}} \text{ OR } \frac{\text{INCH}}{\text{MILLIMETER}}$$

The note indicates the order of the dimensions.

TABULAR DIMENSIONING READOUT CHART

Dual dimensioning often results in drawings which become complicated due to the addition of a second set of dimensions. In such instances, drawings are simplified by using *tabular dimensioning.* The drawing is dimensioned according to the measurement unit for the controlling (design) dimensions. Equivalent dimensions in the secondary measurement system are recorded in table (tabular) form as in figure 18-5A. The corresponding measurements in the table represent a soft conversion of all design sizes on the drawing.

LETTER IDENTIFICATION TABULAR DIMENSIONING

A modification of tabular dimensioning of conversion values is used in some industries by replacing the dimensioning on a drawing with *letters* (figure 18-5B). The actual dimensions for each letter designation are provided in customary inch and SI metric units in a *tabular readout table.* Letter identification tabular dimensioning is often used on parts which are manufactured in many sizes but have identical design features. The tables often include another column for *preferred hard conversion dimensions.* These dimensions represent complete conversion and necessary design, engineering, and dimensional changes to SI metrics.

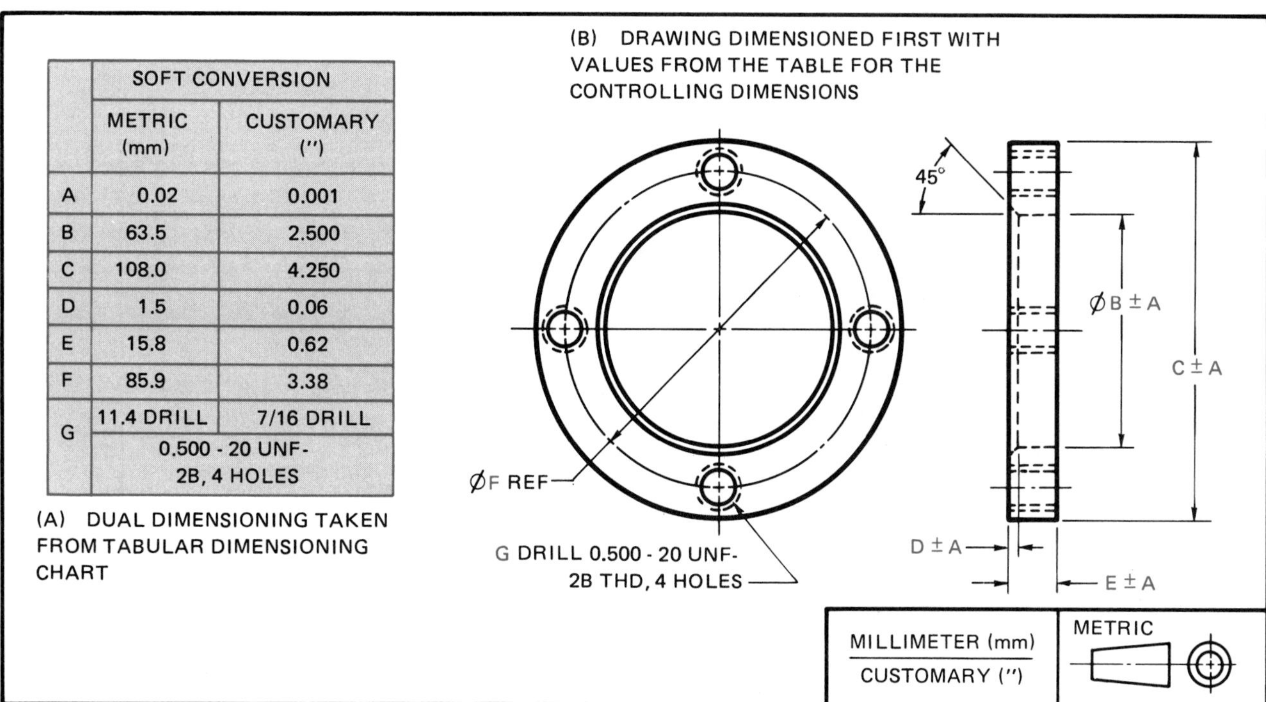

	SOFT CONVERSION	
	METRIC (mm)	CUSTOMARY (")
A	0.02	0.001
B	63.5	2.500
C	108.0	4.250
D	1.5	0.06
E	15.8	0.62
F	85.9	3.38
G	11.4 DRILL	7/16 DRILL
	0.500 - 20 UNF-2B, 4 HOLES	

FIGURE 18-5 LETTER IDENTIFICATION TABULAR (DUAL) DIMENSIONING

NOTE: REFER TO APPENDIX B-1 FOR THE LARGER SCALE DRAWING TO USE WITH THIS ASSIGNMENT.

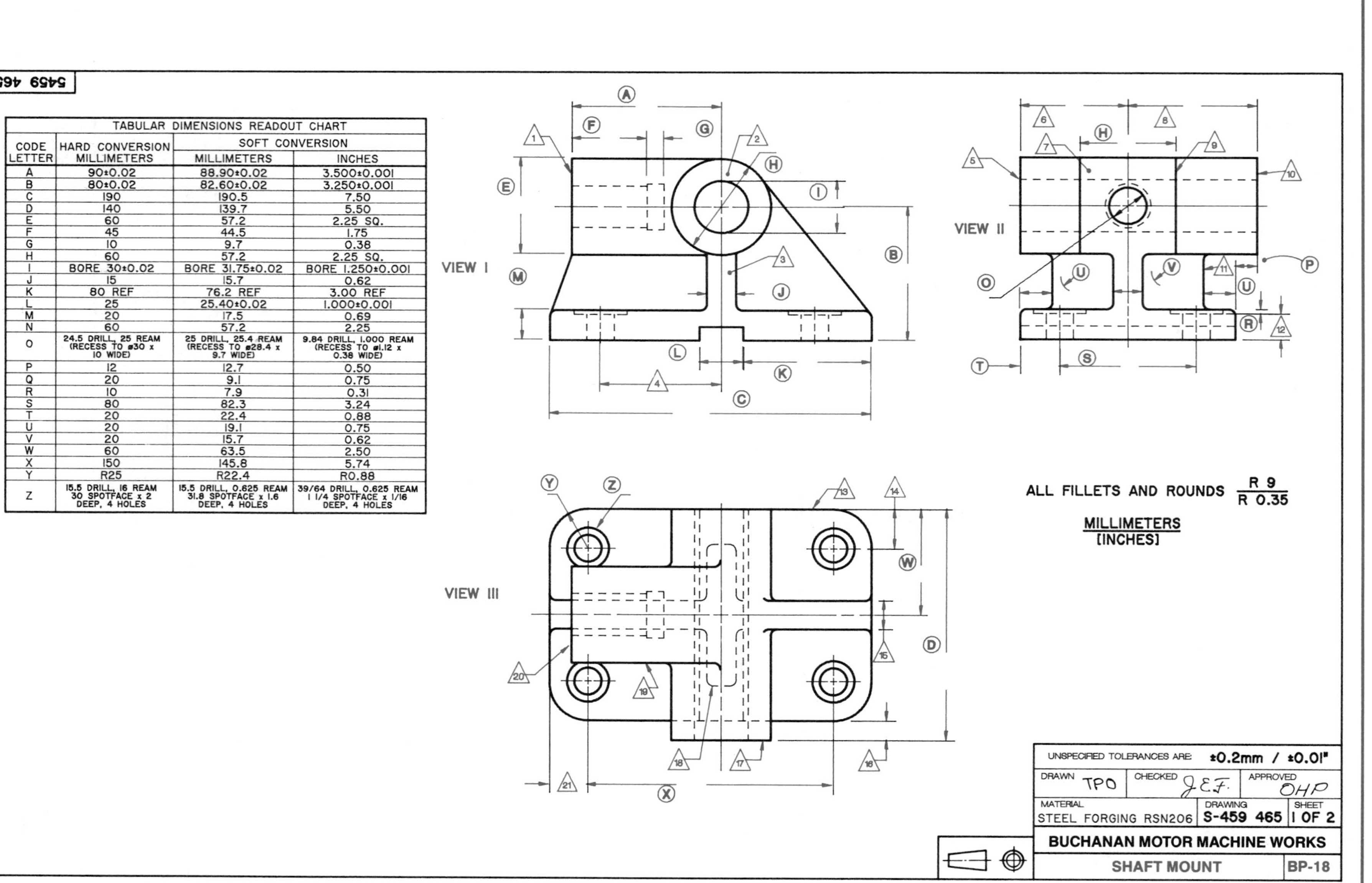

TABULAR DIMENSIONS READOUT CHART

CODE LETTER	HARD CONVERSION MILLIMETERS	SOFT CONVERSION MILLIMETERS	SOFT CONVERSION INCHES
A	90±0.02	88.90±0.02	3.500±0.001
B	80±0.02	82.60±0.02	3.250±0.001
C	190	190.5	7.50
D	140	139.7	5.50
E	60	57.2	2.25 SQ.
F	45	44.5	1.75
G	10	9.7	0.38
H	60	57.2	2.25 SQ.
I	BORE 30±0.02	BORE 31.75±0.02	BORE 1.250±0.001
J	15	15.7	0.62
K	80 REF	76.2 REF	3.00 REF
L	25	25.40±0.02	1.000±0.001
M	20	17.5	0.69
N	60	57.2	2.25
O	24.5 DRILL, 25 REAM (RECESS TO ⌀30 x 10 WIDE)	25 DRILL, 25.4 REAM (RECESS TO ⌀28.4 x 9.7 WIDE)	9.84 DRILL, 1.000 REAM (RECESS TO ⌀1.12 x 0.38 WIDE)
P	12	12.7	0.50
Q	20	9.1	0.75
R	10	7.9	0.31
S	80	82.3	3.24
T	20	22.4	0.88
U	20	19.1	0.75
V	20	15.7	0.62
W	60	63.5	2.50
X	150	145.8	5.74
Y	R25	R22.4	R0.88
Z	15.5 DRILL, 16 REAM 30 SPOTFACE x 2 DEEP, 4 HOLES	15.5 DRILL, 0.625 REAM 31.8 SPOTFACE x 1.6 DEEP, 4 HOLES	39/64 DRILL, 0.625 REAM 1 1/4 SPOTFACE x 1/16 DEEP, 4 HOLES

SHAFT MOUNT (BP-18)

1. a. Identify the projection system used on the print of the Shaft Mount.
   b. Determine the position (name) from which **VIEWS I, II,** and **III** are projected.

2. a. State the three measurement systems in which dimensions are given.
   b. Name the measurement system of the controlling (design) dimensions.

3. a. Give one advantage of using a tabular dimensioning readout chart.
   b. Indicate the effect of dimensional changes according to preferred hard conversion on the production of a part.

4. Locate surfaces △1, △2, and △3 in **VIEW I** in **VIEWS II** and **III**.

5. Locate surfaces △5 and △9 (**VIEW II**) in **VIEW III**.

6. Use brackets and dual dimension features Ⓐ through Ⓜ in **VIEW I**, including Ⓓ in **VIEW III**.

   Note. Use the dual dimensioning values in the conversion table.

7. Indicate the *specified tolerances* for metric and customary inch dimensions Ⓐ, Ⓑ, and Ⓛ.

8. Give the nominal size of (a) all fillets and rounds and (b) the radius of the four corners of the base portion.

9. Compute and record the nominal dual dimensioning for features △4, △6, △8, △12, △14, △16, and △21.

   Note. Use soft conversion values.

10. Give the specifications in customary inch equivalents for features Ⓞ and Ⓩ.

11. a. Indicate the nominal variations between hard conversion dimensions and the controlling metric dimensions for features Ⓐ and Ⓑ.
    b. Tell what possible effect the nominal dimensional variations have on interchangeability of the Shaft Mounts machined according to soft and hard conversion dimensions.

ASSIGNMENT—UNIT 18

Student's Name ______________________

1. a. ______________________
   b. **VIEW I** __________ **VIEW III** __________
   VIEW II __________

2. a. (1) ______________________
   (2) ______________________
   (3) ______________________
   b. ______________________

3. a. ______________________
   ____
   ____
   b. ______________________
   ____
   ____

4.

	VIEW II	VIEW III
△1	________	________
△2	________	________
△3	________	________

5. △5 ________ △9 ________

6. *Note:* Features Ⓐ through Ⓜ are to be dual dimensioned on the Shaft Mount drawing. Brackets are to be used for the equivalent customary units found in the tabular dimensioning chart.

7. ______________________
   ____

8. a. ______________________
   b. ______________________

9.

	Dual Dimensioning		Dual Dimensioning
△4	________	△14	________
△6	________	△16	________
△8	________	△21	________
△12	________		

10. Ⓞ ______________________
    ____
    Ⓩ ______________________
    ____

11.

	Controlling Dimension	Hard Conversion	Nominal Variation
a. Ⓐ	________	________	________
Ⓑ	________	________	________

b. ______________________
____
____
____

HIGH TECHNOLOGY applications

HT-5 INTERLINKAGE OF CADD, CAM, AND CNC MACHINE TOOLS

CADD (computer-aided drafting and design) refers to the following two tasks.

- The *design of a product* (whether it be of a single part, a structure in architecture, an instrumentation control or HVAC system, or any other object or device) by using computer hardware and specially-designed software.
- The *preparation of drawings* (whether they be architectural or structural plans, elevations, sections; perspective views; working engineering drawings with standard views or exploded assemblies; schematic diagrams for electrical/electronics circuitry; or other forms and types of drawings) using computer input.

CADD is used to generate three basic drawing models.

- **Wireform Model**. A drawing created by connecting all line, circle, and arc edges to produce a wireform of the object.
- **Surface model**. A wireform model with surfaces that are covered by using a computer. The model is drawn as it appears when viewed by eye.
- **Solid model**. Representation of an object or part in solid form. A solid model permits cutouts or sectional views that expose internal design features.

COMPUTER-AIDED MANUFACTURING (CAM) AND CNC MACHINES

When CADD is linked with a computer-actuated machine control unit (MCU), the composite system is identified as a CADD/CAM unit or system. CAM depends on CADD-generated input that controls processes, dimensions, finish, cutting speeds, feeds, depth of cut, tool path, and other machining, inspection, or assembly processes.

A chucker and bar turning machine is used to illustrate how CADD, CAM, and computer numerical control (CNC) input are interfaced in high production manufacturing. The *machine control panel* (B) controls machining functions, speeds and feeds, cutting fluid flow, and tool movements.

The *numeric CNC controls* (C) are programmed manually or automated by CADD for a completely automatic cycle. The color screen provides a graphic picture of tool path simulation, dimensional, and other geometric features. At (D), the cutting tools are preset on an interchangeable turret top plate. The tools are sequenced and positioned for cutting through CNC programs and commands of the machine control unit (MCU).

(A) MACHINE SPINDLE AREA

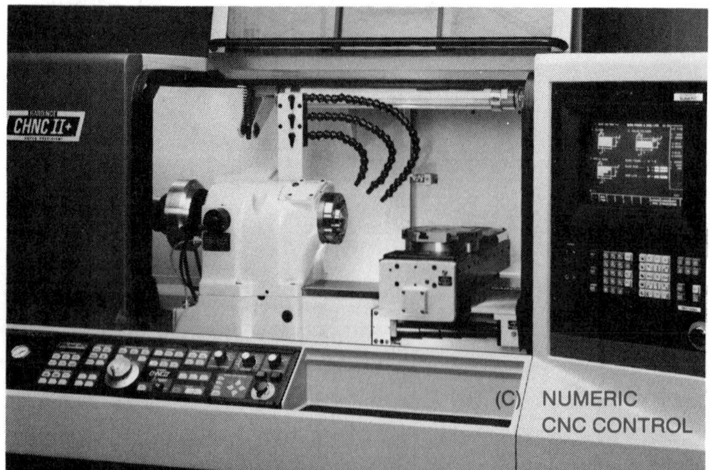

(C) NUMERIC CNC CONTROL

(B) MACHINE CONTROL PANEL

(D) MULTIPLE-TOOLED INTERCHANGEABLE TURRET TOP PLATE

Photos courtesy of HARDINGE BROTHERS INC.

SECTION 6
Machine Elements and Processes

UNIT 19
External Screw Threads and Fasteners

The representation and dimensioning of fasteners, machine elements, and machining processes are governed by ANSI, Canadian Standards Association (CSA), and SI metric standards. This unit deals with external screw threads as one of the most widely used fasteners. The units which follow cover the representation and dimensioning of internal threads, machined surfaces, part features, and other machining processes and products.

Background information follows on design features and functions of screw threads. Representation and dimensioning practices are covered for 29°, 60°, and 90° thread forms in the ANSI, CSA, and SI metric systems.

FUNCTIONS SERVED BY SCREW THREADS

Screw threads serve four major functions:

- Screw threads permit two or more parts to be assembled, disassembled, or held securely together.
- Screw threads control movements on applications of precision lead screws or in measurement, as in the case of a micrometer spindle.
- Screw threads transmit force. The screw jack is a prime example.
- Screw threads change the direction of motion from rotary to linear. For instance, machine handwheels convert rotary motion to longitudinal or vertical movements.

CHARACTERISTICS AND BASIC FEATURES OF SCREW THREADS

Design features of screw threads relate to a continuous ridge of uniform cross-section. This ridge is formed around a cylindrical or cone-shaped surface. The terms which are regularly used for common thread features are illustrated in figure 19-1.

Threads are identified as straight or tapered (as in the case of pipe threads). Threads which are cut on inner surfaces or holes are referred to as *internal threads. External threads*, like a bolt or stud, are cut around the outside surface. Threads that serve to move a part or to tighten two surfaces together when turned clockwise are known as *right-hand threads.* Threads are left-hand if they advance when turned counterclockwise. A

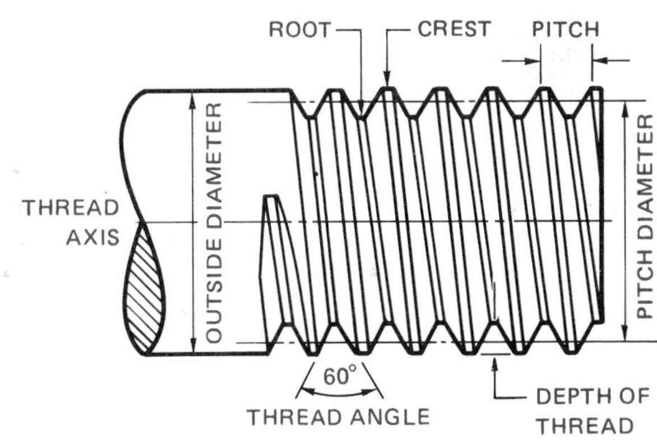

FIGURE 19-1 BASIC 60° THREAD FEATURES AND TERMS

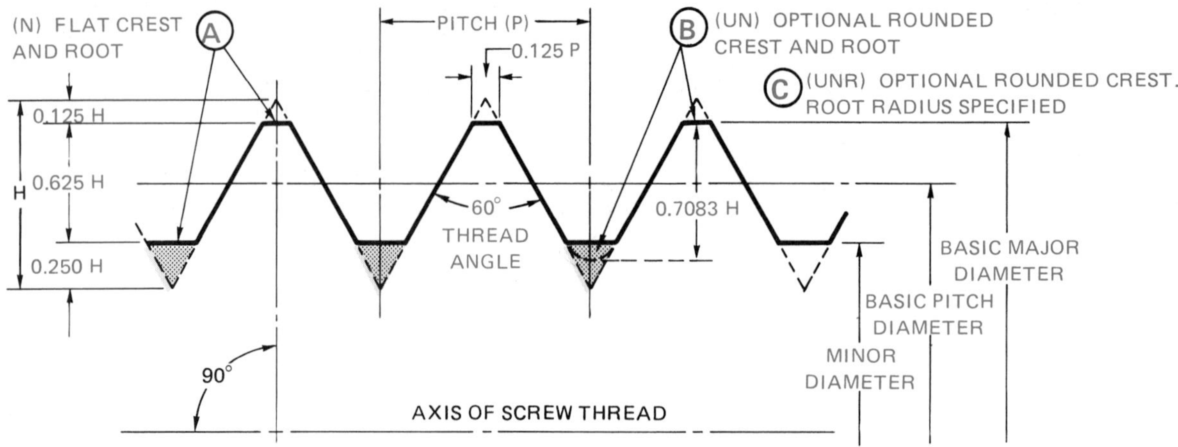

FIGURE 19-2 CHARACTERISTICS OF AMERICAN NATIONAL (N) AND UNIFIED (UN AND UNR) SCREW THREAD PROFILES

thread is assumed to be right-hand unless it is otherwise specified on a drawing by the letters LH.

A thread may have a *single lead*. This means that a part moves a distance equal to the pitch of the thread in one revolution. Some parts require a faster or greater advance movement per revolution. For such applications, the lead is *doubled* (2 × pitch), *tripled* (3 × pitch), or *quadrupled* (4 × pitch). *Lead* means the distance a part advances for each revolution. In cases other than single lead (single pitch) threads, the lead and pitch are given on the drawing as part of the *thread notation*.

Another modified design of the 60° thread form is found in *stub threads*. As the name implies, the threads are of shorter depth. This form is particularly adapted for thin wall sections which cannot accommodate standard depth threads.

THREAD FORM FOR AMERICAN NATIONAL AND UNIFIED SCREW THREADS

Each thread has a shape, or profile, called the *thread form*. The thread form remains constant although the thread dimensions may vary according to the thread size. The most common thread form has an included angle of 60°. This thread form is used in the *American National Thread System*; the *Unified Thread System*, which is standard in the United States, Canada, and the United Kingdom; and the *SI Metric System*. The symbol *N* in a thread designation on a drawing means American National Form; *UN*, Unified form; and *M*, the metric standard thread form. *UN*- and *N*-form threads of the same size are interchangeable.

Figure 19-2 shows the general characteristics of Unified and American National screw thread profiles. The flat crest and root shown at location Ⓐ identify the American National form. The only difference between the *N* form and the Unified (*UN*) thread series form is in the shape of the crest and root. The root on the Unified Thread form may be flat or rounded. The Unified form is shown at position Ⓑ. The designation *UNR* on the drawing of a threaded part means the root is rounded on external threads. The flat or rounded crest is optional.

OTHER COMMON THREAD FORMS

Square and Acme thread forms are used in applications which require either a coarse pitch or a great force. The *square thread* form (figure 19-3A) has sides which are at a right angle to the thread root or axis.

The sides of the *Acme thread* (figure 19-3B) form an included angle of 29°. Square and Acme threads are often *multiple-lead threads* for fast movements. The 29° Brown and Sharpe thread is designed for applications requiring a deep worm thread, as in gearing.

DESIGNATION OF THREAD SERIES

The fineness or coarseness of threads in a particular series is identified in trade handbooks under each *thread series*. This means there is a standard number of threads for each diameter.

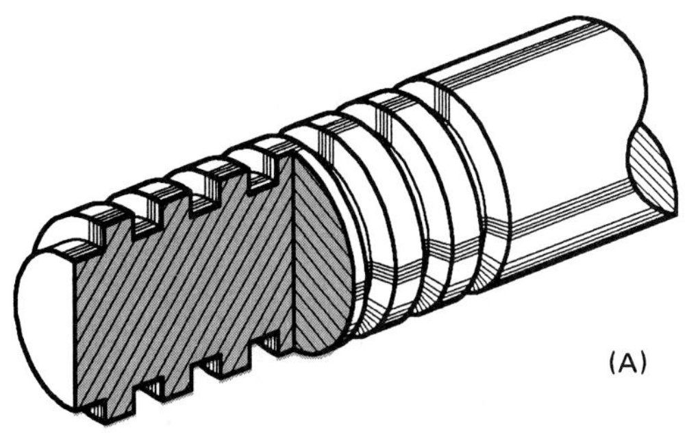

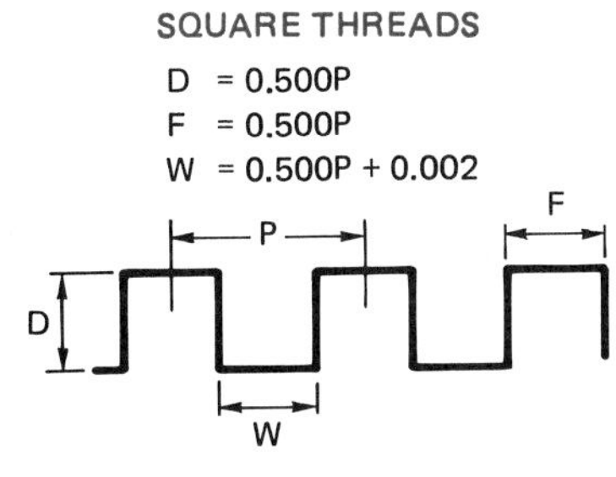

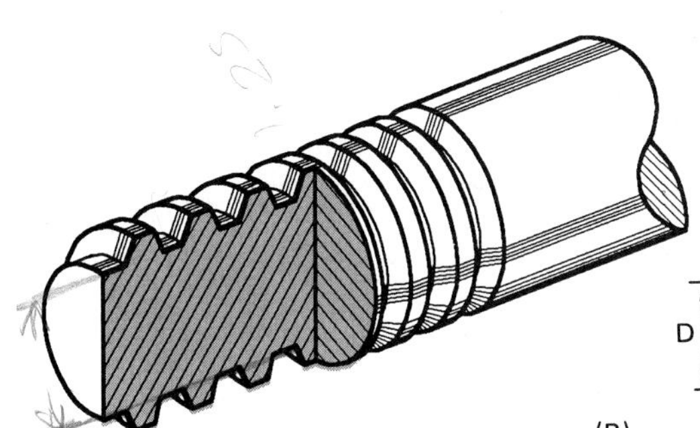

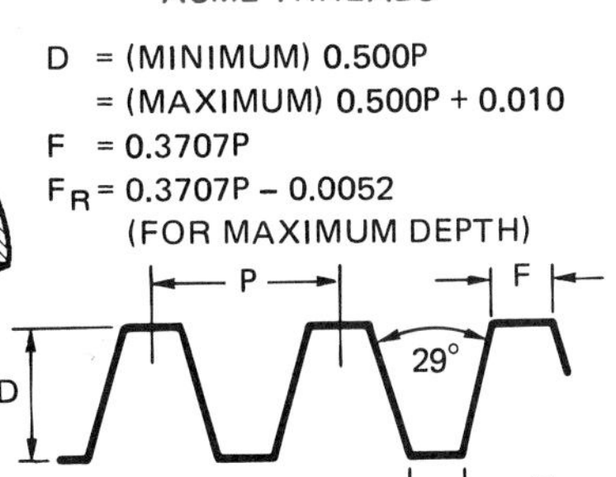

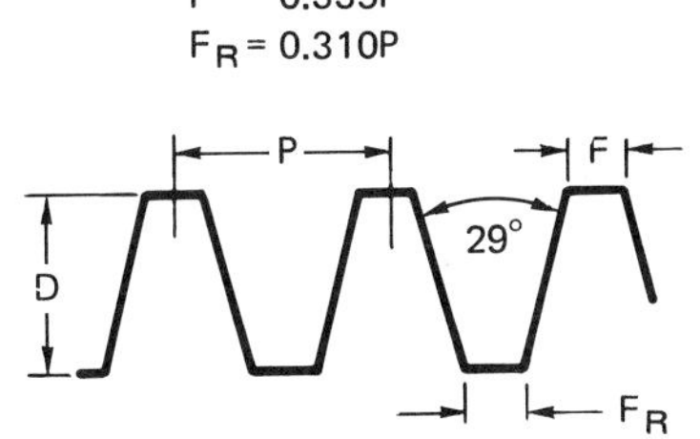

FIGURE 19-3 OTHER COMMON THREAD FORMS (29° AND 90°)

The 60° form threads fall into five categories of thread series. The thread series are listed in table 19-1 in column (a). Designations according to Unified or American National standards, which appear on drawings along with other dimensioning information, are listed in table 19-1 in column (b).

REPRESENTING EXTERNAL SCREW THREADS ON DRAWINGS

Three common methods of representing external screw threads in the ANSI, CSA, and SI metric systems are shown in figure 19-4. Threads may be represented on drawings in pictorial (A), schematic (B), or simplified (C) form.

TABLE 19-1 DESIGNATIONS OF N, UN, AND UNR THREADS

THREAD SERIES (a)	THREAD DESIGNATIONS ON DRAWINGS (b)		
	AMERICAN NATIONAL SYSTEM	UNIFIED SYSTEM	
		*1	*2
Coarse	NC	UNC	UNRC
Fine	NF	UNF	UNRF
Extra Fine	NEF	UNEF	UNREF
Special	NS	UNS	UNRS
Constant Pitch	(*3) N	(*3) UN	(*3) UNR

*1 Rounded crest and/or root are optional
*2 Rounded crest is optional; root radius is specified
*3 Thread pitch precedes the thread designation

DIMENSIONING EXTERNAL SCREW THREADS

An external screw thread in the ANSI series is dimensioned by using a leader and a notation. Figure 19-5 indicates how a thread is designated in order to provide specifications for machining and measuring a particular thread. Information is not always provided for each of the seven categories shown. Sometimes the Class of Fit ⑥ is described by a note. In the example, the length ⑦ may be omitted as this dimension is given on the drawing.

REPRESENTING AND DIMENSIONING PIPE THREADS

The three basic types of ANSI pipe threads include straight, taper, and pressure-tight Dryseal threads.

- *Straight pipe threads*. Where a drawing is dimensioned like $\frac{1}{2}$ **NPS**, the notation indicates (1) that the part contains the

SYSTEM	PICTORIAL REPRESENTATION (A)		SCHEMATIC REPRESENTATION (B)	SIMPLIFIED REPRESENTATION (C)	END VIEW
ANSI SCREW THREADS					
CSA AND ISO SCREW THREADS			INCOMPLETE THREAD RUNOUT		FINE, 270° ROOT DIAMETER

FIGURE 19-4 REPRESENTATIONS OF EXTERNAL SCREW THREADS (ANSI, CSA, AND ISO SYSTEMS)

standard number of threads in the pipe series, and (2) that straight threads are required.

- *Taper pipe thread.* Some parts require a tight, leakproof joint or seal. In such cases, internal and external threads are *cut along a standard taper.* This taper is 0.750″ per foot. An example of a drawing notation for a 1/2″ taper pipe thread is **$\frac{1}{2}$ NPT.**
- *Pressure-tight threads.* Mating threads on parts which are subjected to extremely high pressures require a pressure-tight fit. Dryseal pressure-tight threads are specified as **NPTF** for *the taper series* and **NPSF** for the *Dryseal straight thread series.*

DESIGNATION OF ISO SCREW THREADS

The basic major diameter (nominal diameter) and pitch are used to specify ISO metric screw threads. Figure 19-6 shows how an ISO metric thread in the fine thread series is dimensioned. The pitch for coarse series SI metric threads is often omitted.

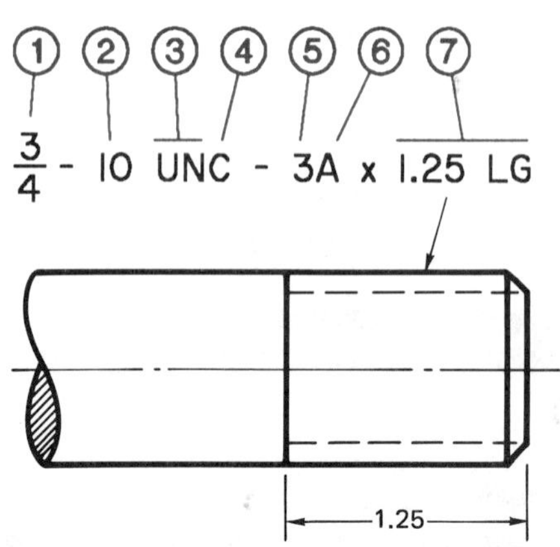

DIMENSIONING CODE

1. OUTSIDE (NOMINAL) DIAMETER OF THREADS
2. NUMBER OF THREADS PER INCH (TPI)
3. THREAD FORM (UNIFIED)
4. THREAD SERIES (COARSE)
5. THREAD CLASS
6. EXTERNAL (OUTSIDE) THREAD
7. LENGTH OF THREAD

FIGURE 19-5 POSITION OF DIMENSIONING INFORMATION FOR AMERICAN NATIONAL AND UNIFIED STANDARD THREADS

The letter **M** preceding the nominal diameter indicates a thread form which is in the SI metric thread system. The letter **x** separates the outside (nominal) diameter from the thread pitch and other information, such as thread length. Basic thread information is followed by a dash. Then, where applicable, other data is given for the *tolerance* which applies to the *pitch diameter* and the *tolerance* of the *crest diameter.* A number designation is used in both instances to indicate the size of each of the two *tolerance grades.* The *callouts* of the grade tolerance numbers in figure 19-6 are represented as items (6) and (8). Each grade is followed by a *positional letter.* Letter (7) indicates the *positional tolerance for the pitch diameter*; (9), *for the crest diameter.*

The lower case letter **e** is used for external threads which require *a large allowance*; **g**, *a small allowance*; and **h**, *no allowance.* Similarly, for internal threads, the capital **G** is used for a *small allowance* and **H**, *no allowance.*

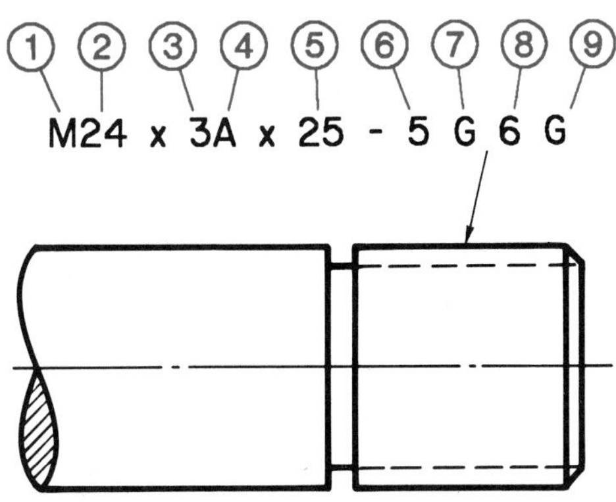

DIMENSIONING CODE

(A) BASIC THREAD CALLOUT

(1) ISO (SI) DESIGNATION
(2) OUTSIDE DIAMETER IN mm
(3) THREAD PITCH IN mm
(4) EXTERNAL THREAD
(5) THREAD LENGTH

(B) ADDITIONAL THREAD CALLOUT

(6) GRADE TOLERANCE NUMBER
(7) POSITIONAL TOLERANCE LETTER
} PITCH DIAMETER TOLERANCE

(8) GRADE TOLERANCE NUMBER
(9) POSITIONAL TOLERANCE LETTER
} CREST DIAMETER TOLERANCE

FIGURE 19-6 POSITION OF DIMENSIONING INFORMATION FOR SI METRIC THREADS

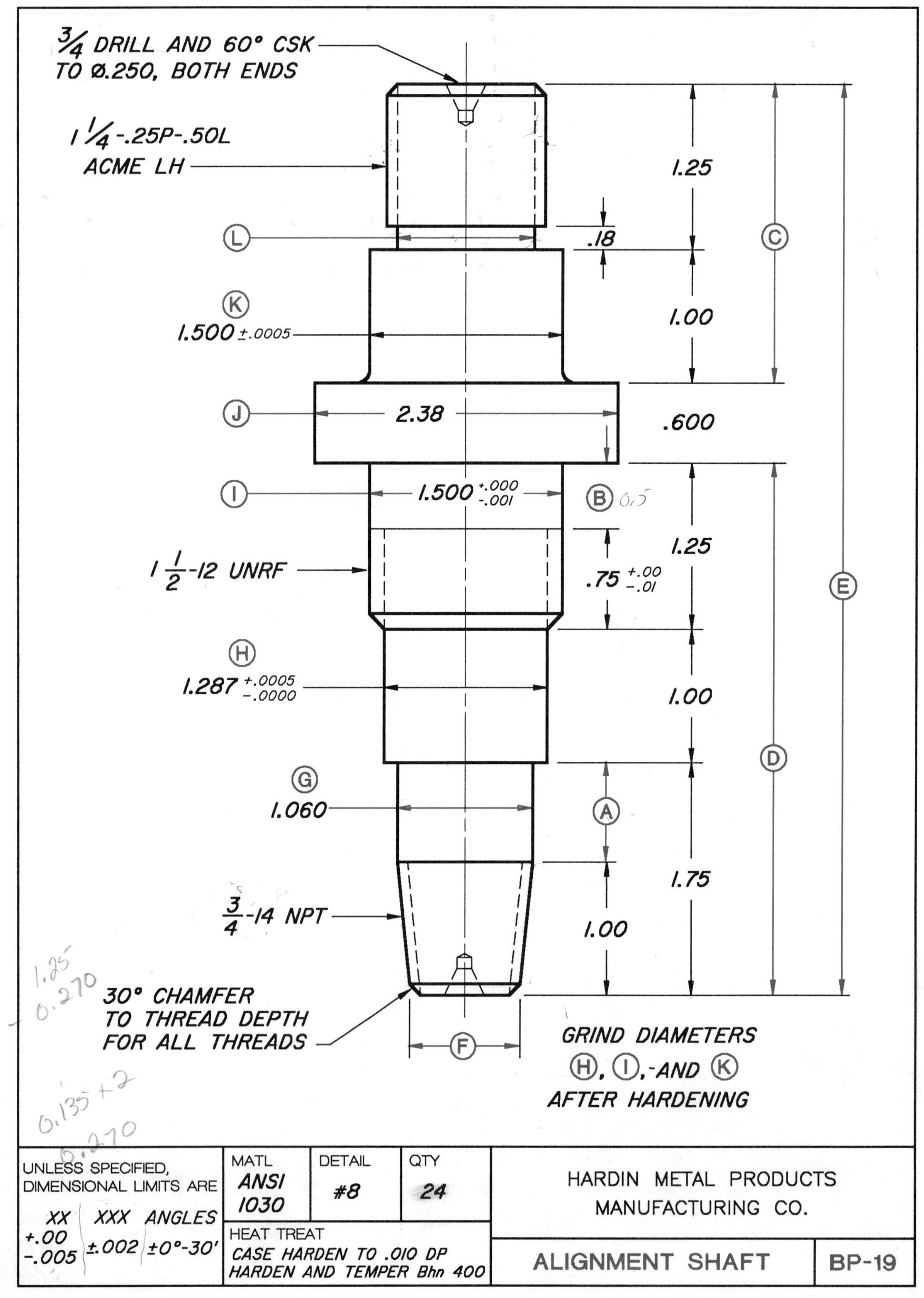
3/4 DRILL AND 60° CSK TO Ø.250, BOTH ENDS
1 1/4-.25P-.50L ACME LH
L
K
1.500 ±.0005
J
2.38
I
1.500 +.000 -.001
1 1/2-12 UNRF
H
1.287 +.0005 -.0000
G
1.060
3/4-14 NPT
30° CHAMFER TO THREAD DEPTH FOR ALL THREADS
F
1.25
.18
C
1.00
.600
B
1.25
.75 +.00 -.01
E
1.00
D
A
1.75
1.00
GRIND DIAMETERS H, I, AND K AFTER HARDENING
UNLESS SPECIFIED, DIMENSIONAL LIMITS ARE
XX +.00 -.005
XXX ±.002
ANGLES ±0°-30'
MATL ANSI 1030
DETAIL #8
QTY 24
HEAT TREAT
CASE HARDEN TO .010 DP HARDEN AND TEMPER Bhn 400
HARDIN METAL PRODUCTS MANUFACTURING CO.
ALIGNMENT SHAFT
BP-19

ALIGNMENT SHAFT (BP-19)

1. a. Identify the kind of steel required to produce the Alignment Shaft.
   b. Give the depth of the hardened case, the heat treatment, and the hardness value.
2. Determine the upper and lower limit dimensions for the following lengths: Ⓐ, Ⓑ, Ⓒ, Ⓓ, and Ⓔ.
3. Give the specification for the center-drilled holes.
4. State the maximum and minimum dimensions to which diameters Ⓗ, Ⓘ, and Ⓚ are to be ground.
5. Indicate the maximum and minimum dimensions to which diameters Ⓖ and Ⓙ are to be turned.
6. a. Identify the system of representation used for the three threads.
   b. Write out the specification for each of the three threads.
7. Identify the following features of the Acme thread:
   a. Nominal outside diameter
   b. Thread pitch
   c. Thread lead
   d. Hand of thread
   e. Nominal root diameter Ⓛ when the single thread depth is 0.135″
8. Provide the following information for the UNRF and/or NPT threads:
   a. Number of threads per inch.
   b. One major difference in the type of threads.
   c. The meaning of the letter **R** in the **UNRF** thread designation.
9. Compute the nominal outside diameter Ⓕ. The NPT has a taper of 0.75″ per foot.
10. Specify how the ends of the 60° threads are to be machined.
11. a. Convert the **1 $\frac{1}{2}$ – 12 UNRF** thread and length to a metric equivalent thread.
    b. Show how the SI metric thread would be specified on a drawing.
    c. Dimension the thread length in metric, including the standard tolerance in mm to two decimal places.
12. State two differences in the manner in which threads are represented in the ANSI and the CSA and SI metric systems (for the end view, where required).

ASSIGNMENT—UNIT 19

Student's Name ____________________

1. a. ____________________
   b. ____________________
   ____
   ____

2.

	Upper Limit	Lower Limit
Ⓐ =	______	______
Ⓑ =	______	______
Ⓒ =	______	______
Ⓓ =	______	______
Ⓔ =	______	______

3. ____________________

4.

	Max. Diameter	Min. Diameter
Ⓗ =	______	______
Ⓘ =	______	______
Ⓚ =	______	______

5.

	Max. Diameter	Min. Diameter
Ⓖ =	______	______
Ⓙ =	______	______

6. a. ____________________
   b. (1) ____________________
   (2) ____________________
   (3) ____________________

7. a. Outside Diameter ______
   b. Pitch ______
   c. Lead ______
   d. Hand ______
   e. (Nominal) Root Diameter ______

8.

	UNRF	NPT
a. TPI	______	______
b. Type	______	______

   c. ____________________

9. Ⓕ = ____________________

10. ____________________

11.

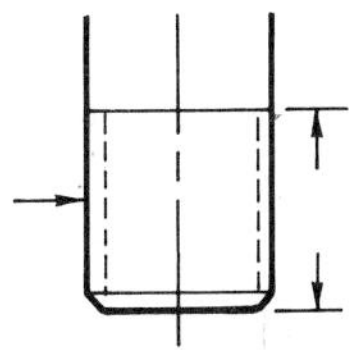

12. a. ____________________
    ____
    b. ____________________
    ____

UNIT 20
Internal Screw Threads

Interchangeable screw threads are generally produced by one of the following five basic production methods:

- Hand or machine cutting, using taps.
- Machine tapping with formed cutters.
- Machine threading with single-point or multiple-point thread chasing tools.
- Machine rolling and forming.
- Casting by such manufacturing processes as sand, die, permanent mold, and shell casting.

FEATURES OF INTERNAL THREADED HOLES

The two most commonly used internal threading processes in jobbing shops include tapping and machining with a single-point cutting tool. In *tapping*, internal threads are formed (*tapped*) by rotating either the tap or the workpiece. The tap cuts the uniform, spiral thread form as it advances into the *tap drill hole.* Some holes are threaded *through* a workpiece (figure 20-1A). Other holes, which do not go through a workpiece, are tapped only part way. Such holes are called *blind-tapped holes* as illustrated in figure 20-1B. A *bottomed-tapped hole* (figure 20-1C) indicates that a blind hole is threaded as deep as practical in a drilled or bored hole.

Holes may be threaded by hand using hand taps. Machine taps are used for machine tapping. The taps are held in and driven by a self-releasing machine attachment like the one shown in figure 20-2. The attachment fits into a machine spindle and is power driven.

TAP DRILL SIZES

As a result of the standardization of

- thread forms (profiles),
- number of threads for each diameter of a thread series,
- classes of thread fits, and
- limits for surface finishes,

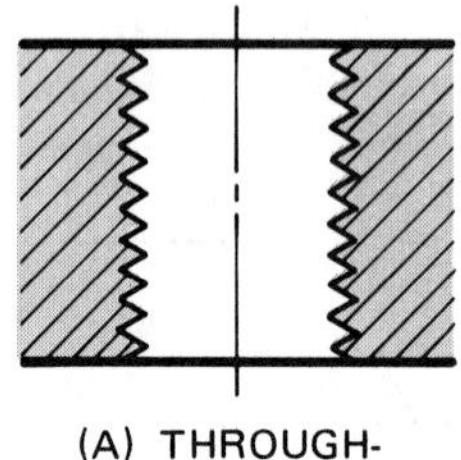

(A) THROUGH-TAPPED HOLE

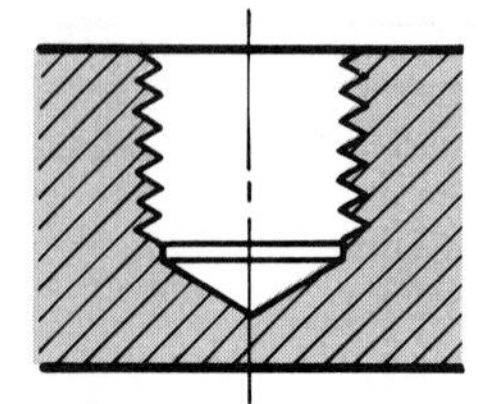

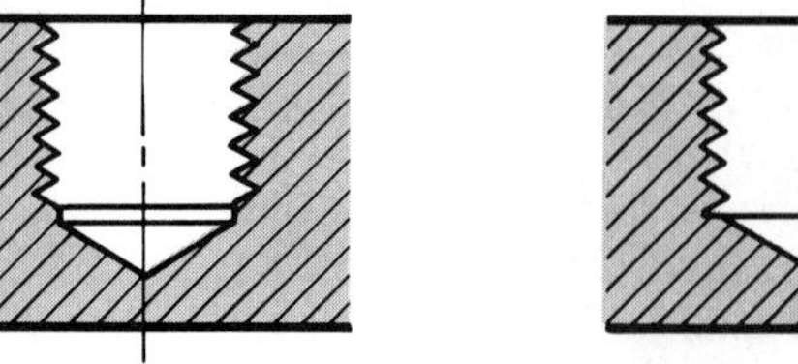

(B) BLIND-TAPPED HOLE

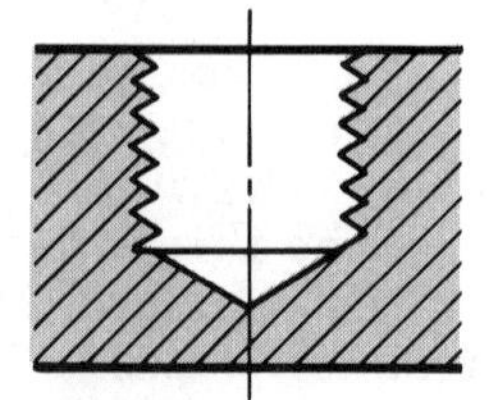

(C) BLIND BOTTOMED-TAPPED HOLE

FIGURE 20-1 THROUGH- AND BLIND-TAPPED HOLES

Courtesy of JARVIS PRODUCTS CORPORATION

FIGURE 20-2 SELF-RELEASING (TORQUE-DRIVEN) TAPPING ATTACHMENT FOR INTERNAL MACHINE THREADING

industrial standards have been developed to govern the diameters of holes (hole sizes) which are to be threaded. Technical tables for hole sizes are provided by manufacturers of fasteners and producers of taps, dies, and other thread-cutting tools. Standards and specifications are also contained in industry handbooks.

Variations of tap hole sizes are based on the required depth of thread from a theoretical 100% full-depth thread, to 75%, 50%, or other practical depth. Drawings generally contain a drill number, letter size, or fractional or decimal dimension for the tap drill for small size taps. A bored or other machined diameter is specified for larger diameter holes.

REPRESENTING INTERNAL THREADS ON DRAWINGS

The three conventional methods of representing internal threads on drawings are illustrated in figure 20-3. Note the variations in the simplified representation according to ANSI standards compared with CSA and SI metric standards.

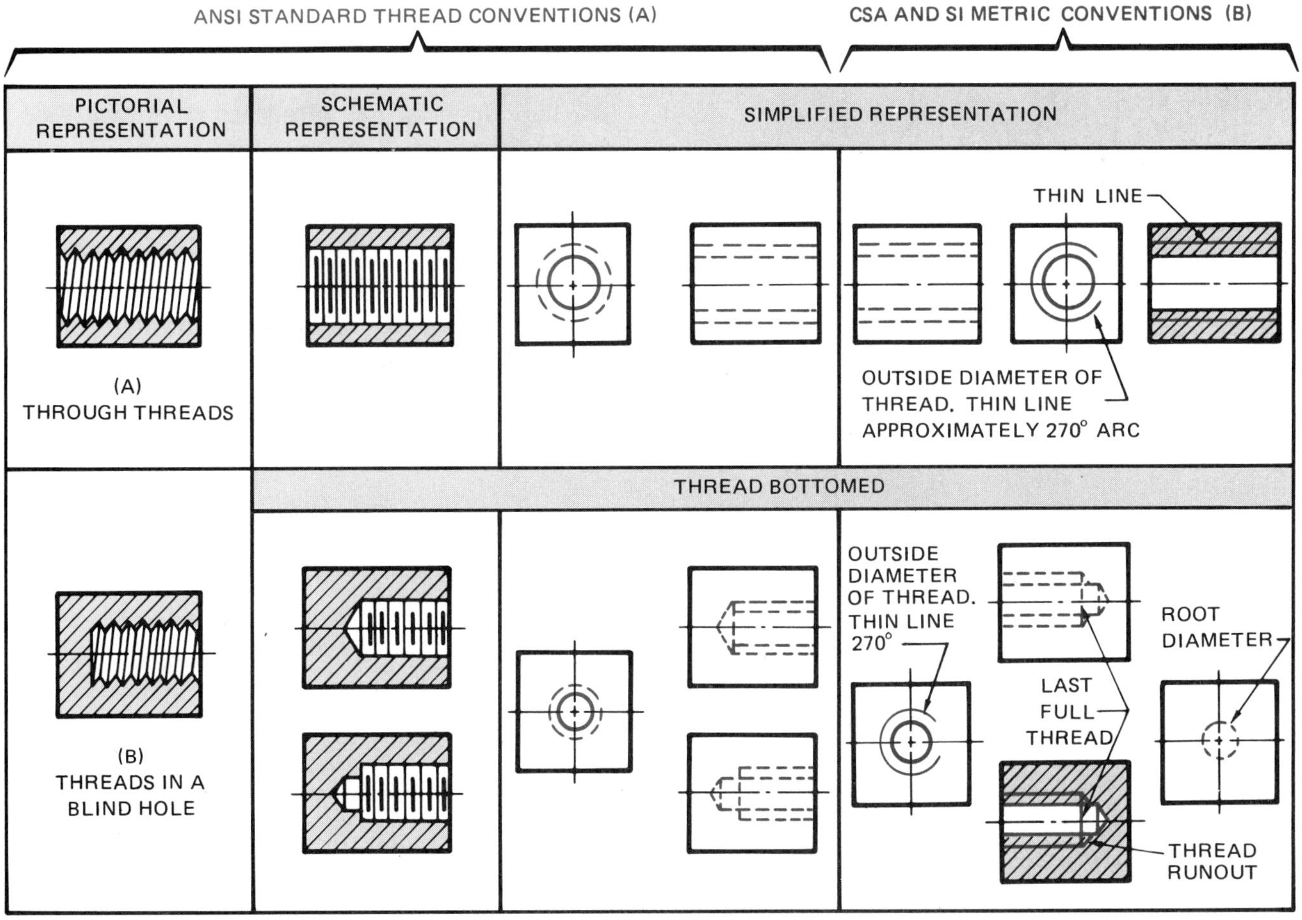

ANSI STANDARD THREAD CONVENTIONS (A)
CSA AND SI METRIC CONVENTIONS (B)
PICTORIAL REPRESENTATION
SCHEMATIC REPRESENTATION
SIMPLIFIED REPRESENTATION
THIN LINE
(A) THROUGH THREADS
OUTSIDE DIAMETER OF THREAD. THIN LINE APPROXIMATELY 270° ARC
THREAD BOTTOMED
(B) THREADS IN A BLIND HOLE
OUTSIDE DIAMETER OF THREAD. THIN LINE 270°
LAST FULL THREAD
ROOT DIAMETER
THREAD RUNOUT

FIGURE 20-3 CONVENTIONS FOR REPRESENTING INTERNAL THREADED HOLES IN THE ANSI, CSA, AND SI METRIC SYSTEMS

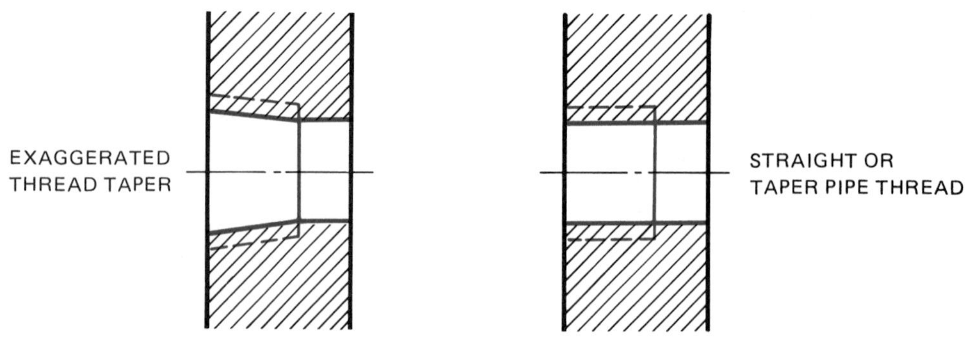

FIGURE 20-4 EXAMPLES OF SIMPLIFIED THREAD REPRESENTATION OF INTERNAL PIPE THREADS

REPRESENTING INTERNAL PIPE THREADS

Standard symbols are used to designate straight pipe threads (NPTS) and tapered pipe threads (NPT). The threads appear on drawings as either straight threads or with an exaggerated taper as shown in figure 20-4.

DIMENSIONING INTERNAL SCREW THREADS

The same dimensioning practices for external ANSI, CSA, and SI metric threads apply to internal threads. The thread depth is specified in the callout or by directly dimensioning the length on the drawing. The standard method for dimensioning an internal thread is demonstrated in figure 20-5.

THREAD CLASSES FOR UNIFIED AND AMERICAN NATIONAL FORM THREADS

A thread class number appears on drawings to indicate the required degree of accuracy between the features of an external and internal thread. The three regular classes of fit are identified as Class 1A or 1B, Class 2A or 2B, and Class 3A or 3B. The letter A, as stated before, indicates an external thread; B, an internal thread.

Each class of fit represents a different amount of clearance between the sides, crests, and roots of the internal and external thread. The new standards require a larger pitch diameter allowance (tolerance) on the external thread.

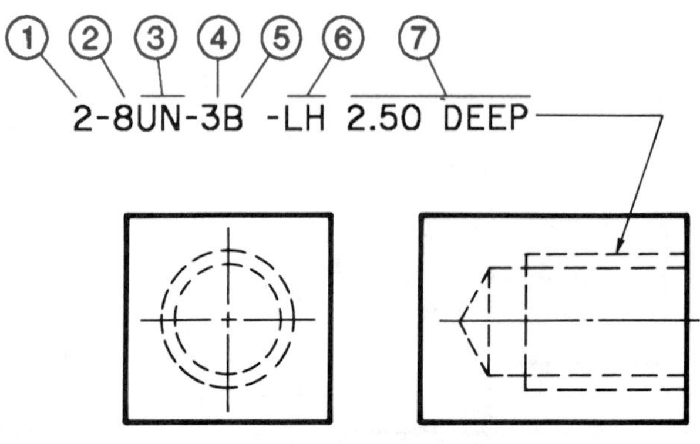

1. 2" – OUTSIDE THREAD DIAMETER
2. 8 – 8 PITCH (CONSTANT) SERIES
3. UN – UNIFIED AMERICAN NATIONAL THREAD FORM
4. 3 – CLASS 3 THREAD FIT
5. B – INTERNAL THREAD
6. LH – LEFT-HAND SCREW THREAD
7. 2.50" – DEPTH OF THREADS

FIGURE 20-5 CONVENTIONAL DIMENSIONING OF A THREADED HOLE

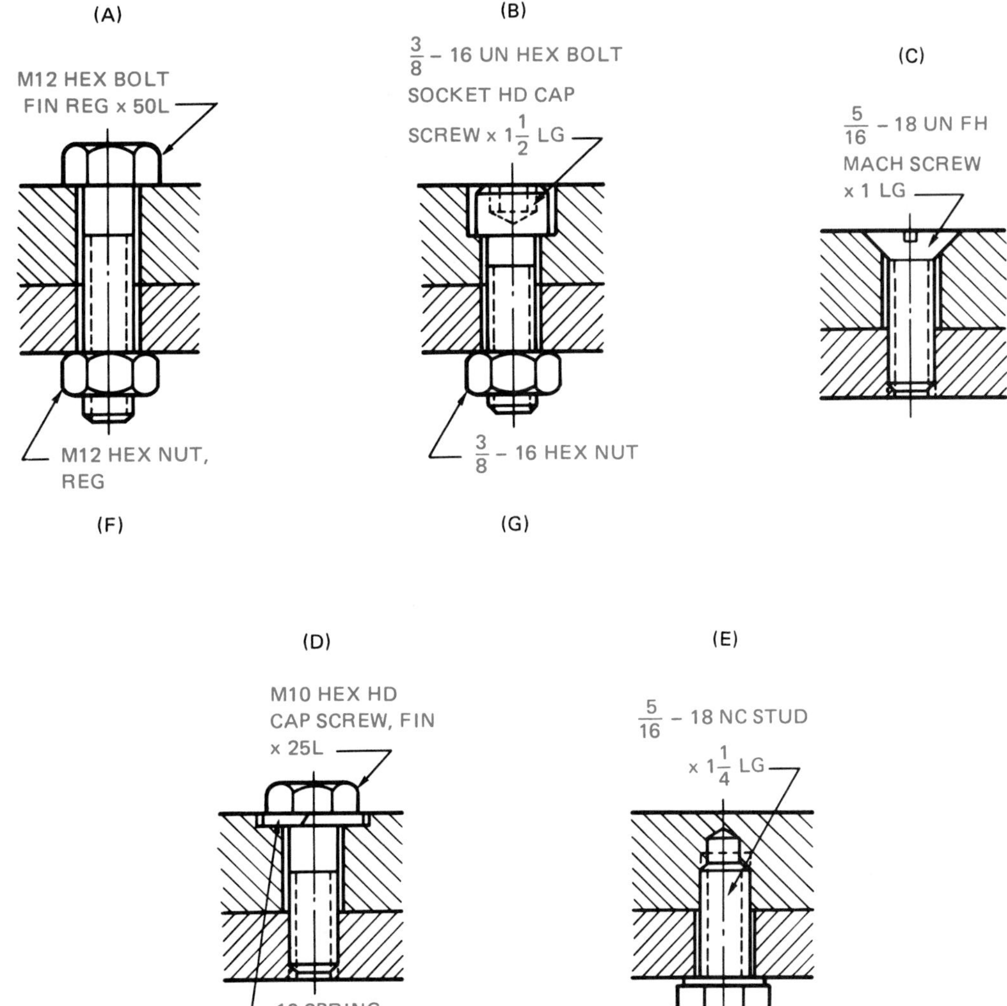

FIGURE 20-6 REPRESENTATION AND SPECIFICATIONS OF SAMPLE THREAD FASTENERS AND WASHERS

- A *Class 1A and 1B fit* provides the greatest allowance (tolerance) between fitted threads.
- A *Class 2A and 2B fit* is the most widely used of the three classes for good quality commercial products.
- A *Class 3A and 3B fit* is used especially for precision applications where a close, snug fit is required.

REPRESENTING AND SPECIFYING COMMON THREADED FASTENERS

When common threaded fasteners are used in the assembly of threaded parts, or parts held together by fasteners, they are generally represented as shown by the examples in figure 20-6. The callouts provide specifications as they appear on drawings for four different screw designs (A, B, C, and D), a conventional stud (E), and three kinds of hexagon nuts (F, G, and H).

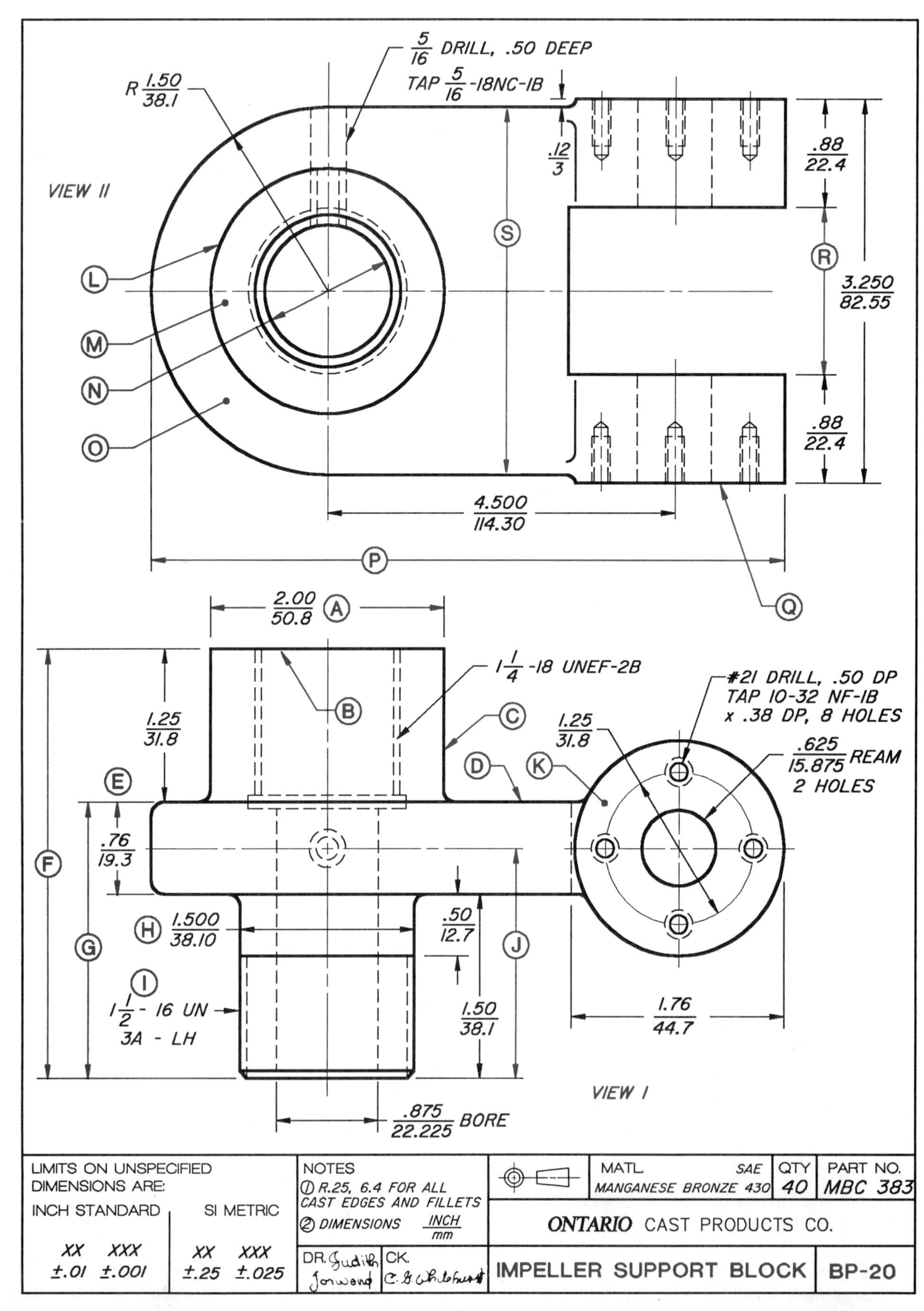

5/16 DRILL, .50 DEEP
TAP 5/16 -18NC-1B
R 1.50/38.1
VIEW II
.12/3
.88/22.4
3.250/82.55
4.500/114.30
L
M
N
O
P
Q
R
S
2.00/50.8 A
B
C
D
E
F
G
H
I
J
K
1 1/4 -18 UNEF-2B
#21 DRILL, .50 DP
TAP 10-32 NF-1B
x .38 DP, 8 HOLES
.625/15.875 REAM
2 HOLES
1.25/31.8
.76/19.3
1.500/38.10
.50/12.7
1 1/2 - 16 UN
3A - LH
1.50/38.1
1.76/44.7
VIEW I
.875/22.225 BORE
LIMITS ON UNSPECIFIED DIMENSIONS ARE:
INCH STANDARD
SI METRIC
XX ±.01
XXX ±.001
XX ±.25
XXX ±.025
NOTES
① R.25, 6.4 FOR ALL CAST EDGES AND FILLETS
② DIMENSIONS INCH/mm
DR. Judith Jorwand
CK. C. G. Whitehurst
MATL. SAE MANGANESE BRONZE 430
QTY 40
PART NO. MBC 383
ONTARIO CAST PRODUCTS CO.
IMPELLER SUPPORT BLOCK
BP-20

ASSIGNMENT—UNIT 20

IMPELLER SUPPORT BLOCK (BP-20)

1. Provide the following information about the part:
   a. The process of manufacturing the original Block
   b. Part number
   c. Material specified
2. State the meaning of each of the two notes in the title block.
3. a. Indicate the system of projection which is used.
   b. Name **VIEWS I** and **II**.
   c. Give the nominal controlling dimensions for Ⓐ and Ⓔ.
4. Identify the following surfaces in **VIEW II**.
   a. Ⓑ c. Ⓓ
   b. Ⓒ d. Ⓚ
5. Give the upper and lower limit dimensions in inch and SI metric measurements for dimensions Ⓕ, Ⓖ, and Ⓟ.
6. Compute the nominal inch/mm dimensions for the following:
   a. Ⓙ c. Ⓡ
   b. Ⓝ d. Ⓢ
7. Give the maximum and minimum inch/mm dimensions for the following features:
   a. Turned surface Ⓗ.
   b. The bored hole.
   c. The reamed holes.
8. Interpret the external thread callout Ⓘ.
   $1\frac{1}{2}$ – 16 UN 3A – LH
9. Interpret the internal thread callout:
   $1\frac{1}{4}$ – 18 UNEF – 2B
10. Describe the meaning of the following items:
    a. Tap drill size
    b. 75% thread depth
11. Provide the following information for the tapped holes:
    a. Tap drill size and depth of the drilled holes.
    b. Thread size and series.
    c. Thread depth.
    d. Number of holes to be tapped.
    e. The class of fit and thread designation.

Student's Name ______________________

1. a. ______________________
   b. Part # ____________ c. ____________
2. a. ______________________
   ____
   b. ______________________
   ____
3. a. ______________________
   b. **VIEW I** ____________ **VIEW II** ____________
   c. Ⓐ = ____________ Ⓔ = ____________
4. a. Ⓑ = ____________ c. Ⓓ = ____________
   b. Ⓒ = ____________ d. Ⓚ = ____________
5.

	Upper Limit (Inch)	Upper Limit (mm)	Lower Limit (Inch)	Lower Limit (mm)
Ⓕ =				
Ⓖ =				
Ⓟ =				

6.

	Inch	mm		Inch	mm
a. Ⓙ =			c. Ⓡ =		
b. Ⓝ =			d. Ⓢ =		

7. a.

	Max. Dimensions inch	Max. Dimensions mm	Min. Dimensions inch	Min. Dimensions mm
a. Ⓗ =				
b.				
c.				

8. 1 1/2″ ______________________
   – 16 ______________________
   U ______________________
   N ______________________
   3 ______________________
   A ______________________
   –LH ______________________
9. 1 1/4″ ______________________
   –18 ______________________
   U ______________________
   NEF ______________________
   –2 ______________________
   B ______________________
10. a. ______________________
    ____
    b. ______________________
    ____
11. a. ____________ ____________
    b. ____________ ____________
    c. ____________ d. ____________
    e. ______________________
    ____

UNIT 21
Dimensioning Machined Surfaces

All machined surfaces, regardless of the degree of production or machining accuracy, have surface irregularities. When the desired quality of surface finish is produced by general manufacturing processes, the part drawing often has a **FINISH ALL OVER** or **FAO** note. Other surface finish symbols and measurements are used in those instances where the machined surface must be held within specified limits of dimensional accuracy and surface quality.

This unit deals with terms, symbols, ratings, and other specifications found on drawings to identify surface texture requirements. These items are described and illustrated for height, width, and direction of surface irregularities and flaws (figure 21-1).

CONTROLLING FACTORS FOR SURFACE TEXTURE

It is common practice to interchangeably use the terms *surface texture*, *surface finish*, *surface roughness*, *surface characteristics*, or just *finish*. *Surface texture* relates to deviations from a nominal profile which form the pattern of the surface. Figure 21-2 shows the nominal profile as represented by a center line. Deviations are produced by roughness, waviness, lay, and flaws.

The engineer or part designer establishes the quality of surface finish on the basis of the following factors:

- Part function, load, speed, and direction of movement as in the case of gears, cams, and moving parts.
- Physical characteristics of contact materials, temperatures, and lubrication. Bearings, journals, pistons, and bushings are examples.
- Surface wear under dry friction machining using cutting tools, dies, punches, and rolls.
- Flatness and smoothness as required for precision micrometer instrument anvils,

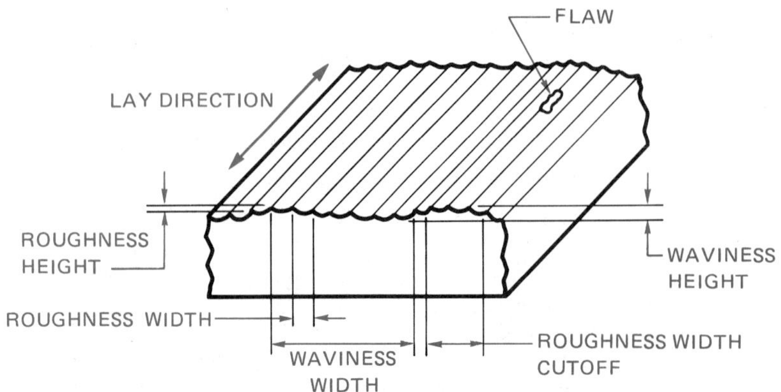

FIGURE 21-1 COMMON SURFACE TEXTURE CHARACTERISTICS

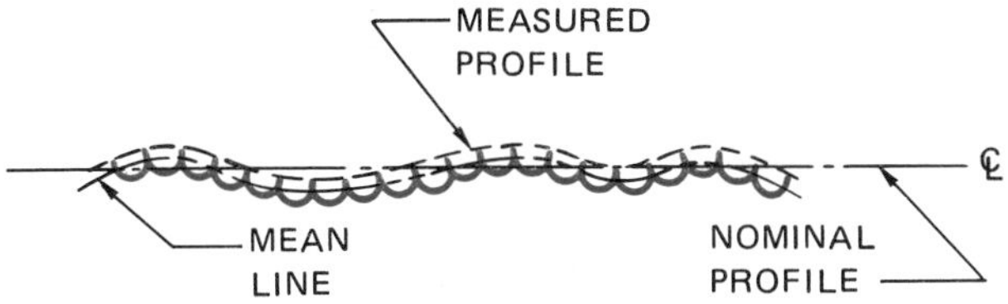

FIGURE 21-2 RELATIONSHIP OF NOMINAL PROFILE, MEASURED PROFILE, AND MEAN LINE (ENLARGED)

gage blocks, and for other extremely high-speed operations.

BASIC SURFACE TEXTURE SYMBOLS

The basic surface texture symbol according to ANSI standards is √‾‾ . The apex of the leg touches the surface which is to be machined or produced by other manufacturing processes. Where it is more functional, the leg may touch an extension line or a leader which points either to the extension line or to the machined surface. The symbol and surface texture information are generally found on one view where size and location dimensions appear. The open symbol √ means the surface may be produced by any manufacturing process. The closed symbol ∇√ indicates that material is to be removed from the surface. The symbol ⊙√ is often used to show that material removal is prohibited. Recommended practices of relating a surface texture symbol to a finished surface are illustrated in figure 21-3.

Identification of Production Methods

Production methods, machining patterns, and dimensional specifications are part of the surface texture symbol. A horizontal line is added to the symbol. A notation often appears above the line to identify the machine process. For example, the symbol √GRIND identifies the production method; √CHROMIUM PLATE , a required surface coating or treatment process.

SURFACE TEXTURE TERMS

- *Surface.* An object, according to ANSI standards, is bounded by a surface. The surface separates one object from another object, substance, or space.
- *Profile.* A profile represents the shape, form, or contour of a machined surface. The profile is viewed on a plane which is perpendicular to the machined surface. A profile (figure 21-2) may be measured by

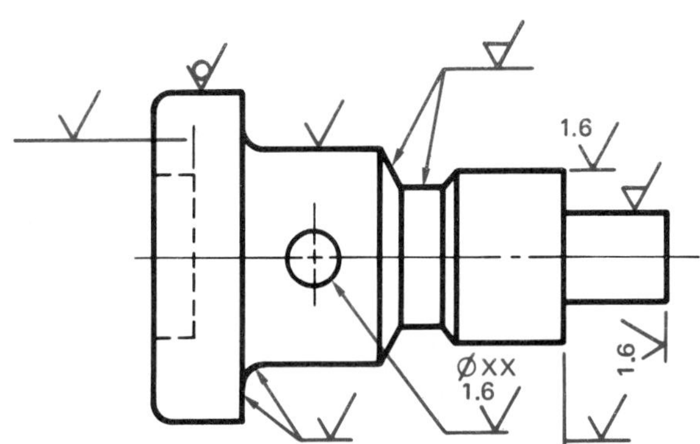

FIGURE 21-3 APPLICATIONS OF ANSI SURFACE FINISH SYMBOLS ON A DRAWING

  a *surface finish analyzer,* a *profilometer,* a *microscope,* or by a comparison with a visual standard, using a *comparator.*
- *Nominal Surface* and *Mean Line.* The nominal surface is theoretically a geometrically perfect surface. The term *mean line,* for practical purposes, represents the average of the surface roughness.

CHARACTERISTICS OF SURFACE ROUGHNESS

Roughness refers to fine surface irregularities which are produced by a cutting tool, feed marks, or production process.

Roughness width cutoff (figure 21-1) relates to the greatest spacing (width) of surface irregularities in a repeating pattern. The placement of the roughness width cutoff value in the surface texture symbol is shown in figure 21-4 as Ⓖ.

Roughness width represents the average groove width from crest to crest (figure 21-1). The value is positioned within the symbol (figure 21-4 as Ⓕ).

Roughness height (figure 21-1) represents the arithmetical average deviation from corresponding crests and roots. Most roughness heights are given as a single value representing the maximum roughness height (figure 21-4 at Ⓐ). When there is a permissible range to the roughness height, the maximum and the minimum values appear above the leg of the symbol (figure 21-4 as shown at Ⓐ and Ⓑ).

Flaws

Surface defects, such as scratches from machining, blow holes produced in casting, and cracks due to hardening, are known as *flaws.* The effect

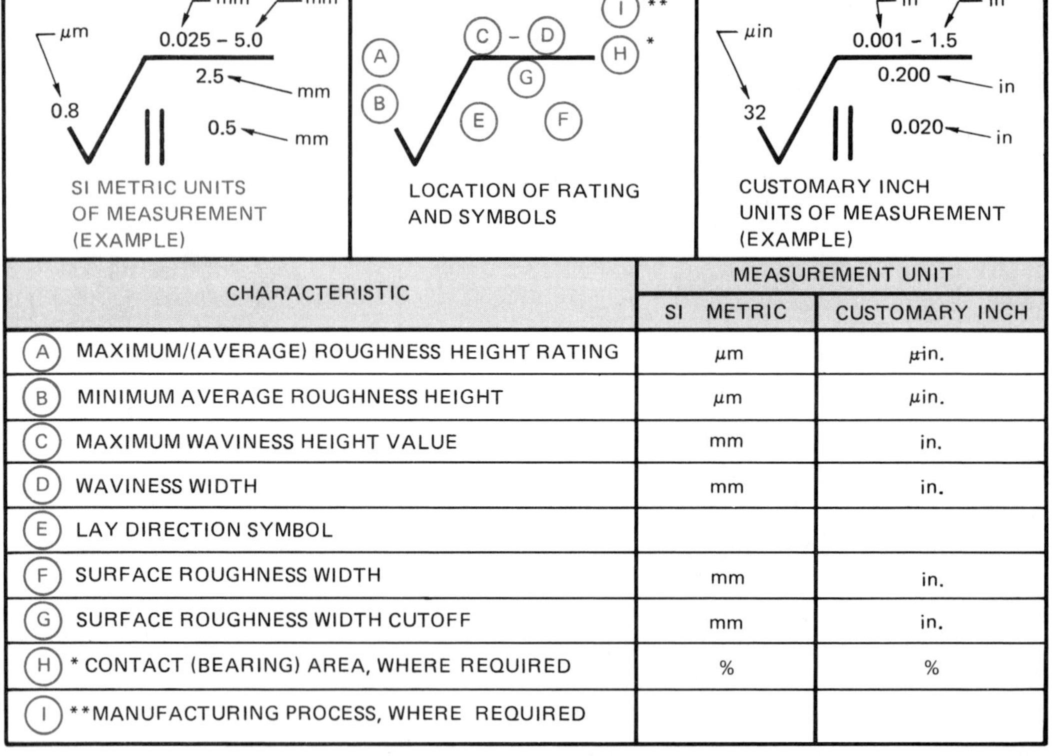

CHARACTERISTIC	MEASUREMENT UNIT	
	SI METRIC	CUSTOMARY INCH
(A) MAXIMUM/(AVERAGE) ROUGHNESS HEIGHT RATING	µm	µin.
(B) MINIMUM AVERAGE ROUGHNESS HEIGHT	µm	µin.
(C) MAXIMUM WAVINESS HEIGHT VALUE	mm	in.
(D) WAVINESS WIDTH	mm	in.
(E) LAY DIRECTION SYMBOL		
(F) SURFACE ROUGHNESS WIDTH	mm	in.
(G) SURFACE ROUGHNESS WIDTH CUTOFF	mm	in.
(H) * CONTACT (BEARING) AREA, WHERE REQUIRED	%	%
(I) **MANUFACTURING PROCESS, WHERE REQUIRED		

FIGURE 21-4 REPRESENTATION OF SURFACE CHARACTERISTICS AND VALUES ON DRAWINGS

of flaws on the roughness height generally is not included in such measurements.

CHARACTERISTICS AND REPRESENTATION OF WAVINESS

Other waviness irregularities are produced by cutting shearing forces, machine feed, strains in hardening, and the working properties of the part itself. The term *waviness* identifies a series of waves of surface variations.

Waviness width refers to the spacing of successive wave peaks or valleys.

Maximum waviness height is the peak-to-valley distance (extending across a series of roughness widths). The placement of maximum waviness height and width values are shown in figure 21-4 at Ⓒ and Ⓓ.

REPRESENTATION OF SURFACE PATTERNS

The direction of the predominant surface pattern as produced in the manufacture of a part is identified as *lay* (figure 21-4 at Ⓔ). Seven symbols commonly used on drawings to indicate the direction of the lay appear in the first column of figure 21-5. The predominant machining or production pattern for each lay symbol is pictured in column (b). Examples are given in column (c) of the placement of each lay symbol within the general surface texture symbol.

RANGES OF SURFACE ROUGHNESS FOR COMMON PRODUCTION METHODS

Dimensions are used with the general surface texture and lay symbols. *Waviness width*, *waviness height*, *surface roughness width*, and *surface roughness height*, are given in either customary inch or SI metric millimeter values. Both maximum and minimum roughness are expressed as microinch (µin.) or micrometer (µm) measurements. Microinch requirements are expressed in millionths of an inch. As just indicated, the symbol µin. is used. In SI metrics, the millionth of a meter surface roughness requirement is followed by the symbol µm.

MICROMETER (SI) AND MICROINCH (CUSTOMARY INCH) RANGES OF SURFACE ROUGHNESS

Product manufacturers and trade handbooks contain tables of information about roughness grades for different manufacturing processes.

Table 21-1 provides a few examples from table A-6, which appears in the appendix. The table associates appropriate surface finishes with different manufacturing processes. For comparison purposes, note the extremely rough surface produced by flame cutting. The roughness ranges facturing processes. In this instance, the solid areas of the table identify the *commercial range.* The dotted areas indicate *coarser or finer surface finishes.* These are produced by changing the method of manufacturing or a particular production process.

TABLE 21-1 EXAMPLES OF MICROINCH AND MICROMETER RANGES OF SURFACE ROUGHNESS

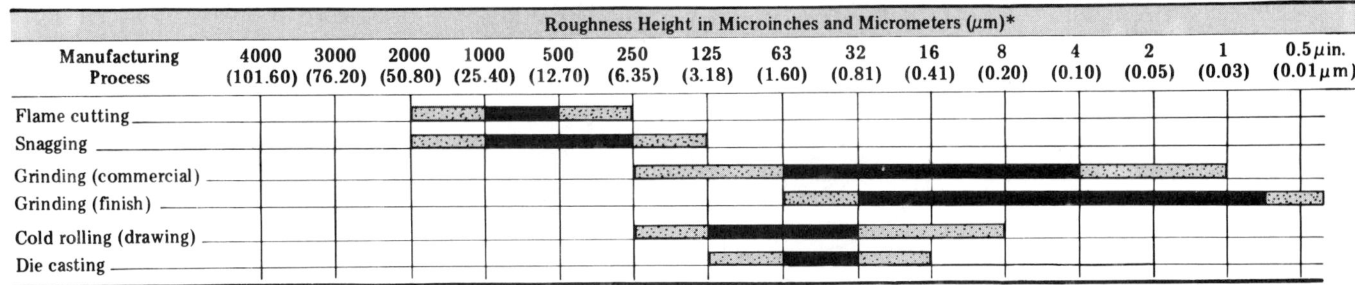

Roughness Height in Microinches and Micrometers (μm)*
Manufacturing Process
4000 (101.60)
3000 (76.20)
2000 (50.80)
1000 (25.40)
500 (12.70)
250 (6.35)
125 (3.18)
63 (1.60)
32 (0.81)
16 (0.41)
8 (0.20)
4 (0.10)
2 (0.05)
1 (0.03)
0.5 μin. (0.01 μm)
Flame cutting
Snagging
Grinding (commercial)
Grinding (finish)
Cold rolling (drawing)
Die casting

Code ▬ General manufacturing (average) surface finish range
▒ Higher or lower range produced by using special processes

*Values rounded to nearest second place μm decimal

from 2000 μin. to 250 μin. (50.8 μm to 6.35 μm). By comparison, precision machined parts may be finish ground in the range between 63 microinches and 0.5 microinches (1.6 μm and 0.01 μm) and finer. Surface roughness tables are coded to show the surface finish range for general manu-

A final example is provided by using a simple drawing (figure 21-6A) to illustrate how surface texture symbols and values are applied. The schematic drawing at (B) assists in identifying and in interpreting each of the required surface texture characteristics.

LAY SYMBOL (A)	PREDOMINANT TOOL OR PRODUCTION PATTERN (B)	APPLICATION OF LAY SYMBOL ON A DRAWING (C)
=	PARALLEL TOOL MARKS	√=
⊥	PERPENDICULAR TOOL MARKS	√⊥
X	ANGULAR DIRECTION TOOL MARKS	√X
M	MULTIDIRECTIONAL OR RANDOM DIRECTION TOOL MARKS	√M
C	CIRCULAR TOOL MARKS	√C
R	FINISH MARKS RADIAL TO THE CENTER OF THE SURFACE TO WHICH THE SYMBOL IS APPLIED	√R
P	PITTED SURFACE	√P

FIGURE 21-5 LAY SYMBOLS, PATTERNS, AND APPLICATIONS

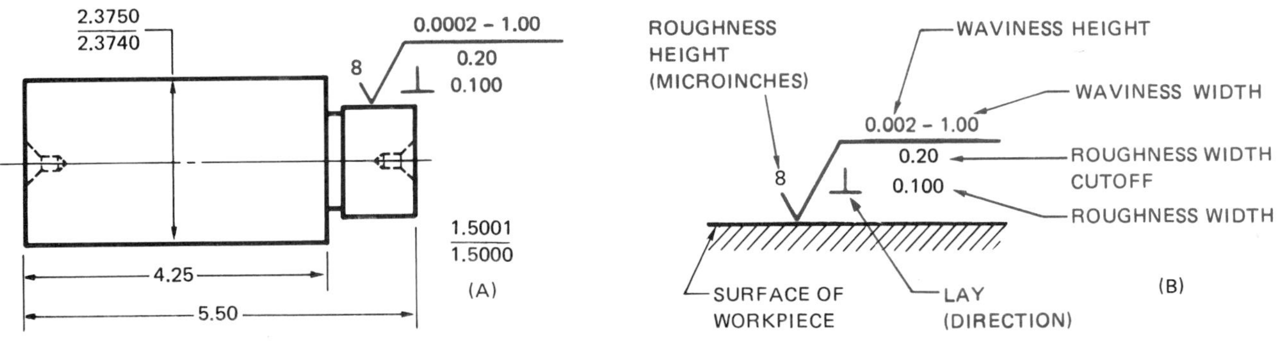

FIGURE 21-6 APPLICATION AND INTERPRETATION OF SURFACE TEXTURE SYMBOLS AND VALUES

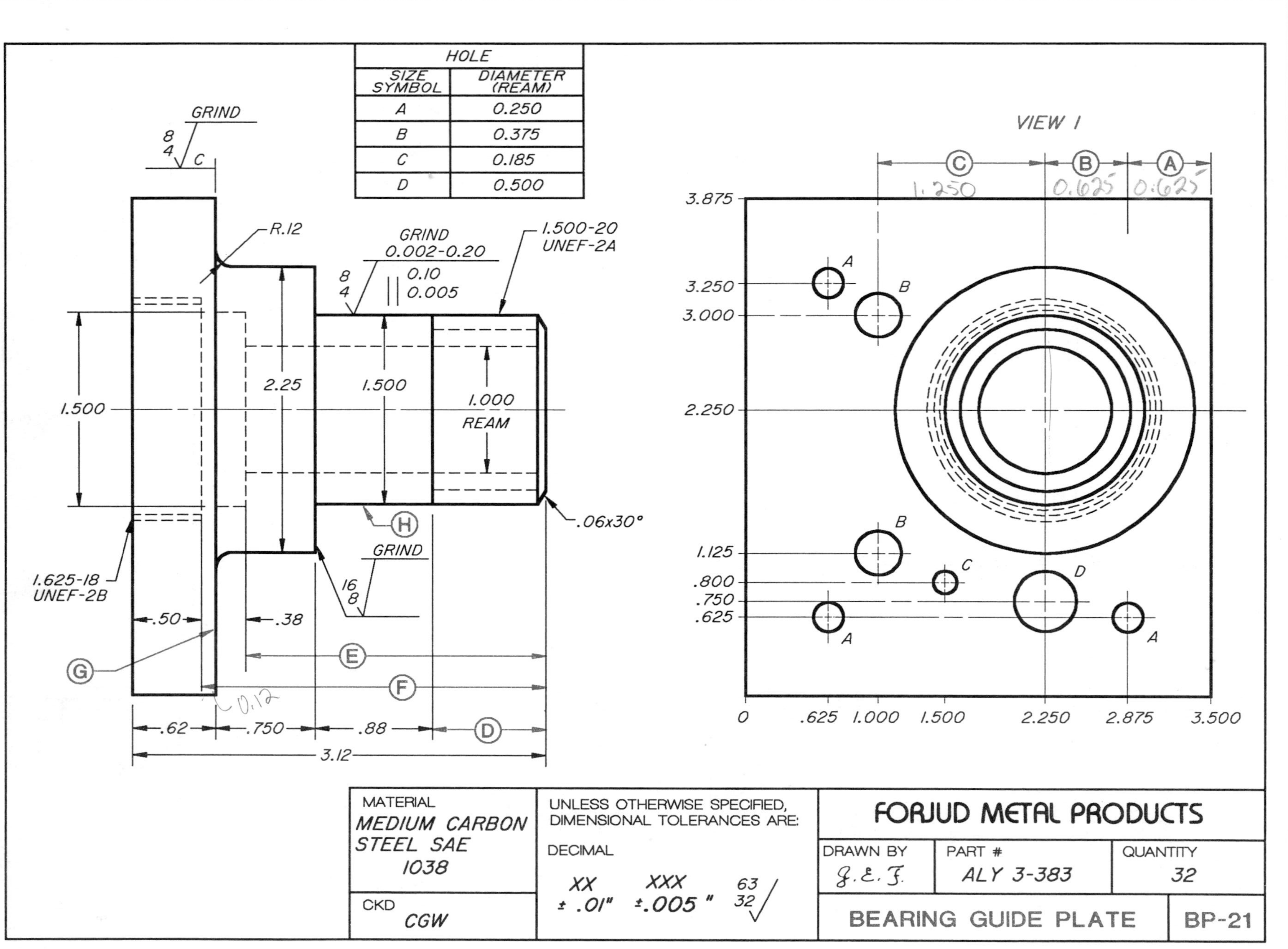

HOLE
SIZE SYMBOL
DIAMETER (REAM)
A 0.250
B 0.375
C 0.185
D 0.500
GRIND
R.12
GRIND 0.002-0.20
0.10
0.005
1.500-20 UNEF-2A
2.25
1.500
1.000 REAM
.06x30°
GRIND
1.625-18 UNEF-2B
.50
.38
.62
.750
.88
3.12
VIEW I
3.875
3.250
3.000
2.250
1.125
.800
.750
.625
0
.625 1.000 1.500 2.250 2.875 3.500
MATERIAL
MEDIUM CARBON STEEL SAE 1038
CKD CGW
UNLESS OTHERWISE SPECIFIED, DIMENSIONAL TOLERANCES ARE:
DECIMAL
XX ±.01"
XXX ±.005"
FORJUD METAL PRODUCTS
DRAWN BY
PART #
ALY 3-383
QUANTITY
32
BEARING GUIDE PLATE
BP-21

ASSIGNMENT—UNIT 21

BEARING GUIDE PLATE (BP-21)

Student's Name ______________________

1. a. Identify the material which is to be used for the Bearing Guide Plate.
   b. Give the part number and the quantity required.
2. a. Identify the kind of dimensioning which is used to locate each hole in **VIEW I**.
   b. Name the dimensioning practice used on the drawing to provide information about holes A, B, C, and D.
   c. State one advantage of using the dimensioning methods followed on the Plate drawing.
3. Give the overall width, height, and depth dimensions.
4. a. State what tolerance is allowed on fractional and decimal dimensions.
   b. Determine the maximum and minimum dimensions for the following features:
   (A) (B) (C)
5. Determine the maximum diameter to which each of the four holes in the Plate is to be reamed.
6. Compute the following nominal dimensions:
   a. (D) b. (E) c. (F)
7. Interpret the thread symbol:
   1 $\frac{1}{2}$ – 20 UNEF – 2A
8. Give the general tolerance for all machined surfaces which are not specified.
   Note. Use reference table A-4 and/or A-5 in the appendix to answer problems 9, 10, and 11. The tables relate to American Standard surface texture values in customary units and SI metric units.
9. Convert the following measurements from inch to millimeters and μin. to μm:
   a. Roughness width cutoff of 0.300″
   b. Roughness waviness height of 0.001″
   c. Roughness height of 8 μin.
10. Determine the roughness height maximum and minimum range in μin. for the following general manufacturing processes:
    a. Drilling b. Finish grinding c. Die casting
11. Give the maximum and minimum ranges in μm for the lower (finer) ranges of surface finish produced by using the following special manufacturing processes:
    a. Finish turning b. Hot rolling
12. Write the following surface texture specification as it would appear on a shop drawing:
    a. Surface to be produced by milling
    b. Maximum roughness height 63 μin.
    c. Minimum roughness height 32 μin.
    d. Roughness width 0.020″
    e. Roughness width cutoff 0.62″
    f. Waviness height 0.002″
    g. Waviness width 1.25″
    h. Lay Multidirectional
13. Provide the following information for surface (G):
    a. Machining process
    b. Maximum roughness height
    c. Lay pattern
14. Interpret the surface texture symbol for feature (H)

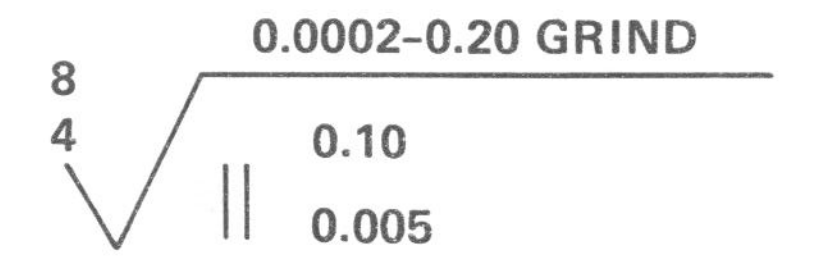

1. a. ______________________
   b. Part # __________ Quantity __________
2. a. ______________________
   b. ______________________
   c. ______________________
3. Width ______ Height ______ Depth ______
4. a. Fractional ________ Decimal ________
   b. Maximum Minimum
   (A) = ________ ________
   (B) = ________ ________
   (C) = ________ ________
5. Hole Max. Dimension Hole Max. Dimension
   A = __________ C = __________
   B = __________ D = __________
6. (D) = ______ (E) = ______ (F) = ______

7. 1$\frac{1}{2}$ ______________________
   –20 ______________________
   UNEF ______________________
   –2A ______________________
8. ______________________
9. SI Metric Equivalent
   a. 0.300″ = ______________
   b. 0.001″ = ______________
   c. 8 μin. = ______________
10. a. __________ c. __________
    b. __________
11. Lower (finer) Range
    Maximum μm Minimum μm
    a. __________ __________
    b. __________ __________

12.

13. a. ______________________
    b. ______________________
    c. ______________________
14. a. Manufacturing Process __________
    b. Min/Max Roughness Height Values = __________
    c. Waviness Height = __________
    d. Waviness Width = __________
    e. Roughness Width Cutoff = __________
    f. Lay __________
    g. Roughness Width = __________

UNIT 22
Dimensioning and Representation of Nonthreaded Fasteners

Keys, pins, rivets, and washers are four common groups of nonthreaded fasteners. In addition, since energy is produced by exerting a push, pull, or twisting force, springs are included as a fifth group. This unit covers the representation of each group of fasteners and general dimensioning practices. Dimensions appear on drawings either in customary inch or SI metric units of measurement. Both systems are represented when dual dimensioning is used on certain drawings.

KEYS, KEYSEATS, AND KEYWAYS

A *key* is a fastener which is used between two parts to prevent them from turning and for one part to drive the other. A *keyseat* is a groove machined into a shaft to "seat" the key. A *keyway* is a groove cut into the hub area of a mating part to receive that portion of the key which extends beyond the shaft. Five common types of keys are illustrated in figure 22-1.

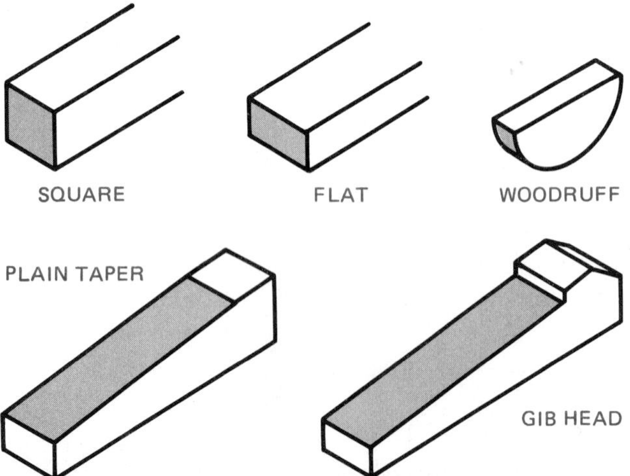

FIGURE 22-1 FIVE COMMON TYPES OF KEYS

Methods of representing square, flat, and Woodruff keys, keyseats, and keyways appear in figure 22-2. Limit dimensions are applied to ensure interchangeability of parts, proper fit, and to meet mass production requirements. A note is used for jobbing shop purposes to specify the width and then the depth of the keyseat and the keyway. The length as well as the key size are dimensioned directly on the drawing. The Woodruff key (figure 22-2B) is semicircular in shape. This key fits into a corresponding semicircular groove in a shaft.

A notation like **#806 AMERICAN STANDARD WOODRUFF KEY** generally appears on drawings. The number indicates the cutter size. The last two digits give the diameter of the round groove and cutter in eighths of an inch. The **−06** means 6/8, or 0.750 diameter. The one or two digits preceding the last two indicate the cutter and groove width in thirty-seconds of an inch (**−8−−**= 8/32, or 0.250″ width). The depth of groove which is indicated on a print permits the parts to be assembled according to the required class of fit.

MACHINE PINS

Machine pins are semipermanent fasteners for parts which are to be assembled and disassembled. Five widely used types of machine pins are illustrated in figure 22-3. Detail working drawings are usually not included for machine pins. Instead, each pin is identified by a detail number. A description of the pin, outside diameter, length,

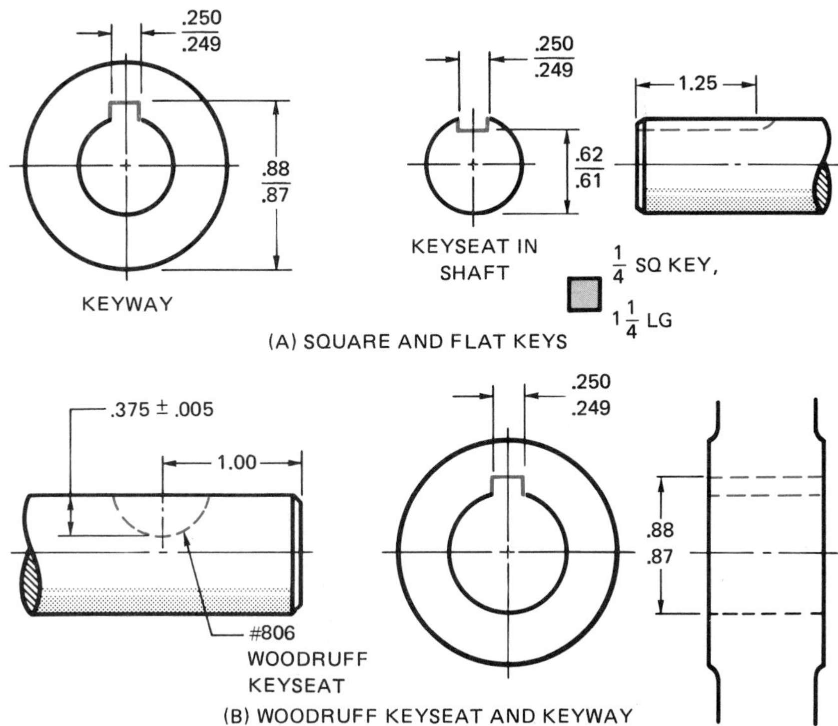

FIGURE 22-2 REPRESENTATION AND DIMENSIONING OF KEYS, KEYSEATS, AND KEYWAYS

and manufacturer's catalog specifications are contained in the parts list on a drawing.

Dowel pins (figure 22-3A) are used primarily to secure parts from moving in an assembly and/or to precisely align them. Dowel pin diameters, lengths, form, hardness, and quality of ground diameter are standardized. Applications of dowel pins for alignment purposes include punches, dies, jigs, fixtures, and other machine components. Holes are generally lapped for precision fit and alignment.

Commercial straight pins serve similar functions to dowel pins. However, the ends of straight pins (figure 22-3B) are usually flat. Diametral sizes and lengths are standard.

Taper pins are designed with standard tapers (figure 22-3C) and pin sizes. A taper pin is driven into parts which are fitted together with tapered holes until the pin is seated. Figure 22-4 shows a general application of a taper pin. The *short-angle taper* aids in aligning the holes and holding the parts together securely. The diameter at the large end is the basic dimension of the taper pin.

Taper pin dimensions and other information as found on drawings include the drill diameter and a note like:

DRILL AND REAM AT ASSEMBLY FOR A #2 TAPER PIN.

Specifications for a required taper pin detail are given in the parts list.

Clevis pins are straight pins with a head on one end and a hole drilled through the body at the other end (figure 22-3D). A cotter pin is inserted through the hole to prevent the clevis pin from working out of position and to hold the parts between the head and the cotter pin in place. Clevis pins provide a movable joint where parts may be readily removed or adjusted.

Cotter pins (figure 22-3E) are designed for general locking applications where precision and strength equal to a solid pin are not required. Cotter pins fit through clearance holes in parts which are held together. For thread applications, the cotter pin also fits through slots in a nut, preventing it from turning or working loose. The cotter pin is held in place by spreading the legs. A typi-

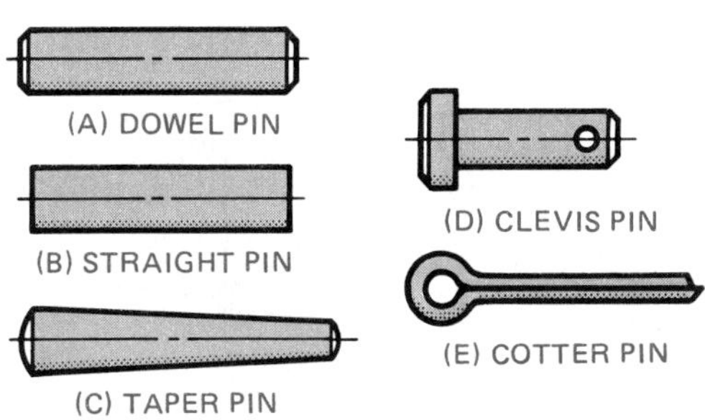

FIGURE 22-3 COMMON TYPES OF MACHINE PINS

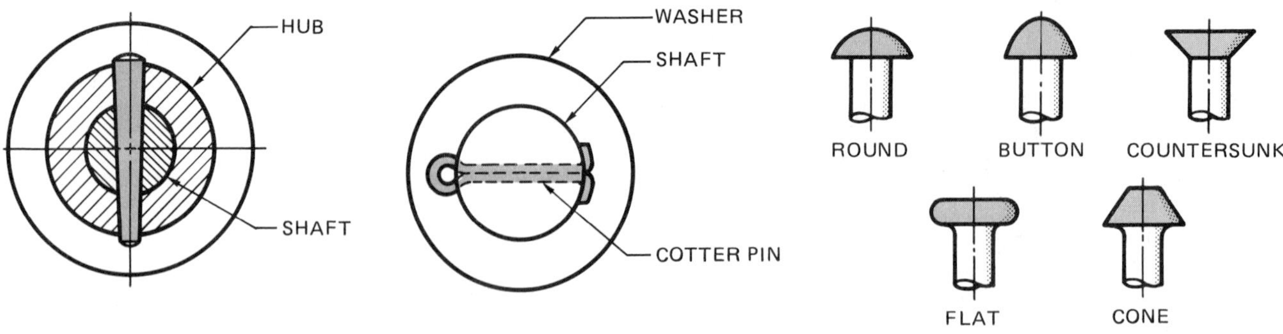

FIGURE 22-4 APPLICATION OF TAPER PIN

FIGURE 22-5 REPRESENTATION OF A WASHER AND COTTER PIN APPLICATION

FIGURE 22-6 SHAPES OF FIVE COMMON RIVET HEADS

cal representation of a cotter pin in an assembly drawing is shown in figure 22-5.

RIVET FASTENERS

Rivets are fasteners whose principal use is to hold sheet metal, plates, and other comparatively thin parts securely together. Common rivet heads include *round*, *button*, *countersunk*, *cone*, and *flat*. Design features of each type appear in figure 22-6. Information is provided on a drawing by a note which gives the body diameter, head type, and length. A typical note reads **ø 8 × 16 ROUND HEAD**. In addition, the center-to-center (pitch) distance is dimensioned directly on the drawing. The pitch is sometimes included with the note. For example:

ø 8 × 16 ROUND HEAD,
25 PITCH.

This indicates the rivet holes are spaced 25 millimeters apart.

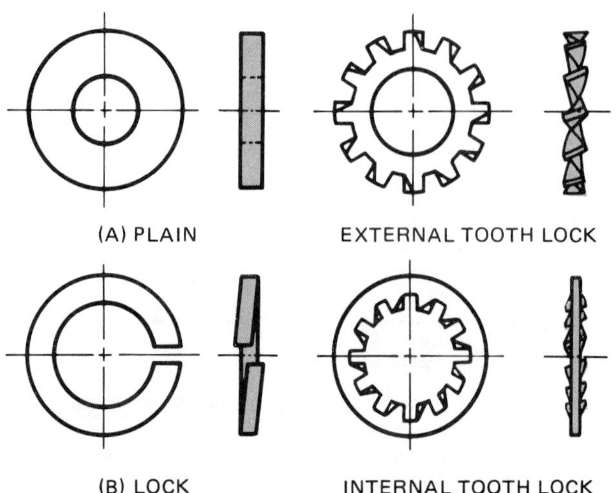

FIGURE 22-7 PLAIN, LOCK, AND TOOTH LOCK WASHER

WASHERS

Washers serve primarily to distribute the load (force) from tightening a bolt, machine screw, or nut over a larger surface area than the threaded fastener. The three basic washer designs include the *plain flat washer*, *lock washer*, and *internal* and *external tooth lock washer* (figure 22-7).

Plain flat washers are identified by a detail number and a simple note reading:

ø $\frac{1}{2}$ × ø 1 × $\frac{1}{16}$ PLAIN WASHER.

In this particular example, the outside diameter is 1″; inside (hole) diameter, 1/2″; thickness, 1/16″.

Lock washers are heat treated, split, and formed as a spiral. The spring tension of the spiral helps to hold the assembled parts securely. The two most widely used designs are for regular and heavy duty. Lock washers are specified on a drawing by a note or a specification in the parts list. The following is a drawing note.

ø $\frac{1}{2}$ REG HELICAL LOCK WASHER

The note gives the nominal hole diameter of 1/2″, series (regular), and type (helical lock washer).

Internal and *external tooth lock washers* are corrugated from near the center to the outer diameter for external washers and just the opposite for internal washers. The washers are hardened. As a bolt head, nut, or other surface is tightened, the teeth-like corrugations of the washer keep the head or screw from loosening. Tooth lock washers are designated on a drawing in the same manner as a regular lock washer except for information about type.

ø $\frac{1}{2}$ INTERNAL TOOTH LOCK WASHER

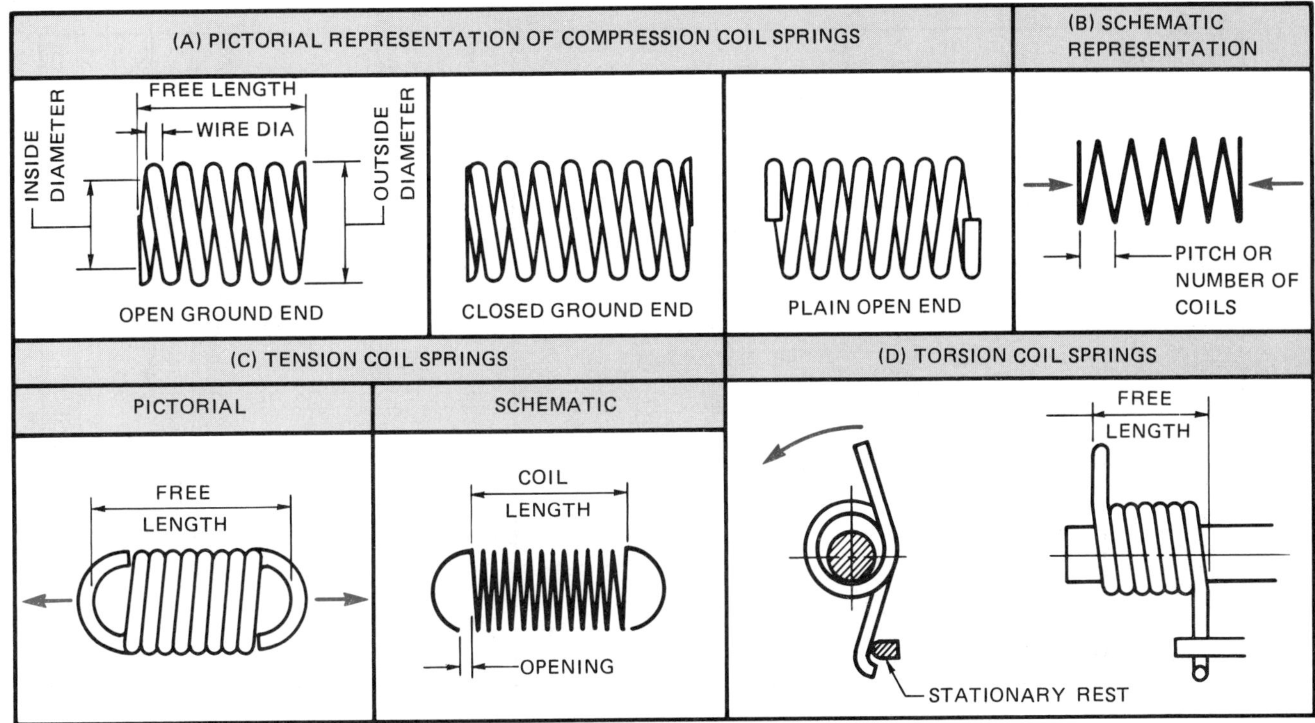

FIGURE 22-8 COMPRESSION, TENSION, AND TORSION COIL SPRINGS

SPRINGS

Coil springs and flat springs are the two basic types. Each is designed to apply energy (force) by a pushing, pulling, or twisting action.

Coil Springs

Coil springs are formed by winding spring steel (or other spring metal) wire into a continuous spiral (figure 22-8A). Coil springs are identified by function as: *compression*, *tension*, or *torsion*. *Compression springs* are designed so that the coils are open. Energy is applied as the coils are compressed. Compression springs are made with *plain open ends*, *plain ground ends*, or *open ground ends*.

Tension coil springs (figure 22-8C) exert a pulling force when the coils are stretched. The coils are wound to touch when in a normal position. The ends are formed to hook or loop around a pin or shaft or to be attached directly to parts which are to be controlled by the spring.

Torsion coil springs are wound in a right- or left-hand direction. Force is applied by the twisting or circular motion produced on the spring (figure 22-8D).

Flat Springs

Flat springs are simply flat pieces of spring steel designed so that one end is secured. Flat springs serve the same functions as coil springs in that they may exert a pushing, pulling, or twisting force.

TERMS APPLIED TO SPRINGS

The information needed by the skilled craft worker or technician relates to wire size (diameter), spring material, inside and outside diameters, pitch, and free, loaded, or solid lengths. Often, the number of coils in a given length is given instead of pitch. Except for the terms relating to length, the other terms are descriptive. *Free length* indicates the length of a spring in its normal, unloaded condition. *Loaded length* represents the overall spring length measured under the actual required load. *Solid length* means the length of a spring when all coils are touching.

DIMENSIONING SPRINGS

Flat springs are not standardized. Flat springs are represented and dimensioned in the same manner as other flat parts. Coil springs are generally represented by schematic drawings as shown in figure 22-8B. The illustrations indicate how each drawing is dimensioned. In cases where dimensioning is omitted, a leader and a note provide the same information.

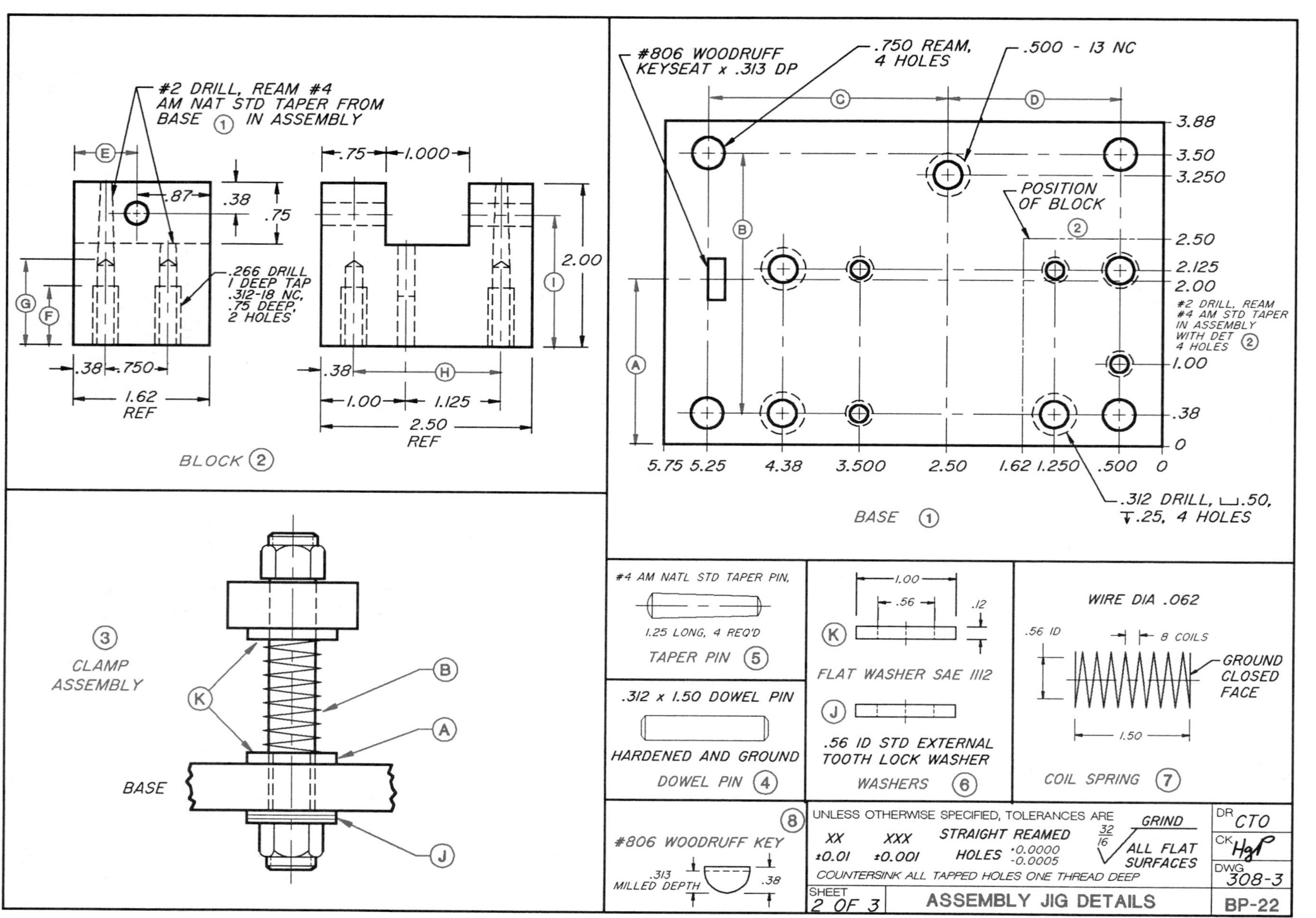

#2 DRILL, REAM #4 AM NAT STD TAPER FROM BASE ① IN ASSEMBLY
.266 DRILL 1 DEEP TAP .312-18 NC, .75 DEEP, 2 HOLES
BLOCK ②
#806 WOODRUFF KEYSEAT x .313 DP
.750 REAM, 4 HOLES
.500 - 13 NC
POSITION OF BLOCK ②
#2 DRILL, REAM #4 AM STD TAPER IN ASSEMBLY WITH DET ② 4 HOLES
.312 DRILL, ⌴.50, ↧.25, 4 HOLES
BASE ①
③ CLAMP ASSEMBLY
BASE
#4 AM NATL STD TAPER PIN.
1.25 LONG, 4 REQ'D
TAPER PIN ⑤
.312 x 1.50 DOWEL PIN
HARDENED AND GROUND
DOWEL PIN ④
FLAT WASHER SAE 1112
.56 ID STD EXTERNAL TOOTH LOCK WASHER
WASHERS ⑥
WIRE DIA .062
8 COILS
GROUND CLOSED FACE
COIL SPRING ⑦
#806 WOODRUFF KEY ⑧
.313 MILLED DEPTH
UNLESS OTHERWISE SPECIFIED, TOLERANCES ARE
XX ±0.01
XXX ±0.001
STRAIGHT REAMED HOLES +0.0000 -0.0005
GRIND ALL FLAT SURFACES
COUNTERSINK ALL TAPPED HOLES ONE THREAD DEEP
DR CTO
CK
DWG 308-3
SHEET 2 OF 3
ASSEMBLY JIG DETAILS
BP-22

ASSIGNMENT—UNIT 22

ASSEMBLY JIG DETAILS (BP-22)

Student's Name ______________________

1. Name each machined detail and nonthreaded part represented on the Jig Detail drawing.
2. Provide the following information:
   a. Kind of dimensioning used for the Base.
   b. Tolerances on two-place decimal dimensions.
   c. Machine process for all flat surfaces.
   d. Countersinking requirement for all tapped holes.
   e. Tolerances on the straight reamed holes.
3. Calculate the following nominal dimensions for the Base:
   a. (A) b. (B) c. (C) d. (D)
4. Determine dimensions (E) through (I) for the Block.
5. Indicate the number of the following kinds of holes in the Base:
   a. Counterbored
   b. Taper reamed
   c. 3/8″ reamed
   d. 1/2″ tapped holes
6. Give the specifications covering all of the holes in the Base.
7. Name the five different nonthreaded fasteners which are detailed on sheet 2 as parts of the Jig.
8. a. Give the specifications for Detail (7).
   b. Explain the meaning of "free length."
   c. Identify the type according to function on the Clamp detail.
9. a. Describe the functions served by parts (J) and (K).
   b. Give the specifications for these two parts.
10. a. State the functions served by the taper pins.
    b. Give the drill size for the taper pin holes.
    c. Interpret the meaning of the note:

    #4 TAPER PIN, 1 $\frac{1}{4}$ LG.

    d. Tell under what conditions the tapered holes in the Block and Base are reamed.
11. a. Name part (4).
    b. Indicate the heat treatment and surface finish process for part (4).
    c. Give the nominal size and length.
12. a. Interpret the dimensioning note for the Woodruff key:

    #806 WOODRUFF KEY.

    b. Give the nominal depth to which the keyseat is milled in the Base.

1. ______________________
____

2. a. __________ d. __________
b. __________ __________
c. __________ e. __________

3. a. (A) = ______ c. (C) = ______
b. (B) = ______ d. (D) = ______

4. a. (E) = ______ d. (H) = ______
b. (F) = ______ e. (I) = ______
c. (G) = ______

5. a. Counterbored Holes ______
b. Taper Reamed Holes ______
c. 3/8″ Reamed Holes ______
d. 1/2″ Tapped Holes ______

6. a. ______________________
b. ______________________
c. ______________________
d. ______________________

7. a. __________ d. __________
b. __________ e. __________
c. __________

8. a. ______________________
____
____
b. ______________________
____
c. ______________________

9. a. (J) ______________________
____
(K) ______________________
____
b. (J) ______________________
(K) ______________________

10. a. ______________________
____
b. Drill Size ______________
c. ______________________
____
d. ______________________
____

11. a. ______________________
____
b. ______________________
____
c. ______________________
____
____

12. a. ______________________
____
b. Nominal Milled Depth ______

UNIT 23
Tapers and Knurls: Design Features and Processes

Standardized procedures are followed for representing and dimensioning tapers and knurled surfaces. Descriptions of each process and information generally contained on shop drawings are covered in this unit.

TAPERS

A part is said to be *tapered* when it uniformly changes in size along its length. The taper may be along a cylindrical part as in the case of a taper shank drill or a taper reamer. The taper may also be flat like a wedge-shaped key or an angle measurement gage.

A taper may be designed with a *steep-angle taper* which permits easy disengagement (self-releasing) as with a milling machine arbor. On the other hand, a *short-angle taper* is particularly suited for applications which require an accurate alignment and a strong frictional (self-holding) force between two tapered parts in order to drive a tool or other mating member. Figure 23-1 provides examples of short- and steep-angle tapers. Tapers may be external or internal.

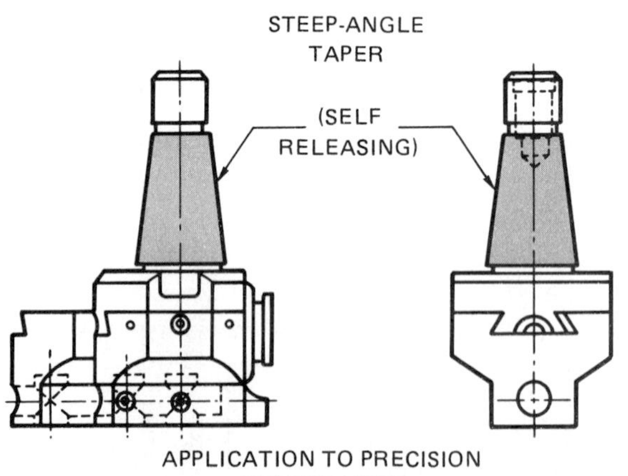

FIGURE 23-1 EXAMPLES OF SHORT-ANGLE AND STEEP-ANGLE TAPERS

STANDARDS FOR TAPERS

Tapered parts may be dimensioned in customary or SI metric units. The most widely used self-holding tapers conform to Brown and Sharpe or Morse standards. Additional taper systems include Jarno, American Standard Machine, British Standard, and ANSI. ANSI standards have been adopted for steep-angle machine tapers.

Standard tapers are often identified on drawings by a taper number. A #6 B & S taper shank or a size 40 American National Standard spindle nose for a milling machine are two examples. With such notations, reference is made to the manufacturer's technical data sheets or trade handbooks for full dimensional and geometric tolerance requirements.

DIMENSIONING AND MEASURING TAPERS

There are three controlling dimensions for tapers (figure 23-2):

- The diameter at the large end
- The diameter at the small end
- The length of the taper, the rate of taper, or the taper angle

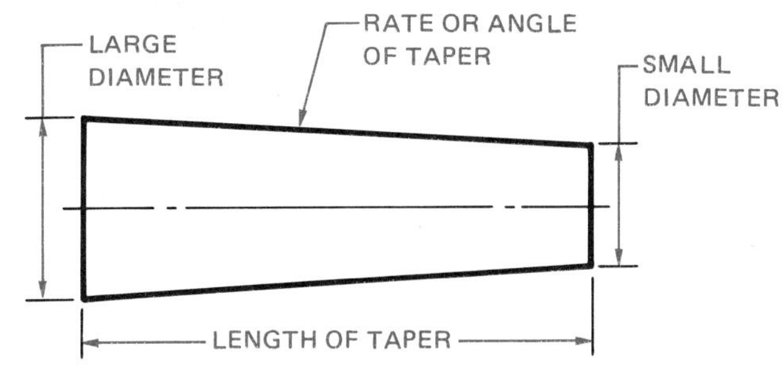

FIGURE 23-2 CONTROLLING DIMENSIONS FOR TAPERS

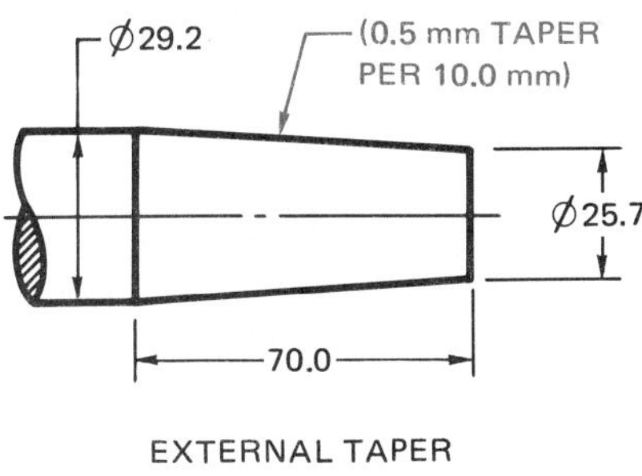

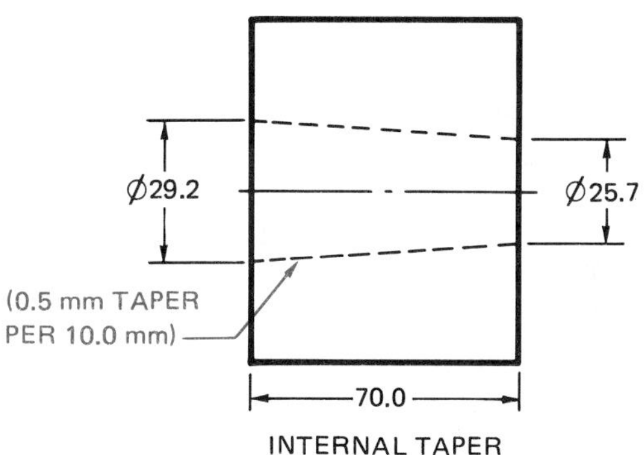

FIGURE 23-3 TECHNIQUE OF DIMENSIONING EXTERNAL AND INTERNAL TAPERS (SI METRIC MM)

Three of the dimensions with tolerances are usually specified on a drawing. The fourth dimension may be given as a *reference* using the letters REF, or as a note contained within parentheses as shown in figure 23-3, or by bracketing the dimension.

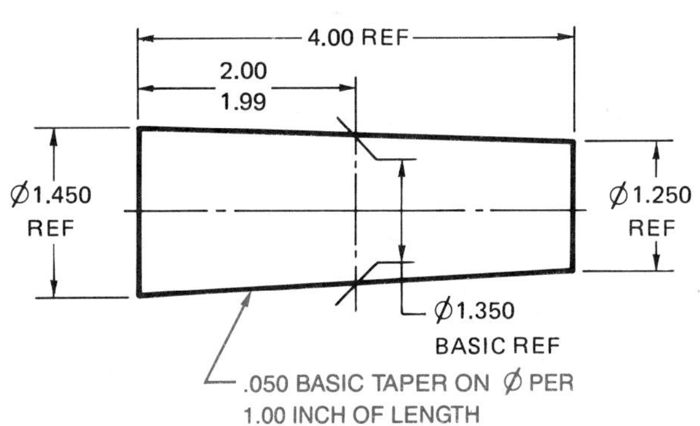

FIGURE 23-4 DRAWING DIMENSIONED TO CONTROL DIAMETER AND POSITION OF TAPER

BASIC DIAMETER METHOD OF DIMENSIONING A TAPER

A dimension is provided on a drawing of a precise taper to indicate the exact size and location of a basic diameter (figure 23-4). By controlling the size and location, the taper is positioned in relation to the taper on the mating part. The basic diameter of 1.350″ controls the position of the taper within the upper limit (2.00″) and the lower limit (1.99″) dimension from the left face of the workpiece.

DIMENSIONING A FLAT TAPER

A flat taper may be dimensioned in the same manner as a cone-shaped taper. Figure 23-5 illustrates this practice.

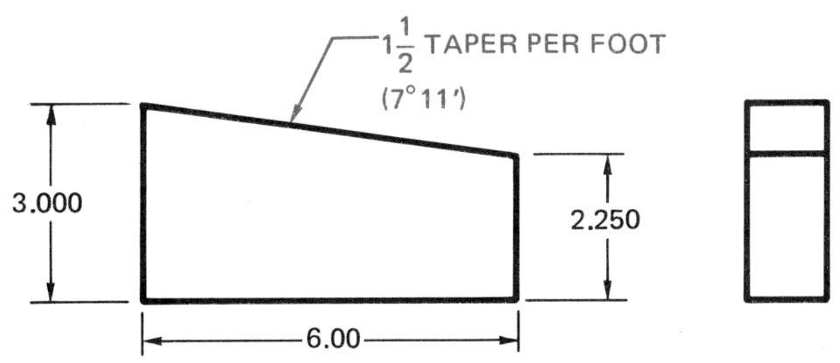

FIGURE 23-5 WORKING DRAWING OF A PART WITH A FLAT TAPER

KNURLS AND KNURLED SURFACES

The term *knurled surface* refers to diamond, straight line, or diagonal pattern raised areas that are impressed upon the surface of a cylinder (figure 23-6). *Knurling* is a material *displacement process* in which a knurled surface is produced by forcing a pair of hardened knurls into a revolving workpiece to impress a pattern.

Purposes for Knurling

Work surfaces are knurled for four general purposes:

- To provide a good gripping surface on tools and parts where force is to be applied.
- To permit ease in handling and fine adjustments of instruments and gages.
- To add to the appearance of a part.
- To increase the diameter of a workpiece in order to press fit two parts or to prevent one part from turning in a mating part.

Basic Patterns of Knurls

There are three basic knurl patterns: *straight-line*, *diamond*, and *diagonal*.

- *Straight-line pattern* (also known as *parallel-ridge*). This pattern is formed by impressing two hardened knurl rolls with parallel cut teeth into a turning workpiece. Grooves are cut parallel to the work axis (figure 23-6A).
- *Right-hand* and *left-hand diagonal patterns*. These two patterns are shown in figure 23-6 at (B). The *right-hand diagonal* (helix) *pattern* is formed by impressing a left-hand diagonal knurl into a workpiece. A *left-hand diagonal knurl pattern* is produced by using a right-hand knurl. The diagonal ridges that are produced are at a 30° angle to the workpiece (figure 23-6B).
- *Diamond pattern*. This pattern is formed by impressing a workpiece with a right-hand and a left-hand diagonal pattern knurl. As the two knurl patterns cross each other, the 30° ridges/grooves form a diamond pattern (figure 23-6C).

  A second method is to use a standard *diamond pattern knurl roll*. A *depressed* diamond pattern is formed in the work surface by using a male diamond point knurl.

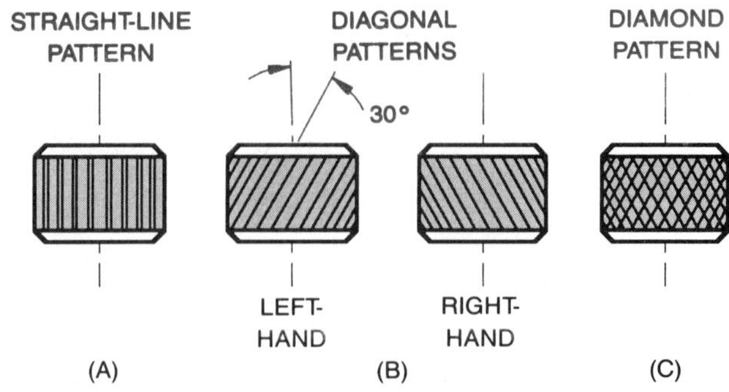

FIGURE 23-6 KNURL PATTERNS

Sizes of Knurl Rolls

The term *pitch* is used to identify the size of a knurl roll. Pitch refers to the number of ridges (teeth) per linear inch. The three most commonly used pitch sizes are *coarse* (14 pitch), *medium* (21 pitch), and *fine* (33 pitch). Knurl rolls are made of high-quality steel, heat treated, and are mounted on hardened pins in knurl holders.

Other knurl sizes range from 12 to 80 *teeth per inch* (TPI) and are available for diamond, straight-line, and diagonal knurling. When the term *normal teeth per inch* is used, it relates to the measurement of the knurl crests along a line of measurement that is *perpendicular to the helix angle of the teeth*.

INCREASE OF WORK DIAMETER PRODUCED BY KNURLING

There are general information tables and specifications established by American Standard (ANSI/ASME) B 94.6 that relate to work blank diameters and the increase in workpiece sizes produced by knurling. Table 23-1 provides an

TABLE 23-1 APPROXIMATE INCREASE OF WORKPIECE DIAMETER PRODUCED BY KNURLING (CIRCULAR PITCH/THREADS PER INCH SYSTEM)
Courtesy of REED ROLLED THREAD DIE COMPANY

NORMAL TEETH PER INCH (TPI) (CIRCULAR PITCH KNURLS)	STRAIGHT AND DIAGONAL KNURLING*	DIAMOND KNURLING* RAISED POINTS (MALE)
12	.034	.038
16	.025	.029
20	.020	.023
25	.016	.018
30	.013	.015
35	.011	.013
40	.009	.010
50	.009	.010
80	.005	.006

*30° Helix Angle

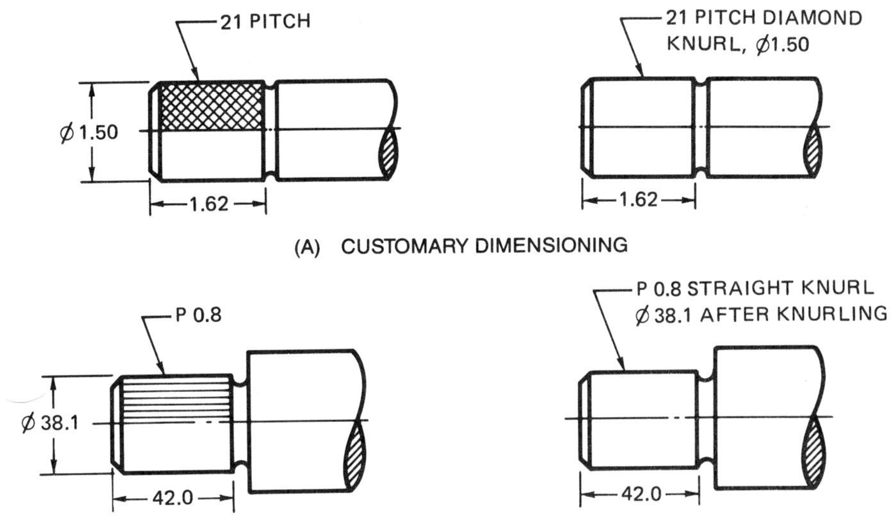

FIGURE 23-7 DIMENSIONS AND NOTES FOR KNURLED SURFACES

example of one manufacturer's data on the effects of straight-line and diagonal knurling and diamond knurling at a 30° helix angle.

The increase in work diameter ranges from .005″ to .034″ (.12 to .86 mm) for straight-line and diagonal knurling and from .006″ to .038″ (.15 to .97 mm) for diamond knurling. These are approximate dimensions and require measuring the first produced part to establish the correct work blank diameter.

DIMENSIONS AND NOTES FOR KNURLED SURFACES

A knurled area is dimensioned for nominal diameter and length. For press fits, the minimum acceptable diameter is given. The size and pattern of knurls appear as a note. The standard pitches of knurls appear as (**14P**), (**21P**), and (**33P**). Comparable metric sizes are available.

In summary, figure 23-7 illustrates the dimensioning and notes for two knurl patterns. *Customary dimensioning* is shown at (A); *ANSI metric dimensioning*, at (B).

USE OF NOTES FOR OPERATIONS AND LIMITS

In preceding units, standard practices were followed to provide information through the use of notes for such processes as drilling, boring, threading, etc. Another practice which is illustrated in figure 23-8 is to include upper and lower limit dimensions in sequence for particular operations.

While abbreviations have not been used, attention is directed to the fact that a whole section of standard *Abbreviations and Symbols for Use on Drawings* has been prepared by ANSI to cover major products and manufacturing processes. Samples are provided in the Appendix.

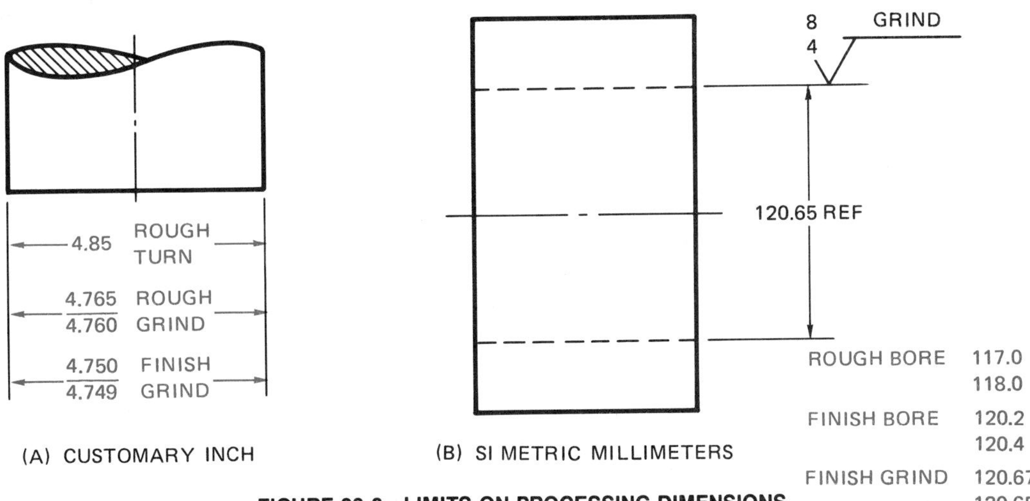

FIGURE 23-8 LIMITS ON PROCESSING DIMENSIONS

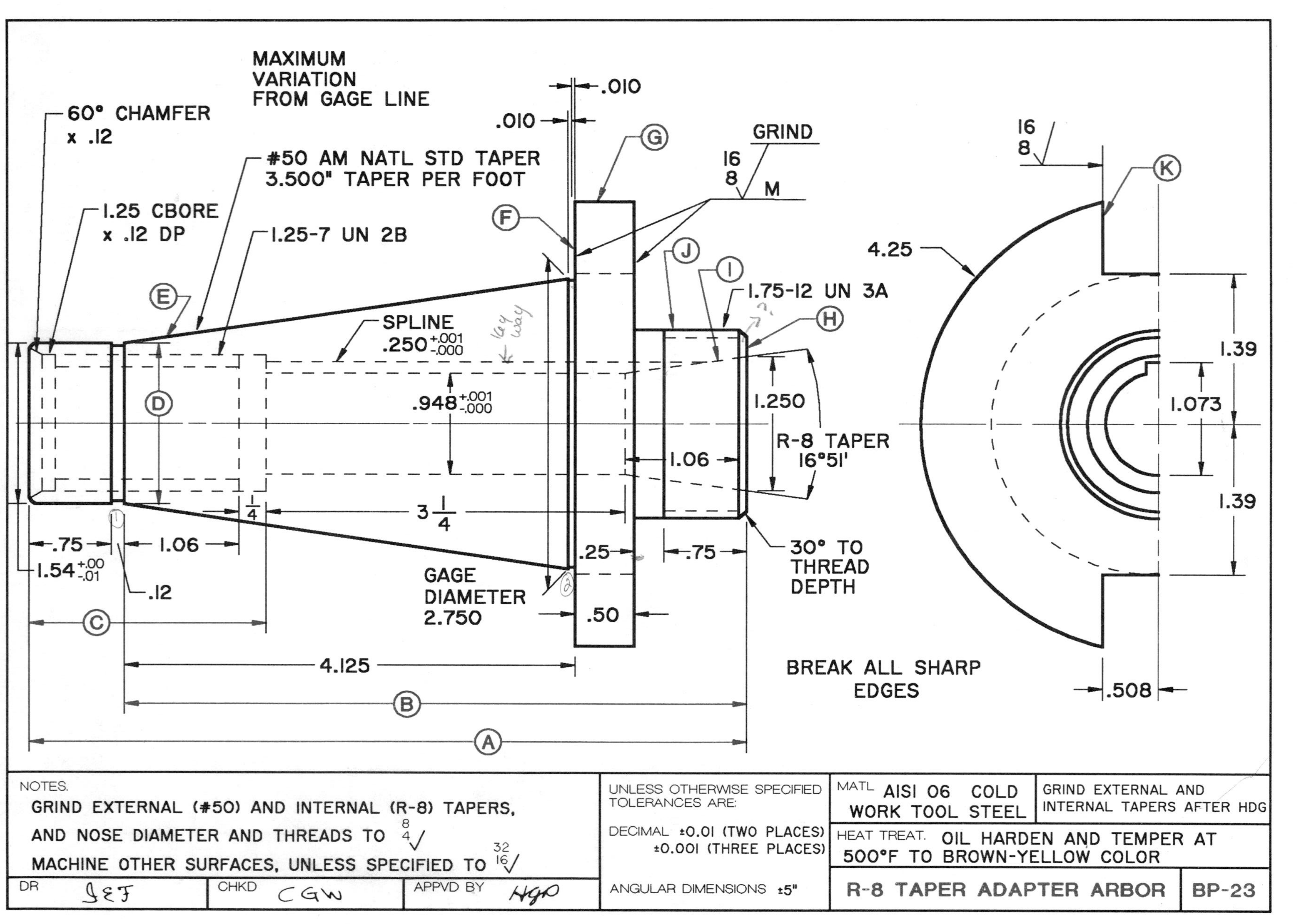
MAXIMUM VARIATION FROM GAGE LINE
.010
.010
60° CHAMFER x .12
#50 AM NATL STD TAPER 3.500" TAPER PER FOOT
1.25 CBORE x .12 DP
1.25-7 UN 2B
GRIND
16 8
M
1.75-12 UN 3A
SPLINE .250 +.001 -.000
.948 +.001 -.000
1.250
R-8 TAPER 16°51'
1.06
3 1/4
1/4
.75
1.06
1.54 +.00 -.01
.12
.25
.75
30° TO THREAD DEPTH
GAGE DIAMETER 2.750
.50
4.125
4.25
1.39
1.073
1.39
.508
BREAK ALL SHARP EDGES
NOTES.
GRIND EXTERNAL (#50) AND INTERNAL (R-8) TAPERS, AND NOSE DIAMETER AND THREADS TO 8/4, MACHINE OTHER SURFACES, UNLESS SPECIFIED TO 32/16
UNLESS OTHERWISE SPECIFIED TOLERANCES ARE:
DECIMAL ±0.01 (TWO PLACES) ±0.001 (THREE PLACES)
ANGULAR DIMENSIONS ±5"
MATL AISI 06 COLD WORK TOOL STEEL
GRIND EXTERNAL AND INTERNAL TAPERS AFTER HDG
HEAT TREAT. OIL HARDEN AND TEMPER AT 500°F TO BROWN-YELLOW COLOR
R-8 TAPER ADAPTER ARBOR
BP-23
DR
CHKD
APPVD BY

ASSIGNMENT — UNIT 23

R-8 TAPER ADAPTER ARBOR (BP-23)

Student's Name ______________________

1. Indicate the tolerance which is allowed on the following dimensions:
   a. Two- and three-place decimals
   b. Fractional
   c. Angular (taper surfaces)
2. Identify the material of which the Adapter is made.
3. State the heat treating processes.
4. a. Give the specifications for (1) the external and (2) the internal threads.
   b. Identify the process for finish machining the nose threads.
   c. Indicate (1) the required quality of surface finish for the tapers and (2) the method of producing the finish.
5. Give the dimensions of:
   a. The counterbored hole
   b. The chamfer angle and length
6. State the shop requirement for all sharp edges.
7. Determine the maximum and the minimum dimensions for machining the two slots.
8. Give the nominal dimensions for cutting (broaching) the keyway (spline).
9. Compute nominal dimensions Ⓐ, Ⓑ, Ⓒ, and Ⓓ.
10. Indicate the required ground finish for surfaces Ⓔ through Ⓚ.
11. State two functions of the Adapter Arbor.
12. Identify the functions of the following features:
    a. Nose threads
    b. Slotted plate
    c. Shank threads
13. a. State two functions of the #50 American National Standard taper body.
    b. Give the effective length of the taper.
    c. Determine the taper per foot and the taper per inch.
    d. Indicate (1) the finish machining process and (2) the required surface texture for the body.
14. a. State the included angle and length of the R-8 taper.
    b. Give the maximum and minimum acceptable finish ground angle.
15. Determine the following maximum and minimum diameters:
    a. Bored hole
    b. Recess (undercut) for the internal thread (as a decimal value)

1. a. XX ________ XXX ________
   b. Fractional ________
   c. Angular ________
2. ________
3. ________
4. a. (1) External (Nose) Thread: ________
   (2) Internal (Shank) Thread: ________
   b. ________
   c. (1) ________ (2) ________
5. a. Counterbore ________
   b. Chamfer ________
6. ________
7.

	Maximum	Minimum
Width	________	________
Distance from Center Line	________	________

8. ________
9. Ⓐ = ________ Ⓒ = ________
   Ⓑ = ________ Ⓓ = ________
10. Ⓔ = ________ Ⓘ = ________
    Ⓕ = ________ Ⓙ = ________
    Ⓖ = ________ Ⓚ = ________
    Ⓗ = ________
11. a. ________
    b. ________
12. a. Nose Threads to thread the inside to attach another tool
    b. Slotted Plates ________
    c. Shank Threads ________
13. a. (1) ________
    (2) ________
    b. ________
    c. Taper Per Foot ________
    Taper Per Inch ________
    d. (1) ________
    (2) ________
14. a. Included Angle ________
    Length ________
    b. Maximum ________
    Minimum ________
15.

	Maximum	Minimum
a.	________	________
b.	________	________

SECTION 7
Sectional Views

UNIT 24
Full Sections: Section Lining

Interior details of simple parts are usually represented on a technical drawing by the use of hidden lines. Other parts often contain a great number of details and/or complicated internal features. The use of hidden lines in such cases results in complicated drawings which are difficult and impractical to interpret.

SECTIONS AND SECTIONAL VIEWS

The shape or inside hidden features of an object, which may not be accurately or clearly represented by regular views, may be drawn in a *sectional view.*

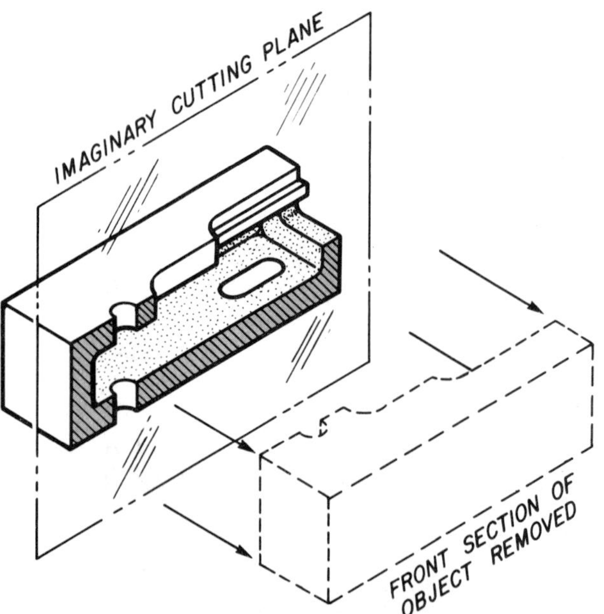

FIGURE 24-1 PASSING AN IMAGINARY CUTTING PLANE THROUGH AN OBJECT AND REMOVING A PORTION (SECTION)

A sectional view is produced by passing an imaginary cutting plane through an object at a selected location. The section which is in front of the cutting plane is theoretically removed. What is then viewed is an exposed section. This section shows the essential details of the shape and sizes of the desired internal features, as illustrated in figure 24-1.

In this unit, a single, flat cutting plane is considered as cutting completely through the object. The sectional view thus produced is called a *full section* (or *full sectional view*). Later units provide additional applications using multi-faced regular and offset cutting planes. *Half section*, *removed section*, *phantom section*, and other types of sections are treated at that time.

LOCATION OF CUTTING PLANES

The location of one or more cutting planes is represented on a drawing by a cutting plane line. To review, a *cutting plane line* is a heavy, broken line. Arrows at the beginning and end of a cutting plane line show the direction from which the object is viewed. Letters are sometimes placed at the corners to identify the section or for reference.

FULL SECTIONAL VIEW

The principles involved in producing a full sectional view are illustrated in figure 24-2. The Link Arm part is shown pictorially at (A). The two views at (B) indicate how the part may be represented by a conventional orthographic projection drawing.

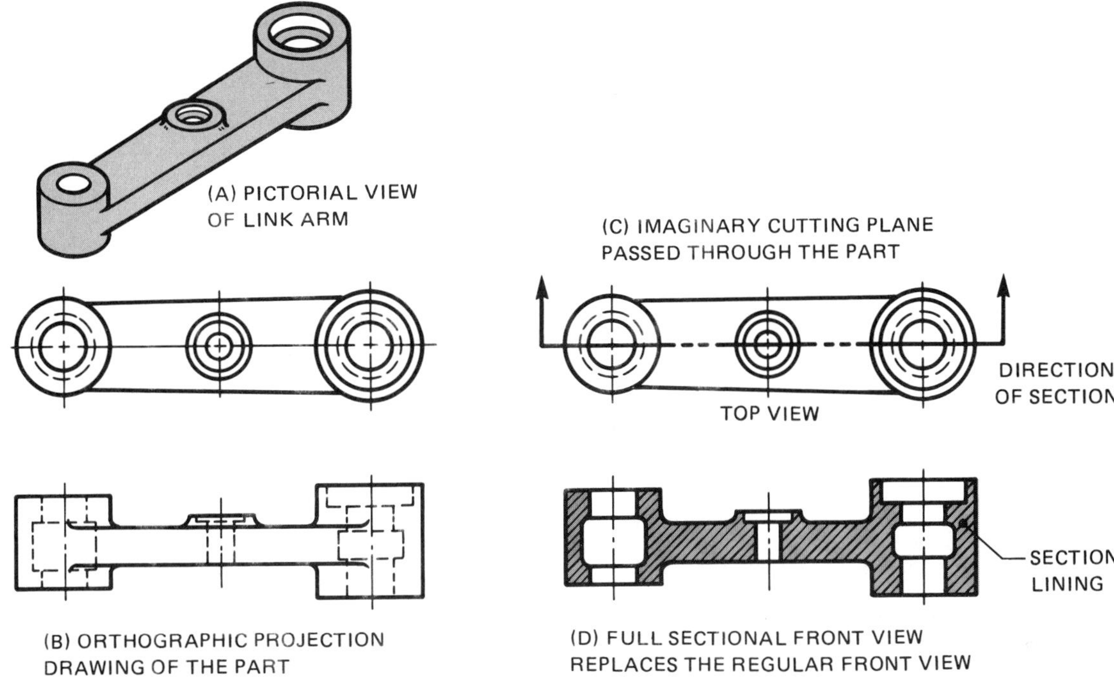

FIGURE 24-2 CONCEPT OF USING A CUTTING PLANE AND SECTION LINING TO CLEARLY REPRESENT INTERNAL DETAILS IN A FULL SECTIONAL VIEW

If a full section is desired, the imaginary cutting plane is located as drawn at (C). The position for viewing is identified in the top view (C) by a cutting plane line. Since the direction of viewing is obvious, no letters are used at the corners of the cutting plane line. The internal details are shown in the front full sectional view. Note that there are no hidden lines in this view.

ANSI SYMBOLS FOR SECTION LINING

Fine, solid, straight lines (called *section lines*) are used (1) to represent the surface which has been cut apart from the object and (2) to clearly identify particular parts within an assembly. Section lines are drawn at an angle, usually 45°. Different combinations of section lines are also used to generally identify specific materials. However, exact material requirements are specified by a note or description in a materials list.

A number of widely used ANSI symbols for section lining are illustrated in figure 24-3. Note that for general sectioning purposes a fine, single section line is used.

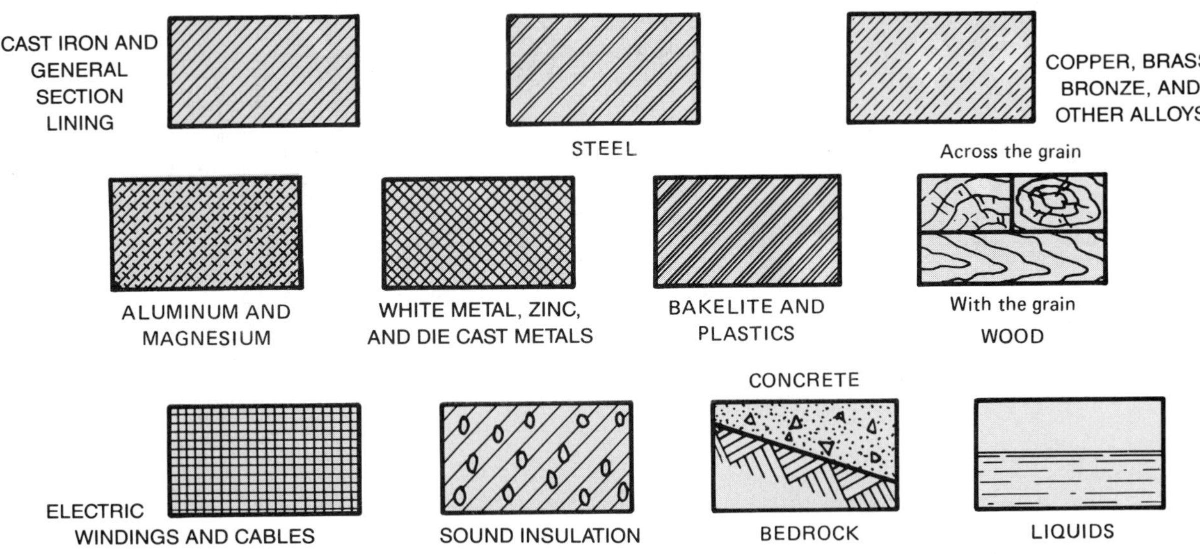

FIGURE 24-3 ANSI SECTION-LINING SYMBOLS

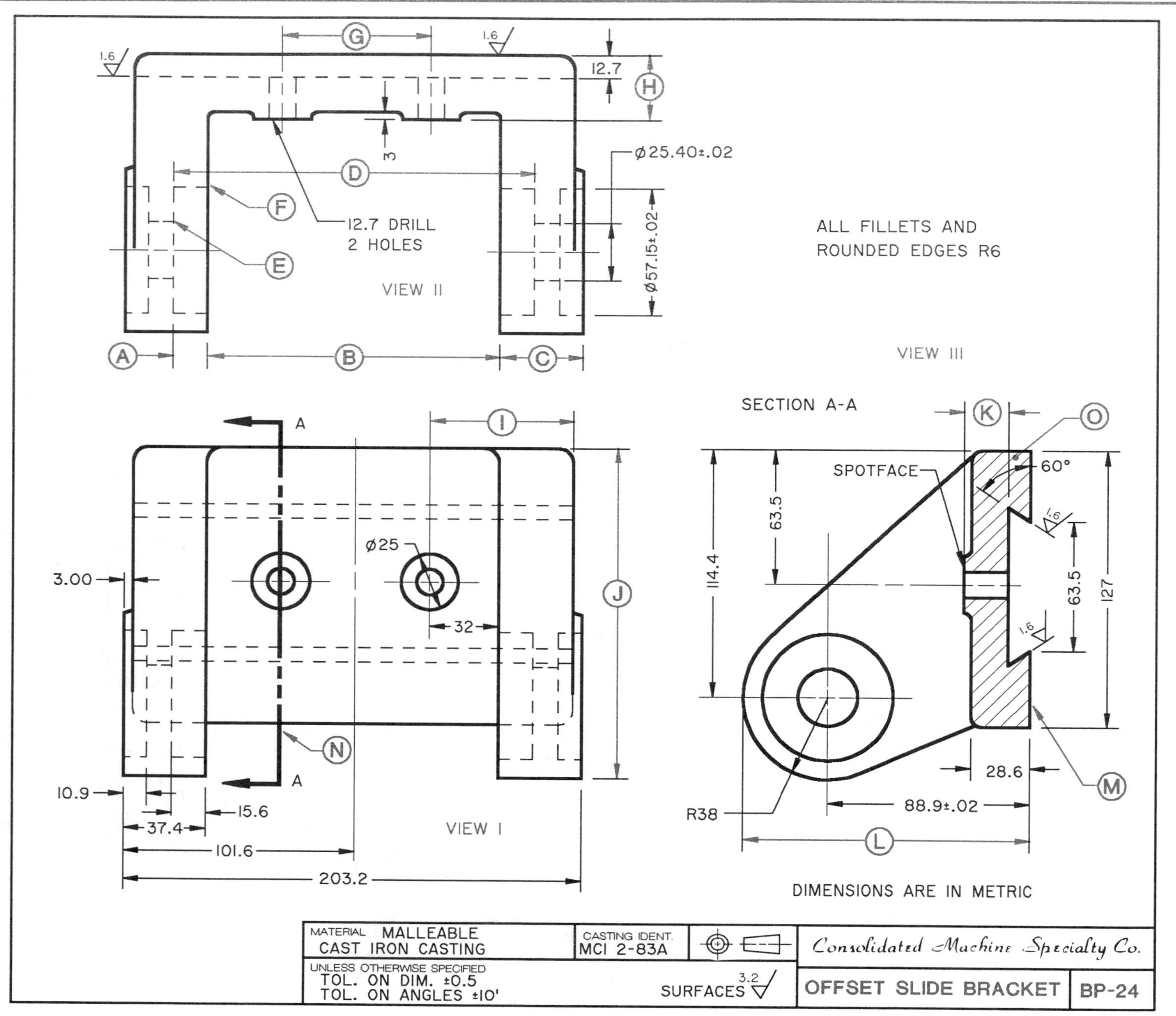
1.6
G
1.6
12.7
H
3
D
Ø25.40±.02
F
12.7 DRILL
2 HOLES
E
Ø57.15±.02
ALL FILLETS AND
ROUNDED EDGES R6
VIEW II
A
B
C
VIEW III
SECTION A-A
K
O
I
A
SPOTFACE
60°
63.5
1.6
Ø25
3.00
J
114.4
63.5
127
32
1.6
N
A
28.6
M
10.9
15.6
37.4
R38
88.9±.02
VIEW I
L
101.6
203.2
DIMENSIONS ARE IN METRIC
MATERIAL MALLEABLE
CAST IRON CASTING
CASTING IDENT.
MCI 2-83A
Consolidated Machine Specialty Co.
UNLESS OTHERWISE SPECIFIED
TOL. ON DIM. ±0.5
TOL. ON ANGLES ±10'
SURFACES 3.2
OFFSET SLIDE BRACKET
BP-24

ASSIGNMENT—UNIT 24

OFFSET SLIDE BRACKET (BP-24)

Student's Name ______________________

1. State (a) the angle of projection and (b) the dimensioning system used on the Bracket drawing.
2. Identify **VIEWS I**, **II**, and **III**.
3. Provide the following information:
   a. The kind of material used for the Bracket
   b. The identification number of the casting
4. Determine the following dimensions:
   a. Size of the base area
   b. Overall thickness of the upright arms
   c. Diameter of the holes drilled in the base pads
   d. Size of all fillets and rounded edges
5. Indicate (a) how the base pads are to be machined and (b) dimension (H).
6. Compute the maximum dimensions for the following measurements:
   (A) (C)
   (B) (D)
7. Determine the upper and lower limit dimensions for (E) and (F).
8. Compute the following minimum dimensions:
   (G) (I) (K)
   (H) (J) (L)
9. Determine the upper and lower limit angle dimension for the dovetail.
10. Indicate the required surface finish for surface (M) and the sides of the dovetail.
11. Give the required surface finish accuracy for all other machined surfaces.
12. a. Name lines (N) and (O).
    b. State briefly the function of these two types of lines.
13. Give one reason for using a sectional view in place of a regular view for a casting with considerable interior details.
14. Draw the section linings for aluminum and brass.

1. a. ______
   b. ______
2. **VIEW I** ______
   VIEW II ______
   VIEW III ______
3. a. ______
   b. ______
4. a. Size of Base ______
   b. Upright Thickness ______
   c. Hole Diameter ______
   d. Fillets/Rounded Edges ______
5. a. ______
   b. (H) = ______

6. (A) = ______ (C) = ______
   (B) = ______ (D) = ______

7.

	Upper Limit	Lower Limit
(E) =	______	______
(F) =	______	______

8. (G) = ______ (J) = ______
   (H) = ______ (K) = ______
   (I) = ______ (L) = ______
9. Upper Limit ______
   Lower Limit ______
10. Surface Finish ______
11. Surface Finish ______

12. a. (N) ______
       (O) ______
    b. Functions
       (N) ______
       ______
       (O) ______
       ______
13. ______
    ______
    ______
14.

ALUMINUM BRASS

UNIT 25
Half Sections, Broken-out Sections, Offset Sections

Principles which deal with cutting planes, cutting plane lines, and section lining are expanded in this unit. With modifications, new applications are made into half sections, broken-out (partial) sections, and offset sections.

SECTIONAL VIEWS OF SYMMETRICAL PARTS

The interior and exterior details of parts or assemblies with symmetrical features are often shown together in a *half section view*. This type of view is developed by using a cutting plane line with the imaginary cutting plane at a right angle or center lines which are 90° to each other.

A half section view combines the advantages of showing general external design features in one-half of a regular view and the internal features as a section view in the second half of the view. This is shown pictorially in figure 25-1A. The cutting plane passes through the symmetrical Flange Plate. One-quarter is removed to expose the internal features. An object such as the Flange Plate is generally represented by one or more regular views (depending on the complexity) and a half-section view as presented at (B).

The cutting plane line, directional arrows, and letters are sometimes omitted from half-section views. As is true in other sectional views, hidden lines are only included in a half section when they are required for dimensioning or to clarify particular features of a part.

BROKEN-OUT (PARTIAL) SECTION DRAWINGS

A *broken-out (partial) section* is used instead of a full or half section to show features contained within only a portion of the interior of an object. The exposed area as represented on a drawing is developed by extending the cutting plane only as far as it is needed. The theoretically cut-away area is then "broken out" to reveal particular internal features. Note in figure 25-2 that the broken-out section is limited by an irregular *break line.* As stated before, the standard break line is a thick, irregular line which is drawn freehand.

OFFSET SECTIONS

Sometimes it is necessary to show a number of features which are not aligned (fall in a straight line) in a section view. The cutting plane in such instances is shifted at one or more places to cut through a desired detail. The section view obtained in this manner is called an *offset section.*

Figure 25-3 provides an example of an offset section. The cutting plane is offset in the top view. The cutting plane line shifts from the major center line to cut through the counterbored holes in the bosses of the casting. The view obtained, showing the internal details along the cutting plane line, is identified at (B) as an *offset sectional view.*

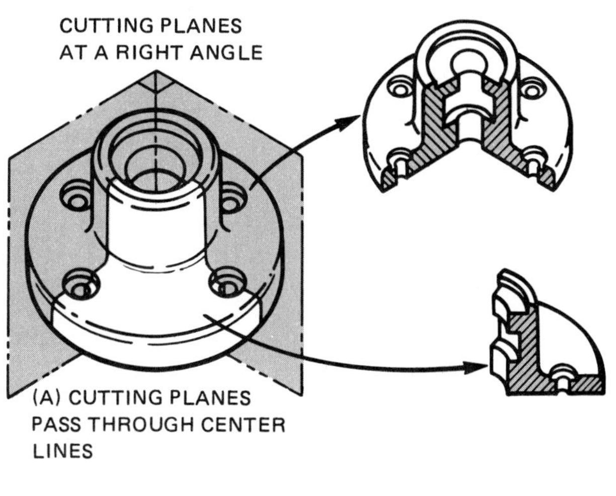

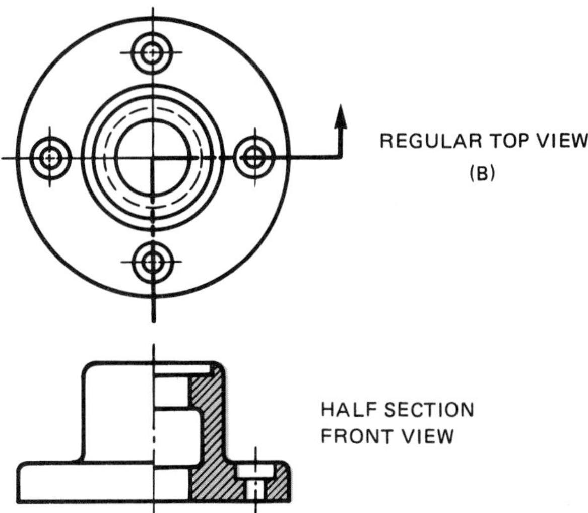

FIGURE 25-1 CUTTING PLANE POSITION AND THE DEVELOPMENT OF A HALF-SECTION VIEW

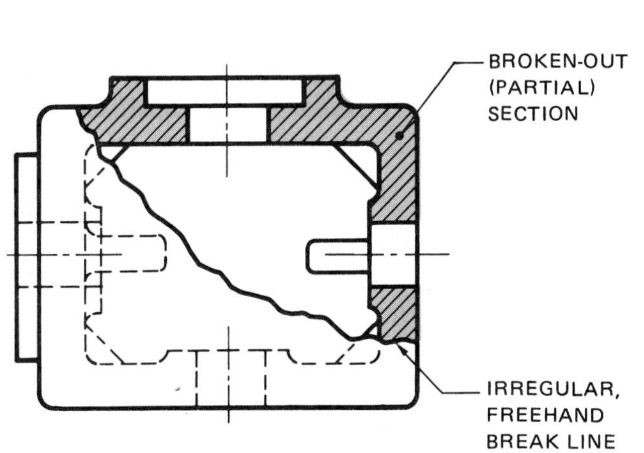

FIGURE 25-2 EXAMPLE OF BROKEN-OUT (PARTIAL) SECTION OF AN OBJECT

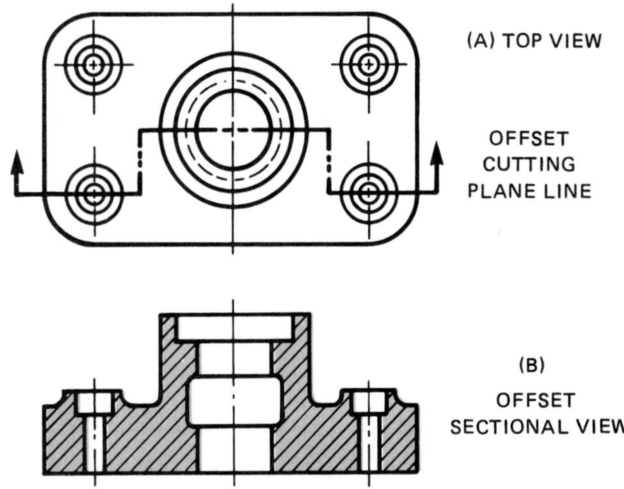

FIGURE 25-3 OFFSET SECTIONAL VIEW OF A MACHINED CASTING

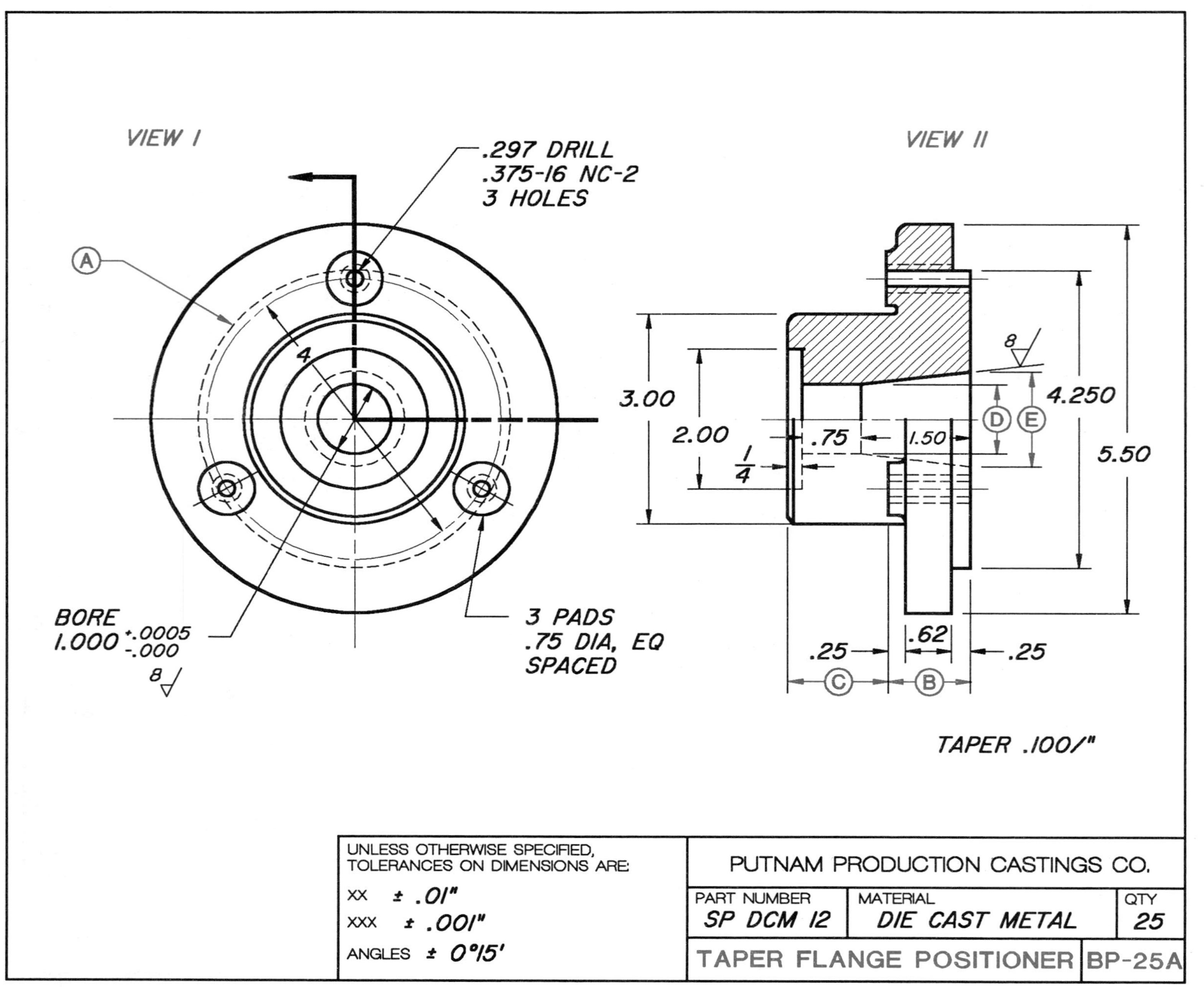

UNLESS OTHERWISE SPECIFIED, TOLERANCES ON DIMENSIONS ARE:	PUTNAM PRODUCTION CASTINGS CO.			
XX ± .01"	PART NUMBER SP DCM 12	MATERIAL DIE CAST METAL		QTY 25
XXX ± .001"				
ANGLES ± 0°15'	TAPER FLANGE POSITIONER		BP-25A	

ASSIGNMENT A—UNIT 25

TAPER FLANGE POSITIONER (BP-25A)

Student's Name ______________________

1. Identify the following items on the drawing of the Positioner:
   a. Part number
   b. Material
   c. Required quantity

2. Name **VIEW I** and **VIEW II**.

3. Cite one advantage of a sectional view over a conventional right-side view of the Positioner.

4. Determine (a) the outside diameter and (b) the overall width (thickness) of the Positioner casting.

5. Provide the following information:
   a. Number of tapped holes
   b. Tap drill size
   c. Tap size
   d. Angle between holes

6. Determine the upper and lower limit dimensions for the following features:
   a. Bored hole
   b. Back flange Ⓐ
   c. Flange thickness Ⓑ
   d. Width Ⓒ
   e. Taper: small diameter Ⓓ
   large diameter Ⓔ
   f. The 120° angle between the tapped holes

7. Change the decimal tolerance limits as follows:

   XX Decimal dimensions $^{+.00''}_{-.01''}$

   XXX Decimal dimensions $^{+.000''}_{-.002''}$

   Then, compute the lower limit dimensions for:
   a. Ⓐ d. Ⓓ
   b. Ⓑ e. Ⓔ
   c. Ⓒ

8. Give the surface finish required for the bored hole and taper.

---

1. a. Part number ______________
   b. Material ______________
   c. Quantity ______________

2. **VIEW I** ______________
   VIEW II ______________

3. ______________
   ____

4. a. Outside Diameter ______________
   b. Width ______________

5. a. Number ______________
   b. Tap Drill Size ______________
   c. Tap Size ______________
   d. Angle ______________

6.

	Upper Limit	Lower Limit
a.	______	______
b. Ⓐ	______	______
c. Ⓑ	______	______
d. Ⓒ	______	______
e. Ⓓ	______	______
Ⓔ	______	______
f.	______	______

7. Lower Limit Dimensions with Changed Tolerance Limits:
   a. Ⓐ ______________
   b. Ⓑ ______________
   c. Ⓒ ______________
   d. Ⓓ ______________
   e. Ⓔ ______________

8. ______________

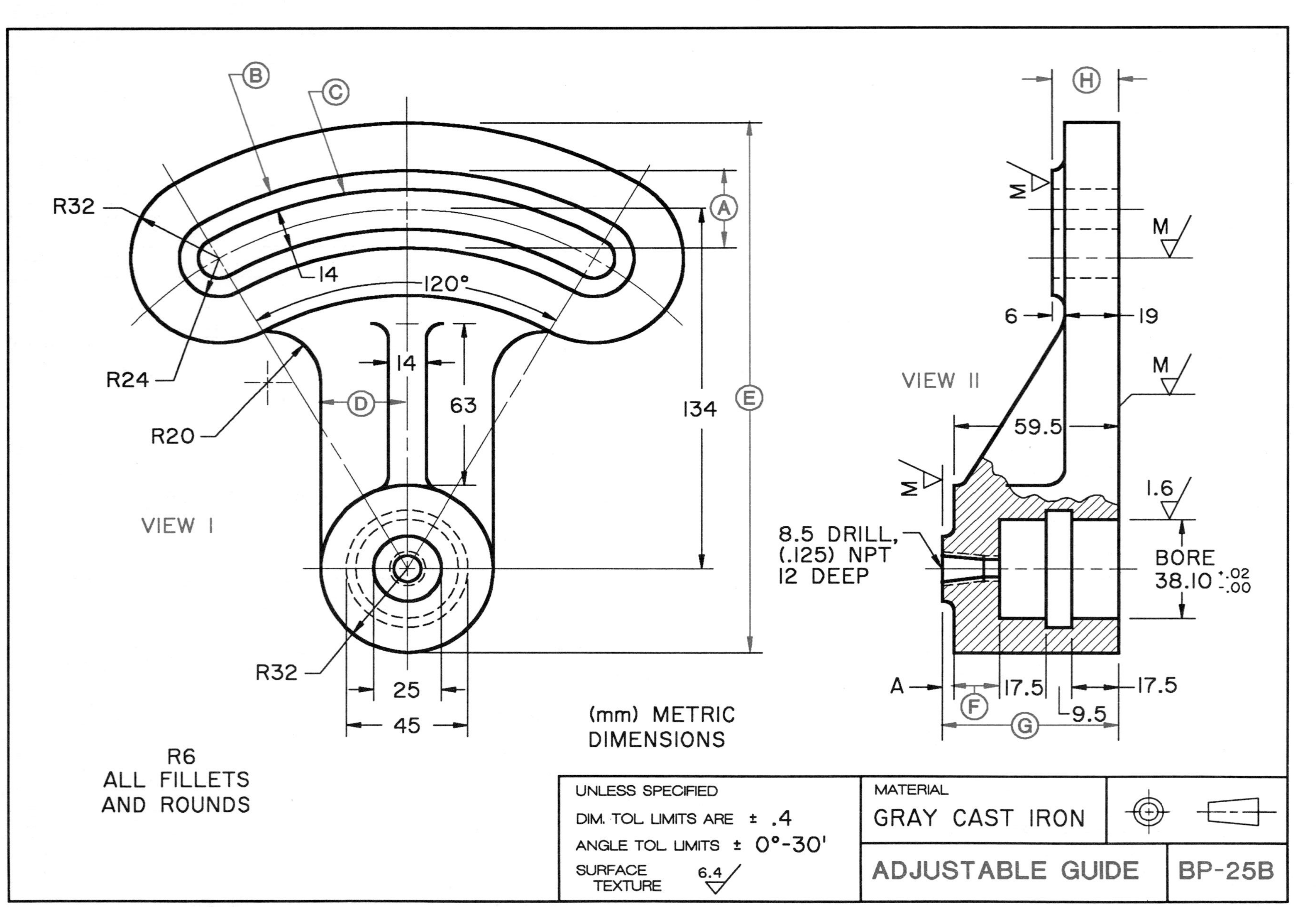
R32
R24
R20
R32
14
120°
14
63
134
A
B
C
D
E
VIEW I
25
45
R6
ALL FILLETS
AND ROUNDS
VIEW II
H
M
M
6
19
M
59.5
M
1.6
8.5 DRILL,
(.125) NPT
12 DEEP
BORE
38.10 +.02 -.00
A
17.5
17.5
9.5
F
G
(mm) METRIC
DIMENSIONS
UNLESS SPECIFIED
DIM. TOL LIMITS ARE ± .4
ANGLE TOL LIMITS ± 0°-30'
SURFACE
TEXTURE
6.4
MATERIAL
GRAY CAST IRON
ADJUSTABLE GUIDE
BP-25B

ASSIGNMENT B—UNIT 25

ADJUSTABLE GUIDE (BP-25B)

Student's Name ____________________

1. Name **VIEW I** and **VIEW II**.
2. State (a) the projection angle and (b) the measurement system used on the Guide drawing.
3. Identify four features in the broken-out portion of **VIEW II** which are not cross hatched.
4. a. Give the total number of surfaces which are to be machined.
   b. Indicate the general surface finish requirement.
   c. Identify the required quality of surface finish of the bored ∅ 38.10 hole.
5. Explain the note (**VIEW II**):

   8.5 DRILL, ($\frac{1}{8}$") NPT

   12 DEEP
6. Refer to the pad and rib in **VIEW I**. Determine the dimensions of the following features:

   a. Ⓐ d. Ⓓ
   b. Ⓑ e. Ⓔ
   c. Ⓒ
7. Determine the upper and lower limit dimensions of the following features:

   a. Bored hole

   b. Center distance between the bored hole and the elongated slot

   c. Dimension Ⓕ

   d. Depth of the bored hole

   e. Overall widths Ⓖ and Ⓗ

1. **VIEW I** ____________________

   VIEW II ____________________

2. a. ____________________

   b. ____________________

3. a. ____________________

   b. ____________________

   c. ____________________

   d. ____________________

4. a. Machined Surfaces ____________

   b. General Surface Finish ____________

   c. Bore Surface Finish ____________

5. **8.5 DRILL** ____________

   ($\frac{1}{8}$") NPT ____________

   12 DEEP ____________

6. a. Ⓐ ________

   b. Ⓑ ________

   c. Ⓒ ________

   d. Ⓓ ________

   e. Ⓔ ________

7.

	Upper Limit	Lower Limit
a.	________	________
b.	________	________
c. Ⓕ	________	________
d.	________	________
e. Ⓖ	________	________
Ⓗ	________	________

UNIT 26
Revolved and Removed Sections: Rotated Features

An orthographic projection view of an object with many details is often difficult to draw and to interpret. Under these conditions, drawings are sometimes simplified by *violating the principles of true projection.* Certain features are drawn as though they were revolved or moved slightly out of position. Hidden lines that otherwise would coincide and become a single line with another line are moved just enough to display the line.

REVOLVED SECTIONS

One of the easiest techniques of representing a rib, spoke, or other irregular shape is to include a *revolved section* as part of a regular view. A revolved section is one where a cutting plane passes through an object at a particular location. The cutting plane is then revolved 90°. The shape of the object appears on location on the drawing or blueprint as a *revolved section.*

Figure 26-1 shows an application of a revolved section. Standard views are drawn at (A). A study of the right-side view shows that it is difficult to establish the shape of the rib area. The part drawing is simplified by passing a cutting plane through the part and rotating it 90°. The front view obtained in this manner shows the shape clearly. Dimensions may be easily added. The drawing is simplified and easy to read, as shown at (B).

The outline of a revolved section is represented by a fine, continuous line. If the revolved section is *broken out* as illustrated at (C), the outline of the section is represented by a thick, continuous line.

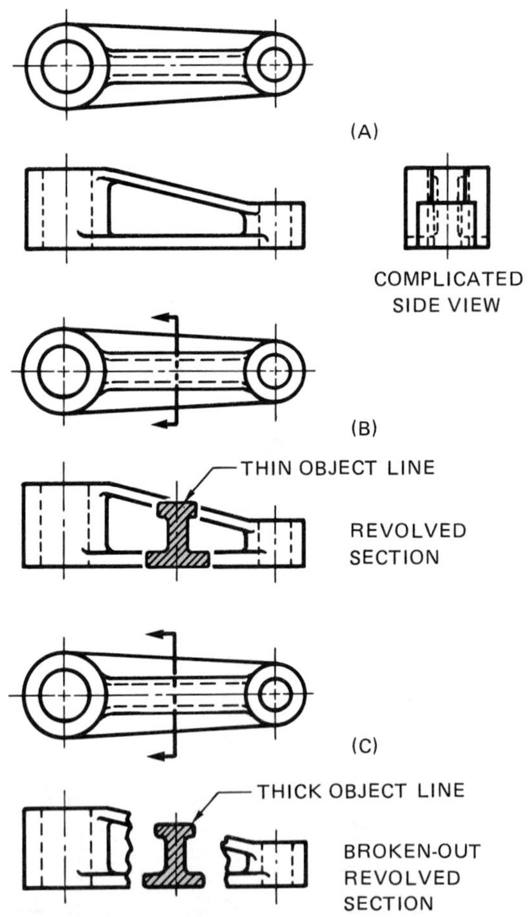

FIGURE 26-1 REVOLVED (ROTATED) SECTION USED TO CLEARLY DESCRIBE THE WEB SHAPE

REMOVED SECTIONS

Another technique of accurately describing the shape of a specific area of an object is to completely remove one or more sections from a regular view. This kind of drawing representation is called a

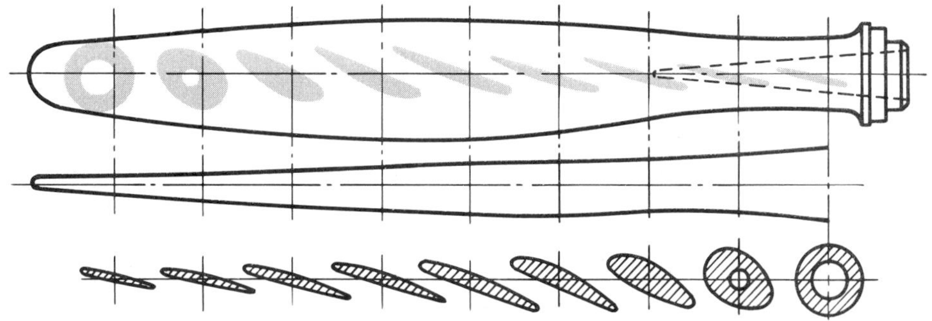

FIGURE 26-2 REMOVED SECTIONS DRAWN ABOVE EACH NATURAL PROJECTED POSITION

removed section or a *removed section view.* Removed sections are practical when a number of cross-sectional views are needed to show the changing shape (sections or areas) of a part.

Removed sections also permit drawing a section of a part to a larger scale in order to magnify particular details or for ease in dimensioning. Removed sections are drawn on the same drawing sheet, away from the regular views. The section is identified to show the position of each cutting plane unless the location of the section is apparent from one other view. This is the case in figure 26-2 in which nine removed sections of a fan blade are included. It should be noted that the removed sections show the gradually changing shape of the blade at nine successive stages. Whenever practical, ANSI standards suggest that a removed section be drawn in its natural projected position.

The reading and drawing of a sectional view require an understanding of certain part features which are not cross-sectioned. Among these features are gear teeth, shafts, keyways, holes, ribs, spokes, partitions, and others. Figure 26-3 illustrates how some of these features are represented on a drawing.

SECTIONS OF ROTATED FEATURES

It is common practice in drafting to *rotate features* in order to clearly show the true shape. This condition is illustrated in figure 26-4. The drawing of the Housing Cover shows two sectional views. Note the difficulty in interpreting the right ribs and the left lug in the true projection view at (A). The preferred method of representation is to revolve the lugs and ribs. It is assumed the cutting plane line is revolved until the ribs and lug are parallel to a regular plane. The revolved section at (B) is preferred. Note that the bore, lugs, and ribs are not cross-sectioned.

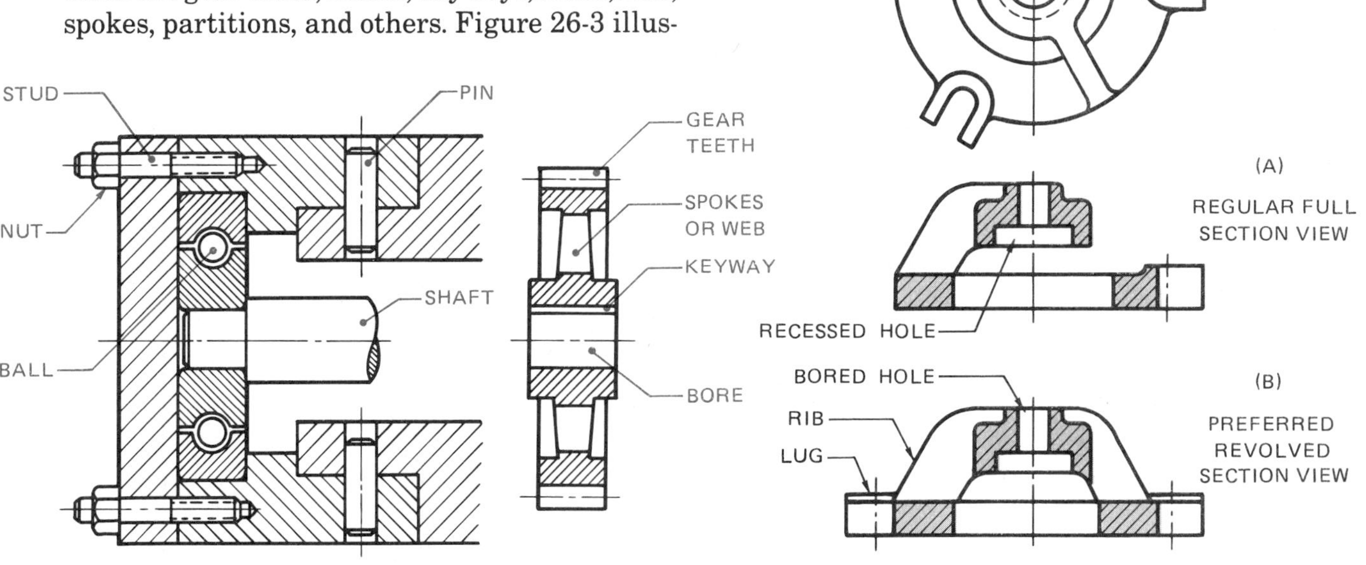

FIGURE 26-3 EXAMPLES OF FEATURES WHICH ARE NOT CROSS-SECTIONED IN SECTIONAL DRAWINGS

FIGURE 26-4 ROTATED FEATURES REPRESENTED BY A REVOLVED SECTION VIEW

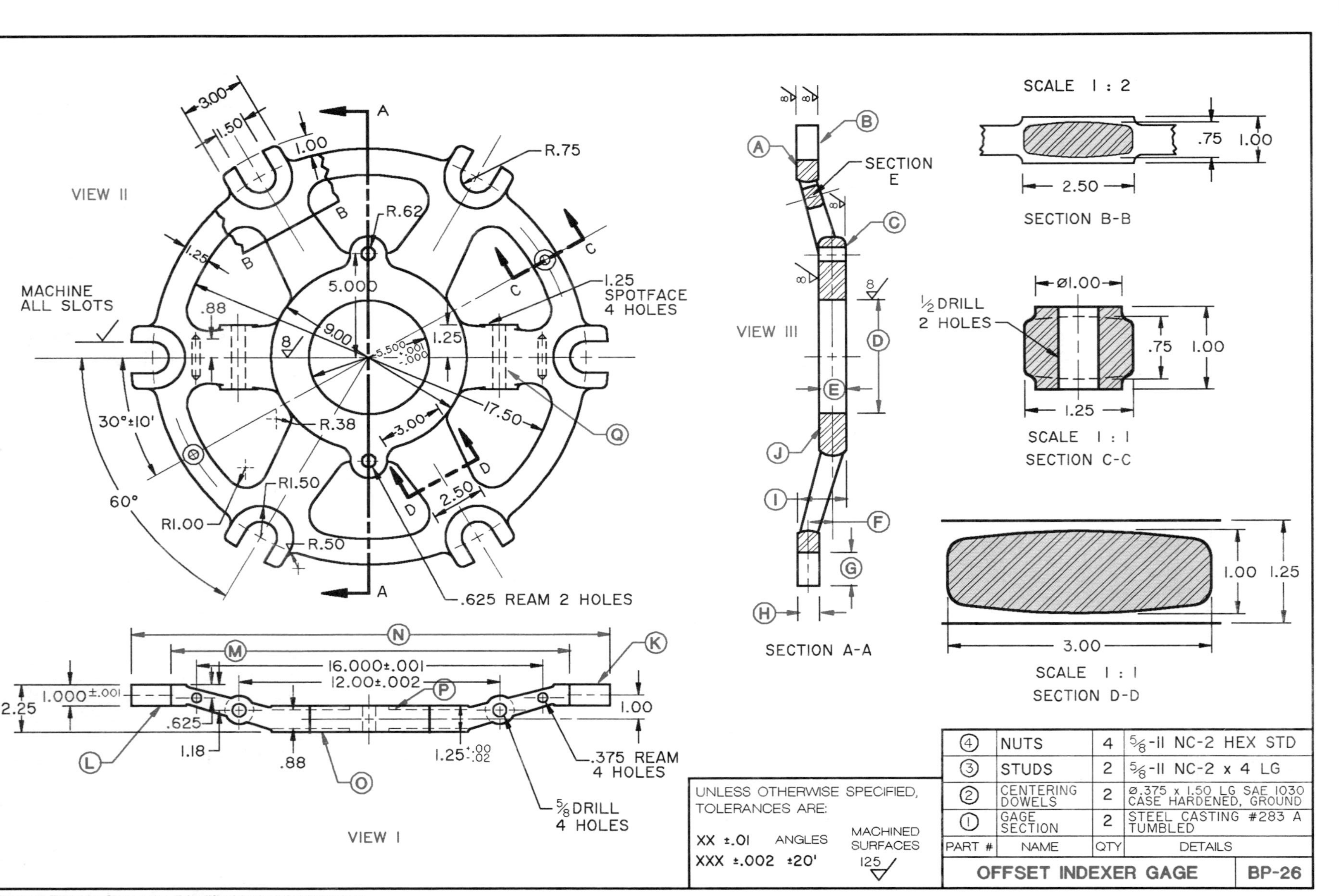

PART #	NAME	QTY	DETAILS
4	NUTS	4	⅝-11 NC-2 HEX STD
3	STUDS	2	⅝-11 NC-2 x 4 LG
2	CENTERING DOWELS	2	Ø.375 x 1.50 LG SAE 1030 CASE HARDENED, GROUND
1	GAGE SECTION	2	STEEL CASTING #283 A TUMBLED

OFFSET INDEXER GAGE — BP-26

General Electric Company (Modified Industrial Drawing)

ASSIGNMENT—UNIT 26

OFFSET INDEXER CAGE (BP-26)

Student's Name ______________________

1. Name each of the different parts of the assembled Cage.
2. Give the sizes of (a) the centering dowels and (b) the studs used to secure the Cage sections together.
3. Identify each of the following views or sections:
   a. **VIEW I**
   b. **VIEW II**
   c. Section A–A
   d. Section B–B
   e. Section C–C
   f. Section D–D
   g. Section E
4. Describe briefly the difference between sectional views A–A and B–B.
5. Tell why hidden lines are used in section C–C when in principle such lines are omitted.
6. Identify three features of the Cage assembly which are not cross-sectioned on a sectional view.
7. Provide the following information:
   a. Number of surfaces to be machined as identified in Section A–A
   b. Required degree of accuracy to which the surfaces are to be machined
8. a. Determine the center distances between each of the following features.
   b. Specify the tolerance in each case:
   (1) Centering dowels
   (2) Studs
   (3) Offset dimension (F)
   (4) 1/2″ drilled holes in the outer rim
   (5) .625″ reamed holes
9. Identify the letter of the feature in **VIEW I** which corresponds with each of the following surfaces in Section A–A:
   a. (A) b. (B) c. (C) d. (J)
10. Determine the upper and lower dimensional limits of the following features:
    a. Bored hole (D)
    b. Flange thickness (E)
    c. Center distance (F)
    d. Lug thickness (H)
    e. Overall thickness (width) (I)
    f. Diameter (M)
    g. Outside diameter (N)
11. Give the width and depth of slots (G).
12. Determine (a) the number and (b) the drill diameters of holes (Q).

1. ______________________
____

2. a. Centering Dowels ______________________
b. Studs ______________________

3. a. ______________________
b. ______________________
c. ______________________
d. ______________________
e. ______________________
f. ______________________
g. ______________________

4. ______________________
____
____
____

5. ______________________
____
____

6. a. ______________________
b. ______________________
c. ______________________

7. a. Machined Surfaces ______________________
b. Surface Finish Accuracy ______________________

8.

	Center Distances (a)	Tolerances (b)
(1)		
(2)		
(3)		
(4)		
(5)		

9. a. (A) = ________ c. (C) = ________
b. (B) = ________ d. (J) = ________

10.

	Upper Limits	Lower Limits
a. (D)		
b. (E)		
c. (F)		
d. (H)		
e. (I)		
f. (M)		
g. (N)		

11. (G) Width ________ Depth ________

12. a. Number of Holes ______________________
b. Drill Diameter ______________________

UNIT 27 Phantom, Auxiliary, and Assembly Drawings

PHANTOM OR HIDDEN SECTIONS

A phantom (hidden) section is used to represent in one view the interior and exterior features of an object which is not symmetrical. A phantom view is generally a sectional view. The phantom portion is *superimposed* on (included within) a regular view. Light, evenly spaced broken lines of the material are used as section lining for phantom sections. Figure 27-1 provides an example of a nonsymmetrical part. Note that the view shows interior details in section on an exterior view without removing the front portion of the object.

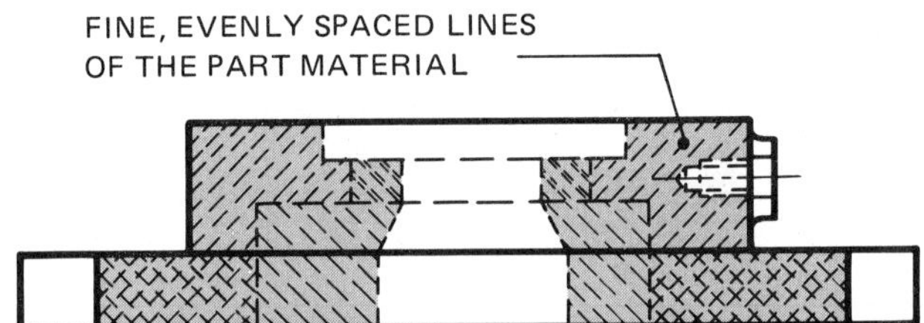

FIGURE 27-1 PHANTOM (HIDDEN) SECTIONAL VIEW SUPERIMPOSED ON A REGULAR VIEW

AUXILIARY SECTION VIEWS

Auxiliary section views serve two purposes which are similar to those of regular auxiliary views:

- To simplify interior and exterior details on a drawing which has angular features or areas of a part.
- To represent features on angular surfaces in their true size and shape.

An *auxiliary section view* represents a section of an object as viewed at an angle or on an auxiliary plane. An auxiliary section view may be a complete view or a partial view, using cross-hatch lines. Figure 27-2 provides an example of a partial auxiliary section view along cutting plane lines D–D. While the auxiliary view is projected from a standard view, unnecessary details are omitted. This practice simplifies the drawing and adds to the ease and accuracy in reading a print.

REPRESENTATION OF PARTS WITH COMPLICATED INTERIOR FEATURES

A sectional view may be considered a regular orthographic view in which internal features are clearly exposed by removing a portion of the object. A number of section views are often provided for parts which have complicated internal features. If a single view were used, the maze of required hidden lines needed to represent each feature would make the print difficult to read. When a number of sectional views is used, each such view is considered by itself without reference to any details which may have been removed from any other view.

PHANTOM REFERENCE OUTLINE USED WITH A SECTIONAL VIEW

Some sectional view drawings are simplified by using phantom lines. Fine, long lines, followed by two short dashes, are used as phantom lines to

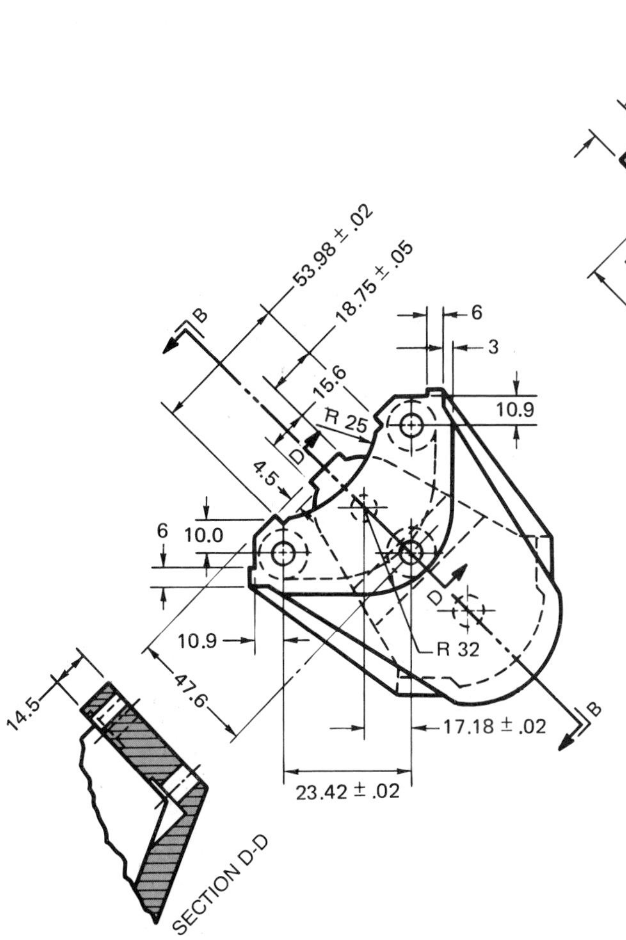

FIGURE 27-2 AUXILIARY SECTION VIEWS REPRESENT TRUE SIZES AND SHAPES OF SURFACES

represent *reference surfaces*. Figure 27-3A shows *phantom outline* surfaces which are spread across considerable distances. In order to assist in interpreting a drawing so each feature may be machined and measured in relation to other reference surfaces, the section view (Figure 27-3B) contains a phantom outline of the reference surfaces Ⓐ, Ⓑ, Ⓒ, and measuring roll Ⓓ. The 38.2 ±.04 mm dimension and the 25.4 ±.02 mm ball diameter provide dimensional information relative to the reference surfaces and the two machined faces Ⓔ and Ⓕ.

ASSEMBLY SECTIONAL DRAWINGS

An assembly sectional drawing, like a regular assembly drawing or a pictorial assembly drawing, shows within one view the relationship of the separate parts of a machine, structure, device, or electrical, pneumatic, instrumentation system, or other unit. As stated before, in addition to providing complete information about an assembled unit, assembly sectional drawings are often combined with detail drawings to provide detailed specifications for both the assembly and each separate part. Assembly sectional drawings are widely applied in operating manuals, descriptive promotional catalogs, technical product information, and for production needs.

Individual parts, as represented by sectional assembly drawings, are identified by different cross-hatching. The exception is in the case of shafts, bearings, spokes, etc., where cross-hatch lines are omitted.

OUTLINE SECTIONING

Outline sectioning is a useful technique for indicating a large sectioned area. In outline sectioning, widely-spaced lines are drawn or sketched only along the border surfaces. This technique of sectioning focuses attention on internal details as contrasted with looking over an object that is entirely cross hatched.

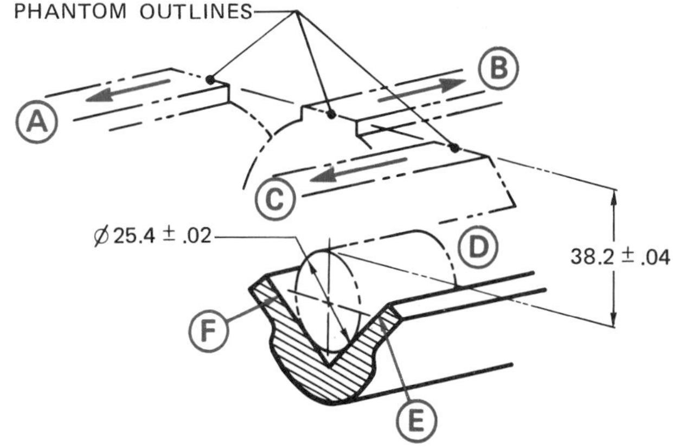

FIGURE 27-3 APPLICATION OF PHANTOM OUTLINES AND A SECTION IN ESTABLISHING A MACHINING AND MEASUREMENT DIMENSION

NOTE: REFER TO APPENDIX B-2 FOR THE LARGER SCALE DRAWING TO USE WITH THIS ASSIGNMENT.

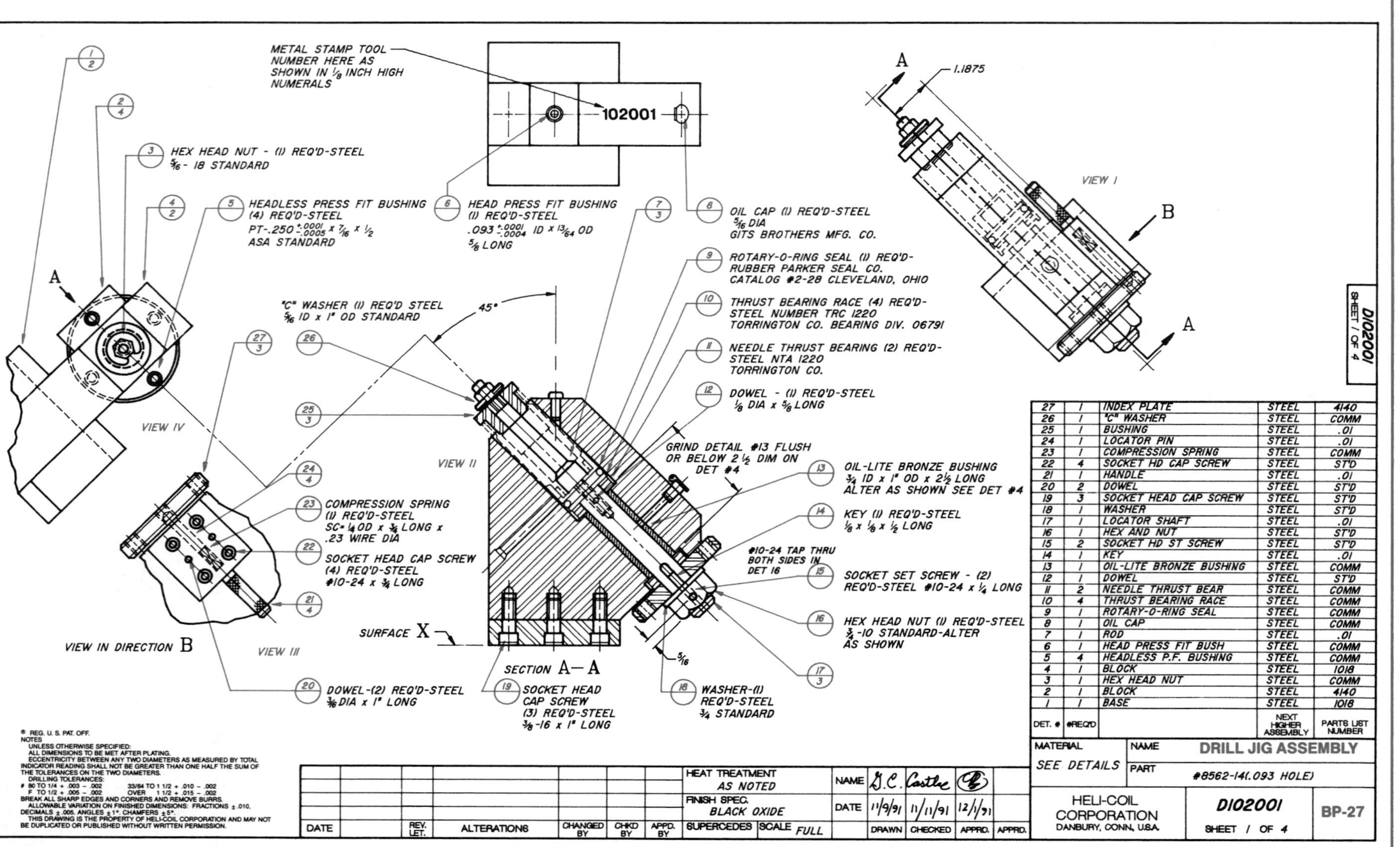

Courtesy of Heli-Coil Products, division of Mite Corporation

DRILL JIG ASSEMBLY (BP-27)

1. Identify (a) the part number, (b) the tool number, and (c) the number of sheets in the drawing set.
2. Refer to the parts list of the title block.
   a. Give the number of *different parts* used in constructing the Drill Jig.
   b. Determine the number of parts which are purchased commercially.
   c. Indicate the number of *different* standard fasteners which are required.
   d. Identify (1) the detail number and (2) the classification of steel for the (a) base (b) blocks and (c) index plate.
3. Explain briefly what the symbol (17/3) means.
4. Name each of the following views:
   a. **VIEW I** c. **VIEW III**
   b. **VIEW II** d. **VIEW IV**
5. a. Identify one view in which a phantom outline is used.
   b. State the purpose for using the phantom outline.
   c. Tell how a part as represented by phantom outline is identified.
6. State the function of the phantom lines which extend from **VIEW I** to **VIEW II** and **VIEW II** to **VIEW IV**.
7. Give (a) the detail number and (b) the name of each of five parts in **VIEW I** which do not require cross-hatching.
8. Identify three different materials which are required for constructing the Drill Jig:
   a. Give the part name
   b. State the material in the part
9. Provide the specifications for the following:
   a. The drill bushing (6)
   b. The locator pin bushing (5)
10. Give (a) the detail numbers and (b) the specifications of the following parts:
    (1) Oil-Lite Bronze bushing
    (2) Rotary seal
11. State (a) the size and (b) the quantity required for the following parts.
    (1) (12) **Dowel** (2) (14) **Key**
12. Give three reasons for using the auxiliary assembly section view instead of a regular front view.

ASSIGNMENT—UNIT 27

Student's Name ______________________

1. a. Part # ______________________
   b. Tool # __________ c. Sheets in Set __________
2. a. __________ b. __________ c. __________
   d.

	Detail Number (1)	Steel Class'n (2)
(a) Base	________	________
(b) Block	________	________
(c) Block	________	________
(d) Index Plate	________	________

3. ______________________
   ____
   ____
4. a. __________ c. __________
   b. __________ d. __________
5. a. ______________________
   b. ______________________
   ____
   c. ______________________
   ____
6. ______________________
   ____
   ____
7.

	Detail Number (a)	Part Name (b)
(1)	________	________
(2)	________	________
(3)	________	________
(4)	________	________
(5)	________	________

8.

	Part Name (a)	Material (b)
(1)	________	________
(2)	________	________
(3)	________	________

9. a. ______________________
   ____
   b. ______________________
   ____
10.

	Detail Number (a)	Specification (b)
(1) Oil-Lite Bronze Bushing	________	________
(2) Rotary Seal	________	________

11.

	Part Size (a)	Number Required (b)
(1)	________	________
(2)	________	________

12. a. ______________________
    ____
    b. ______________________
    ____
    c. ______________________
    ____

SECTION 8
Tolerancing

UNIT 28
Linear and Angular Measurement Tolerances

The production of interchangeable parts depends on specifying limits and fits for different applications according to uniform standards. The designer and the producer of a part, a component of many parts, or a complete mechanism are guided by the following principles:

- The same ANSI, SI metric, and Canadian National Standard definitions of terms for limits and fits are followed.
- Dimensional limits are established from a series of preferred tolerances and allowances which are standard.
- Tolerances are applied according to a uniform system.
- Preferred basic sizes are selected, where practical, to reduce material, tooling, and other machining costs.

While recommended ANSI standards for limits and fits are widely used, adaptations are needed. Compensation must be made for such factors as operating temperatures, humidity, lubrication, bearing load, speed, and part size.

TERMS APPLIED TO ANSI STANDARD LIMITS AND FITS

A number of the principal terms which apply to limits and fits are defined at this time. Additional terms are introduced throughout this section.

Nominal Size. This designation is for general identification purposes. A nominal size represents a particular dimension or unit of length but without specified limits of accuracy. For example, a 1″-diameter bar of steel provides a general description without stating acceptable maximum or minimum diameters.

Reference Dimension. A dimension without tolerance used for reference and checking purposes. For example, a length, diameter, or center distance which is given on a drawing for reference and checking but not for layout, machining, or inspection.

Basic Size. The exact theoretical size from which all limiting allowances and tolerances are made.

Actual Size. A measured dimension (size).

Design Size. A part size after clearance allowance is made and tolerances are applied.

Allowance. A prescribed (intentional) difference in the dimensions of mating parts. *Positive allowance* represents the minimum clearance between two moving, mating parts. *Negative allowance* (such as a press fit of a gear on a shaft) provides the maximum interference between the parts.

Tolerance. The total permissible variation in the size of a part. In other words, tolerance is the dimensional range within which a feature or part will perform a required function and may be interchangeable. A tolerance provides the craftsperson with information for machining and fitting mating parts. Tolerance is the difference between the upper and lower limits of a dimension. For example, if the upper limit dimension is **4.587″** and the lower limit is **4.572″**, the tolerance is **4.587″ − 4.572″** or **.015″**.

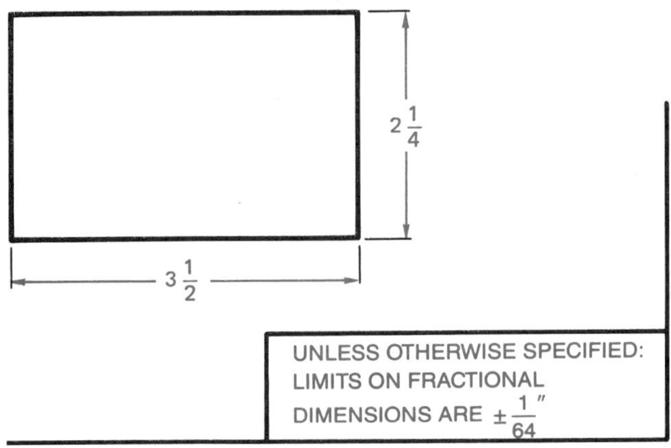

FIGURE 28-1 APPLICATION OF FRACTIONAL TOLERANCES

Limits of Size. The maximum and minimum sizes of a feature. The extreme permissible sizes of acceptance.

Tolerance Limit. The positive or negative variation a dimension (size) may depart from the design size.

LINEAR (TOLERANCE) LIMITS

Tolerance limits are generally specified on a drawing as a note, or are included in the title block or added to the dimension. The two most widely used basic tolerance limits relate to linear and angular measurements.

When customary units of measure are used, and the part does not need to be machined to a high degree of accuracy, the limits may be specified as a fraction. More precise measurement limits are given as decimals of customary or SI metric units of measure.

The drawing of the rectangular block and the note (as shown in figure 28-1) specify that the

LIMITS ON PRACTICAL DIMENSIONS ARE $\pm\frac{1}{64}''$.

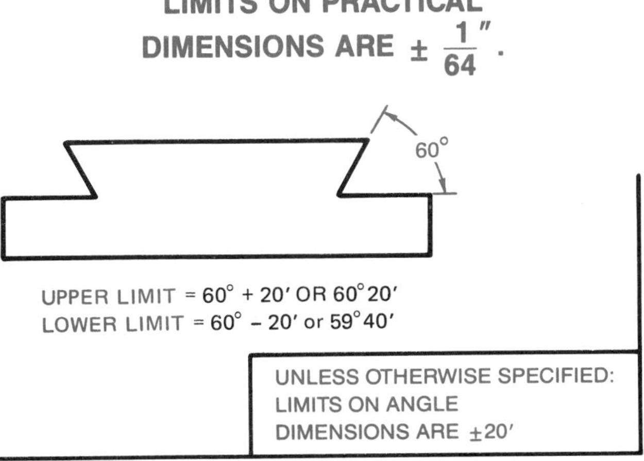

FIGURE 28-2 SPECIFYING ANGLE LIMITS IN THE TITLE BLOCK

part may be machined one 64th-inch smaller or larger than the nominal sizes of $3\frac{1}{2}''$ and $2\frac{1}{4}''$.

The *maximum acceptable sizes,* which permit the part to be interchangeable are:

$$3\frac{1}{2}'' + \frac{1}{64}'' = 3\frac{33}{64}'' \text{ and}$$

$$2\frac{1}{4}'' + \frac{1}{64}'' = 2\frac{17}{64}''.$$

The *minimum acceptable* machined sizes are:

$$3\frac{1}{2}'' - \frac{1}{64}'' = 3\frac{31}{64}'' \text{ and}$$

$$2\frac{1}{4}'' - \frac{1}{64}'' = 2\frac{15}{64}''.$$

LIMITS OF SIZE

The maximum sizes of $3\frac{33}{64}''$ and $2\frac{17}{64}''$ are called the *upper limits.* The minimum sizes of $3\frac{31}{64}''$ and $2\frac{15}{64}''$ are the *lower limits* of the two dimensions. The *actual size* as measured with a measuring instrument (or steel rule in this case) must be between the upper and lower limit of each dimension:

$$3\frac{31}{64}'' \text{ to } 3\frac{33}{64}'' \text{ and } 2\frac{15}{64}'' \text{ to } 2\frac{17}{64}''.$$

ANGULAR DIMENSIONS AND TOLERANCES

Surfaces which are not parallel to each other are dimensioned in degrees (°), minutes (′), and seconds (″). A numerical value is used with these symbols to specify the angular measurement. The tolerance may be included with the dimension or may appear in a title block or as a note near the dimension. In figure 28-2, the required angular limits for the 60° angle is given as a note.

The angle limits of ±20 are shown in figure 28-3 as part of the angle dimension.

FIGURE 28-3 ANGULAR LIMITS INCLUDED WITH DIMENSION OF ANGLE

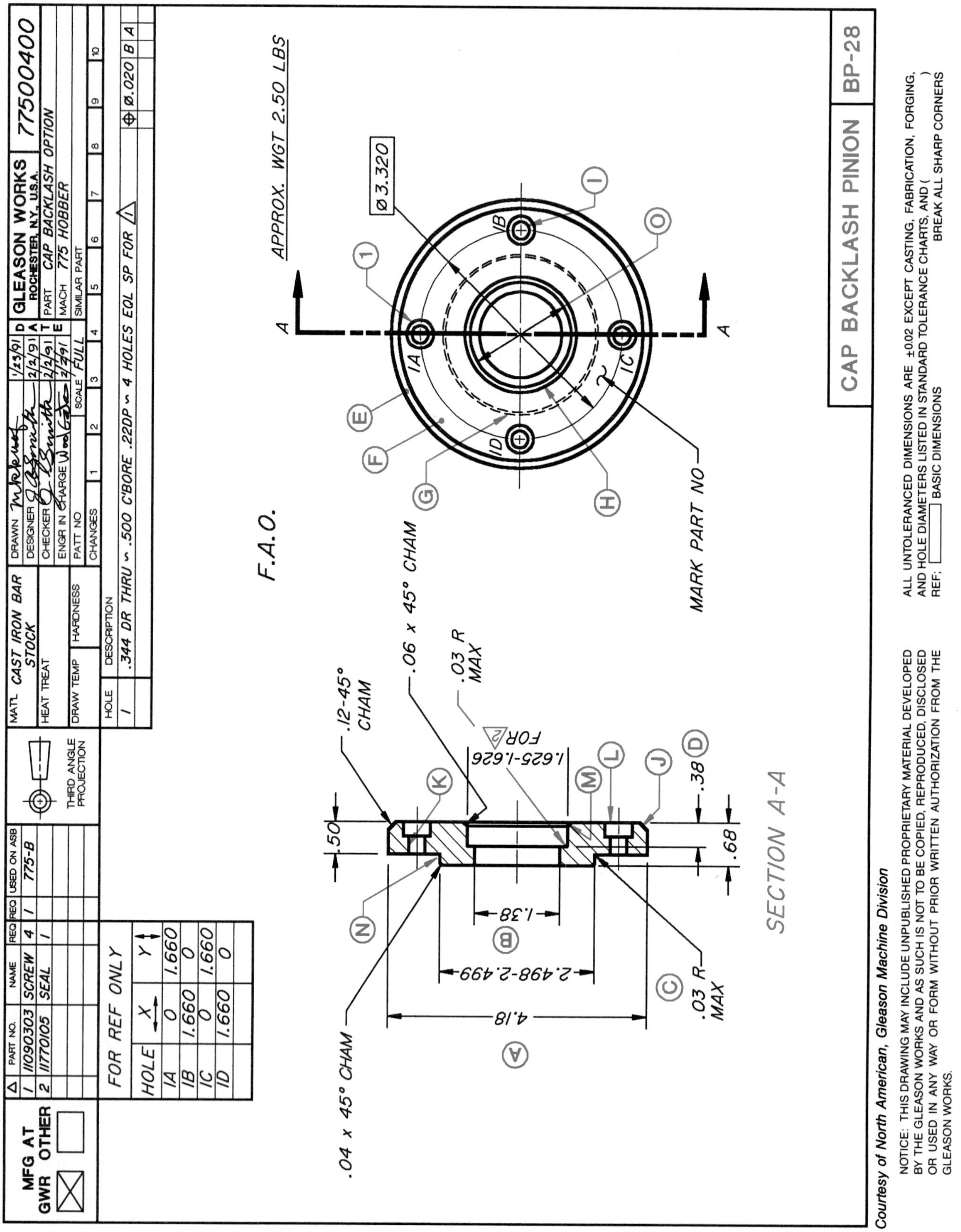

MFG AT		Δ	PART NO.	NAME	REQ	REQ	USED ON ASB
GWR	OTHER	1	11090303	SCREW	4	1	775-B
☒	☐	2	11770105	SEAL	1		

THIRD ANGLE PROJECTION

MAT'L	CAST IRON BAR STOCK
HEAT TREAT	
DRAW TEMP	HARDNESS

		DATE
DRAWN	M. Kekuof	1/23/91
DESIGNER	J. A. Smith	2/2/91
CHECKER	J. L. Smith	2/2/91
ENGR IN CHARGE	Wood Gates	2/2/91
PATT NO	SCALE FULL	

CHANGES 1 2 3 4 5 6 7 8 9 10

GLEASON WORKS
ROCHESTER, N.Y., U.S.A.
77500400
PART CAP BACKLASH OPTION
MACH 775 HOBBER
SIMILAR PART

HOLE	DESCRIPTION	
1	.344 DR THRU ⌵ .500 C'BORE .22DP ⌵ 4 HOLES EQL SP FOR △1	⌖ Ø.020 B A

FOR REF ONLY

HOLE	X	Y
1A	0	1.660
1B	1.660	0
1C	0	1.660
1D	1.660	0

F.A.O.

APPROX. WGT 2.50 LBS

.04 x 45° CHAM
.50
.12-45° CHAM
.06 x 45° CHAM
.03 R MAX
1.625-1.626 FOR △2
4.18
2.498-2.499
1.38
.03 R MAX
.38
.68
A B C D J K L M N
SECTION A-A

A
Ø 3.320
E F G H I O 1
1A 1B 1C 1D
MARK PART NO
A

CAP BACKLASH PINION | BP-28

Courtesy of North American, Gleason Machine Division

NOTICE: THIS DRAWING MAY INCLUDE UNPUBLISHED PROPRIETARY MATERIAL DEVELOPED BY THE GLEASON WORKS AND AS SUCH IS NOT TO BE COPIED, REPRODUCED, DISCLOSED OR USED IN ANY WAY OR FORM WITHOUT PRIOR WRITTEN AUTHORIZATION FROM THE GLEASON WORKS.

ALL UNTOLERANCED DIMENSIONS ARE ±0.02 EXCEPT CASTING, FABRICATION, FORGING, AND HOLE DIAMETERS LISTED IN STANDARD TOLERANCE CHARTS, AND ( )
REF; ☐ BASIC DIMENSIONS
BREAK ALL SHARP CORNERS

ASSIGNMENT—UNIT 28

CAP BACKLASH PINION (BP-28)

Student's Name ____________________

1. Identify the kind of material which is used for the Pinion.
2. Tell why a full-section A–A view is used instead of a left-side view.
3. Give the maximum dimensions for the following features:
   a. (A) c. (C)
   b. (B) d. (D)
4. State what the specifications are for holes (1).
5. Name (a) parts △1 and △2, (b) give the required quantity of each part, and (c) identify the part numbers.
6. Give (a) the nominal size and (b) the angle of each beveled corner.
7. Assume the angular tolerance is ±15′.
   a. Give the upper and lower limit angular dimension for the equally spaced holes.
   b. Give the allowable angular tolerance between the holes.
8. Identify the following features in the front view by the letter of the corresponding feature in the section view:
   a. (E) d. (H)
   b. (F) e. (I)
   c. (G)
9. Indicate (a) the distance holes 1 A and 1 C are from the horizontal (X) axis and (b) 1 B and 1 D are from the vertical (Y) axis.
10. Tell what the symbol △2 in the section view means.
11. Give the minimum and the maximum diameters for (G) and (O).
12. Identify the company's practice with untoleranced dimensions as related to the production of part blanks prior to machining.

1. ____________________
2. ____________________

____

____

	Max. Dimension
3. a. (A)	______
b. (B)	______
c. (C)	______
d. (D)	______

4. ____________________

____

5.	Part Name (a)	Req'd. Quantity (b)	Part Number (c)
△1	______	______	______
△2	______	______	______
6. Size (a)	______	______	______
Angle (b)	______	______	______

7. a. Upper Limit ______ Lower ______

b. Tolerance ____________________

8. a. (E) = ______ d. (H) = ______

b. (F) = ______ e. (I) = ______

c. (G) = ______

9. a. ______ b. ______

10. ____________________

____

11.	Minimum	Maximum
(G)	______	______
(O)	______	______

12. ____________________

____

____

____

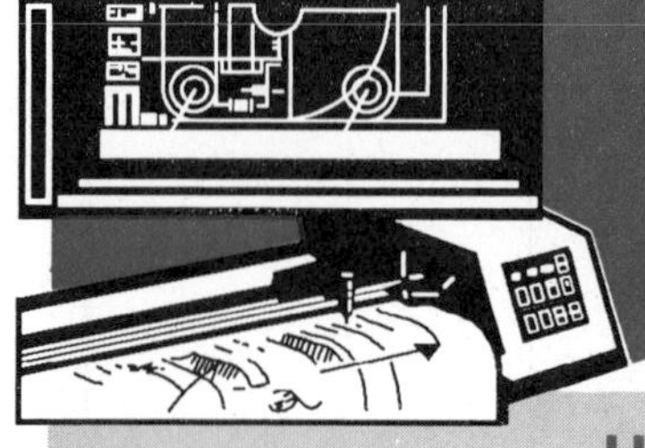

HIGH TECHNOLOGY applications

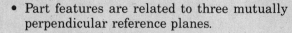

HT-6 CADD: GEOMETRIC TOLERANCING

CADD: GEOMETRIC TOLERANCING SYMBOL LIBRARY

ANSI Y 14.5M, as revised, provides standards for form size, shape, and other characteristics of symbols that are used in geometric tolerancing, related dimensioning, and guidelines for applying symbols on regular and CADD drawings.

Geometric tolerancing and other engineering drawing symbols may be hand drawn or computer generated, using a menu that includes a predrawn symbol library. CADD systems include *graphic tablets* on which plastic symbol library *master* and *subtemplates* are overlayed. The templates are used for digitizing, menu picking, and manipulating images; and for accurately tracing and drawing symbols. Two reduced size symbol templates are illustrated: geometric dimensioning and tolerancing (MMDT), and mechanical fasteners (MMFAS). Graphic tablets with master template and subtemplate overlays and CADD software are available to meet a wide variety of occupational design and drawing requirements.

CADD: GUIDELINES USED IN GEOMETRIC TOLERANCING

CADD geometric tolerancing symbols and dimensioning are applied to drawings using drafting room practices that are based on the following ANSI Y 14.5 standards.

- The major features of an object are used to establish the basic coordinate system.
- Features of an object on a drawing are oriented to subcoordinates of major coordinates.
- Part features are related to three mutually perpendicular reference planes.
- Datums are related according to part function and the sequential order used in CAM.
- Uniform coordinate dimensioning is used for a complete drawing.
- Coordinate or tabular dimensioning on a CADD drawing identify approximate dimensions of an arbitrary profile.
- Regular profile (contour) changes are defined at the point of tangency; an irregular profile, by at least four points.
- Polar coordinate dimensioning is used to define circular hole patterns.
- Angles are dimensioned in degrees and decimal parts of a degree.
- Numerical control (NC) programming dimensions generally split a tolerance and are based on the mean.
- Similarly, CADD dimensions are based at the mean of the tolerance.
- Geometric tolerancing, symbols, and dimensioning are used on drawings to control particular geometric form, location, and other part characteristics.

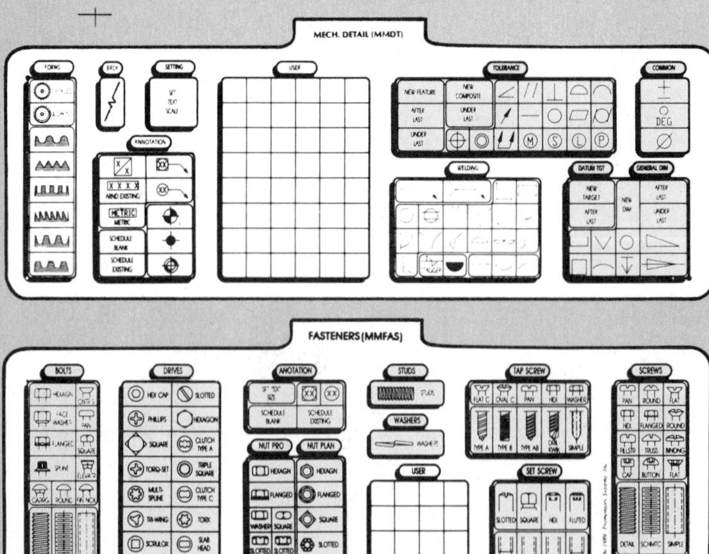

Drawing courtesy of SUMMAGRAPHICS CORPORATION

UNIT 29
Unilateral and Bilateral Tolerances

Since a drawing provides the worker with full specifications of a part, the tolerances on dimensions indicate the degree of accuracy to which each feature is to be machined. Three- and four-place decimal tolerances are generally used with customary units of measure for a precision part to function correctly in relation to a mating part. Two-place decimal tolerances and fractional tolerances are common for general machining accuracy.

UNILATERAL TOLERANCES

A *unilateral tolerance* is a single-direction tolerance. Such a tolerance is specified on a drawing as either above (+) or below (−) a basic size dimension. Three examples of the total tolerance in one direction are given in figure 29-1.

The unilateral tolerance on part (A) indicates a lower limit dimension of **4″** and an upper limit of **$4\frac{1}{64}$″**. The acceptable machining range is between **4″** and **$4\frac{1}{64}$″**. The + **$\frac{1}{64}$″** tolerance indicates the largest machineable dimension. On part (B), the (**−1 mm**) unilateral tolerance requires the part to be machined between **87 mm** and **88 mm**. The minus tolerance (**−15′**) on the angular surface (C) means that the adjacent surfaces are to be machined to a dimensional accuracy of between **45°15′** and **45°30′**.

BILATERAL TOLERANCES

Bilateral limits include two-direction limits. A tolerance with bidirectional limits permits dimensional variation in a range above (+) and below (−) a basic dimension. In other words, the dimension of a particular feature may vary in both (+) and (−) directions.

Bilateral limits may be *equal* or *unequal*. If the plus and minus limits for a basic size (dimension) are the same, the limits are said to be equal. However, there are some parts which require a different plus and minus limit. In such cases, each

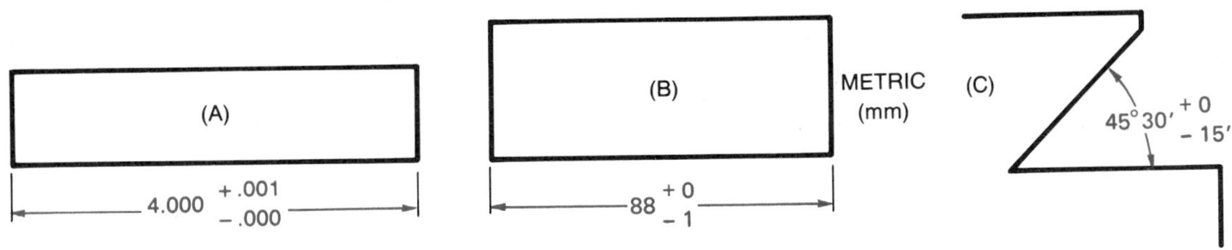

FIGURE 29-1 EXAMPLES OF UPPER AND LOWER LIMITS FOR UNILATERAL TOLERANCES

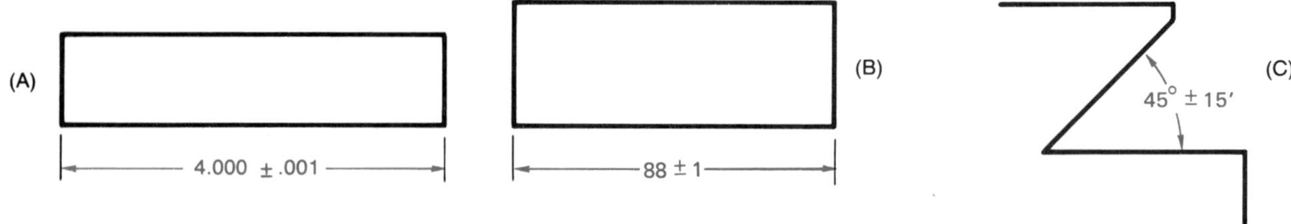

FIGURE 29-2 APPLICATION OF UPPER AND LOWER LIMITS ON BILATERAL TOLERANCES

limit is unequal. The high limit is placed above the low limit with the dimension. When a dimension is given as a note, the low limit appears first.

Limits are given from the smaller (lower limit) to the larger (upper limit) dimension. The bilateral limits ±.001″ and ±1 mm for parts (A) and (B) in figure 29-2 indicate the upper and lower limits for producing each part. The dimensional range for part (A) is **3.999″** to **4.001″**; part (B), from **87 mm** to **89 mm**. The acceptable angle measurement for part (C) ranges from **44°45′** to **45°15′** The respective tolerances are .002″, 2 mm, and 0°30′.

TOLERANCES ON DECIMAL DIMENSIONS

Measurements in decimal units of measure permit quick, accurate dimensional readings using precision instruments. Readings in thousandths ($\frac{1}{1,000}$″; 0.001″) and ten-thousandths ($\frac{1}{10,000}$″; 0.0001″) are common. Decimal tolerances are used with decimal inch and SI metric units of measurement.

There are four common methods of showing decimal tolerances on drawings. These are illustrated in figures 29-3 through 29-6. Figure 29-3 indicates the practice of including decimal tolerances as one of the notes in a title block.

The size range for machining the **2.62″** features is from **2.62″ − 0.01″** (=**2.61″**) to **2.62″ + 0.01″** (=**2.63″**). The acceptable dimensional range for the **1.84″** feature is from **1.84″ − 0.01″** (=**1.83″**) to **1.84″ + 0.01″** (=**1.85″**). The tolerance for each dimension = .02″.

Another drafting room practice is to include the decimal limits with the dimension (figure 29-4). The acceptable dimensional range for the **74.2 mm** dimension is from the lower limit of **74.15 mm** to the upper limit of **74.25 mm**; a tolerance of .10 mm.

Upper and lower limit sizes are often simplified by dimensioning a feature on a drawing as shown in figure 29-5.

Unequal limits are given in terms of the (+) and (−) variations in relation to a basic size. For example, in figure 29-6, the upper limit for the basic size of **4.750″** is **4.750″ + 0.001″**, or **4.751″**. The lower limit equals **4.750″ − 0.002″**, or **4.748″**.

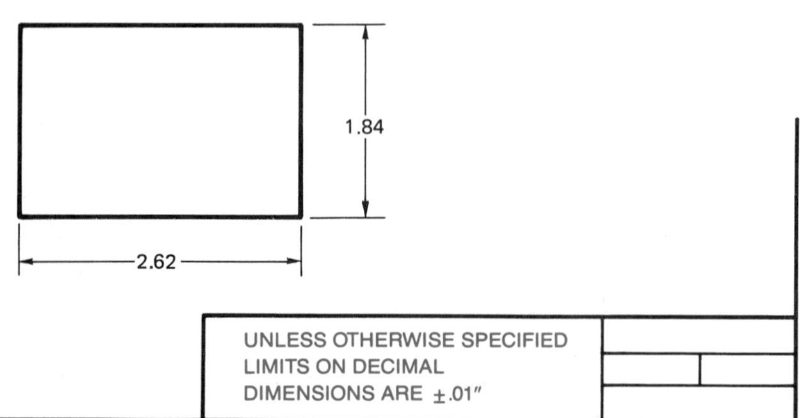

FIGURE 29-3 LIMITS ON DECIMAL DIMENSIONS NOTED IN TITLE BLOCK

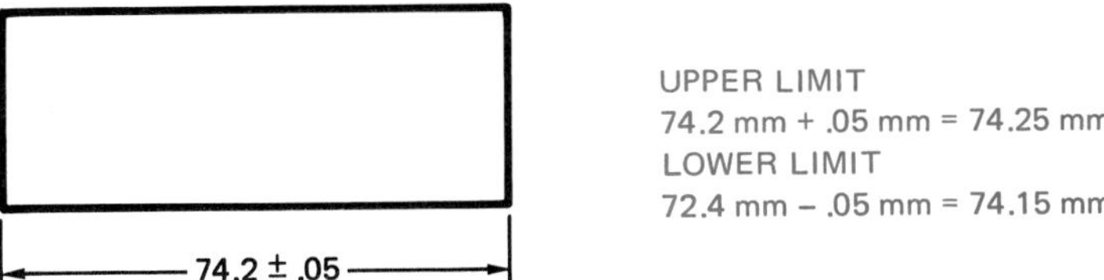

FIGURE 29-4 DECIMAL LIMITS AS PART OF A DIMENSION

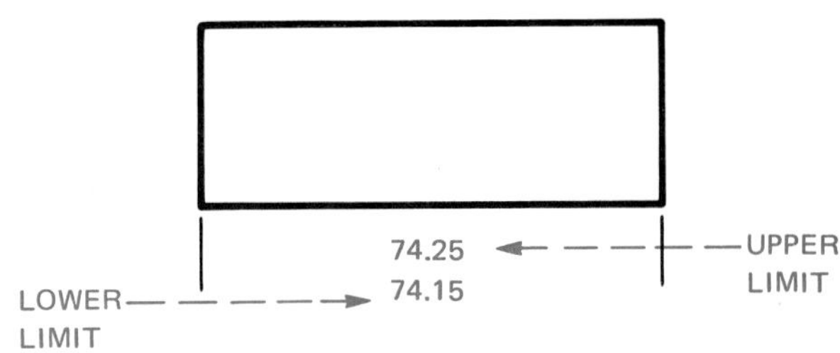

FIGURE 29-5 DIMENSION INCLUDES UPPER AND LOWER LIMITS

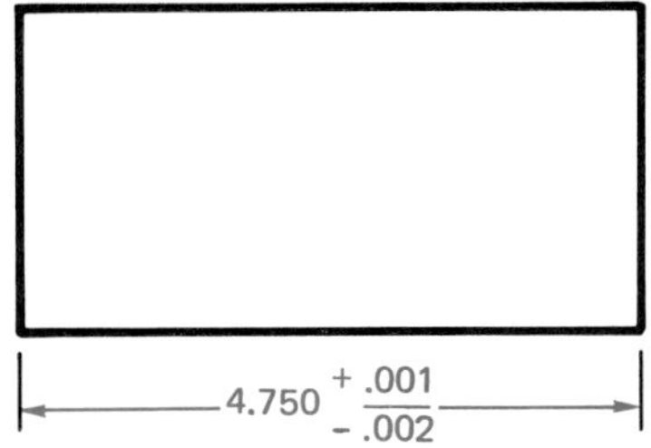

FIGURE 29-6 DIMENSIONING FOR UNEQUAL LIMITS

SPECIFYING GENERAL TOLERANCES IN A TITLE BLOCK

Different features often require machining or fabricating a part to different degrees of dimensional accuracy. Fractions, two-place and/or three-place decimals for customary units, one- or two-place decimals for SI metric dimensions, and angle tolerances sometimes appear on a blueprint in a title block as *general tolerances.* Specific tolerances for a particular dimension are usually given as part of the dimension. Figure 29-7 provides an example of general limits incorporated in a title block.

<table>
<tr><td colspan="5">UNLESS OTHERWISE SPECIFIED:
ALL UNTOLERANCED DIMENSIONS ARE ± .02″</td></tr>
<tr><td rowspan="2">FRACTIONAL LIMITS
$\pm\frac{1}{64}''$</td><td colspan="3">DECIMAL LIMITS</td><td rowspan="2">ANGLE LIMITS
± 45′</td></tr>
<tr><td>ONE PLACE
± .1″</td><td>TWO PLACES
± .01″</td><td>THREE PLACES
± .001″</td></tr>
</table>

FIGURE 29-7 EXAMPLE OF GENERAL LIMITS INCLUDED IN A TITLE BLOCK

NOTE: REFER TO APPENDIX B-3 FOR THE LARGER SCALE DRAWING TO USE WITH THIS ASSIGNMENT.

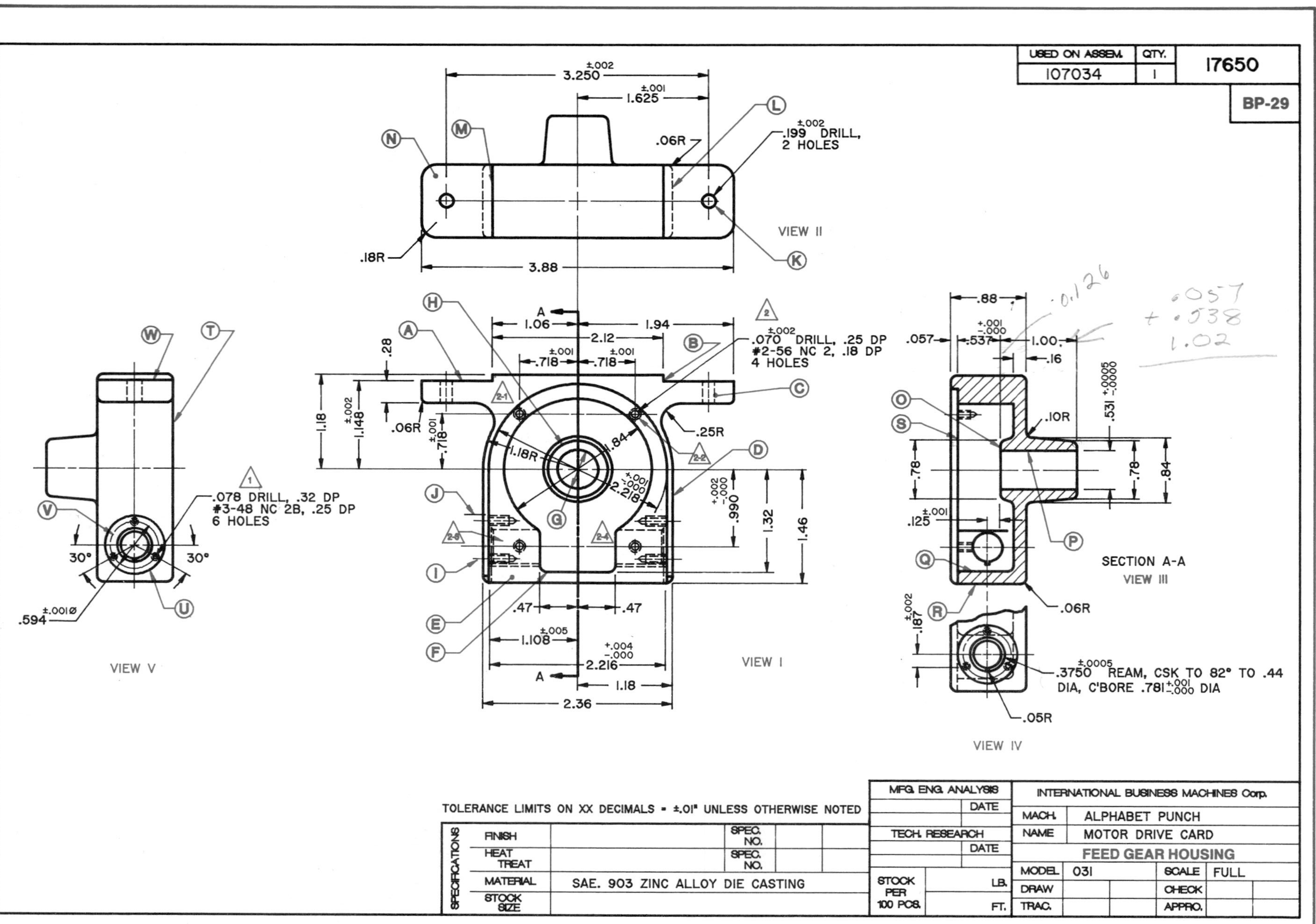

ASSIGNMENT—UNIT 29

FEED GEAR HOUSING (BP-29)

Student's Name ______________

1. Name **VIEWS I** through **V**.
2. State (a) the material and (b) the process for manufacturing the Housing.
3. Identify the letters in **VIEW II** which correspond with the following features in **VIEW I**.
   a. Ⓐ c. Ⓒ
   b. Ⓑ d. Ⓓ
4. Determine the letter of the feature in **VIEW III** which corresponds with the following features in **VIEW I**.
   a. Ⓔ c. Ⓖ
   b. Ⓕ d. Ⓗ
5. Give the letter of the feature in **VIEW V** which corresponds with the following features in **VIEW I**.
   a. Ⓘ b. Ⓙ c. Ⓐ
6. Give the maximum overall dimension for (a) the width, (b) height, and (c) depth of the Housing.
7. Indicate the kinds of tolerances which apply to (a) fractional and (b) decimal dimensions.
8. Give the specifications for the △1 tapped holes.
9. Indicate the upper and lower limit dimensions for the drill diameter of the #2–56 tapped holes (△2).
10. Determine the minimum and maximum acceptable measurements from the centers of holes △2-1 and △2-2 and △2-3 and △2-4 from the horizontal axis.
11. Provide the same information for holes △2-1 and △2-3, and △2-2 and △2-4 from the vertical (Y) axis.
12. Determine the upper and lower limit angle dimensions between the center lines of each pair of #3–48 tapped holes. Use a tolerance of ±15″ for angular dimensions.
13. Give the maximum and minimum acceptable measurements for the following features:
    a. The drilled holes in **VIEW II**.
    b. The diameter of the bored hole in the hub.
    c. The through reamed hole.
    d. The counterbored holes.
    e. Surface Ⓐ to the center line of the hub bored hole.

1. **VIEW I** ______________
   VIEW II ______________
   VIEW III ______________
   VIEW IV ______________
   VIEW V ______________
2. a. ______________
   b. ______________
3. a. Ⓐ = ________ c. Ⓒ = ________
   b. Ⓑ = ________ d. Ⓓ = ________
4. a. Ⓔ = ________ c. Ⓖ = ________
   b. Ⓕ = ________ d. Ⓗ = ________
5. a. Ⓘ = ________ c. Ⓐ = ________
   b. Ⓙ = ________
6. a. Width ______________
   b. Height ______________
   c. Depth ______________
7. a. Fractional ______________
   b. Decimal ______________
8. ______________
   ____
9. Upper Limit ________ Lower Limit ________

10. From Axis X	Min.	Max.
△2-1, △2-2	________	________
△2-3, △2-4	________	________

11. From Axis Y	Min.	Max.
△2-1, △2-3	________	________
△2-2, △2-4	________	________

12. Upper Limit ________ Lower ________

13.	Maximum	Minimum
a.	________	________
b.	________	________
c.	________	________
d.	________	________
e.	________	________

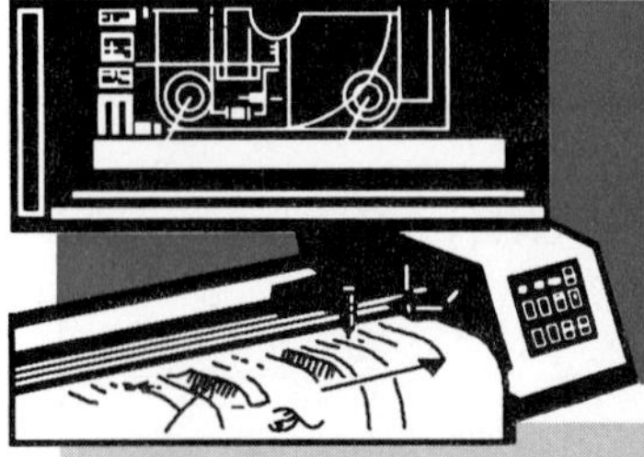

HIGH TECHNOLOGY applications

HT-7 NC/CNC COMPUTER-ASSISTED APT PROGRAMMING

Machine tool manufacturers use several different program-oriented languages to generate computer-assisted programs (CAP) for the manufacture, inspection, and assembly of parts and components. The selection of the computer language depends on the complexity of mathematical computations, tooling setups, and machining processes.

Symbols, modified English word spelling, and simplified language are used to reduce the entries in computer program statements. One of the most powerful and widely used program-oriented computer processor languages is known as APT (*Automatically Programmed Tools*).

APT AND CAP

The application of APT to computer-assisted programming may be illustrated by using the part drawing and part programming layout of a workpiece that is to be manufactured on a numerical control lathe. Incremental coordinates and dual dimensioning are used.

The part programming layout drawing uses **Z** for the longitudinal (spindle) axis and **X** for transverse (cross slide) movements. Note the + and − coordinates for the tool movements and feeds. The letters indicate operations, sequences, and directions of cut.

Once the programming layout drawing is completed, the part programmer prepares a series of APT language statements.

The first program manuscript statements are postprocessor and auxiliary statements—for example:

```
PART NØ PIN NØ6          INTØL/.ØØ1
MACHIN HARDINGE 6        ØUTØL/.ØØ1
```

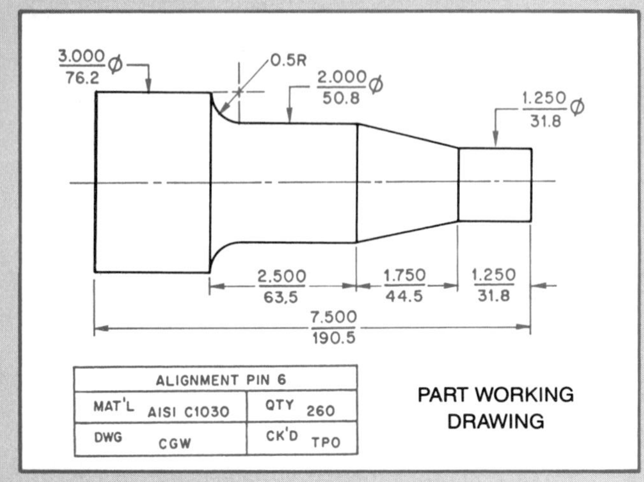

PART WORKING DRAWING

```
CUTTER/.Ø94         FEDRAT/4,IPM
CLPRNT              SPINDL/4ØØ,RPM
```

The second set of statements in APT programming relate to geometry—for example:

```
SP=PØINT/7.75, -1.65  C1=CIRCLE/2.5, -1.5, .5
PI=PØINT/7.75, -.613  L4=LINE/P4, P5
```

The geometry statements are followed by motion statements—for example:

```
FRØM/SP             GØFWD/C1, PAST, L4
GØ TØ, L1           GØ TØ/SP
```

The program is concluded by two closing statements:

```
CØØLNT/ØFF          FINI
```

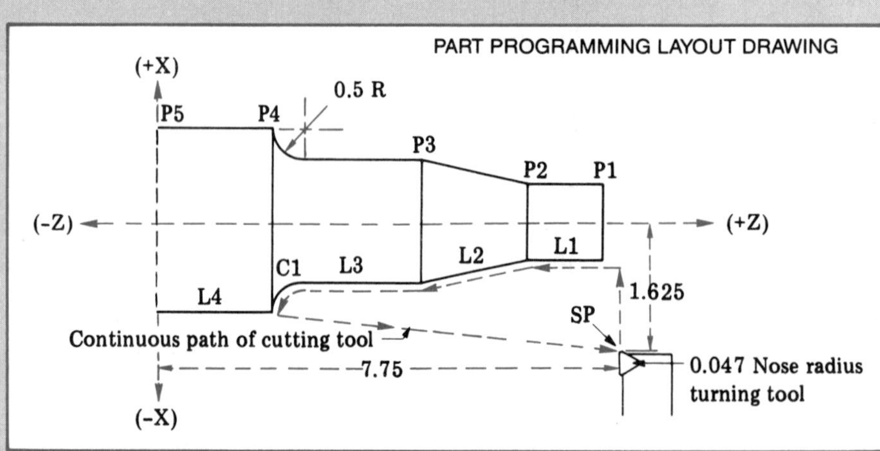

PART PROGRAMMING LAYOUT DRAWING

Adapted from ADVANCED MACHINE TOOL TECHNOLOGY AND MANUFACTURING PROCESSES *Courtesy of C. THOMAS OLIVO ASSOCIATES*

UNIT 30
Geometric Tolerancing: Form, Location, Runout

Geometric tolerancing is used on many industrial drawings to specify tolerances which apply to shape, form, or the position of particular features of a part. Special tolerance symbols such as those shown in figure 30-1A relate to geometric characteristics.

GEOMETRIC CONTROL SYMBOLS

ANSI standard geometric characteristic symbols are grouped in figure 30-1A according to:

- Form tolerances (including surface relationships),

TYPE OF TOLERANCE		GEOMETRIC CHARACTERISTICS	SYMBOL
FORM TOLERANCES		STRAIGHTNESS	—
		FLATNESS	⏥
		ROUNDNESS (CIRCULARITY)	○
		CYLINDRICITY	⌭
		PROFILE OF A LINE	⌒
		PROFILE OF A SURFACE	⌓
	RELATIONSHIP	ANGULARITY	∠
		PERPENDICULARITY (SQUARENESS)	⊥
		PARALLELISM	//
LOCATION TOLERANCES		POSITION	⌖
		CONCENTRICITY	◎
		SYMMETRY	⌯
RUNOUT TOLERANCES		CIRCULAR RUNOUT	↗
		TOTAL RUNOUT	⌰

FIGURE 30-1A ANSI GEOMETRIC CHARACTERISTIC SYMBOLS FOR ENGINEERING DRAWINGS

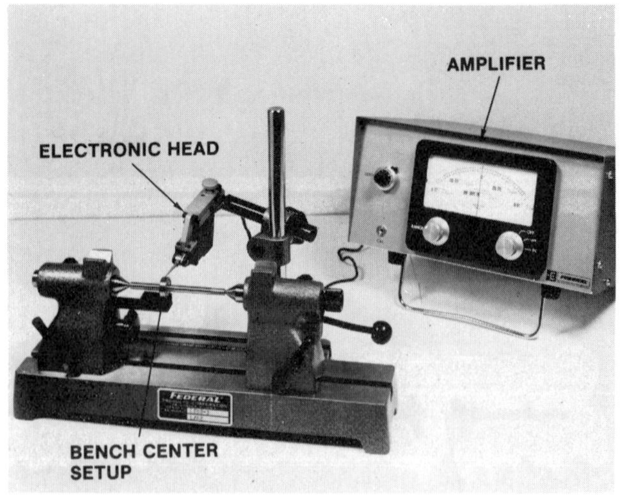

FIGURE 30-1B INSPECTION AND MEASURING SETUP FOR FORM, LOCATION, AND RUNOUT TOLERANCES

- Location tolerances, and
- Runout tolerances.

A typical measuring and inspection setup for form, location, and runout tolerance is illustrated in figure 30-1B.

FEATURE CONTROL SYMBOLS

Tolerance information appears on blueprints as a *feature control symbol* which is enclosed in a rectangular block. The symbol is a simplified form for providing tolerance information that might otherwise be given in a longer note. The makeup of a feature control symbol is illustrated in figure 30-2.

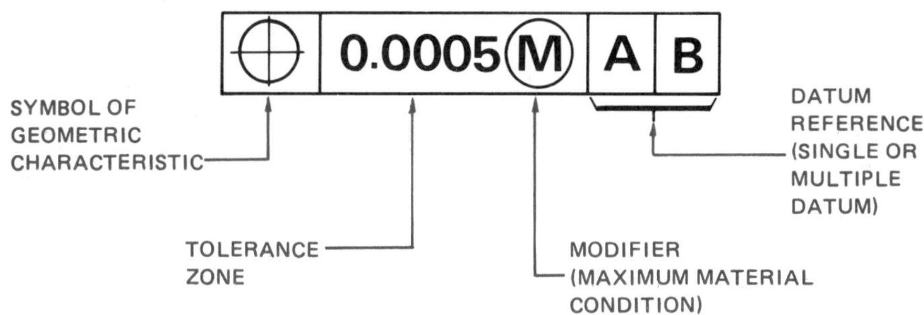

FIGURE 30-2 EXAMPLE OF SPECIFICATIONS WITHIN A FEATURE CONTROL SYMBOL

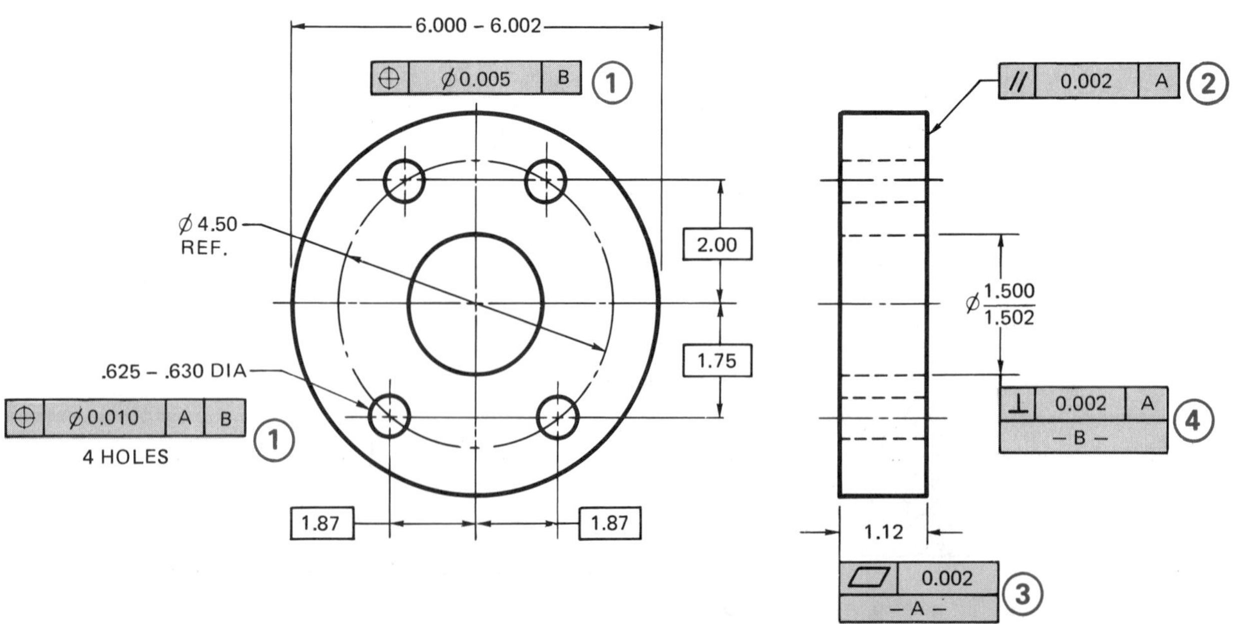

FIGURE 30-3 PLACEMENT OF FEATURE CONTROL SYMBOLS

PLACEMENT OF FEATURE CONTROL SYMBOLS

Feature control symbols are generally placed on engineering drawings as shown in figure 30-3 at (1), (2), (3), and (4).

- The symbol is added to a note or a dimension. (1)
- A leader is run from the symbol block to the feature. (2)
- A corner, side, or end of the symbol block is added to an extension line of the feature. (3)
- A symbol block is attached to the dimension line of the feature. (4)

DATUM REFERENCE SYMBOL

A datum reference symbol is used with form, location, and runout tolerances. A *tolerance zone* is given in the symbol as either a linear or a diametral measurement. A *modifier* may be included in the feature control symbol. The modifier may indicate a *maximum material condition* ((M)) or a tolerance application *regardless of feature size* ((S)).

DATUM SPECIFICATION

Datum identification and *datum targets* are illustrated at (A) and (B) in figure 30-4. A *reference letter* (in this instance -A-) is preceded and followed by a dash and is contained in a block or

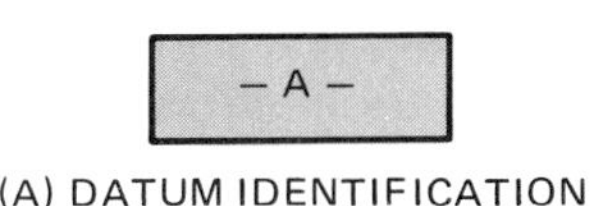

(A) DATUM IDENTIFICATION

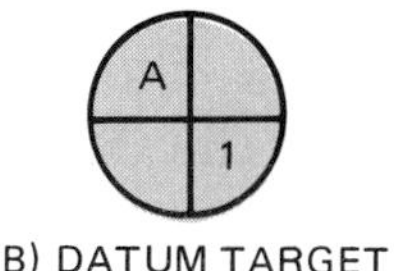

(B) DATUM TARGET

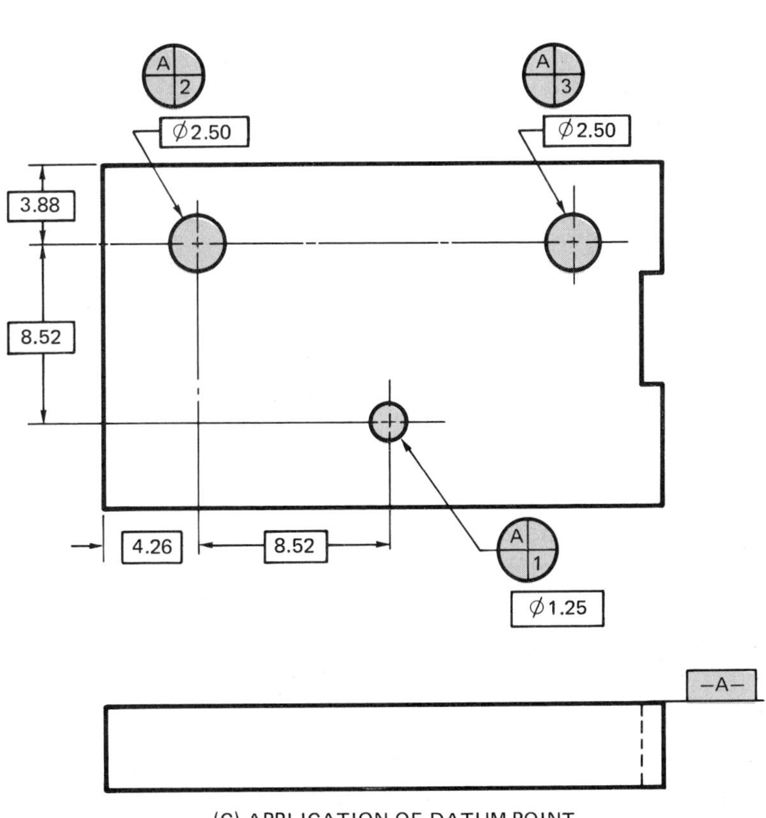

(C) APPLICATION OF DATUM POINT AND AREA TARGETS

FIGURE 30-4 DATUM IDENTIFICATION, DATUM TARGET SYMBOLS, AND APPLICATION OF FEATURE CONTROLS

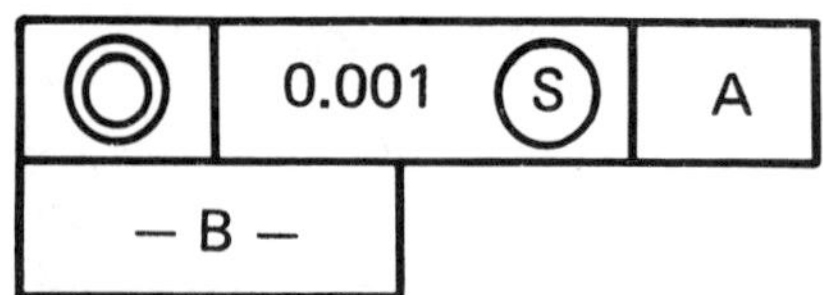

FIGURE 30-5 DATUM CONTROLLED BY ANOTHER FEATURE CONTROL SYMBOL

frame. This datum identifies a theoretically exact point, line, or surface from which other features are specified.

The datum targets (A/1), (A/2), and (A/3) in the illustration at (C) provide specifications for the three holes.

When a datum is controlled by another feature control symbol, this information appears as shown in figure 30-5. Information in this form, which is connected with a feature of a part, is called a *drawing callout*. The drawing callout in figure 30-5 is interpreted to mean that regardless of the feature size ((S)), the concentricity ((◎)) of feature -B- is governed by the 0.001″ tolerance of feature -A-.

LOCATION TOLERANCING

Location tolerances relate to position (⌖), concentricity (◎), and symmetry (⌯). Location tolerances control the following relationships:

- Center-to-center distances of holes, slots, grooves, sections, etc.
- The location of part features from such datums as planes, lines, and cylindrical surfaces.
- The specifications of features with a common axis.

Common drawing callouts using a feature control symbol for *positional tolerancing* are provided in table 30-1. An interpretation of each general application is given in the right column.

FORM TOLERANCING

Form tolerances are applied to the geometric shape or the condition of a feature. Geometric characteristics such as straightness (——), flatness (▱), roundness (○) (circularity), profile (⌓), etc., and corresponding symbols used within a feature control symbol block appear in table 30-2. Again, interpretations are given in the right column for each drawing callout example used for form tolerancing.

TOLERANCES OF RELATIONSHIP

Tolerances of relationship deal with angularity (∠), perpendicularity (⊥), and parallelism (//). The symbol of each geometric characteristic, followed by a given tolerance, indicates the condition and quantity of tolerance within an acceptable limit. In other words, the accuracy to which holes, diameters, slots, cut-away sections, or plane surfaces must be produced is specified by a tolerance for the characteristic being measured.

Angularity (∠), perpendicularity (⊥), and parallelism (//) tolerances relate to the orientation of a surface element or axis to a datum plane or axis.

Angularity tolerance (∠) refers to a tolerance zone established around a basic angle. A drawing callout specifies a required angle tolerance for a surface element or an axis. In figure 30-6, the angular surface must be machined within a 0.005″ tolerance zone at the basic angle of 45°.

Perpendicularity tolerance (⊥) refers to a tolerance zone for a surface element or axis which is at 90° to a particular datum plane or axis.

Parallelism tolerance (//) refers to a tolerance zone wherein a surface element or axis is equidistant at all points from a datum plane or axis. The tolerance zone specifies the allowable variation between two parallel lines, planes, or cylindrical surfaces. A drawing callout like | // | .005 | A | indicates that the machined surface must be parallel and lie within a tolerance zone of **0.005″** in relation to surface (plane) **A**.

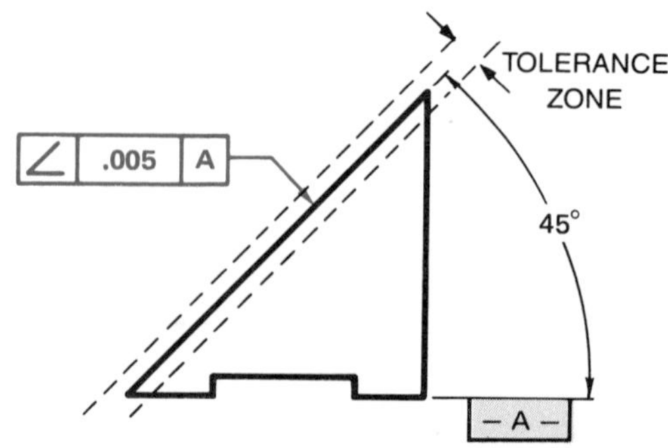

FIGURE 30-6 APPLICATION OF ANGULARITY TOLERANCE

RUNOUT TOLERANCES

Circular runout, (↗) when applied to a cylindrical feature and its datum axis, relates to roundness and concentricity. In the case of surfaces at a right angle to a datum axis, circular runout controls the surface elements of *perpendicularity.*

Total runout tolerance is illustrated by the drawing callout in figure 30-9. The callout specifies the acceptable roundness and concentricity runout tolerance of the cylindrical feature. The second feature control symbol indicates that the end must be machined flat and perpendicular to the axis within 0.001″. The total runout clearance relates to the control of all surfaces.

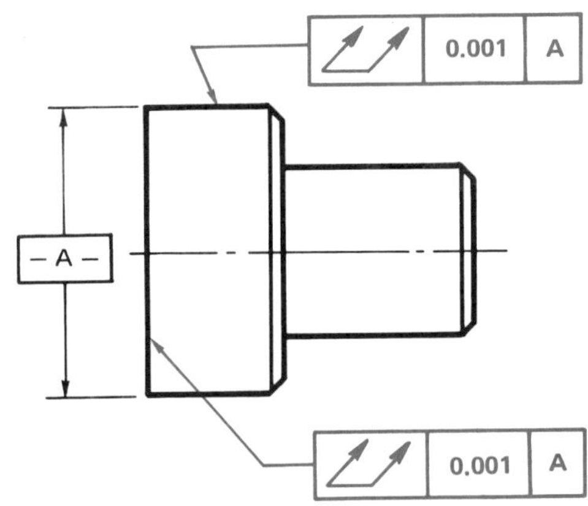

FIGURE 30-7 DRAWING CALLOUT OF TOTAL RUNOUT TOLERANCE

APPLICATION OF GEOMETRIC TOLERANCING CONDITIONS TO CADD/CAM

General geometric tolerancing and dimensioning rules require some modification in their application in CADD/CAM systems. Selected dimensioning and tolerancing guidelines, as established by ANSI Y14 M standards, follow as examples.

- The major features of the part are used to establish the basic coordinate dimensioning system. However, the major features are not necessarily defined as datum.
- Subcoordinated systems related to the major coordinates are used to locate and orient the features of the part.
- Three mutually perpendicular reference planes (to which the part features are related) are used to define the part features.
- The function of the part establishes the relationship of the datums. Datum features are related in the order of precedence in which they are used for CAM (computer aided manufacturing).
- Complete and accurate dimensioning of geometric shapes is required. Regular geometric shapes may be defined by mathematical formulas. However, coordinate dimensions are not used to define a profile feature unless required for inspection or reference.
- Coordinate and tabular dimensions are used to identify approximate dimensions on an arbitrary profile.
- The same coordinate dimensioning system is used on the entire drawing.
- A continuity of profile is required in CADD. Changes in contour are clearly defined at the change point or point of tangency. At least four points are defined along an irregular profile.
- Circular hole patterns may be defined by using polar coordinates.
- Angles are dimensioned in degrees and decimal parts of a degree, when possible. Example: **60°45′** is dimensioned **60.75°**.
- Geometric tolerancing controls specific form and location for all part features.
- Dimensions are based on the *mean* of a tolerance, permitting the numerical control programmer to split a tolerance. Dimensions without limits should conform to the capabilities of the NC machine tool and part function. In such cases, bilateral profile tolerances may be recommended.

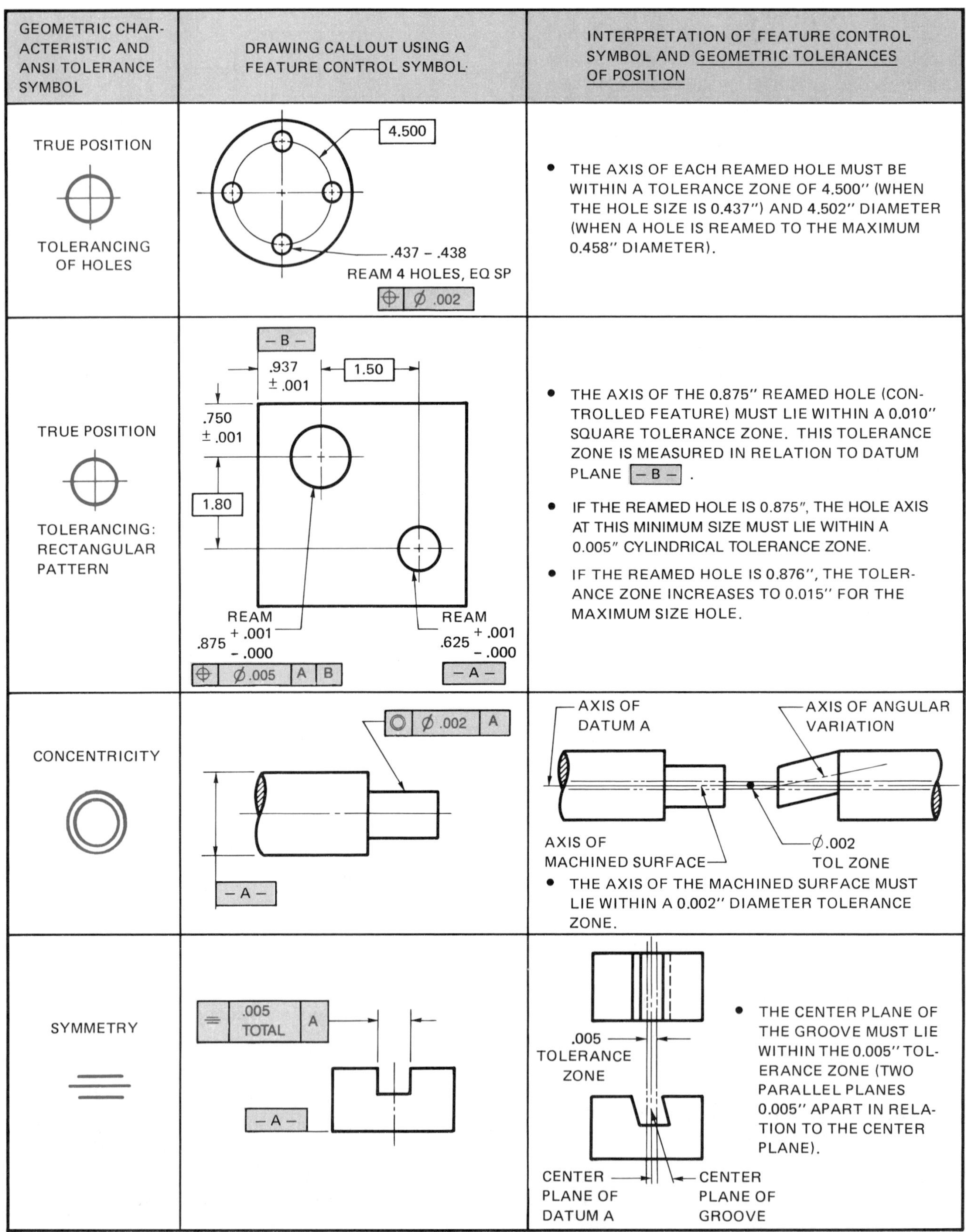

GEOMETRIC CHARACTERISTIC AND ANSI TOLERANCE SYMBOL	DRAWING CALLOUT USING A FEATURE CONTROL SYMBOL	INTERPRETATION OF FEATURE CONTROL SYMBOL AND GEOMETRIC TOLERANCES OF POSITION
TRUE POSITION ⌖ TOLERANCING OF HOLES	4.500 .437 – .438 REAM 4 HOLES, EQ SP ⌖ ∅ .002	• THE AXIS OF EACH REAMED HOLE MUST BE WITHIN A TOLERANCE ZONE OF 4.500" (WHEN THE HOLE SIZE IS 0.437") AND 4.502" DIAMETER (WHEN A HOLE IS REAMED TO THE MAXIMUM 0.458" DIAMETER).
TRUE POSITION ⌖ TOLERANCING: RECTANGULAR PATTERN	– B – .937 ± .001 1.50 .750 ± .001 1.80 REAM .875 + .001 – .000 ⌖ ∅ .005 A B REAM .625 + .001 – .000 – A –	• THE AXIS OF THE 0.875" REAMED HOLE (CONTROLLED FEATURE) MUST LIE WITHIN A 0.010" SQUARE TOLERANCE ZONE. THIS TOLERANCE ZONE IS MEASURED IN RELATION TO DATUM PLANE – B – . • IF THE REAMED HOLE IS 0.875", THE HOLE AXIS AT THIS MINIMUM SIZE MUST LIE WITHIN A 0.005" CYLINDRICAL TOLERANCE ZONE. • IF THE REAMED HOLE IS 0.876", THE TOLERANCE ZONE INCREASES TO 0.015" FOR THE MAXIMUM SIZE HOLE.
CONCENTRICITY ◎	◎ ∅ .002 A – A –	AXIS OF DATUM A AXIS OF ANGULAR VARIATION AXIS OF MACHINED SURFACE ∅.002 TOL ZONE • THE AXIS OF THE MACHINED SURFACE MUST LIE WITHIN A 0.002" DIAMETER TOLERANCE ZONE.
SYMMETRY ≡	≡ .005 TOTAL A – A –	.005 TOLERANCE ZONE CENTER PLANE OF DATUM A CENTER PLANE OF GROOVE • THE CENTER PLANE OF THE GROOVE MUST LIE WITHIN THE 0.005" TOLERANCE ZONE (TWO PARALLEL PLANES 0.005" APART IN RELATION TO THE CENTER PLANE).

TABLE 30-1 SAMPLE APPLICATIONS WITH INTERPRETATIONS OF ANSI FEATURE CONTROL SYMBOLS FOR POSITIONAL TOLERANCING

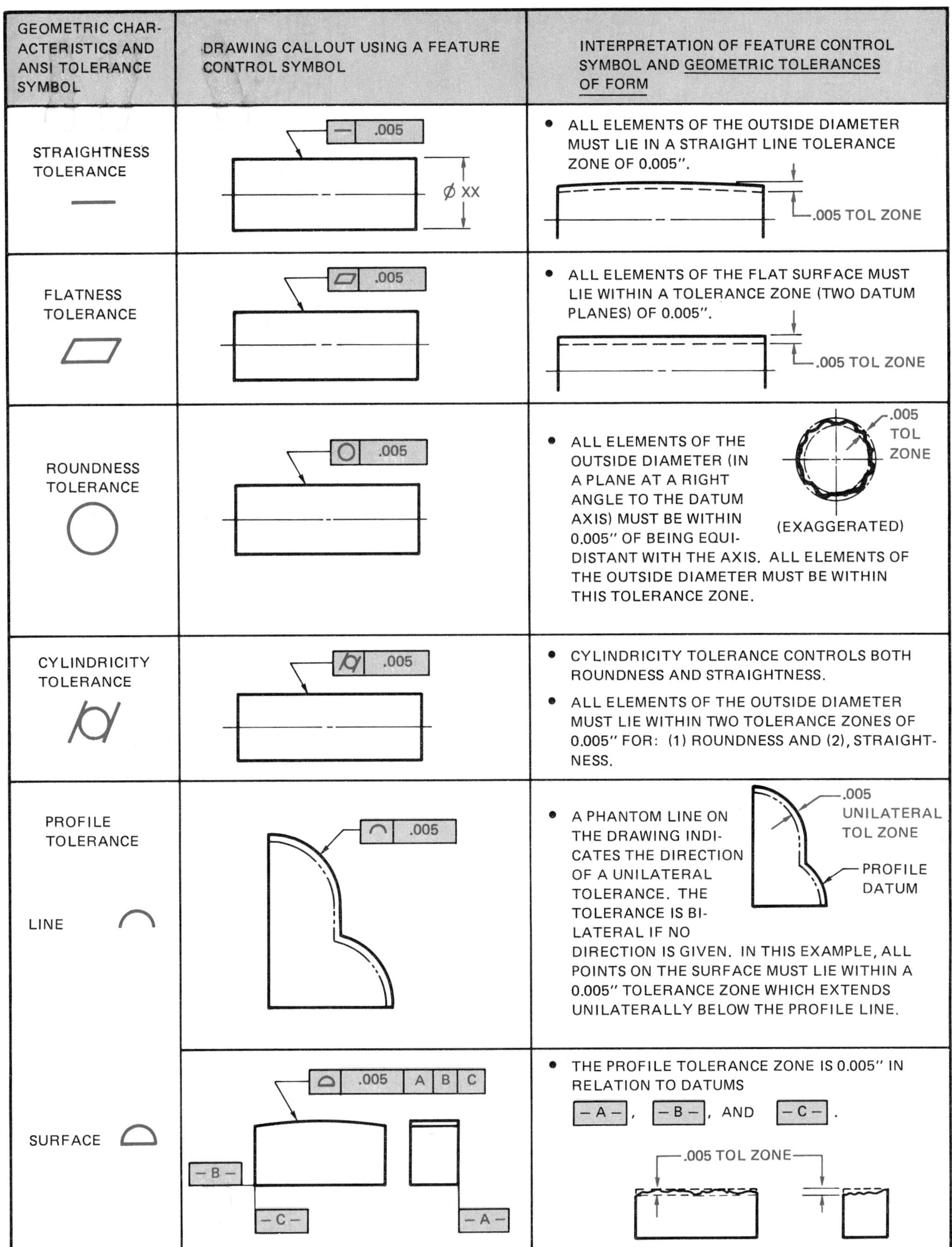

GEOMETRIC CHARACTERISTICS AND ANSI TOLERANCE SYMBOL	DRAWING CALLOUT USING A FEATURE CONTROL SYMBOL	INTERPRETATION OF FEATURE CONTROL SYMBOL AND GEOMETRIC TOLERANCES OF FORM
STRAIGHTNESS TOLERANCE —	— .005; Ø XX	• ALL ELEMENTS OF THE OUTSIDE DIAMETER MUST LIE IN A STRAIGHT LINE TOLERANCE ZONE OF 0.005". .005 TOL ZONE
FLATNESS TOLERANCE ▱	▱ .005	• ALL ELEMENTS OF THE FLAT SURFACE MUST LIE WITHIN A TOLERANCE ZONE (TWO DATUM PLANES) OF 0.005". .005 TOL ZONE
ROUNDNESS TOLERANCE ○	○ .005	• ALL ELEMENTS OF THE OUTSIDE DIAMETER (IN A PLANE AT A RIGHT ANGLE TO THE DATUM AXIS) MUST BE WITHIN 0.005" OF BEING EQUIDISTANT WITH THE AXIS. ALL ELEMENTS OF THE OUTSIDE DIAMETER MUST BE WITHIN THIS TOLERANCE ZONE. .005 TOL ZONE (EXAGGERATED)
CYLINDRICITY TOLERANCE ⌭	⌭ .005	• CYLINDRICITY TOLERANCE CONTROLS BOTH ROUNDNESS AND STRAIGHTNESS. • ALL ELEMENTS OF THE OUTSIDE DIAMETER MUST LIE WITHIN TWO TOLERANCE ZONES OF 0.005" FOR: (1) ROUNDNESS AND (2), STRAIGHTNESS.
PROFILE TOLERANCE LINE ⌒	⌒ .005	• A PHANTOM LINE ON THE DRAWING INDICATES THE DIRECTION OF A UNILATERAL TOLERANCE. THE TOLERANCE IS BILATERAL IF NO DIRECTION IS GIVEN. IN THIS EXAMPLE, ALL POINTS ON THE SURFACE MUST LIE WITHIN A 0.005" TOLERANCE ZONE WHICH EXTENDS UNILATERALLY BELOW THE PROFILE LINE. .005 UNILATERAL TOL ZONE; PROFILE DATUM
SURFACE ⌓	⌓ .005 A B C; –B–; –C–; –A–	• THE PROFILE TOLERANCE ZONE IS 0.005" IN RELATION TO DATUMS –A–, –B–, AND –C–. .005 TOL ZONE

TABLE 30-2 SAMPLE APPLICATIONS WITH INTERPRETATIONS OF ANSI FEATURE CONTROL SYMBOLS FOR FORM TOLERANCING

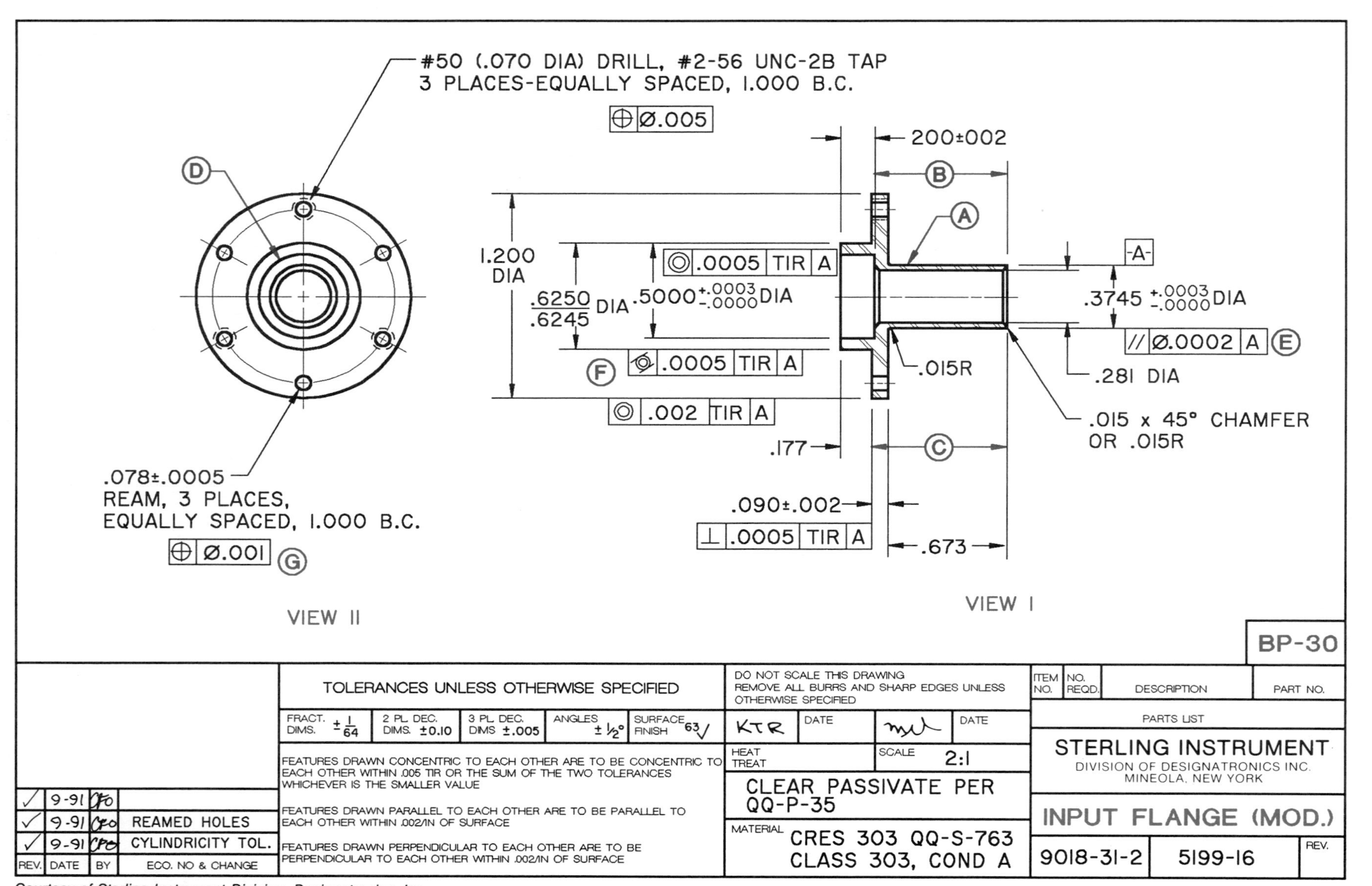
#50 (.070 DIA) DRILL, #2-56 UNC-2B TAP
3 PLACES-EQUALLY SPACED, 1.000 B.C.
⊕ Ø.005
200±002
B
A
D
1.200 DIA
.6250/.6245 DIA
.5000 +.0003/-.0000 DIA
◎ .0005 TIR A
F
⌭ .0005 TIR A
◎ .002 TIR A
.015R
-A-
.3745 +.0003/-.0000 DIA
// Ø.0002 A E
.281 DIA
.015 x 45° CHAMFER
OR .015R
.177
C
.090±.002
⊥ .0005 TIR A
.673
.078±.0005
REAM, 3 PLACES,
EQUALLY SPACED, 1.000 B.C.
⊕ Ø.001 G
VIEW II
VIEW I
BP-30
TOLERANCES UNLESS OTHERWISE SPECIFIED
FRACT. DIMS. ± 1/64
2 PL. DEC. DIMS. ±0.10
3 PL. DEC. DIMS ±.005
ANGLES ± 1/2°
SURFACE FINISH 63
FEATURES DRAWN CONCENTRIC TO EACH OTHER ARE TO BE CONCENTRIC TO EACH OTHER WITHIN .005 TIR OR THE SUM OF THE TWO TOLERANCES WHICHEVER IS THE SMALLER VALUE
FEATURES DRAWN PARALLEL TO EACH OTHER ARE TO BE PARALLEL TO EACH OTHER WITHIN .002/IN OF SURFACE
FEATURES DRAWN PERPENDICULAR TO EACH OTHER ARE TO BE PERPENDICULAR TO EACH OTHER WITHIN .002/IN OF SURFACE
DO NOT SCALE THIS DRAWING
REMOVE ALL BURRS AND SHARP EDGES UNLESS OTHERWISE SPECIFIED
KTR
DATE
DATE
HEAT TREAT
SCALE 2:1
CLEAR PASSIVATE PER QQ-P-35
MATERIAL CRES 303 QQ-S-763 CLASS 303, COND A
ITEM NO.
NO. REQD.
DESCRIPTION
PART NO.
PARTS LIST
STERLING INSTRUMENT
DIVISION OF DESIGNATRONICS INC.
MINEOLA, NEW YORK
INPUT FLANGE (MOD.)
9018-31-2
5199-16
REV.
9-91
9-91 REAMED HOLES
9-91 CYLINDRICITY TOL.
REV. DATE BY ECO. NO & CHANGE

Courtesy of Sterling Instrument Division, Designatronics, Inc.
(Modified Industry Blueprint)

ASSIGNMENT—UNIT 30

INPUT FLANGE (MOD.) (BP-30)

Student's Name ______________________

1. Specify the material for the Flange.
2. Indicate the required finish.
3. a. Name the two views.
   b. Identify by letter a missing feature from **VIEW II**.
   c. State why **VIEW I** is used instead of a conventional view.
4. Give the following unspecified tolerances:
   a. Decimals: XX places, XXX places
   b. Angles
   c. Surface finish
5. State what the unspecified tolerance is for each of the following:
   a. Parallel features
   b. Perpendicular features
   c. Concentric features (total indicator runout)
6. Indicate the tolerancing systems which are used for general dimensioning on the drawing.
7. Determine the nominal dimensions for the following features:
   a. Outside diameter of the Flange
   b. Machined outside diameter of the hub
   c. Overall length (width) of the Flange
8. Write the specifications for the following features:
   a. The tapped holes
   b. The reamed holes
9. Compute nominal dimensions Ⓑ, Ⓒ, and Ⓓ.
10. Interpret the symbol [-A-].
11. a. Identify five different geometric characteristic symbols used on the drawing.
    b. State (1) the characteristic to be measured and (2) what type of tolerance is represented by each symbol.
12. a. Interpret the specifications within feature control symbol Ⓔ.
    b. Give the nominal, maximum, and minimum diameters.
13. a. Interpret the specifications within feature control symbol Ⓕ.
    b. Give the minimum and maximum acceptable dimensions for this machined hole.
14. Interpret feature control symbol Ⓖ.

1. ______________________
2. ______________________
3. a. ______________________
   b. ______________________
   c. ______________________
   ____
4. a. Decimal: XX ________ XXX ________
   b. Angles ______________________
   c. Surface Finish ______________________
5. a. [//] Features // to ______________________
   b. [⊥] inch ⊥ to ______________________
   c. [◎] inch ◎ to ______________________
   ____

6. ______________________
7. a. Outside Diameter ______________________
   b. Hub Outside Diameter ______________________
   c. Length ______________________
8. a. ______________________
   ____
   b. ______________________
   ____
9. Ⓑ = ________
   Ⓒ = ________
   Ⓓ = ________
10. ______________________
    ____
    ____

11.

Symbol a.	Characteristic b. (1)	Type of Tolerance b. (2)

12. a. ______________________
    ____
    b. Nominal ________ Max. ________
    Min. ________
13. a. ______________________
    ____
    ____
    b. ______________________
    ____
14. ______________________
    ____

SECTION 9
Machine Elements (Actuators)

UNIT 31
Spur Gearing: Rack and Pinion Gears

Gears and cams are classified as *machine actuators*. Any movement of a gear tooth or a cam is transmitted from the driver to actuate a mating part.

The reading and interpreting of a shop or laboratory drawing of a gear or cam requires a knowledge of design terms. In turn, dimensional measurements for setups, indexing, machining, and measuring gear and cam features involve computations and the use of handbooks and other technical reference data.

This section provides technical information which the craftsperson needs to know in order to accurately interpret drawings of the following actuators:

- Spur gears, including rack and pinion gears.
- Helical, worm, and worm gears.
- Bevel and miter bevel gears.

The last unit in the section deals with descriptions, principles, and applications of three common cam motions. *Constant velocity* (uniform), *parabolic*, and *harmonic motion* cams are described and applied.

METHODS OF PRODUCING GEARS

The demands for positive and uniform driving mechanisms, which started during the Industrial Revolution, accelerated the development of gear-cutting machines. The form milling of individual gear teeth gave way to gear hobbing, gear shaping, broaching, and shaving processes for producing gears. In addition, gears are cast, rolled on hot or cold forming machines, extruded, stamped, or formed by powder metallurgy. Precision gear tooth forms and high quality surface textures are produced by shearing and burnishing. Hardened gear teeth are generally finished by grinding and lapping.

SPUR GEARING TERMINOLOGY AND MEASUREMENTS

The design features and measurement terms which provide the craftsperson and technician with all the information needed for machine setup, cutting, and measuring gear teeth and gears are illustrated in figure 31-1. The designations on the drawing and the definitions which follow conform to ANSI standards.

Pitch Circle. An imaginary circle representing the point at which the teeth of two mating gears are tangent. The diameter of the pitch circle indicates the gear size. Most gear dimensions are computed with respect to the pitch circle.

Pitch Diameter. The diameter of the pitch circle.

Addendum. The vertical distance between the pitch circle and the outside diameter *(addendum circle)*. The adjustable tongue on the vertical beam of a gear tooth vernier caliper is set to a corrected addendum measurement (known as the chordal addendum) in order to measure the *chordal thickness*.

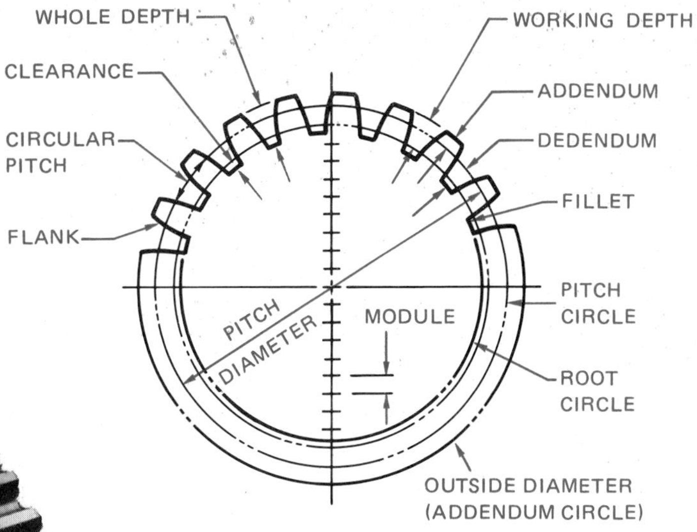

WHOLE DEPTH
WORKING DEPTH
CLEARANCE
ADDENDUM
CIRCULAR PITCH
DEDENDUM
FILLET
FLANK
PITCH DIAMETER
MODULE
PITCH CIRCLE
ROOT CIRCLE
OUTSIDE DIAMETER (ADDENDUM CIRCLE)

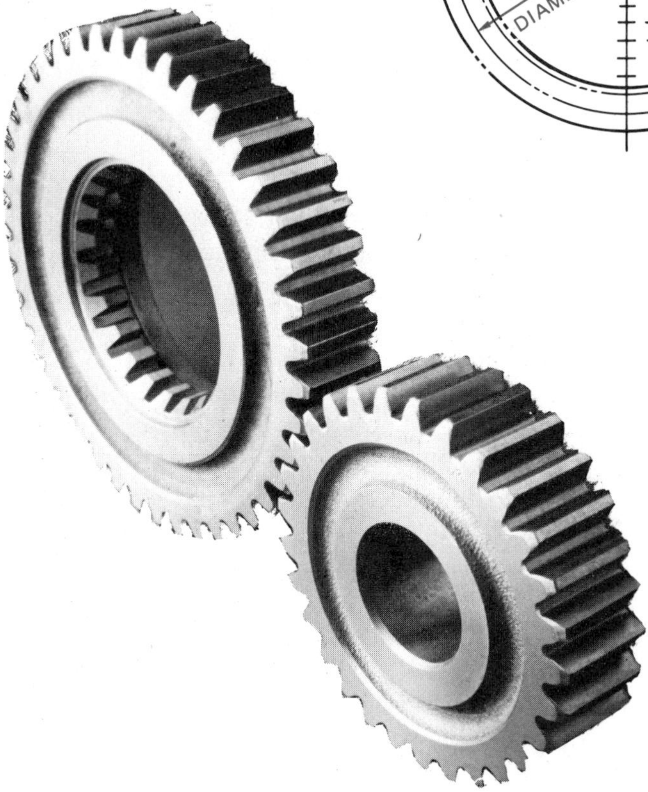

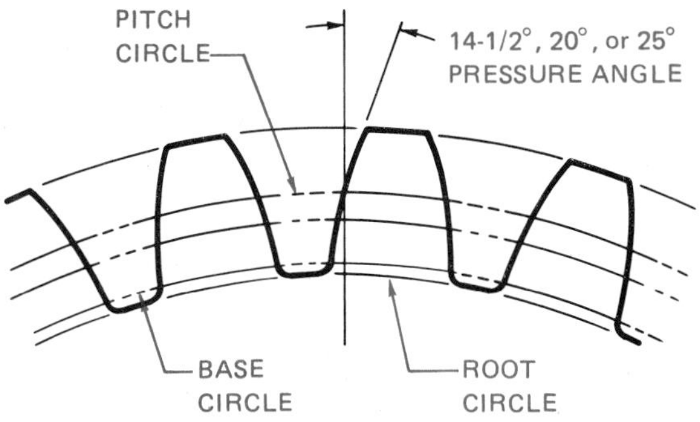

PITCH CIRCLE
14-1/2°, 20°, or 25° PRESSURE ANGLE
BASE CIRCLE
ROOT CIRCLE

CIRCULAR THICKNESS
CIRCULAR PITCH
CHORDAL THICKNESS
CHORDAL ADDENDUM
FILLET

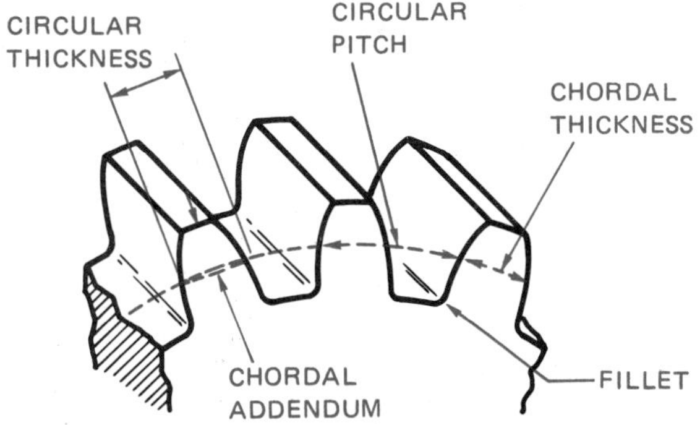

FIGURE 31-1 COMMON SPUR GEAR TERMS AND MEASUREMENTS (PHOTOGRAPH COURTESY OF GLEASON WORKS, MACHINE DIVISION.)

Addendum Circle. The outside diameter of the gear.

Dedendum. The vertical distance between the pitch circle and the root circle. The dedendum includes clearance between the outside diameter of the mating gear or rack and the root circle.

Root Circle. The diameter at the root of the gear.

Base Circle. The circle (diameter) from which the tooth profile is generated.

Pressure Angle. The angle at which forces from the tooth of one gear are transmitted to the mating tooth of another gear or rack. ANSI pressure angles of 14 1/2°, 20°, and 25° are commonly used with regular and stub (shorter) length gear teeth.

Circular Pitch. The circular (curved) distance along the pitch circle between a point on one tooth to a corresponding point on the next tooth.

Circular Thickness. The thickness of a gear tooth as measured along the arc on the pitch circle.

Chordal Thickness. The tooth thickness measured along the straight line (chord) of the pitch circle. The horizontal beam of the gear tooth vernier is used to take the chordal thickness (tooth) measurement.

Chordal Addendum. The distance between the outside diameter and the chord for the chordal thickness. This dimension is read on the vertical beam of a gear tooth vernier caliper.

Whole Depth. The overall depth to which each gear tooth is cut.

Working Depth. The depth to which one tooth extends into the corresponding gear tooth space on a mating gear or rack.

Clearance. The distance between the outside diameter of one gear and the root circle of

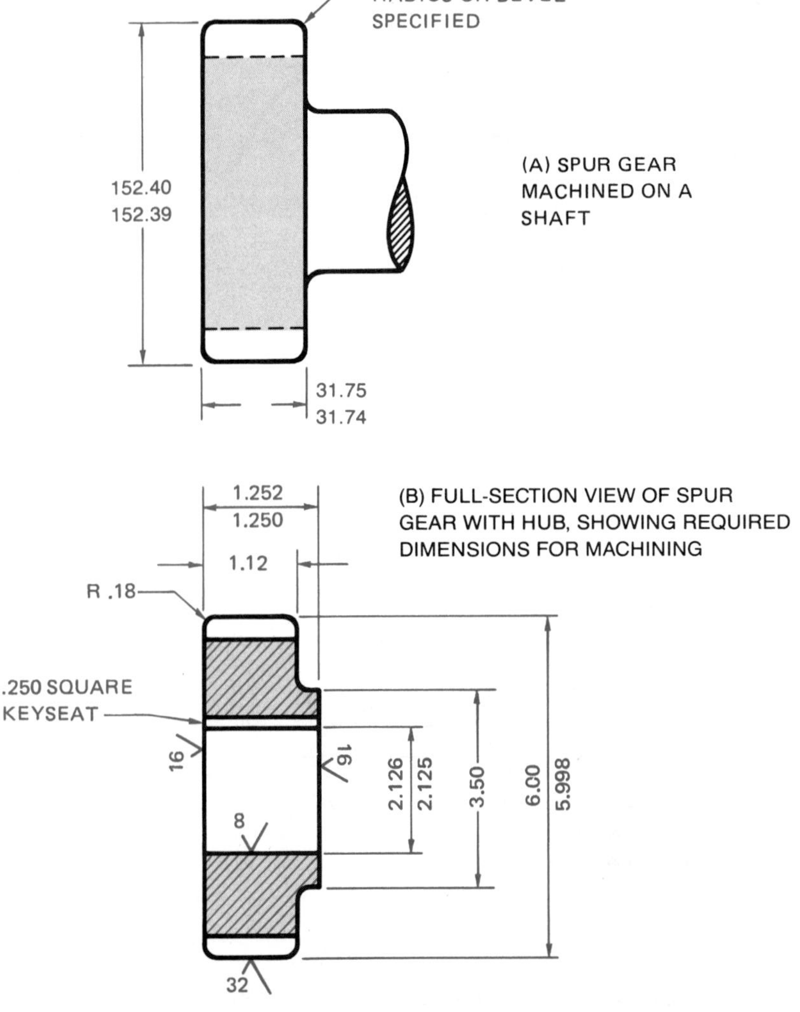

FIGURE 31-2 PREFERRED REPRESENTATION AND DIMENSIONING OF SPUR GEARS

the mating gear, or the base circle and root circle of a gear.

Face Width. The distance between the two sides (faces) of a gear tooth.

Number of Teeth. The number of teeth to machine in a gear.

Diametral Pitch. The number of teeth per inch of pitch diameter.

Diametral Pitch System. The standard by which most gearing is designed in the United States. A *module* which represents the ratio between pitch diameter in inches and the number of teeth. In other words, the pitch diameter equals the number of teeth per inch of pitch diameter.

SI Metric Module. An actual dimension which is equal to the pitch diameter in millimeters divided by the number of teeth in the gear.

Metric Module Equivalent. A metric measurement equivalent of a diametral pitch. A metric module equivalent is computed by dividing 25.4 by the diametral pitch.

REPRESENTATION, DIMENSIONING, AND TOOTH DATA FOR SPUR GEARING

The preferred representation of spur gears or pinions which are machined as part of a shaft is illustrated in figure 31-2A. A gear that is designed with a hub or holes is represented by a section view which cuts through the axis (figure 31-2B). The one view may be a full or partial section.

Another common practice (when details need to be clarified) is to use a full top view with a few teeth and a section view, as illustrated in figure 31-3. Dimensions and notes for *finish-machining* the gear blank are included on both views. Other information relating to the gear tooth form and cutter setup is included as *gear tooth data. Reference dimensions,* such as pitch diameter (figure 31-3) are included as data from which other values may be computed.

SPUR GEAR RACK AND PINION TERMINOLOGY AND REPRESENTATION

A *spur gear rack* (figure 31-4) is a straight bar, usually of round, square, or rectangular cross-section, with teeth cut on one side. A rack transforms rotary motion into linear motion. The pitch line of the rack teeth and the pitch circle of the gear intersect.

The term *pinion* is generally used to denote the smaller of two gears. In the case of a rack and pinion, the pinion is the circular gear. Selected design features of a rack are identified in figure 31-5. Note that a *lineal pitch* is used in place of circular pitch. Also, a *lineal thickness* is specified as being equal to the circular thickness of the mating pinion.

A rack is generally represented on a drawing as illustrated in figure 31-6. The example provides dimensions for machining and measuring the rack. Additional rack tooth data is provided in table form.

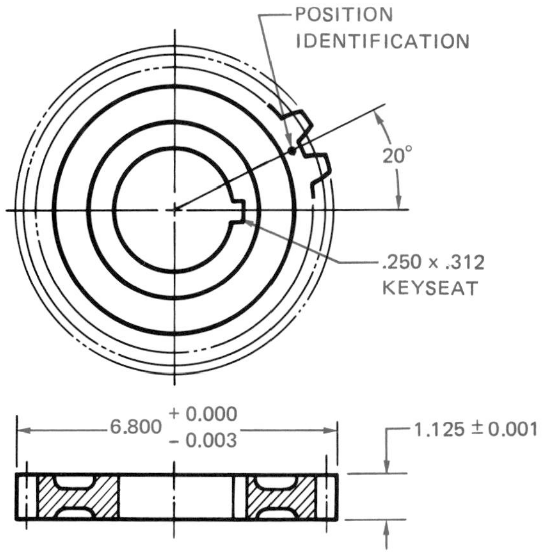

SPUR GEAR TOOTH DATA	
NUMBER OF TEETH	32
DIAMETRAL PITCH	5
PRESSURE ANGLE	20°
PITCH DIAMETER (REF)	6.400
CIRCULAR THICKNESS	0.314
CIRCULAR PITCH	0.628
ADDENDUM	0.200
WHOLE DEPTH	0.450
WORKING DEPTH	0.400
MEASURING DIAMETER	0.346
MEASUREMENT OVER PINS	6.883

FIGURE 31-3 REPRESENTATION OF A SPUR GEAR WITH DIMENSIONS AND MACHINING AND MEASUREMENT DATA

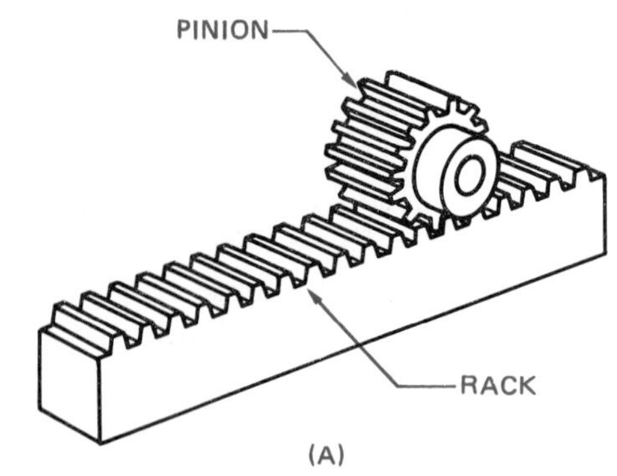

FIGURE 31-4 EXAMPLE OF COMMON SPUR GEAR RACK AND PINION

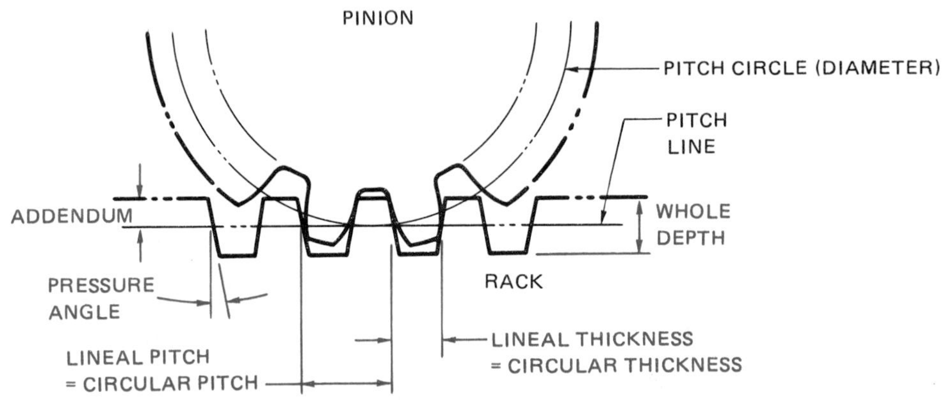

FIGURE 31-5 BASIC DESIGN FEATURES OF RACK TEETH

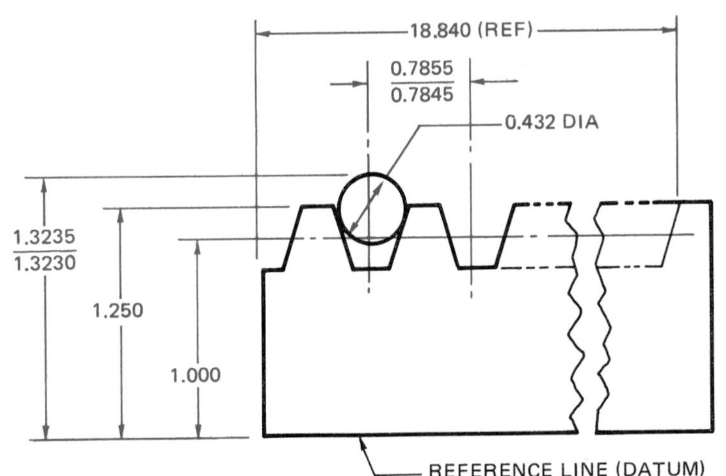

RACK TOOTH DATA	
PRESSURE ANGLE	20° 0′
NUMBER OF TEETH	24
DIAMETRAL PITCH	4
TOOTH THICKNESS AT PITCH LINE	0.3927
NOMINAL LINEAR PITCH	0.785
WHOLE DEPTH	0.5393
MAXIMUM ALLOWABLE PITCH VARIATION	0.001
ACCUMULATED PITCH ERROR, MAXIMUM	0.024

FIGURE 31-6 MACHINING AND MEASURING DIMENSIONS AND RACK TOOTH DATA

USE OF HANDBOOK FORMULAS OF STANDARD SPUR GEARS

Setup, machining, and measurement dimensions are not always furnished on a drawing. Handbook formulas or values are used according to information which is given in a note or a data table. The drawing values are substituted in formulas in order to compute required dimensions.

Standard symbols and notations are used. For instance, **P** denotes diametral pitch; **N**, number of teeth; **O**, outside diameter, and **A**, addendum. The general formulas require simple mathematical processes to compute a particular dimension. Two examples are given to indicate the nature of spur gear formulas.

$$N = (P) \times (D)$$

N = Number of Teeth
P = Diametral Pitch
D = Diameter of Gear
P_c = Circular Pitch

$$P_c = \frac{(\pi) \times (D)}{(N)}$$

Handbook tables provide rules and formulas for standard spur and other tooth forms for 14 1/2°, 20°, and 25° pressure angles for fine pitch and coarse pitch series. Additional specifications on classes of fit and other design features are available from the American Gear Manufacturer's Association (AGMA).

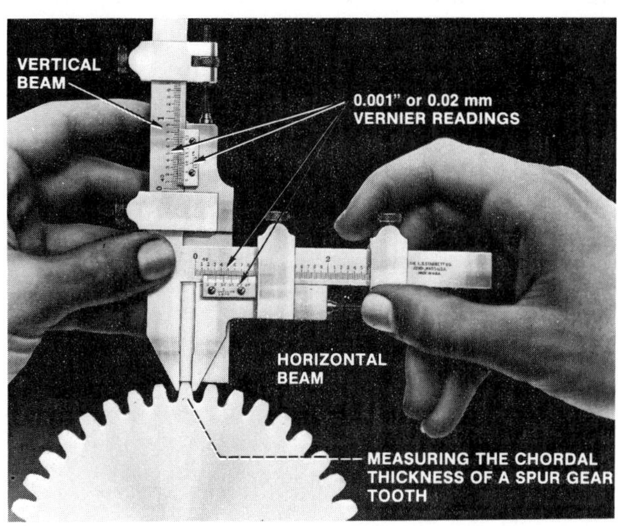

FIGURE 31-7 FEATURES AND APPLICATION OF A GEAR TOOTH VERNIER CALIPER (COURTESY OF THE L.S. STARRETT COMPANY)

GEAR TOOTH MEASUREMENT

Gear teeth are usually measured mechanically with a gear tooth vernier caliper (figure 31-7) or an outside micrometer. The vernier caliper is used to measure the tooth thickness (chordal thickness) at a depth known as the *chordal addendum*. In the case of measurement with a micrometer (figure 31-8), a dimension is given in a table (or may be computed) for an overall measurement with a given diameter *pin* (wire size). Often, the measurement information is given on a drawing. Precise inspection of gear tooth forms, dimensional measurements, and surface texture accuracies are also established by using optical comparators.

Gears are measured or tested for the following geometric characteristics and surface conditions, in addition to general dimensional measurements:

- Concentricity, perpendicularity, and parallelism
- Tooth size and form
- Surface texture
- Lead (on helical gears) and tooth angles (on bevel gears)

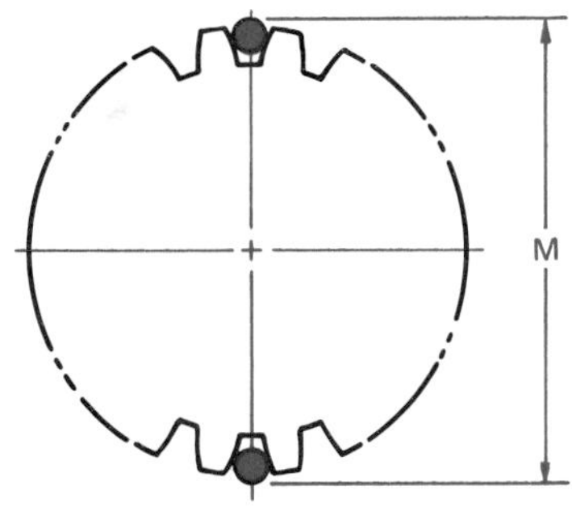

(A) EVEN NUMBER OF TEETH

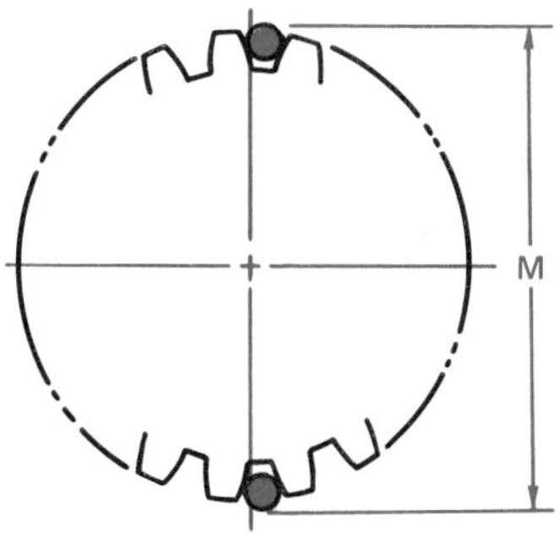

(B) ODD NUMBER OF TEETH

M = MICROMETER MEASUREMENT OVER WIRES

FIGURE 31-8 MICROMETER (TWO-WIRE) METHOD OF MEASURING THE DEPTH AND ACCURACY OF SPUR GEAR TEETH

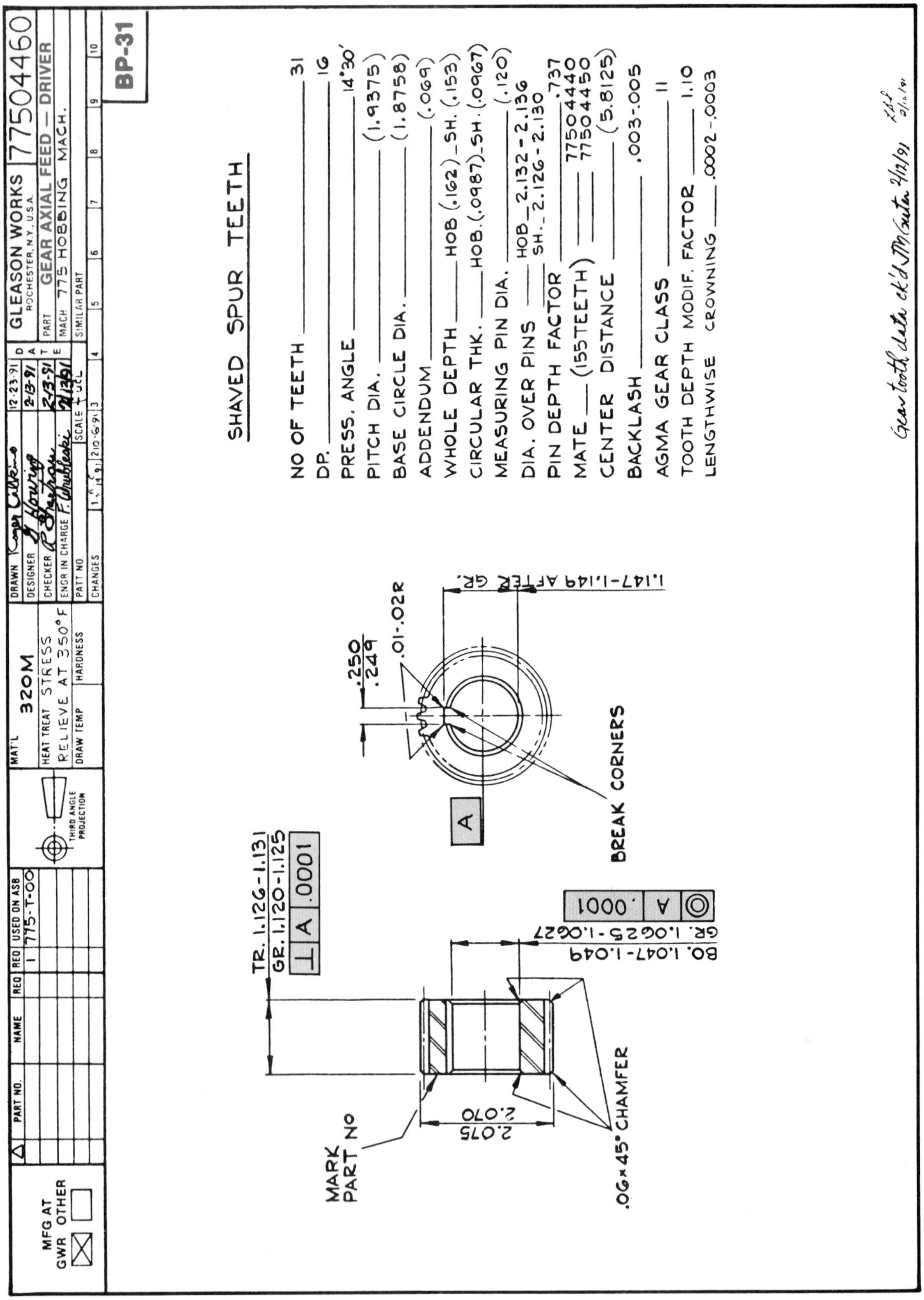

MFG AT	PART NO.	NAME	REQ	REQ	USED ON ASB
GWR ☒ OTHER ☐				1	775-T-00

THIRD ANGLE PROJECTION

MAT'L	320M
HEAT TREAT	STRESS RELIEVE AT 350°F
DRAW TEMP	HARDNESS

		DATE
DRAWN	Roger Calkins	12-23-91
DESIGNER	[illegible] Howing	2-13-91
CHECKER	R. Sherman	2-13-91
ENGR IN CHARGE	F. Wrubleski	2/13/91
PATT NO	SCALE	
CHANGES	1	2 10-6-91

GLEASON WORKS 77504460
ROCHESTER, N.Y., U.S.A.

PART GEAR AXIAL FEED — DRIVER
MACH 775 HOBBING MACH.
SIMILAR PART

BP-31

TR. 1.126-1.131
GR. 1.120-1.125
⊥ | A | .0001

MARK PART NO

2.075
2.070

.06×45° CHAMFER

BO. 1.047-1.049
GR. 1.0625-1.0627
◎ | A | .0001

.250
.249

.01-.02R

A

BREAK CORNERS

1.147-1.149 AFTER GR.

SHAVED SPUR TEETH

NO OF TEETH	31
DP.	16
PRESS. ANGLE	14°30′
PITCH DIA.	(1.9375)
BASE CIRCLE DIA.	(1.8758)
ADDENDUM	(.069)
WHOLE DEPTH	HOB (.162) SH. (.153)
CIRCULAR THK.	HOB. (.0987) SH. (.0967)
MEASURING PIN DIA.	(.120)
DIA. OVER PINS	HOB 2.132-2.136 SH. 2.126-2.130
PIN DEPTH FACTOR	.737
MATE (155TEETH)	77504440 77504450
CENTER DISTANCE	(5.8125)
BACKLASH	.003-.005
AGMA GEAR CLASS	11
TOOTH DEPTH MODIF. FACTOR	1.10
LENGTHWISE CROWNING	.0002-.0003

Gear tooth data ck'd JM Guten 2/12/91 ZBS 2/1·/91

Courtesy of North American, Gleason Machine Division

ALL UNTOLERANCED DIMENSIONS ARE ±0.2 EXCEPT CASTING, FABRICATION, FORGING, AND HOLE DIAMETERS LISTED IN STANDARD TOLERANCE CHARTS, AND ( )
REF; ☐ BASIC DIMENSIONS
BREAK ALL SHARP CORNERS

ASSIGNMENT—UNIT 31

GEAR AXIAL FEED – DRIVER (BP-31)

Student's Name ______________________

1. Name the manufacturing process used to machine the gear teeth.
2. Identify (a) the material used and (b) the specifications for heat treating.
3. a. Determine the upper and lower limits for machining the outside diameter of the gear blank.
   b. Indicate the size of the chamfer for the teeth and hole.
4. Give the minimum and maximum diameters for boring the hole.
5. Give the minimum and maximum diameters for grinding the bored hole.
6. Determine the tolerance for the ground hole.
7. State (a) the maximum width of the keyseat and (b) the maximum overall height after grinding.
8. Indicate (a) the perpendicularity tolerance zone of the sides and (b) the concentricity tolerance zone of the hole.
9. Give (a) the diametral pitch, (b) the number of gear teeth, and (c) the pressure angle of the gear teeth.
10. Give (a) the diameter of the measuring pins and (b) the maximum diameter measurement over the pins for (1) the hobbed teeth and (2) the shaved teeth.
11. State the base circle reference diameter.
12. Give (a) the center-to-center reference distance between the mating gears and (b) the number of teeth in the mating gear.
13. Determine the upper and lower limit whole depth dimensions to which the teeth are hobbed. Apply a tolerance of $\pm 0.0005''$.
14. Compute the dimensions for setting a gear tooth vernier caliper to measure (a) the nominal chordal addendum and (b) the nominal chordal thickness of the shaved gear teeth.

    Note. Refer to a handbook table for formulas for standard outside diameters.
15. Assume the spur gear is to mesh with a rack.
    a. Compute the nominal lineal pitch of the rack teeth.
    b. Give the lineal (basic) tooth thickness of the hobbed teeth.
16. Use the reference table in the appendix for SI Metric Spur Gears. Calculate the following reference dimensions for an 8-module, 32-tooth SI metric gear.
    a. Pitch diameter
    b. Outside diameter
    c. Whole depth
    d. Tooth (chordal) thickness

1. ______________________
2. a. Material ______________________
   b. ______________________
3. a. Upper ______________________
   Lower ______________________
   b. ______________________
4. Boring Min. ________ Max. ________
5. Grinding Min. ________ Max. ________
6. Tolerance ______________________
7. a. Max. Keyseat Width ______________________
   b. Max. Overall Height ______________________
8. a. Perpendicularity Tolerance Zone ________
   b. Concentricity Tolerance Zone ________
9. a. Diametral Pitch ______________________
   b. Number of Teeth ______________________
   c. Pressure Angle ______________________
10. a. Diameter Measuring Pins ________
    b. Max. Diameter (1) hobbed teeth ________
    (2) shaved teeth ________
11. Base Circle Reference Diameter ________
12. a. Reference Center Distance ________
    b. Mating Gear Teeth ________
13. Whole Depth Min. ________ Max. ________
14. a. Vertical Blade, Chordal Addendum ______________________
    b. Horizontal Beam, Chordal Thickness ______________________
15. a. Nominal Lineal Pitch ______________________
    b. Lineal Tooth Thickness ______________________
16. a. Pitch Dia. ______________________
    b. Outside Dia. ______________________
    c. Whole Depth ______________________
    d. Chordal Thickness ______________________

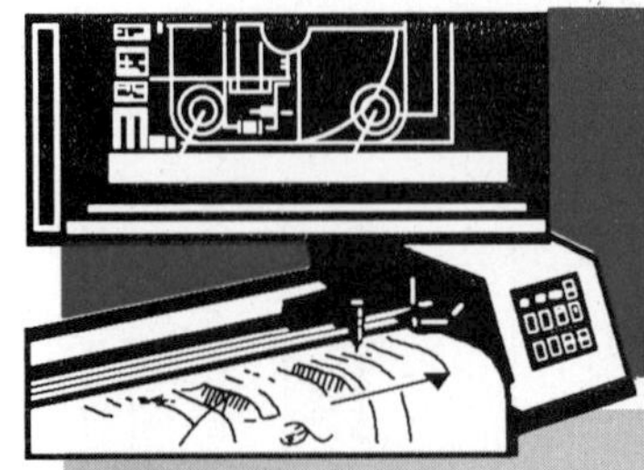

HIGH TECHNOLOGY applications

HT-8 CADD/CIM AND CELLULAR MANUFACTURING

GROUP TECHNOLOGY (GT): CLASSIFICATION AND CODING

A *group technology (GT) system* depends on a very large manufacturing data base for a great number of machined parts. Parts are grouped according to *families of workpieces* (parts) that have common design features and require similar manufacturing processes. Each family of parts is subdivided into *part family cells* which consist of parts that have minor differences in design but require common tooling, machine tools, and manufacturing processes.

In group technology, parts are assigned according to a numeric or numeric and alphanumeric (number and letter) designation. The coded part designation depends on part geometry, workpiece material, manufacturing processes, surface finish, dimensional accuracy, geometric tolerances, and other design features.

The basic chart for concentric parts illustrates a group technology classification and coding system matrix and an example of a coded part (**13188075**). Particular attributes (features) of the product are identified in each vertical column. A variation of each attribute appears in each horizontal column.

The classification and coded information is computer processed in what is called cellular manufacturing. Cells or groups of NC (and conventional) machine tools are clustered for a required battery of manufacturing processes involving common tooling, according to a specific group technology classification.

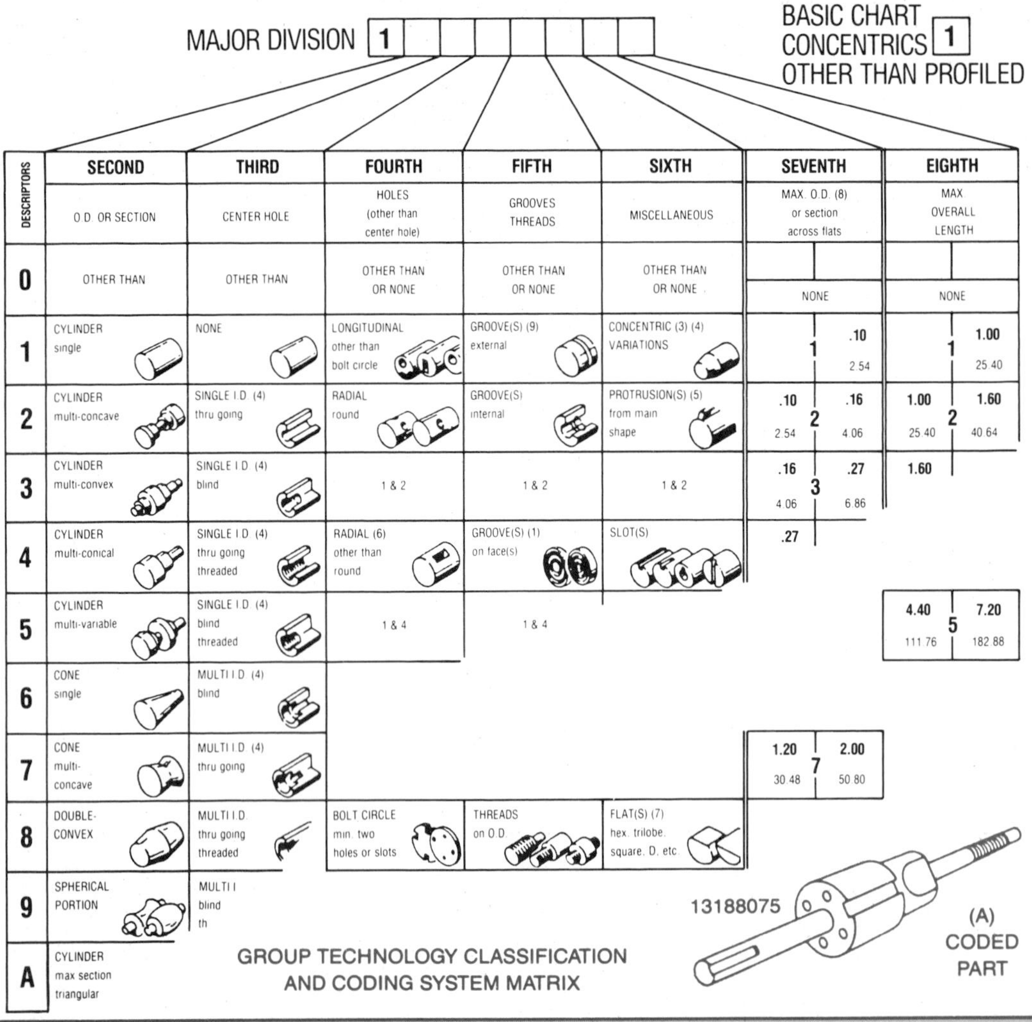

MAJOR DIVISION [1]

BASIC CHART CONCENTRICS [1] OTHER THAN PROFILED

DESCRIPTORS	SECOND	THIRD	FOURTH	FIFTH	SIXTH	SEVENTH	EIGHTH
	O.D. OR SECTION	CENTER HOLE	HOLES (other than center hole)	GROOVES THREADS	MISCELLANEOUS	MAX. O.D. (8) or section across flats	MAX OVERALL LENGTH
0	OTHER THAN	OTHER THAN	OTHER THAN OR NONE	OTHER THAN OR NONE	OTHER THAN OR NONE	NONE	NONE
1	CYLINDER single	NONE	LONGITUDINAL other than bolt circle	GROOVE(S) (9) external	CONCENTRIC (3) (4) VARIATIONS	1 .10 2.54	1 1.00 25.40
2	CYLINDER multi-concave	SINGLE I.D. (4) thru going	RADIAL round	GROOVE(S) internal	PROTRUSION(S) (5) from main shape	.10 2.54 2 .16 4.06	1.00 25.40 2 1.60 40.64
3	CYLINDER multi-convex	SINGLE I.D. (4) blind	1 & 2	1 & 2	1 & 2	.16 4.06 3 .27 6.86	1.60
4	CYLINDER multi-conical	SINGLE I.D. (4) thru going threaded	RADIAL (6) other than round	GROOVE(S) (1) on face(s)	SLOT(S)	.27	
5	CYLINDER multi-variable	SINGLE I.D. (4) blind threaded	1 & 4	1 & 4			4.40 111.76 5 7.20 182.88
6	CONE single	MULTI I.D. (4) blind					
7	CONE multi-concave	MULTI I.D. (4) thru going				1.20 30.48 7 2.00 50.80	
8	DOUBLE-CONVEX	MULTI I.D. thru going threaded	BOLT CIRCLE min. two holes or slots	THREADS on O.D.	FLAT(S) (7) hex, trilobe, square, D, etc.		
9	SPHERICAL PORTION	MULTI I blind th					
A	CYLINDER max section triangular						

13188075

(A) CODED PART

GROUP TECHNOLOGY CLASSIFICATION AND CODING SYSTEM MATRIX

Drawing courtesy of SCHLUMBERGER CAD/CAM DIVISION

UNIT 32
Helical and Worm Gearing

Helical gears serve similar functions to spur gears in transmitting rotary motion and load between two shafts. The axes of the shafts may be parallel and lie in the same plane, as shown in figure 32-1A. Helical gears permit application to shaft axes which lie in different planes (figure 32-1B). Generally, helical gears are designed for use on parallel shafts, shafts with axes at 45°, and shafts at 90° to each other.

In operation, helical gears are smooth, quiet, and efficient. These conditions are due to the gradual overlapping engagement of the gear teeth and continuous action. The strength of a helical gear tooth is greater than a comparable size spur gear tooth on which the load is suddenly transferred from one tooth to the next.

However, helical gears produce end thrust when used alone and for nonparallel applications. The use of double helical gears on parallel shaft movements equalizes end thrust and removes the need for thrust bearings. The combination of a duplicate set of right- and left-hand helical gears, mounted on a common shaft, is called a *herringbone gear.*

REPRESENTATION AND MEASUREMENT OF HELICAL GEAR TEETH

Similar terminology is used for both spur and helical gearing. Again, handbook tables provide technical information related to symbols for main design features. Formulas are provided for computing each required dimension and measurement.

(A) HELICAL GEARING — PARALLEL AXES

(B) GEAR AXES AT RIGHT ANGLE

FIGURE 32-1 COMMON APPLICATIONS OF HELICAL GEARING (FIG. 32-1A COURTESY OF GLEASON WORKS, MACHINE DIVISION, AND FIG. 32-1B COURTESY OF BOSTON GEAR WORKS)

There are, however, two additional terms.

Normal diametral pitch (P_{nc}). The value for the diametral pitch of the helical gear tooth cutter.

Normal circular pitch (T_n). The distance between a reference point on one gear tooth and a corresponding point on the next tooth. The gear tooth measurement is taken on the pitch circle on a reference plane that is at 90° to the helix angle (figure 32-2).

Since a helix angle is to be considered, many machining and measuring dimensions must be computed. Trigonometric function tables and angle values are used with simple formulas to obtain essential dimensions. Technical tables of helical gear symbols, rules, and formulas are contained in handbooks and other technical literature. A chart is available to establish the milling cutter size, as a *normal diametral pitch* is used to cut a helical gear instead of a diametral pitch.

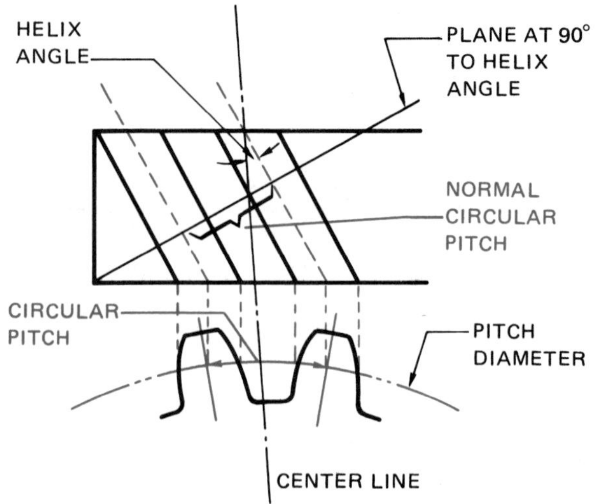

FIGURE 32-2 DIFFERENCE BETWEEN NORMAL CIRCULAR PITCH AND CIRCULAR PITCH MEASUREMENT

WORM GEARING

Industrial worm gearing is used to (1) transmit power efficiently and (2) to obtain a large reduction in velocity between two nonintersecting shafts. The shafts are generally at a right angle to each other. Two gears are used in worm gearing. A *worm*, with teeth similar to those of an Acme screw thread form, serves as the driver. The larger driven gear is the *worm gear*. Figure 32-3 identifies common terminology. Four specific terms which apply to worm gearing follow.

Axial Pitch of a Worm. This equals the circular pitch of the mating worm gear. Axial pitch represents the distance between corresponding reference points on adjacent worm threads.

Axial Advance (Lead). The distance a worm thread advances in one revolution. This movement is equal to the thread pitch for a single lead thread, twice the thread pitch for a double lead thread, etc.

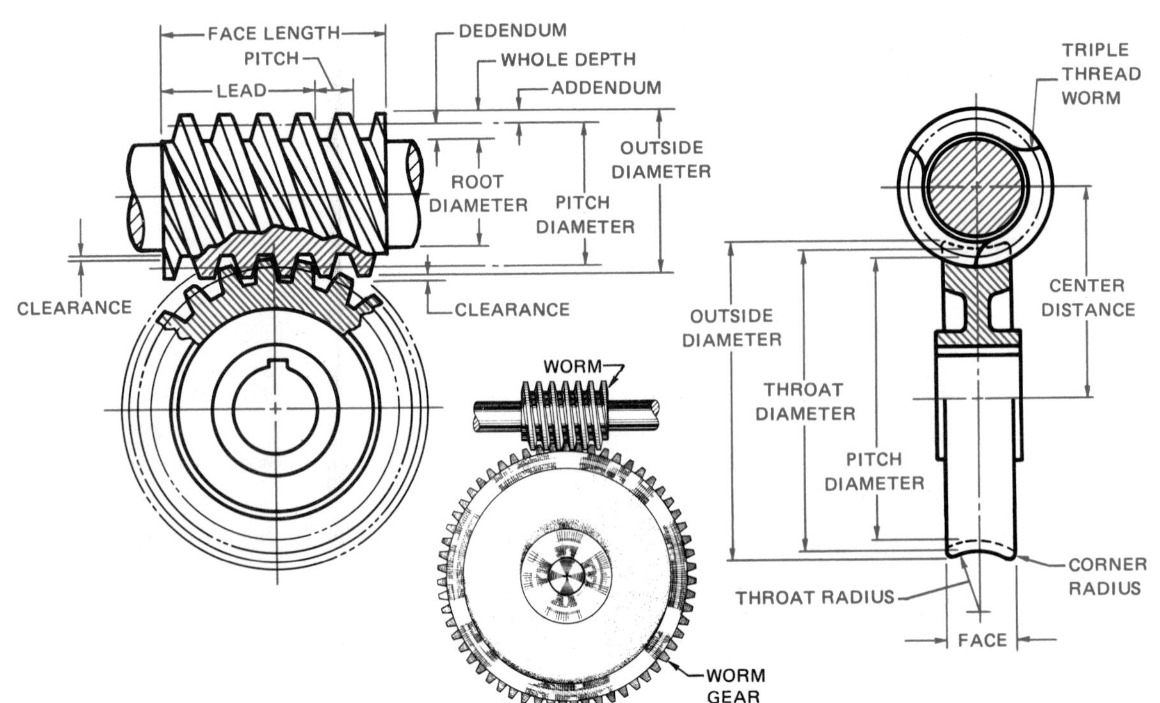

FIGURE 32-3 COMMON WORM AND WORM GEAR TERMS RELATING TO MACHINING AND MEASUREMENT REQUIREMENTS

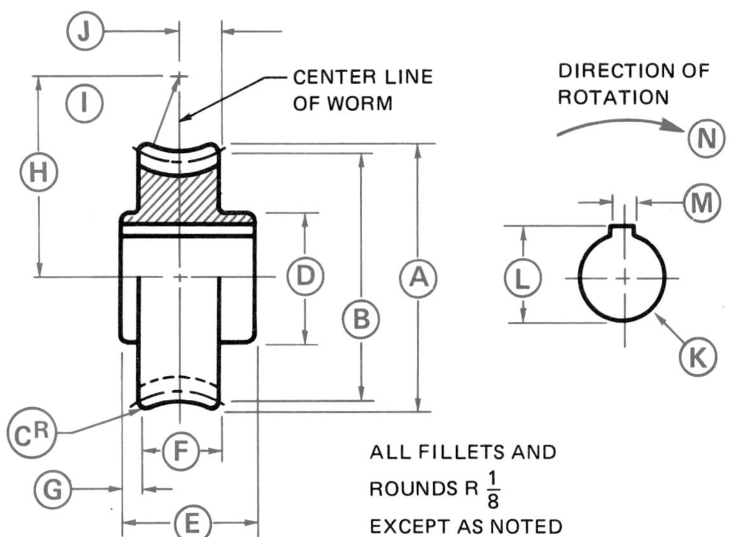

(A) DIMENSIONS FOR MACHINING THE GEAR BLANK

(B) REQUIRED GEAR TOOTH DATA	
NUMBER OF TEETH	
PITCH DIAMETER	
ADDENDUM	
WHOLE DEPTH (APPROX.)	
WORM PART NUMBER	
BACKLASH ASSEMBLED	
HOB NUMBER	

FIGURE 32-4 ANSI REPRESENTATION AND GEAR TOOTH DATA FOR A STANDARD WORM GEAR

Lead Angle. The angle formed by (1) the tangent to the helix of the thread at the pitch diameter and (2) a plane perpendicular to the axis of the worm.

Throat Diameter. A diametral measurement of the worm gear. A measurement at the bottom of the tooth arc which is equal to the pitch diameter plus twice the addendum.

REPRESENTATION OF WORMS AND WORM GEARS

A standard worm gear is usually represented in one view by a full or half section as shown in figure 32-4A. Dimensional and surface texture information for machine setup, machining, and measuring are either read directly from a working drawing (figure 32-4A, items (A) through (N)) or are found in an accompanying table of gear tooth data as shown in figure 32-4B. If no drawing is provided for the worm, reference dimensions such as those identified at (B) are given in a table.

ANSI techniques of representing and dimensioning a worm are illustrated in figure 32-5A. The required machining dimensions which normally appear on a front view or front and end views are identified by the letters (A) through (I). Other required data for cutter selection, machining, and measurement often appear as supplementary thread specifications (figure 32-5B).

ANSI standards, terms, rules, and formulas (which are often used with dimensional and other information as provided on a blueprint) are published in trade handbooks and manufacturers' technical bulletins.

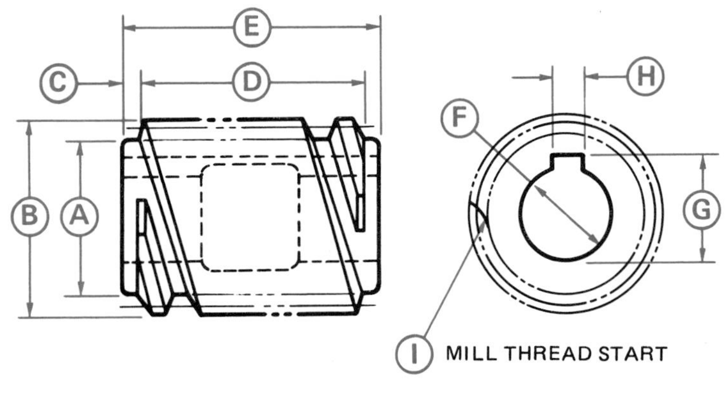

(A) MACHINING DIMENSIONS

(B) REQUIRED WORM THREAD DATA	
NUMBER OF THREADS	
PITCH DIAMETER (NOMINAL)	
AXIAL PITCH	
LEAD RIGHT (OR LEFT) HAND	
LEAD ANGLE	
NORMAL PRESSURE ANGLE (NOMINAL)	
ADDENDUM	
WHOLE DEPTH (APPROX.)	
NORMAL CHORDAL ADDENDUM	
NORMAL CHORDAL THICKNESS	
WORM GEAR PART NUMBER	
MANUFACTURING PROCESS	

FIGURE 32-5 ANSI REPRESENTATION AND DIMENSIONING OF A STANDARD WORM GEAR

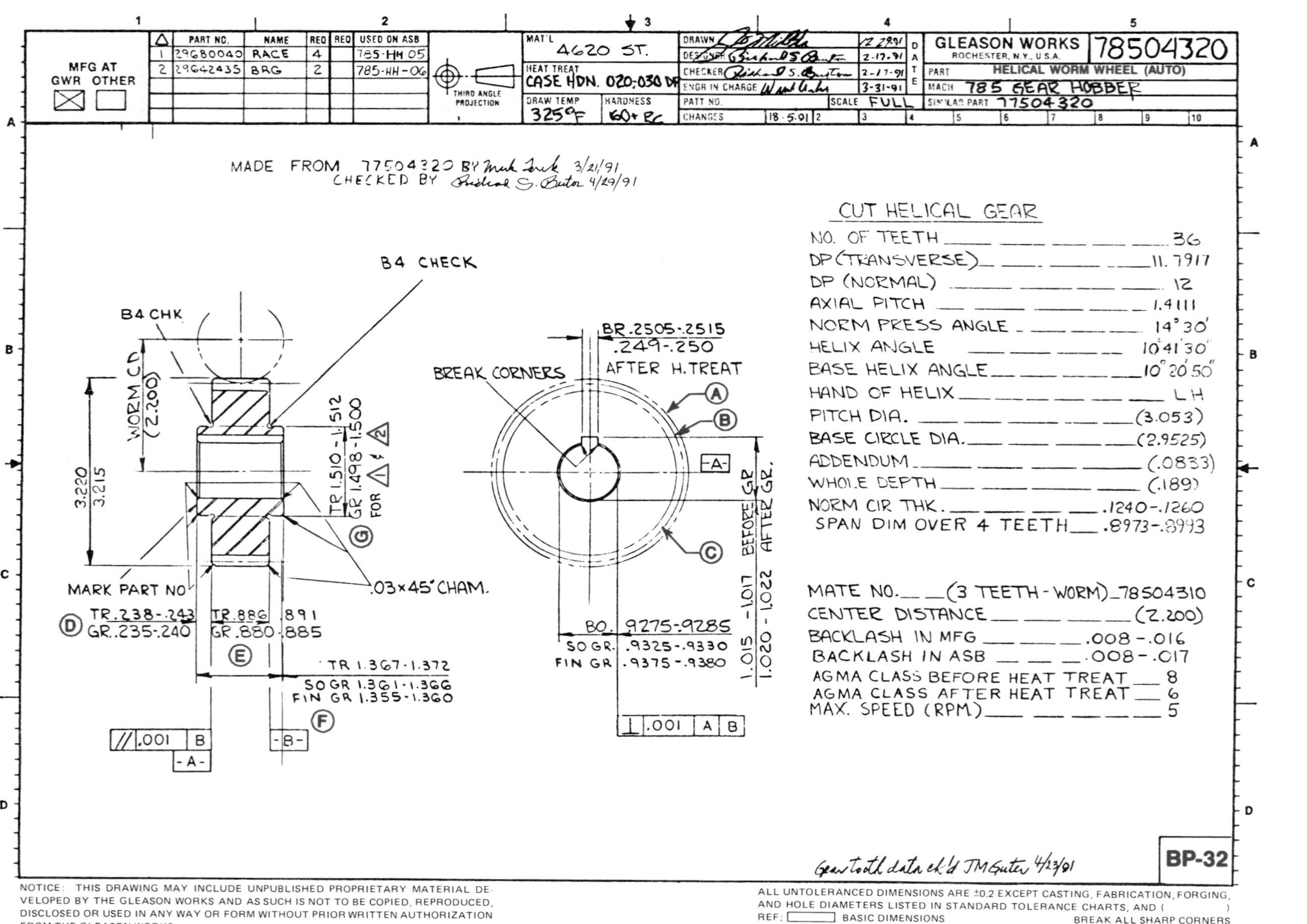
GLEASON WORKS
ROCHESTER, N.Y., U.S.A.
78504320
HELICAL WORM WHEEL (AUTO)
785 GEAR HOBBER
77504320
MAT'L 4620 ST.
HEAT TREAT CASE HDN. .020-.030 DP
DRAW TEMP 325°F
HARDNESS 60+ Rc
THIRD ANGLE PROJECTION
SCALE FULL
MFG AT GWR OTHER
PART NO. NAME REQ USED ON ASB
1 29680040 RACE 4 785-HH-05
2 29642435 BRG 2 785-HH-06
MADE FROM 77504320 BY
CHECKED BY
CUT HELICAL GEAR
NO. OF TEETH 36
DP (TRANSVERSE) 11.7917
DP (NORMAL) 12
AXIAL PITCH 1.4111
NORM PRESS ANGLE 14°30'
HELIX ANGLE 10°41'30"
BASE HELIX ANGLE 10°20'50"
HAND OF HELIX LH
PITCH DIA. (3.053)
BASE CIRCLE DIA. (2.9525)
ADDENDUM (.0833)
WHOLE DEPTH (.189)
NORM CIR THK. .1240-.1260
SPAN DIM OVER 4 TEETH .8973-.8993
MATE NO. (3 TEETH-WORM) 78504310
CENTER DISTANCE (2.200)
BACKLASH IN MFG .008-.016
BACKLASH IN ASB .008-.017
AGMA CLASS BEFORE HEAT TREAT 8
AGMA CLASS AFTER HEAT TREAT 6
MAX. SPEED (RPM) 5
B4 CHECK
B4 CHK
BREAK CORNERS
BR .2505-.2515
.249-.250
AFTER H.TREAT
BO .9275-.9285
SO GR .9325-.9330
FIN GR .9375-.9380
1.015-1.017 BEFORE GR
1.020-1.022 AFTER GR.
.03×45° CHAM.
TR 1.510-1.512
GR 1.498-1.500
TR 1.367-1.372
SO GR 1.361-1.366
FIN GR 1.355-1.360
TR .886
GR .880
.891
.885
TR .238-.243
GR .235-.240
MARK PART NO
3.220
3.215
WORM ℄ (2.200)
BP-32
NOTICE: THIS DRAWING MAY INCLUDE UNPUBLISHED PROPRIETARY MATERIAL DEVELOPED BY THE GLEASON WORKS AND AS SUCH IS NOT TO BE COPIED, REPRODUCED, DISCLOSED OR USED IN ANY WAY OR FORM WITHOUT PRIOR WRITTEN AUTHORIZATION FROM THE GLEASON WORKS.
ALL UNTOLERANCED DIMENSIONS ARE ±0.2 EXCEPT CASTING, FABRICATION, FORGING, AND HOLE DIAMETERS LISTED IN STANDARD TOLERANCE CHARTS, AND ( )
REF: BASIC DIMENSIONS
BREAK ALL SHARP CORNERS

ASSIGNMENT—UNIT 32

HELICAL WORM WHEEL (AUTO) (BP-32)

Student's Name ____________________

1. a. Name the two different types of gears and state the number of teeth in each gear.
   b. Give the identification number of the mating gear.
2. a. Identify the material of which the part is made.
   b. Indicate the required heat treatment.
3. Interpret the symbols ⚠1 and ⚠2 by giving (a) the part name, (b) the required quantity, and (c) the assembly (ASB) on which used.
4. a. Identify lines Ⓐ, Ⓑ, and Ⓒ by type.
   b. List the design feature which is represented by each line.
5. a. Give the tolerance range for untoleranced dimensions.
   b. Indicate the upper and the lower dimensional limits for the outside diameter.
   c. Determine the nominal chamfer dimensions for the bored hole and corners of the gear teeth.
6. a. Identify the two datums which are used.
   b. Write out and interpret the meaning of the two feature control symbols.
7. a. Determine the maximum height of the keyway (1) before and (2) after grinding.
   b. Give the machining tolerances for (1) the width and (2) the depth (height) of the keyway.
8. a. State the minimum dimensions for ground features Ⓓ, Ⓔ, Ⓕ (FIN GR), and Ⓖ.
   b. Give the dimensional tolerances for these features.
9. Determine the following nominal dimensions of the helical gear:
   a. Center-to-center distance between the helical gear and the worm
   b. Dedendum + Clearance
   c. 2 × Whole Depth
   d. Pitch diameter (Outside diameter − 2 × addendum)
10. Provide the following specifications:
    a. Pressure angle
    b. Tooth helix angle at the base
    c. Tolerance for the normal circular tooth thickness

1. a. (1) ____________________
   (2) ____________________
   b. Mate Number ____________________

2. a. ____________________
   b. ____________________

3.

	Part Name (a)	Qty (b)	ASB (c)
⚠1	______	___	______
⚠2	______	___	______

4.

	(a) Type Line	(b) Feature
Ⓐ	______	______
Ⓑ	______	______
Ⓒ	______	______

5. a. ____________
   b. Max. ________ Min. ________
   c. ____________

6. a. ________, ________
   b. Feature Control Symbol — Interpretation
   ____
   ____
   ____
   ____

7. a. Maximum (1) Before ________
   (2) After ________
   b. Machining Tolerance (1) ________
   (2) ________

8. a. Ⓓ = ________ Ⓔ = ________
   Ⓕ = ________ Ⓖ = ________
   b. ____________________
   ____

9. a. ________ c. ________
   b. ________ d. ________

10. a. ________ c. ________
    b. ________

UNIT 33
Bevel Gearing

Bevel gears are conical in shape. Bevel gears are used to connect gears and transmit power and motion between two shafts which are at an angle to each other. There are two basic types of bevel gears: *straight tooth* and *hypoid tooth*. The tooth form of bevel gears is the same profile as found on spur gear teeth.

APPLICATIONS OF BEVEL GEARS

Straight-tooth bevel gears are widely used when the axes of two gears intersect, as shown in figure 33-1. The term *miter bevel gears* identifies a pair of mating gears of the same size which transmit motion at 90° to the axis of the driver gear. The driver gear (which is usually the smaller gear in the set) is called the *pinion gear*; the driven gear, simply a gear.

Hypoid bevel gears are of the curved tooth type. These gears are used to transmit power and motion between shafts at practically any angle. The two most common curved tooth type bevel gears are *spiral bevel gears* and *hypoid gears*. A hypoid gear resembles a spiral bevel gear except that the pinion axis is offset in relation to the gear axis (figure 33-2). The spiral form, like helical gear teeth, provides for smoother operation, particularly at high speeds.

GENERATION OF BEVEL GEAR TEETH

A correctly formed bevel gear tooth is of uniform cross-section throughout its length. The size diminishes uniformly from the large end to the small end. The production of an accurate tooth form requires machining on a generating type of gear cutting machine. However, there are applications where it is desirable to cut straight bevel gears by milling.

Rules and formulas for calculating dimensions for milling, as well as other generating processes for machining bevel gears, are provided by manufacturers and in trade handbooks. While the 20° pressure angle is common, bevel gears are also designed with the older 14 1/2° pressure angle and a 25° form. Tables are used by the crafts-

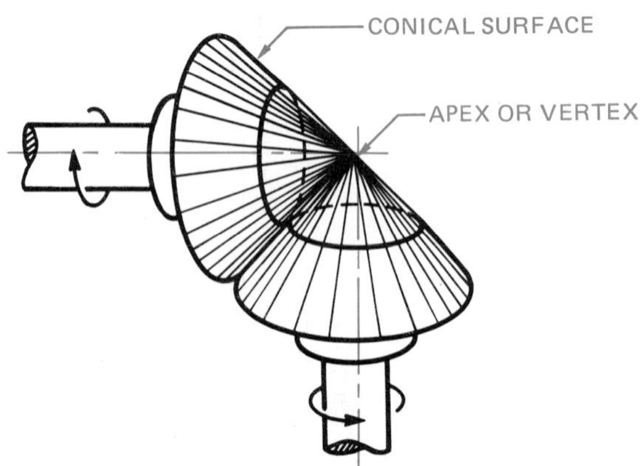

FIGURE 33-1 AXES OF STRAIGHT TOOTH BEVEL GEARS

FIGURE 33-2 PAIR OF HYPOID GEARS (COURTESY OF GLEASON WORKS, MACHINE DIVISION)

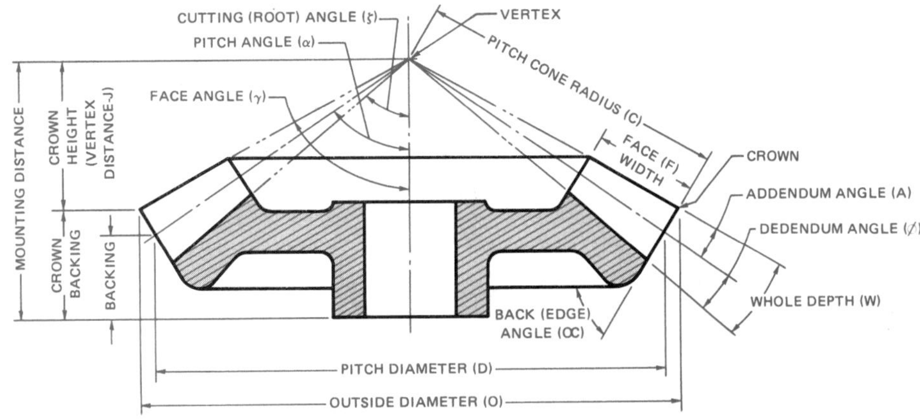

FIGURE 33-3 DESIGN FEATURES, TERMS, AND SYMBOLS APPLIED TO BEVEL GEARS

person to establish dimensions, cutter sizes, and amount of cutter offset, when such data are not provided on a drawing.

FEATURES, TERMS, AND SYMBOLS

The general features of bevel gears which appear on blueprints are illustrated in figure 33-3. The section view gives the terms and symbols which identify each feature.

Formulas, similar to those used for computing spur gear dimensions, apply to the following milled straight-tooth bevel gear features:

- Addendum
- Dedendum
- Clearance
- Circular pitch
- Diametral pitch
- Pitch diameter
- Pressure angle
- Whole tooth depth
- Module (for metric teeth)
- Chordal thickness (large end)
- Circular thickness (large end)

Dimensions that are not provided on a drawing, including the amount of gear tooth offset, are generally computed using handbook formulas.

DRAWINGS OF BEVEL GEARS

A single bevel gear or a pair of bevel gears are usually represented by a one-view section drawing. Gear teeth, while not represented on a working drawing, are often shown on an assembly drawing.

The section view generally includes dimensions for machining the bevel gear blank. A complementary table provides specifications for cutter selection, machine setups, and dimensions for measuring the gear teeth. The ANSI representation and dimensioning of a bevel gear, with accompanying tooth data, are shown in figure 33-4. ANSI standards are used for uniformly detailing gear drawings.

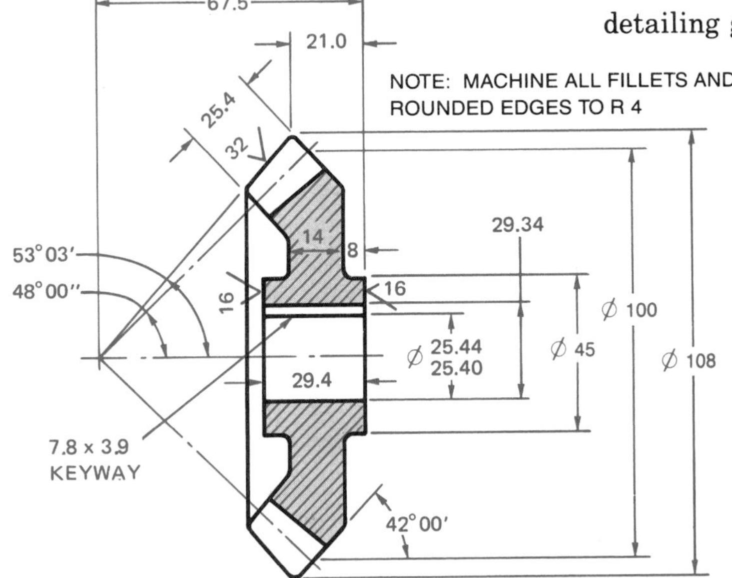

BEVELED GEAR TOOTH DATA	
NUMBER OF TEETH	20
PRESSURE ANGLE	20°
DIAMETRAL PITCH	5
MODULE	5.08
CUTTING ANGLE	41°25′
WHOLE DEPTH	11.13
CIRCULAR PITCH	15.96
CHORDAL ADDENDUM	5.18
CHORDAL THICKNESS	7.98

FIGURE 33-4 ANSI REPRESENTATION, DIMENSIONS, AND TOOTH DATA FOR A BEVEL GEAR

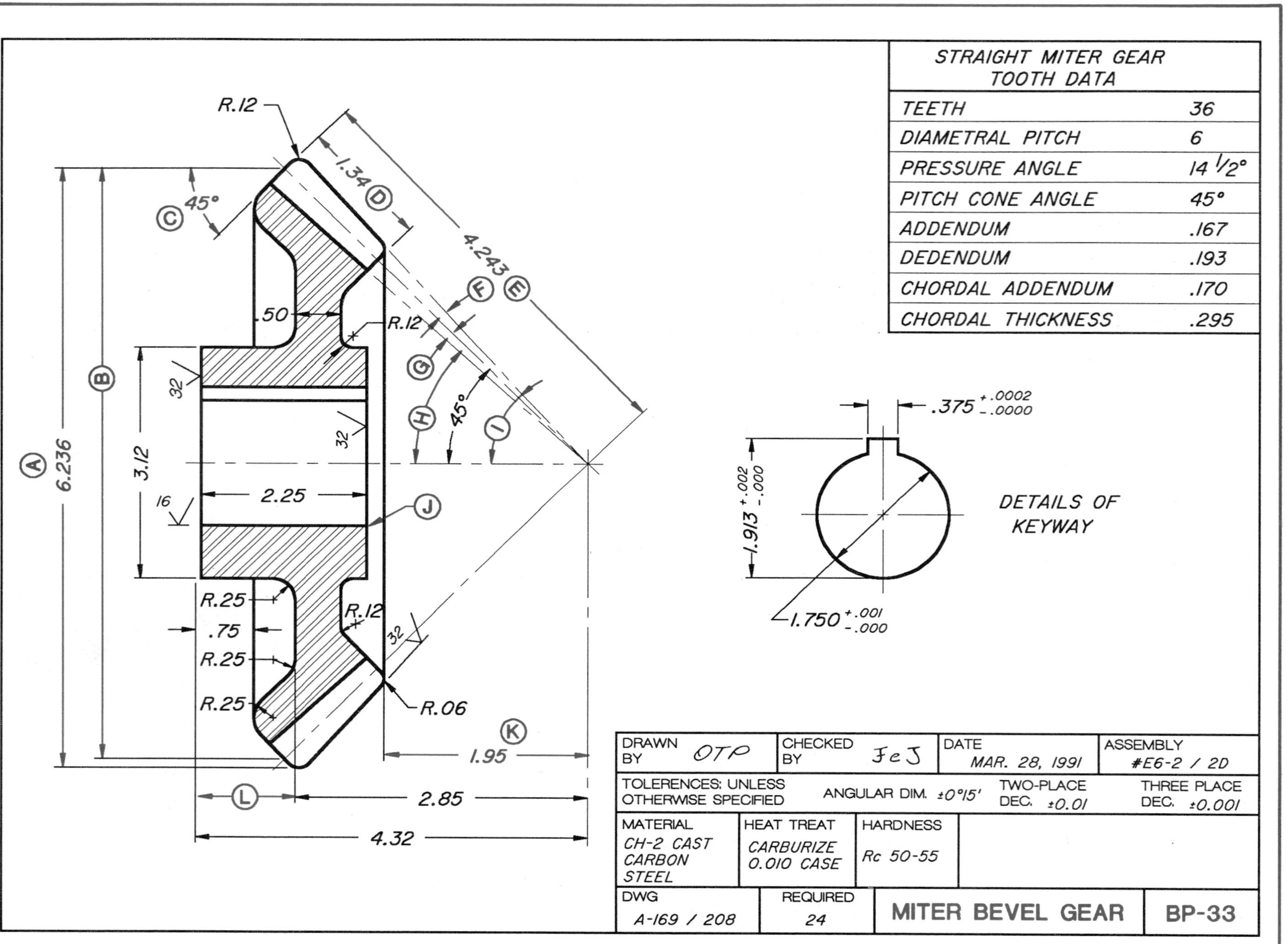
STRAIGHT MITER GEAR TOOTH DATA
TEETH 36
DIAMETRAL PITCH 6
PRESSURE ANGLE 14 1/2°
PITCH CONE ANGLE 45°
ADDENDUM .167
DEDENDUM .193
CHORDAL ADDENDUM .170
CHORDAL THICKNESS .295
DETAILS OF KEYWAY
.375 +.0002 -.0000
1.750 +.001 -.000
1.913 +.002 -.000
DRAWN BY OTP
CHECKED BY FeJ
DATE MAR. 28, 1991
ASSEMBLY #E6-2 / 2D
TOLERENCES: UNLESS OTHERWISE SPECIFIED
ANGULAR DIM. ±0°15'
TWO-PLACE DEC. ±0.01
THREE PLACE DEC. ±0.001
MATERIAL CH-2 CAST CARBON STEEL
HEAT TREAT CARBURIZE 0.010 CASE
HARDNESS Rc 50-55
DWG A-169 / 208
REQUIRED 24
MITER BEVEL GEAR
BP-33
R.12
1.34 D
4.243 E
45° C
F
G
H
I
45°
.50
R.12
A 6.236
B
3.12
2.25
J
32
16
R.25
.75
R.25
R.25
R.12
R.06
1.95 K
L
2.85
4.32

ASSIGNMENT—UNIT 33

MITER BEVEL GEAR (BP-33)

Student's Name ______________________

1. Name the view which represents the miter bevel gear.
2. Describe straight miter bevel gears.
3. State two general methods of machining bevel gear teeth.
4. Give the specifications for the:
   a. material
   b. heat treatment
   c. hardness testing
5. Give the surface finish dimensions for the:
   a. Bored hole
   b. Sides of the hub section
   c. Tooth face
6. Determine the following upper and lower limit dimensions for the keyway:
   a. Width
   b. Overall height
7. Determine the upper and lower limit dimensions for machining the following bevel gear blank features:
   a. Width of hub
   b. Bore diameter
   c. Recessed bore face (J)
   d. Back edge angle
   e. Face width
   f. Outside diameter
8. Give the letter on the drawing which corresponds to each of the following bevel gear features:
   a. Pitch cone radius
   b. Crown height
   c. Outside diameter
   d. Face width
   e. Back edge angle
   f. Addendum angle
   g. Dedendum angle
9. Compute the following nominal dimensions. Use handbook tables, as required:
   a. Pitch diameter
   b. Addendum angle
   c. Dedendum angle
   d. Cutting angle
   e. Face angle
   f. Whole depth of tooth
   g. Crown backing
10. Give the upper and lower limit dimensions for the (a) cutting angle and (b) face angle.

1. ______________________
2. ______________________
____
____
3. (a) ______________________
(b) ______________________
4. a. ______________________
b. ______________________
c. ______________________
5. a. ________ b. ________ c. ________
6. a. Width = ______________________
b. Overall Height = ______________________

7.

	Upper Limit	Lower Limit
a. Width of Hub =	________	________
b. Bore Diameter =	________	________
c. Recessed Face =	________	________
d. Back Edge Angle =	________	________
e. Face Width =	________	________
f. Outside Diameter =	________	________

8. a. ____________ e. ____________
b. ____________ f. ____________
c. ____________ g. ____________
d. ____________

9. a. ______________________
b. ______________________
c. ______________________
d. ______________________
e. ______________________
f. ______________________
g. ______________________

10.

	Upper Limit	Lower Limit
(a) Cutting Angle	________	________
(b) Face Angle	________	________

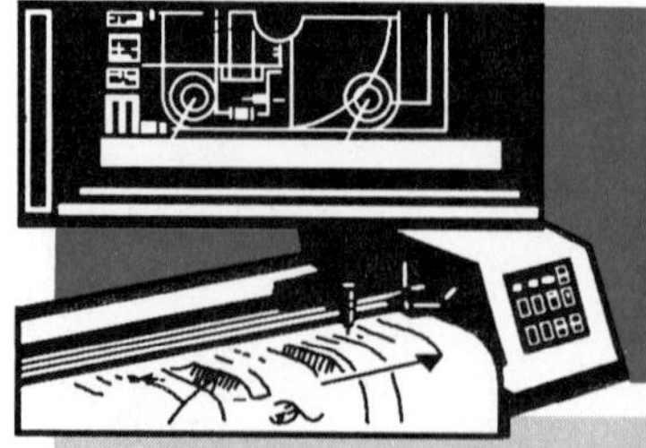

HIGH TECHNOLOGY applications

HT-9 CADD/CIM AND INDUSTRIAL ROBOTS

PARABOLIC-SHAPE ROBOTIC ENVELOPE MOVEMENTS

Robots serve two major computer-programmed functions in industry: handling and processing. *Handling applications* relate to the movement of materials, parts, tooling; loading and unloading work stations; storage, retrieval, and conveyor systems; and associated machine tool handling requirements. *Processing applications* deal with the actual manipulation of tools and equipment to perform specific computer programmed tasks such as: welding, spraying, assembling, and others.

Movements of robots are linear and rotary along a number of different axes. The photo of an industrial robot illustrates six basic axes of movement. The two line art views show the parabolic-shape *work envelope.* This envelope identifies the range, reach, and degrees of freedom of the robot.

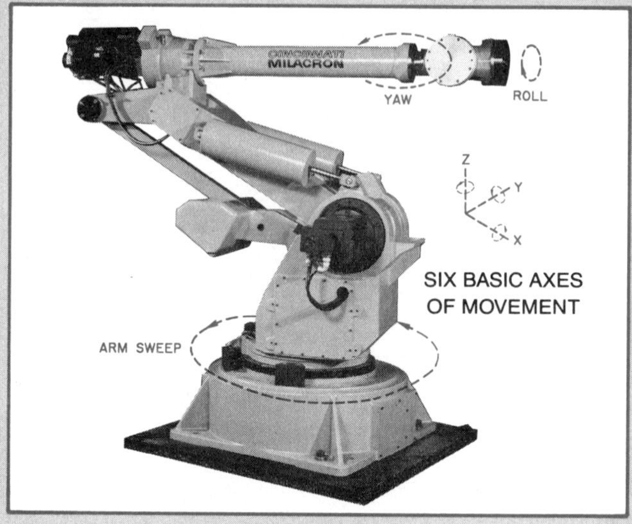

PROGRAMMING FOR TEACH AND AUTOMATIC ROBOTIC MODES

Robots are computer controlled and are generally programmed through a two-mode control system that includes a *teach mode* and an *automatic mode.* A robot is programmed in a teach mode for position, function, and speed, for a sequence of points. The program is then viewed on a computer screen and modified, as needed.

Once it is established that the verified teach functions and movements can be performed safely, they are computer programmed for the automatic mode. As a part of a computer-aided manufacturing network, sequences of computer operations and changes may be entered into computer memory.

Generally there are two sets of robot *grippers (fingers).* One set is used to load parts with unmachined surfaces. A second set of grippers for the robot hand (*end effector*) is designed to protect finished surfaces during removal.

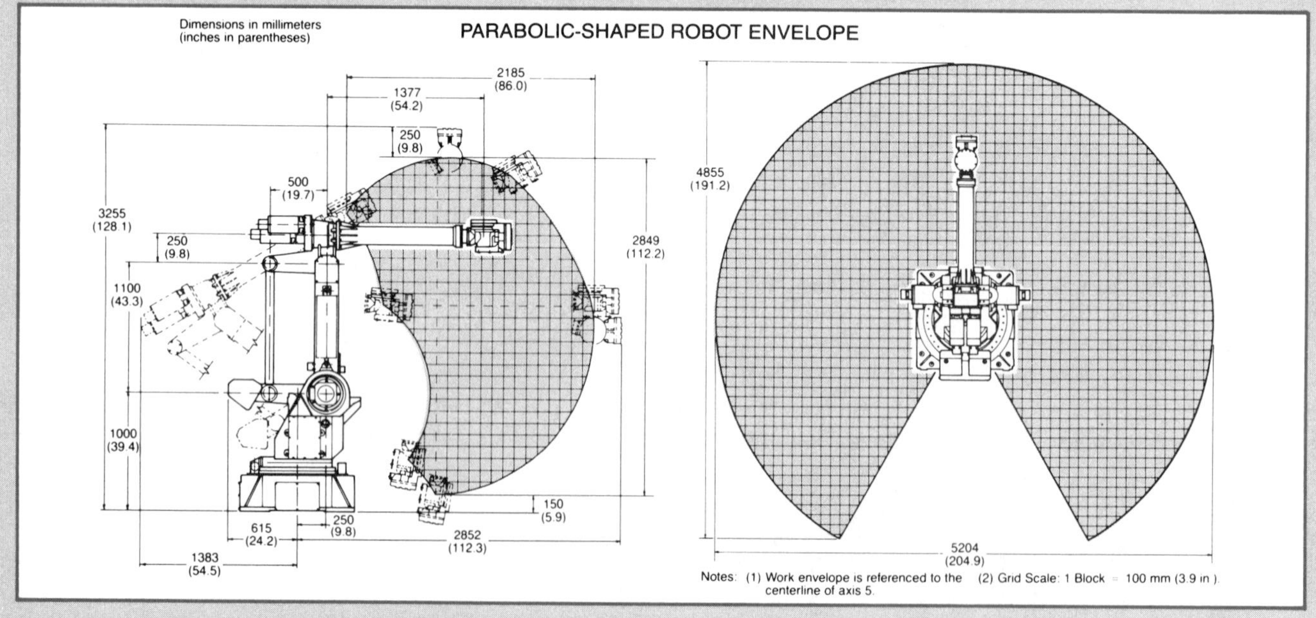

Drawings courtesy of CINCINNATI MILACRON INC.

UNIT 34
Cam Motions and Cam Drawings

A *cam* is a machine element which transmits rotating, oscillating, or reciprocal motion through another machine element known as a *follower.* The motion may be continuous for a full cycle or interrupted for part of a cycle. *Motion* relates to the *rate of movement.* The distance a rotating cam moves a follower is called *displacement.*

FUNCTIONS OF CAMS AND FOLLOWERS

Two common cam designs are known as *radial* (or plate, or disc) and *cylindrical (drum)* (figure 34-1). These cams transform rotary motion to linear motion. A cam is said to be of the *positive type* when the follower is continuously engaged. A pin or roller on the follower rides in a machined groove in the cam. The two line drawings (figure 34-1) show a tapered and straight side roller follower for a cylindrical cam and a recessed plate cam.

Figure 34-2 illustrates two *nonpositive types* of followers for plate and toe and wiper cams. A nonpositive follower requires gravity, a spring, or other additional force to hold the follower against a cam.

Line drawings of three other basic cam follower designs appear in figure 34-3. The knife-edge follower at (A) permits precise movements and acceleration, particularly where motion must be transmitted corresponding to sharp contour features of the mating cam.

CAM MOTIONS AND DEFINITIONS

There are three common cam and follower motions:

- Uniform (constant velocity)
- Parabolic
- Harmonic

Cams are designed to transmit a single motion, two of these motions, or a combination of all three motions.

TAPERED ROLLER FOLLOWER

CYLINDRICAL (DRUM) CAM

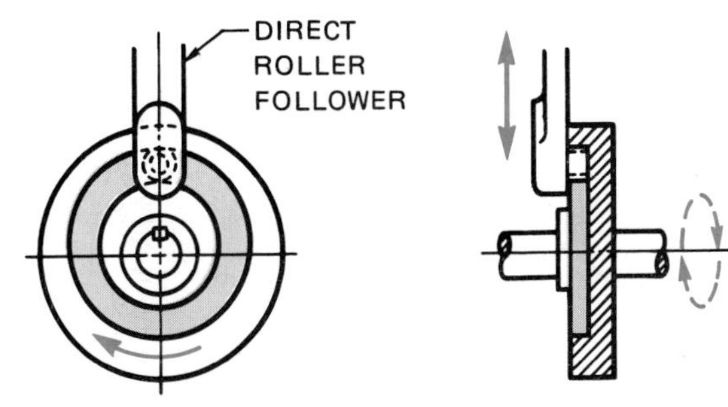

ECCENTRIC RECESSED PLATE CAM

FIGURE 34-1 EXAMPLES OF POSITIVE-TYPE CYLINDRICAL AND PLATE (RADIAL) CAMS

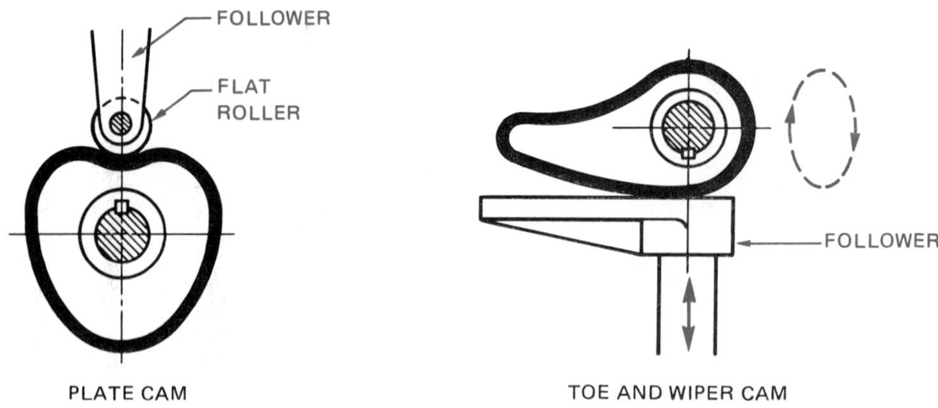

FIGURE 34-2 NONPOSITIVE TYPES OF FOLLOWERS

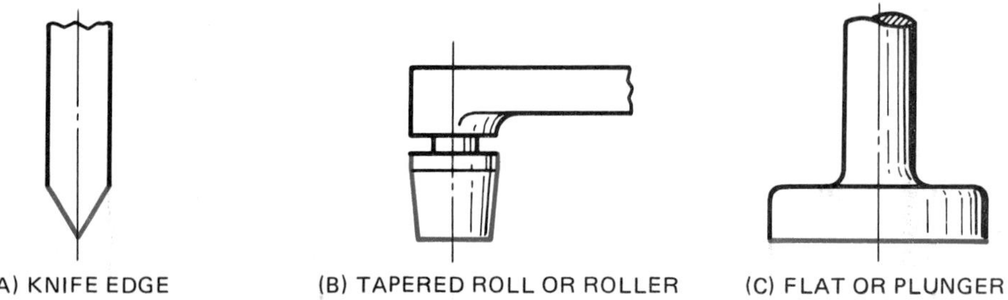

FIGURE 34-3 COMMON CAM FOLLOWER DESIGNS

Uniform (constant velocity) Cam and Follower Motion. A motion which causes a cutting tool or other mechanical device to rise and fall at a constant rate of speed from the start to the finish of a stroke or cycle.

Parabolic Motion. A uniformly accelerated or decelerated motion.

Harmonic Motion. Cam and follower movements which require a constant velocity, but not necessarily a uniformity of motion. High-speed mechanisms often depend on harmonic motion.

The constant speed rise and fall on a uniform motion cam and follower, as an example of movement through 90°, is illustrated at (A) in figure 34-4. Usually, straight-line uniform motion is designed with arcs near the peaks. The arcs relieve the shock which otherwise would be produced by rapidly starting and stopping the follower.

Figure 34-4B represents harmonic motion during another 90° phase of a cycle. Figure 34-4C shows how uniform accelerated-decelerated motion is represented on a blueprint. The acceleration and deceleration occur at a constant speed.

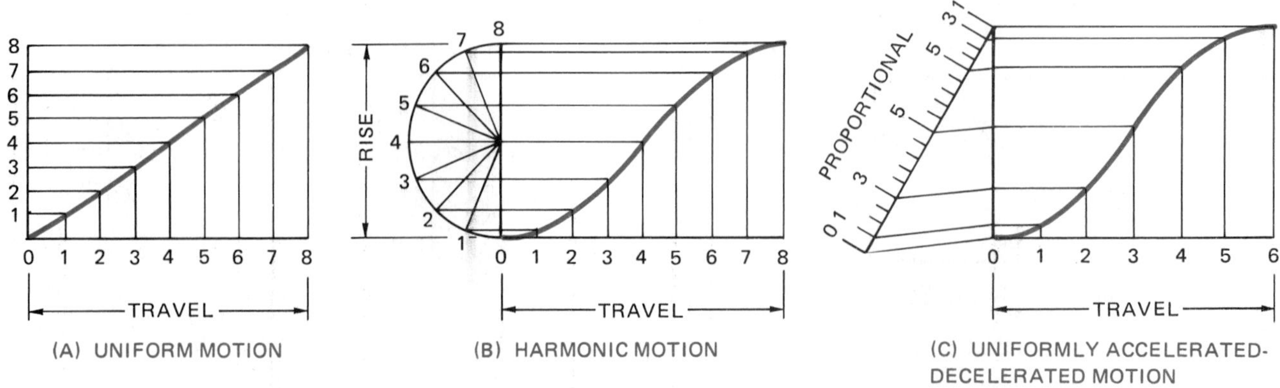

FIGURE 34-4 DRAFTING ROOM AND LAYOUT TECHNIQUES OF PRODUCING THREE KINDS OF MOTION

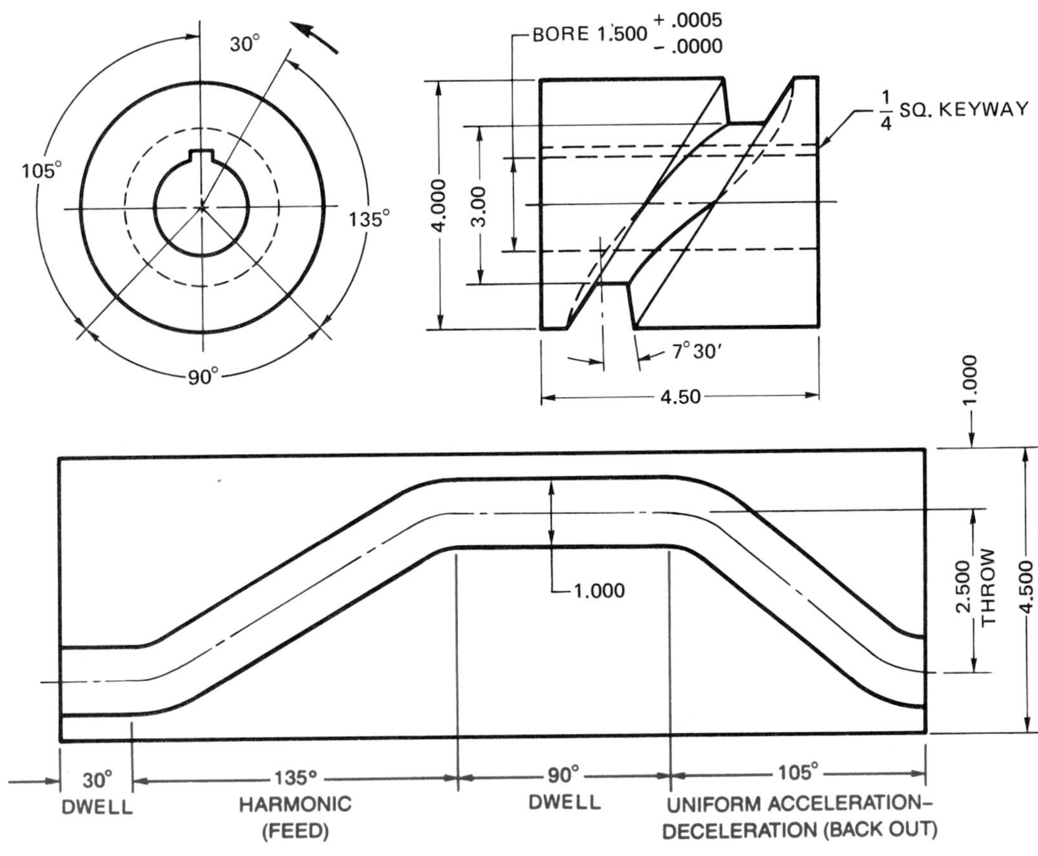

FIGURE 34-5 WORKING DRAWING WITH DIMENSIONS FOR MACHINING A CYLINDRICAL CAM

BASIC CAM TERMS

Six terms are generally applied in the shop relative to cams. The terms are *lobe, rise and fall, lead, profile,* and *dwell.*

Lobe. A projecting area of a plate or disc cam. The lobe imparts a reciprocal motion to the follower. Single, double, and triplicate lobes indicate the number of times the follower is caused to rise or fall in one cam revolution.

Rise and Fall. The distance a cam lobe (or other feature) raises or lowers the follower.

Lead. The total distance a 360° uniform rise cam (with one lobe) moves a follower in one revolution.

Cam Profile. The actual form (profile) of the working surface of a cam.

Dwell. The time interval when a moving cam produces no movement in a follower.

REPRESENTATION OF CAM AND FOLLOWER MOTIONS ON DISPLACEMENT DIAGRAMS

Cam motions are visually represented by a *displacement diagram.* Such a diagram provides information about the kind of motion and the extent to which a cam causes a follower to rise or fall during a part or for a total working cycle.

A displacement diagram (figure 34-5) shows a *working circle (cycle)* of 360° divided into a uniform number of degrees or parts. In this case, the length of the displacement diagram is equal to the circumference of the cylindrical cam. The follower movement is plotted as a rise, fall, or dwell.

The displacement diagram shows that through 135° the cam movement is harmonic motion. The follower is lifted gradually from its starting position to a halfway point in the full rise. The speed then gradually decreases until the follower reaches the full rise. After a dwell where there is no feeding of the follower for 90°, the cam motion for the next 105° is uniform accelerated-decelerated. The cutting tool, through the uniform motion of the follower, is backed out of the cut. As the cam continues to rotate, there is a dwell for the remaining 30° of the cycle.

A displacement diagram is used for layout, machining, and measuring the cam profile for cylindrical (drum) as well as plate (disc) type cams.

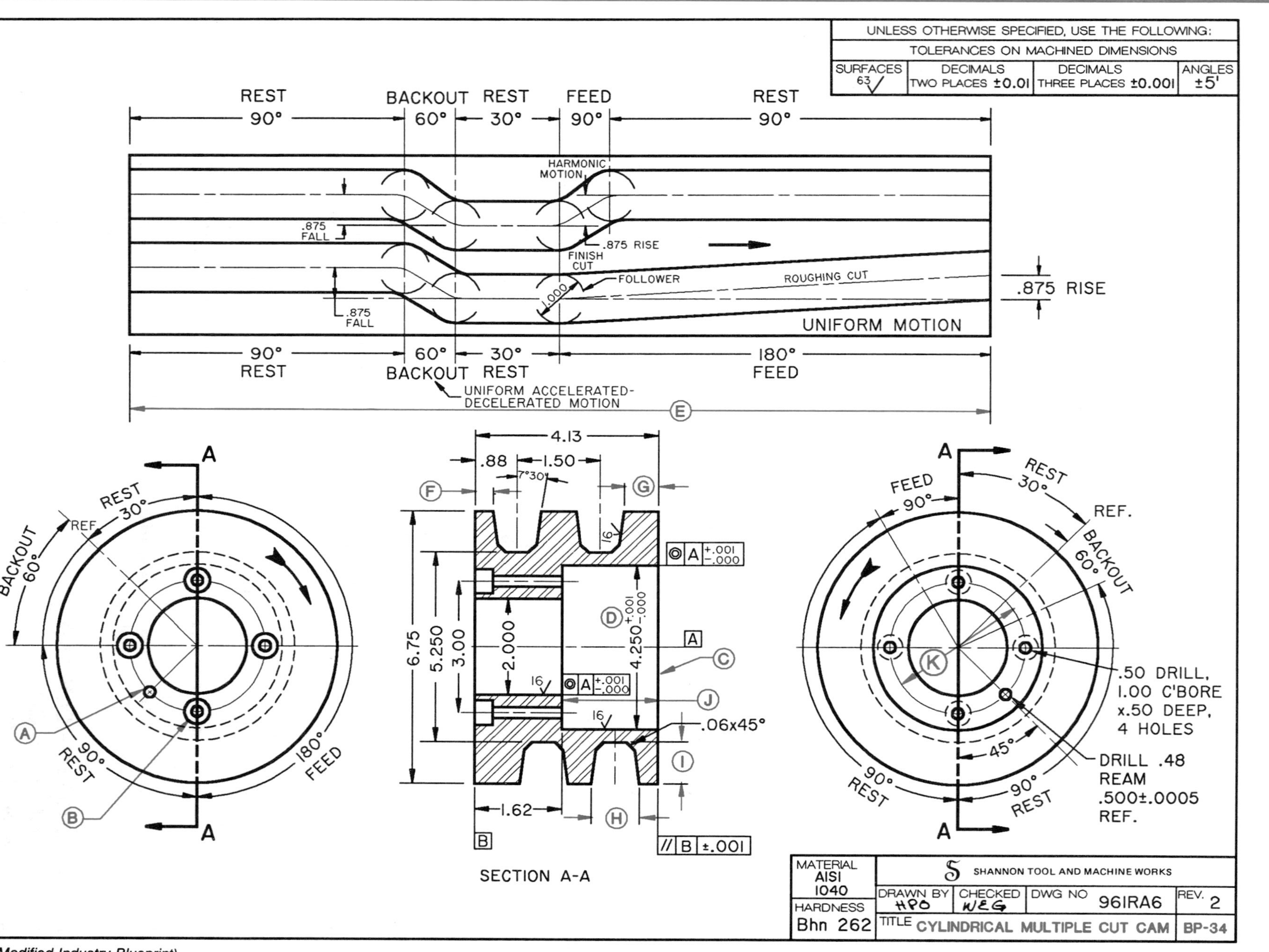

(Modified Industry Blueprint)

ASSIGNMENT—UNIT 34

CYLINDRICAL MULTIPLE CUT CAM (BP-34)

Student's Name ____________

1. Name the three views and the layout drawing.
2. State the purpose served by hole (A).
3. Give the upper and lower dimensional limits of hole (A).
4. State the specifications for holes (B).
5. Describe the geometric tolerancing which applies to face (C).
6. a. Give the upper and lower limits for dimension (D).
   b. Translate the geometric tolerancing symbol used with dimension (D).
7. Give (a) the material and (b) the hardness specifications.
8. Specify the accuracy of the surface texture for the groove walls and the two bored holes for the shaft.
9. Determine dimensions (E), (F), (G), (H), (I), (J), and (K).
10. State (a) the cam motion used for the roughing cut and (b) the degrees of arc for the cutting and tool return areas of the cam.
11. State (a) the cam motions used for the finish cut and (b) the degrees of arc for the cut and tool return.
12. Give the degrees of arc for the rest areas between the end and beginning of each cut.
13. Determine the throw of the cam for (a) the roughing and (b) the finish cuts.
14. Give the nominal root diameter of the cam grooves.
15. Determine the (a) nominal, (b) upper limit, and (c) lower limit of the included angle for the follower.
16. Compute the nominal depth of the cam grooves.

1. a. ____________
   b. ____________
   c. ____________
   d. ____________
2. ____________
   ____
3. Upper Limit ______ Lower Limit ______
4. ____________
   ____
5. ____________
   ____
6. a. Upper Limit ______ Lower Limit ______
   b. ____________
   ____
7. a. Material ____________
   b. Hardness ____________
8. ____________
9. (E) = ______ (I) = ______
   (F) = ______ (J) = ______
   (G) = ______ (K) = ______
   (H) = ______
10. a. Motion ______ b. Degrees of Arc ______
11. a. Motion ______ b. Degrees of Arc ______
12. ____________
    ____
13. a. Roughing Cut ______ Finish Cut ______
14. Nominal Root Diameter ____________
15. a. Nominal Included Angle ____________
    b. Upper Limit ____________
    c. Lower Limit ____________
16. Nominal Groove Depth = ____________

UNIT 35
Bearings: Features, Representation, and Applications

SLIDING CONTACT, PLAIN JOURNAL, AND THRUST BEARINGS

The two major classifications of bearings are *sliding contact* and *rolling contact*. Sliding contact bearings permit movement between mating surfaces according to three basic designs:

- *Radial sleeve bearings* support rotating shafts or journals. The most common designs are the full journal bearing with 360° contact and the partial journal bearing with 180° contact.
- *Thrust bearings* support axial loads on rotating members.
- *Guide or slipper bearings* guide moving parts along a straight path.

Motion between parts of plain, sliding contact bearings takes place under one or a combination of the following conditions:

- *Dry operation* without gaseous or liquid lubrication, as in the case of plastics;
- *Hydrodynamic condition* in which the bearing surfaces are separated by a wedge (film buildup) of a lubricating medium; or
- *Hydrostatic condition* wherein the lubricating medium, under pressure, produces a force opposite to the applied load. This action causes the rotating element to lift away from the bearing surface.

ROLLING CONTACT (ANTI-FRICTION) BEARINGS

Balls or rollers are substituted in rolling contact bearings for the hydrodynamic or hydrostatic conditions required in sliding contact bearings to carry the load. The precision of rolling contact bearings accounts for their durability, limited runout, accurate radial and axial clearances, and smooth operation. Due to the fact that balls and rollers have a greatly reduced starting friction, roller contact bearings are designated as *anti-friction bearings*.

Precision anti-friction bearings have a low torque, a minimum of vibration, extreme accuracy of rotation and position, and excellent endurance operating at high speeds and temperatures.

STANDARDS FOR DIMENSIONS, TOLERANCES, AND FITS

Standards, as specified by the Anti-Friction Bearing Manufacturer's Association (AFBMA), are used in the production of interchangeable deep-groove ball and roller bearings. Such bearings conform to standard dimensions, tolerances, and fits.

The AFBMA has also adopted a uniform designation code to assist manufacturers and users. Letter and number symbols identify such characteristics of bearings as series, type, and size; closures (open, shielded, sealed); radial run-out and play; material; and cage design, etc. Tech-

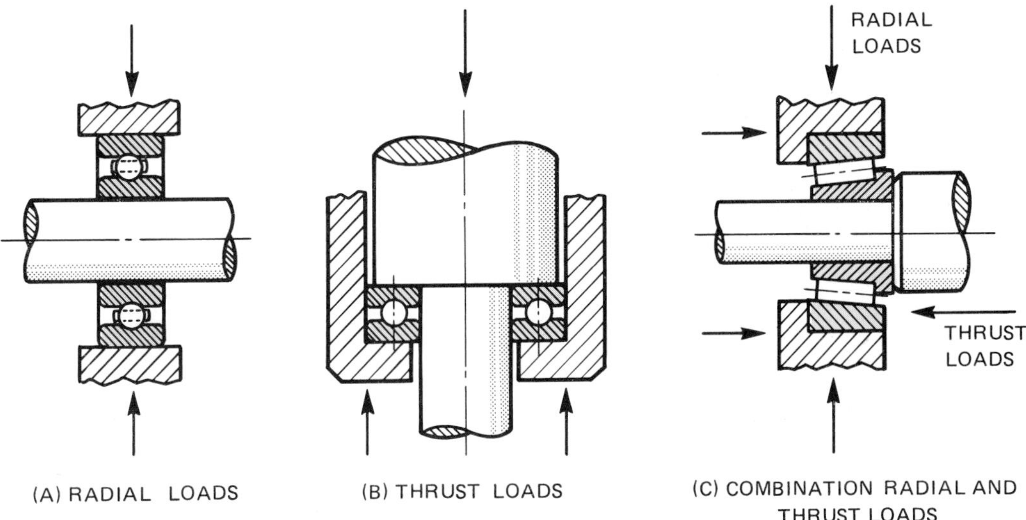

FIGURE 35-1 BEARING LOADS SUPPORTED BY BALL AND ROLLER BEARINGS

niques of representing bearings on drawings have also been standardized.

In addition, AFBMA ball bearing standards have been accepted by the ANSI. In turn, through international agreements, the AFBMA/ANSI standards conform with the standards for the manufacturing of standard grade ball bearings developed by the International Organization for Standardization (ISO).

BEARING LOADS

Ball and roller bearings support (a) radial, (b) thrust, and (c) combination radial and thrust loads. Figure 35-1 shows radial and thrust applications of ball bearings at (A) and (B) and the use of such bearings for combination loads at (C). Since roller bearings have a higher load capacity than ball bearings, they are applied to meet slower speed, heavy duty requirements.

GENERAL BALL BEARING DESIGNS

There are three general categories of ball bearings: (1) *deep groove*, (2) *angular contact*, and (3) *ball thrust*. Features of each are shown graphically in figure 35-2. Each category of ball bearings is designed for a different set of applications.

Deep Groove Ball Bearings

This design (figure 35-2A) meets specifications for a bearing to sustain a radial and a thrust load in either direction, operating up to very high speeds.

Angular Contact Ball Bearings

Angular contact bearings (figure 35-2B) are suited for high-speed applications, to support axial

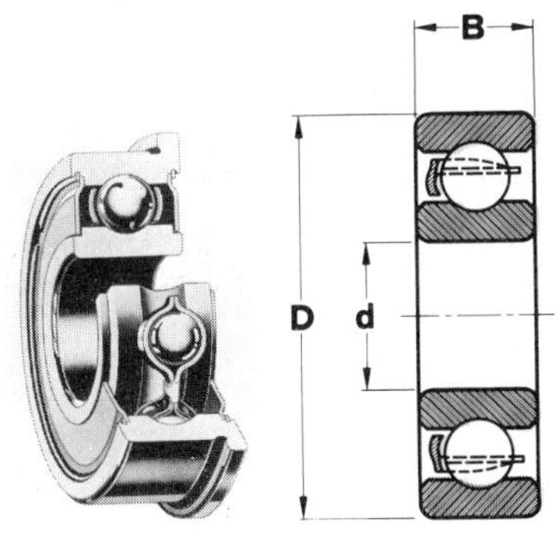

(A) DEEP GROOVE BEARING

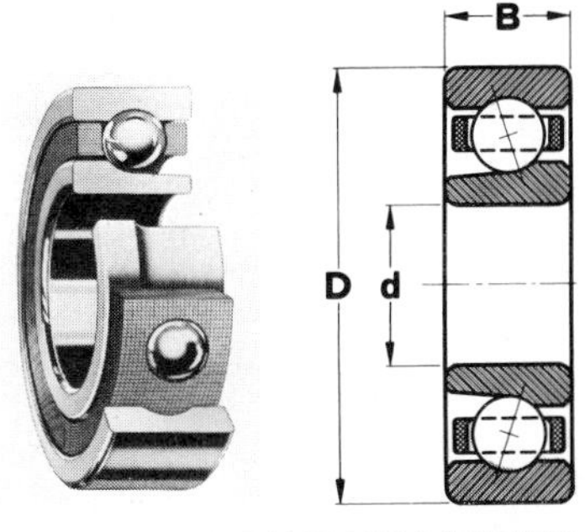

(B) ANGULAR CONTACT BEARING

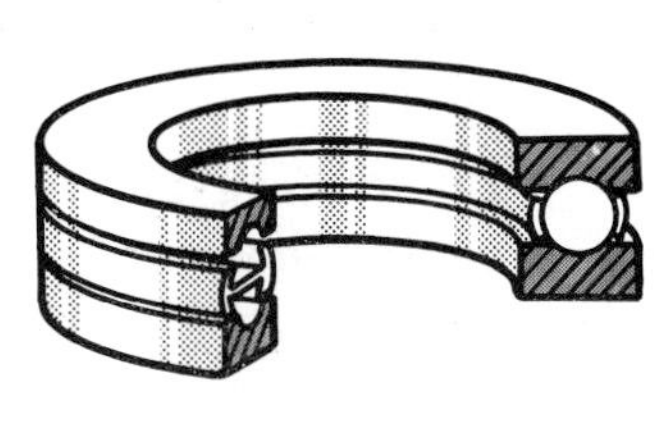

(C) BALL THRUST BEARING

FIGURE 35-2 GENERAL BALL BEARING DESIGNS (COURTESY OF THE BARDEN CORPORATION)

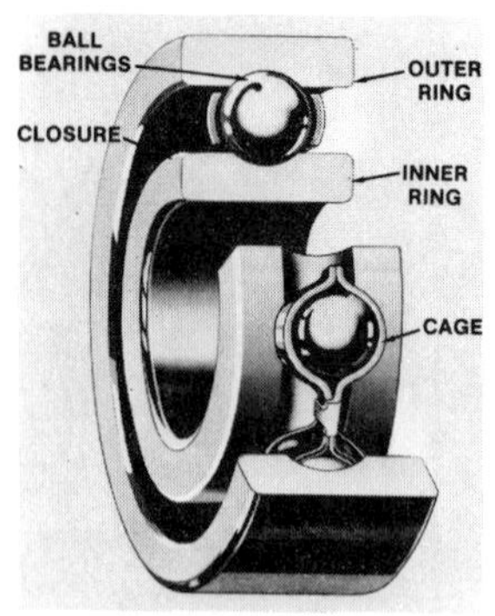

MAIN PARTS OF ANTI-FRICTION BEARINGS

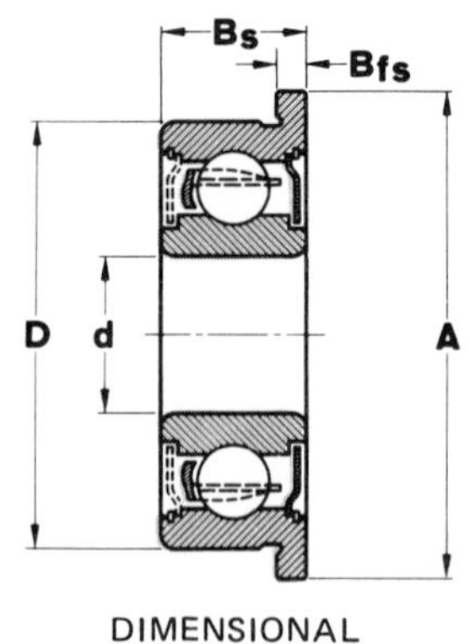

DIMENSIONAL SPECIFICATIONS

FIGURE 35-3 REPRESENTATION OF BEARING PARTS AND SPECIFICATIONS (EXAMPLE: DEEP-GROOVE BALL BEARING WITH SHIELDED, FLANGED OUTER RING) (COURTESY OF THE BARDEN CORPORATION)

and radial loads, and where the rigidity of the system requires preloading. Thrust loads are limited to one direction. *Separable and nonseparable* types of angular control bearings are available. The separable type of bearing permits the assembly of rotating members when the bearing is located in a blind hole.

Ball Thrust Bearings

This type of bearing (figure 35-2C) is widely used in machine tool, automotive, aerospace, and other industries to accommodate thrust loads in one direction. Ball thrust bearings have high thrust capacity and minimum axial deflection.

FEATURES OF BALL AND ROLLER BEARINGS

There are five principal parts to ball and roller bearings (figure 35-3):

- *Outer Ring.* This member is generally designed for either a press fit into a bored hole and/or to shoulder against a flanged surface.
- *Inner Ring* for deep groove or angular contact-bearing designs and to permit mounting on shafts.
- *Cage* for accurately spacing and holding the balls. Cages are made of metal, phenolic, and other plastics. A few selected cage designs are shown in figure 35-4.
- *Bearings* of ball, roller, or needle design. The bearings are heat treated, ground, and lapped to extremely precise dimensions, and tolerances for geometric features.

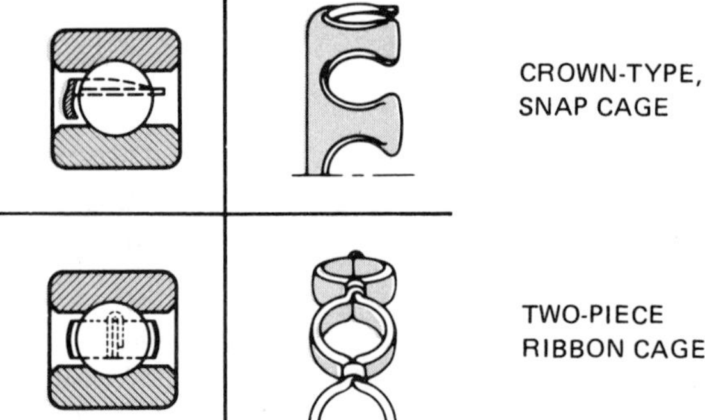

DEEP GROOVE BEARINGS

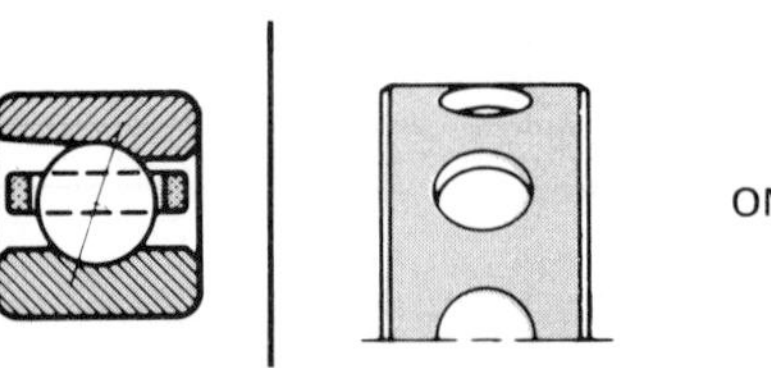

ANGULAR CONTACT BEARING

FIGURE 35-4 CROSS SECTIONS AND ILLUSTRATIONS OF SELECTED CAGES (COURTESY OF THE BARDEN CORPORATION)

- *Bearing Closures (shields and seals).* Closures exclude contaminants, contain the lubricant, and prevent damage to the inner parts during assembly. Bearings may be *open* or *shielded* with a bearing closure.

PRELOADED BEARINGS

Some bearing applications require that a permanent load thrust be placed on the bearings. This condition is called *preloading.* Preloading is used to:

- Eliminate all radial and end play,
- Increase the rigidity of the system,
- Prevent ball skidding under very high acceleration,
- Limit any change in the inner and outer ring contact angle at very high speeds, and
- Reduce nonrepetitive runout.

Figure 35-5A shows a set of angular contact bearings designed for preloading. Note that the inside faces are relieved by an amount called the *preload offset.* At position (B) the *duplexed (set) bearings* are secured in operating position. The faces of the inner rings are in contact. In cases where bearings are seated against internal shoulders, the total preload offset is designed into the outer ring.

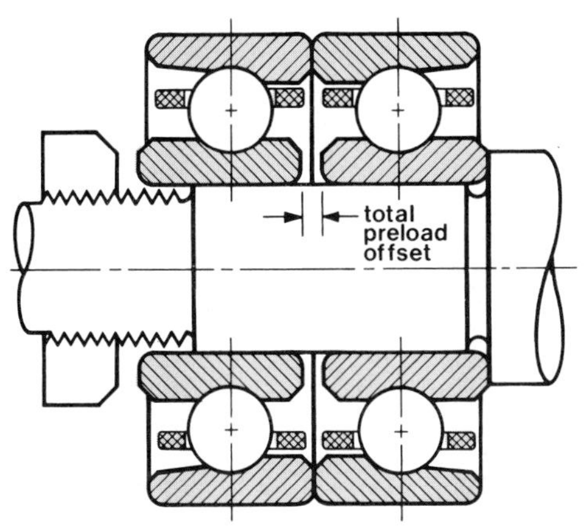

(A) CONDITION PRIOR TO FINAL ASSEMBLY

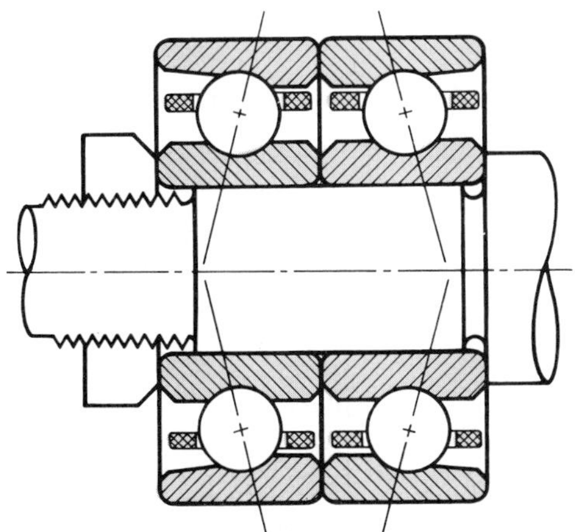

(B) OPERATING POSITION (FINAL ASSEMBLY)

FIGURE 35-5 APPLICATION OF PRELOADED ANGULAR CONTACT DUPLEXED BEARINGS (COURTESY OF THE BARDEN CORPORATION)

NOTE: REFER TO APPENDIX B-4 FOR THE LARGER SCALE DRAWING TO USE WITH THIS ASSIGNMENT.

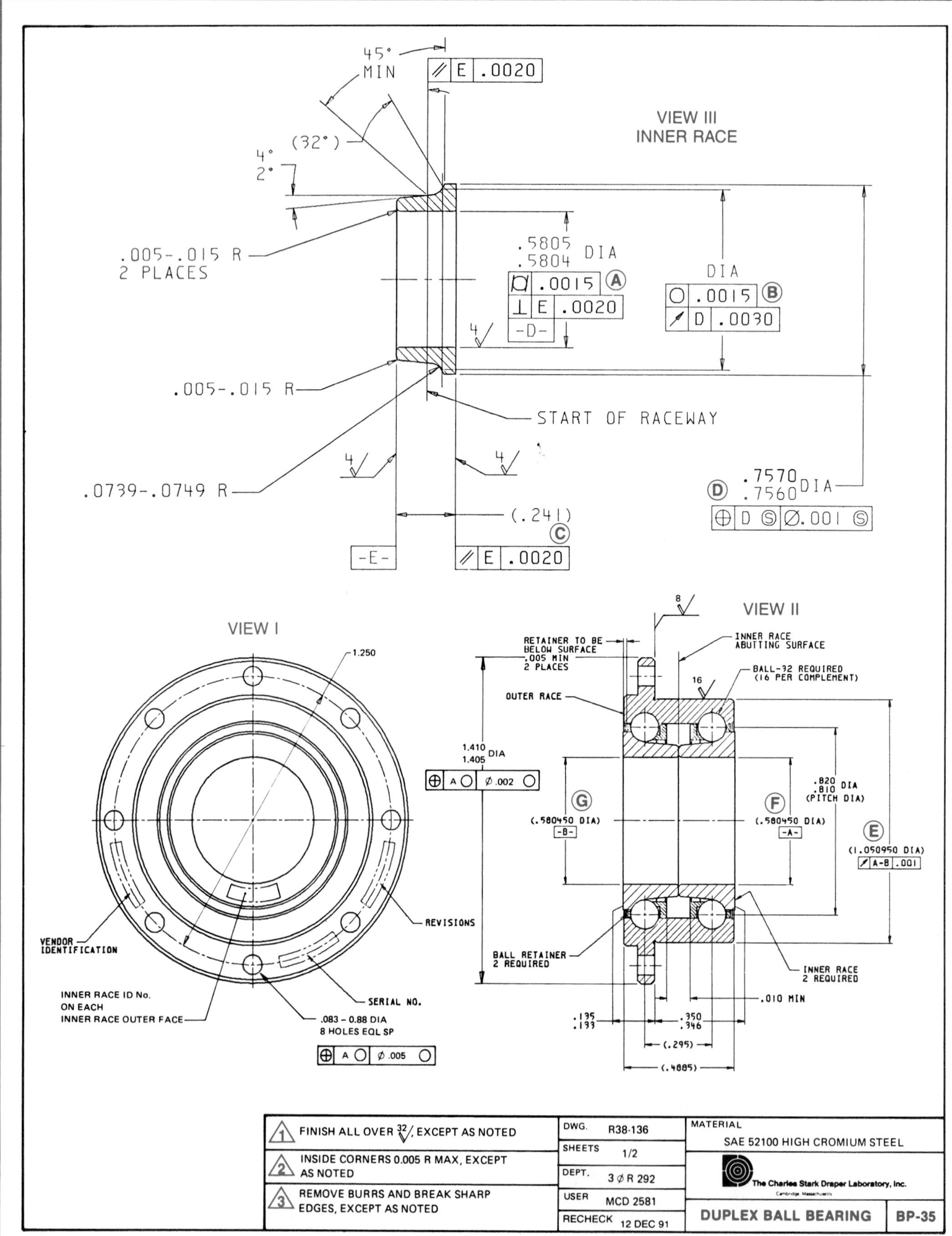

Courtesy of The Charles Stark Draper Laboratory, Inc.
(Computer-aided design (CAD) drawing)

ASSIGNMENT—UNIT 35

DUPLEX BALL BEARING (BP-35)

Student's Name ____________

1. Name **VIEWS I**, **II**, and **III**.
2. a. Identify the basic type of ball bearing as represented by the drawing.
   b. State a principal use for this type of ball bearing.
3. State two functions for each of the following anti-friction ball bearing parts:
   a. Cage
   b. Bearing closures (shields/seals)
4. Indicate the material from which the balls and rings are to be machined.
5. List the identifying information provided on the outer face of (a) the outer race and (b) the inner race.
6. Establish the following information about the holes in the outer flange:
   a. Nominal pitch diameter
   b. Machining specifications
7. Provide the following dimensions for the outer race:
   a. Upper and lower limits for the outside diameter of the flange.
   b. Overall width (nominal width).
   c. Center-to-center distance between the two raceways (reference dimension).
   d. Upper and lower limits for the pitch diameter of the raceways.
8. Determine the upper and lower limit dimensions of the following features of the inner races:
   a. Bore diameter
   b. Ground angle of the body
   c. Reference width
9. Indicate the surface specifications for the following surfaces:
   a. Outer Race
      (1) Inside flange face
      (2) Body outside diameter
   b. Inner Race
      (3) Bore
      (4) Outside faces
10. Interpret each of the following geometric tolerancing symbols:
    (A) | ⌭ | .0015 | / | ⊥ | E | .0020 | / | -D- |
    (B) | ○ | .0015 | / | ↗ | D | .0030 |
    (C) | // | E | .0020 |
    (D) | ⊕ | D Ⓢ | ϕ .001 Ⓢ |
    (E) | ↗ | A-B | .001 |
11. Explain the meaning of the following supplemental notes:
    △1 △2 △3

1. **VIEW I** ____________
   VIEW II ____________
   VIEW III ____________
2. a. ____________
   ____
   b. ____________
   ____
3. a. Cage (1) ____________
   ____
   (2) ____________
   ____
   b. Closures (1) ____________
   (2) ____________
4. ____________
5. a. Outer Race ____________
   ____
   ____
   b. Inner Race ____________
   ____
6. a. ____________
   b. ____________
   ____
7. a. Flange O.D. Max. ____________
   Min. ____________
   b. Overall Nominal Width ____________
   c. C-to-C Distance ____________
   d. Pitch Diameter Max. ____________
   Min. ____________
8. Upper Lower
   a. Inside Diameter ______ ______
   b. Body Angle ______ ______
   c. Width (Reference) ____________
9. a. Outer Race
   (1) Flange Face ____________
   (2) Body Outside Diameter ____________
   b. Inner Race
   (3) Bore ____________
   (4) Outside Faces ____________
10. (A) ____________
    (B) ____________
    (C) ____________
    (D) ____________
    (E) ____________
11. △1 ____________
    △2 ____________
    △3 ____________

UNIT 36
Splines: Functions and Design Features

SPLINES

The term *splines* relates to a series of uniform, parallel surfaces (teeth) which are machined in a shaft or are cut into a mating hub or other fitting. Shafts, hubs of sliding gears and drives, and other members are splined to serve three main functions:

- To couple shafts to transmit heavy loads without slippage (press fit).
- To transmit power between a shaft and a sliding machine member, as in the case of movable gears in a quick-change gear box (sliding fit).
- To hold parts securely that may later require removal in order to change an angular position relationship (sliding fit).

American National Standard Involute Splines

There are two common forms of splines:

- Parallel side splines having straight sides, as shown in figure 36-1.
- ANSI involute splines which are similar in form to a 30° pressure angle internal and external involute gear (figure 36-2).

The involute spline has three advantages over parallel side splines.

- Involute splines have more strength and greater torque-transmitting ability.
- Involute splines may be produced by gear manufacturing processes and may be fitted accurately.
- Involute splines are self-centering under load to equalize stresses.

DESIGN FEATURES AND DRAWING REPRESENTATION

Generally, involute splines are designed at 30°, 37 1/2°, and 45° pressure angles. There are two types of fits for ANSI splines: *side fit* and *major diameter fit*. The tooth sides of side fit splines act as drivers and to centralize the splines. The

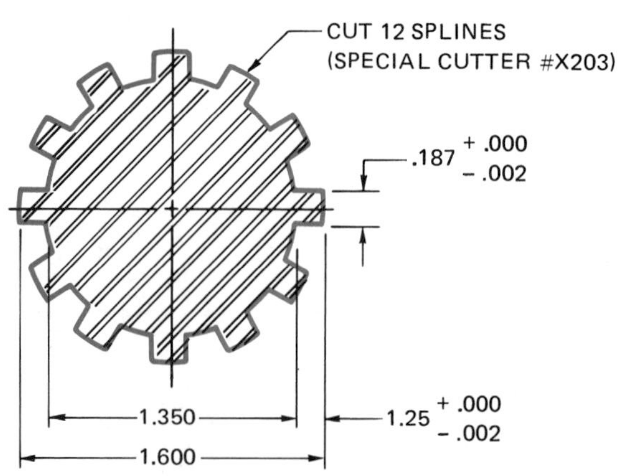

FIGURE 36-1 EXAMPLE OF PARALLEL SIDE SPLINE

INVOLUTE SPLINE DRAWING DATA (FLAT ROOT, MAJOR DIAMETER FIT)

	(B) INTERNAL	(C) EXTERNAL
Number of Teeth	XX	XX
Pitch	XX/XX	XX/XX
Pressure Angle	30°	30°
Base Diameter (REF)	X.XXXXX	X.XXXXX
Pitch Diameter (REF)	X.XXXXX	X.XXXXX
Major Diameter	X.XXXX/X.XXXX	X.XXXX/X.XXXX
Form Diameter	X.XXX	X.XXX
Minor Diameter	X.XXX/X.XXX	X.XXX minimum
*Circular Space Width	*Maximum Actual X.XXXX *Minimum Effective X.XXXX	

	(B) INTERNAL	(C) EXTERNAL
*Circular Tooth Thickness		*Maximum Effective X.XXXX *Minimum Actual X.XXXX

*NOTE: The major diameter and effective spline must be concentric at maximum material conditions.

The following information may be added as required:

	(B) INTERNAL	(C) EXTERNAL
Pin Diameter	X.XXXX	X.XXXX
Pin Measurements (REF)		
Maximum Between Pins	X.XXX	
Minimum Over Pins		X.XXXX
Corner Clearance	X.XXX/X.XXX	
Chamfer Height		X.XXX/X.XXX

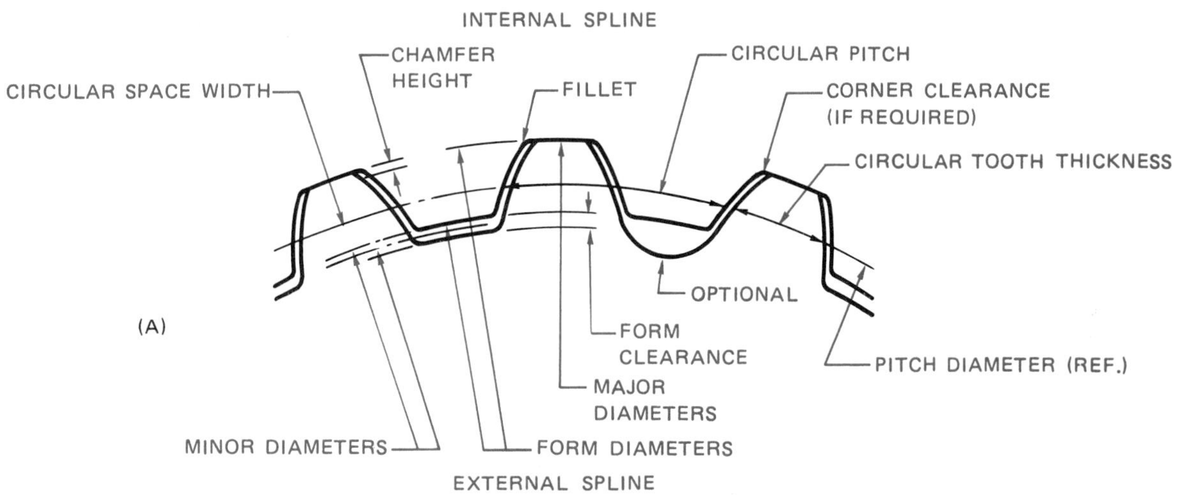

FIGURE 36-2 ANSI INVOLUTE SPLINE TERMS, SYMBOLS, AND DRAWING DATA (30° PRESSURE ANGLE, FLAT ROOT, MAJOR DIAMETER FIT)

mating parts of the major diameter fit contact at the major diameters for centralizing. The teeth sides serve as drivers.

Some of the major design features of an involute spline are represented in figure 36-2A. Oftentimes, instead of using a graphic illustration for spline teeth, the designer provides specifications only. In still other cases, a simplified drawing with accompanying data is provided.

The manner in which typical data is given appears in figure 36-2B for internal involute splines and at (C) for external involute splines. Similar information is provided for parallel side splines. Note that the number of X's following a dimension indicates the number of decimal places required in the design of each feature and/or machining and measuring.

Handbook tables provide formulas and dimensional specifications for basic dimensioning, classes of tolerances, flat root or rounded fillet root, key form measurements, and other engineering data. Within the shop, a splined shaft is often referred to as a *multiple-spline shaft*. The mating member is identified as a *fluted hole or fluted member.*

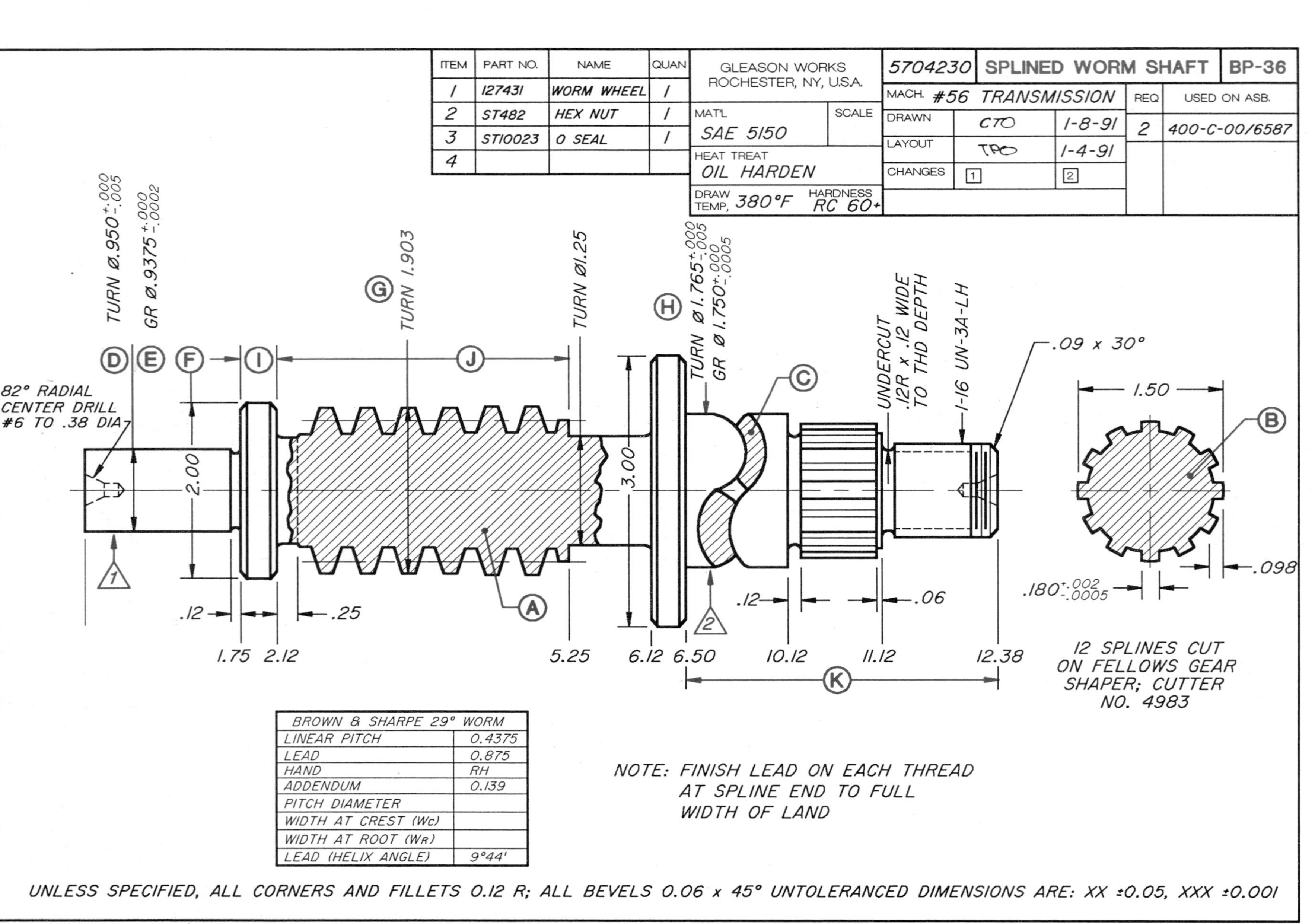
ITEM
PART NO.
NAME
QUAN
1 127431 WORM WHEEL 1
2 ST482 HEX NUT 1
3 ST10023 O SEAL 1
4
GLEASON WORKS
ROCHESTER, NY, U.S.A.
5704230
SPLINED WORM SHAFT
BP-36
MACH. #56 TRANSMISSION
REQ
USED ON ASB.
MAT'L SAE 5150
SCALE
DRAWN CTO 1-8-91
2
400-C-00/6587
LAYOUT TRO 1-4-91
HEAT TREAT OIL HARDEN
CHANGES 1 2
DRAW TEMP. 380°F
HARDNESS RC 60+
TURN Ø.950 +.000 -.005
GR Ø.9375 +.000 -.0002
G
TURN 1.903
TURN Ø1.25
H
TURN Ø1.765 +.000 -.005
GR Ø1.750 +.000 -.0005
UNDERCUT .12R x .12 WIDE TO THD DEPTH
1-16 UN-3A-LH
.09 x 30°
D
E
F
I
J
C
82° RADIAL CENTER DRILL #6 TO .38 DIA
2.00
3.00
1.50
B
1
A
2
.12
.25
.12
.06
.180 +.002 -.0005
.098
1.75 2.12
5.25
6.12 6.50
10.12
11.12
12.38
K
12 SPLINES CUT ON FELLOWS GEAR SHAPER; CUTTER NO. 4983
BROWN & SHARPE 29° WORM
LINEAR PITCH 0.4375
LEAD 0.875
HAND RH
ADDENDUM 0.139
PITCH DIAMETER
WIDTH AT CREST (WC)
WIDTH AT ROOT (WR)
LEAD (HELIX ANGLE) 9°44'
NOTE: FINISH LEAD ON EACH THREAD AT SPLINE END TO FULL WIDTH OF LAND
UNLESS SPECIFIED, ALL CORNERS AND FILLETS 0.12 R; ALL BEVELS 0.06 x 45° UNTOLERANCED DIMENSIONS ARE: XX ±0.05, XXX ±0.001

ASSIGNMENT—UNIT 36

SPLINED WORM SHAFT (BP-36)

Student's Name ____________________

1. a. Name the Ⓐ and Ⓑ sections of the Splined Worm Shaft.
   b. Explain briefly what function is served by design feature Ⓒ.
   c. Identify the material represented by section lining.
2. a. Identify the three types of dimensioning used on the drawing.
   b. State one advantage of using arrowless dimensioning.
3. Indicate the required material.
4. Give the specifications of the center-drilled holes.
5. Determine the upper and lower limit dimensions for diameters Ⓓ through Ⓗ.
6. Identify the processes for machining surfaces △1 and △2.
7. a. Give the specifications for the 60° V-form thread and interpret each symbol or value.
   b. Interpret the meaning of the dimension *.09×30°*.
   c. State the dimensions of the thread undercut.
8. Compute maximum lengths Ⓘ, Ⓙ, and Ⓚ.
9. a. Identify the major gear thread form.
   b. Translate the meaning of linear pitch.
   c. Tell what the different measurements (ratio) for linear pitch and lead mean.
   d. State how the end of each thread is to be machine finished.
10. Compute the following nominal dimensions:
    a. Pitch diameter, when $P_d = OD - 2A$
    b. Width (W_c) of flat at crest, when $W_c = 0.355 \times$ linear pitch
    c. Width (W_r) of flat at root, when $W_r = 0.310 \times$ linear pitch
11. a. Identify two common design forms for splines.
    b. Name the design form of the splines on the Shaft.
    c. Determine the maximum and minimum sizes for the following features:
       (1) Width of spline tooth
       (2) Depth of spline tooth

1. a. Ⓐ ____________________
   Ⓑ ____________________
   b. Ⓒ ____________________
   ____
   c. ____________________
2. a. (1) ____________________
   (2) ____________________
   (3) ____________________
   b. ____________________
   ____
   ____
3. ____________________
4. ____________________
   ____
   ____

5.

	Upper Limit	Lower Limit
Ⓓ	________	________
Ⓔ	________	________
Ⓕ	________	________
Ⓖ	________	________
Ⓗ	________	________

6. ____________________
7. a. ____________________
   ____
   ____
   b. ____________________
   c. ____________________
8. Ⓘ = ________ Ⓚ = ________
   Ⓙ = ________

9. a. ____________________
   b. ____________________
   ____
   ____
   c. ____________________
   ____
   d. ____________________
   ____
10. a. P_d = ________ c. W_r = ________
    b. W_c = ________
11. a. (1) ____________________
    (2) ____________________
    b. ____________________
    c.

	Maximum	Minimum
(1)	________	________
(2)	________	________

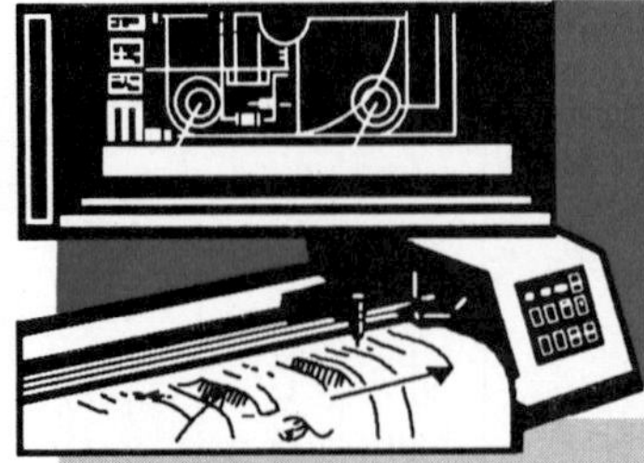

HIGH TECHNOLOGY applications

HT-10 CIM/CELLULAR MANUFACTURING: AUTOMATED FLOW OF PARTS

Work loading and unloading of CNC/CIM individual machine tools, machining centers, group technology machine cells, and totally flexible manufacturing systems require the use of work holding, positioning, delivery and off-loading equipment. Commercially available *palletizing systems* and *components* are used to provide for the accurate, continuous flow of materials and parts to vertical and horizontal machine tools.

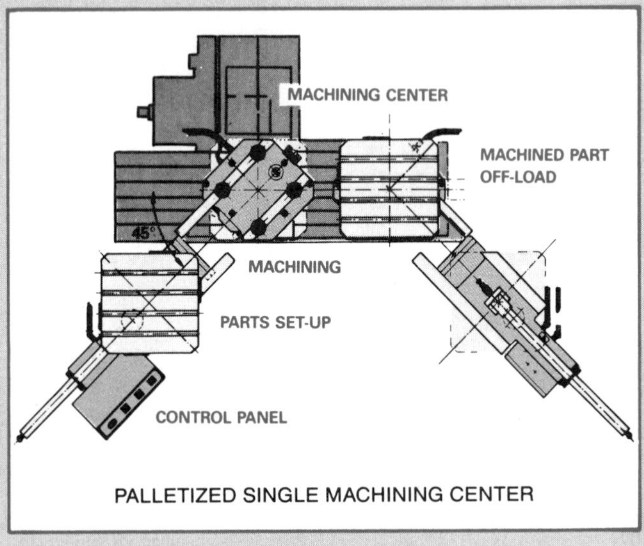

PALLETIZED SINGLE MACHINING CENTER

PALLETIZING COMPONENTS AND SYSTEMS

A *palletizing system* includes round, square, and rectangular mounting plates that are known as *pallets.* Parts are accurately positioned and secured on a pallet in relation to operations to be performed and for straight or rotational alignment at a load or unload station. The setup is then moved onto a *pallet receiver* and placed on a motorized *pallet transfer mechanism.*

A pallet transport system may be designed for a simple on-loading and off-loading machining center work station, as illustrated. Other more complex applications (such as the integration of two or more similar or different machining stations) may require the addition of a number of *rail guided vehicles* and *pallet loaders* to meet automatic on-load and off-load needs of machining cells.

A fully automated four machining center cell and guided palletized system is displayed in the illustration below.

Also available for palletized systems are dividing heads, horizontal and vertical rotary tables, coordinate inspection tables and other accessories. These serve as workholding fixtures and permit other auxiliary machine functions.

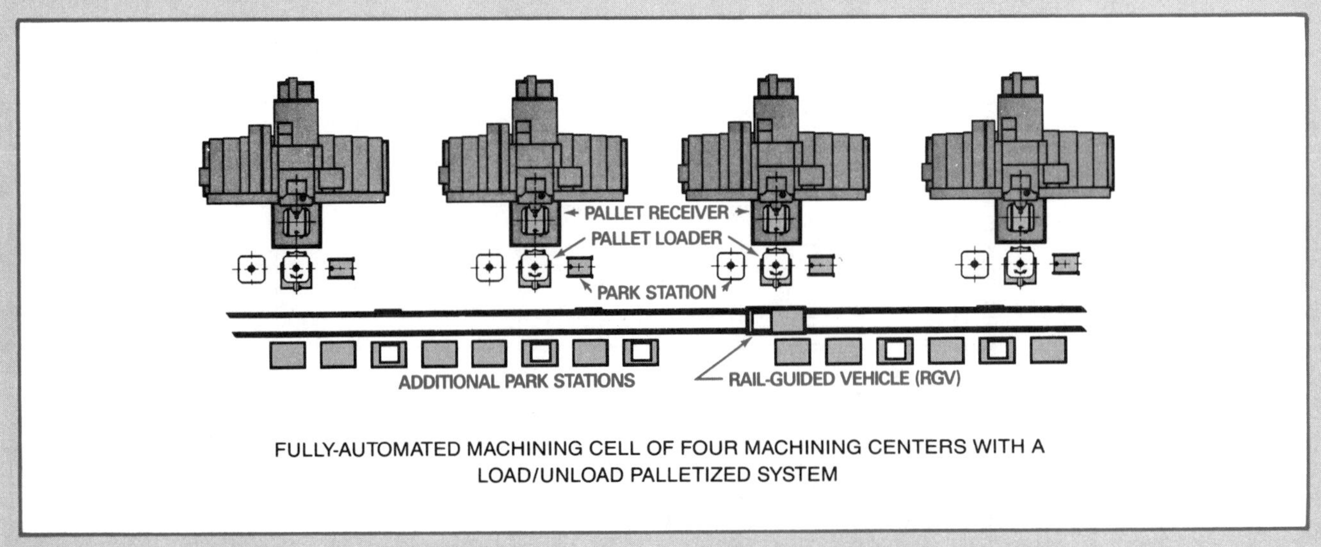

FULLY-AUTOMATED MACHINING CELL OF FOUR MACHINING CENTERS WITH A LOAD/UNLOAD PALLETIZED SYSTEM

Courtesy of EIMELDINGEN CORPORATION

SECTION 10
Pictorial Drawings

UNIT 37
Pictorial Assembly Drawings and Applications

Pictorial drawings are used in industry, in architecture, and in other businesses to illustrate a part, an assembly of many parts, or a building or construction project. While pictorial drawings primarily show external features, sections are sometimes removed to expose interior details. Pictorial drawings provide a quick, visual method of identifying design features and information about the assembly and functioning of a part or assembly.

CHARACTERISTICS OF PICTORIAL DRAWINGS

A pictorial drawing is a one-view drawing. The several faces of the object look almost the way they appear visually. Pictorial drawings do not show surfaces, details, and measurements in true size and shape. Often a photograph of a part or mechanism is air-brushed to bring out certain features and to serve the same function as a pictorial line drawing. The addition of a photo or pictorial drawing such as figure 37-1 simplifies seeing a part or a mechanism and aids in interpreting the function of an assembled unit.

In this particular example, the drawing is often referred to as a *phantom pictorial drawing*. The drawing shows the functions, location of parts, and principles of operation. This information is important in manufacturing, for marketing the device, for assembly and disassembly purposes, and for testing and development.

In actual practice, a set of drawings for this device may consist of the pictorial assembly drawing, a mechanical assembly drawing, working drawing, and other production drawings used in programming. The separate parts are fabricated, machined, heat treated, assembled, tested, etc. according to specific information contained on the set of drawings.

FIGURE 37-1 PHANTOM PICTORIAL DRAWING USED BY MANUFACTURER TO SHOW AND DESCRIBE OPERATIONAL FEATURES OF A TURBINE-TYPE PNEUMATIC (AIR-DRIVEN) GRINDING HEAD (COURTESY OF MOORE SPECIAL TOOL COMPANY, INC.)

RATCHET SCREW
RATCHET STOP
RATCHET PLUNGER
RATCHET BODY
RATCHET SPRING
ADJUSTING NUT
SPINDLE NUT
BARREL SPRING
THIMBLE
SLEEVE
LOCK NUT
SPINDLE
ANVIL
FRAME

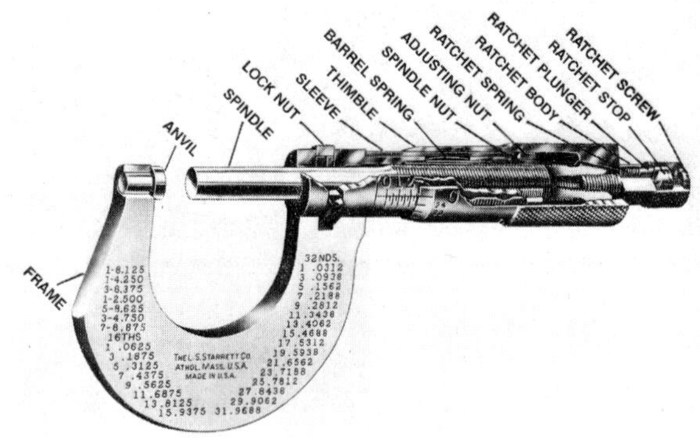

FIGURE 37-2 CUT-AWAY PICTORIAL ASSEMBLY DRAWING OF A MICROMETER WITH EXPOSED PARTS NAMED (COURTESY OF THE L.S. STARRETT COMPANY)

GENERAL TYPES OF PICTORIAL ASSEMBLY DRAWINGS

Two types of pictorial assembly drawings are in common use. In the first type, the object is drawn pictorially with interior parts exposed. Each part is generally named for easy reference, to show operations, and other characteristics. The cut-away pictorial assembly drawing of a micrometer (figure 37-2) with part names labeled is an example.

Parts are often shaded to highlight certain features. Highlighting makes a drawing look more like a photographic picture except that it focuses attention on particular parts or areas. Pictorial drawings may be precisely drawn with instruments. Simple parts or mechanisms are often sketched freehand. Techniques of freehand sketching are described in a later section.

Another drafting room practice of representing an assembled unit is to use an *exploded drawing.* In this second type, each part of a device is shown pictorially in its location and in the order of assembly. Some manufacturers provide a descriptive part name. Guide lines are used to show the sequence in assembly. Part numbers are often cross-referenced to a parts list on a drawing. Parts lists provide other specifications and related technical information. An example of an exploded pictorial drawing is provided in figure 37-3. Note the use of guide lines for assembly, disassembly, maintenance, and servicing.

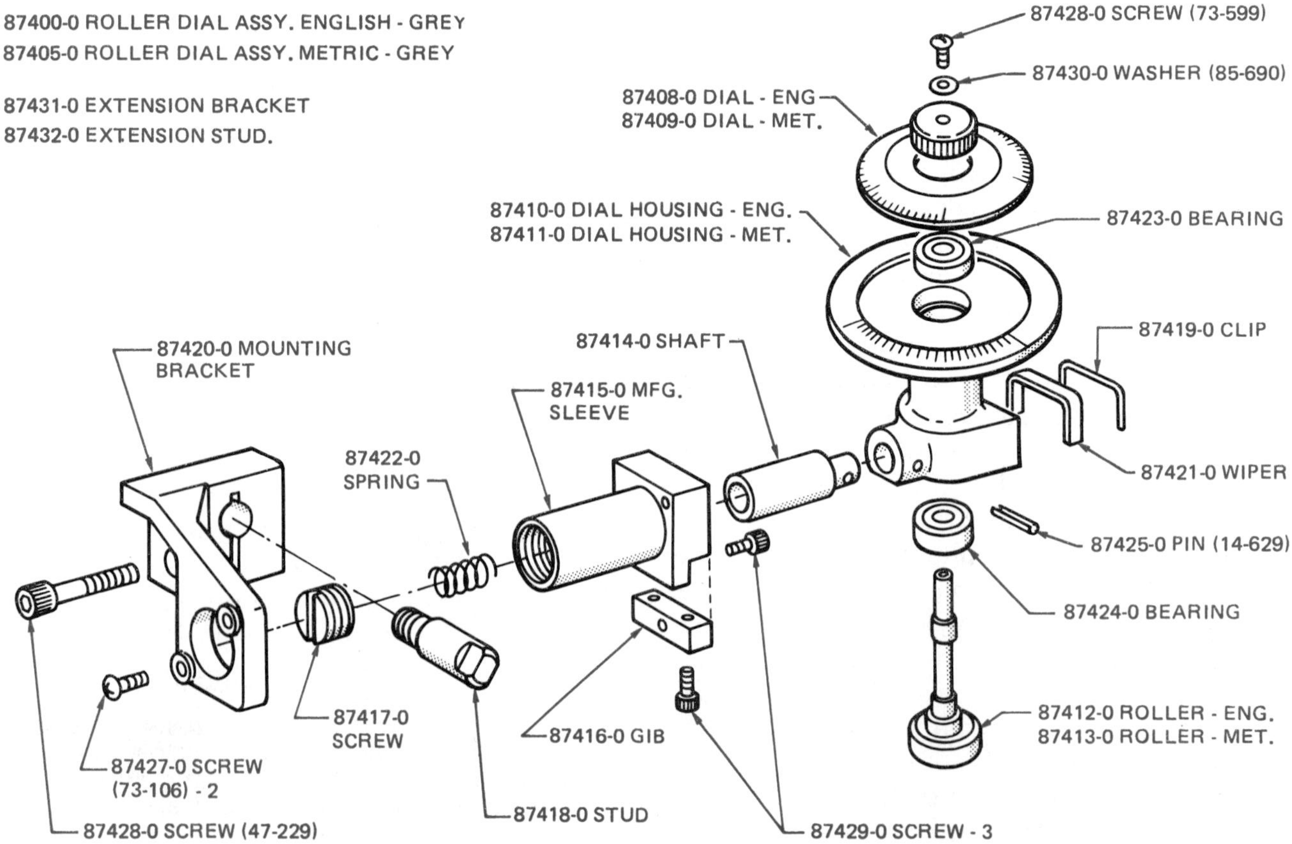

87400-0 ROLLER DIAL ASSY. ENGLISH - GREY
87405-0 ROLLER DIAL ASSY. METRIC - GREY
87431-0 EXTENSION BRACKET
87432-0 EXTENSION STUD.
87428-0 SCREW (73-599)
87430-0 WASHER (85-690)
87408-0 DIAL - ENG
87409-0 DIAL - MET.
87410-0 DIAL HOUSING - ENG.
87411-0 DIAL HOUSING - MET.
87423-0 BEARING
87419-0 CLIP
87414-0 SHAFT
87420-0 MOUNTING BRACKET
87415-0 MFG. SLEEVE
87422-0 SPRING
87421-0 WIPER
87425-0 PIN (14-629)
87424-0 BEARING
87412-0 ROLLER - ENG.
87413-0 ROLLER - MET.
87417-0 SCREW
87416-0 GIB
87427-0 SCREW (73-106) - 2
87418-0 STUD
87428-0 SCREW (47-229)
87429-0 SCREW - 3

FIGURE 37-3 EXPLODED PICTORIAL ASSEMBLY DRAWING WITH NUMBERED PARTS AND PART NAMES (COURTESY OF THE CLAUSING MACHINE TOOL CORPORATION)

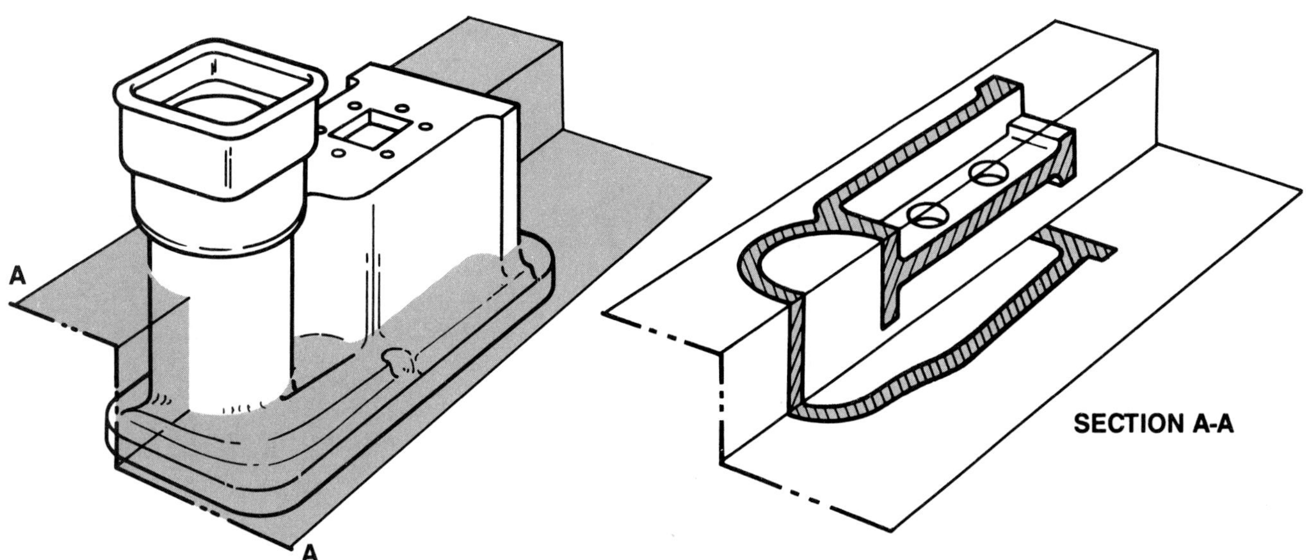

FIGURE 37-4 OFFSET CUTTING PLANE USED WITH PICTORIAL DRAWING TO SHOW COMPLICATED INTERNAL DETAILS

APPLICATIONS OF PICTORIAL REPRESENTATION DRAWINGS

Design Assembly Drawings

Designers often prepare a pictorial drawing for developmental projects. This kind of drawing provides the framework around which a finished product is developed. Subassemblies may be broken out to show major components and to further study movements, functions, and design elements. Later, precise assembly, detail, working, and other production drawings are developed.

Machine Maintenance and Servicing Pictorial Drawings

Cut-away assembly or part pictorial drawings are used to expose details. Full, half, and partial sections are removed in order to see internal design features. Figure 37-4 is a typical example of combining cutting planes with a pictorial drawing. In this instance, **SECTION A-A** shows the internal web design and other construction details that are not visible in the pictorial drawing.

Pictorial Drawings for Technical Manuals and Marketing Materials

Pictorial assembly drawings are included in many manufacturers' technical manuals to serve three prime functions:

- To show internal details of construction and operation.
- To quickly identify parts and mechanisms requiring replacement, servicing, or maintenance work.
- To provide the sequential steps for assembly, disassembly, and checking a machine unit, electronic circuit, piping system, or other component.

NOTE: REFER TO APPENDIX B-5 FOR THE LARGER SCALE DRAWING TO USE WITH THIS ASSIGNMENT.

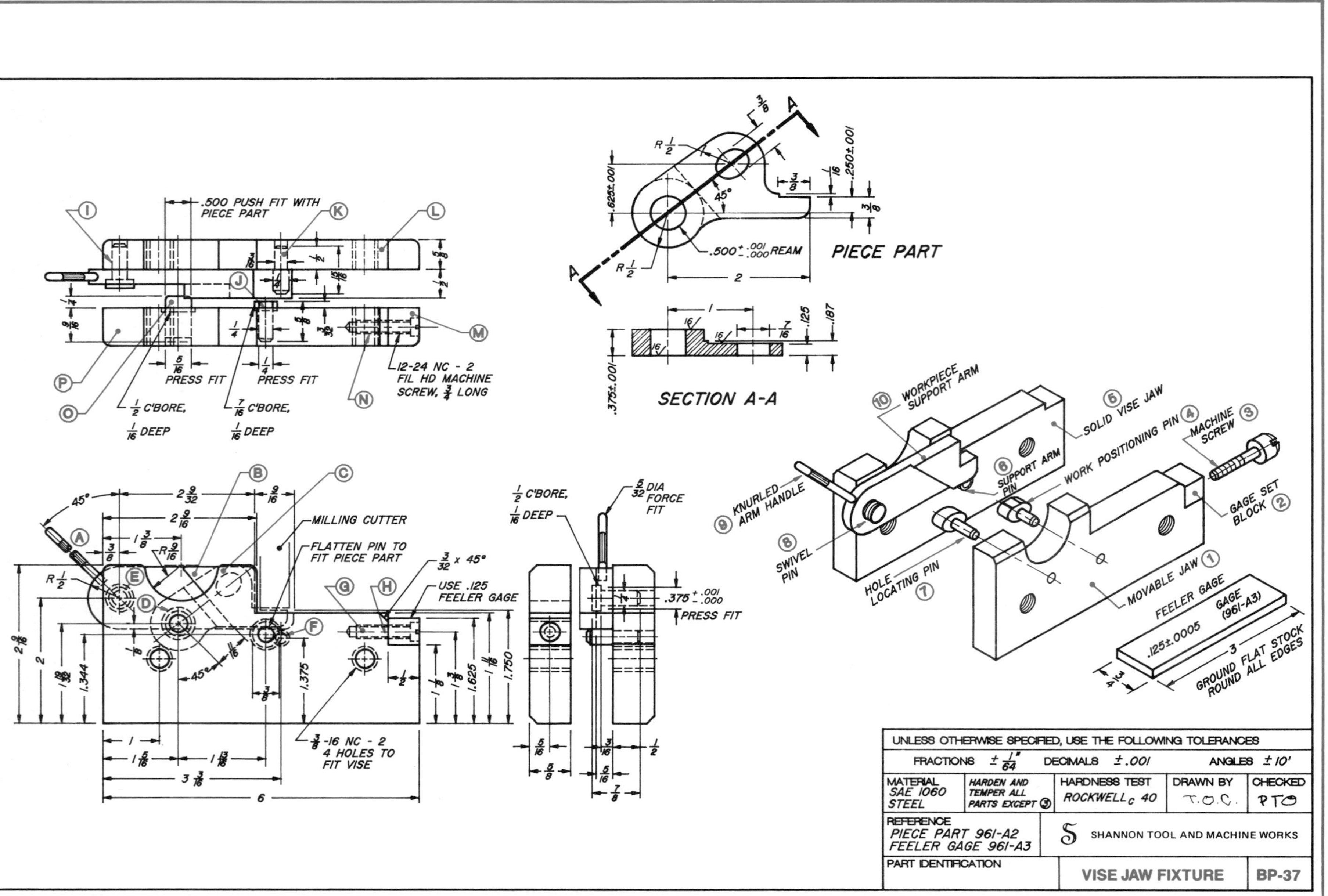

ASSIGNMENT—UNIT 37

VISE JAW FIXTURE (BP-37)

1. State three different drafting room techniques used to represent the Vise Jaw Fixture.
2. Give the view and letter which identifies the piece part.
3. Identify the letters in the top view which represent parts ① through ⑤.
4. Identify the letters in the front view which represent parts ⑥ through ⑩.
5. State two functions served by the Vise Jaw Fixture.
6. State the function of each of the following parts:
   a. Gage set block ②
   b. Swivel pin ⑧
   c. Hole locating pin ⑦
   d. Support arm pin ⑥
7. Give the specification of the steel to be used for each part of the fixture except ③.
8. Specify the required Rockwell C reading for all hardened and tempered parts.
9. Determine the following upper and lower limit measurements from the base of the movable jaw:
   a. To the finish depth of cut
   b. To the top surface of the gage set block
   c. To the center of the pin
10. Give the nominal measurement from the base of the vise jaws to the horizontal center line of pins
    Ⓙ
    Ⓚ
    Ⓞ
    Ⓘ
11. Determine the nominal overall length, width (height), and thickness (depth) of the
    a. Workpiece support arm
    b. Piece part
12. Identify the (a) part number, (b) finished dimensions, and (c) material for the feeder gage.

Student's Name ____________________

1. a. ____________________
   b. ____________________
   c. ____________________
2. ____________________
3. ① ________ ④ ________
   ② ________ ⑤ ________
   ③ ________
4. ⑥ ________ ⑨ ________
   ⑦ ________ ⑩ ________
   ⑧ ________
5. a. ____________________
   b. ____________________
6. a. ____________________
   ____
   b. ____________________
   ____
   c. ____________________
   ____
   d. ____________________
   ____
7. ____________________
8. ____________________

9.	Upper	Lower
a.	________	________
b.	________	________
c.	________	________

10. Ⓙ ____________________
    Ⓚ ____________________
    Ⓞ ____________________
    Ⓘ ____________________

11.	Length	Width	Thickness
a.	________	________	________
b.	________	________	________

12. a. ____________________
    b. ____________________
    c. ____________________

UNIT 38
Axonometric, Oblique, and Perspective Drawings

There are three basic types of pictorial drawings:

- *Axonometric drawings.* This classification includes *isometric, dimetric,* and *trimetric projections.* The object is inclined at an angle to the plane of projection.
- *Oblique drawings.* These include *cavalier* and *cabinet* projection. One face of the object is parallel to the plane of projection.
- *Perspective drawings.* In this instance, the features of an object appear to converge toward a point in the distance.

PRINCIPLES OF AXONOMETRIC PROJECTION

Isometric Projection Drawings

Isometric drawings (figure 38-1A) are built around three axes. The axes are spaced 120° apart; isometric meaning *equal measure.* A number of faces of the object are seen regardless of the position (view). Principal measurements of an object are represented on drawings in terms of width, height, and depth (figure 38-1A).

Edges and features of an object which are parallel to an axis are drawn parallel to the axis.

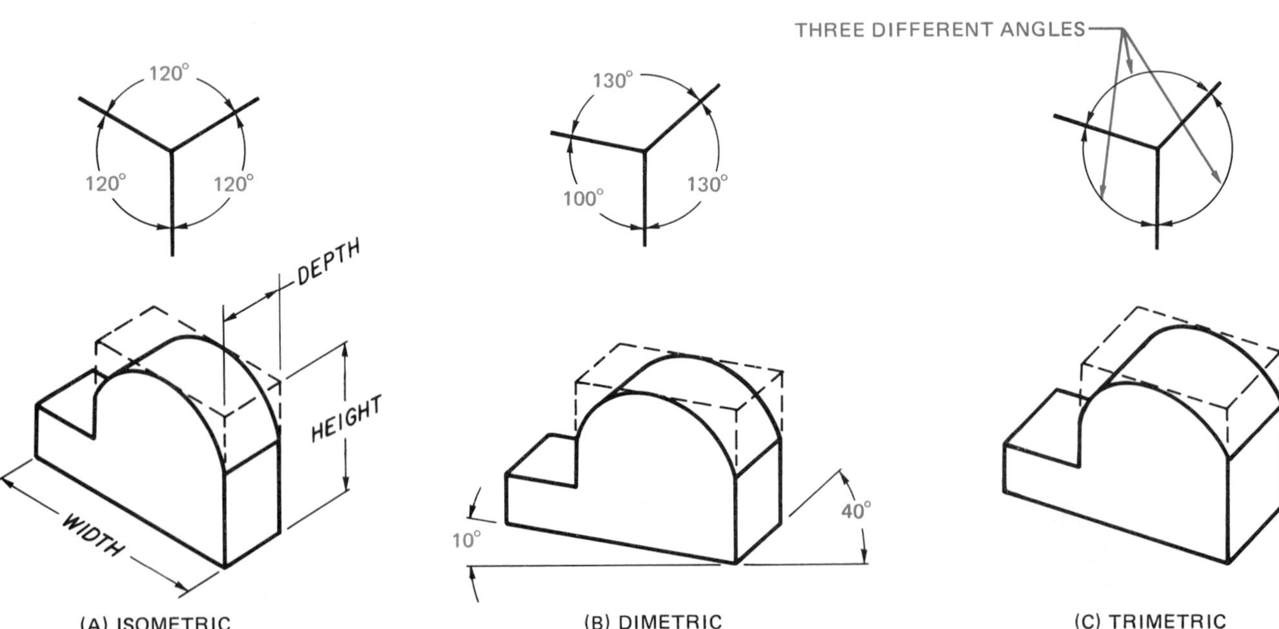

FIGURE 38-1 AXONOMETRIC PICTORIAL DRAWING

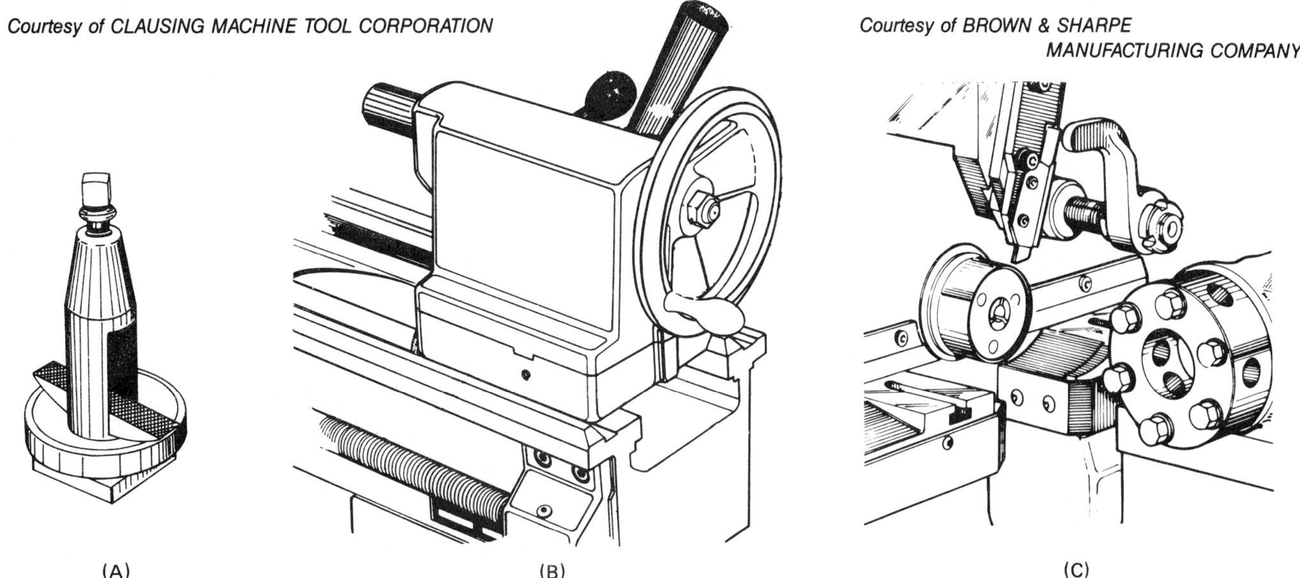

FIGURE 38-2 INDUSTRIAL APPLICATIONS OF ISOMETRIC DRAWINGS [(A) and (B), AND DIMETRIC DRAWING (C)]

Non-isometric lines (such as nonparallel, angular, and circular lines) do not appear in their true length and shape. Holes appear as ellipses; parts of circular features as elliptical arcs. Hidden lines are normally not found on isometric drawings. Figure 38-2 shows how parallel, nonparallel, and circular features of a tool post (A), tailstock assembly (B), and machine tool components (C) are represented on an industrial isometric and dimetric drawing.

While isometric drawings are generally outside views, it is often necessary to show internal features to clarify construction and other details. For this purpose, full, half, and broken-out sections are used with isometric drawings.

Dimetric Projection Drawings

A *dimetric drawing* is a form of axonometric drawing. The object is turned so that two axes make the same angle with the plane of projection. The two axes in figure 38-1B are 130° apart. The third axis makes an angle of 100°. A dimetric drawing (figure 38-2C) displays a part graphically with less distortion than an isometric drawing.

Trimetric Projection Drawings

The object of a *trimetric drawing* is turned so that each of the three axes is drawn at a different angle to the plane of projection (figure 38-1C). Lines on each axis are drawn to a different scale. While difficult to make, trimetric drawings show an object with less distortion than either isometric or dimetric drawings.

OBLIQUE DRAWINGS

Oblique drawings are a modification of orthographic and axonometric projection. Most descriptive features of a part appear on the principal face which is parallel to the plane of projection. This means the lines representing the face of the object are seen in true size and shape. Two of the axes used in oblique drawings are at a right angle to each other. The third axis, which recedes from the face, is drawn at any convenient angle to a horizontal plane. Receding edges are drawn parallel to the third axis.

Cavalier Oblique Drawings

Cavalier oblique drawings are usually made with the receding axis at 30° or 45°. The same drawing scale is used for height, width, and depth measurements of all features on the principal face and receding faces (figure 38-3A).

Cabinet Oblique Drawings

Cabinet oblique drawings have a receding axis which may be drawn at any angle to the horizontal. The depth for design features along a reced-

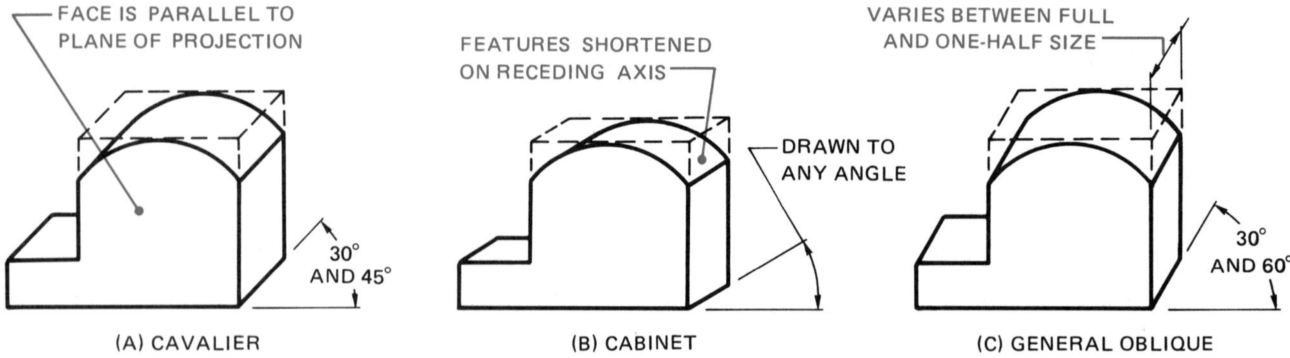

FIGURE 38-3 BASIC TYPES OF OBLIQUE DRAWINGS

ing axis is shortened as shown in figure 38-3B. Usually, a one-half scale is used. Cabinet oblique drawings provide a more natural-looking proportioned object.

General Oblique Drawings

General oblique drawings are made with variable scales for measurements of part features along the receding axes. Common axis angles are 30° and 60°. The purpose of the variable scale is to graphically show a part as naturally as possible (figure 38-3C). Figure 38-2C illustrates parts on an automatic screw machine as they appear on a general oblique drawing.

PERSPECTIVE DRAWINGS

A *perspective drawing* is the most realistic pictorial form for representing parts, units, or whole projects. Examples include assembly lines, buildings, and structures which need to be viewed as extending into the distance. Perspective drawings may appear on blueprints with other regular orthographic or section views in order to simplify the reading of a print and to show how a part actually looks. A perspective drawing shows the receding edges and lines as ultimately meeting in a *fixed (station) point* (VP).

Perspective drawings differ from other types of pictorial drawings in which features are drawn to receding axes. The plane of projection from which the object is viewed is called a *picture plane.* The object rests on a *ground plane.* The picture plane and a *horizontal plane* intersect at eye level at the *horizontal line.* The term *vanishing point* identifies the place where lines which represent an object meet, if extended.

There are four kinds of perspective drawings:

- Parallel
- Angular
- Oblique
- True

Parallel (Single-Point) Perspective Drawings

One face of the object appears on a parallel perspective drawing as parallel to the picture plane. *One-point parallel perspective* means that all lines from the face recede toward a single vanishing point, as illustrated in figure 38-4. The object looks distorted in this kind of perspective drawing.

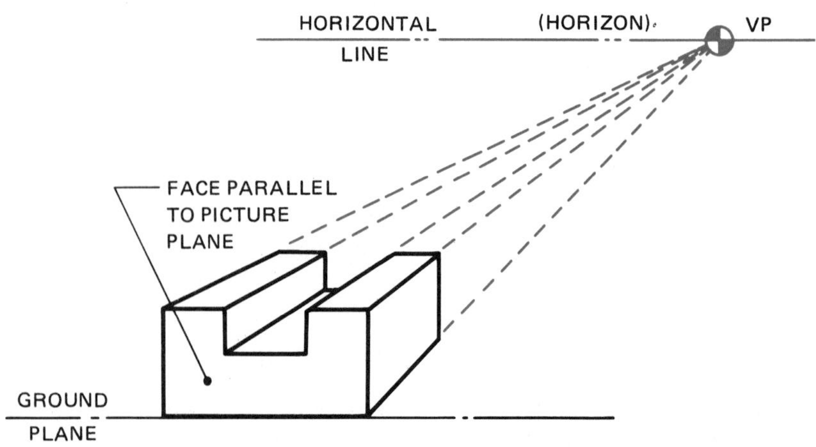

FIGURE 38-4 PARALLEL (SINGLE-POINT) PERSPECTIVE

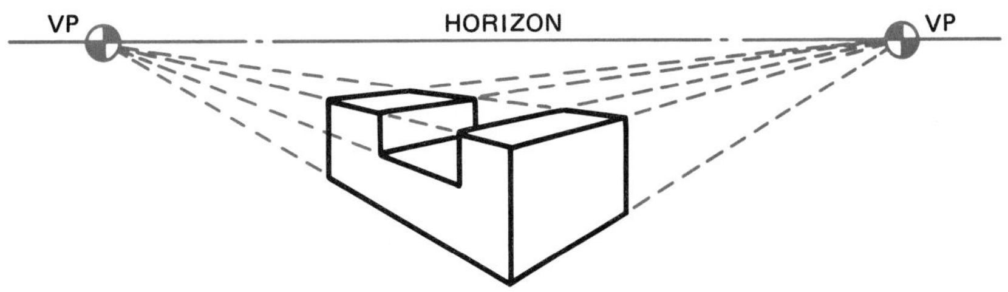

FIGURE 38-5 ANGULAR (TWO-POINT) PERSPECTIVE

Angular (Two-Point) Perspective Drawings

An object drawn in *angular (two-point) perspective* is seen as being placed at an angle to the picture plane (figure 38-5). The angle may be 15°, 30°, or 45°. The 30° angle is used most frequently. The lines for different features and edges of a part are drawn to two vanishing points as seen in the example of a two-point perspective drawing. Circles and arcs (radii) appear as ellipses or as part of an ellipse.

Oblique (Three-Point) Perspective Drawings

None of the principal edges of an object is parallel to the picture plane in an *oblique (three-point) perspective drawing*. Each line for a parallel edge recedes toward a separate vanishing point. There are three vanishing points, as illustrated in figure 38-6. Oblique perspective drawings are difficult to make. Their principal use is for architectural drawings.

True Perspective Drawings

A *true perspective drawing* shows the object with the accuracy of a photographic picture. The width, height, and depth are each foreshortened. Foreshortening makes the object look more realistic.

True perspective is widely applied to structural and architectural drawings. Such drawings are generally prepared using a *perspective drawing board* or *drafting machine.* A finely printed sheet, called a *perspective grid sheet* is available for drawing a part mechanically and to simplify the making of a freehand sketch of an object.

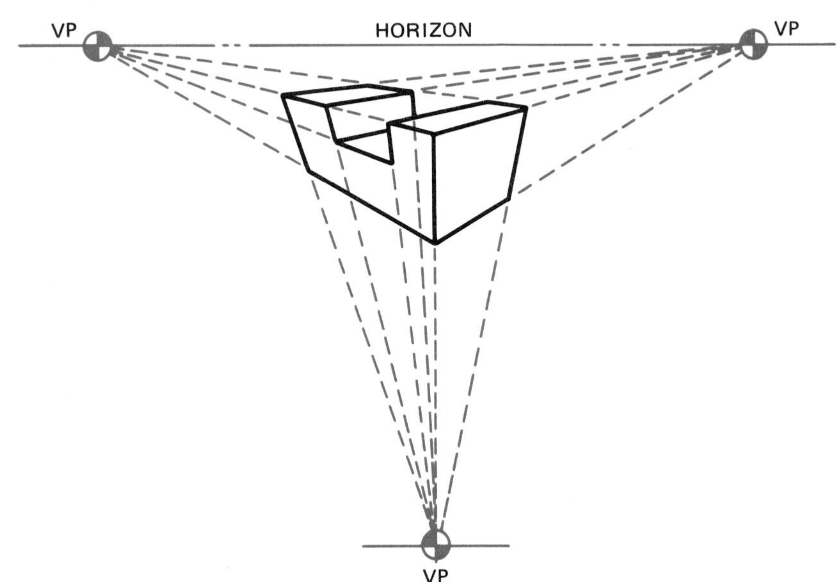

FIGURE 38-6 OBLIQUE (THREE-POINT) PERSPECTIVE

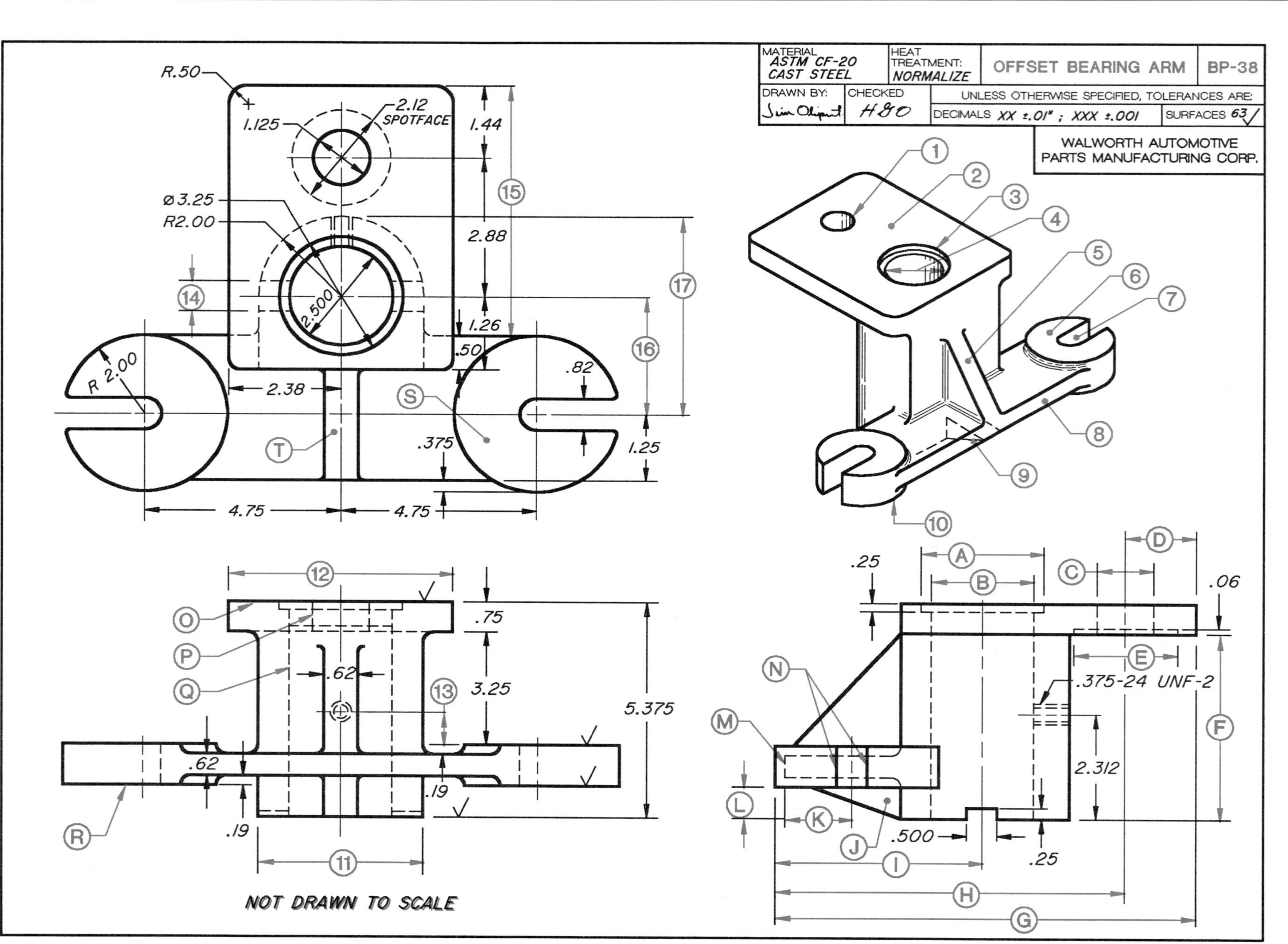
MATERIAL
ASTM CF-20
CAST STEEL
HEAT TREATMENT:
NORMALIZE
OFFSET BEARING ARM
BP-38
DRAWN BY:
CHECKED
UNLESS OTHERWISE SPECIFIED, TOLERANCES ARE:
DECIMALS XX ±.01" ; XXX ±.001
SURFACES 63
WALWORTH AUTOMOTIVE
PARTS MANUFACTURING CORP.
R.50
1.125
2.12
SPOTFACE
1.44
Ø3.25
R2.00
2.88
2.500
1.26
.50
.82
R 2.00
2.38
.375
1.25
4.75
4.75
.75
.62
3.25
5.375
.62
.19
.19
.25
.06
.375-24 UNF-2
2.312
.500
.25
NOT DRAWN TO SCALE

ASSIGNMENT — UNIT 38

OFFSET BEARING ARM (BP-38)

Student's Name ______________________

1. Identify (a) the material in the part and (b) the required heat treatment.
2. Give the letter of the detail in the front view which represents features (1), (2), (4), and (10) of the Offset Bearing Arm.
3. Locate the detail letters in the top view which represent features (5) and (6).
4. Locate the detail letters in the right-side view which represent features (3), (7), (8), and (9).
5. Determine the nominal and maximum diameters for holes (B) and (C).
6. Give the nominal diameter and the depth for spotfacing area (E).
7. Determine heights (F) and (L) and the overall width (G).
8. Determine the upper and lower limit measurements to the center lines for (H), (I), and (K).
9. Give the letters which identify the ribs.
10. Indicate the thickness of the ribs and the pads for the elongated slots.
11. Compute casting measurements (11) and (12).
12. Determine dimension (13) for the center line location.
13. Give the thread specification.
14. Identify the machining dimensions for slot (14).
15. Give the upper and lower limit dimensions for the overall height of the part.
16. Determine dimensions (15), (16), and (17).

1. a. ______________________
   b. ______________________
2. (1) = __________
   (2) = __________
   (4) = __________
   (10) = __________
3. (5) = __________
   (6) = __________
4. (3) = __________
   (7) = __________
   (8) = __________
   (9) = __________

5.

	Nominal Diameter	Maximum Diameter
(B) =	__________	__________
(C) =	__________	__________

6. ______________________
7. (F) = ______________________
   (L) = ______________________
   (G) = ______________________

8.

	Upper Limit	Lower Limit
(H) =	__________	__________
(I) =	__________	__________
(K) =	__________	__________

9. ______________________
10. Ribs __________ Pads __________
11. (11) = __________ (12) = __________
12. (13) = __________
13. ______________________
14. Slot (14) ______________________
15. Upper Limit __________ Lower Limit __________
16. (15) = __________
    (16) = __________
    (17) = __________

UNIT 39
Dimensioning Pictorial Drawings

Pictorial drawings are dimensioned according to the same basic rules as for standard multiview drawings. Pictorial drawings may be used as working drawings when all required dimensions and other specifications are placed on the drawing. *Aligned* and *unidirectional systems of dimensioning* are used.

ALIGNED (PICTORIAL PLANE) DIMENSIONING

In the *aligned system* of dimensioning pictorial drawings, extension lines, dimension lines, and dimensions are placed parallel to the pictorial planes (pictorial drawing axes). Figure 39-1 shows

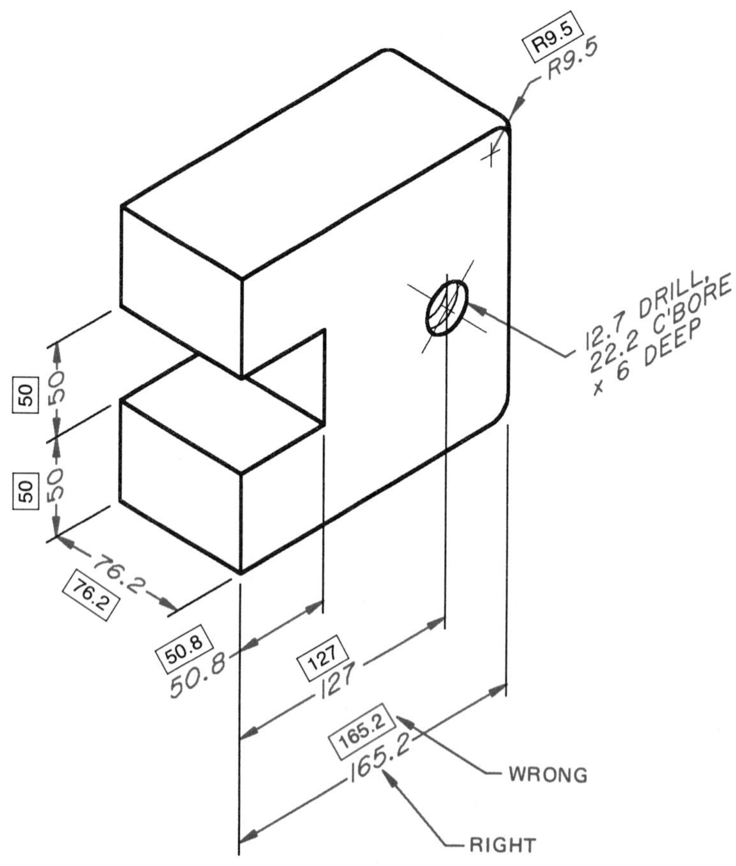

FIGURE 39-1 CORRECT AND INCORRECT ALIGNED DIMENSIONING OF PICTORIAL DRAWINGS

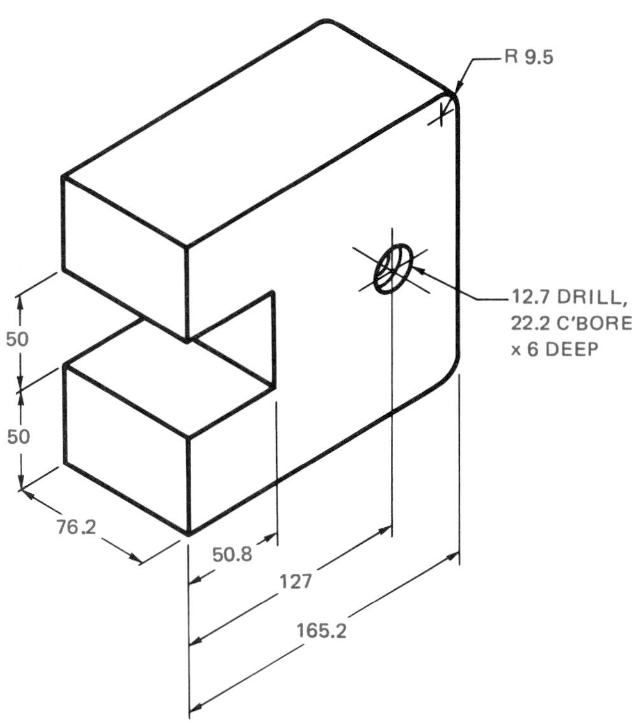

FIGURE 39-2 UNIDIRECTIONAL DIMENSIONING APPLIED TO A PICTORIAL DRAWING

correct and incorrect dimensioning in the aligned system. SI metric dimensions are used in the example and assignment.

UNIDIRECTIONAL DIMENSIONING

The *unidirectional system of dimensioning* pictorial drawings is easier to use. Dimensions, notes, and other technical information are lettered in a vertical position. Figure 39-2 shows the dimensions used in the preceding example as they appear in unidirectional dimensioning.

BASIC DIMENSIONING PRACTICES

In general, rules for dimensioning multi-view drawings apply to pictorial drawings. More specific rules relating to aligned and unidirectional dimensioning follow:

- Extension and dimension lines are drawn parallel to the picture planes.
- Dimensions are lettered parallel to the extension lines (aligned dimensioning) and vertically in relation to the horizontal plane for unidirectional dimensioning.
- Dimensions are given for visible features. Hidden lines are not generally given.
- Notes and other specifications are lettered parallel with the horizontal plane.
- Dimensioning across hidden surfaces and dimension lines should be avoided if possible. This is to prevent errors in reading dimensions which may be confusing in interpreting specific features of the part.

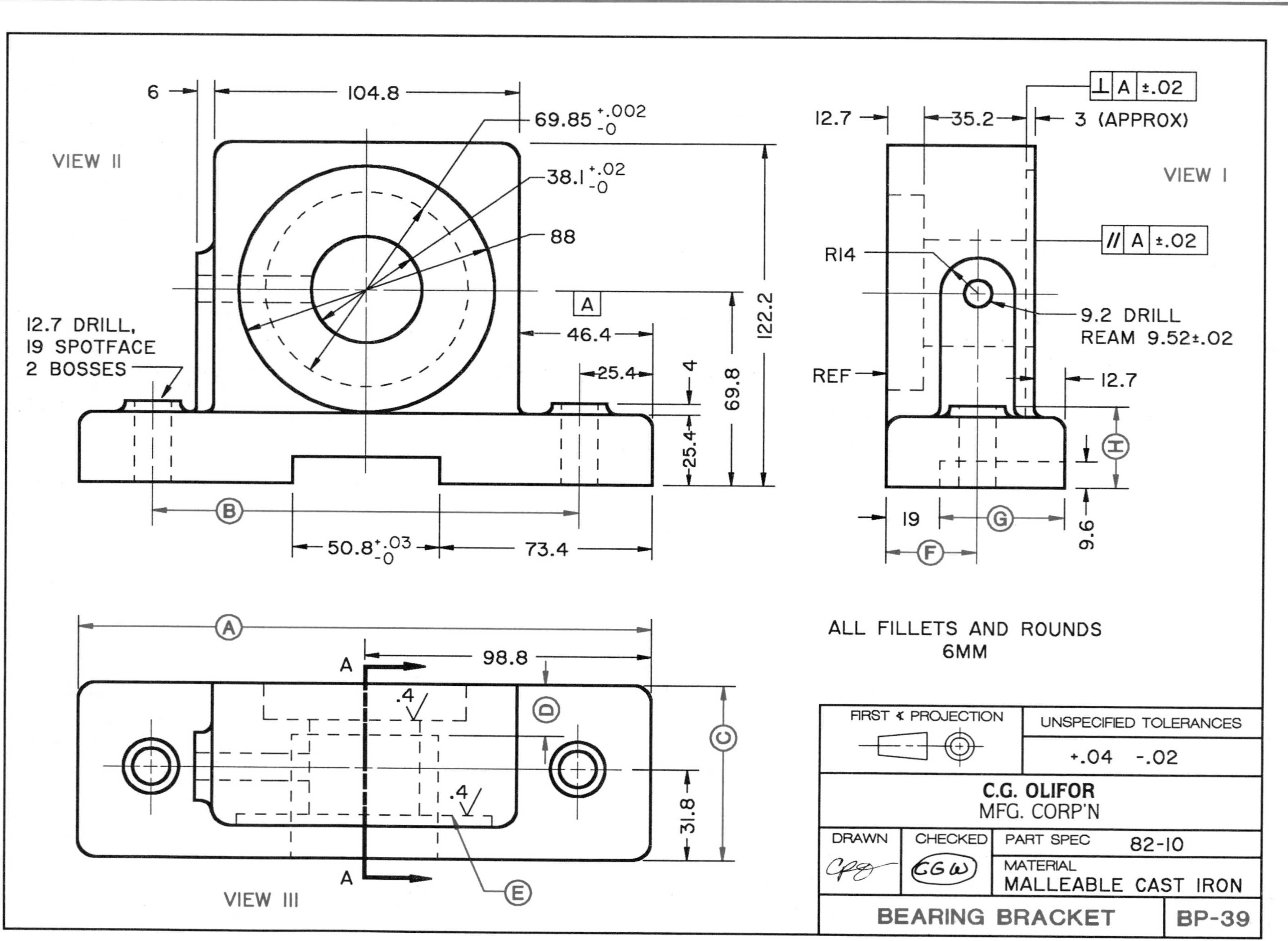
VIEW II
6
104.8
69.85 +.002 -0
38.1 +.02 -0
88
A
46.4
25.4
4
25.4
69.8
122.2
12.7 DRILL,
19 SPOTFACE
2 BOSSES
B
50.8 +.03 -0
73.4
VIEW I
12.7
35.2
3 (APPROX)
⊥ A ±.02
// A ±.02
R14
9.2 DRILL
REAM 9.52±.02
REF
12.7
H
19
G
F
9.6
A
98.8
.4
D
C
31.8
.4
E
VIEW III
ALL FILLETS AND ROUNDS
6MM
FIRST ∢ PROJECTION
UNSPECIFIED TOLERANCES
+.04 -.02
C.G. OLIFOR
MFG. CORP'N
DRAWN
CHECKED
PART SPEC
82-10
MATERIAL
MALLEABLE CAST IRON
BEARING BRACKET
BP-39

ASSIGNMENT—UNIT 39

BEARING BRACKET (BP-39)

Student's Name ______________________

1. Identify the system of projection.
2. Name **VIEWS I**, **II**, and **III**.
3. Determine the nominal overall length Ⓐ.
4. Compute the nominal center-to-center distance Ⓑ.
5. Compute the maximum and minimum overall width Ⓒ.
6. Give the (a) nominal, (b) maximum, and (c) minimum dimensions for measurement Ⓓ.
7. Indicate the nominal overall height to which the two bosses are to be spotfaced.
8. Give the surface finish requirement for the faces of the recessed holes.
9. State the (a) upper and (b) lower limits of the through-bored hole.
10. Interpret the geometric tolerancing symbol for the through-bored hole.
11. Give the approximate amount of material to be machined from the casting for surface Ⓔ.

12. Interpret the geometric tolerancing symbol for surface Ⓔ.
13. Determine nominal dimensions Ⓕ, Ⓖ, and Ⓗ.
14. Add extension lines, dimension lines, and the nominal dimensions for features Ⓘ through Ⓟ on the full-section (**A–A**) isometric sketch of the Bearing Bracket.

    Note. Dimension according to the unidirectional dimensioning system.

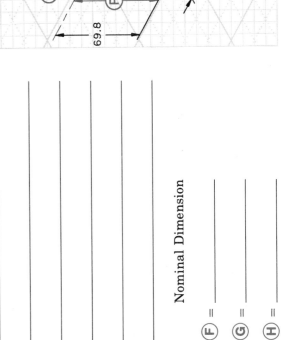

1. ______________________

2. **VIEW I** ______________________

   VIEW II ______________________

   VIEW III ______________________

3. Ⓐ = ______________________

4. Ⓑ = ______________________

5. Maximum Minimum

   Ⓒ = ______________________

6. a. Nominal Ⓓ = ______________________

   b. Maximum = ______________________

   c. Minimum = ______________________

7. ______________________

8. ______________________

9. a. Upper Limit = ______________________

   b. Lower Limit = ______________________

10. ______________________

11. ______________________

12. ______________________

13. Nominal Dimension

    Ⓕ = ______________________

    Ⓖ = ______________________

    Ⓗ = ______________________

14.

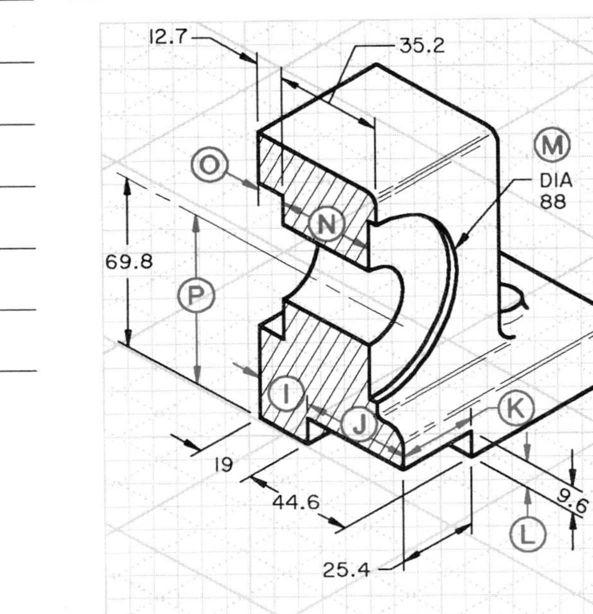

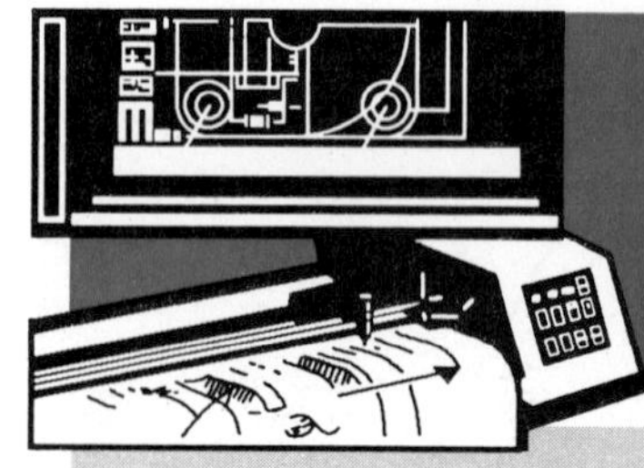

HIGH TECHNOLOGY applications

HT-11 CADD/CIM/FMS (FLEXIBLE MANUFACTURING): SYSTEM CONFIGURATIONS

Equipment floor plans and layouts for integrated, automated manufacturing depend on such factors as: the complexity, machining accuracy, and number of work processes; subsystems for work positioning and delivery devices; robotic processes, and the like. CADD/CIM/and other linkages to tooling, manufacturing, and inspection are additional considerations.

There are four typical shop layouts used in FMS: *in-line, loop, ladder,* and *open-field.* The layouts are referred to as *configurations* and are illustrated by line art at (A), (B), (C), and (D).

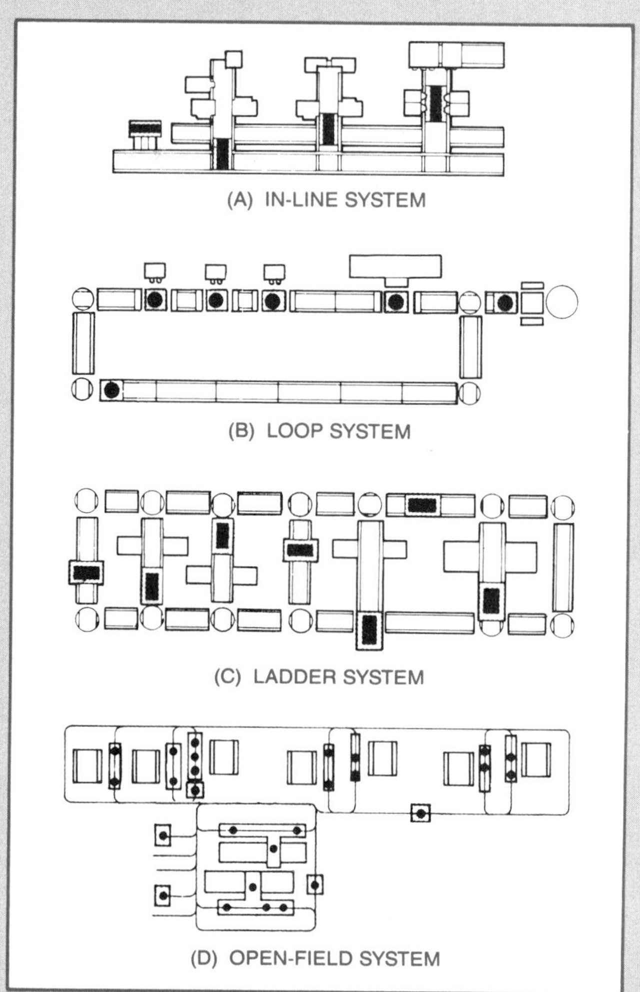

(A) IN-LINE SYSTEM

(B) LOOP SYSTEM

(C) LADDER SYSTEM

(D) OPEN-FIELD SYSTEM

IN-LINE EQUIPMENT CONFIGURATION FOR FMS

The *in-line configuration* (A) is used for sequential machining processes. The workpiece is loaded and the machining processes are started at one end of a battery of machines. After the machining processes are completed at this first stage, the workpiece is picked up and moved to the next machine tool (stage) in the line. The workpiece is moved through each successive stage until all processes are completed.

LOOP EQUIPMENT CONFIGURATION FOR FMS

In a *loop configuration,* the equipment that is required for the manufacture of a part or component is placed around a loop pattern (B). The illustration shows a directional table located at the right end of the loop. The directional table controls the routing of palletized workpieces. Rail-guided vehicles sequentially feed workpieces from any machining stage to the next. All workpiece movements, tooling changes, machining processes, and robotic input are computer controlled.

LADDER FMS CONFIGURATION

The *ladder layout configuration* (C) derives its name from the combination of cross rungs that run between the outer legs of a battery of machine tools and other FMS equipment. Programmable computers provide commands for routing a workpiece among each machine center that is tooled for a series of machining operations. The routing is programmed for whatever sequence is required to manufacture a specific part.

OPEN-FIELD FMS CONFIGURATION

An *open-field configuration* is the most complex of the four basic FMS configurations. The open-field floor plan for equipment placing and work sequencing is best adapted to *random-order* manufacturing. The equipment and tooling are located according to *machining sequences.* Wire-guided carts are used for on- and off-loading of workpieces that are positioned on workholding tables and accessories for auxiliary computerized functions.

Generally, in an open-field configuration, the machine control unit (MCU) at each manufacturing stage interfaces with a mainframe computer for the entire FMS operation.

Drawing courtesy of C. THOMAS OLIVO ASSOCIATES

SECTION 11
Industrial Production Working Drawings

UNIT 40
Title Blocks: Specifications, Change Notes

Practically all industrial drawings contain details and specifications which complement the information represented on the drawing itself. This added information is usually contained within a block which is bounded by the bottom and right drawing border lines. The term *title block* is used for this area and for the contents within the block. A title block serves two main functions:

- To provide a system for cataloging each technical drawing and sketch for easy reference and storage.
- To provide a record of design, development, manufacturing, and other pertinent data relating to a complete assembly, or mechanism, or a single part.

COMMON TITLE BLOCK ELEMENTS

A common title block is illustrated in figure 40-1. In its simplest form, practically all title blocks include a company name Ⓐ, a descriptive, short title or name of a part or assembly Ⓑ, and a part identification number Ⓒ. Places are usually provided for dates and signatures of persons responsible for any phase of engineering, design, manufacturing, or inspection, and for final approval Ⓓ.

General tolerance limits on linear and angular dimensions are specified Ⓔ, including surface finish requirements. When a series of drawings is required to fully describe a part or mechanism, a notation about the sequence and number of drawings is included Ⓕ.

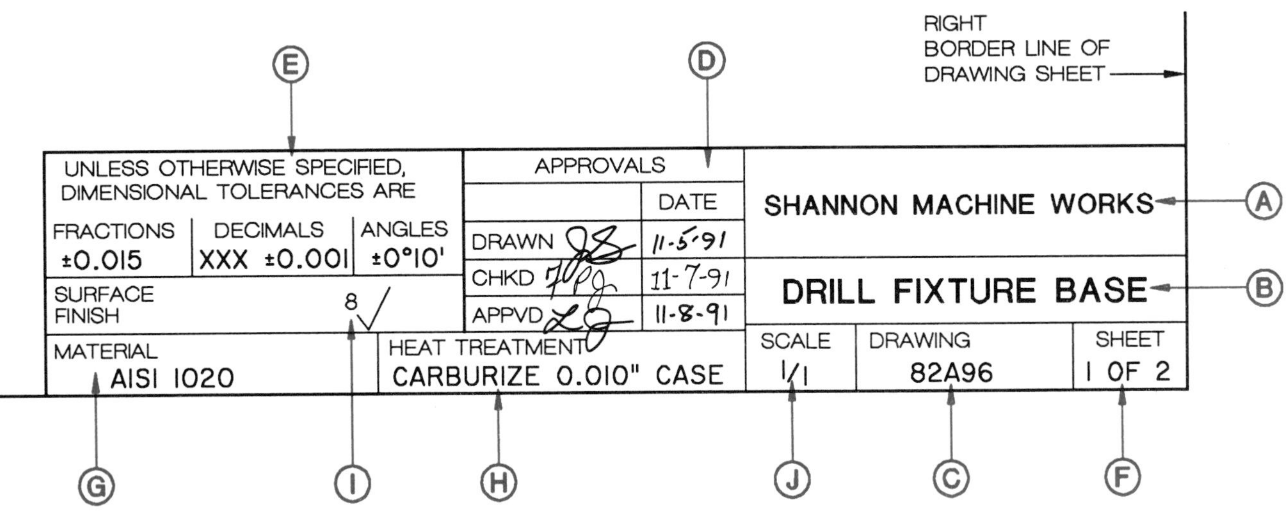

FIGURE 40-1 COMMMON TITLE BLOCK ELEMENTS

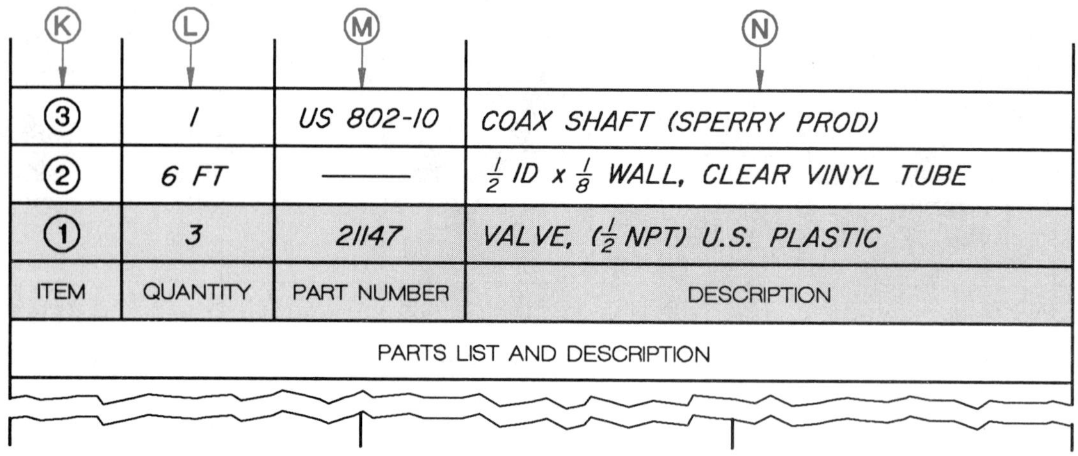

ITEM	QUANTITY	PART NUMBER	DESCRIPTION
(3)	1	US 802-10	COAX SHAFT (SPERRY PROD)
(2)	6 FT	———	$\frac{1}{2}$ ID x $\frac{1}{8}$ WALL, CLEAR VINYL TUBE
(1)	3	21147	VALVE, ($\frac{1}{2}$ NPT) U.S. PLASTIC

PARTS LIST AND DESCRIPTION

FIGURE 40-2 EXAMPLE OF TITLE BLOCK EXTENSION TO INCLUDE PARTS AND DESCRIPTIONS

Sometimes, the material for a simple part is given on the drawing near one of the views. In other cases, the title block contains spaces for recording the material (G), required heat treatment (H), and surface texture or surface processing (I). If the scale is not included on the drawing, it may appear in the title block (J). Where the weight of a part is important, this information may also be given in the title block.

MATERIALS IDENTIFICATION AND SPECIFICATIONS

Title block information may be expanded beyond the common elements (figure 40-2). Additional spaces may be planned to list each item (K). On assembly drawings, the item number generally corresponds with the sequence function of the part in relation to the assembly drawing and not the part identifying number. The following information is then provided for each item: quantity (L), identification number of the part (M), and a description (N), which may cover material, size, and source from which it may be secured.

Other Specification Information

Other columns may be added in table form to the title block or in a separate area near it. For instance, items may be listed relating to processing among successive work stations, next assembly, final assembly, inspection checks, and sign-off points. Figure 40-3 provides an example.

Each specification (O) provides the worker with additional information about the part requirements. If such specifications were to appear on the drawing, the drawing would become unnecessarily crowded and complicated.

PART REVISIONS AND DRAWING CHANGES

The mechanic and technician often work with individual parts or assemblies on which design

SPECIFICATIONS
FIN. RACE R'DNESS-0.0003"
FIN. RACE R.O.C. - 4.0
FIN "X" - 0.0005"
ECC. FIN. RACE TO FIN. HUB
O.D. - 0.0004" REF.

(O)

FIGURE 40-3 ADDITION OF SPECIFICATIONS TO TITLE BLOCK INFORMATION

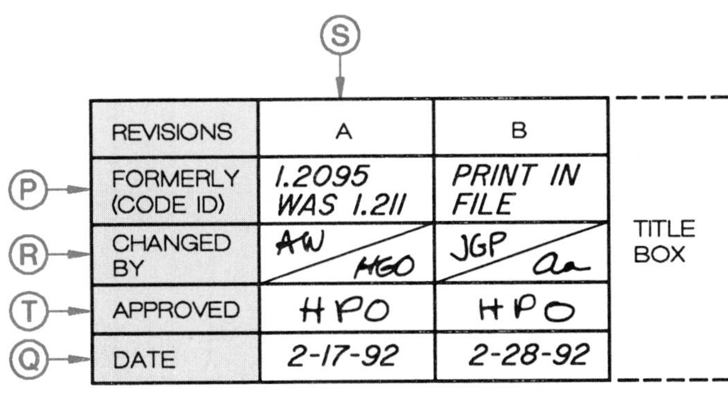

FIGURE 40-4 ELEMENTS OF CHANGE BLOCKS FOR REVISIONS

changes have been made. Sometimes, these changes appear on an *engineering work order.* The order is separate from the original drawing and provides permanent change information. The changes may relate to dimensional variations, modified feature design, quality of finish, and the like. The worker must relate the engineering work order to the companion technical drawing.

A drawing change may also be called a *change notice, revision,* or *alteration.* Change information is usually recorded in a *change block* like the one illustrated in figure 40-4. It is important that the change block contain an identifying number or code for each changed feature (P), date of change (Q), the person making the change (R), a brief, general description (S), and the approval of the change (T).

STANDARDS FOR TITLE BLOCKS AND ACCOMPANYING TABLES

Title block sizes, shape, and content are prepared according to standards for such government agencies as the Department of Defense or as a subcontract. Such standards are intended to ensure the interchangeability of contents and to maintain accurate systems control. However, in general shop and laboratory practice, the design of the title block and any accompanying tables are determined by what an industry or company decides to be most functional and cost-effective for their specific needs.

SYMBOLS USED TO REPRESENT DRAWING CHANGES

When a product, part, or detail is to be changed, the original drawing and/or computer program for manufacturing the part is altered to reflect the change. A revision symbol and/or number is often placed next to the changed value or feature to identify a change. The first change is numbered 1 or R1, the second; 2 and R2, etc. A circle or triangle symbol around the revision number helps to make the identification stand out from other drawing notations. The changes (R1) and (R2) in figure 40-5A quickly identify dimension changes on the drawing.

Note in each instance that a heavyweight line is placed under each dimension. This line tells that only the dimension on the drawing has been changed. In other words, the heavyweight line is the symbol for **NOT TO SCALE**.

After a change is made and correctly labeled, it is noted as an *engineering change notice* (ECN) in a title/block or in a corner on the drawing (figure 40-5B).

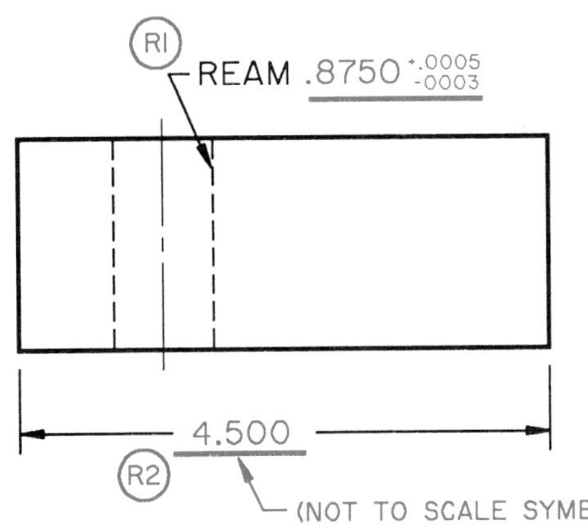

(A) APPLICATION OF SYMBOLS FOR DRAWING CHANGES

REVISION	DATE	ECN #	DESCRIPTION	APPROVED
RI	9-2-91	000372	.8750 +.0005 -.0003 WAS .750±.001	HGO
R2	12-5-91	000395	4.5000 WAS 4.375	SJB

(B) ENGINEERING CHANGES RECORDED IN TITLE/CHANGE BLOCK

FIGURE 40-5 USE OF SYMBOLS IN REPRESENTING DRAWING CHANGES AND RECORDING CHANGES

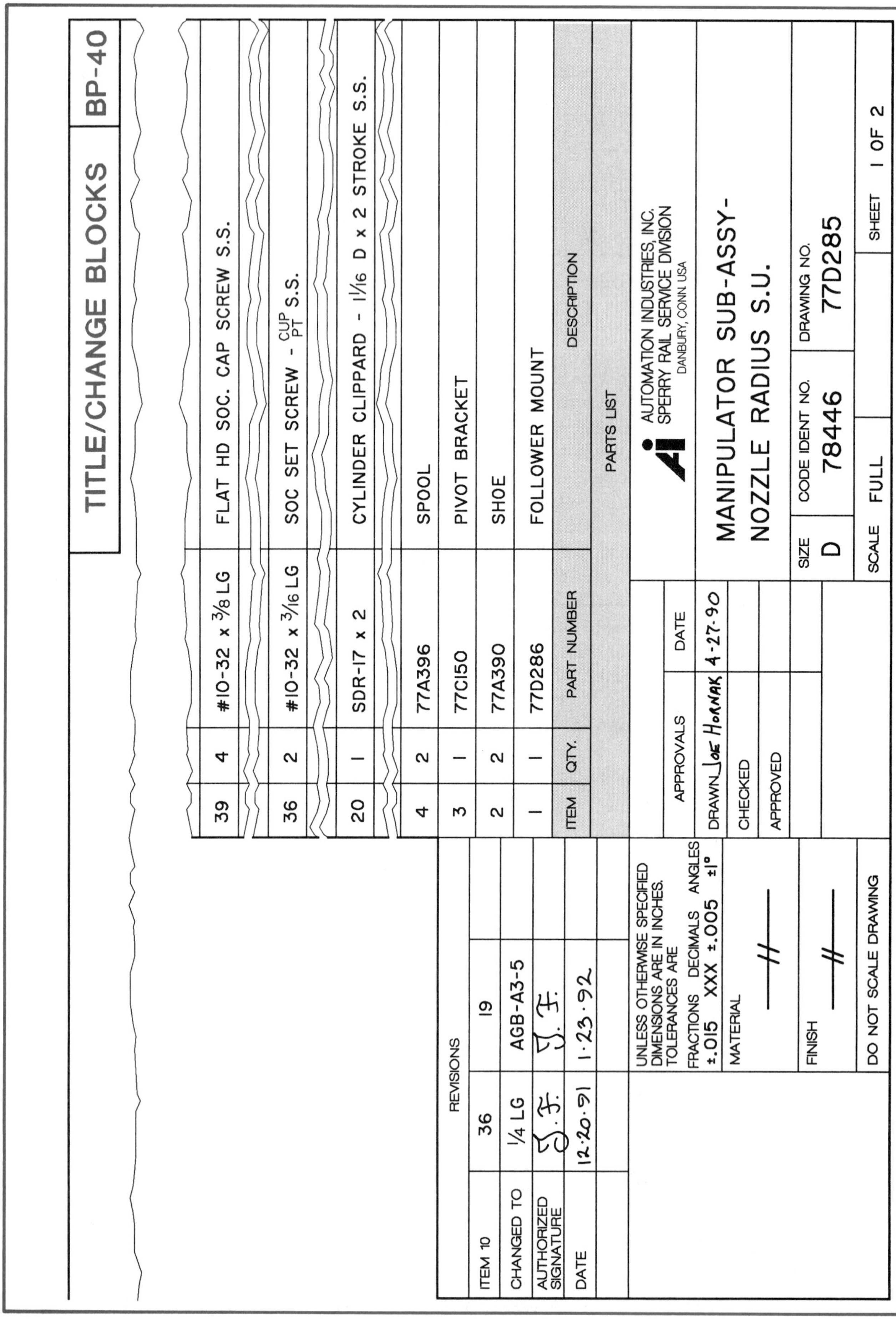

TITLE/CHANGE BLOCKS — BP-40

ITEM	QTY.	PART NUMBER	DESCRIPTION
39	4	#10-32 x 3/8 LG	FLAT HD SOC. CAP SCREW S.S.
36	2	#10-32 x 3/16 LG	SOC SET SCREW - CUP PT S.S.
20	1	SDR-17 x 2	CYLINDER CLIPPARD - 1 1/16 D x 2 STROKE S.S.
4	2	77A396	SPOOL
3	1	77C150	PIVOT BRACKET
2	2	77A390	SHOE
1	1	77D286	FOLLOWER MOUNT

PARTS LIST

REVISIONS

ITEM 10	36	19	
CHANGED TO	1/4 LG	AGB-A3-5	
AUTHORIZED SIGNATURE	J. F.	J. F.	
DATE	12·20·91	1·23·92	

UNLESS OTHERWISE SPECIFIED DIMENSIONS ARE IN INCHES. TOLERANCES ARE

FRACTIONS	DECIMALS	ANGLES
±.015	XXX ±.005	±1°

MATERIAL —//—

FINISH —//—

DO NOT SCALE DRAWING

APPROVALS	DATE
DRAWN Joe Hornak	4-27-90
CHECKED	
APPROVED	

AUTOMATION INDUSTRIES, INC.
SPERRY RAIL SERVICE DIVISION
DANBURY, CONN USA

MANIPULATOR SUB-ASSY-
NOZZLE RADIUS S.U.

SIZE	CODE IDENT NO.	DRAWING NO.
D	78446	77D285

SCALE FULL — SHEET 1 OF 2

Courtesy of Automation Industries, Inc., Sperry Rail Service Division (Adaptation of Industrial Drawing)

ASSIGNMENT—UNIT 40

TITLE/CHANGE BLOCKS (BP-40)

Student's Name ______________________

1. Identify the name of the part or assembly.
2. Give (a) the drawing number and (b) the number of sheets in the set.
3. State (a) the specific code identification and (b) drawing size.
4. Indicate the scale to which the drawing is made.
5. List the drafting room control (approval) procedure.
6. Give the unspecified drawing tolerance for:
   a. fractional dimensions,
   b. three-place decimals, and
   c. angular dimensions.
7. Provide the following information for the Spool part.
   a. Item number on the subassembly drawing
   b. Required quantity
   c. Reference number
8. Give the specifications for item (20).
9. List the requirements for item (39).
10. State one difference between an engineering work order and a drawing change order.
11. Give the full specifications of part (36) according to the latest revision.
12. Specify the latest identification of part (19).

1. ______________________
2. a. __________ b. __________
3. a. __________ b. __________
4. ______________________
5. ______________________
____
____
6. a. Fractional Dimensions: __________
b. Decimals: __________
c. Angles: __________

7. a. Item Number __________
b. Quantity Required __________
c. Ref. # __________
8. Item (20) __________
____
____
9. Item (39) __________
____
____

10. ______________________
____
____
____
11. Part (36) __________
____
12. Part (19) __________
____

UNIT 41
Detail Production Drawings

A *detail drawing* provides a complete description of a part. The drawing includes as many principal, auxiliary, section, phantom, and other views as are needed to accurately describe the object and its features. Included on the drawing are dimensions, notes, material and part specifications, tolerances, heat treatment and surface finishes, scale, and the like. All of this information is essential to producing, inspecting, measuring, and assembling and testing a part.

TYPES OF DETAIL DRAWINGS

There are many types of detail drawings. Simple parts may be adequately described by one-, two-, or three-view working drawings. One or more working drawings may be included on the same sheet, depending on the number of parts which are assembled as a unit.

Separate detail drawings are required in some industries, particularly where the part is large and/or complicated. A detail drawing may also be used for a specific manufacturing process. Patternmaking, forging, welding, machining, and stamping detail drawings are examples.

Pattern Detail Drawings

A cast part may require a separate *pattern* (or *patternmaking) detail drawing.* This type of drawing provides shape and dimensional information for making a pattern with or without cores for holes and hollow areas in a casting. Allowances are provided for metal shrinkage, removal of the pattern from a mold, and for machining.

Machining Detail Drawings

This type of detail drawing usually contains considerably more information about each surface, and dimensional, specification, and other machining requirements than is found on a pattern detail drawing. Precision reference surfaces, complete dimensioning, tolerancing, surface finish, and other details necessary in the production of a part are shown on the drawing. Figure 41-1 provides an example of a two-view machine detail drawing.

Forging Detail Drawings

Forging detail drawings are generally prepared with just the essential information for forming a part by forging processes. A forging detail drawing is sometimes placed side by side with a machine detail drawing. Figure 41-2 shows such a *combination drawing.* Sectional views are widely used in forging drawings.

A single drawing which provides full information for forging and machining is referred to as a *composite drawing.* On a composite drawing, the outline of the finished part is represented by solid lines. The outline of the forging is often shown by phantom lines. Dimensions of finished sizes appear on a composite drawing. Allowances for surface irregularities, shrinkage, and machining are added to the forging sizes.

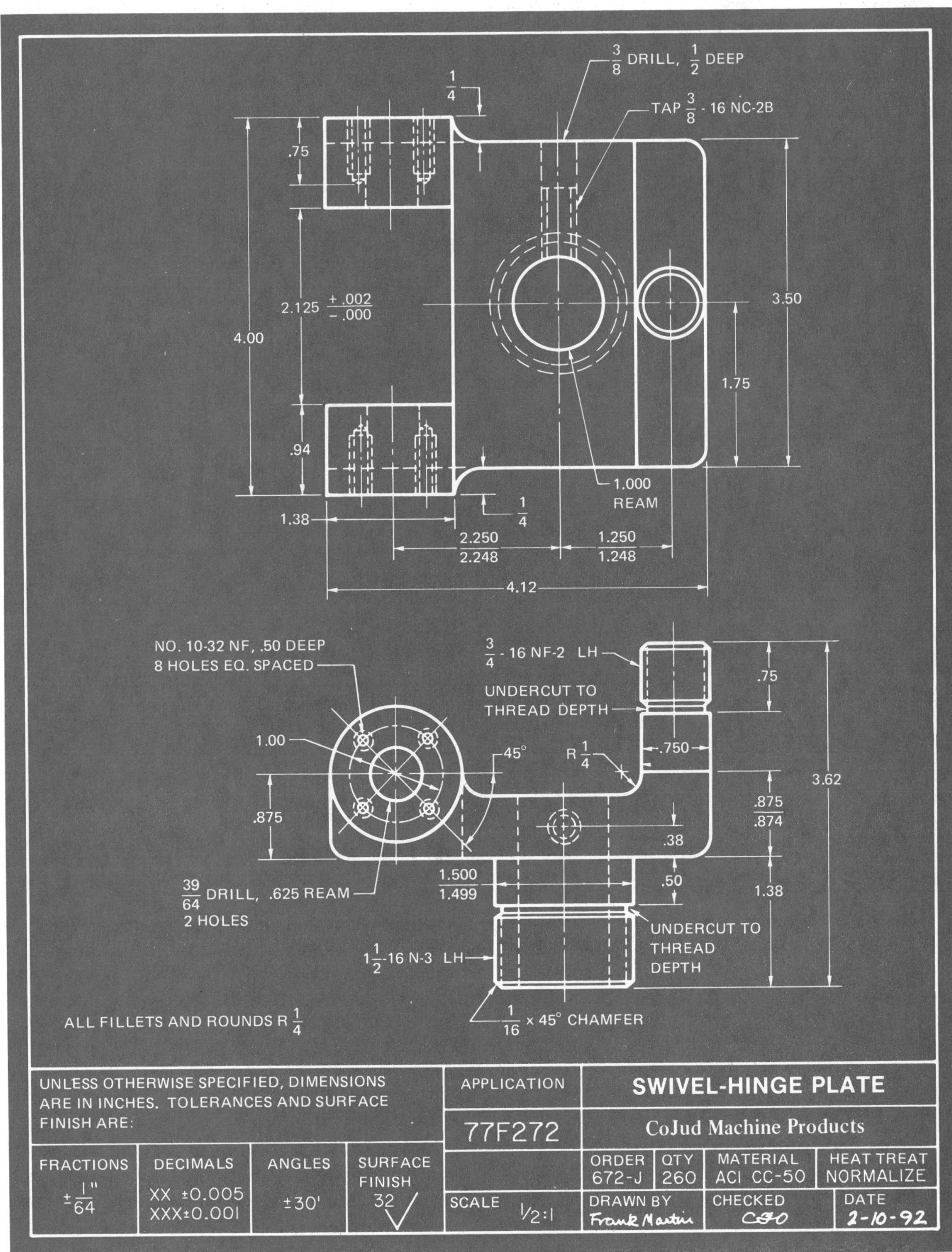

FIGURE 41-1 EXAMPLE OF A TWO-VIEW MACHINE DETAIL DRAWING

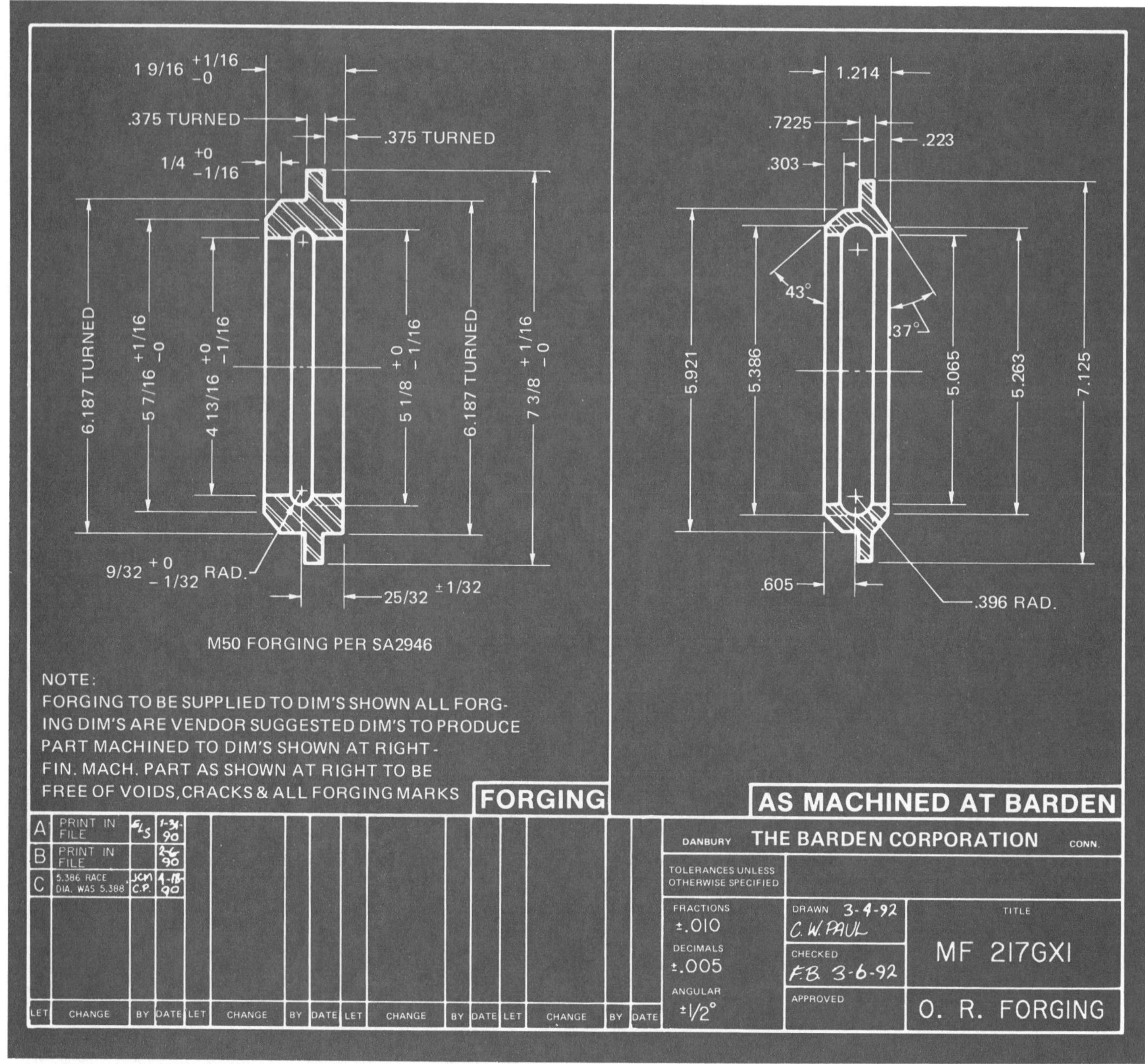

FIGURE 41-2 TWO-PART DETAIL DRAWING WITH SEPARATE INFORMATION FOR FORGING AND MACHINING PROCESSES (COURTESY OF THE BARDEN CORPORATION)

Welding Detail Drawings

A *welding detail drawing* represents a part which is formed by a number of separate pieces of metal when they are welded together. Each piece is dimensioned. The drawing shows the part as it appears before welding. A welding detail drawing usually has each numbered part with its specifications in a parts list.

Metal Stamping Detail Drawings

Metal stamping detail drawings are really multiview working drawings. The selection of views, dimensioning, tolerancing, and other specifications on stamping drawings follow basic principles and conventional standards governing technical drawings. A typical metal stamping detail drawing is illustrated in figure 41-3.

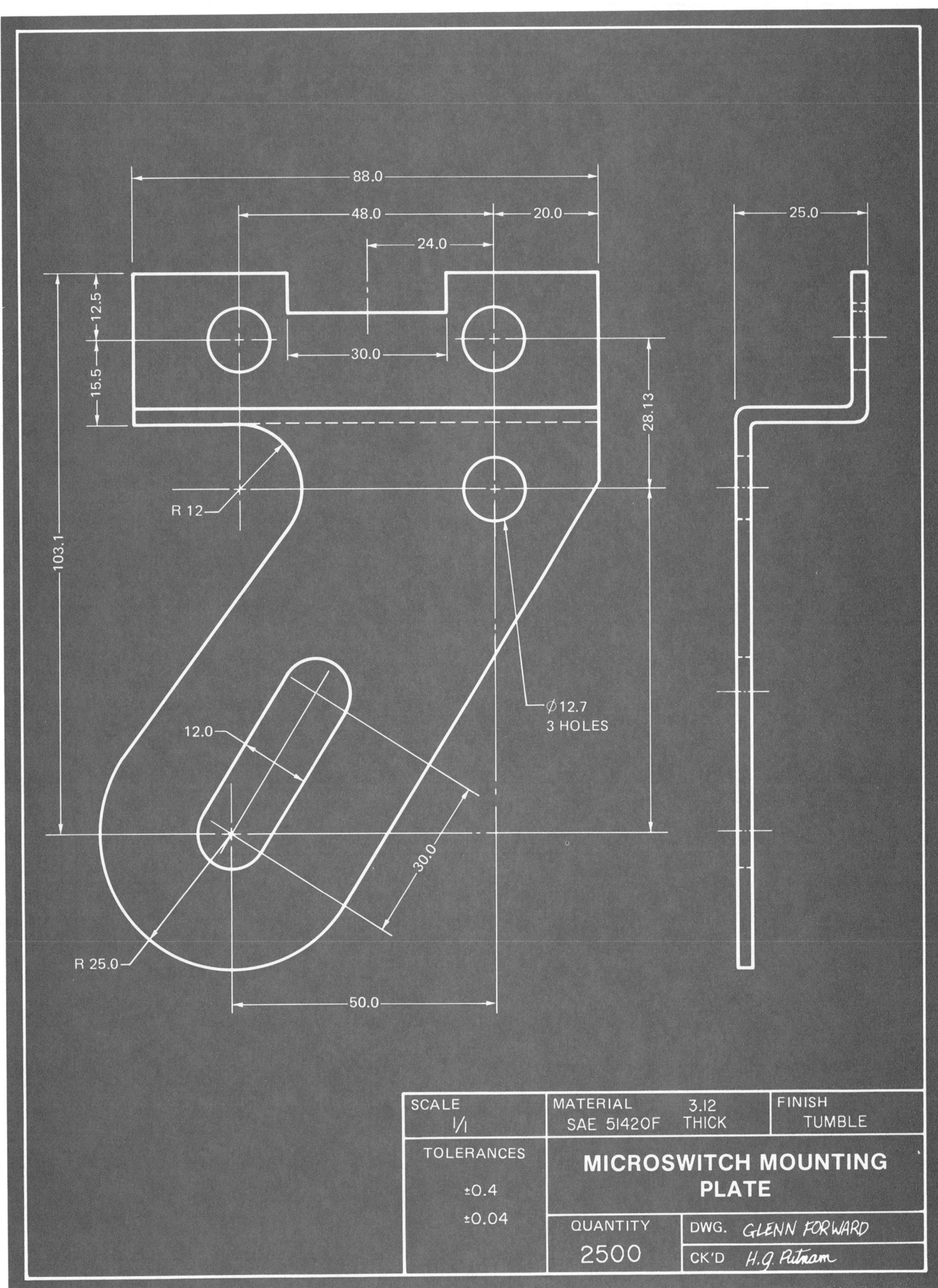

FIGURE 41-3 EXAMPLE OF METAL STAMPING DETAIL DRAWING

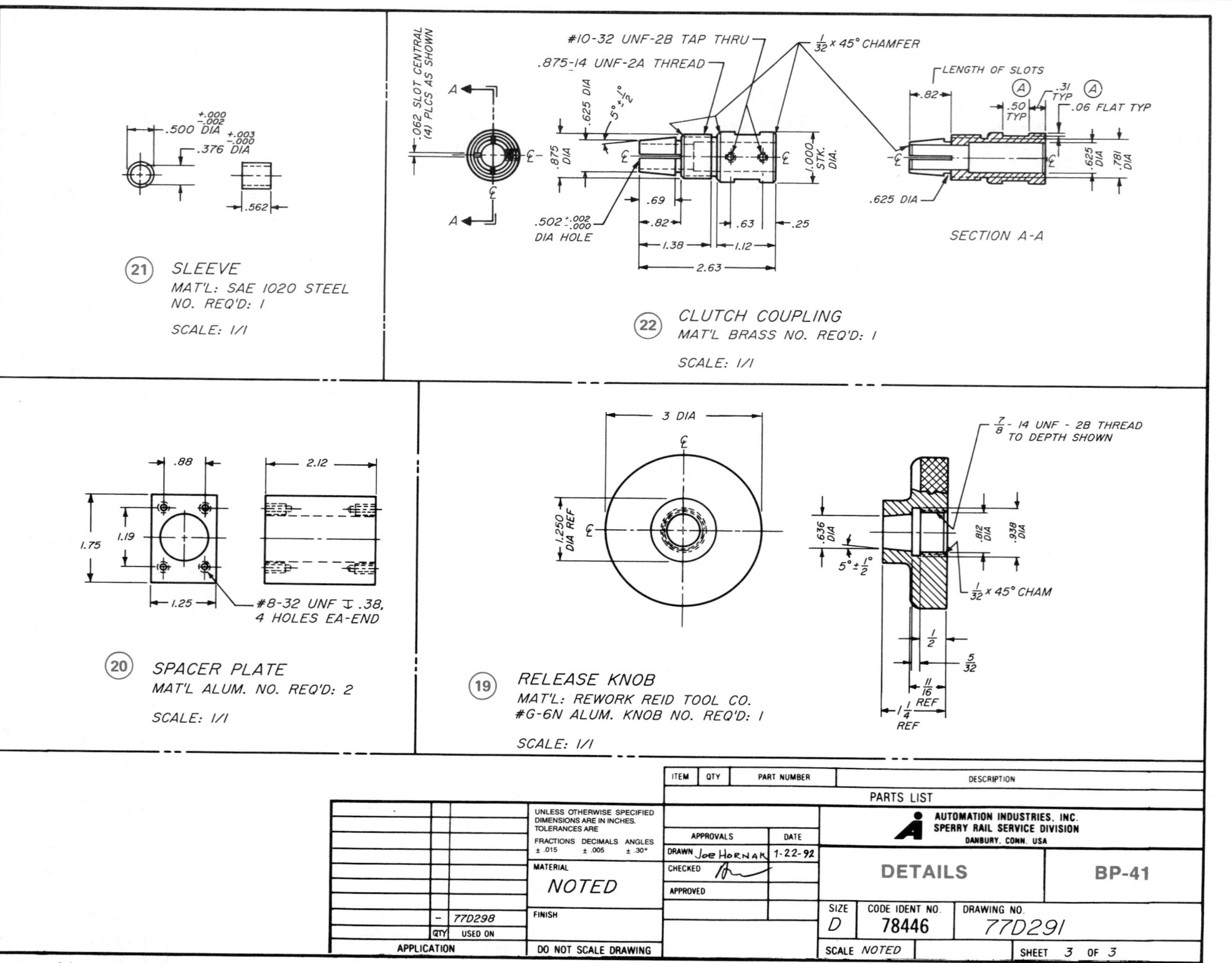

Courtesy of Automation Industries, Inc., Sperry Rail Service Division (Modified Industrial Drawing)

ASSIGNMENT—UNIT 41

DETAIL DRAWINGS (BP-41)

Student's Name ______________________

1. Describe briefly the function of detail drawings.
2. List three different types of detail drawings for particular manufacturing processes.
3. Supply the following information:
   a. Drawing number c. Sheet #
   a. Code identification d. Scale
4. Identify the part name which represents the function of detail items (19), (20), (21), and (22).
5. List the material required for each of the four parts.
6. Determine the upper and lower limits and nominal size of:
   a. Each reference dimension and
   b. The taper angle of part (19).
7. Identify
   a. The supplier of part (19) and
   b. The specification
8. Give the specifications for the tapped holes in part (20).
9. Determine the following tolerances for part (21):
   a. Outside diameter c. Width (length)
   b. Inside diameter
10. Give the following nominal, upper, and lower limit dimensions for part (21):
    a. Outside diameter c. Width (length)
    b. Inside diameter
11. Describe two functions served by Section A–A of part (22).
12. Give the standard chamfer to which the holes and edges are to be machined.
13. Establish the nominal, upper, and lower limit dimensions (part (22)) for the following:
    a. Two inside diameters
    b. The taper angle
    c. The small diameter end of the taper
14. Give the following information for the threaded features on part (22):
    a. Thread sizes
    b. Thread series
    c. Fit
    d. Meaning of "A" and "B" designation
15. State briefly the processes for producing the threaded features.
16. Give the specifications for slotting the tapered nose portion.
17. Describe (a) what is meant by the note FLAT TYP. Section A–A, and (b) the nominal dimension for this feature.
18. Explain the assembly and function of parts (19) and (22) in relation to part (21), including the tapped holes.

1. ______________________

2. a. ______ b. ______ c. ______

3. a. ______ c. ______
   b. ______ d. ______

4. a. (19) ______ c. (21) ______
   b. (20) ______ d. (22) ______

5. (19) ______ (21) ______
   (20) ______ (22) ______

6.

	Ref.	Upper Limit	Lower Limit
a.			
b.			

7. a. ______ b. ______

8. ______

9. a. ______ b. ______ c. ______

10.

	Nominal	Upper Limit	Lower Limit
a.			
b.			
c.			

11. a. ______ b. ______

12. ______

13.

	Nominal	Upper Limit	Lower Limit
a.			
b.			
c.			

14. a. Sizes ______ ______
    b. Series ______ ______
    c. Fit ______ ______
    d. Designation ______ ______

15. a. ______ b. ______

16. ______

17. a. ______ b. ______

18. ______

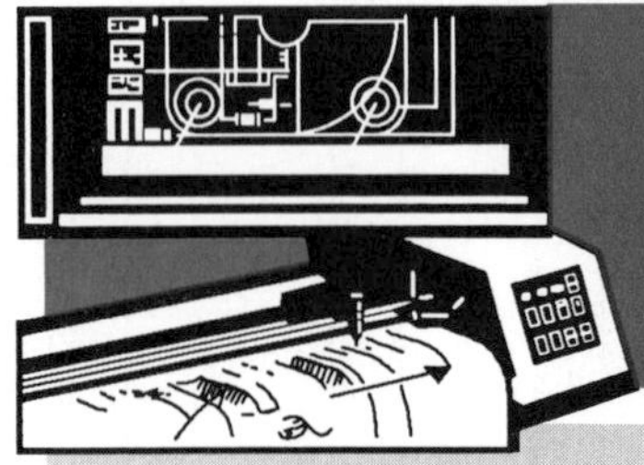

HIGH TECHNOLOGY applications

HT-12 CADD/FMS: AUTOMATED SUBSYSTEMS

A flexible manufacturing system (FMS) is defined as a computer-controlled, random-order, highly-automated manufacturing system. A FMS interfaces the input of CADD/CAM and the computer control of machine control units (MCU) that command NC/CNC/DNC and other subsystems.

Four other subsystems that are equally important to FMS relate to the following:

- Materials handling.
- Tool handling.
- In-process and post-process gaging.
- Precision surface sensing probing.

To review, materials-handling subsystems relate to such functions as: work positioning (fixturing) and work holding; transporting workpieces in a controlled path and on-line loading into precise positions at each different stage of manufacturing; and off-loading and work transfer.

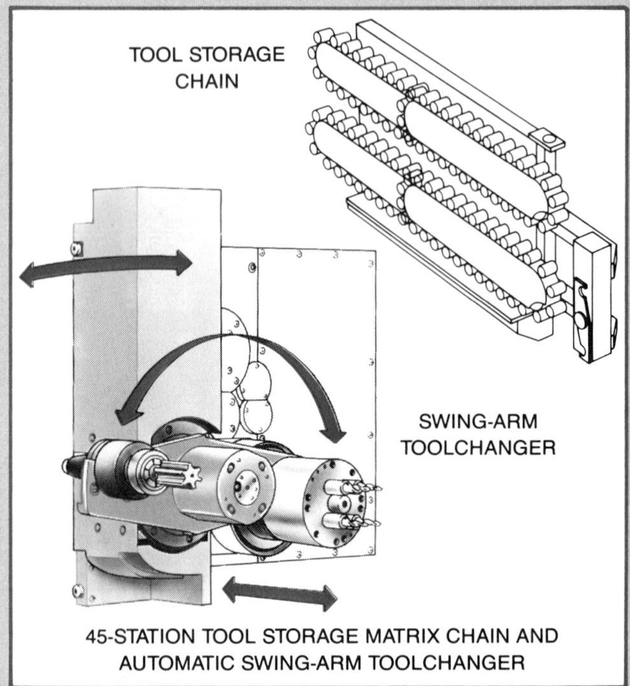

45-STATION TOOL STORAGE MATRIX CHAIN AND AUTOMATIC SWING-ARM TOOLCHANGER

NC AUTOMATED TOOL-HANDLING SYSTEMS

Numerical control functions (as related to tool handling for CNC machine tools, CADD, and FMS requirements) provide computerized commands for the following tasks:

- Determining required cutting tool and machining setups.
- Establishing operation sequences and time intervals.
- Setting cutting speeds and feeds.
- Controlling tool changer movements for positioning, mounting, removing, and storing tools.

Cutting tools, instruments, and other devices are mounted in automatic tool drums and/or tool chains. The illustration shows two 45-station tool storage matrix chains and an automatic swing-arm tool changer.

NC IN-PROCESS AND POST-PROCESS GAGING

CADD/CAM/FMS depend on *in-process and post-process gaging. In-process gaging* means that form, dimensional measurements, and inspection processes are performed prior to the removal of a workpiece from a machining station.

Post-process gaging takes place after a part is machined. While all part features may be checked, out-of-tolerance conditions cannot be compensated for during machining.

In-process gaging permits corrections to be made during machining through commands of the MCU. In-process gaging requires the linkage of a computer-controlled measuring machine (CCMM) into a group of NC machines that are remotely controlled (by DNC).

CADD/CAM SURFACE SENSING PROBES

Electronically controlled *surface sensing probes* are used in CNC to locate, measure, check, and compensate for surface defects, dimensional variations from allowable tolerances, and machine drift away from normal conditions. A precision surface sensing probe consists of four major components, as follows:

- An *interchangeable stylus* (sensing probe).
- A *probe body* whose shank is designed to fit a machine spindle taper.
- An *inductive machine-mounted module* that inter-connects to an electronic control board.
- An *electronic printed circuit (control) board* that interfaces with the inductive module to instantaneously record dimensional, geometric, and other surface measurements beyond an acceptable quality level (AQL).

Courtesy of CINCINNATI MILACRON INC.

UNIT 42
Assembly Production Drawings

An *assembly drawing* contains all the parts of a structure or machine. Each part is drawn in the relative position in which it functions. Assembly drawings serve the following purposes:

- To represent a complete unit so that each part may be easily identified and recorded on a parts list.
- To develop specifications which relate to geometric tolerancing, class of fit, wear, surface texture, material; as well as to shape, size, dimensional measurements, purchasing, and other information.
- To provide the detailer with basic data from which separate detail drawings for each part may be prepared.
- To visually illustrate internal features and/or the completed product. Modified spin-off drawings may be suggested to accommodate different needs for assembly drawings for production, testing, operating, marketing, etc.

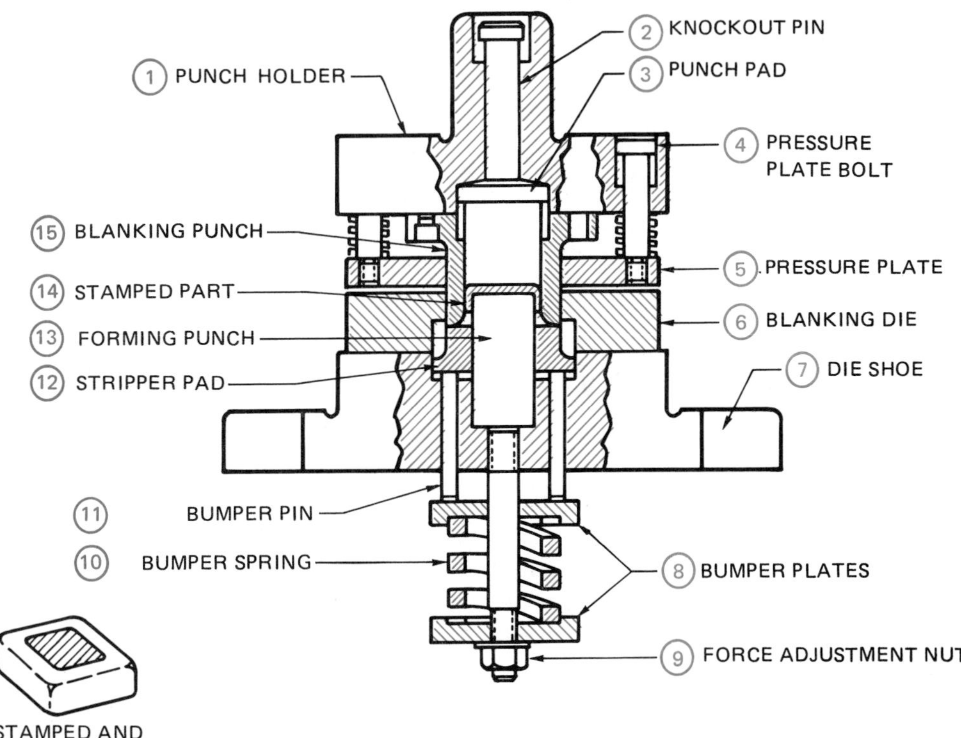

FIGURE 42-1 SINGLE-VIEW ASSEMBLY DRAWING USING FULL AND CUT-AWAY SECTIONS

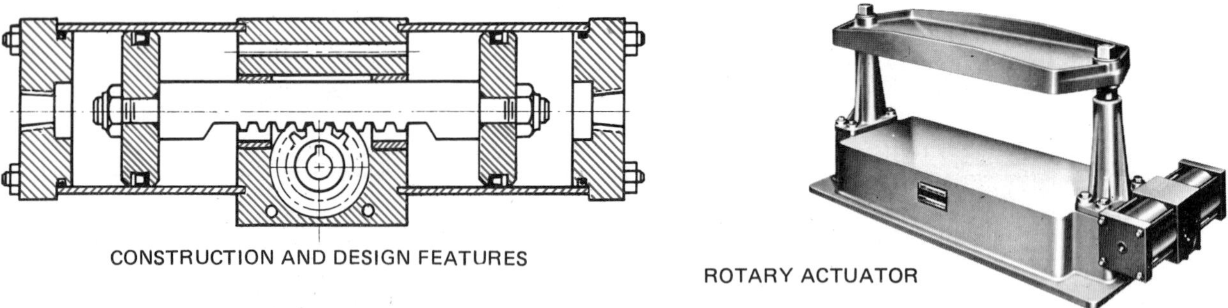

FIGURE 42-2 COMBINATION FULL-SECTION ASSEMBLY DRAWING AND PHOTOGRAPH OF COMPLETE MECHANISM (COURTESY OF UNIVERSAL VISE AND TOOL COMPANY, SWARTZ FIXTURE DIVISION)

SELECTION OF VIEWS FOR ASSEMBLY DRAWINGS

The number of views on an assembly drawing depends on the complexity of the unit and the purpose to be made of the drawing. For example, figure 42-1 is a single-view assembly drawing of a combination die. Note that certain parts appear in full section and other part features are shown within cut-away sections. This assembly drawing is adequate for identifying all of the parts and for illustrating the steps in blanking and forming the shell.

The *combination assembly drawing* illustrated in figure 42-2 includes a mechanical drawing of the actuating mechanism and a photograph of the complete pneumatic fixture. The full-section assembly drawing shows the design and construction features of the rotary actuator. The actual form and parts of the pneumatic fixture appear in the accompanying photograph. Although it is common practice to identify and list the separate parts, these are not included on this particular assembly drawing.

Omitting Features on Assembly Drawings

Details about certain features are often omitted in order to improve the clarity of an assembly drawing. This information generally appears on the detail drawing of each part.

A common type of assembly drawing is pictured in figure 42-3. There are three standard views. The piece part appears in phantom outline in the front view. The principal cutter-locator dimensions are also given. On some assembly drawings, detail dimensions are omitted. Only center distances, special movement, and overall dimensions are given. The supplemental two-view working drawing provides full information for producing the Feeler Gage.

Note that each different part is identified by a leader and a number. Specifications for each numbered part appear in the material list which usually includes part names.

PARTIALLY DETAILED ASSEMBLY DRAWINGS

There are many exceptions to dimensioning, particularly in relation to partially detailed assembly drawings. Figure 42-4 illustrates the use of such a drawing for a drilling and tapping fixture. There is a top view, partial right-side view, and an auxiliary section view. The piece part is shown in locked position in phantom outline. The supplemental two-view working drawing describes the piece part.

Partially detailed assembly drawings provide sufficient information for producing and assembling many parts. This practice eliminates the preparation of a number of detail drawings. The craftsperson handles fewer drawings, has essential detail information, and easily visualizes each part.

TYPES OF ASSEMBLY DRAWINGS

- *General assembly drawings* show the outlines of parts and the relationship of movements. Figure 42-5 illustrates the main spindle assembly of a precision jig grinder. This drawing focuses on the relationship of the main spindle parts, which are shown in section. Other details of the downfeed hand wheel and the grinding head are displayed by the outline of each unit.

  Where there are a great many parts or a unit is very large, the general assembly drawing is broken down into smaller functional drawings. These are called *subassembly drawings.*

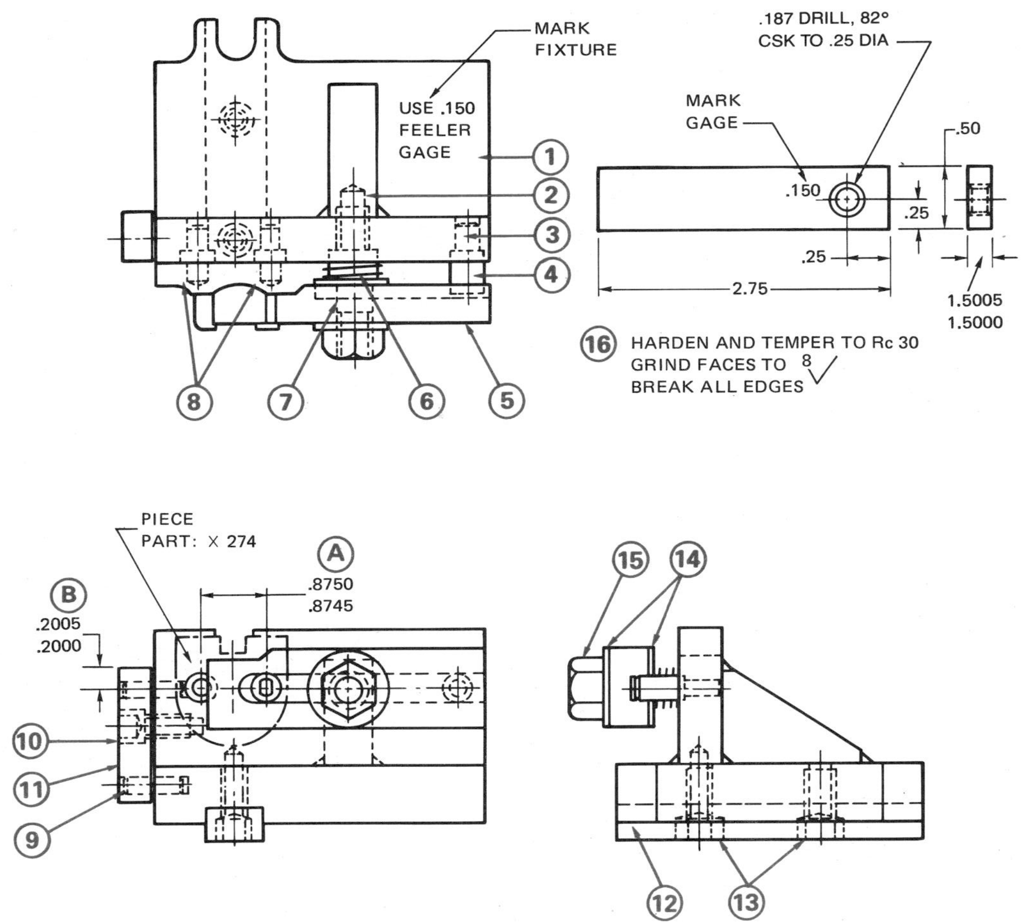

MARK FIXTURE

USE .150 FEELER GAGE

.187 DRILL, 82° CSK TO .25 DIA

MARK GAGE

.150 .25 .25 2.75 .50 1.5005 1.5000

(16) HARDEN AND TEMPER TO Rc 30
GRIND FACES TO 8
BREAK ALL EDGES

PIECE PART: X 274

(A) .8750 .8745

(B) .2005 .2000

PARTS LIST				
DET. #	PART IDENTITY	QTY	SIZES	MAT'L
1	BASE	1	$\frac{5}{8}$ x 3 x 4	AISI 1020
2	SUPPORT	1	$\frac{5}{8}$ x 2 x $1\frac{1}{4}$	AISI 1020
3	VERTICAL PLATE	1	$\frac{5}{8}$ x $1\frac{1}{2}$ x 4	AISI 1020
4	GUIDE PIN	1	$\frac{3}{8}$ DIA x 1	DRILL ROD
5	CLAMP PLATE	1	$\frac{1}{2}$ x 1 x $3\frac{1}{4}$	AISI 1020
6	SPRING	1	.025 DIA	SPG WIRE
7	CLAMP SCREW	1	$\frac{1}{2}$ DIA x $1\frac{5}{8}$ LG	AISI 1020
8	LOCATOR PINS	2	$\frac{1}{2}$ x 1	DRILL ROD
9	DOWEL PINS	2	$\frac{3}{16}$ DIA x $\frac{3}{4}$ LG	STD
10	SCREW	1	$\frac{5}{16}$-18NC x $\frac{1}{2}$ SOC. HD CAP SCR	STD
11	SETUP PLATE	1	$\frac{11}{16}$ x $\frac{5}{8}$ x $1\frac{3}{4}$	AISI 1090
12	TONGUE	1	$\frac{5}{8}$ x $\frac{3}{8}$ x 4	AISI 1020
13	SCREWS	2	$\frac{5}{16}$ x18NC x $\frac{7}{16}$ SOC. HD CAP SCR	STD
14	WASHERS	2	$\frac{3}{8}$ ID-1OD x $\frac{3}{32}$ OD	AISI 1020
15	CLAMP NUT	1	$\frac{3}{8}$-16NC HEX. FIN.	STD
16	FEELER GAGE	1	$\frac{11}{64}$ x $\frac{1}{2}$ x 3 GROUND STOCK	B & S

DET. #	CHANGE	DATE	APPVD.
B	WAS .1875 / .1870	10-20-90	JMR
A	WAS .860 / .861	10-20-90	JMR

CHANGE NOTES

FIXTURE ASSEMBLY

MACHINE TOOL IDENT.	MILLING MACHINE MM 22		DRAWN BY JOJ
TOOLING IDENT.	FIXTURE X274		CHECKED BY HGP
WORK STATION		3 MM	APPVD BY TPO
DESCRIPTION SPACER PLATE		SHEET 1 OF 3	DATE 1-8-92

FIGURE 42-3 TYPICAL ASSEMBLY DRAWING

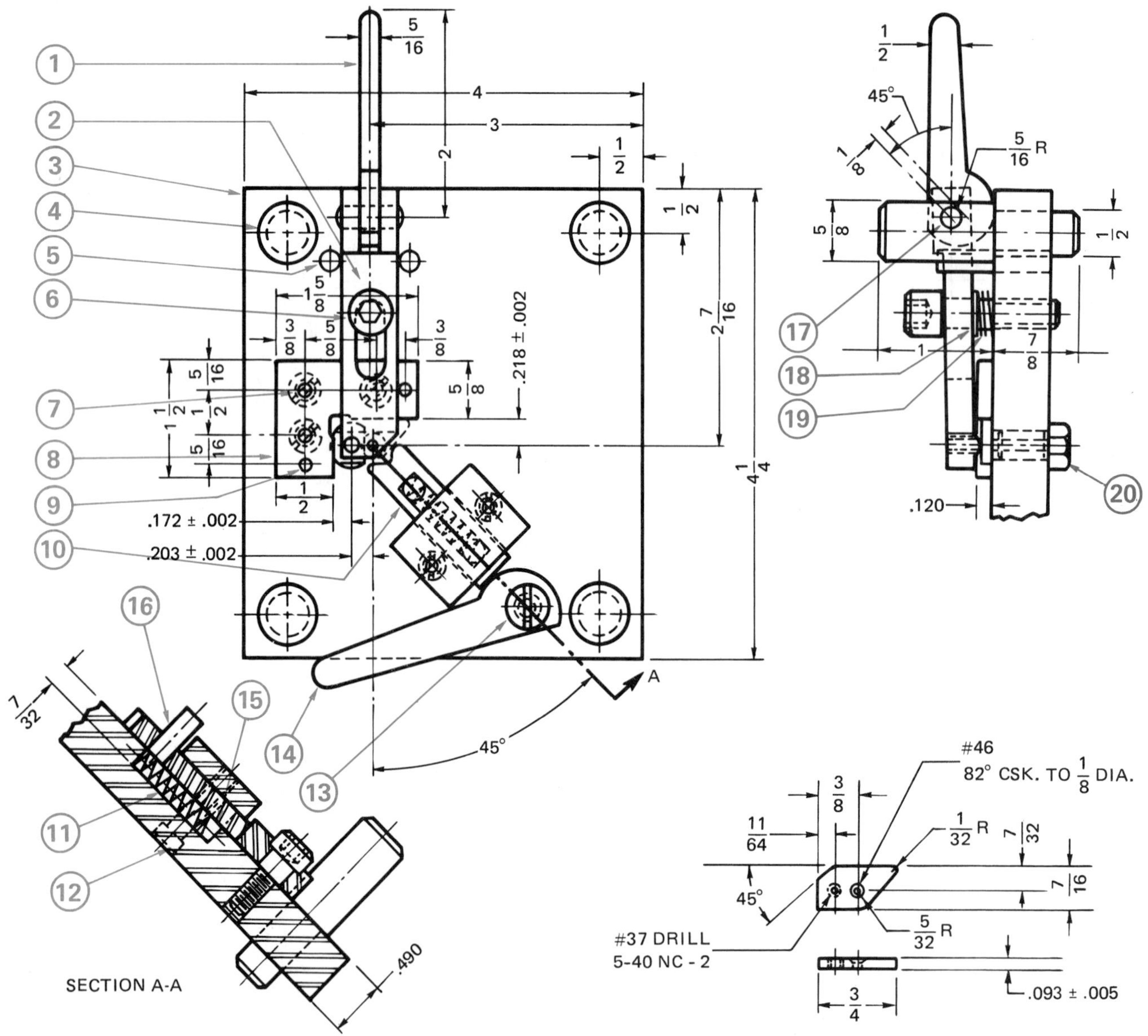

FIGURE 42-4 PARTIALLY DETAILED ASSEMBLY DRAWING WITH WORKING DRAWING OF MANUFACTURED PART

- A *working assembly drawing (partially detailed assembly drawing)*, as stated earlier, is often used for simple mechanisms. The single drawing incorporates the function of a detail drawing and an assembly drawing.
- *Pictorial assembly drawings* (prepared either mechanically or freehand) provide another common type of assembly drawing. Cutaway sections are used to expose parts and features of parts.
- *Installation assembly drawings* provide details necessary for shop layouts in placing equipment and establishing overall work area dimensions under maximum operating conditions. In addition to dimensional data, specifications are included for utility and other energy requirements. Related *maintenance assembly* and *subassembly drawings* show clearly the operation of each component. Such drawings are used for servicing, lubricating, and operating manuals.
- *Catalog assembly drawings* provide design features of parts and mechanisms. Replacement information about sizes, identification numbers of parts, manufacturers, etc., usually appear in a table which accompanies the assembly drawing.

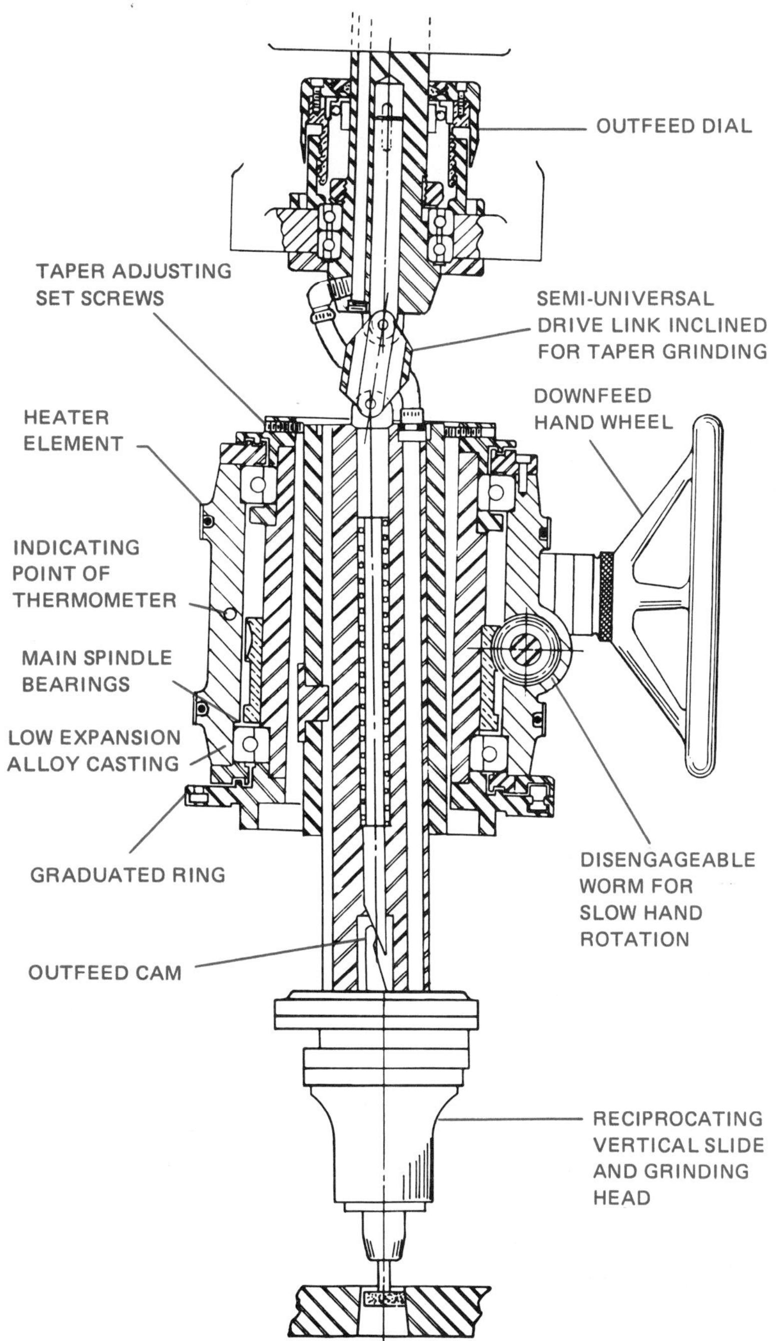

FIGURE 42-5 GENERAL ASSEMBLY DRAWING CONSISTING OF COMBINATION SECTION AND FULL OUTLINE OF A MAIN SPINDLE ASSEMBLY (COURTESY OF MOORE SPECIAL TOOL COMPANY, INC.)

NOTE: REFER TO APPENDIX B-6 FOR THE LARGER SCALE DRAWING TO USE WITH THIS ASSIGNMENT.

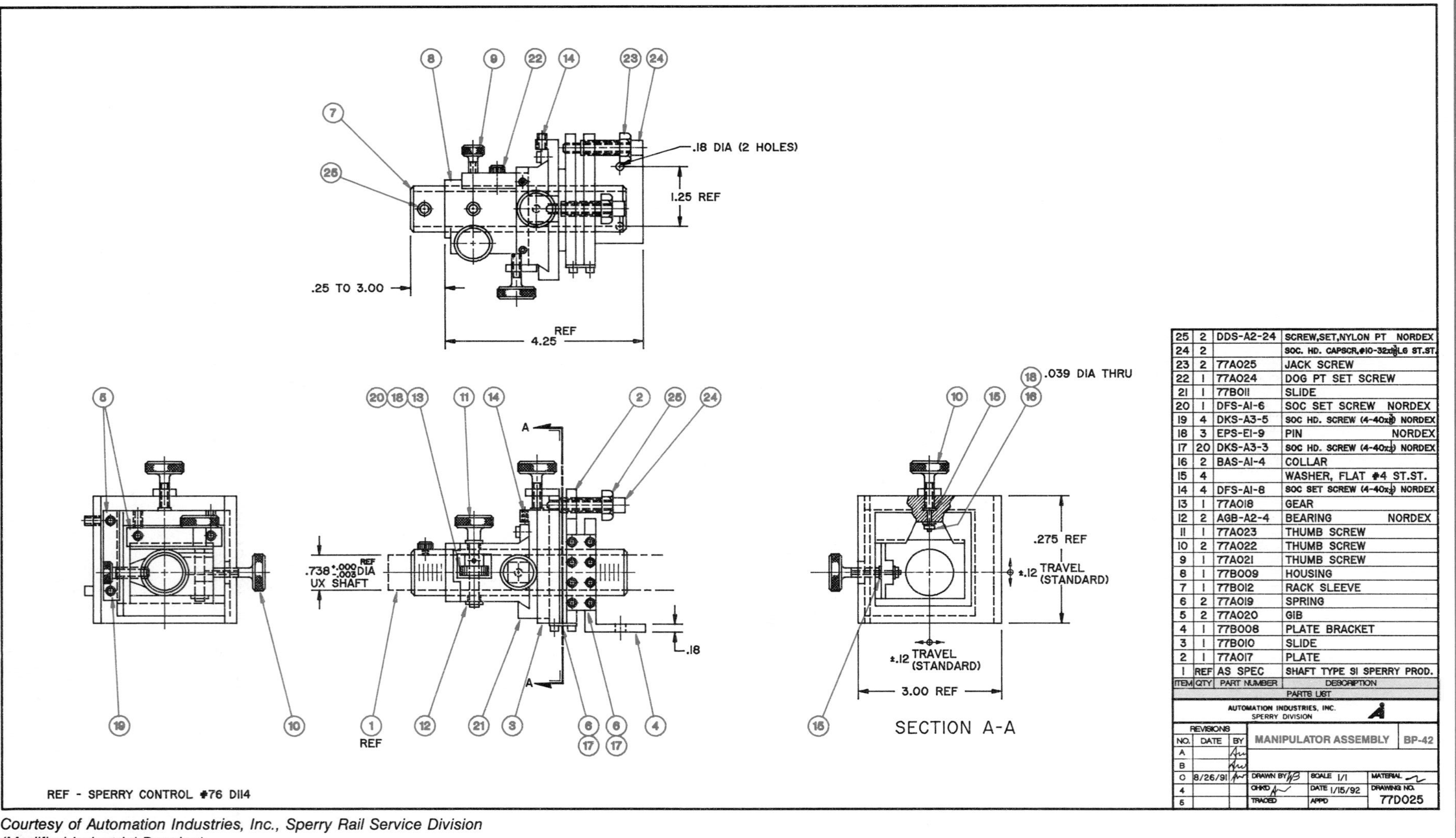

Courtesy of Automation Industries, Inc., Sperry Rail Service Division (Modified Industrial Drawing)

MANIPULATOR ASSEMBLY (BP-42)

1. a. Identify two different types of assembly drawings (other than a general assembly drawing).
   b. State the prime function served by each type of assembly drawing.
2. Name each of the four views used on the Manipulator drawing.
3. Identify the type of drawing represented by the four views.
4. Describe briefly the purpose of using a cut-away section in the Section A–A view.
5. Provide the following information:
   a. The number of different items (parts) that are represented on the drawing.
   b. The detail numbers, part numbers, and the quantity required of each thumb screw.
   c. The detail numbers, names, and parts in the assembly which actuate the rack sleeve (7).
6. Identify the part names and item numbers of parts other than washers, collars, screws, pins, and springs for which detail drawings are needed.
7. State the function of each of the following items:
   a. (25) Nylon point set screws
   b. (2) Plate
   c. (4) Plate bracket
   d. (5) Gibs
   e. (10) (Two) knobs
   f. (11) Thumb screw
   g. (19) Socket head screws
8. Establish the following Reference dimensions:
   a. Width
   b. Height
   c. Depth
   d. Center distance between the two holes
   e. Part (1)
9. a. Explain the meaning of
   (18)
   (16) .039 DIA THRU
   b. Give the fastener name, detail number, and part number.
10. Give the following travel ranges:
   a. Slide assembly—vertically
   b. Slide assembly—horizontally
   c. Rack Sleeve—longitudinally

ASSIGNMENT—UNIT 42

Student's Name ______________________

1. a. (1) ____________ (2) ____________
   b. (1) ____________ (2) ____________
2. a. ______________________
   b. ______________________
   c. ______________________
   d. ______________________
3. ______________________
4. ______________________
   ____
5. a. ______________________
   b.

Det. No.	Pt. No.	Qty. Req'd.

   c.

Det. No.	Pt. Name	Pt. No.

6.

Pt. Names	Item Nos.	Pt. Names	Item Nos.

7. a. (25) ______________________
   ____
   b. (2) ______________________
   ____
   (4) ______________________
   ____
   (5) ______________________
   ____
   (10) ______________________
   ____
   (11) ______________________
   (19) ______________________

8.

	Reference Dimension
a. Width	
b. Height	
c. Depth	
d. Center Distance	
e. Part (1)	

9. a. ______________________
   ____
   b. Name __________ Detail Number __________
   Part Number __________

10.

Movement	Travel Range
a. Vertically	
b. Horizontally	
c. Longitudinally	

UNIT 43
Simplified (Functional) Industrial Drawings

The nature and extent to which technical drawings are simplified depend primarily on one of three possible uses:

- *General, functional technical drawings.* These drawings follow ANSI standards and are used across a number of industrial fields. The technical drawings are applied in engineering, design, production, inspection, assembly, and other processing requirements.
- *Technical production drawings within a specific field.* Aerospace, automotive, machine tool, electrical/electronic, and instrumentation are examples of specific

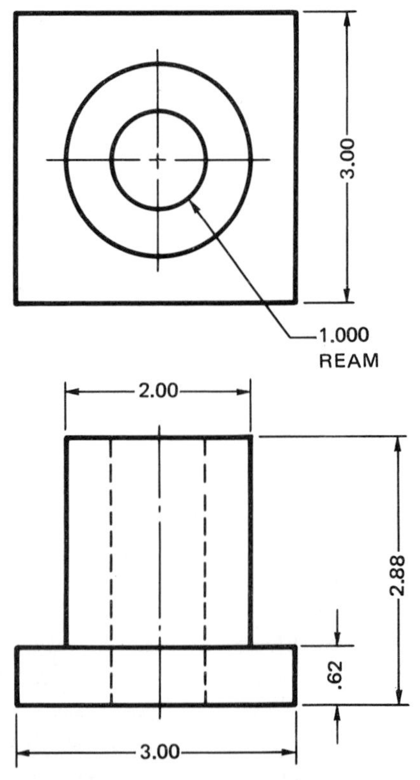

(A) CONVENTIONAL TWO-VIEW DRAWING

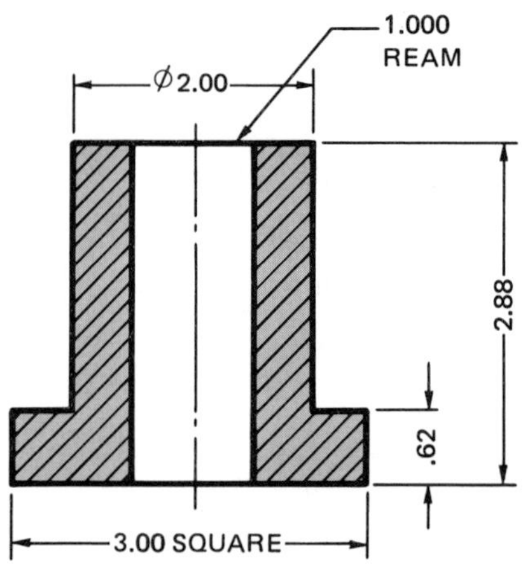

(B) PREFERRED SIMPLIFIED REPRESENTATION

FIGURE 43-1 APPLICATION OF SIMPLIFIED DRAFTING PRACTICES ELIMINATES ONE VIEW

fields. Drawings for use within such fields may include simplified drafting techniques accepted as standards by each particular industry.

- *Local or in-company technical drawings.* These are functional drawings used by workers within a plant who are familiar with a particular product line. Simplified drawings may be used to provide limited details and other information as required in a specific company.

Simplified drawings are used to replace a number of former conventional drafting room practices. Older technical drawings often contained more details than were required, duplicated information, and occasionally made the reading of a print difficult or confusing.

CHARACTERISTICS OF SIMPLIFIED INDUSTRIAL DRAWINGS

Today, the designer and detailer give consideration to the following factors when deciding whether a conventional or a simplified drawing is to be prepared:

- *Readability, clarity,* and *ease of interpretation.* Figure 43-1 provides an example.
- *Single meaning.* All details, data, and other required information must be *uniformly and accurately interpreted* by all workers involved in the production of a part, structure, or mechanism.
- *Complete information.* Only essential lines, views, details, notes, and dimensions are included on a simplified drawing.
- *Economy.* This factor relates to the efficient use of time for design and drafting as well as economy in the interpretation of the drawing.

GUIDELINES FOR INTERPRETING BLUEPRINTS OF SIMPLIFIED DRAWINGS

The reading of a blueprint of a simplified industrial drawing requires an understanding of changed drafting room practices. The changes are broadly grouped in relation to: (a) types and number of views, (b) dimensioning techniques, (c) lines, (d) symbols, notes, and other written specifications, and (e) different combinations of techniques of representation. Blueprints often include combinations of mechanically prepared technical drawings, freehand sketches, photographic drawings, and other types of drawings. Simplified drawings usually involve conventional and modified drafting room practices.

The reader of a simplified drawing needs to recognize the following changes:

- *Unnecessary views are eliminated.* Notations on one view often permit the replacement of one or more views. Symbols may be used in place of notes. In the previous example (Figure 43-1B), the addition of the word **SQUARE** to the 3.00″ dimension provides complete information about the shape and size of the base and the remaining features. This functional one-view drawing is preferred to the two views (Figure 43-1A).
- *Symbols and abbreviations* like **THD** (thread), **THK** (thick or thickness), **HEX** (hexagon), **ϕ** (diameter), or **℄** (center line) replace the written word where possible.
- *Partial views* of *symmetrical objects* are used instead of full views.
- *Repetitive details and unnecessary hidden lines* are omitted. Details are not shown of hardware and other general stock items. Specifications in parts lists often substitute for simple detail drawings.
- *Unimportant details are removed* from a conventional drawing. Figure 43-2A is a conventional drawing of a rectangular plate with four holes. The functional drawing at (B) provides the same information.
- *Parts, features, notes, symbols,* and *repetitive line patterns* are produced on original drawings from preprinted pressure-sensitive overlays.
- *Tables accompany the drawing* when there is a series of holes or repetitive details. Position and size dimensions are contained in a table and are identified by a number or symbol relating to each particular feature.
- *Elaborate pictorial or repeating details* are omitted.

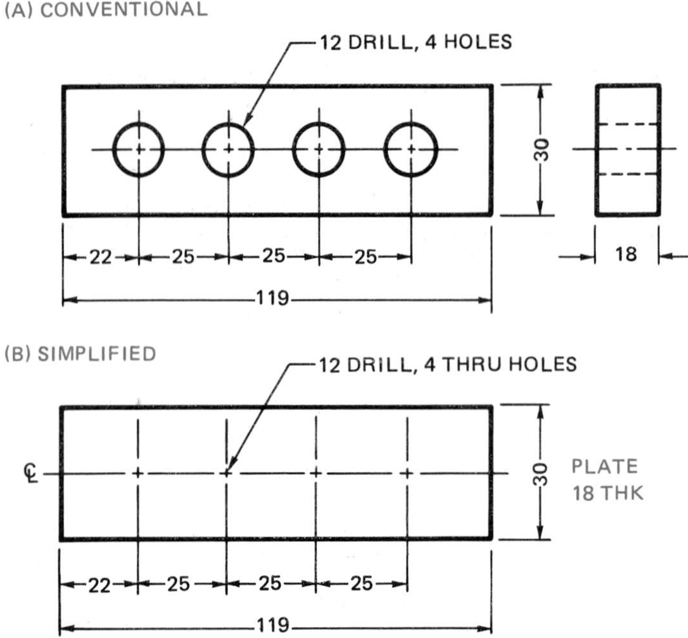

FIGURE 43-2 ELIMINATION OF UNNECESSARY DETAILS SIMPLIFIES REPRESENTATION OF OBJECT

OTHER SIMPLIFIED REPRESENTATION TECHNIQUES

Base-line dimensioning may be considered a simplified technique. Holes of varying sizes are represented by easily identified and interpreted hole symbols. Figure 43-3A shows conventional symbols for representing a hole and a counterbored hole. Other forms of representing holes on simplified drawings are presented at (B).

Where hole locations are symmetrical, only one set of dimensions may be found on one-half of a view. Figure 43-4 combines the use of the following simplified representation practices:

- Base-line dimensioning with dimensioning of certain symmetrical features.
- The use of conventional hole symbols for four counterbored holes and the two .500″ reamed holes. Two other symbols are used (and) for the two 3/8″ drilled holes and the five .375″ reamed holes, respectively.
- The identification of the part as detail ④ followed by the part thickness abbreviation (**THK**) and the material note (**AISI 1090 TOOL STEEL**).

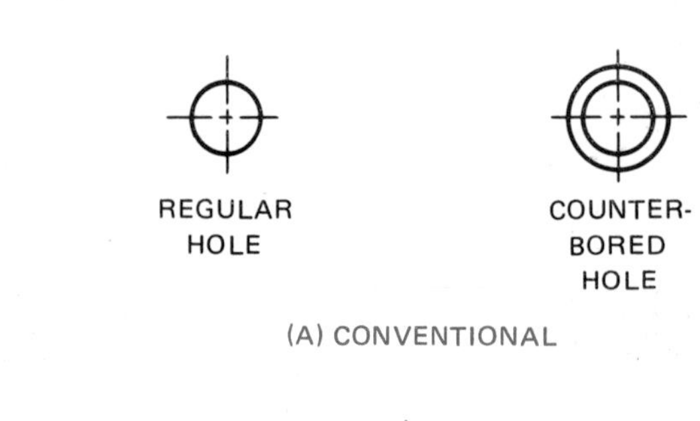

(B) SYMBOLS USED FOR IDENTIFYING DIFFERENT HOLE SIZES ON THE SAME DRAWING

FIGURE 43-3 SYMBOLS USED TO REPRESENT HOLES ON CONVENTIONAL AND SIMPLIFIED INDUSTRIAL DRAWINGS

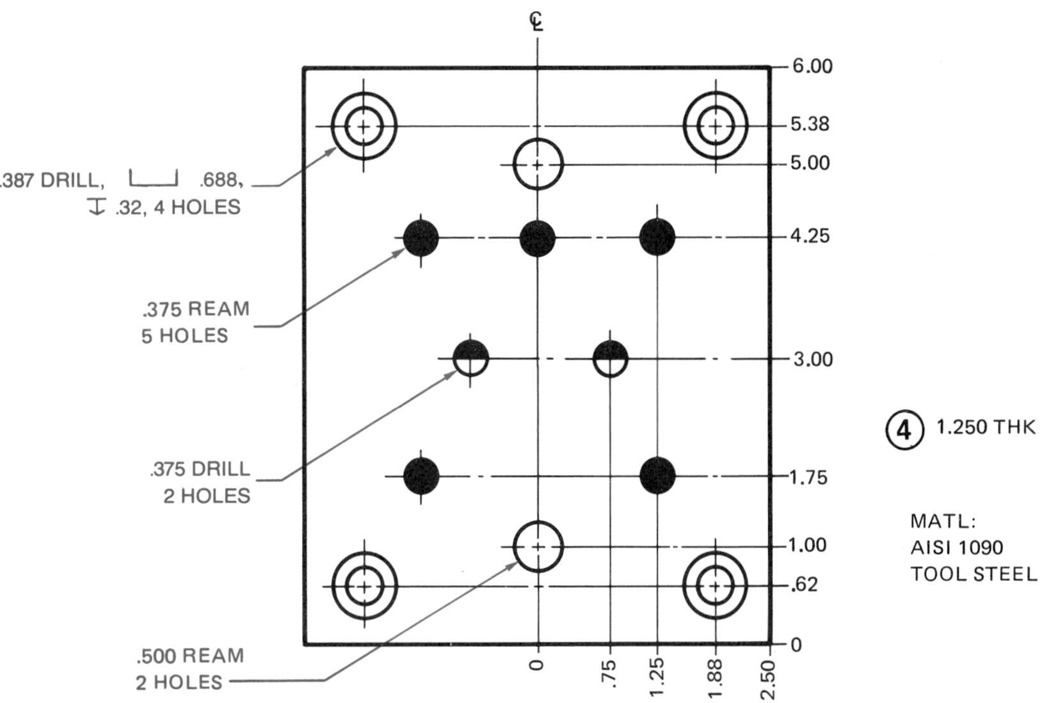

FIGURE 43-4 APPLICATION OF SIMPLIFIED HOLE REPRESENTATION AND COORDINATE DIMENSIONING PRACTICES

SIMPLIFIED HALF AND PARTIAL VIEW DRAWINGS

Half views provide a practical technique for simplifying industrial drawings of parts that have symmetrical features. *Partial views* permit parts to be represented when unnecessary hidden features and details may be omitted.

Two examples are provided in figure 43-5 to illustrate the use of simplified half view and partial view drawings. Time and drawing space are saved at (A) by using a *symmetry line* (center line) and half view to represent the equally-spaced holes and part features.

The drawing at (B) uses a *viewing plane line* (A—A) to indicate the direction of viewing and the partial view showing the elliptical form as viewed. When used in combination with the least number of views and full dimensioning, the partial view gives size, shape, and details of the elliptical portion of the part. The partial view generally replaces a full conventional view.

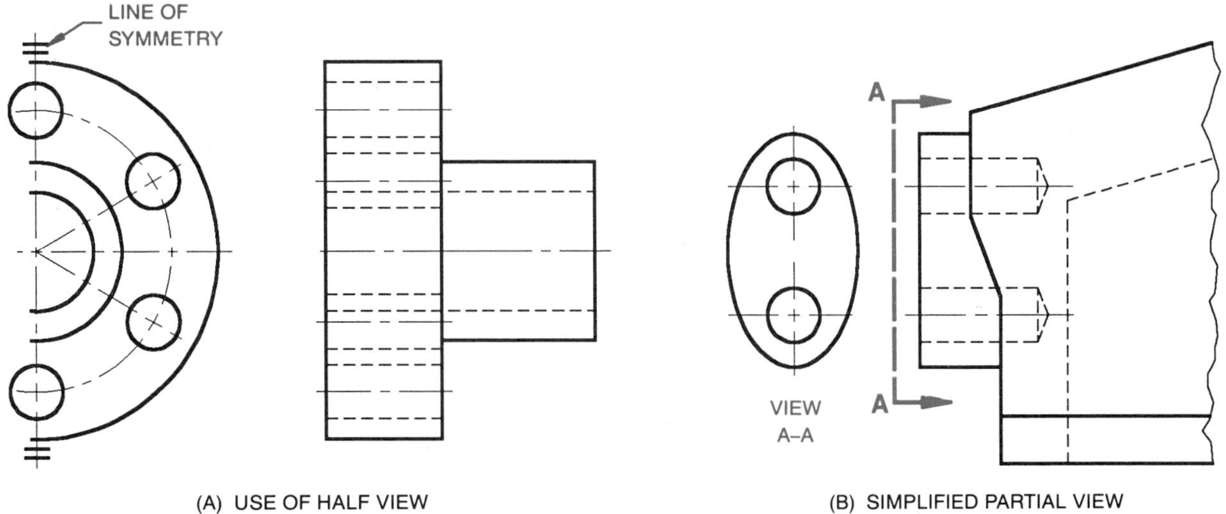

(A) USE OF HALF VIEW

(B) SIMPLIFIED PARTIAL VIEW

FIGURE 43-5 SIMPLIFIED HALF AND PARTIAL VIEW DRAWINGS

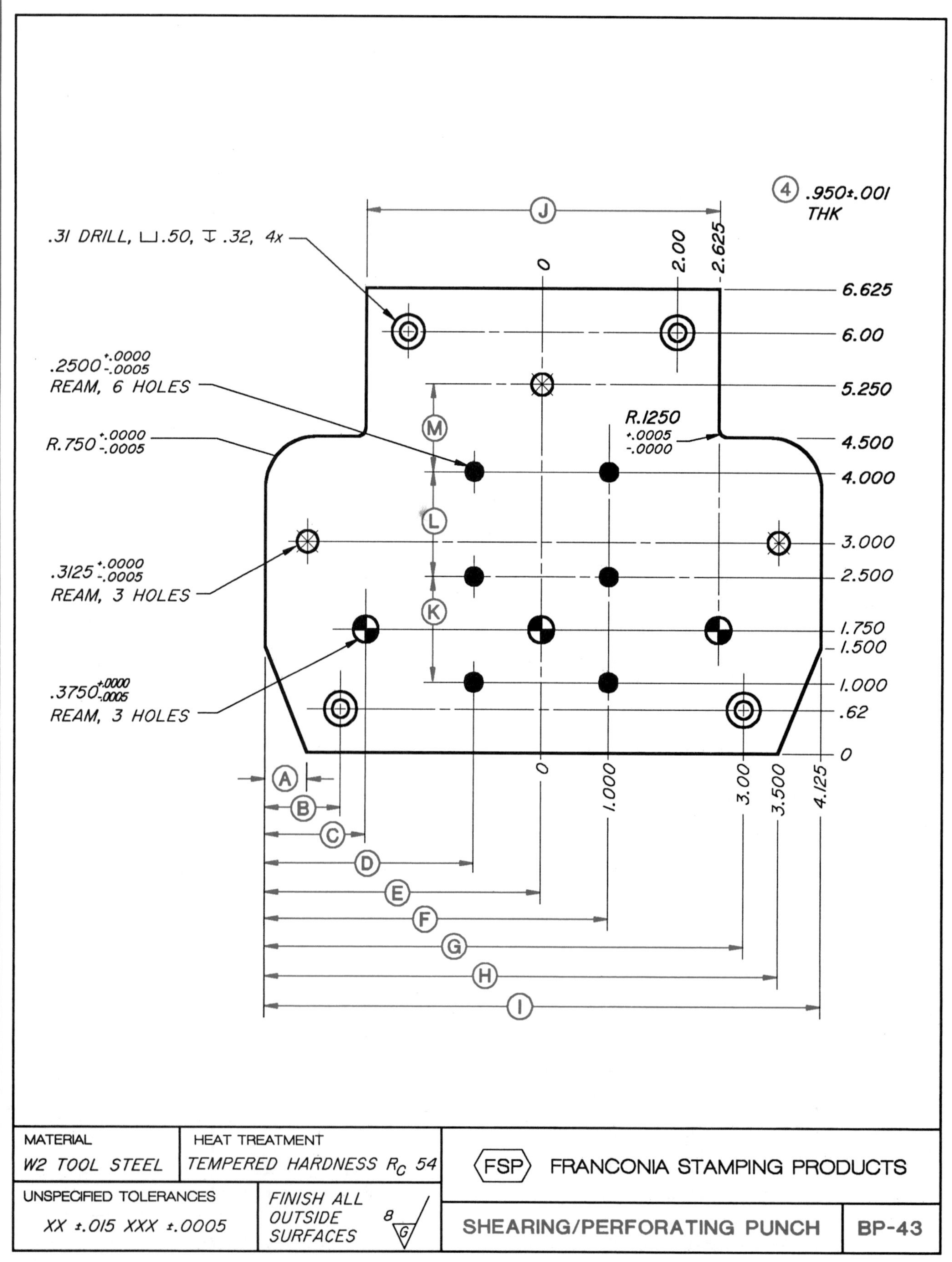

4 .950±.001 THK
.31 DRILL, ⌴.50, ↧.32, 4x
J
0
2.00
2.625
6.625
6.00
5.250
.2500 +.0000 -.0005 REAM, 6 HOLES
M
R.1250 +.0005 -.0000
R.750 +.0000 -.0005
4.500
4.000
L
3.000
.3125 +.0000 -.0005 REAM, 3 HOLES
2.500
K
1.750
1.500
.3750 +.0000 -.0005 REAM, 3 HOLES
1.000
.62
0
A
B
C
D
E
F
G
H
I
0
1.000
3.00
3.500
4.125
MATERIAL
W2 TOOL STEEL
HEAT TREATMENT
TEMPERED HARDNESS R_C 54
FSP
FRANCONIA STAMPING PRODUCTS
UNSPECIFIED TOLERANCES
XX ±.015 XXX ±.0005
FINISH ALL OUTSIDE SURFACES
8
G
SHEARING/PERFORATING PUNCH
BP-43

ASSIGNMENT—UNIT 43

SHEARING/PERFORATING PUNCH (BP-43)

1. State two advantages of simplified industrial drawings over conventional drawings.
2. Cite two factors which are considered in establishing the nature and extent to which a drawing may be simplified.
3. Name three requirements of simplified industrial drawings.
4. Refer to the steel section of a trade handbook.
   a. Identify the classification (type) of metal used for the Punch.
   b. Give the carbon content.
   c. State two physical properties of the metal.
5. Give the hardness of the Punch after tempering.
6. Identify the type of dimensioning used on the drawing of the Punch.
7. Describe the meaning of the ℄ symbol.
8. Draw freehand the symbols for the following features as used on the drawing of the Punch:
   a. Reamed holes
   b. Drilled and counterbored holes
   c. Thickness of Punch
9. Indicate (a) the machining process for finishing all external surfaces and (b) the required surface texture tolerance.
10. Determine the upper and lower limit dimensions of the following features:
    a. Overall width
    b. (1) (drill diameter)
    (2), (3), (4) (reamer diameters)

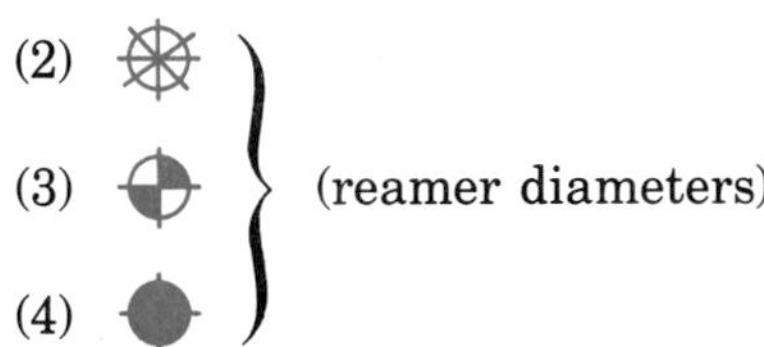

    c. Large and small radii
    d. Punch thickness
    e. Dimensions **.62** and **6.00**
11. Compute nominal dimensions Ⓐ through Ⓙ.
12. Determine center-to-center distances Ⓚ, Ⓛ, and Ⓜ.
13. Identify two drafting techniques for simplifying the drawing of an object with symmetrical features or when details and hidden features may be omitted.

Student's Name ____________________

1. a. ____________________
   b. ____________________
2. a. ____________________
   ____
   b. ____________________
   ____
3. a. ____________________
   b. ____________________
   c. ____________________
   ____
4. a. ____________________
   ____
   b. ____________________
   c. Properties. (1) ____________________
   (2) ____________________
5. ____________
6. ____________________
7. ____________________
   ____
8. a. ____________________
   b. ____________ c. ____________
9. a. ____________________
   b. ____________________

10.	Upper Limit	Lower Limit
a.		
b. (1)		
(2)		
(3)		
(4)		
c. Large		
Small		
d.		
e.		

11. Ⓐ ________ Ⓕ ________
    Ⓑ ________ Ⓖ ________
    Ⓒ ________ Ⓗ ________
    Ⓓ ________ Ⓘ ________
    Ⓔ ________ Ⓙ ________
12. Ⓚ ________ Ⓛ ________ Ⓜ ________
13. a. ____________ b. ____________

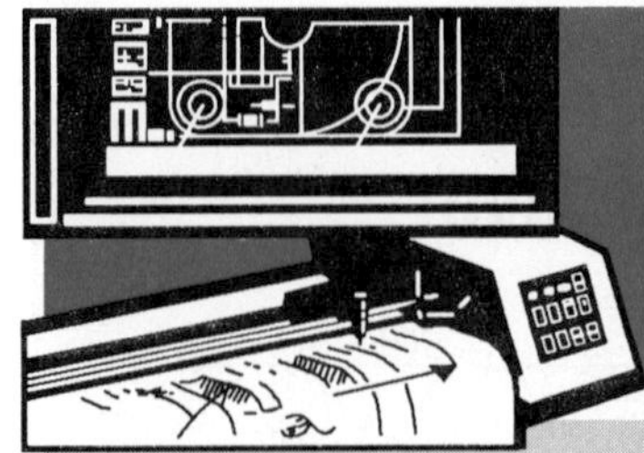

HIGH TECHNOLOGY applications

HT-13 NEW MATERIALS: PARTICLE METALLURGY STEELS (PM) AND POWDERED METALS

Two comparatively new processes and materials that are greatly impacting on CADD and CAM technology and machining relate to particle metallurgy (PM) steels and powdered metals and nonmetallic materials.

PARTICLE METALLURGY (PM) STEELS

Particle metallurgy (PM) steels have uniform steel particles that are evenly distributed throughout a cross-section. There is limited alloy segregation as compared with conventional steels. Particle metallurgy (PM) steels are produced by passing fluid molten alloy steel through a high-pressure atomizer to form small metal droplets. Upon solidification, the droplets form into fine steel powder particles. The particles are screened for size and then loaded into a steel canister (from which all gasses are evacuated). The canister is sealed, heated, and pressure is applied to compact the particles. The particle metallurgy steel mass that is produced is then forged, rolled, and formed by other methods into standard forms and sizes for use in manufacturing.

Advantages of PM steels relate to such properties as: machinability in an unhardened state; grindability in a hardened state; toughness and wear resistance (resulting in greater productivity); less distortion in heat treating; better shock load resistance and tool deflection; and other desirable characteristics.

POWDER METALLURGY: PRODUCT CHARACTERISTICS

Simply defined, powder metallurgy (related to metal cutting tools) deals with selecting and blending steel powders to form a specific composition and the addition of a binder. This composition is flowed into forming dies for compacting blanks under heavy pressure. The blanks are then sintered in atmospherically-controlled furnaces (as shown in the flow chart at (A)).

Second finishing operations (shown at B) are performed on powdered metal parts. A *repressing* operation produces a cutting tool that has greater density and strength. *Sizing* processes produce cutting tools of greater dimensional accuracy and closer geometric tolerances; *coining* produces a higher quality surface finish. Other second process examples include: heat treating, precision grinding, brazing, and surface coatings.

Advantages of manufacturing powdered metal and nonmetallic parts include the following: economical manufacturing costs resulting from reducing second machining operations; versatility in forming parts of intricate design; quality control of product material; high degree of dimensional tolerance ($\pm 0.001''/\pm 0.02$ mm) and quality of surface finish; controlled porosity from fine grain to porous textures (for impregnating with nonmetallic materials); and adaptability to the manufacture of metallic and nonmetallic parts.

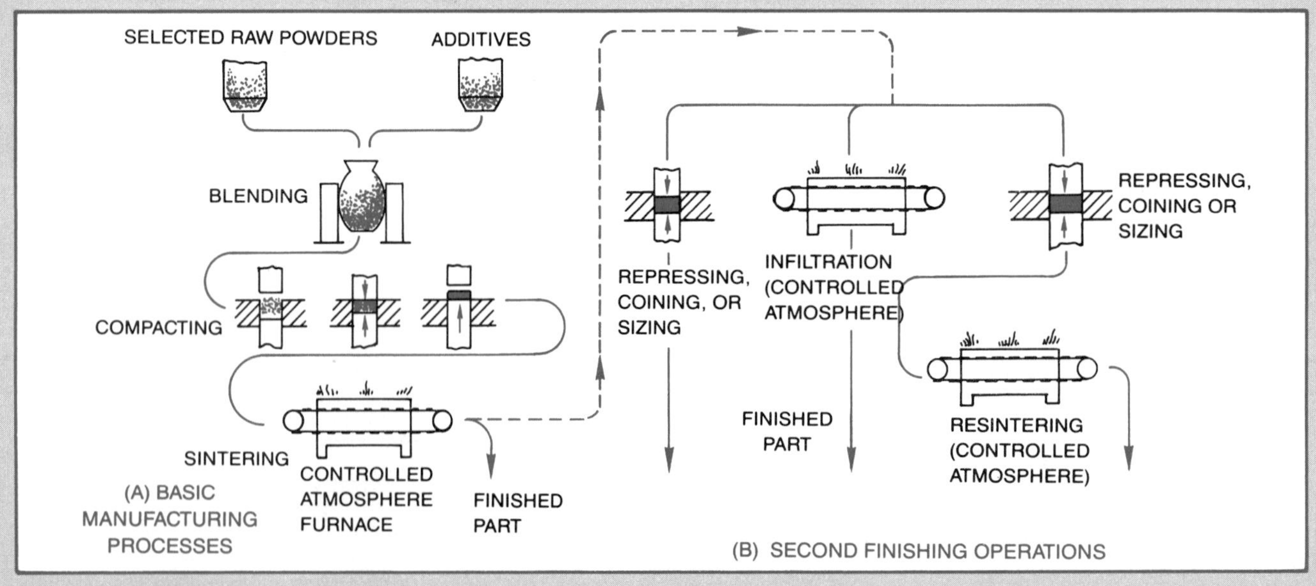

Drawing from Advanced Machine Tool Technology and Manufacturing Technology — *Courtesy of C. THOMAS OLIVO ASSOCIATES*

SECTION 12
Machine and Tool Drawings

UNIT 44
Properties, Identification, and Heat Treatment of Metals

Materials for parts, machines, and different structures are selected according to chemical, physical, and other properties. If a part serves as a bearing, the properties of the selected material must permit movement with limited friction, heating, and wear. Metals like bronze and bearing alloys are considered for such applications. By contrast, cutting dies for high production require a tool steel. Such a steel is capable of being hardened to a considerable depth, maintains a cutting edge, and may be tempered to withstand shock.

Mechanical properties relate to hardenability, malleability, strength, toughness, brittleness, ductility, and plasticity. Other properties are corrosion resistance, thermal conductivity (ability to conduct heat or cold), and expansion. These properties and other factors are considered by the designer.

The worker who constructs, assembles, or tests a part usually identifies the kind of material from a note or specification which is included in the title block of a drawing or on a parts list. This information provides a key to machining and nonmachining processes which may be required to change the physical properties of a part.

Additional title block information may indicate a specific *heat treatment* and a required degree of hardness. Other mechanical processing to produce a special surface finish (which does not alter physical properties) may also be given.

This unit provides basic information which persons who interpret prints need to know in relation to the following:

- General classifications and applications of steels, alloys, and aluminum,
- Common heat treating processes, and
- Hardness testing.

GENERAL CLASSIFICATION OF STEELS

The ability to use trade and engineering handbooks and manufacturers' technical data about the composition and properties of materials is important. A drawing may merely indicate under MATERIAL that an **SAE** or **AISI 4020 STEEL** is to be used. The handbook information identifies the steel as a molybdenum alloy steel with certain chemical composition and mechanical properties.

There are two basic classifications of steel: *plain* and *alloy. Plain carbon steels* (with 0.02% to 0.30% carbon content) may not be hardened unless more carbon is added to the outer surface. Generally, low carbon steels are known as machine, cold-drawn, or cold-rolled steels. Common usage is for shafts, bolts, nuts, and general purposes.

Medium carbon steels (0.30% to 0.60% carbon content) are hardenable and are used for hammers, wrenches, and similar parts. *High carbon steels* (0.60% to 1.50% carbon content) may be hardened to a considerable depth and degree of hardness. High carbon steels are widely used for edge-cutting tools such as drills, reamers, milling cutters, punches, dies, and taps.

Alloy steels contain such alloying elements as nickel, chromium, molybdenum, tungsten, and vanadium. These elements change the mechanical properties of steel. Tensile strength, toughness, hardenability, and wear resistance are increased. Other properties are improved.

SAE/AISI STEEL CLASSIFICATIONS ON DRAWINGS

The Society of Automotive Engineers (SAE) and the American Iron and Steel Institute (AISI) system of classifying steels is a basic four-digit number code. The first digit consists of one of nine numbers. Each number indicates the general type of steel, as follows.

- 1 Carbon
- 2 Nickel
- 3 Nickel-chrome
- 4 Molybdenum
- 5 Chromium
- 6 Chrome-vanadium
- 7 Tungsten
- 8 Nickel-chromium-molybdenum
- 9 Silicon-manganese

The second digit identifies the series classification of the basic type of steel. For instance, the **SAE/AISI** "**10**" series carbon steels are *nonsulphurized*. An **SAE/AISI** "**23**" (**23XX**) series alloy steel has a 3.50% nickel content. It is at this point that the worker refers to a handbook for further information about the material.

The third and fourth digits indicate the midpoint of the carbon content range. The percent of carbon is called *points*. The carbon content in the example (figure 44-1) is 0.30% or *30 points*.

Sometimes a material is specified on a drawing as **SAE/AISI 71560**. The **7** identifies a tungsten steel which has 15% tungsten (**15**) and 60% or (**60**) points of carbon. When the letter H follows a code number, it refers to a specific hardenability standard. Details about composition, properties, and applications for H designated steels are found in handbooks.

TOOL AND DIE STEELS

A letter code is often used to designate tool and die steels. The letter **W** designates a *water-hardening tool steel*; **O**, *oil-hardening tool steel*; **H-12**, a hot-worked tool steel of chromium (**12**) type, etc.

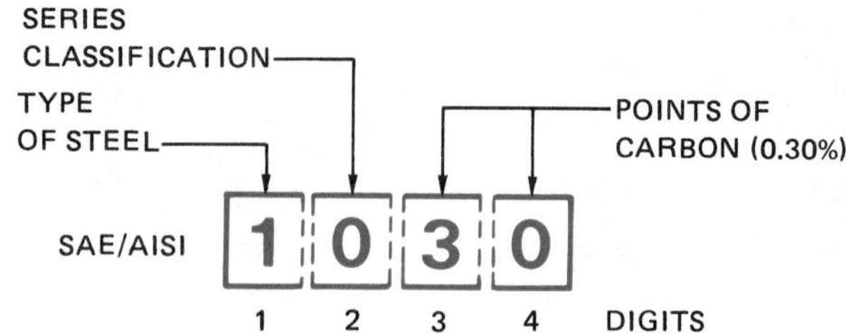

FIGURE 44-1 SAE/AISI BASIC CODE SYSTEM FOR DESIGNATING STEELS

UNIFIED NUMBER SYSTEM

The *Unified Number System (UNS)* represents a joint effort of the SAE, the American Society for Testing and Materials (ASTM), and others to provide a single correlated identification system. Sixteen letters and a five numeral code (00001 to 99999) are used. Eight letters (**D,F,G,H,J,K,S**, and **T**,) denote ferrous (iron and steel) metals and alloys. Eight Letters (**A,C,E,L,M,P,R**, and **Z**) identify nonferrous metals and alloys.

By comparison, an **SAE/AISI 1030** plain carbon steel has a corresponding UNS designation of **G10300**. A tool steel like **AISI T2** has a UNS code of **T12002**.

ALUMINUM ASSOCIATION DESIGNATION SYSTEM

Parts constructed of wrought aluminum and aluminum alloys are usually identified by a four-digit numerical code. The first-digit numerals **1** through **8** are assigned to almost pure aluminum (**1**), and various alloying elements like (**2**) for copper-type alloy through (**7**). Numeral (**8**) relates to any other alloying element. Numeral (**9**) is not assigned at present.

The second-digit code numeral indicates the degree of control of the impurities. Numerals in the third and fourth digits indicate to the nearest hundredth of one percent the purity of aluminum (in series **1**) or the percent of different alloys in the group.

Temper and hardness are identified on a drawing by a code letter, as follows:

- F Hardness as fabricated
- H Strain hardened (for wrought alloys only)
- O Annealed (for wrought alloys only)
- T Thermally (heat) treated
- W Solution heat treated

A one-, two-, or three-digit numeral code may be used with a *temper designation*. Each digit provides additional information about temper. A temper designation such as **H12** shows that the part is strain hardened (**H1**) and then annealed (**H12**).

OTHER COMMON METALS

Cast irons are specified on drawings under six general classes:

- Gray cast iron
- White cast iron
- Chilled cast iron
- Alloy cast iron
- Malleable cast iron
- Ductile (nodular) cast iron

The physical properties of cast iron vary according to composition and heat treatment. *Gray cast irons* have excellent machinability, low elasticity, high damping capacity, and are easy to manufacture. *White cast irons* are adapted to parts which require high-wear and abrasive resistance properties. *Chilled cast irons* are used for wear-resistant surfaces.

The addition of alloys produces alloy cast irons with increased strength, corrosion and heat resistance, and other property changes. *Alloy cast irons* are ideal for automotive engine, brake, and other systems; machine tool beds, and other machine parts.

Some cast irons are heat treated so the grain structure is changed and the metal becomes malleable. *Malleable cast irons* are used where great strength, ductility, and resistance to shock are required properties.

Ductile or nodular cast iron is a new type of cast iron. Its properties of toughness, shock and wear resistance, hardenability, and machinability approach regular grades of carbon steel. Nodular cast irons are adaptable to automotive, aerospace, heavy equipment, machine tool, and other industrial, structural, and civil applications.

HEAT TREATING PROCESSES

Heat treating deals with the heating and cooling of a metal or an alloy. The purpose of heat treating is to alter physical properties. Variations in hardness, for example, result from changing the cell structure through heating and cooling at a specified rate within certain critical points (temperatures).

FIGURE 44-2 ELECTRIC TOOLROOM HEAT-TREATING FURNACE WITH AUTOMATIC TEMPERATURE CONTROL AND DIGITAL READOUT (COURTESY OF THE THERMOLYNE CORPORATION)

NORMALIZING

A drawing may contain a note to **NORMALIZE**. This means the part requires heat treating to produce a uniform grain structure. A *normalized* steel part is then able to receive further heat treatment.

HARDENING

A hard grain structure is formed in steel by quenching a part from its hardening temperature at its *critical cooling rate.* Parts made of tool steels, high-speed steels, and other alloy steels are machined and formed while the material is in a machinable condition. Punches, dies, cutters, and other edge-cutting and forming tools are hardened by heating and cooling. Drawings often indicate temperatures and rate of cooling according to the geometry of the part, processes to be performed, and other requirements. Figure 44-2 illustrates a typical toolroom furnace. The heat is automatically controlled and the temperature is recorded on a digital readout instrument.

TEMPERING

A hardened part is under severe internal stresses which cause distortion and even crack-

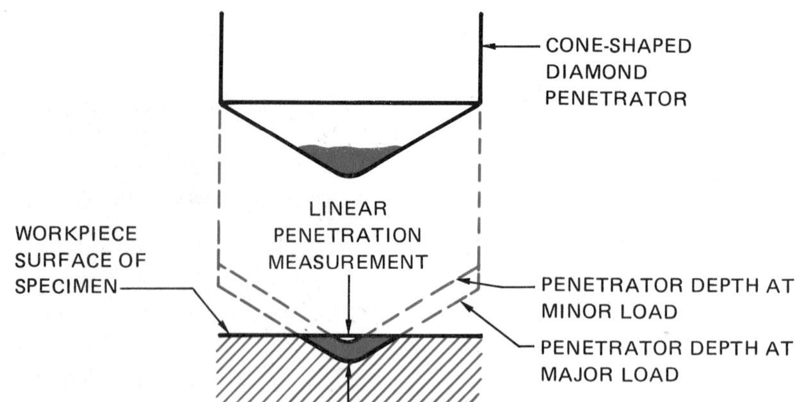

FIGURE 44-3 LINEAR PENETRATION—BASIS OF ROCKWELL HARDNESS TEST READINGS

ing. These problems are overcome by *tempering.* A hardened part may be tempered by reheating to a temperature that is called a *transformation temperature range* and then cooled at a specified rate.

In jobbing shops, tempering is controlled by heating to a particular color and then cooling. The colors are produced at certain temperatures. The colors on polished steel range from a very light yellow at 380 °F (193 °C) to a light blue at 640 °F (338 °C). The colors in between range from light to dark shades of straw, brown, purple, and blue. Maximum hardness is produced by tempering at 380 °F; minimum hardness at 640 °F.

CASE HARDENING

Many parts require just a hardenable outer case, leaving the core soft and retaining the properties of a low-carbon steel. Additional carbon, carbon-nitrogen, or nitrogen is taken into the outer case.

A drawing note such as **CASEHARDEN .030″** indicates the depth of penetration. In this case, the part is brought up to and held at the case-hardening temperature in the presence of a carbonaceous bath, with air excluded. The carbon absorbing/penetrating process continues until the required depth is reached. Carburized parts are reheated to hardening temperature and quenched, followed by tempering.

HARDNESS TESTING

Drawings contain notes which indicate required degrees of hardness according to a hardness measurement system used in that particular industry or plant. The symbol **R** identifies the Rockwell hardness testing system. **R_C** indicates a **C** scale measurement is required. A reading like **R_C40** means a "40" hardness on the Rockwell C scale. Other hardness measurement systems and scales include Vickers, Shore, Scleroscope, and Brinell instruments. A hardness value in one system may be converted to an equivalent value in any other system by referring to tables of conversion numbers. For comparison purposes, an **R_C60** value on a Rockwell C scale is the equivalent of **R_A81.2** (Rockwell A scale), **697** Vickers, **81** Shore, and **BHN 613** on a Brinell scale.

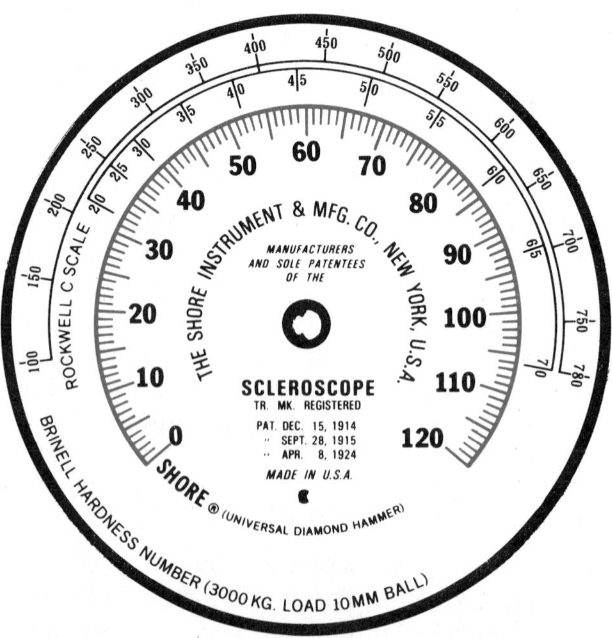

FIGURE 44-4 DIRECT-READING DIAL ON SCLEROSCOPE WITH EQUIVALENT ROCKWELL C SCALE AND BRINELL HARDNESS NUMBERS (COURTESY OF SHORE INSTRUMENT AND MANUFACTURING COMPANY)

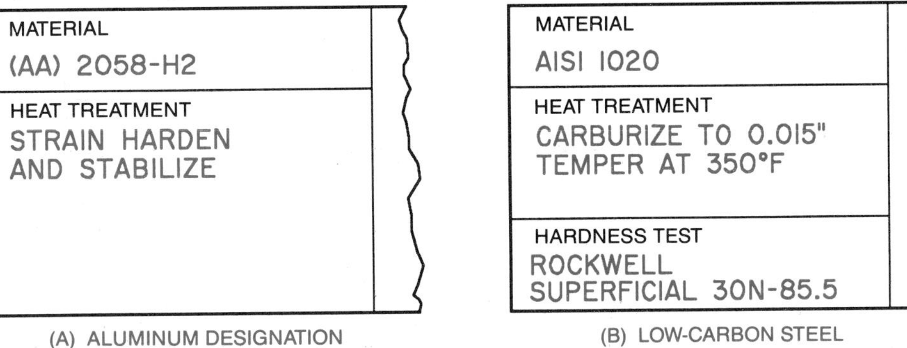
MATERIAL
(AA) 2058-H2
HEAT TREATMENT
STRAIN HARDEN
AND STABILIZE

(A) ALUMINUM DESIGNATION

MATERIAL
AISI 1020
HEAT TREATMENT
CARBURIZE TO 0.015"
TEMPER AT 350°F
HARDNESS TEST
ROCKWELL
SUPERFICIAL 30N-85.5

(B) LOW-CARBON STEEL

MATERIAL
SAE W1080
HEAT TREATMENT
ANNEAL BEFORE
MACHINING TO BHN
247-25B
HARDEN AFTER MACHINING
HARDNESS TEST
BHN 481-508

(C) WATER-HARDENING TOOL STEEL

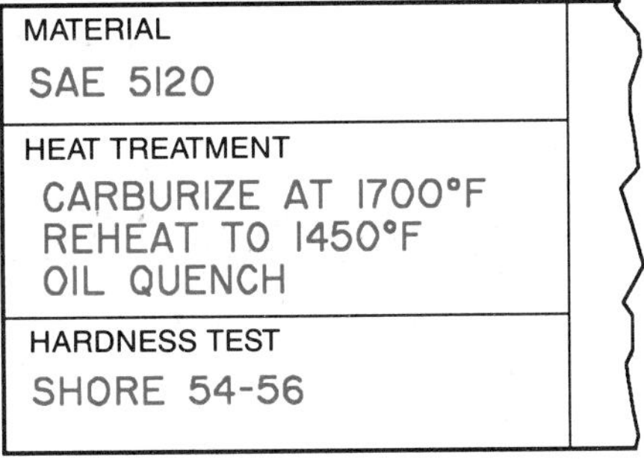
MATERIAL
SAE 5120
HEAT TREATMENT
CARBURIZE AT 1700°F
REHEAT TO 1450°F
OIL QUENCH
HARDNESS TEST
SHORE 54-56

(D) CHROMIUM ALLOY STEEL

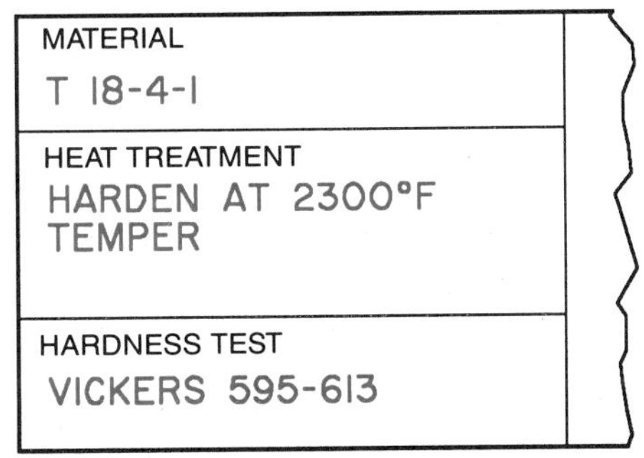
MATERIAL
T 18-4-1
HEAT TREATMENT
HARDEN AT 2300°F
TEMPER
HARDNESS TEST
VICKERS 595-613

(E) TUNGSTEN HIGH-SPEED STEEL

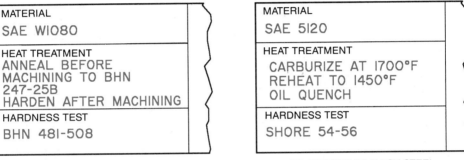

FIGURE 44-5 SAMPLE TITLE BLOCK INFORMATION ABOUT MATERIAL, HEAT TREATMENT, AND HARDNESS TESTING

Hardness testing is carried on to define the capacity of a material to resist wear, maintain a cutting edge, or to establish other properties. Instruments for hardness testing are designed to measure either the depth of penetration for a given load or the height a special hammer rebounds.

Figure 44-3 illustrates the shape of the *penetrator*, the initial load, and the penetration at the major load. Regardless of the hardness testing system, a shallow penetration indicates a high degree of hardness. The deeper the penetration, the softer the material. The numbers in each testing system range from the highest like **R_C68** to the lowest of **R_C20** or **BHN739** to **BHN226**.

The depth of penetration is translated into a hardness number. The hardness is read on a direct-reading scale. One model is displayed in figure 44-4. Note the inner scale provides Scleroscope hardness numbers. The middle scale numbers are R_C values. The outer scale gives equivalent BHN values.

METHODS OF SPECIFYING MATERIALS, HEAT TREATING, AND HARDNESS TESTING ON DRAWINGS

Techniques of specifying the required material, heat treatment processes, and hardness test values are illustrated in figure 44-5. Sample specifications are provided for four different types of steels and alloys and an aluminum part.

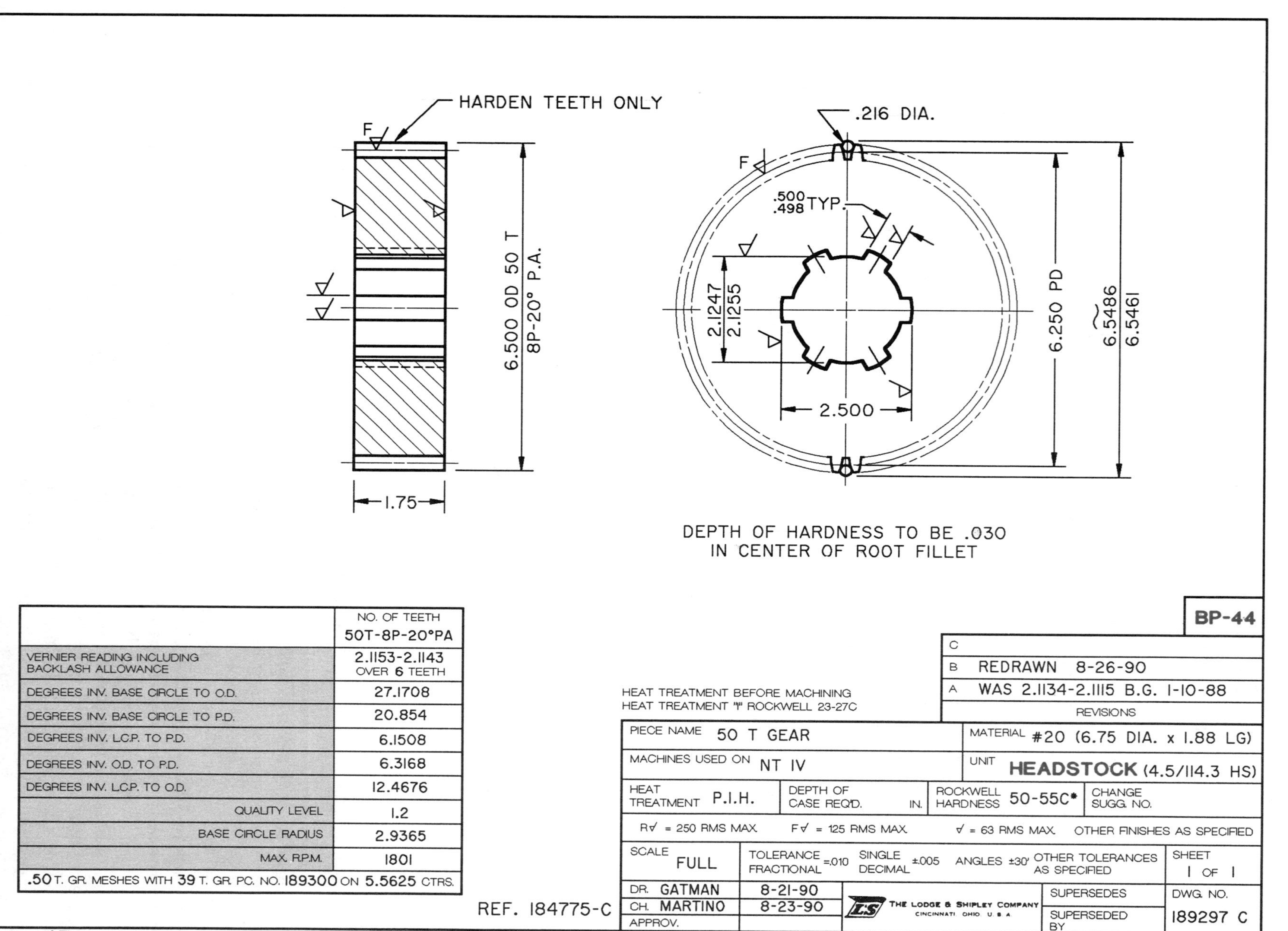
HARDEN TEETH ONLY
F
6.500 OD 50 T
8P-20° P.A.
1.75
.216 DIA.
.500
.498 TYP.
2.1247
2.1255
2.500
6.250 PD
6.5486
6.5461
DEPTH OF HARDNESS TO BE .030
IN CENTER OF ROOT FILLET
NO. OF TEETH 50T-8P-20°PA
VERNIER READING INCLUDING BACKLASH ALLOWANCE 2.1153-2.1143 OVER 6 TEETH
DEGREES INV. BASE CIRCLE TO O.D. 27.1708
DEGREES INV. BASE CIRCLE TO P.D. 20.854
DEGREES INV. L.C.P. TO P.D. 6.1508
DEGREES INV. O.D. TO P.D. 6.3168
DEGREES INV. L.C.P. TO O.D. 12.4676
QUALITY LEVEL 1.2
BASE CIRCLE RADIUS 2.9365
MAX. R.P.M. 1801
.50 T. GR. MESHES WITH 39 T. GR. PC. NO. 189300 ON 5.5625 CTRS.
REF. 184775-C
BP-44
C
B REDRAWN 8-26-90
A WAS 2.1134-2.1115 B.G. 1-10-88
REVISIONS
HEAT TREATMENT BEFORE MACHINING
HEAT TREATMENT "P" ROCKWELL 23-27C
PIECE NAME 50 T GEAR
MATERIAL #20 (6.75 DIA. x 1.88 LG)
MACHINES USED ON NT IV
UNIT HEADSTOCK (4.5/114.3 HS)
HEAT TREATMENT P.I.H.
DEPTH OF CASE REQ'D. IN.
ROCKWELL HARDNESS 50-55C*
CHANGE SUGG. NO.
R√ = 250 RMS MAX
F√ = 125 RMS MAX
√ = 63 RMS MAX
OTHER FINISHES AS SPECIFIED
SCALE FULL
TOLERANCE FRACTIONAL ±.010
SINGLE DECIMAL ±.005
ANGLES ±30'
OTHER TOLERANCES AS SPECIFIED
SHEET 1 OF 1
DR. GATMAN 8-21-90
CH. MARTINO 8-23-90
APPROV.
THE LODGE & SHIPLEY COMPANY CINCINNATI OHIO U.S.A.
SUPERSEDES
SUPERSEDED BY
DWG. NO. 189297 C

Courtesy of The Lodge & Shipley Company (Modified Industrial Drawing)

ASSIGNMENT—UNIT 44

HEADSTOCK GEAR (BP-44)

Student's Name ______________________

1. a. Identify (1) the machine unit of which the gear is a part and (2) the gear classification.
   b. Give the basic specifications which identify the gear size and design.
2. Note:
   a. The number of teeth in the meshing gear,
   b. The meshing gear part number, and
   c. The center distance between the two mating gears.
3. a. State the current maximum and minimum vernier measurement (including backlash allowance) over six teeth.
   b. Give the old, revised readings.
4. a. Identify the kind of line which is used to represent the outside diameter and the base circle.
   b. Indicate the meaning of the irregular line (∿).
   c. Name the system used to measure the pitch diameter.
   d. Give the nominal diameter of the measuring rolls.
   e. State the measurement range over the rolls.
5. Give the maximum and minimum dimensions for the following features:
   a. Outside diameter
   b. Pitch diameter (using $^{+0.000}_{-0.001''}$ tolerance)
   c. Nominal radius of the base circle
   d. Pressure angle of teeth
   e. Thickness of gear
6. a. Give the number of splines to be cut in the gear hub.
   b. Determine the included angle between each set of splines.
   c. Give the angle between the center line of the first spline and the vertical center line of the teeth.
   d. State the angular tolerance.
   e. Give the minimum and maximum diameter of the bore.
   f. Establish what the nominal width (diameter) is to which opposite splines are cut.
   g. Tell what the maximum and minimum width is of each spline.
7. a. Identify the different surface texture symbols used on this drawing.
   b. State the maximum surface texture values.
8. a. Name the hardness testing system for the Gear.
   b. Specify the required hardness scale reading and range (1) before, and (2) after machining.
   c. Identify the features which are to be hardened.
   d. Give the specification for the depth and condition of heat treating.

1. a. (1) ______________________
   (2) ______________________
   b. ______________________
   ____
2. a. ______________________
   b. ______________________
   c. ______________________
3. a. Maximum ________ Minimum ________
   b. Maximum ________ Minimum ________
4. a. ______________________
   b. ______________________
   c. ______________________
   d. Roll Diameter ______________________
   e. ______________________

5. Maximum | Minimum
   a. ____________ ____________
   b. ____________ ____________
   c. Nominal ______________________
   d. ____________ ____________
   e. ____________ ____________

6. a. ____________ c. ____________
   b. ____________ d. ____________
   e. Min. ____________ Max. ____________
   f. ____________
   g. Max. ____________ Min. ____________
7. a. ______________________
   b. ______________________
8. a. ______________________
   b. (1) ______________________
   (2) ______________________
   c. ______________________
   d. ______________________

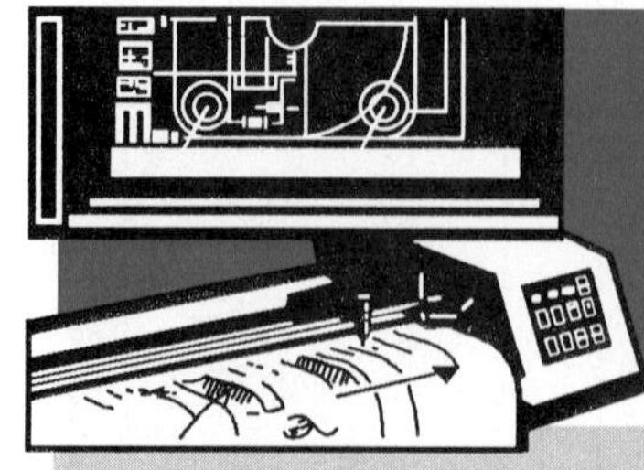

HIGH TECHNOLOGY applications

HT-14 METALLURGY: CUTTING TOOL COATINGS

TITANIUM NITRIDE (TiN) COATINGS

Titanium nitride (TiN) coatings are being applied to part surfaces such as: cutting, forming, and drawing dies to permit materials to flow more easily; end mills and other cutting tool edges; and to improve wear resistance of surfaces that are in moving contact.

Extremely hard titanium coatings (TiN) significantly reduce the tendency of materials to adhere to the edges and flutes of cutting tools. This property of *lubricity* reduces the cutting forces between a workpiece and cutter and also reduces heat that is generated and dissipated from the cutting area and cutting tool. Lubricity permits higher machining operating speeds and greater tool wear life, reduced tool deflection and cutting tool down time, improved surface finish, greater control of tolerances, and (importantly) greater cost effectiveness.

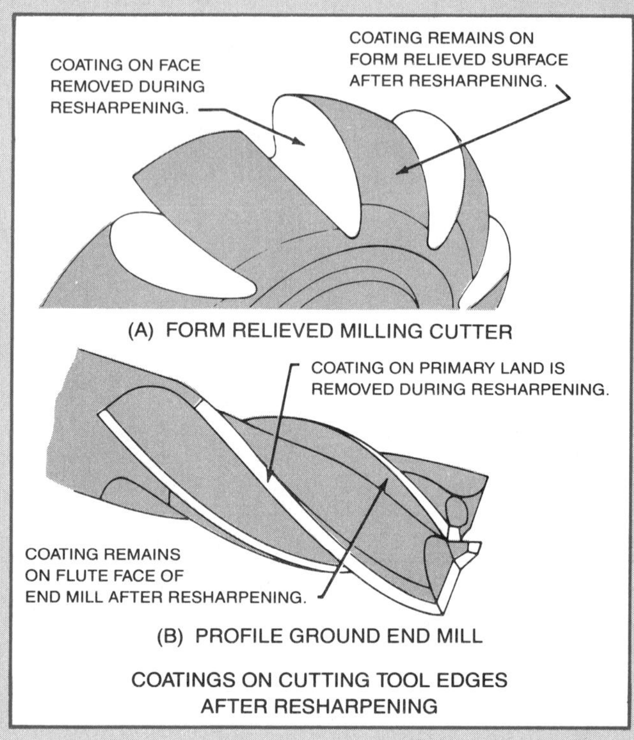

COATING ON FACE REMOVED DURING RESHARPENING.
COATING REMAINS ON FORM RELIEVED SURFACE AFTER RESHARPENING.
COATING ON PRIMARY LAND IS REMOVED DURING RESHARPENING.
COATING REMAINS ON FLUTE FACE OF END MILL AFTER RESHARPENING.

(A) FORM RELIEVED MILLING CUTTER

(B) PROFILE GROUND END MILL

COATINGS ON CUTTING TOOL EDGES AFTER RESHARPENING

APPLICATION OF TiN SURFACE COATINGS

Cutting tools and the surfaces of other parts may be coated by titanium nitride (TiN) by a *Physical Vapor Deposition Process*, as shown in the schematic drawing. Physically and chemically clean parts to be coated are secured in tool holders in a vacuum chamber. The tool holders are connected to a power supply and become cathodes in a high voltage circuit.

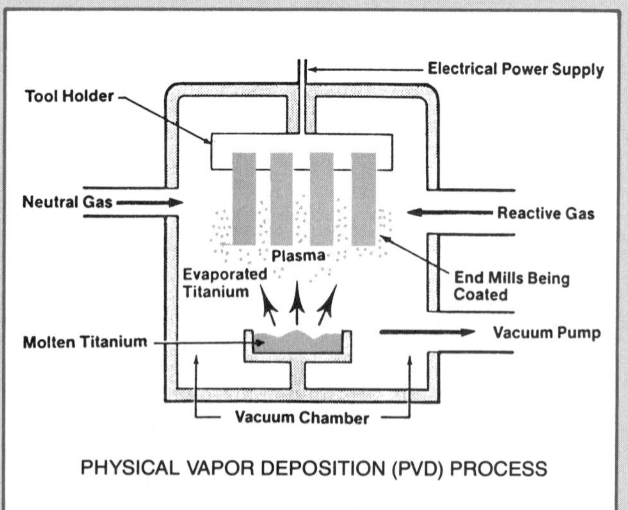

Electrical Power Supply
Tool Holder
Neutral Gas
Reactive Gas
Plasma
Evaporated Titanium
End Mills Being Coated
Molten Titanium
Vacuum Pump
Vacuum Chamber

PHYSICAL VAPOR DEPOSITION (PVD) PROCESS

A charged argon gas is introduced into the gas evacuation chamber. Positive argon ions in the highly-charged, high voltage field bombard against each tool to produce atomically-clean surfaces.

At the same time, the titanium in the reaction chamber is heated by an electron beam until it evaporates into titanium ions. A nitrogen gas is introduced to combine with the titanium ions that are accelerating toward the cutting tools. At a 900° F temperature range, the titanium bombardment forms a securely bonded, torsion-, deflection-, and stress-resistant TiN coating. Generally, a coating thickness of 0.0001″ (0.0025 mm) is adequate to affect the characteristics of the cutting tool and productivity.

TiN COATED CUTTING TOOLS: MAINTENANCE

A form-relieved milling cutter and a profile ground end mill are used to illustrate cutter maintenance. A TiN coated cutter may be reground and sharpened and still retain high cutting efficiency, provided there are no craters on the teeth or peripheral, or flank wear.

Drawings courtesy of BALZERS TOOL COATING INC.

UNIT 45
Jig and Fixture Design Features and Drawings

A high percent of the world's progress since the early manufacture of interchangeable parts may be attributed to the use of jigs and fixtures, punches and dies, formed cutting tools, and measurement gages. While some of the work-holding, positioning, and tool-guiding functions which were formerly performed with the aid of jigs and/or fixtures are now automated, these devices are still widely used to manufacture parts to highly precise limits of accuracy.

This unit examines certain basic design principles of jigs and fixtures. Then, the functions and representation of major components are presented so a worker may readily interpret assembly, subassembly, and detail drawings.

FUNCTIONS OF JIGS AND FIXTURES

A *jig* is a mechanical device. It serves three prime functions:

- To nest and/or hold a piece part (workpiece) securely so it may be machined accurately.
- To position a piece part in a fixed, exact dimensional and positional relationship to one or more cutting tools.
- To guide cutting tools.

A *fixture* serves as a work-holding and positioning device. The piece part is positioned accurately and secured in relation to machining, testing, or inspection processes.

The designation of a jig or fixture on a drawing provides an insight into the process and/or the machine tool. Jigs are commonly used for such hole-forming operations as drilling, reaming, and tapping. Fixtures are widely applied to such machines as milling, grinding, and turning, and to inspection processes. Other types of fixtures include broaching, bending, assembling, welding, and testing. Fixtures are further classified for processes such as face milling, boring, slotting, form grinding, and so forth.

CONSIDERATIONS IN THE DESIGN OF JIGS AND FIXTURES

One or more reference planes are involved in the design of jigs and fixtures. The machine table or a face plate are regarded as reference planes. Consideration in design is also given to the shape and surface condition of the workpiece (piece part). The part may have straight and square surfaces, or they may be at an angle to each other, circular, or contour in shape. The surface condition may be plane, curved, or rough. Geometric features and size also affect design.

MAJOR COMPONENTS OF JIGS AND FIXTURES

A jig or a fixture is constructed from a number of parts. Many of these parts, which are called *components*, are of standard size, shape, and material and are generally purchased. Specifications and catalog reference are sometimes included in a parts list or are identified in a callout on an assembly drawing. Working detail drawings are prepared for components which are

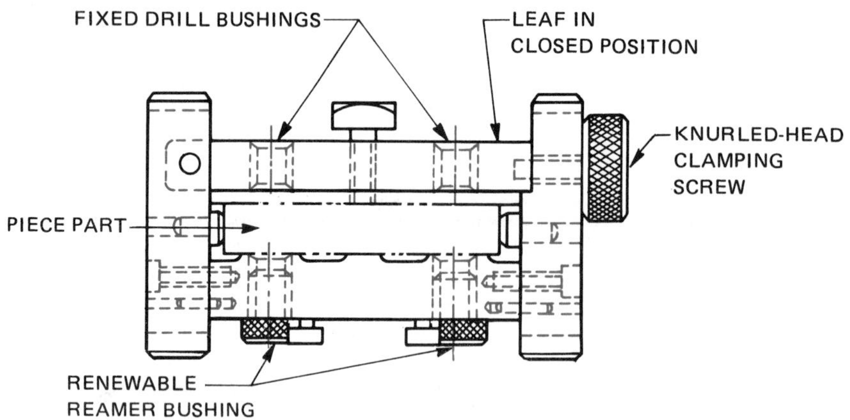

FIGURE 45-1 EXAMPLE OF LEAF JIG FOR DRILLING AND REAMING

of special design, giving dimensional, shape, construction, and other details.

The components are broadly grouped into seven major categories:

- Jig and fixture bodies
- Locating devices and work supports
- Clamping devices
- Fasteners and standard parts
- Jig leaves
- Drill jig bushings
- Ejectors

JIG AND FIXTURE BODIES

The *leaf*, *box*, *open-plate*, and *indexing* drill jigs are four common types. A *leaf jig* permits loading and unloading by swinging a hinged cover over a nested piece part, locking the leaf in position, performing the operation, and then unloading. Figure 45-1 is an example of a typical leaf drill jig. The drawing shows the piece part (in phantom outline) nested in the jig. The leaf is illustrated in "locked up" position. Holes may be drilled using the drill bushings in the leaf plate to guide the drill. The jig is then turned over. The piece part is reamed using renewable reamer bushings in the base. These bushings serve positioning and hole-sizing functions.

Another design of this same leaf jig may include renewable drill and reamer bushings in the leaf. Once the holes are drilled, the drill bushing is replaced by a bushing for reaming.

The form of this leaf jig is *box shape*. Thus, this name is used to categorize jigs which are formed as a box. Bushings which guide cutting tools may be included in one, two, or more sides. The box form may be built by using fasteners to hold the sides and the base or top, by casting, or by welding.

The main feature of an *open-plate jig* is a plate. Flush-mounting, press-fit bushings are included in the design. Sometimes, other permanent or

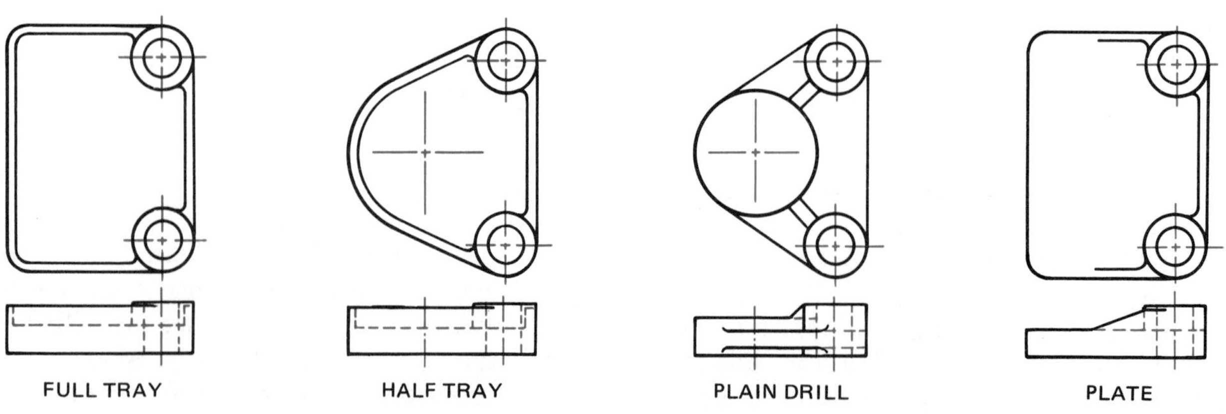

FIGURE 45-2 BASIC DESIGNS OF FIXTURE HEADS (COURTESY OF UNIVERSAL VISE AND TOOL COMPANY, SWARTZ FIXTURE DIVISION)

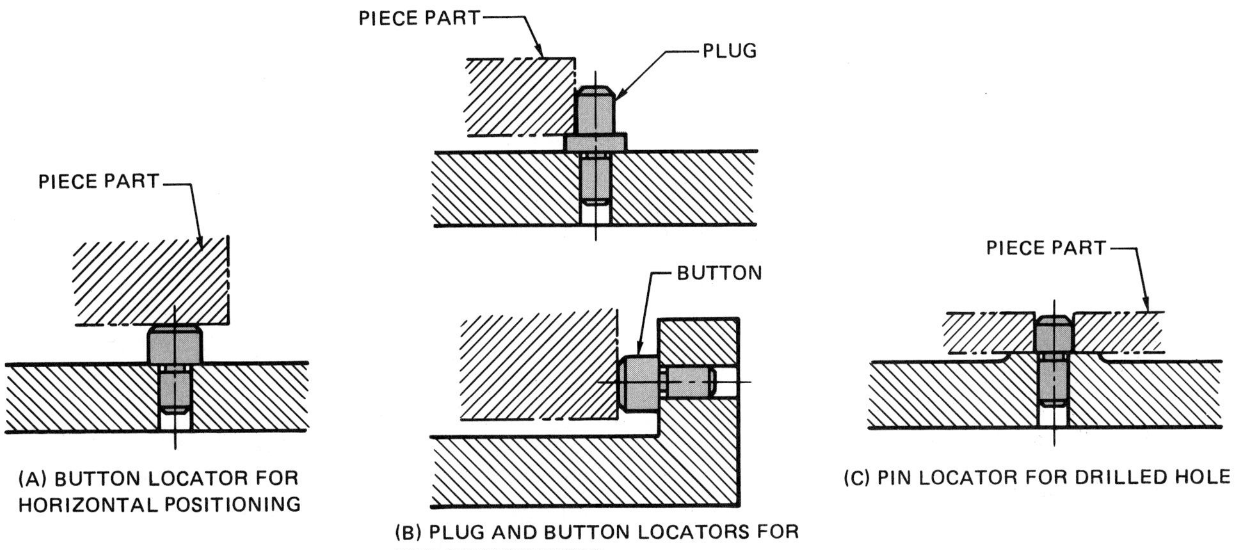

FIGURE 45-3 DESIGN FEATURES AND APPLICATIONS OF COMMON JIG AND FIXTURE LOCATORS

renewable replacement bushings are also used. This type of jig is especially adapted for carrying on a single machining process like drilling, reaming, or tapping at one loading.

Parts requiring the machining of angular surfaces involve an *indexing jig*. Successive holes and surfaces are positioned at a specific angle by using an indexing mechanism or a divider plate from which angle positions are located.

Feet (legs) may be cast, machined, or fitted into the body of a jig or fixture. The reduced surface area of the feet permits the jig or fixture to seat better on the reference surface.

UNIVERSAL FIXTURES

The *universal fixture* is designed so that a top plate may be moved to clamp or release a positioned piece part. The top plate is controlled by a rack and pinion or a pneumatic actuator. The *actuator design* is particularly adapted to automated and numerical control applications. Four of the basic types of heads around which jigs and fixtures are designed are illustrated in figure 45-2.

LOCATING DEVICES AND WORK SUPPORTS

Locators serve to uniformly position a piece part. Position relates to the reference plane (such as the machine table or a face plate) and the longitudinal and the cross-feed movement of the table or a carriage.

Pins, plugs, and *buttons* are three common designs of round locators. General shape, position, and application are shown in figure 45-3. When a hole serves as a reference, the pin type shown at (C) is either rounded or beveled to permit the part to center easily.

Rest plates and pads are used when the surface of the piece part is to rest solidly. The drawing at (A) in figure 45-4 shows both a supporting and a positioning pad. The plate at (B) provides only support for the part.

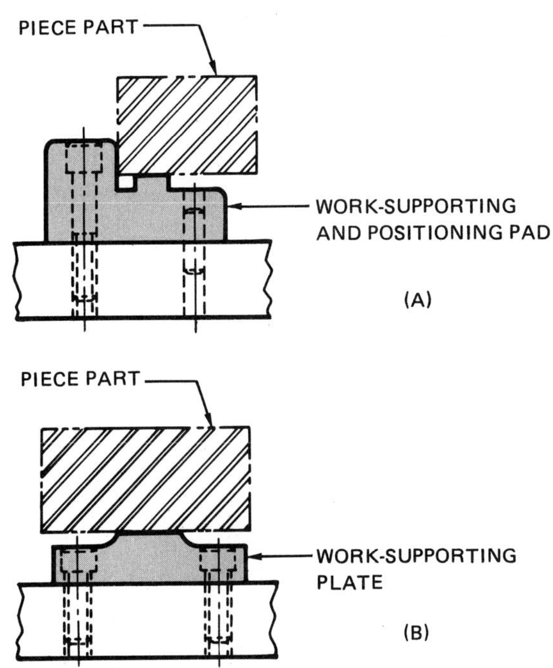

FIGURE 45-4 APPLICATIONS OF REST PLATE AND WORK SUPPORTING PAD

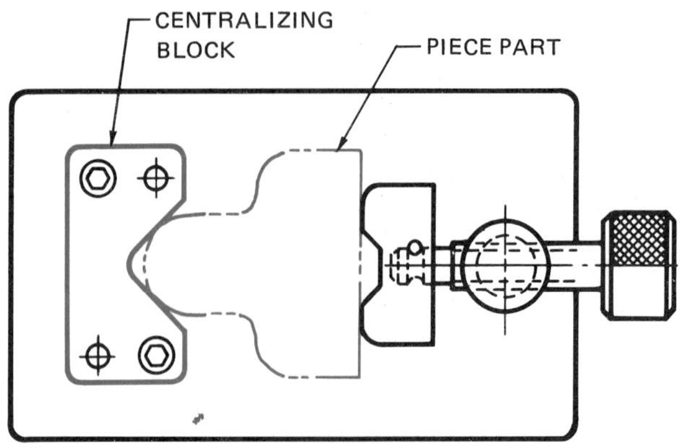

FIGURE 45-5 USE OF CENTRALIZING BLOCK TO POSITION A PIECE PART

The term *nest, nesting,* or *locating nest* refers to an area of a jig or fixture which receives and positions a piece part. Nests are designed to serve three functions:

- To accommodate size variations.
- To permit easy loading and unloading.
- To provide adequate space for chip removal.

A *centralizing block* (figure 45-5) is another method of positioning either a regular or an irregularly shaped part.

WORK-HOLDING (CLAMPING) DEVICES

Since jigs and fixtures are production tooling, it is necessary to provide a quick-acting component for efficient loading and unloading. Once loaded, the work-holding forces must be distributed to hold the workpiece securely, clamping it without distortion or damage. The clamping device must also be able to resist vibration and chatter. The clamping forces in the design and operation of jigs and fixtures are directed toward a pad or area of a piece part which can safely handle the holding and cutting forces.

Strap, quick-acting, and *latch clamps* are used as *top clamping devices.* The simple sliding strap clamp serves as a lever. One end rests on a pin or block. The opposite end applies force on the piece part when a *nut, knob,* or *knurled handle* is turned against the strap clamp.

Quick-acting swing and sliding strap clamps permit the clamp to pivot out of position. Once loaded, the clamp (which is held by springs at a fixed height) is swung over the piece part. The eccentric cam is then moved on the rest block to apply force to clamp the piece part. Figure 45-6 illustrates the design and operating features of a quick-acting swing and cam-clamping device.

Side-clamping devices are used when the *top surface* is to be machined without obstruction. One of the simplest side clamps is a cone-pointed socket head screw with a hardened point. For other applications, a *hinged toe clamp* directs the work-holding forces down on the reference and locator surfaces.

JIG LEAF

A *jig leaf* in simplest form is a hinged plate. Its function is to provide an easy, open loading and unloading area. If the plate contains drill bushings or locators, it also serves to position and guide drills, reamers, and taps.

Once loaded, the leaf is locked in a closed position. Force is applied on the piece part by a quarter-turn, thumb, or knurled head screw.

FASTENERS AND STANDARD PARTS

Screw fasteners, nuts, and pins are standard, commercially available parts. These are identified on parts lists by trade names, manufacturer's catalog designations, and other specifications.

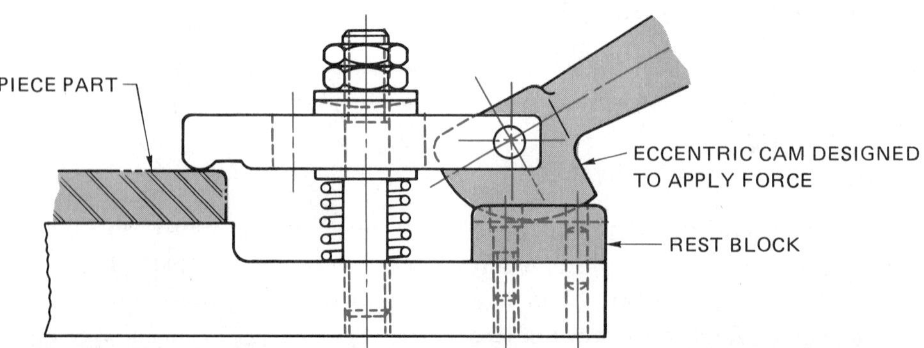

FIGURE 45-6 EXAMPLE OF QUICK-ACTING SWING AND CAM-CLAMPING DEVICE

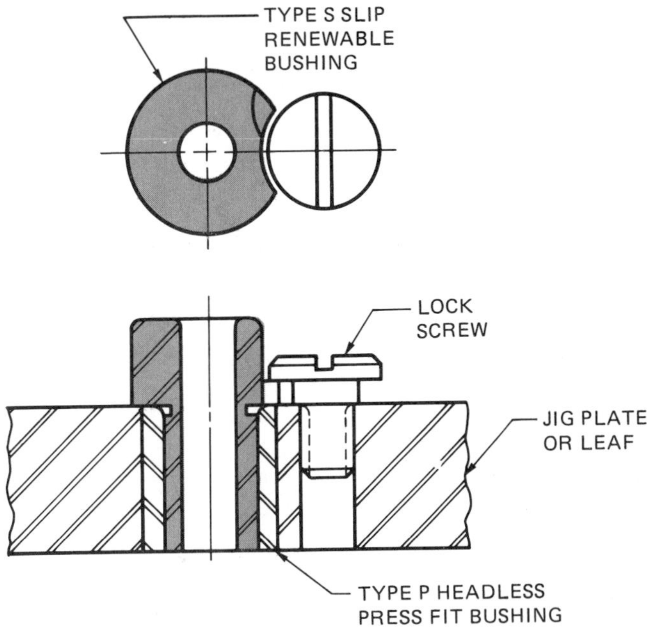

FIGURE 45-7 SECTION OF BUSHING ASSEMBLY

Fillister head, socket head, and *hexagon head screws; flat, cup, cone, oval,* and *half-dog point set screws;* and *flathead* and *shoulder screws* are common. *Wing, thumb, knurled head,* and *quarter-turn screws* are used on parts which are secured by hand in a jig or fixture.

Dowel pins and *taper pins* are employed for accurately positioning mating parts. *Jig feet* with press fit or threaded fit and in round, hexagon, or flat form are commercially available. *Plain, "U" design, spring,* and *lock washers* are common. *Round, tension,* and *compression springs* and *flat springs* are used to hold parts at an original position when forces are released.

DRILL JIG BUSHINGS

Bushings are designed to position and to guide such cutting tools as drills, reamers, and taps. Bushings are hardened and tempered to resist wear and to maintain accuracy. Precision bushings are ground externally and internally and are lapped and concentric to within 0.0003″ (0.007 mm). Sizes and forms have been standardized by producers.

Two basic designs of bushings include *permanent press fit* and *renewable.* The headless (Type P) press fit bushing is used for flush mounting; the head type (H), for resisting the effects of heavy axial loads.

Renewable bushings are used with *liners* for production runs which require bushing changes because more than one operation is to be performed in each hole. A *Type S slip renewable bushing* accurately fits into a permanent liner (master) bushing. The bushing is locked in position by turning the knurled head. Figure 45-7 shows how a Type S drill bushing, headless Type P press fit bushing, and lock screw are represented as a section on a drawing.

PIECE PART EJECTORS

Ejectors are included in the design of jigs and fixtures when it is difficult to remove a finish-machined part. If just a slight force is needed, one or more *ejector pins* (which are activated by springs) are included. An *ejector lever* is sometimes used with ejector pins to apply a greater lifting force to move a finished piece part out of a jig or fixture.

VARIATIONS IN FIXTURE DESIGN

While fixtures have been considered using primarily a horizontal reference plane, other reference planes are used for vertical and angular mountings on milling, grinding, boring, and other production machines. Also, fixtures are designed using magnet and vacuum chucks.

Progressive stations on a rotary fixture on automated and numerically controlled machining centers often combine mechanically, hydraulically, electrical (magnetic), and fluid- (vacuum-) actuated jigs and fixtures.

NOTE: REFER TO APPENDIX B-7 FOR THE LARGER SCALE DRAWING TO USE WITH THIS ASSIGNMENT.

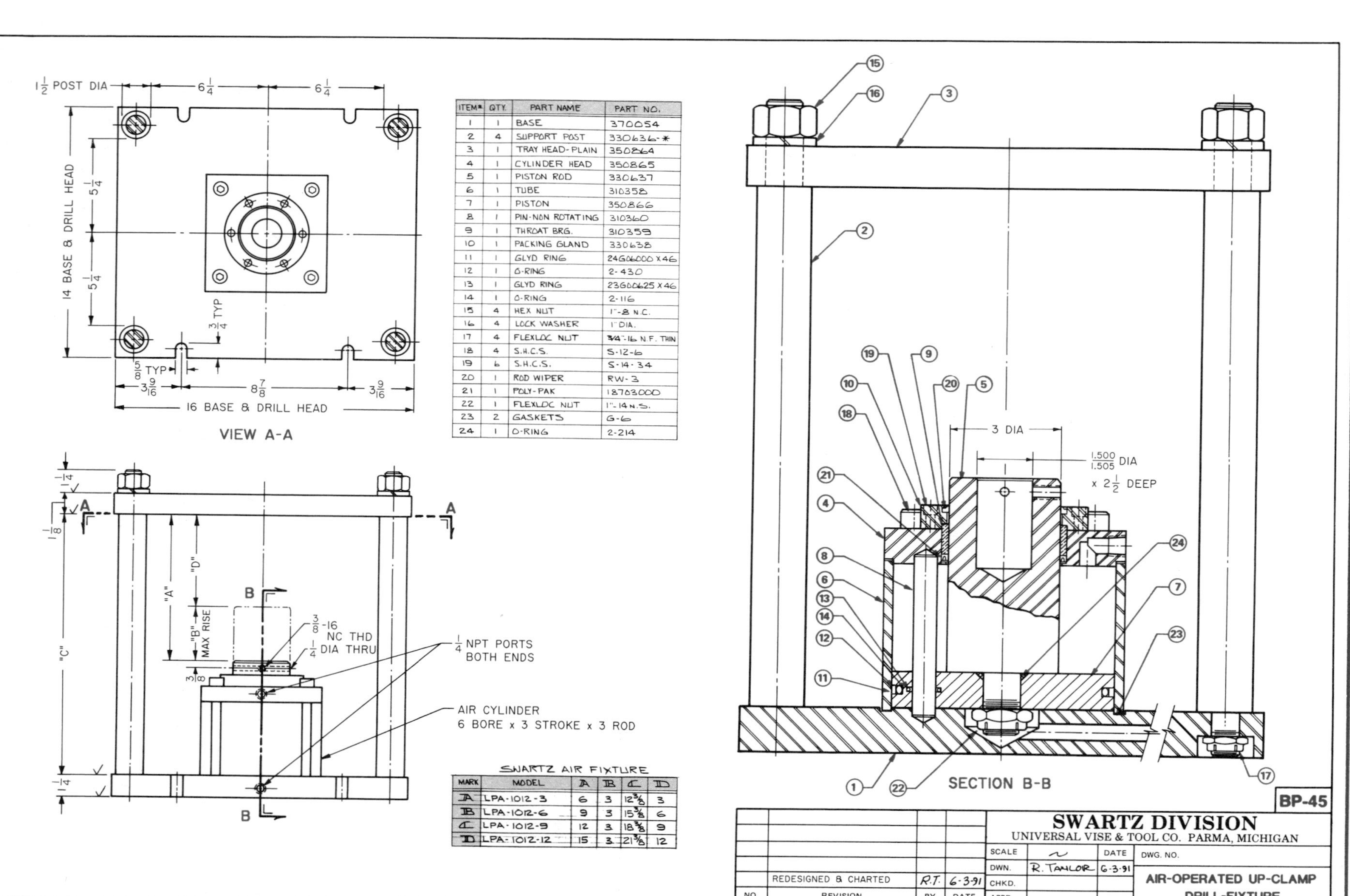

ITEM#	QTY.	PART NAME	PART NO.
1	1	BASE	370054
2	4	SUPPORT POST	330636-*
3	1	TRAY HEAD-PLAIN	350864
4	1	CYLINDER HEAD	350865
5	1	PISTON ROD	330637
6	1	TUBE	310358
7	1	PISTON	350866
8	1	PIN-NON ROTATING	310360
9	1	THROAT BRG.	310359
10	1	PACKING GLAND	330638
11	1	GLYD RING	24G06000 X46
12	1	O-RING	2-430
13	1	GLYD RING	23G00625 X46
14	1	O-RING	2-116
15	4	HEX NUT	1"-8 N.C.
16	4	LOCK WASHER	1" DIA.
17	4	FLEXLOC NUT	3/4"-16 N.F. THIN
18	4	S.H.C.S.	S-12-6
19	6	S.H.C.S.	S-14-34
20	1	ROD WIPER	RW-3
21	1	POLY-PAK	18703000
22	1	FLEXLOC NUT	1"-14 N.S.
23	2	GASKETS	G-6
24	1	O-RING	2-214

SWARTZ AIR FIXTURE

MARK	MODEL	A	B	C	D
A	LPA-1012-3	6	3	12 3/8	3
B	LPA-1012-6	9	3	15 3/8	6
C	LPA-1012-9	12	3	18 3/8	9
D	LPA-1012-12	15	3	21 3/8	12

Courtesy of Universal Vise and Tool Company, Swartz Fixture Division (Modified Industrial Drawing)

UP-CLAMP FIXTURE (BP-45)

NOTE: This drawing represents the basic design of a particular type of fixture. Other components are added to produce a functional jig or fixture.

1. Explain each of the following conditions:
   a. Why **VIEW A–A** is not a regular top view.
   b. Why **section B–B** is used instead of a side view.
   c. Why **section B–B** is drawn to a different scale than the front view.
   d. What the phantom outline in the front view shows.
2. a. Determine the maximum upward rise (stroke) of models **LPA 1012-3** through **12**.
   b. Give the overall distance "A" for each of the four models.
   c. Explain briefly the importance of dimension "A."
3. a. Give the maximum (1) width and (2) depth of jig or fixture components which can be accommodated in a 16″-size base and tray head.
   b. State (1) the center-to-center distance between the bolt holes for strapping the fixture to a table and (2) the width of the bolt slots.
   c. Determine the thickness (height) of (1) the base and (2) the tray head.
4. Study the sectional area for **B–B**.
   a. Identify the energy system used to operate the clamping mechanism.
   b. Give (1) the cylinder bore, (2) the length of stroke, and (3) the diameter of the piston rod.
   c. Explain briefly how the fixture operates.
5. List (a) the thread specifications and (b) give the required quantity of fasteners or threaded holes for the following parts:
   (1) Socket head cap screws holding the cylinder head to the base
   (2) Regular hexagon nuts
   (3) Socket head cap screws Det. (19)
   (4) Flexloc nuts Det. (17)
   (5) Port threads
   (6) Piston rod tapped hole
6. Identify the nonfastener parts in the upper cylinder design according to (a) item number, (b) part name, (c) identification, and (d) quantity required.
7. a. Give the drawing detail number of the different seal parts used between the piston, the nonrotating pin, and the inside wall of the tube.
   b. Name each of the seal parts
   c. List each identification number

ASSIGNMENT—UNIT 45

Student's Name ______________________

1. a. ______________________
   ____
   b. ______________________
   ____
   c. ______________________
   ____
   d. ______________________
   ____
2. a. ______________________
   b. (1) __________ (3) __________
   (2) __________ (4) __________
   c. ______________________
   ____
3. a. (1) Maximum Width __________
   (2) Maximum Depth __________
   b. (1) Center-to-Center Distance= __________
   (2) Width of Bolt Slots= __________
   c. (1) Base Thickness= __________
   (2) Tray Head Thickness= __________
4. a. ______________________
   b. Cylinder Bore __________
   Length of Stroke __________
   Diameter of Piston Rod __________
   c. ______________________
   ____
   ____
   ____

5.

	Thread Size or Specification (a)	Quantity (b)
(1)		
(2)		
(3)		
(4)		
(5)		
(6)		

6.

Item No. (a)	Part Name (b)	Identification (c)	Quantity (d)

7.

Detail No. (a)	Part Name (b)	Identification Number (c)

UNIT 46
Cutting Tools, Gages, and Measurement Setups

Machining processes require a machine tool, a control system, and a cutting tool. Cutting action takes place when the workpiece is held rigidly and the form of the cutting tool permits material to "flow" over the cutting edges as cutting force is applied.

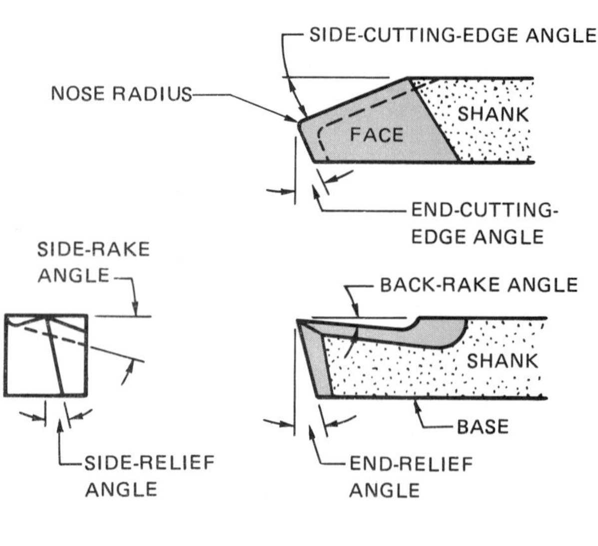

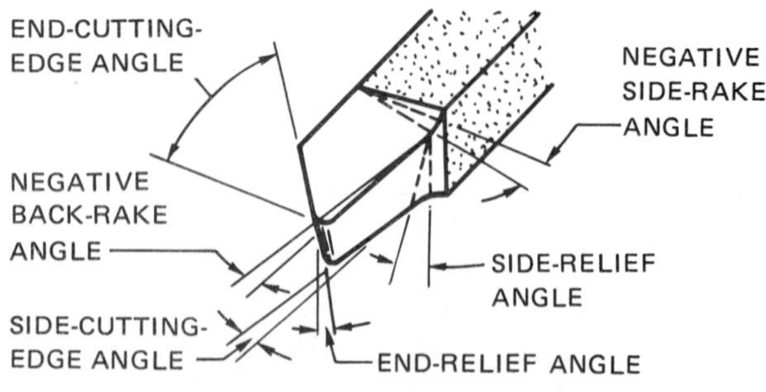

FIGURE 46-1 COMMON DESIGN FEATURES AND TERMS FOR SINGLE-POINT CUTTING TOOLS

FUNCTIONS OF CUTTING TOOLS

Cutting tools are designed to produce three types of chips: continuous, discontinuous, and segmental. Chips are controlled for worker safety and to prevent damage to a workpiece. Parallel, angular, and grooved *chip breakers* control the flow of chips.

DESIGN FEATURES OF SINGLE-POINT CUTTING TOOLS

Cutting tools may be single-point or have multiple cutting edges. A lathe tool bit is a good example of a *single-point cutting tool;* a milling cutter, a *multiple-edge cutting tool.* In each case, the designer, the machine worker, and the parts inspector use common terminology. Drawings and handbooks provide dimensional information, usually in terms of angles. Figure 46-1 illustrates that the cutting edge of a single-point tool bit is ground (1) with a *side-cutting* and an *end-cutting edge angle* and has (2) a *side-relief* and *end-relief angles,* and (3) a *side-rake angle* and a positive *back-rake angle.* Design, production planning, and sometimes working drawings provide information about the best angles to grind the cutter for particular machining processes. Cutting speeds and feeds are often suggested.

DESIGN FEATURES OF MULTIPLE-POINT CUTTING TOOLS

Multiple-fluted cutters such as reamers, taps, drills, dies, and end mills have cutting edges

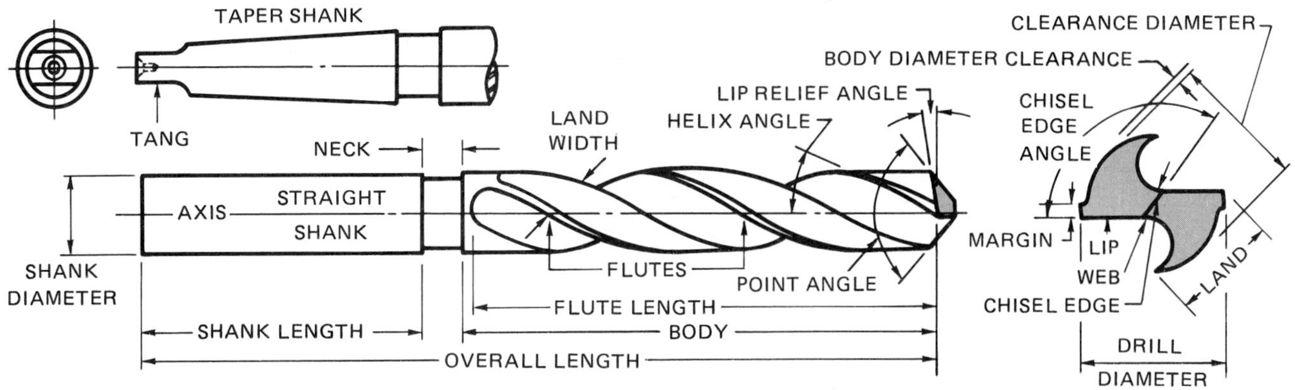

(A) STANDARD FEATURES AND TERMINOLOGY FOR A TWIST DRILL

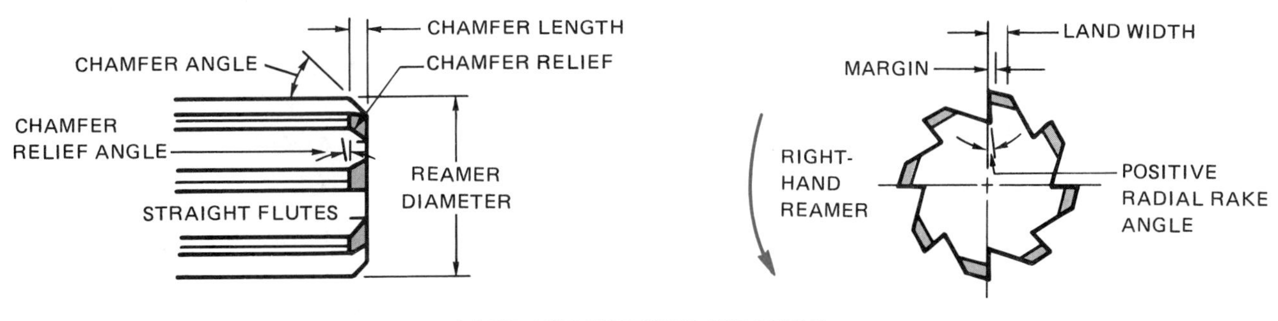

(B) REAMER FEATURES AND TERMS

FIGURE 46-2 COMMON FEATURES AND TERMINOLOGY FOR MULTIPLE-EDGE, HOLE-FORMING CUTTING TOOLS

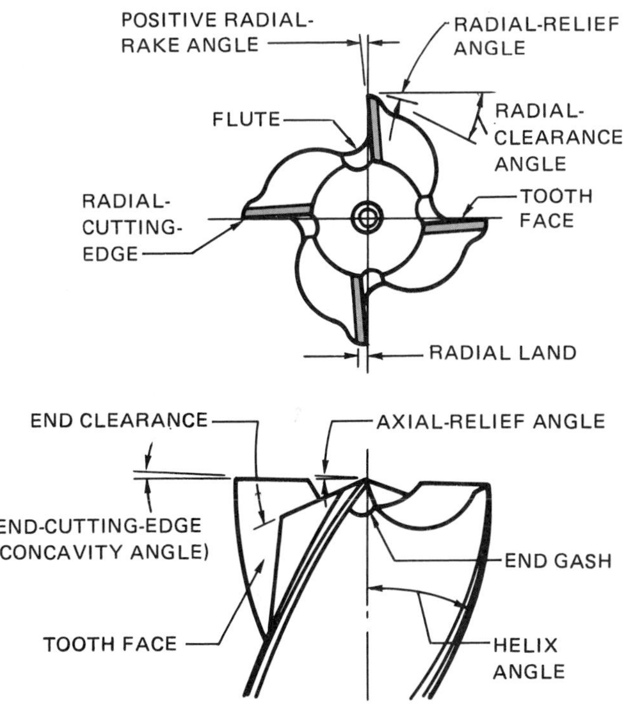

FIGURE 46-3 DESIGN FEATURES AND COMMON TERMINOLOGY FOR END MILLS

which are described by other terms. Figure 46-2A provides standard terminology for the main features of a twist drill. The design features of the cutting teeth on a machine reamer are shown at (B).

Since end mills are widely used, design features of fluted end mills are pictured in figure 46-3.

CUTTING TOOL HOLDERS

Solid cutting tools are generally held by a taper with a tang (milled flat end area) which fits into a slot in an adapter. The shank of a drill, reamer, tap, or other tool or holder which is straight is held in a chuck. Cutting tools designed with a bored hole are usually positioned and driven by an *arbor, adapter,* or *collet.* Figure 46-4 is a section view of an assembly for mounting a taper shank end mill. Note that the cutting tool is held in a collet, the collet in an adapter, and the adapter in the tapered machine spindle.

Many of the dimensions and design features found on drawings of cutting tools deal with cut-

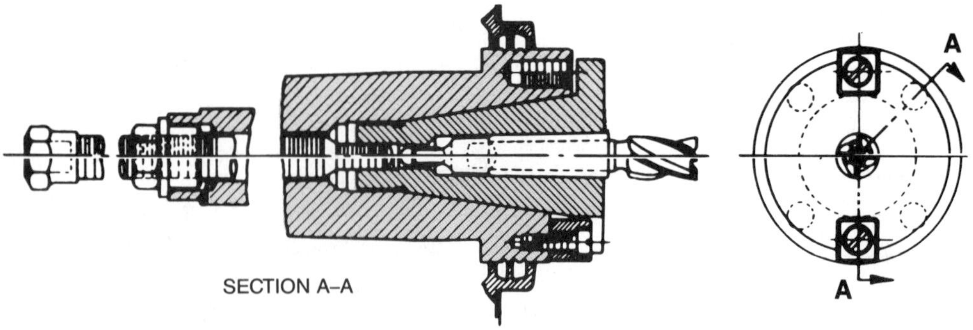

SECTION A–A

FIGURE 46-4 SECTION VIEW OF CUTTER ASSEMBLY

ting action and holding. *Angles* provide clearance and relief for cutting and for flowing chips away from the workpiece and cutter. *Tapers* are used for centering and holding; *flutes* and *lands* for forming cutting edges.

Other information found on production plans, which complement detail working drawings, deal with tool setups and machine stations, feeds and speeds, and cutting lubricants. The plans indicate the position (station) of each cutting tool in the sequence of operations. The cutting edges of tools in a high production or numerical control setup are often positioned so that the cutting edge of each tool lies in the same tool plane. This reduces machining time as each cutting edge is at a fixed distance from the workpiece.

DESIGN FEATURES AND FUNCTIONS OF GAGES

Gaging is defined by industry and the ANSI as a process of *measuring manufactured materials to ensure specified uniformity of size and other geometric features, as required.* Gaging provides for quality control and dimensional accuracy within the limits specified for the product.

Gages used in the gaging process may be grouped in three broad categories: (1) fixed, (2) indicating,

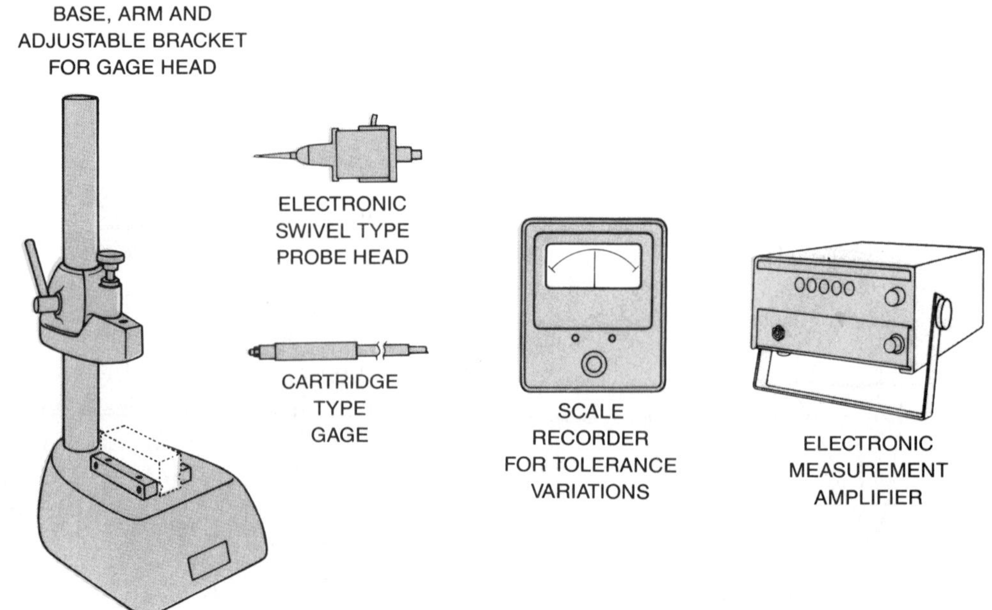

FIGURE 46-5 EXAMPLES OF COMPONENTS IN AN ELECTRONIC GAGING SYSTEM (COURTESY OF FEDERAL PRODUCTS CORPORATION)

and (3) combination. *Fixed gages* are designed for parts inspection to establish dimensional accuracy within prescribed high and low limits of accuracy. Some of the common fixed gages are *plug* (for inside diameters), *ring* (for outside diameters), *pin*, *thread*, *angle* and *taper*, and *flatness*. These hand gages are used for checking processes such as milling, drilling, grinding, broaching, and others.

SPECIAL GAGES: COMPARATORS

Special gages are also designed for automatically controlling continuous process machines. Special gages are often used in combination with *air* (pneumatic), *mechanical*, *electrical/electronic*, and *optical comparators*. These instruments provide high amplification of surface texture, diametral measurements, and other geometric features of the workpiece. The geometric characteristics which are measured deal with the following:

- Parallelism
- Flatness
- Squareness
- Roll and pitch
- Angularity
- Concentricity
- Alignment
- Profile (form)

Optical Comparator

The *optical comparator*, often called a *contour projector* (figure 46-6), is widely used in measurement, parts inspection for geometrically toleranced parts, and for accurate comparison of irregularly-shaped surfaces against a master chart (form of the part). Applications of contour surface measurements include: thread forms; angles, radii, arcs, and combinations; special form gages; flat and circular cutting and forming tools; and others.

The range of tolerancing measurements of the model comparator illustrated is from ±0.0001″ to ±0.0005″ (±0.002 mm to ±0.01 mm). Comparators may be used for machined steel and other nonferrous parts, plastics, and soft materials, without physical contact.

The optical comparator (figure 46-6) has a series of accessories including a precision rotary indexing vise, adjustable centers, and special parts positioning fixtures.

COMBINATION MEASUREMENT AND GAGING SYSTEMS

Some gaging systems are designed with an electronic or spring-type *gaging head* and an *amplifier*. The amplifier converts minute quantities of electrical energy into a movement which records product variations on a particular scale. In other applications, electronic circuits translate dimensional and surface variations between a mounted gaging head and a workpiece to either an *electronic recorder* or a *digital display readout unit*. Figure 46-5 shows the major components of an electronic gaging system.

Drawings reflect the increasing use of geometric tolerancing. The part producer needs to translate tolerancing information specified on a drawing by the designer into actual measurements. It is at this point that gages and other measuring instruments are used. Gaging and measuring ensure that machined surfaces are controlled within specified dimensional and geometric tolerances.

Courtesy of JONES & LAMSON DIVISION; TEXTRON INC.

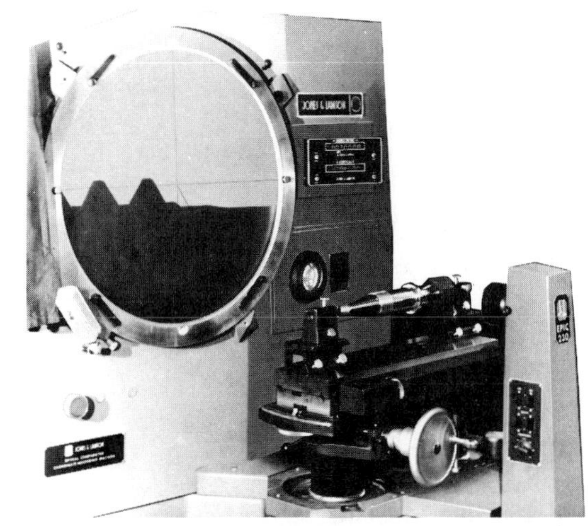

FIGURE 46-6 OPTICAL COMPARATOR SETUP PROVIDES DIGITAL READOUTS FOR DIMENSIONAL MEASUREMENTS AND ACCURATE CONTOUR COMPARISON

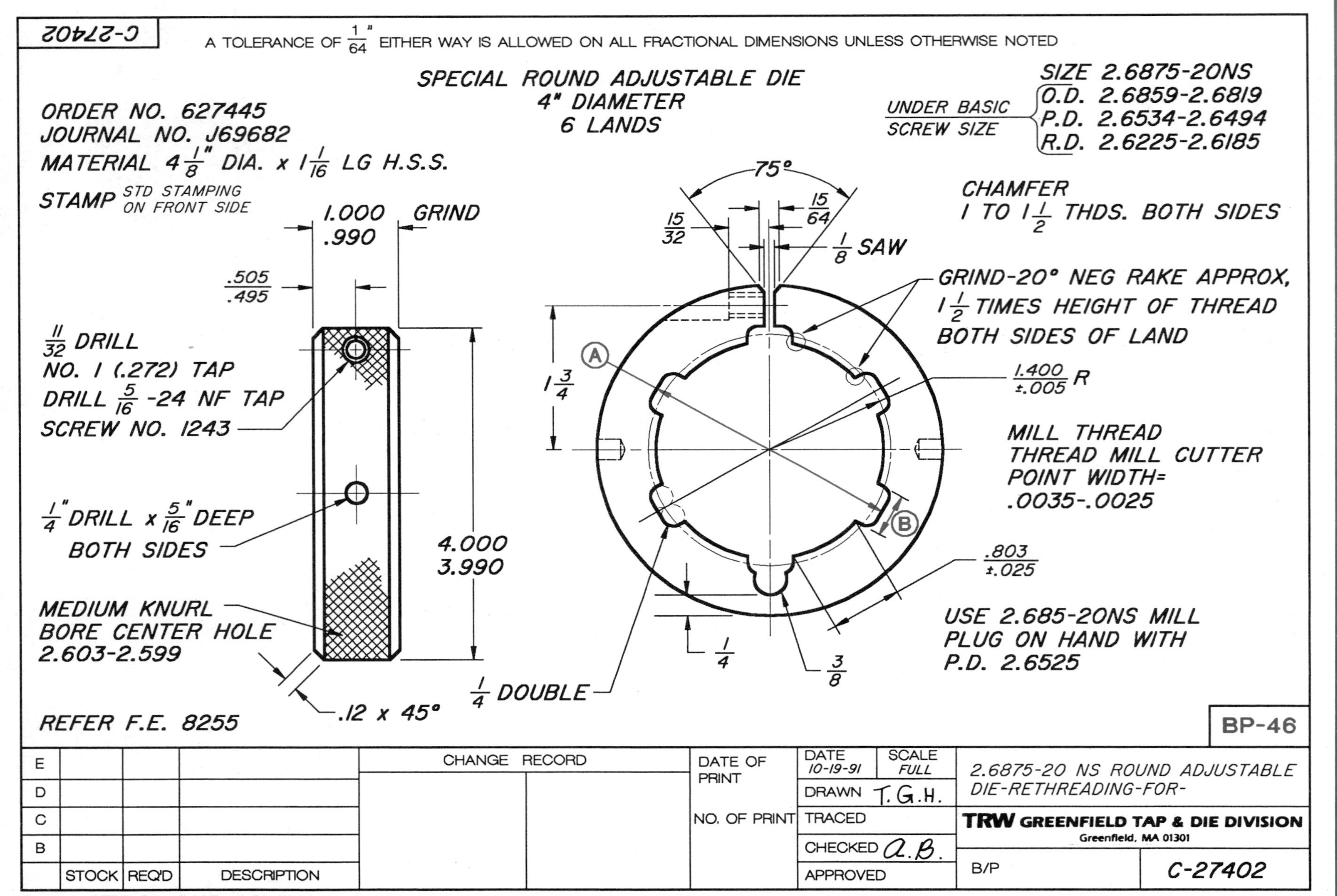
C-27402
A TOLERANCE OF $\frac{1}{64}$" EITHER WAY IS ALLOWED ON ALL FRACTIONAL DIMENSIONS UNLESS OTHERWISE NOTED
SPECIAL ROUND ADJUSTABLE DIE
4" DIAMETER
6 LANDS
ORDER NO. 627445
JOURNAL NO. J69682
MATERIAL $4\frac{1}{8}$" DIA. x $1\frac{1}{16}$ LG H.S.S.
STAMP STD STAMPING ON FRONT SIDE
UNDER BASIC SCREW SIZE
SIZE 2.6875-20NS
O.D. 2.6859-2.6819
P.D. 2.6534-2.6494
R.D. 2.6225-2.6185
CHAMFER
1 TO $1\frac{1}{2}$ THDS. BOTH SIDES
GRIND-20° NEG RAKE APPROX,
$1\frac{1}{2}$ TIMES HEIGHT OF THREAD
BOTH SIDES OF LAND
$\frac{1.400}{\pm.005}$ R
MILL THREAD
THREAD MILL CUTTER
POINT WIDTH=
.0035-.0025
$\frac{.803}{\pm.025}$
USE 2.685-20NS MILL
PLUG ON HAND WITH
P.D. 2.6525
.505
.495
1.000
.990
GRIND
4.000
3.990
75°
$\frac{15}{32}$
$\frac{15}{64}$
$\frac{1}{8}$ SAW
$1\frac{3}{4}$
$\frac{1}{4}$
$\frac{3}{8}$
A
B
$\frac{11}{32}$ DRILL
NO. 1 (.272) TAP
DRILL $\frac{5}{16}$-24 NF TAP
SCREW NO. 1243
$\frac{1}{4}$" DRILL x $\frac{5}{16}$" DEEP
BOTH SIDES
MEDIUM KNURL
BORE CENTER HOLE
2.603-2.599
.12 x 45°
$\frac{1}{4}$ DOUBLE
REFER F.E. 8255
BP-46
E
D
C
B
STOCK
REQ'D
DESCRIPTION
CHANGE RECORD
DATE OF PRINT
NO. OF PRINT
DATE 10-19-91
SCALE FULL
DRAWN T.G.H.
TRACED
CHECKED Q.B.
APPROVED
2.6875-20 NS ROUND ADJUSTABLE
DIE-RETHREADING-FOR-
TRW GREENFIELD TAP & DIE DIVISION
Greenfield, MA 01301
B/P
C-27402

Courtesy of TRW, Greenfield Tap & Die Division
(Modified Industrial Drawing)

ASSIGNMENT—UNIT 46

SPECIAL ADJUSTABLE DIE (BP-46)

Student's Name ______________________________

1. a. Identify the kind of material used for the Die.
   b. Give the stock size before machining.
   c. State how the outside diameter is machined.
2. a. Indicate the system of dimensional tolerancing used on the drawing.
   b. Give the unspecified tolerance on fractional dimensions.
3. Determine the maximum and minimum dimensions of the following features:
   a. Outside diameter of Die.
   b. Die width (thickness).
   c. Distance from the center axis to the center of the threaded hole.
   d. Diameter of the drilled hole.
   e. Vertical center line and the end of the tapped hole.
   f. Width of the saw slot.
4. a. List the specification for the tapped hole.
   b. Identify the screw thread form and series.
5. Give the nominal dimensions for (a) the outside chamfered edges and (b) the included angle of the slot.
6. a. Identify the thread size of the Die.
   b. Determine the following nominal diameters, using a tolerance of $\pm0.002''$ in each case:
      (1) Bored center hole
      (2) Pitch diameter
      (3) Outside diameter
   c. Interpret the callout

      CHAMFER 1 TO $1\frac{1}{2}$ THREADS, BOTH SIDES
7. Compute the upper and the lower limit dimensions for Ⓐ and Ⓑ.
8. Give the maximum and minimum width of each land.
9. Tell how the thread size is to be gaged.
10. Indicate (a) how the Die threads are to be formed and (b) what information is given about the cutter.
11. Identify the following:
    a. Kind and angle of rake
    b. Process by which the rake is produced
    c. Position of rake
    d. Height of rake
12. State the design function served by the 3/8″ recessed groove.

1. a. ______________________________
   b. ______________________________
   c. ______________________________
2. a. ______________________________
   b. ______________________________

3.

	Maximum	Minimum
a. Diameter	________	________
b. Width	________	________
c. Distance	________	________
d. Diameter	________	________
e. Depth	________	________
f. Width	________	________

4. a. ______________________________
   b. ______________________________
5. a. ______________________________
   b. ______________________________
6. a. ______________________________
   b. (1) Bored Center Hole ____________
      (2) Pitch Diameter ____________
      (3) Outside Diameter ____________
   c. ______________________________
   ____

7.

	Upper Limit	Lower Limit
Ⓐ =	________	________
Ⓑ =	________	________

8. Max. ____________ Min. ____________
9. ______________________________
10. a. ______________________________
    b. ______________________________
11. a. Rake kind and angle ____________
    b. Machine process ____________
    c. Position ____________
    d. Height ____________
12. ______________________________
____
____

UNIT 47
Punch and Die Principles and Features

Punches and dies are used to produce interchangeable parts at high production levels. Regular and irregularly shaped parts may be stamped to precise internal and/or external shapes and sizes. Die-cutting processes include *blanking, piercing, lancing, bending, forming, cutting off, shaving, trimming,* and *drawing.*

Each die is usually identified by the cutting and/or forming processes. One or a combination of processes may be required to blank out, or form, or assemble a simple part. Punches and dies are mounted on and operated from machine tools called *punch presses, power presses,* or just *presses.*

BASIC METAL DIE-CUTTING AND FORMING PROCESSES

Blanking is the process of stamping (cutting out) a part to a specified shape and size. A combination punch and die set is used. As the punch forces against a roll, strip, or sheet, the metal is sheared by the die, and a required part is produced. The die size establishes the dimensional size of the part.

The die is relieved (cut at a slight angle) a short distance from the cutting edges. The enlarged area produced by the angle permits the blanked parts to pass through the die. A *stripper plate* and a *stop pin* are two other important design features. Another factor to be considered is *die clearance.* Clearance represents the difference between the cutting edges of the punch and die. Die clearance is determined by the thickness of the material and the type of cutting process.

Returning to the stripper plate, this feature serves to strip the stock from the punch. The stop pin provides a constant reference point for positioning the stock for successive stamping strokes. Figure 47-1 shows how stock is positioned; the relationship of the punch, die block, stripper plate and stop pin; and the final blanked part.

Piercing refers to the stamping of holes as design features of a part.

Cutting off refers to separating or cutting away the workpiece from the strip or roll of stock.

Notching refers to the removal of material which otherwise becomes surplus and affects the forming of a part. Notching permits metal to flow as it is drawn or formed to shape.

Trimming refers to the removal of irregular or surplus material in order to produce smooth, accurate surfaces on a stamped part.

Shaving refers to a thin, shaving cut. Shaving produces a dimensionally accurate and smooth finished edge.

POWER PRESSES

Stamped parts vary in size, form, and degree of accuracy. Fine, small stamped parts which are a fractional part of a square inch or square centimeter may be produced on high-speed automated presses. Large panels, bodies, and heavy parts may also be stamped using heavy-duty hydraulic power presses of 35,000 tons capacity or more.

Die drawings include such terms and dimensions as *stroke, shut height,* and *die area. Stroke*

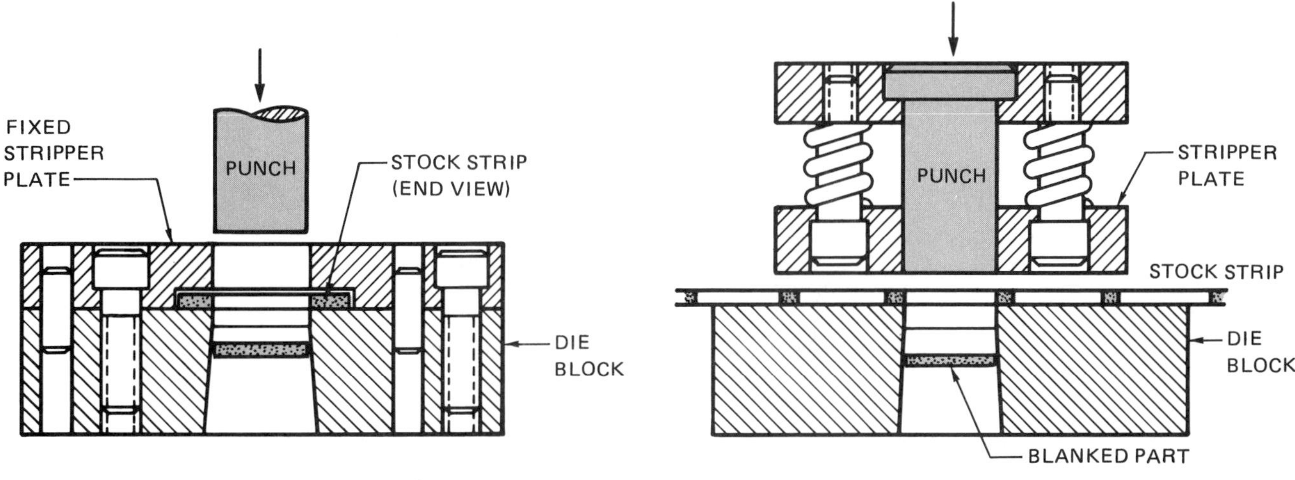

FIGURE 47-1 MAJOR COMPONENTS OF A SIMPLE BLANKING PUNCH AND DIE SET

indicates the maximum movement of the ram of a power press. *Shut height* is the distance between the bottom of the press stroke and the up stroke adjustment. *Die area* represents the space available on a punch and die set for all components. Figure 47-2 is a commercially available two-post die set with the major parts listed.

CONVENTIONAL AND COMPOUND DIES

Inverted Dies

Many stamping and forming processes require the punch and die components to be inverted. *Inverted dies* employ a *pressure pad* around the punch. In this type of die, the punch is secured to the die shoe. The pressure pad forces the stock off the punch at the end of each cutting stroke. As the blank is formed, it is forced into the cavity of the die which is mounted on the punch holder. An *ejector* and *knockout rod* force the blank out of the die. Thus, each stamped part is removed instead of being forced through the die.

Progressive Dies

Multiple stamping and forming processes require *progressive dies.* This means that two or

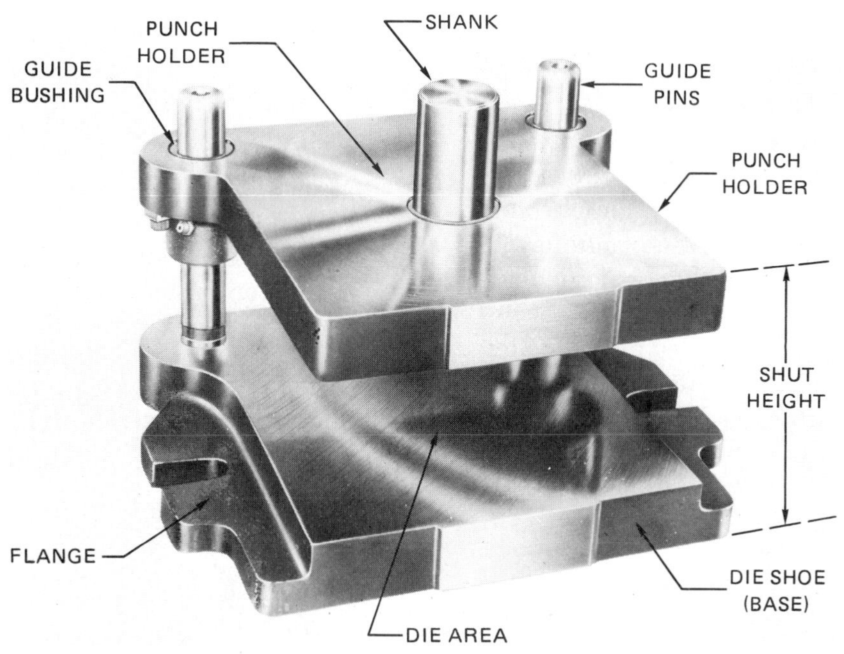

FIGURE 47-2 STANDARD TWO-POST DIE SET WITH PARTS AND DESIGN FEATURES LABELED (COURTESY OF THE DANLY MACHINE CORPORATION)

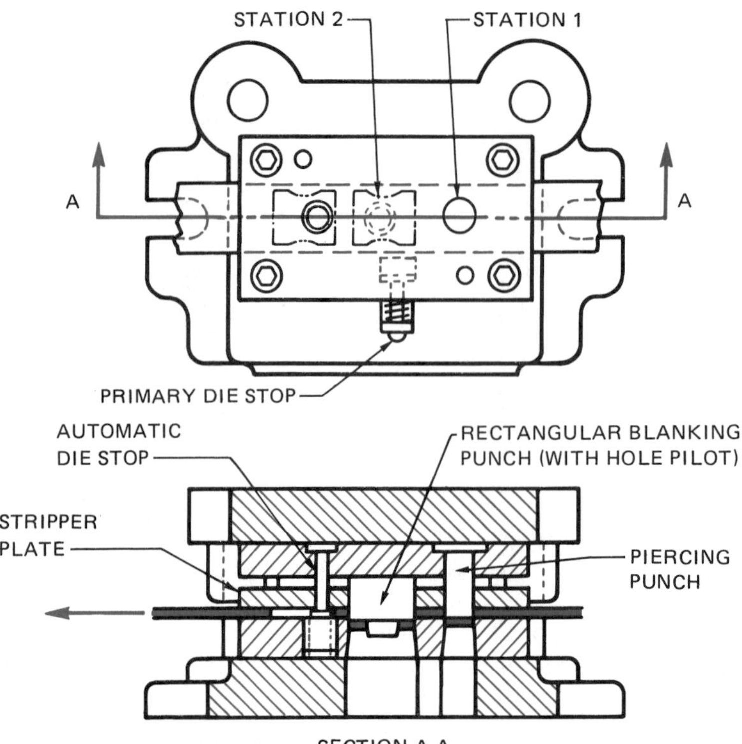

FIGURE 47-3 TYPICAL DRAWING WITH A REGULAR TOP VIEW OF DIE AND A SECTIONAL VIEW (WITH CALLOUTS AND PARTS LIST OMITTED)

more press operations are performed for each stroke. Progressive dies often incorporate piercing, forming, blanking, and trimming operations. The strip or roll is fed progressively from one station where a particular stamping process is performed to the next process *(station)*. After the strip is once cycled through each station, a finished part is produced at each stroke of the press. Figure 47-3 is an example of a typical punch and die drawing, except that leaders and an accompanying parts list are omitted. Drawings usually contain a regular top view and a full section of the die.

A simple two-station progressive die is illustrated. As the die is being set up for production, the center hole in the part is pierced at station 1. The strip stock is then advanced and positioned with the *primary stop*. The rectangular punch at station 2 centers in the punched hole and blanks out the completely formed part. Once the die is set up, one pierced and stamped part is produced at each stroke of the press.

Compound Dies

Compound dies produce two or more operations at one station. For example, a rectangular part with a centrally located hole may be produced with a compound die. The rectangular blanking punch serves also as the piercing die for the center hole. This punch/die component is mounted on and secured to the die block. The mating stamping die and piercing punch are attached to the punch plate.

Combination Dies

Combination dies are designed to perform cutting and noncutting operations during a single complete stroke of a power press. The cutaway section (figure 47-4) shows the internal design features of a combination die that is used to produce the stamped and formed part.

A rectangular slug is first blanked during the downward stroke of the power press ram. The cup is formed during the remaining part of the downward stroke. The forming punch (that is mounted on the die shoe) forces the blanked slug into the hollowed blanking punch (in the punch section) as the die sections come together.

The shaped/formed part is ejected out of the die during the upward stroke by action of the *punch pad* and the *stripper pad.*

DRAWINGS OF PUNCHES AND DIES

Drawings of punches and dies generally include assembly drawings using standard views, sectional views, and auxiliary views. Working drawings are provided for components which are not commercially available. Parts are identified by leaders. Material and part requirements are specified on a parts list.

Noncommercial parts such as die blocks and punches are usually represented by standard views. The use of base-line, ordinate, and tabular dimensioning is widespread practice. Dimensional tolerances on critical punch and die components are generally more precise tolerances.

Base-line and ordinate dimensioning are practical because they indicate precise horizontal (X axis), transverse (Y axis), and vertical (Z) movements of a table, knee, or machine spindle. Also, numerically controlled jig borers and jig grinders may be programmed easily and precisely for tool and work positions. Where a series of details is close together, the extension line from the center location of a feature may be moved out of position. Also, instead of placing all dimensions in one line, some may be staggered. In still other instances, some of the ordinate dimensions are given within the drawing itself. These variations of ordinate dimensioning are applied in the drawing of the Die Block assignment for this unit.

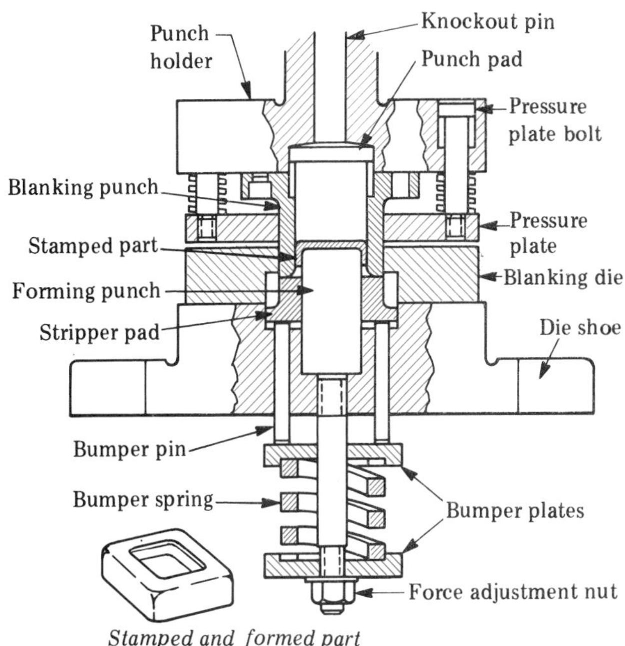

FIGURE 47-4 SECTION VIEW OF COMBINATION DIE

SHEET METAL BENDING, FORMING, AND DRAWING DIES

In sheet metal work, *bending* refers to the shaping of metal across its length so that different areas of the workpiece are in more than one plane. *Forming* refers to the flowing of metal so a finished object is produced which has the form of the punch or die. *Drawing* is generally associated with producing cylindrical, square, or rectangularly shaped shell parts.

Different bends and forms are produced on a *press brake*. *Brake bending* and *forming dies* are illustrated in figure 47-5. Where required, *pressure pads* are used to control the position of the workpiece and to permit the uniform flow of metal. The models in figure 47-5 show the use of pressure pads in forming a *corrugated form* at (D). The setup at (E) is for producing *rounded (coiled) edges*.

Coiled edges are widely used in sheet metal work for safety, appearance, and strength. Curling dies, consisting of highly polished curling surfaces on the punch and die, are used.

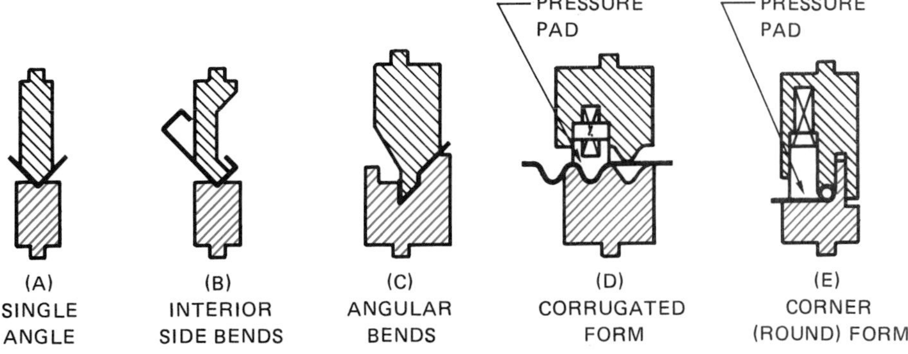

FIGURE 47-5 STANDARD SHEET METAL BENDS AND FORMS PRODUCED ON A PRESS BRAKE

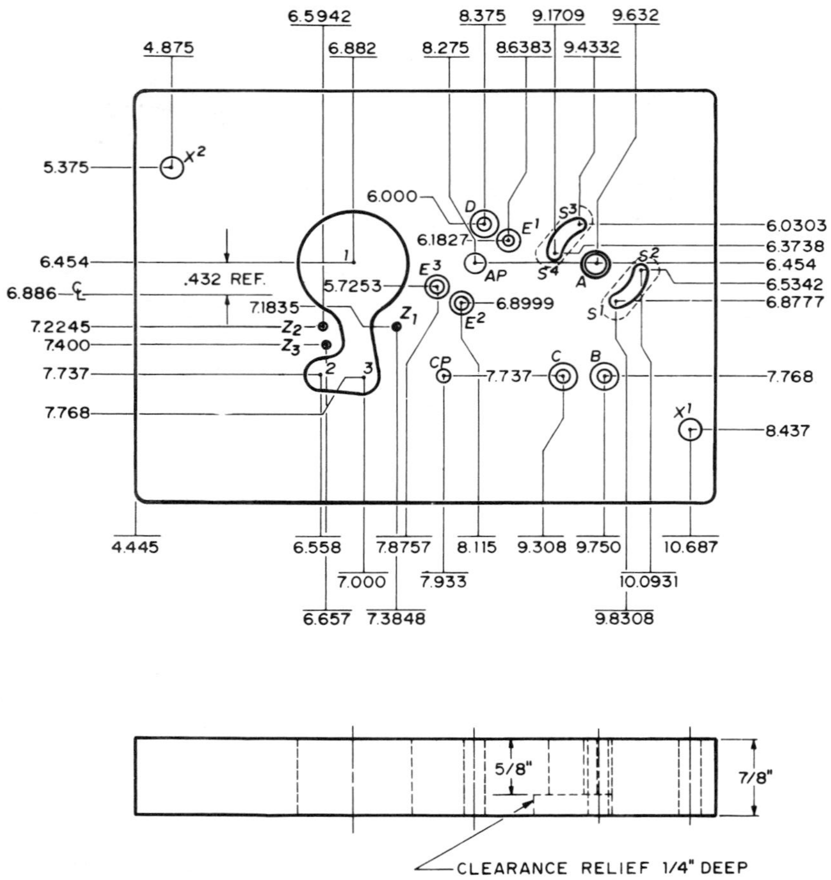

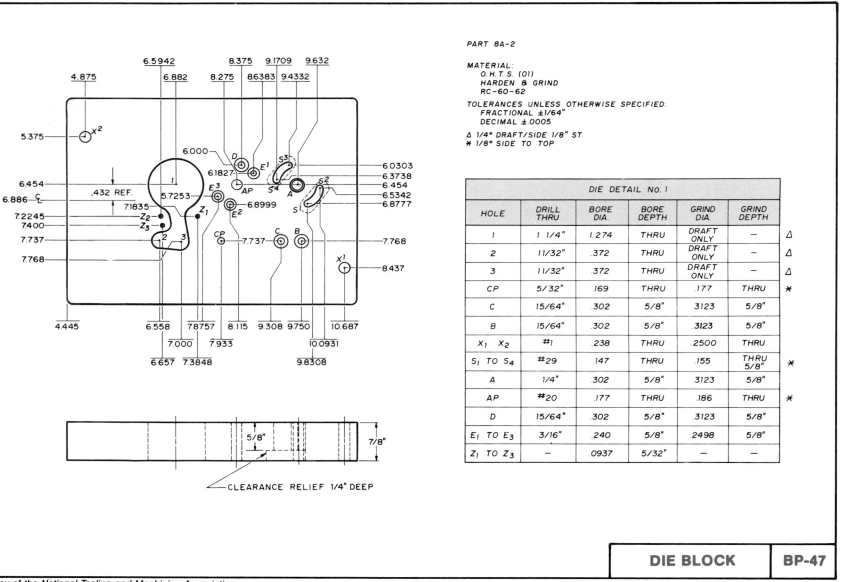

PART 8A-2

MATERIAL:
O. H. T. S. (01)
HARDEN & GRIND
RC-60-62

TOLERANCES UNLESS OTHERWISE SPECIFIED:
FRACTIONAL ±1/64"
DECIMAL ±.0005

Δ 1/4° DRAFT/SIDE 1/8" ST.
✱ 1/8° SIDE TO TOP

CLEARANCE RELIEF 1/4" DEEP

DIE DETAIL NO. 1						
HOLE	DRILL THRU	BORE DIA.	BORE DEPTH	GRIND DIA.	GRIND DEPTH	
1	1 1/4"	1.274	THRU	DRAFT ONLY	–	Δ
2	11/32"	.372	THRU	DRAFT ONLY	–	Δ
3	11/32"	.372	THRU	DRAFT ONLY	–	Δ
CP	5/32"	.169	THRU	.177	THRU	✱
C	15/64"	.302	5/8"	.3123	5/8"	
B	15/64"	.302	5/8"	.3123	5/8"	
X_1 X_2	#1	.238	THRU	.2500	THRU	
S_1 TO S_4	#29	.147	THRU	.155	THRU 5/8"	✱
A	1/4"	.302	5/8"	.3123	5/8"	
AP	#20	.177	THRU	.186	THRU	✱
D	15/64"	.302	5/8"	.3123	5/8"	
E_1 TO E_3	3/16"	.240	5/8"	.2498	5/8"	
Z_1 TO Z_3	–	.0937	5/32"	–	–	

DIE BLOCK | BP-47

Courtesy of the National Tooling and Machining Association

ASSIGNMENT—UNIT 47

DIE BLOCK (BP-47)

Student's Name ______________________

1. a. Identify the kind of steel which is to be used for the Die Block.
   b. Name the machine finishing process for all of the Die cutting features.
   c. Give the hardness testing range.
2. Indicate the unspecified tolerances for (a) fractional and (b) decimal dimensions.
3. Identify three variations in the dimensioning on this drawing from conventional ordinate dimensioning.
4. Interpret the following note.

   1/4° DRAFT/SIDE 1/8″ ST
5. List each group of holes which are to be drilled, bored, and ground to the same specification.
6. Calculate the amount of material which is left for finish grinding the following holes (correct to three decimal places).
   a. **C**
   b. X_1 and X_2
   c. S_1 to S_4
   d. **AP**
7. Determine the upper and lower limit dimensions for the following features from datum **Y** (vertical dimensions).
   a. The horizontal center line.
   b. The center distance of hole **1**
   c. Hole Z_2
   d. Hole Z_3
   e. Holes **CP** and **C**.
   f. Hole E_3
8. Determine the upper and lower limit dimensions for the following features from the **X** (horizontal) datum:
   a. Hole **D**   c. Hole S_2
   b. Hole E_1
9. Give the upper and lower limit diameters of the following ground holes:
   a. S_1 to S_4 b. E_1 to E_3
10. Determine the minimum bore or depth (ground) for each of the following features:
    a. Holes **1**, **2**, and **3**.
    b. Holes **A**, **C**, and **D**.
    c. Holes **B** and E_1 to E_3
11. Compute the following maximum and minimum overall center distances:
    a. Along the **X** axis.   b. Along the **Y** axis.
    (1) E_1 to E_3 (1) S_1 to S_3
    (2) X_1 to X_2 (2) S_3 to S_4

1. a. ______________________
   b. ______________________
   c. ______________________
2. a. Fractional ______________________
   b. Decimal ______________________
3. a. ______________________
   ____
   b. ______________________
   ____
   c. ______________________
   ____
4. ______________________
   ____
5. ____________ ____________
   ____________ ____________
   ____________ ____________
6. a. ____________ c. ____________
   b. ____________ d. ____________

7.	Upper Limit	Lower Limit
a.		
b.		
c.		
d.		
e.		
f.		

8.	Upper Limit	Lower Limit
a.		
b.		
c.		

9.	Upper Limit	Lower Limit
a.		
b.		

10. a. ______________________
    b. ____________ c. ____________

11.	Maximum	Minimum
a. **X** Axis		
(1)		
(2)		
b. **Y** Axis		
(1)		
(2)		

SECTION 13
Computerized Systems: NC, CNC, CAD, and CAM

UNIT 48
Numerical Control (NC)/Computer Numerical Control (CNC) Symbols, Programming, and Systems

Numerical control is a complete system of numbers and letters which provides recorded and other computerized instructions. Numerical control (NC) provides a method of compiling a series of operating instructions into a medium. The instructions may be utilized a number of times and at any time. The instructions for control are in some form of a permanent process control record. The medium containing the coded commands may be a perforated tape, tabulating card, magnetic tape, remote computer storage, diskette, or memory within the unit itself.

BASIC FUNCTIONS OF NUMERICAL CONTROL

There are eight basic functions of a numerical control system:

- *Controlling position movements.* NC is used to position a cutting tool, a forming tool, a recorder, a plotter on a drafting machine, and so forth, in relation to a fixed reference point or datum.
- *Controlling tool movements*, as when setting up and machining.
- *Establishing operation sequences*, including time intervals.
- *Controlling speeds and feeds.*
- *Monitoring* tool performance and machining processes.
- *Providing* machining accuracy *readouts.*
- *Changing process sequences.*
- *Controlling shutdown or recycling.*

Machine tools, other production equipment, instruments, and measuring devices may be programmed. *On command*, electrical energy impulses may be transmitted through an electronics controller to a machine. At that point, the form of energy is converted to mechanical or fluid (pneumatic) energy. This energy is used to actuate motors, switches, clutches, and brakes. These devices control, guide, and time each movement of a machine component and/or a cutting tool or other process or measuring device.

The major advantages of NC and CNC machines include the following:

- *Repeatability.* Successive parts may be reproduced within extremely fine limits of accuracy.
- *Productivity.* Once programmed accurately, operator error is decreased and a fixed productivity level can be maintained.
- *Reduced Setup Time.* NC and CNC programming permit the machining of a single workpiece or quantity production with minimum tooling and workpiece setup time.
- *Eliminates Production Tooling.* NC and CNC often eliminate the need for additional production tooling such as jigs and fixtures due to the flexibility to quickly and accurately position and control machining processes.

- *Efficiency of Tool Changes.* Tool storage drum and carriage installations on NC and CNC machine tools permit efficient storage, retrieval, and tool positioning.

SPECIAL FEATURES OF NC DRAWINGS

Drawings of parts which are to be produced by the application of NC to a machine tool or machining center usually contain complementary technical information and other *documentation*. For instance, in addition to the design features of the part, the programmer must be able to identify the positions and sizes of workholding clamps and devices. Thus, it is important that part drawings are drawn accurately to scale. The exact relationship between tooling and the workpiece needs to be established.

- Working drawings are generally prepared with the plan view drawn in the position from which the workpiece is viewed *on the machine table.*
- The drawing of the workpiece is fully dimensioned. Scaling a part for noncritical dimensions is not recommended.
- Cutter sizes are often noted on drawings so there is a record of every item which is related to programming for producing the part.
- The number of views and sections must be adequate to show cutter outlines. Such drawings permit visual inspection to check clearances with clamps, bosses, and other obstructions.
- The operator's work copy of the NC drawing generally shows the cutter path. This provides information about the general direction of travel.
- Special notes, including quality control and inspection records, appear on NC drawings. These permit uniform communications among the designer, programmer, tooling shop, and operator.
- It is important that all changes in design, tooling, and production be incorporated on the NC drawing.

CATEGORIES OF NC PRODUCED WORKPIECES

There are nine basic categories of NC workpieces ranging from two-axis to five-axis positioning.

Group One. Straight Cuts. Point-to-Point Programming. The machining of parts in this group is controlled by positioning commands along two axes (**X** and **Y**, or **XZ**, or **YZ**). There may also be an additional callout for depth of cut. Figure 48-1 shows point-to-point drilling operations at (**A**), straight milling in one plane at (**B**), and rectangular pattern operations at (**C**).

Group Two. Single Plane: Contour Profiling. This group includes straight lines other than those which are parallel to the **X** or **Y** axes. Straight lines are tangent to circles and arcs; circles intersect circles, and straight lines intersect circles. Figure 48-2 provides examples of contour profiling in a single plane.

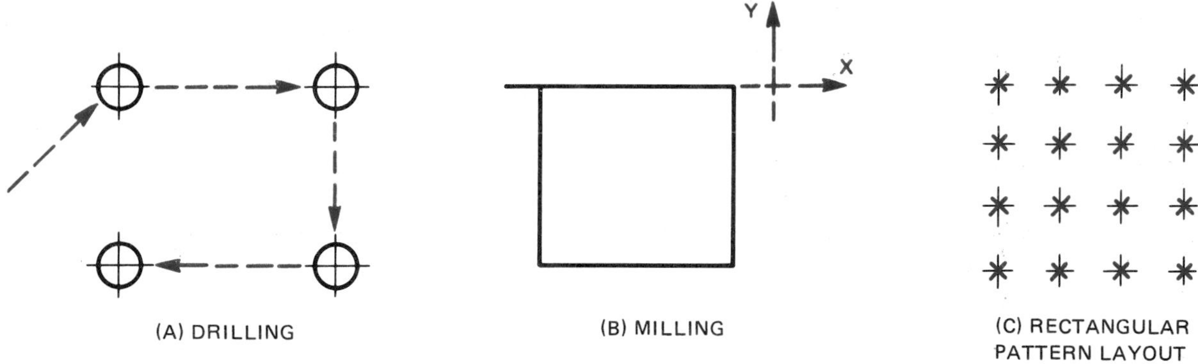

FIGURE 48-1 POINT-TO-POINT OPERATIONS

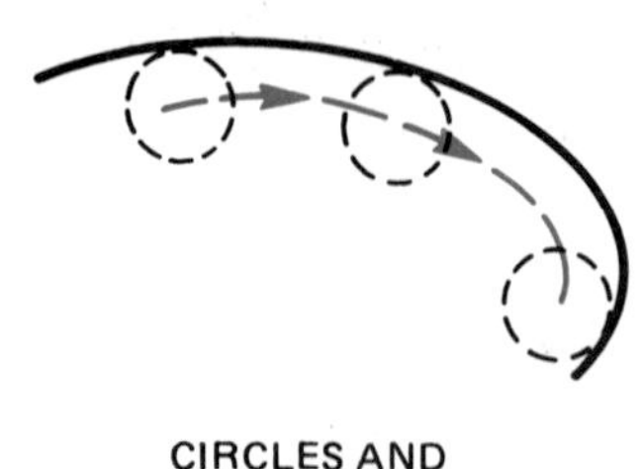

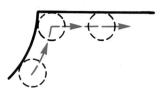

CIRCLES AND ARCS

STRAIGHT LINES INTERSECTING CIRCLES

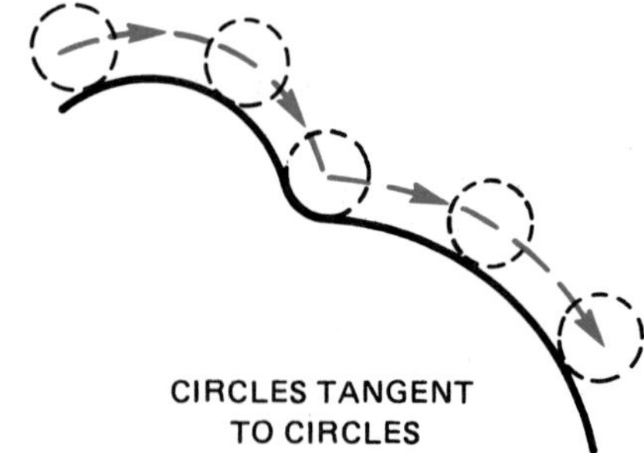

CIRCLES TANGENT TO CIRCLES

FIGURE 48-2 CONTOUR PROFILING

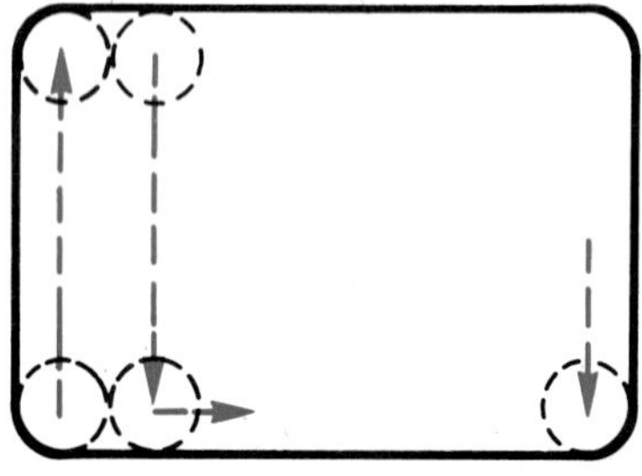

(A) STRAIGHT REPETITIVE CUTS

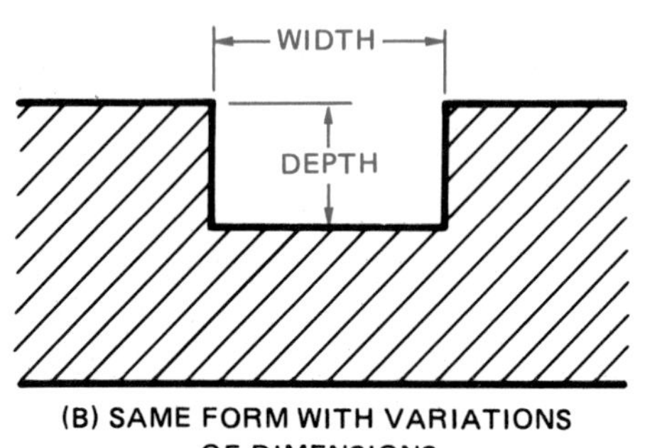

(B) SAME FORM WITH VARIATIONS OF DIMENSIONS

FIGURE 48-3 REPETITIVE CUTS

Group Three. Repetitive Cuts. Drawings such as figure 48-3A show the position of straight, repetitive cuts to mill the recessed rectangular area. Figure 48-3B illustrates a family of parts. While the form is the same, the dimensions for width and depth vary.

Group Four. Complex Curves and Straight Lines. Elliptical (formula) curves, nonparallel straight lines, and irregularly spaced operations around a circle may be programmed with two-axis programming.

Group Five. Controlled Rotary Motion. A rotary table is generally used to program the machining of cams and other workpieces of circular form. The axis of rotation is either parallel or perpendicular to the spindle axis.

Group Six. Three-Dimension Continuing Motion. Parts such as contour dies and complex aerospace and automotive components require the simultaneous motion of three or more spindle axes. Computers are often used to calculate the shape and size of the part and the necessary cutter path.

Group Seven. Sculptured Surfaces. Three to five axes of machine tool motion are programmed. All surfaces are broken down into a mathematical definition. The positioning and cutter movements result in machining the required sculptured surface.

Group Eight. Formula-Defined and Regular Surfaces. Aircraft wings and turbine engine blades are examples of workpieces which may be programmed and machined.

Group Nine. Multi-axis, Simultaneous, Five-Axis Motion. Controls are required in this work group for *simultaneous commands* for positioning and cutting motions.

NUMERICAL CONTROL DRAWINGS: INTERPRETATION

Numerical control drawings of parts are based on a *rectangular (Cartesian)* coordinate system. Figure 48-4 shows the four quadrants with X and Y axis in the horizontal plane and Z dimensions in the vertical plane. The + and − values, representing movements of the machine table (X and Y) and spindle (Z) are shown in figure 48-5. The Z axis is associated with the machine spindle, regardless of its position.

The *datum* or *point of origin (zero reference point)* may be located on or away from the workpiece. *Rotational numerical control motion* may also be associated with X, Y, and Z axes as illus-

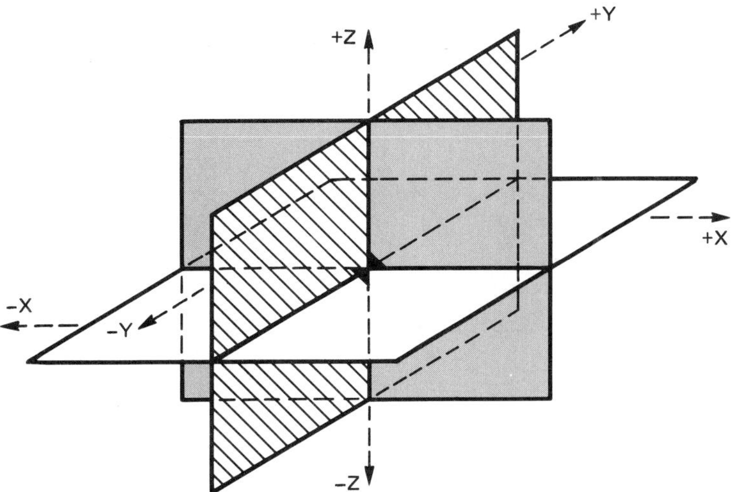

FIGURE 48-4 RECTANGULAR (CARTESIAN) COORDINATE SYSTEM

trated in figure 48-6. Lowercase letters are used to indicate the rotational axes. Numerical control drawings should provide information indicating the direction and basic axis around which motion occurs.

Point Values

Point values are used for axis movements to indicate distance and direction. Once programmed, the commands control these functions. Machining processes involve either *point-to-point (positioning)* or *continuous path (contouring) movements. Positioning* is associated with the spindle movement. Actually, it is the table and workpiece which are moved to a specified location to perform a programmed sequence of events. Such processes as drilling, tapping, welding, or riveting are completed at the programmed location. After each operation, the spindle is *retracted* (brought out of the workpiece and clear of all obstructions) and moved to the next location.

Continuous path operations require the tool to remain in constant contact with the workpiece as coordinate movements take place. The path of the cutting tool is programmed to produce a particular *profile* or *sculptured surface* (such as an embossed die).

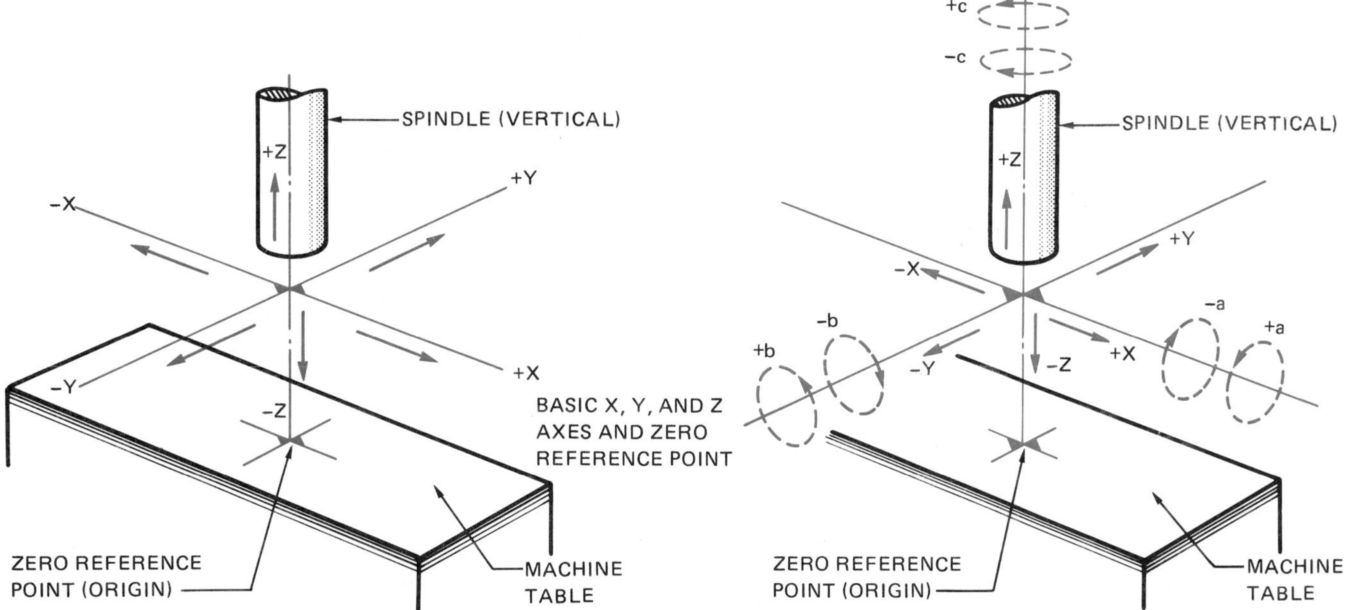

FIGURE 48-5 THREE-DIMENSIONAL COORDINATES

FIGURE 48-6 ROTATIONAL NUMERICAL CONTROL AXES

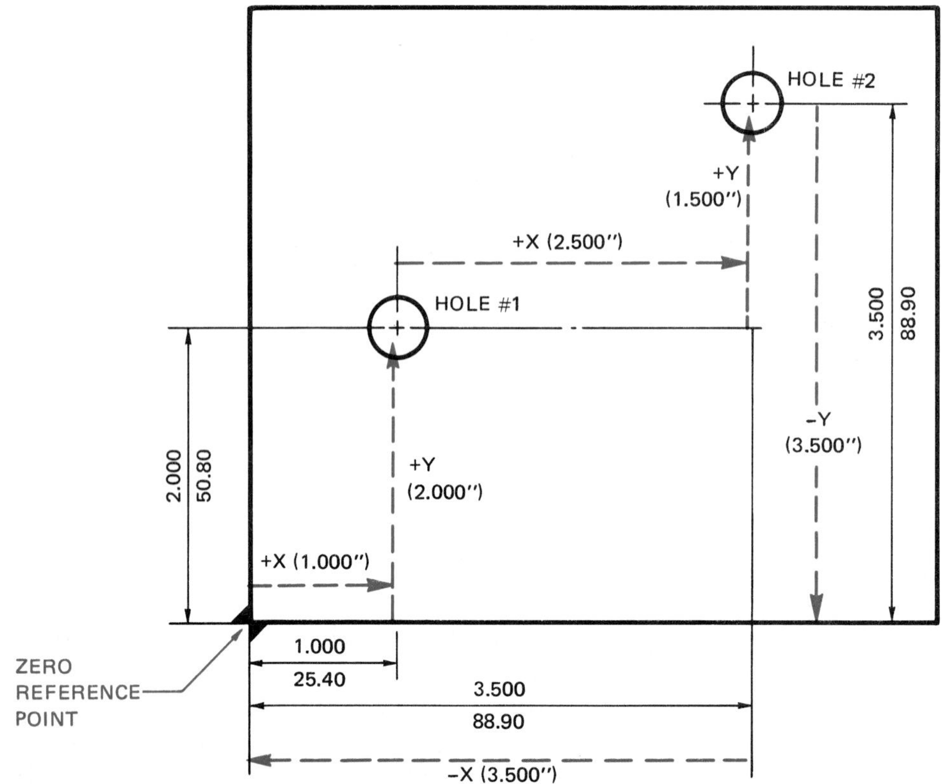

FIGURE 48-7 POSITIONING SPINDLE USING INCREMENTAL DIMENSIONS

PROGRAMMING A NUMERICALLY CONTROLLED MACHINE

In addition to positioning, the NC programmer lists tools, feed rates, spindle speeds, table indexing; starting, stopping, changing, and retracting; and other *auxiliary functions*. Spindle direction (**02**), tool selection (**05**), and coolant control (**07**) are examples.

Incremental and Absolute (Coordinate) Programming

Incremental programming (as illustrated in figure 48-7) refers to the movement of the spindle from one position to the next. In this case, the spindle is positioned for hole **#1** by moving the machine table and workpiece + 1.000″ (25.4 mm) along the **X** axis and + 2.000″ (50.8 mm) along the **Y** axis from the zero reference point. Hole **#2** is positioned by retracting the drill from hole **#1** to clear the workpiece and any work-holding straps or bolts. The table is then moved + 2.500″ (63.5 mm) along the **X** axis and + 1.500″ (38.1 mm) along the **Y** axis.

By contrast, in *absolute programming* table and machine movements are all taken from the *same fixed reference* or *floating reference point*. Absolute measurements are similar to coordinate measurements. Dimensional errors, which accumulate from incremental positioning, are eliminated.

A *floating zero* may be established at any position that will make programming easier. For example, the absolute zero in figure 48-8 is floated to the intersection of the **X** and **Y** axes. The floated zero permits positioning the four holes to be drilled by plus (+) and minus (−) movements (as required) in each of the four quadrants.

PROGRAMMING FORMAT

A numerical control *manuscript* lists all the work positions, cutting tool, and machine tool functions in a logical sequence. The program which is written *manually* or with the aid of a computer (CNC) is generally processed by a tape preparation unit. The program may be written by a *letter address system* using a word as a command. The word (letter) is followed by designated Arabic numerals. When completed, the program is fed through a machine control unit (MCU). The MCU produces specific instructions (motion commands)

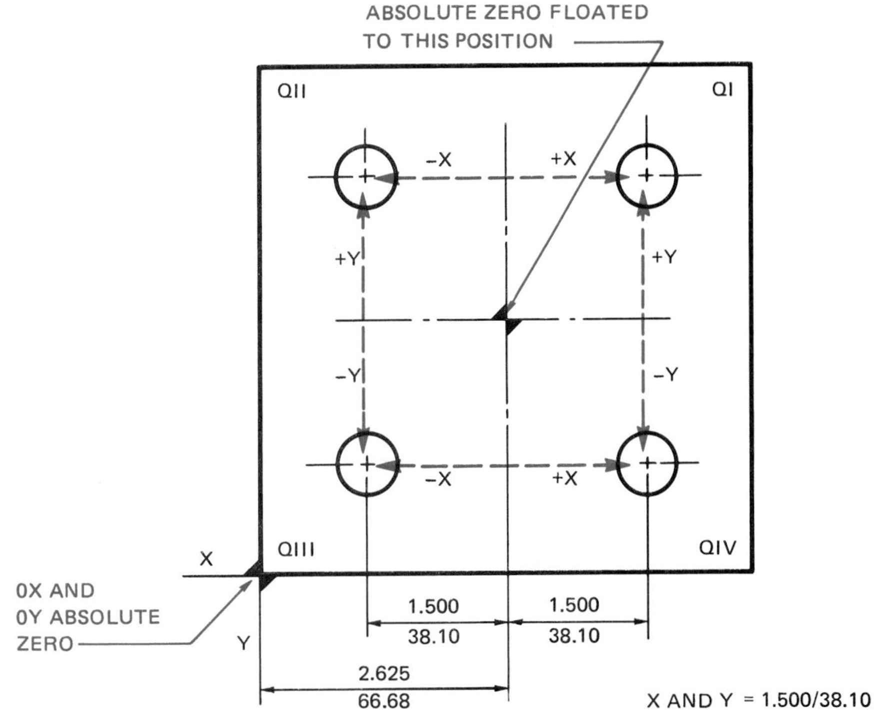

FIGURE 48-8 SPINDLE (CENTERS) POSITIONED IN RELATION TO FLOATED ZERO

for a *servo drive unit*. Each *servo motor (drive motor)* actuates a lead screw, feed mechanisms, speed control, or other motion-controlling device. *Sensors* on the servo motor feed back signals through an *encoder* to the MCU. The signals tell that a particular command (distance, position, or process) is completed. The MCU controls and commands all functions.

The coded input information for a process, machine function, or particular task is sectioned on a tape as a complete entity *block*. Each block is separated from the next one by an *end-of-block (EOB) code*. Where tapes are used, the tape consists of channels. Holes are punched in the channels using a *two-word binary* vocabulary to control all functions. NC word language contains *dimension* and *nondimension words*.

SAMPLE COMMAND ENTRIES

A few sample commands which the MCU accepts from one numerical control letter address continuing system follow:

- **G** *Function*. This preparatory command sets up the control for a specific type of operation. For example, **G01** is a *slope command*. **G04** is a *dwell command*. Such a command is used when tapping, breaking chips, bar feeding, and so forth. The command is followed by a time period, such as **G04 × 25000** This command results in a dwell of 2.5 seconds. A **G33** command relates to *thread cutting;* **G90** *absolute programming;* **G91**, *incremental programming;* and **G94** *feed rate* in inches or millimeters per minute.
- **X**, **Y**, and **Z** *functions* are distance (linear) commands usually of the cross slide, carriage, and spindle, respectively.
- **S** *Functions*. These commands relate to *spindle speed*. The **T** *function* selects the *turret station and offset*.
- **M** *Functions*, followed by two digits, command machine actions. For example, on one manufacturer's NC machine tool, the **M00** *Program Stop* command stops the program, stops the spindle, turns the coolant off, and permits the use of collet switches. An **M20** *Feed Stock* command provides the signal controlling the airflow to the stock feed mechanism. **M21** relates to *open collet;* **M23** *vertical cutoff*. An **M43** *High-Clutch Forward* command causes the spindle to turn clockwise in the 600- to 3,000-RPM speed range.

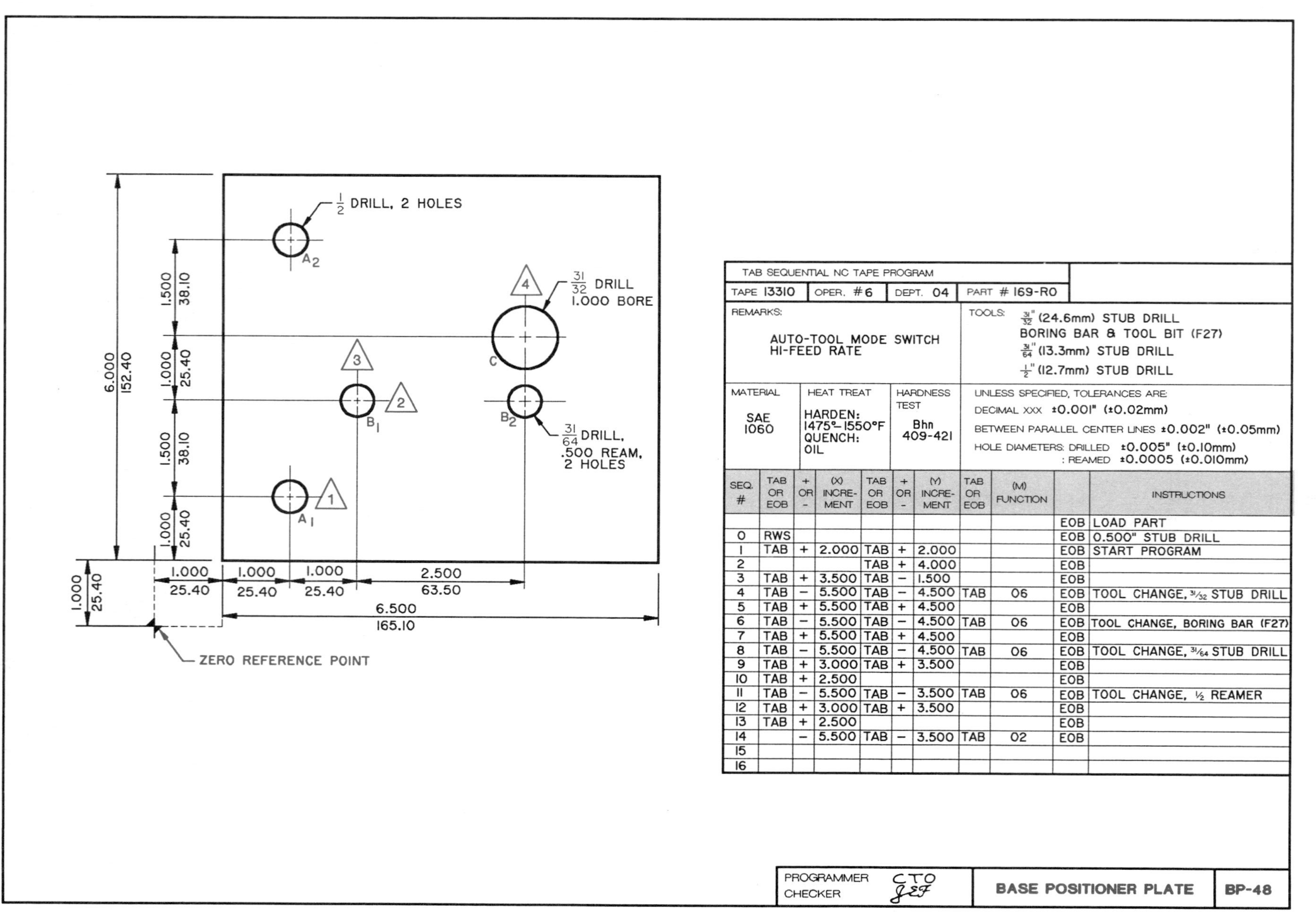

TAB SEQUENTIAL NC TAPE PROGRAM

TAPE 13310 | OPER. #6 | DEPT. 04 | PART # 169-RO

REMARKS: AUTO-TOOL MODE SWITCH
HI-FEED RATE

TOOLS: $\frac{31}{32}$" (24.6mm) STUB DRILL
BORING BAR & TOOL BIT (F27)
$\frac{31}{64}$" (13.3mm) STUB DRILL
$\frac{1}{2}$" (12.7mm) STUB DRILL

MATERIAL	HEAT TREAT	HARDNESS TEST	UNLESS SPECIFIED, TOLERANCES ARE:
SAE 1060	HARDEN: 1475°–1550°F QUENCH: OIL	Bhn 409-421	DECIMAL XXX ±0.001" (±0.02mm) BETWEEN PARALLEL CENTER LINES ±0.002" (±0.05mm) HOLE DIAMETERS: DRILLED ±0.005" (±0.10mm) : REAMED ±0.0005 (±0.010mm)

SEQ. #	TAB OR EOB	+ OR -	(X) INCRE-MENT	TAB OR EOB	+ OR -	(Y) INCRE-MENT	TAB OR EOB	(M) FUNCTION		INSTRUCTIONS
									EOB	LOAD PART
0	RWS								EOB	0.500" STUB DRILL
1	TAB	+	2.000	TAB	+	2.000			EOB	START PROGRAM
2				TAB	+	4.000			EOB	
3	TAB	+	3.500	TAB	–	1.500			EOB	
4	TAB	–	5.500	TAB	–	4.500	TAB	06	EOB	TOOL CHANGE, 31/32 STUB DRILL
5	TAB	+	5.500	TAB	+	4.500			EOB	
6	TAB	–	5.500	TAB	–	4.500	TAB	06	EOB	TOOL CHANGE, BORING BAR (F27)
7	TAB	+	5.500	TAB	+	4.500			EOB	
8	TAB	–	5.500	TAB	–	4.500	TAB	06	EOB	TOOL CHANGE, 31/64 STUB DRILL
9	TAB	+	3.000	TAB	+	3.500			EOB	
10	TAB	+	2.500						EOB	
11	TAB	–	5.500	TAB	–	3.500	TAB	06	EOB	TOOL CHANGE, ½ REAMER
12	TAB	+	3.000	TAB	+	3.500			EOB	
13	TAB	+	2.500						EOB	
14		–	5.500	TAB	–	3.500	TAB	02	EOB	
15										
16										

PROGRAMMER CTO
CHECKER JEF

BASE POSITIONER PLATE | BP-48

ASSIGNMENT—UNIT 48

BASE POSITIONER PLATE (BP-48)

Student's Name ____________________

1. Cite three features of numerical control drawings which differ from conventional drawings.
2. List three basic functions of a numerical (NC) control system on a machine tool.
3. a. Make a simple sketch showing the following machine table and spindle movements from the fixed reference point:
   (1) **X 1750, −X 1750**
   (2) **Y 500, −Y 350**
   (3) **Z 250, −Z 150**
   b. Letter each axis and give the value of each movement.
4. a. State briefly what the difference is between point-to-point (incremental) and continuous path positioning.
   b. Define "zero reference point."
   c. Indicate the difference between incremental programming and absolute programming.
5. Determine the upper and lower limit center-to-center customary inch standard dimensions between the following holes.
   a. A_1 and A_2 (Y axis)
   b. A_1 and B_1 (Y axis)
   c. A_1 and C (Y axis)
   d. A_2 and C (X axis)
6. Indicate (a) the machining operation(s) and (b) the maximum diameters for holes A, B, and C.
7. Establish the maximum and minimum metric dimensions between the following features:
   a. △1 and △2 (Y axis)
   b. △3 and △4 (X axis)
8. a. Identify the NC programming system used on the Plate.
   b. Give the number of tool changes and the sequence numbers at which each change occurs.
9. Identify the location of the X and Y distances of the zero reference point for the tool changes.
10. Determine the X and Y increments for positioning the spindle for the following holes:
   a. A_2 b. B_1 c. C
11. State the functions which are represented by (a) code **M06** and (b) code **M02**.
12. Give the distances required to move the spindle back (retract) to the zero reference point for the following tool changes:
   a. Sequence #4
   b. Sequence #6
   c. Sequence #11
13. Interpret the following sequences:
   a. #1 b. #2 c. #8

1. a. ____________________
   b. ____________________
   c. ____________________
2. a. ____________________
   b. ____________________
   c. ____________________
3. [sketch grid]
4. a. ____________________
   b. ____________________
   c. ____________________

5.

	Upper Limit	Lower Limit
a.		
b.		
c.		
d.		

6.

	Operation (a)	Max. Diameter (b)
A		
B		
C		

7.

	Maximum	Minimum
a.		
b.		

8. a. ____________________
   b. ____________________
9. X = ________ Y = ________

10.

	X Increment	Y Increment
a. A_2		
b. B_1		
c. C		

11. a. ____________________
    b. ____________________

12.

	Seq. #	X Axis	Y Axis
a.	4		
b.	6		
c.	11		

13. a. Seq. #1. ____________________
    b. Seq. #2. ____________________
    c. Seq. #8. ____________________

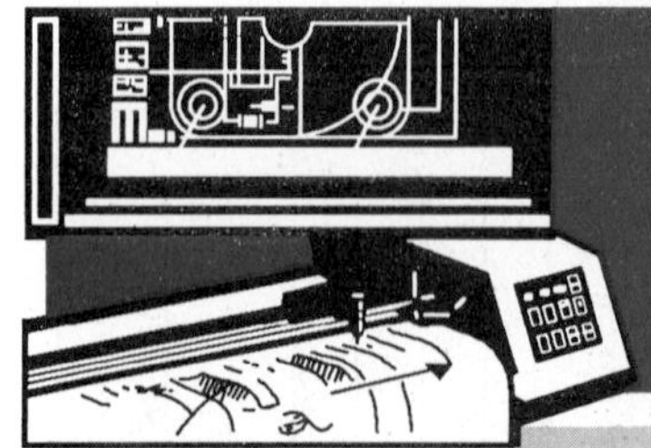

HIGH TECHNOLOGY applications

HT-15 CADD/CNC/CAM-NONTRADITIONAL MACHINING: LASER BEAM MACHINING

Laser beam equipment and machining processes are compatible with CADD/CAM and CNC programmable multi-purpose, multi-axes nontraditional carbon dioxide (CO_2) laser beam machining (LBM). CO_2 laser beam hard surfacing, surface hardening, and machining processes are described briefly.

LASER BEAM HARD SURFACING (CLADDING)

Hard surfacing (cladding) relates to the process of alloying and strongly binding a hard surface alloy material to a softer metal substrate. The purpose is to improve surface hardness and wear and corrosion resistance.

A hard surface is produced by concentrating a high-energy carbon dioxide (CO_2) laser beam across an *interaction zone* on the workpiece that is to be hard coated. Upon heating to melting temperatures, the preplaced hard surface alloying material (powder or chips) and the substrate layer (core) in the interaction zone flow, blend, and bond to produce a uniform texture hard surface facing. Due to the fact that hard surfacing deposits may be accurately controlled to within ±0.005″ to ±0.010″ (±0.127 mm to ±0.254 mm), a minimum amount of machining is required to produce dimensionally and geometrically accurate finished parts.

CO_2 LASER BEAM SURFACE HARDENING

The area, path, rate of spot transformation, and depth of penetration of a laser beam across a localized surface may be accurately controlled. *Surface* hardening takes place when a highly concentrated laser beam (controlled by an *optical integrator*) focuses an intense heat on a specific area of a hardenable workpiece that is rapidly cooled to room temperature.

Laser positioning, intensity, and working speed are computer controlled. Laser beam surface hardening is uniform in depth, workpiece distortion and warping are minimal, quenching equipment is eliminated, and the process is easily controlled and stable.

MULTI-AXIS, WRIST FOCUSING, UNIVERSAL LASER BEAM HEAD

LASER BEAM MACHINING (LBM)

A high-intensity, high-temperature, controlled, and directed laser light energy beam produces the temperature needed to produce a narrow, geometrically straight *kerf.* Accurate cutting takes place without workpiece distortion and in a wide range of difficult-to-cut ferrous and nonferrous metals and other materials. Cutting forces, workpiece fixtures, and tooling setups are minimal.

In *CO_2 laser beam machining,* a high-energy laser beam is focused on a starting *interaction point.* The intense heat produces a cavity that is extended to penetrate a workpiece. A high-velocity inert gas (like argon or helium) is used to blow the vaporized particles clear of the kerf and workpiece and to speed the removal of oxides from the cut surfaces.

A laser cutting center containing a multi-axis movement directional laser head (as shown in the photo) is CNC programmed to machine a contoured (sculptured) surface, to cut holes and slots, and for other machining processes.

Courtesy of TRUMPF INDUSTRIAL LASERS INC.

UNIT 49
Computer Numerical Control: Systems and Drawings

Up to this point, principles and applications of numerical control to machines, setups, and processes have related to manual programming. These same functions, plus time-consuming calculations of the programmer in preparing and editing the manuscript and other process/product/machine tool information, may be computer programmed. *Computer-assisted programs (CAP)* may be used to provide for automatic, multiple-axis, fixed machining cycles, buffer storage, programming slide directions, establishing linear and circular coordinate data, and other functions. Computer-assisted programs may be written in many different processor languages.

COMPUTER-ASSISTED PROGRAMMING LANGUAGES

Symbols and precise meaning English words are employed in CAP. The computer has computational capacity to generate numerical part program data in a form which may be fed through the MCU to the machine.

Computers for CNC use *buffer storage* in the memory of the NC system. When required, output signals from buffer storage are generated from the advance-stored information. The output signals provide the commands for the control of a machine tool. Computer programming uses two *binary digits* to control the operation of electronic components. The digits relate to on or off, charge or discharge, positive or negative, and conductor or nonconductor.

All circuitry controlling the processing and executing of instructions is contained in a *central processing unit (CPU)*. The circuits deal with *basic memory* and *storage, retrieval*, and *logic*. Punched tape, magnetic tape, diskettes, and signals from other computers provide *input information*.

COMPUTER-ASSISTED PART PROGRAMMING

Programming begins with a sketch or conventional drawing. The part is then viewed in relation to axes, the nature and paths of the cutting tools, and the characteristics of the NC machine tool.

The *Automatically Programmed Tools (APT)* system is a very widely used processor language. This system is described by studying a part which requires straight, taper, and radius turning processes. A conventional part drawing is shown in figure 49-1. The **Z** dimensions (for longitudinal carriage movement) are given as *incremental coordinates*. The programmer prepares a programming layout (figure 49-2). The drawing provides the following information:

- Each turning (machining) process (**L1** through **L4**)
- Radius turning information (**C1**)
- Length of each cut (**P1** through **P4**)
- Starting point and nose radius of the cutting tool
- Reference point or plane
- **Z** (longitudinal travel) and **X** (transverse) movement of the machine cross slide
- Continuous path of the cutting tool

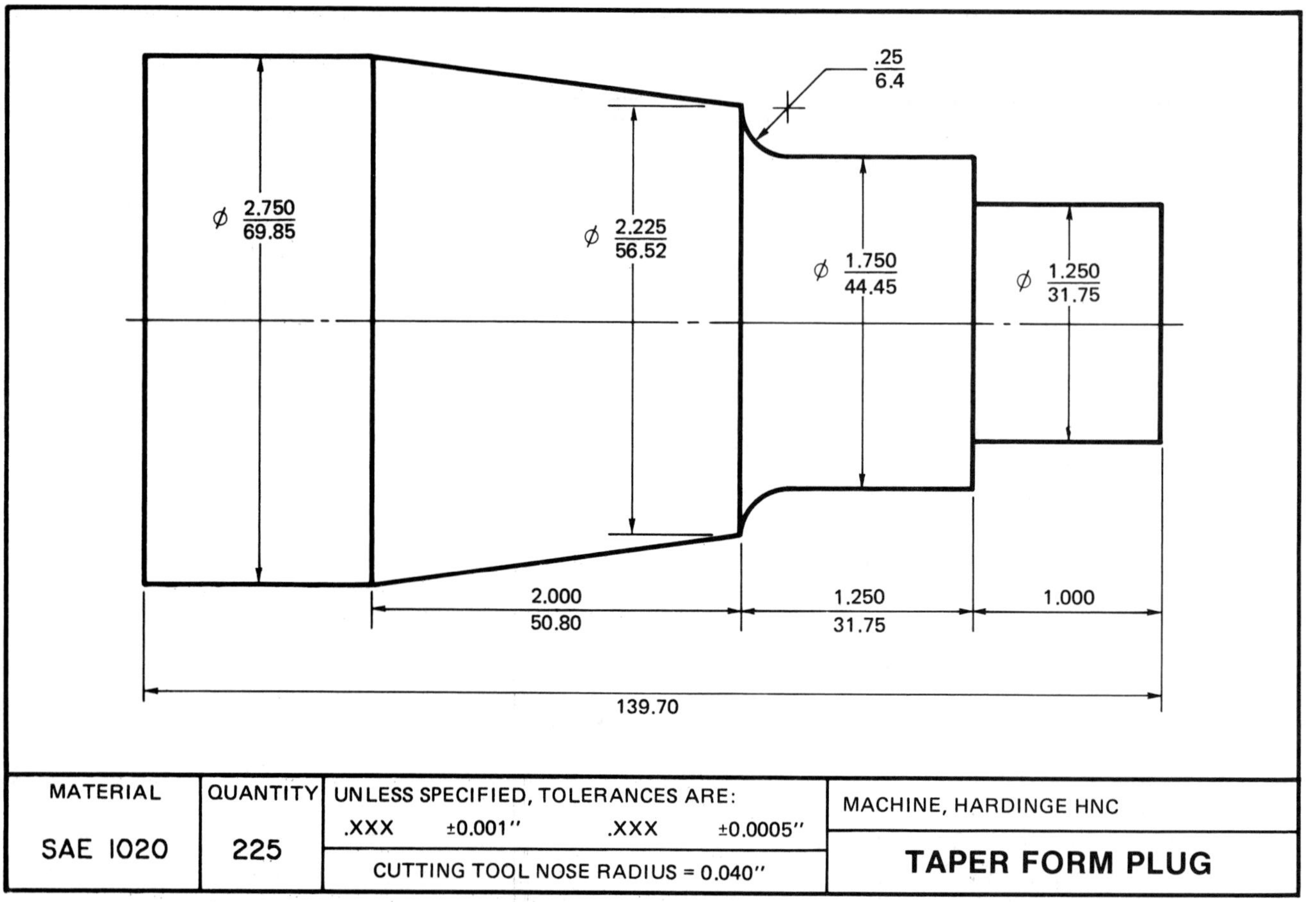

.25
6.4
Ø 2.750
69.85
Ø 2.225
56.52
Ø 1.750
44.45
Ø 1.250
31.75
2.000
50.80
1.250
31.75
1.000
139.70
MATERIAL
SAE 1020
QUANTITY
225
UNLESS SPECIFIED, TOLERANCES ARE:
.XXX ±0.001″ .XXX ±0.0005″
CUTTING TOOL NOSE RADIUS = 0.040″
MACHINE, HARDINGE HNC
TAPER FORM PLUG

FIGURE 49-1 CONVENTIONAL ONE-VIEW DRAWING OF PART TO BE COMPUTER PROGRAMMED FOR NC MACHINING

(+X)
P5
P4
P3
R 0.25
P2
P1
(–Z)
(+Z)
1.500
L1
C1
L2
L4
L3
CONTINUOUS PATH
OF CUTTING TOOL
.040 NOSE RADIUS OF
CUTTING TOOL
SP
5.750
(–X)

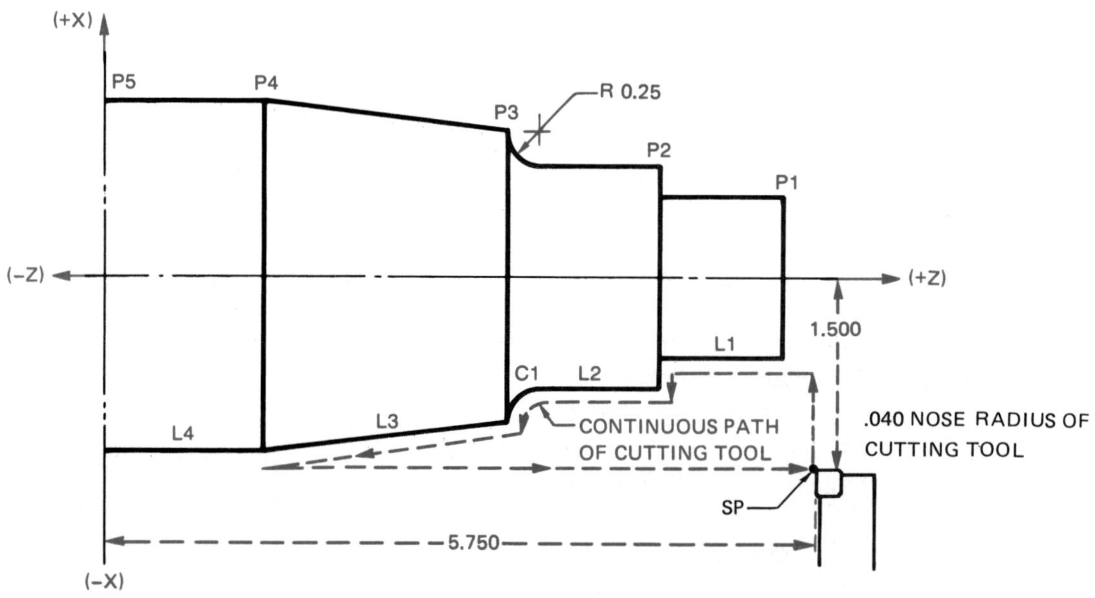

FIGURE 49-2 PROGRAMMING LAYOUT DRAWING OF CONTOURING OPERATIONS

① POSTPROCESSOR AND AUXILIARY STATEMENTS	② GEOMETRY STATEMENTS	③ MOTION STATEMENTS	④ AUXILIARY STATEMENTS
PART NØ TAPER FORM PLUG	SP = PØINT/5.750, – 1.500	FRØM SP	CØØLNT ØFF
MACHIN/HARDINGE/HNC	P1 = PØINT/5.500, – .612	GØ/TØ,L1	FINI
INTØL/.001	P2 = PØINT/4.500, – .875	GØLFT/L1,TØ,L2	
ØUTØL/.002	P3 = PØINT/3.250, – 1.125	GØRGT/L2,TANTØ,C1	
CUTTER/.080	P4 = PØINT/1.250, – 1.375	GØFWD/C1,PAST,L3	
CØØLNT/ØN	P5 = PØINT/0, – 1.375	GØLFT/L3,PAST L4	
CLPRNT	L1 = LINE/P1,P2	GØTØSP	
FEDRAT/3,IPM	L2 = LINE/P2,RIGHT,TANTØ,C1		
SPINDL/500,RPM	C1 = CIRCLE/1.250, – 1.112,.25		
	L3 = LINE/P3,P4		
	L4 = LINE/P4,P5		

FIGURE 49-3 PREPARATION OF COMPUTER-ASSISTED PROGRAM IN APT PROCESSOR LANGUAGE

PREPARING THE PROGRAM MANUSCRIPT

The programmer next determines the sequence of each APT word in order to prepare *statements* for the computer-assisted program. Four APT types of statements are used in the following sequence:

- *Postprocessor statements* to control the machine tool and cutting tool positions, machining setups, and machining processes. Additional computations may be made by a postprocessor computer program to produce a control tape or other input information.
- *Geometry statements* to provide information about part features.
- *Motion statements* to describe positions of cutting tools.
- *Auxiliary statements* to furnish additional information which is not provided in other statements.

An *APT system* dictionary is used to establish the correct spelling of the word for each process. The computer accepts only specific allowable words for programming.

For example, in figure 49-2, the cutting tool radius is **0.040″**. The cutter diameter is **0.080″**. The APT word **CUTTER**, followed by the nose diameter, provides the statement **CUTTER/.080**. An APT statement like **ØUTØL/.002** means that the computer is to stay within **0.002″** on the outside of curves. The APT program manuscript for producing the machined part in the example (figure 49-1) is shown in figure 49-3.

COMPUTERIZED COMPENSATION FEATURE (CDC) AND ELECTRONIC GAGING

Computer-assisted machining systems are expanded by the storage of *compensation information*. Compensation features on NC machines permit command adjustments for cutter diameter and tool length variations and workpiece irregularities. For example, a cutter diameter computerized compensation (CDC) feature permits the NC program to be varied in increments of 0.0001″ (0.002 mm) to compensate for undersize or oversize cutter diameters.

Dial indicator fixtures and electronic tool gages are used for off-line gaging of lengths and diameters. The electronic tool gage provides a visual digital readout and also input into the storage of length and diameter, and other tool data in the memory of the CNC system.

COMPUTER LANGUAGE FLEXIBILITY

CNC machine tools, instruments, and other measurement devices are designed for programming using English or metric decimal point measurements and direct feeds per minute or revolution. Editing can also be done in any block of data stored in the control's memory.

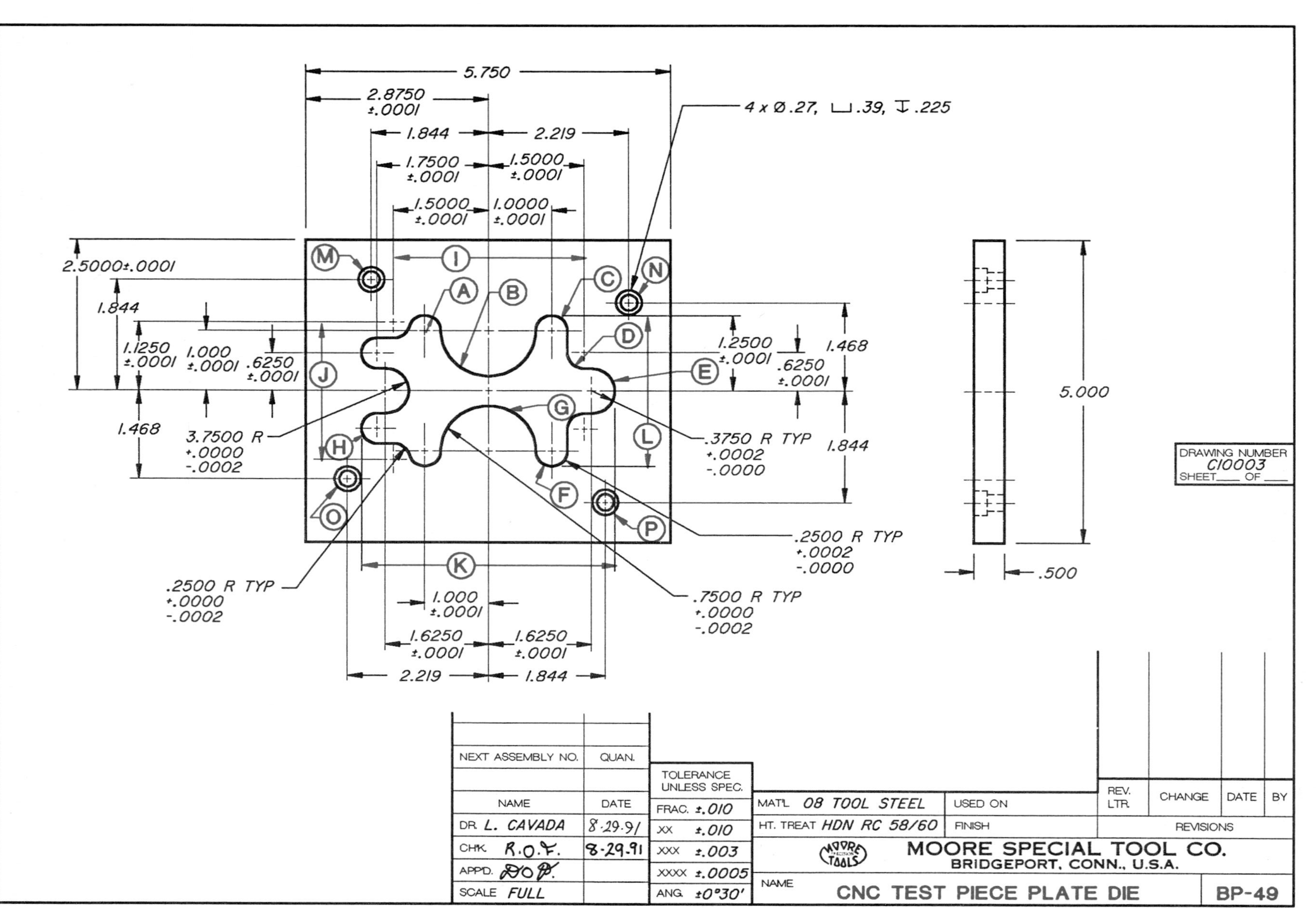

Courtesy of Moore Special Tool Company (Modified Industrial Drawing)

ASSIGNMENT—UNIT 49

CNC TEST PIECE PLATE DIE (BP-49)

Student's Name ______________________

1. Cite two functions of computers which are applied in computer-assisted programming.
2. Identify three machine, setup, or tool processes which may be programmed for a computer numerical control system.
3. Describe briefly what purpose is served by each of the following computer programming items:
   a. Conventional part drawing
   b. Programming layout drawing
   c. Manuscript preparation
4. Give the sequence in which automatically programmed tools (APT) statements are written into a program.
5. Refer to an APT dictionary and interpret the following words and values:
   a. **INTØL/.002**
   b. **FEDRAT/2IPM**
   c. **TANTØ, A2**
6. Explain briefly what function is served by an electronic tool gage in a CNC system.
7. Refer to the computer numerical control drawing of the Plate Die.
   a. Identify the required material.
   b. Give the hardness testing system, scale, and range.
   c. State the unspecified tolerances for three- and four-place decimal dimensions.
8. Give the specifications for the counterbored holes.
9. Determine the maximum and minimum radii of circular features (A), (B), and (D).
10. Give the maximum and minimum center-to-center distances along the **X** axis for counterbored holes (M) and (N) and (O) and (P).
11. Establish the nominal table positioning movements along **X** and **Y** axes for the following machined surfaces: (A), (C), (E), (F), (G), and (H)
12. Compute the maximum and minimum overall center-to-center dimensions for features (I) and (J).
13. Calculate the nominal overall dimensions (K) and (L).
14. Determine the tool positioning movements along the **X** and **Y** axes for counterbored holes (M) and (O).

1. a. ______________________
   ____
   b. ______________________
   ____
2. a. ______________________
   b. ______________________
   c. ______________________
3. a. ______________________
   ____
   b. ______________________
   ____
   c. ______________________
   ____
4. a. ______________________
   b. ______________________
   c. ______________________
   d. ______________________
5. a. ______________________
   b. ______________________
   c. ______________________
6. ______________________
   ____
   ____
7. a. ______________________
   b. ______________________
   c. .XXX = ________ .XXXX = ________
8. ______________________
9.

	Maximum Radius	Minimum Radius
(A)	________	________
(B)	________	________
(D)	________	________

10.

	Maximum	Minimum
(M) and (N)	________	________
(O) and (P)	________	________

11.

	X Axis	Y Axis
(A)	________	________
(C)	________	________
(E)	________	________
(F)	________	________
(G)	________	________
(H)	________	________

12.

	Maximum	Minimum
(I)	________	________
(J)	________	________

13. (K) ________ (L) ________

14.

	X Axis	Y Axis
(M)	________	________
(O)	________	________

UNIT 50
Computer-aided Design (CAD) and Computer-Aided Manufacturing (CAM) Systems

The world's first electronic computer, which provided control input into a single machine tool for selected basic processes, ushered in a second industrial revolution in 1945. The capacities of computers and the range of applications have spread from the basic machine tools to complex machining centers within industry. Important spin-offs have expanded computer-aided applications to agriculture, business, commerce, health, and all other sectors of the economy.

The two preceding units treated the output of computers in relation to numerically controlled (NC) and computer numerically controlled (CNC) machine tools. At first, applications were made to a single piece of equipment like a drilling machine. More sophisticated applications were made later of computer input and controls to combinations of machine tools and feeder systems within CNC machining centers.

COMPLETE COMPUTER-AIDED DESIGN/COMPUTER-AIDED MANUFACTURING SYSTEMS

The term *complete manufacturing* is inclusive of preliminary marketing analyses, management collection data, product design, production scheduling, inspection and quality control, selective assembly, inventory and materials control, and business administration functions.

Significant breakthroughs in complete manufacturing are now possible by combining CAD and CAM systems. Computer-aided design software systems involve a *hierarchical order* of computers. This means that computers are designed and are capable of interlocking the functions and products of what formerly were single computers or a simple family into complete entities.

For example, the manufacture of a part or component depends on the quantity needed for sale or use within a particular market. In tooling up for production, individual processes are computer programmed for a machine tool or a machining center. The computer provides control input to the machine control unit (MCU) concerning cutting tools, speeds and feeds, dimensional and surface finish tolerances, etc. Manufacturing is also dependent on inspection and quality control, distribution, inventory, and other controls. A CAD system contains all the entities which cut across each of these separate functions.

HIERARCHICAL LEVELS OF COMPUTERS

CAD provides the capacity beyond information development and output to evaluate and to accurately control action from information received from one source and to transmit it to an affected activity when required. A CAD system may be considered as a system of *minicomputers* (with limited input) which control basic movements and processes. These are interlocked with *intermediary computers* and, ultimately, a *mainframe computer* for a completely computer-aided manufacturing program. Figure 50-1 illustrates the concept of a completely interlocked CAD/CAM program. Note under *Production* that part of the CAD system

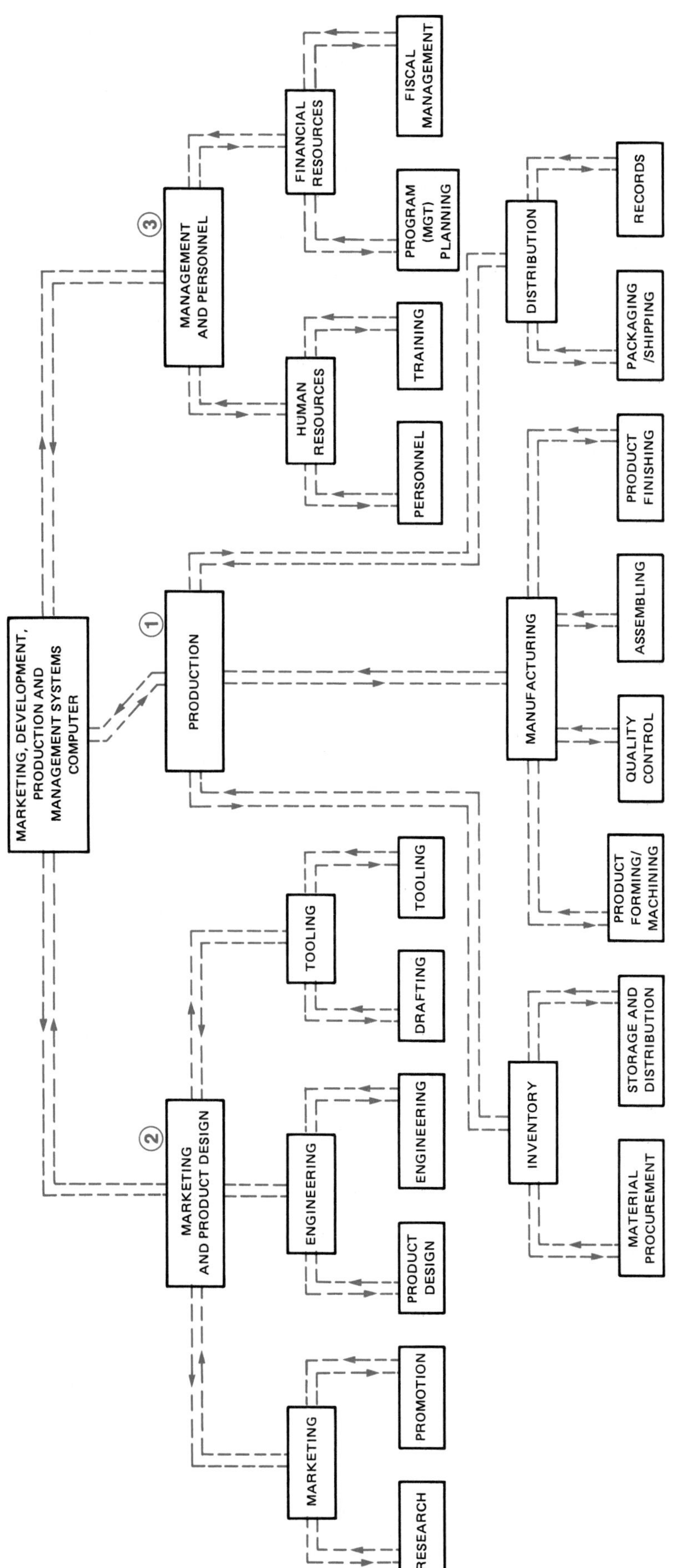

FIGURE 50-1 CONCEPT OF HIERARCHIES OF COMPUTERS IN A COMPLETELY COMPUTER-AIDED (PRODUCT) DESIGN (CAD), COMPUTER-AIDED MANUFACTURING SYSTEM (CAM)

embraces functions in *Manufacturing* such as machining or product forming, inspection (quality control), assembling, and finishing. Each of these functions is computer controlled through intermediate computer capability. Other functions under *Production* relate to inventory and distribution. Similarly, engineering and drafting and tooling are an integral part of *Product Design*. In summary, the hierarchies of computers are identified by the three computer banks for (1) production, (2) marketing and product design, and (3) management.

SELECTED COMPONENTS OF A COMPLETE CAD/CAM SYSTEM

In the overall computer-aided design system as illustrated, the CAD system facilitates information and controls from one area within the system to another. The CAD system completely interlocks with every function and process in CAM.

Product Design

The computer serves to store, recall, compute, manipulate, and process design information. In a CAD system, it is recognized that the same mathematical data base for product design engineering may be applied to numerically controlled (NC) manufacturing and to inspection, quality control, and assembly.

Interactive Graphics Terminal

The work of the product designer, programmer, and planner may now be visually displayed on a cathode ray tube. New data may be entered and referred to the computer for processing by an *interactive graphics terminal.*

Machine Operation Compensation

Computer data may be used within a CAD/CAM system to automatically compensate for any out-of-tolerance movement of the machine tool and for tool wear. In other words, all machine movements and cutting tool actions are instantly controlled and maintained.

Production Control and Scheduling

Selected, current data collection about events at any machining or processing station affect production scheduling and control. As this information is fed to the computer, the quantity of materials and parts is controlled and made available as a programmed computer calls for more material.

Inspection and Quality Control

Printouts or visual displays of parts which are inspected by computers are related to an automatic gaging system. The computer may be programmed to produce an output signal whenever there is a shift in a dimensional tolerance. Tooling or machine changes can be programmed and made automatically for any out-of-tolerance deviation. The computer may also be programmed for selective dimensional and surface texture inspection and control. A production process may be stopped immediately and corrected when a tolerance is exceeded.

Selective Mating Components

The computer may be used as an automatic assembly device to control the measurements, specifications, and selection of mating components from a production lot. Selection from many mass-produced parts with wide tolerance variations often results in lower manufacturing costs, as parts held to finer tolerance limits are more expensive to produce.

Materials and Just-in-time (JIT) Inventory Control

Accurate inventory, lead time in purchasing materials in relation to need, and the efficient movement of materials and parts are important factors in reducing overhead costs. Computer Just-in-time (JIT) inventory and materials control programs are usually provided by microcomputer inputs within a CAD/CAM system.

MAINFRAME COMPUTERS, MINICOMPUTERS, AND MICROPROCESSORS

The hardware for a computer-aided design (CAD) system for a complete CAM system may include one or more mainframe computers, minicomputers, and microprocessors.

Mainframe Computers

This type of computer is a composite computer system with central memory or logic and all required input and output devices. A *mainframe computer* accepts information, partitions its memory or logic, processes the data, and distributes it to one or more input or output devices simultaneously.

A mainframe computer in a total CAM system may serve any number of functions in production, manufacturing, management, etc., at the same time.

Minicomputers

A *minicomputer* has central logic (or memory) and specific input and output devices. These function in a manner similar to a mainframe computer. However, instead of working simultaneously with input and output devices, the minicomputer is slower due to the fact that it switches back and forth between devices.

Microprocessors

A *microprocessor* is the logic element of a computer with a permanent input and output connection. For example, microprocessors are used in programming terminals and machine control units (MCU). A microprocessor provides computational capability to other equipment.

WORDS AND BITS

Each *word* used in the three types of computers just introduced is composed of nine *binary bits* (figure 50-2). Each bit has a positive or negative polarity. The combination of the binary notations within *bits* in a word establishes the numeric, alphabetic, or symbol (comma, slash, period, etc.) value.

A mainframe computer may have a word capacity ranging up to the millions. By comparison, a mini-computer range may extend to 4,000 words; a microprocessor, from 1,000 to 4,000 words.

COMPUTER-AIDED DRAFTING (GRAPHICS)

Two-Dimensional Engineering Drawings

Many engineering drawings are generated by automated drafting machines which are controlled by computer-aided input. Drawings may thus be prepared which combine features of detail, section, assembly, installation, and other types of drawings. Each engineering drawing must still provide complete information about materials, design features, dimensioning, tolerances, and other part or component specifications.

A CAD system may be used to generate drawings and other programming information. This information may be fed manually to the MCU to control the numerically controlled equipment. Also, the information may be stored in memory and later fed electronically to the MCU.

Drafting machines convert computer command words from a MCU into movements of a *carriage* which moves over a plane or cylindrical drum-type surface and a *drawing head* or *plotter device* for producing a finished engineering drawing.

Three-Dimensional Pictorial Drawings

Computerized drafting machines are capable of producing isometric, dimetric, and trimetric drawings from orthographic drawings. Machine capability is extended to include the drawing of one-, two-, or three-point perspective drawings by adding a *perspective module* to the equipment.

Computer-aided drafting and design (CADD) drawings depend on **X**, **Y**, and **Z** axes and the input of three-dimensional and other mathematical data to provide a computer graphics system. Once a three-dimensional drawing is produced through CAD, it is possible to program different sections, surfaces, and features, to rotate them to produce a better view, and to cut away and observe internal details.

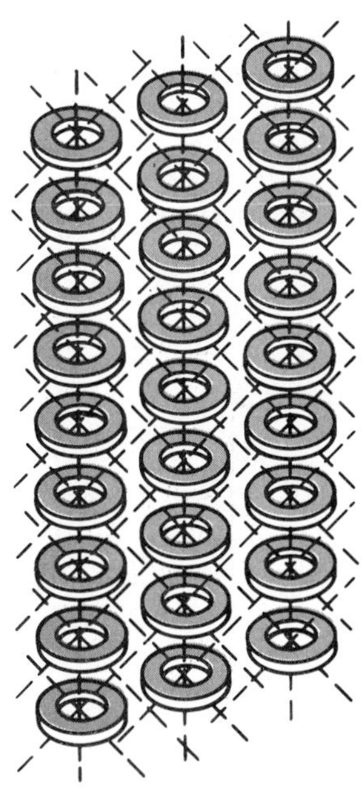

FIGURE 50-2 CONCEPT OF BINARY BITS CAPABLE OF HOLDING (EXISTING IN) A PLUS OR MINUS ELECTRICAL POLARITY

WIRE-FORM AND SOLID-FORM PICTORIAL DRAWINGS

Pictorial line drawings may be generated in *wire form* or in *solid form.* A wire-form drawing consists of a series of lines (resembling *wires*) that describe external and internal features of a part or mechanism. A wire-form drawing permits all edges to be viewed at the same time. Wire-form drawings are not functional for complex parts as the great number of lines that are generated for the drawing are confusing to read.

A solid-form pictorial drawing is preferred for parts that have complicated details. A solid-form drawing shows the surfaces as they are actually seen. CADD systems make it possible for design, programming, manufacturing, and other personnel to cut into an object at any position and to rotate the view or section in order to generate the best possible drawing of particular part features.

Figure 50-3 provides an example of a CAD-generated three-dimensional solid model at (A) and a three-dimensional cutaway surface model at (B). The cross-sectioned area exposes internal features of the part. In CAD, drawings are displayed graphically on the video screen for checking or modification. Once an accurate drawing is completed, a black and white or multicolor print may be produced by *pen* or *electrostatic plotters, laser,* or other form of *printer.*

CADD/CIM SYSTEMS

CADD input of original drawings and mathematical and other input are used in computer-aided manufacturing to produce a nondimensional drawing prior to generating a *tool path display* for machining a part. The tool path display drawing on the computer screen permits the operator to determine whether the programmed machining will produce the required part.

Courtesy of HOUSTON INSTRUMENT DIVISION, AMETEK, INC.

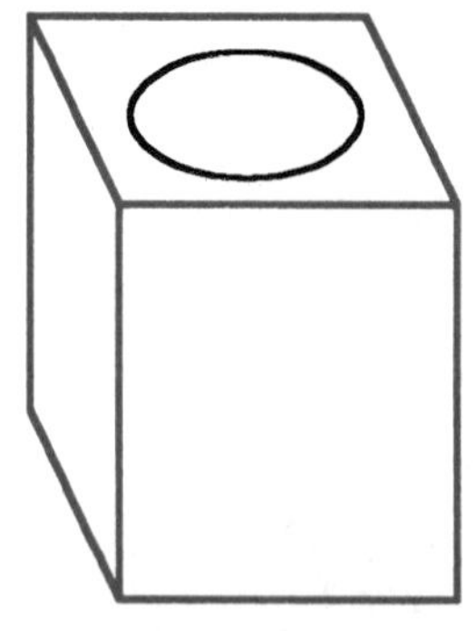

(A) THREE-DIMENSIONAL SOLID MODEL

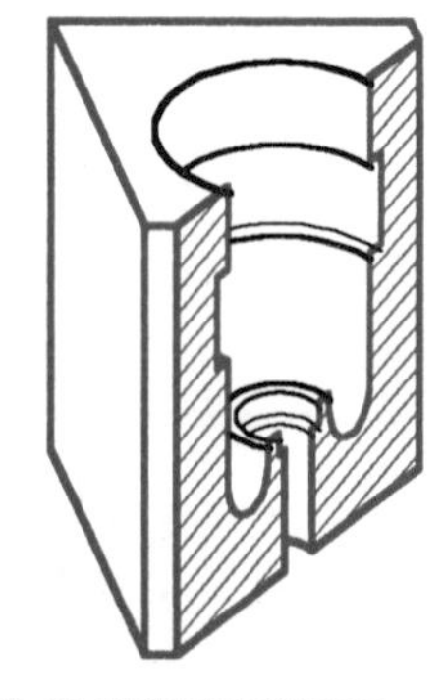

(B) CUT-AWAY THREE-DIMENSIONAL SURFACE SHOWING INTERNAL DETAILS

FIGURE 50-3 CAD-GENERATED THREE-DIMENSIONAL SOLID AND CUT-AWAY MODELS

Courtesy of MCDONNELL DOUGLAS MFG. AND ENGINEERING SYSTEMS CO.

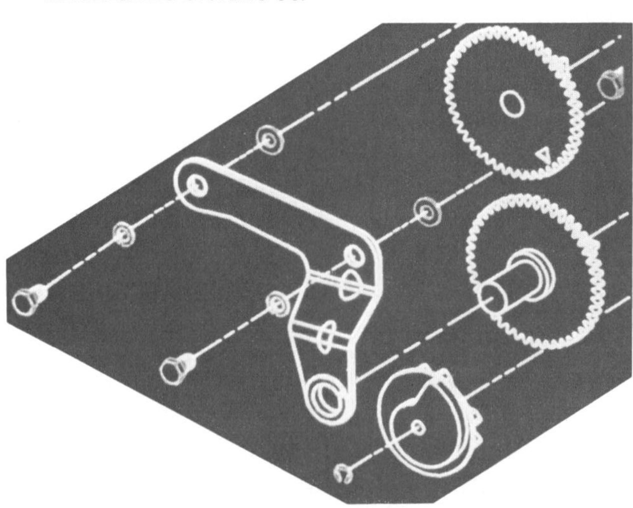

FIGURE 50-4 CAD THREE-DIMENSIONAL EXPLODED ASSEMBLY DRAWING GENERATED FROM SEPARATELY DRAWN PARTS

The drawing on the screen (together with dimensioning and other computer-generated data) may be sent in CIM directly to the numerical control machine tool without the need to generate hard copy drawings.

Another general application of CAD to pictorial drawings is illustrated in figure 50-4. An exploded three-dimensional assembly drawing may be generated from the separately prepared parts drawings.

COMPLEMENTARY PICTORIAL DRAWINGS

Increasingly, greater use is being made of CAD for the preparation of pictorial three-dimensional drawings. These drawings complement standard view working drawings by displaying how a part or mechanism actually looks (figure 50-5).

COMPUTER GRAPHICS: GRAPHS AND MAPS

Computer-aided drafting machines are also designed to produce graphs, maps, and other charts.

Courtesy of MCDONNELL DOUGLAS UNIGRAPHICS

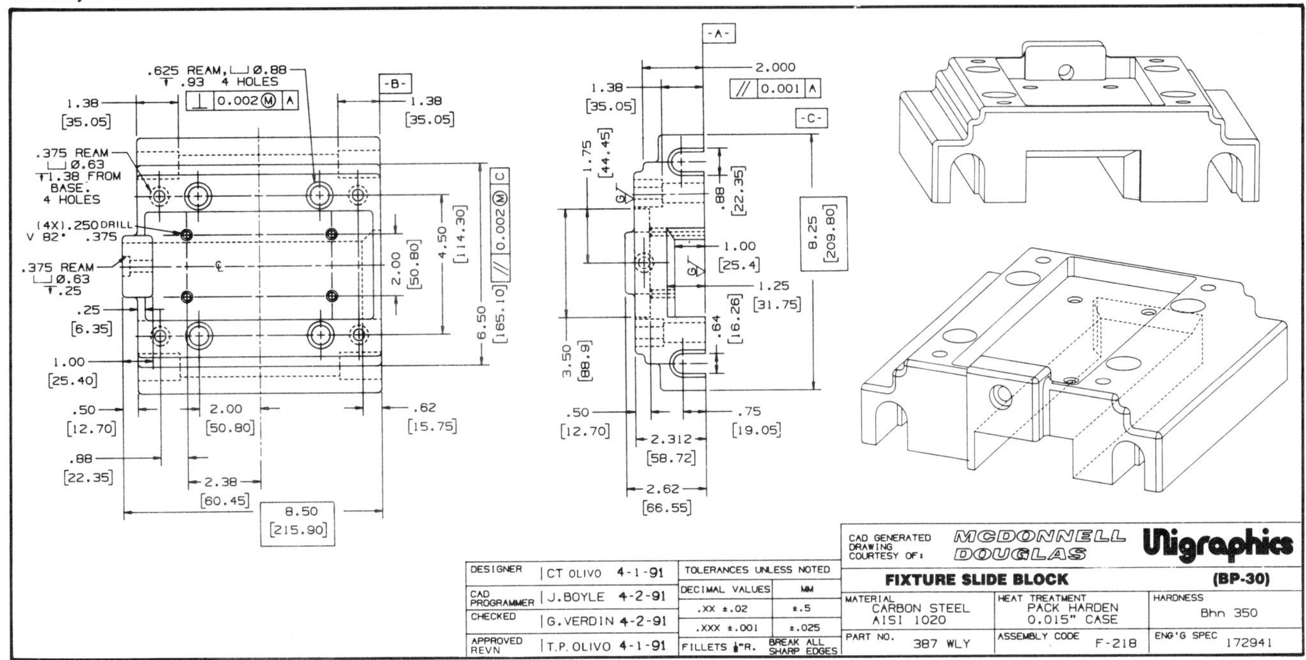

FIGURE 50-5 EXAMPLES OF CAD-GENERATED PICTORIAL DRAWINGS THAT COMPLEMENT STANDARD VIEWS

Graphing involves the drawing of the *grid* or *coordinate system*, *straight* and *curved profile lines*, and adding *calibrations*, *legends*, and other specifications. Mathematical computations may be computer programmed. Figure 50-6 is a simple example of a line graph produced by feeding computer-generated mathematical information and other stored data into the MCU to control the end product of the drafting machine.

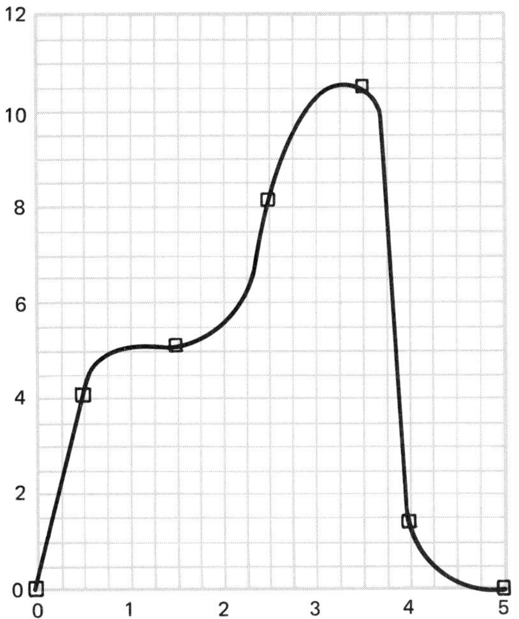

FIGURE 50-6 LINE GRAPH PRODUCED FROM CAD-GENERATED MATHEMATICAL INFORMATION

CAD-GENERATED DRAWINGS AND MACHINING COMMANDS

The drawing in the assignment which follows provides an application of CAD in design and in generating machine commands or NC tapes. This input is fed to the NC machine.

Four different levels of machining are listed on the selected CAD generated drawing as **20**, **25**, **30**, and **35**. The end mill (cutting tool) sizes are also given: **0.275″**, **0.525″**, **0.1875″**, **0.250″**, **0.094″**, and **0.190″**. The tool paths are shown on different CAD drawings for particular levels.

The CAD drawings permit "machining" the part on the CAD tube before it is actually machined. Commands for the drilling, reaming, tapping, and hole-forming processes were generated by other CAD processes. The Dust Cover is finish machined by cutting to length, thus separating the part from the excess stock which is needed for positioning and securing the workpiece for machining.

In this CAD application, form, size, position and location dimensions, geometric tolerancing, surface finish requirements, and machining processes are all derived from a working drawing of the **DUST COVER**.

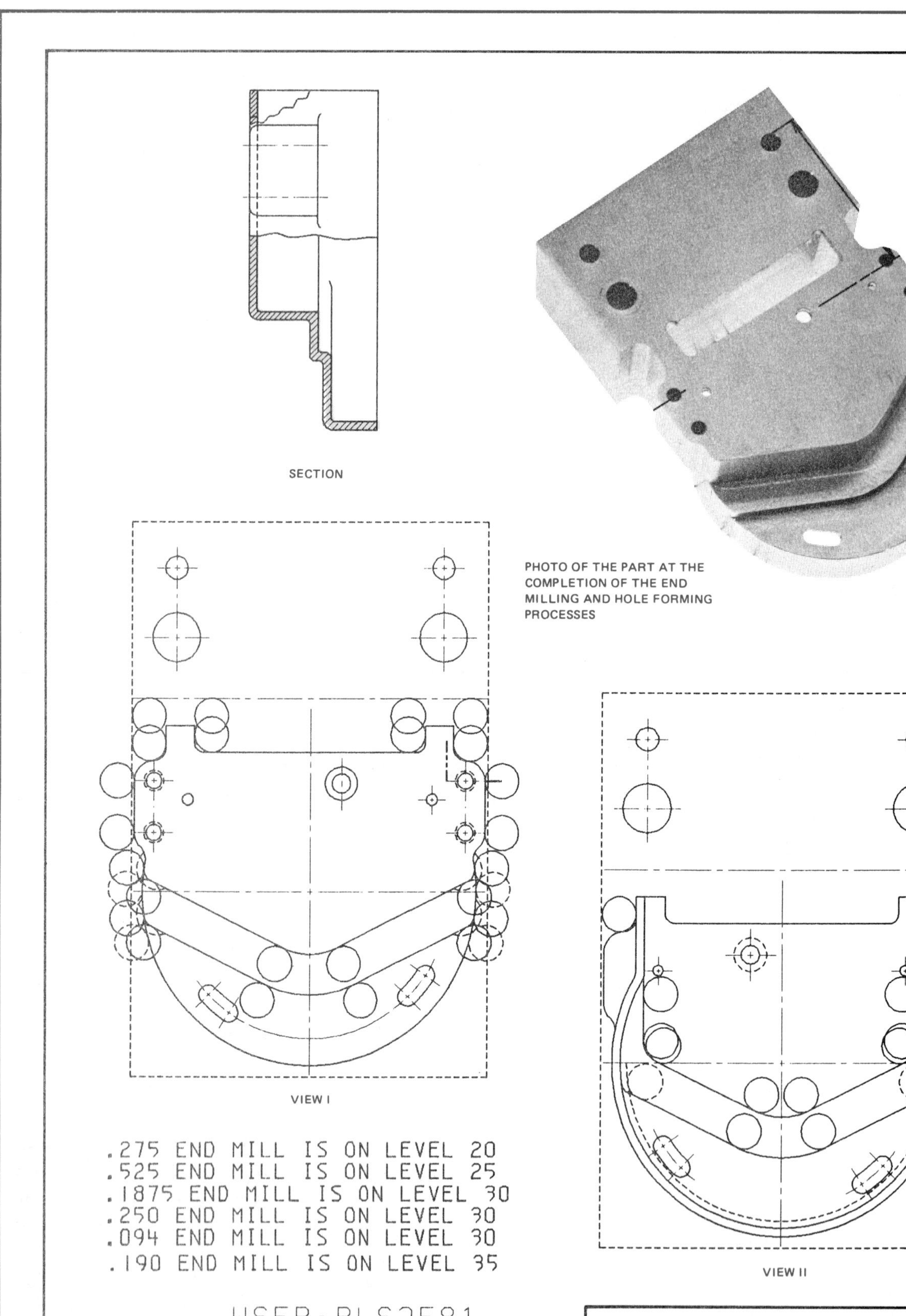

NOTE. THIS CAD GENERATED DRAWING REPRESENTS END MILLING AT ONLY THE 35 LEVEL

DUST COVER	BP-50

Courtesy of The Charles Stark Draper Laboratory, Inc. *Drawing Adaptation*

ASSIGNMENT—UNIT 50

DUST COVER (BP-50)

Student's Name ______________________

1. Describe briefly the function of each of the following systems or pieces of computer hardware:
   a. CAD system
   b. Machine control unit (MCU)
   c. Tool compensation mechanism
   d. Complete computer-aided manufacturing system.
2. Identify three major types of hardware which may be incorporated into a CAD system.
3. Cite one function of an interactive graphics terminal.
4. Give an example of a complete CAD/CAM system.
   a. Identify the product or part to be manufactured.
   b. Prepare a block diagram (similar to figure 50-1) to show the major categories of activities in the total program.
   c. Letter each category.
5. Cite one advantage of the selective mating of components which are mass produced to a wide tolerance.
6. List three different types of graphics products which are produced on an automated drafting machine. The machine is controlled by a computer in a CAD system.
7. Name three functions served by the CAD-generated drawing of the **Dust Cover (BP-50)**.
8. Name **VIEW I**, **VIEW II**, and the section view.
9. Identify the geometric form of the metal blank from which the part is to be machined.
10. Determine the number of holes used for positioning and securing the workpiece.
11. State the levels at which the different end mills are to be positioned for depth of cut.
12. Give the diameters of the cutting tools which are to be used to end mill the different contours on Level 30.
13. Identify the level on which end milling with the 0.525″ cutter takes place.

1. a. ______________________
   ____
   ____
   b. ______________________
   ____
   ____
   c. ______________________
   ____
   ____
   d. ______________________
   ____
   ____
2. a. ______________________
   b. ______________________
   c. ______________________
3. ______________________
   ____
4. ______________________
   ____
   a. ______________________
   b. Prepare separate block diagram.
   c. Letter categories on the block diagram.
5. ______________________
   ____
   ____
6. a. ______________________
   b. ______________________
   c. ______________________
7. a. ______________________
   b. ______________________
   c. ______________________
   ____
8. **VIEW I** ______________________
   VIEW II ______________________
   Section View ______________________
9. ______________________
10. ______________________
11. Levels ______________________
12. ______________________
13. ______________________

SECTION 14
Manufacturing and Occupational Drafting Systems

UNIT 51
Welding Symbols, Standards, and Drawings

Welding drawings are generally assembly drawings. Welding involves the fabricating of a unit from a number of parts. In addition to standard information about features, dimensions, processes, etc., welding drawings include the use of symbols and code letters. These are the shorthand language adopted by the American Welding Society (AWS) and ANSI as standards for accurately interpreting size, type, design, and other requirements of welded structures.

STANDARD WELDING PROCESSES AND CUTTING METHODS

There are six categories of *welding processes* and two for *cutting methods*. The processes and methods, together with the appropriate letter code are listed in figure 51-1.

WELDING PROCESSES				CUTTING PROCESSES	
GAS WELDING	**LETTER CODE**	**SOLID STATE WELDING**	**LETTER CODE**	**ARC CUTTING (AC)**	**LETTER CODE**
Oxyacetylene Welding	OAW	Ultrasonic Welding	USW		
Oxyhydrogen Welding	OHW	Forge Welding	FOW	Air Carbon–Arc	AAC
Pressure Gas Welding	PGW	Cold Welding	CW	Metal–Arc	MAC
ARC WELDING		**BRAZING**		Oxygen–Arc	AOC
Plasma–Arc Welding	PAW	Infrared Brazing	IRB		
Gas Tungsten–Arc Welding	GTAW	Furnace Brazing	FB		
Flux Cored Arc Welding	FCAW	Induction Brazing	IB		
Shielded Metal–Arc Welding	SMAW	Resistance Brazing	RB		
RESISTANCE WELDING		**OTHER PROCESSES**		Plasma–Arc	PAC
Resistance–Spot Welding	RSW	Thermit Welding	TW	Oxygen Cutting	OC
Flash Welding	FW	Laser Beam Welding	LBW	Chemical Flux	FOC
Percussion Welding	PEW	Electron Beam Welding	EBW	Metal Powder	POC

FIGURE 51-1 EXAMPLES OF WELDING AND CUTTING PROCESSES AND LETTER CODES (DESIGNATIONS) (COURTESY OF THE AMERICAN WELDING SOCIETY)

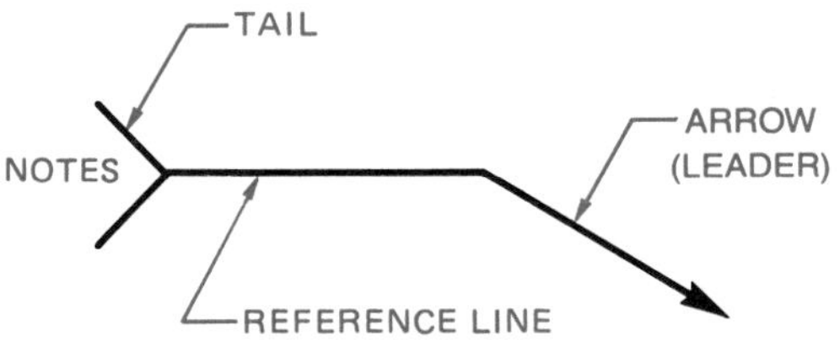

FIGURE 51-2 BASIC WELDING SYMBOL

STANDARD WELDING SYMBOL AND ELEMENTS

The three parts of a basic welding symbol include a *reference line*, *arrow*, and, where needed, a *tail* (figure 51-2). Other symbols, code letters, dimensions, and notes are added. Collectively, the symbol and notations convey design and all other information needed by the craftsperson or technician to fabricate the part.

AWS and ANSI standards for the location of the elements of a welding symbol are illustrated in figure 51-3. The function of each basic element of the symbol is identified by a number from ① through ⑮.

MULTIPLE REFERENCE LINES

Multiple welding processes are represented on drawings by the addition of reference lines with symbols and/or notes. Figure 51-4 shows the use of three reference lines to identify three operations.

BASIC WELDING JOINTS

Butt, *corner*, *T-*, *lap* and *edge joints* are the most widely used welding joints. Each type is pictured in figure 51-5. Note the location of the welding symbol arrow, *arrow side* of each joint, and the *other side*.

BASIC WELDING SYMBOLS

A symbol denoting the type of weld may appear on a drawing on the *arrow side*, *other side*, or *both sides*. For some welds, no arrow side or other side is used. Basic welding symbols are shown in figure 51-6 on the arrow side for illustrative purposes.

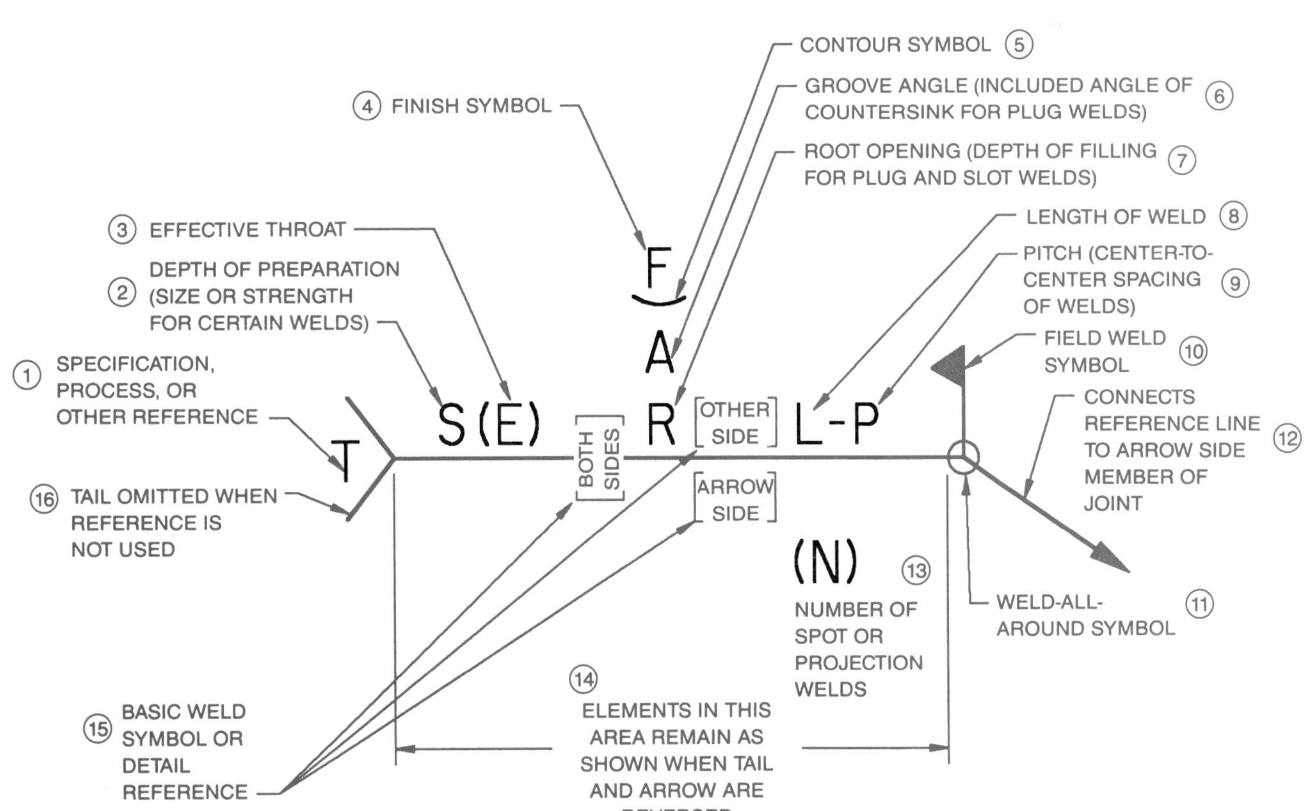

FIGURE 51-3 AWS/ANSI STANDARDS FOR THE LOCATION OF ELEMENTS OF A WELDING SYMBOL

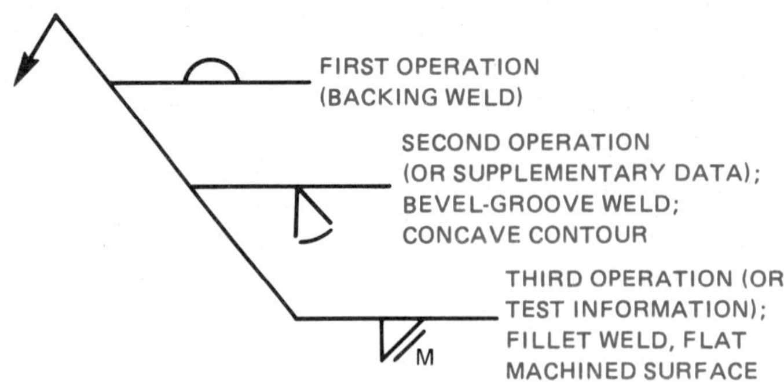

FIGURE 51-4 USE OF MULTIPLE REFERENCE LINES

SUPPLEMENTARY WELDING SYMBOLS USED WITH OTHER WELDING SYMBOLS

When a weld extends completely around a joint, the weld-all-around symbol ( ) is used.

If the weld is made at a place other than that of initial construction, it is known as a *field weld*. The field weld symbol is a flag ( ) attached to the reference line.

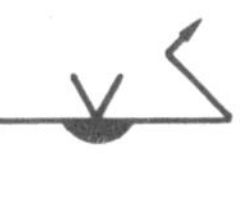

A *melt-through* symbol may be used with any applicable weld symbol. Melt-through symbols are dimensioned for height only.

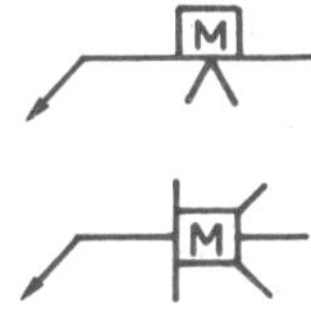

A *joint with backing symbol* is shown with a V groove; a *joint with spacer*, with a double-bevel groove. The materials and dimensions of backing are specified in the tail area.

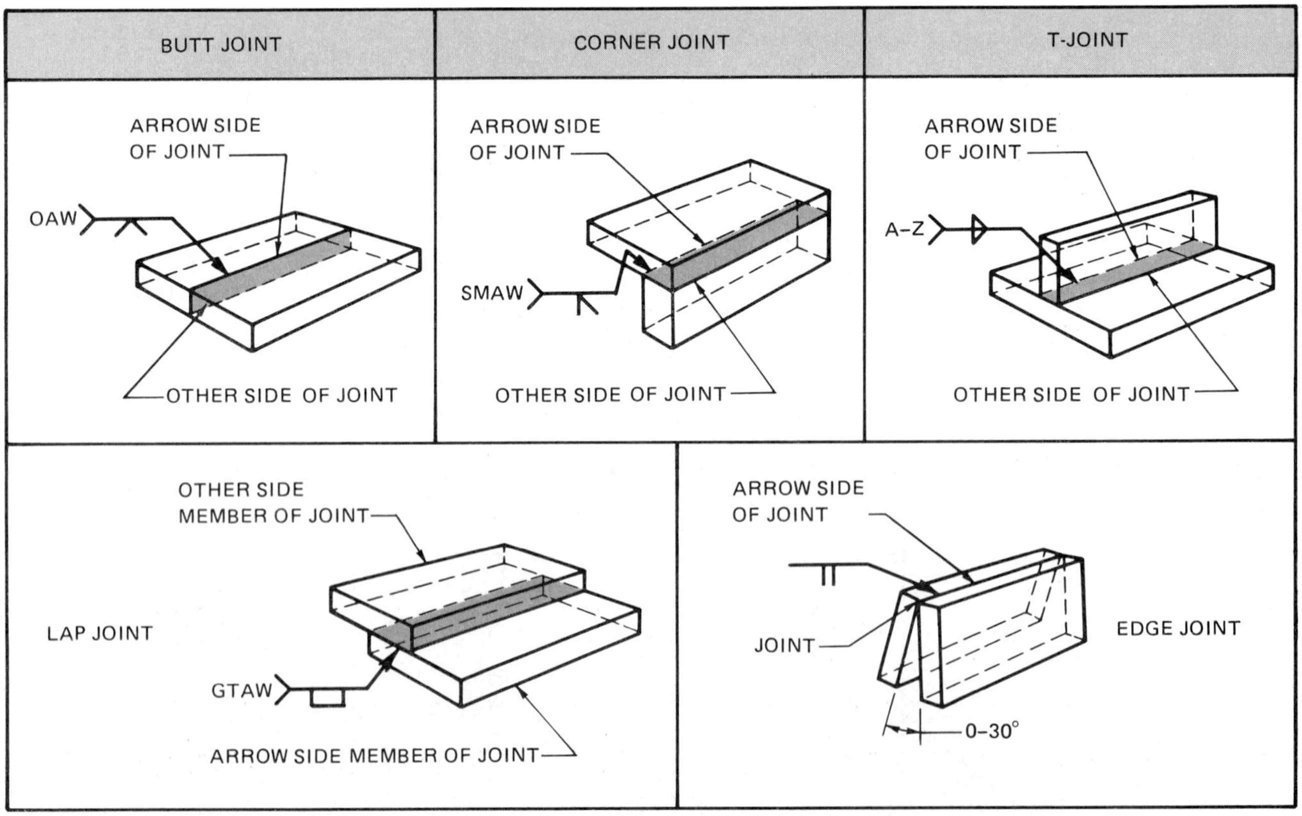

BUTT JOINT	CORNER JOINT	T-JOINT
ARROW SIDE OF JOINT; OAW; OTHER SIDE OF JOINT	ARROW SIDE OF JOINT; SMAW; OTHER SIDE OF JOINT	ARROW SIDE OF JOINT; A-Z; OTHER SIDE OF JOINT
LAP JOINT: OTHER SIDE MEMBER OF JOINT; GTAW; ARROW SIDE MEMBER OF JOINT	EDGE JOINT: ARROW SIDE OF JOINT; JOINT; 0–30°	

FIGURE 51-5 BASIC WELDING JOINTS WITH IDENTIFICATION OF ARROW SIDE AND OTHER SIDE AND EXAMPLES OF WELDING SYMBOL APPLICATIONS (COURTESY OF THE AMERICAN WELDING SOCIETY)

FILLET	PLUG OR SLOT	SPOT OR PROJECTION	SEAM	BACK OR BACKING	SURFACING	SCARF FOR BRAZED JOINT	FLANGE EDGE
				GROOVE WELD SYMBOL			

GROOVE							FLANGE
SQUARE	V	BEVEL	U	J	FLARE-V	FLARE-BEVEL	CORNER

FIGURE 51-6 BASIC WELDING SYMBOLS REFERENCED TO THE ARROW SIDE FOR LOCATION SIGNIFICANCE (COURTESY OF THE AMERICAN WELDING SOCIETY)

A *convex* ( ) or *concave* ( ) *contour symbol* indicates which weld face to finish to a convex (or concave) contour. The code letter indicates the method of producing a specified contour, but not the surface finish requirement.

M

A *flush contour* symbol (——) shows which face of the weld is to be made flush. If a finish symbol is used as illustrated, the letter symbol indicates the method of obtaining the required contour, but not the degree of finish.

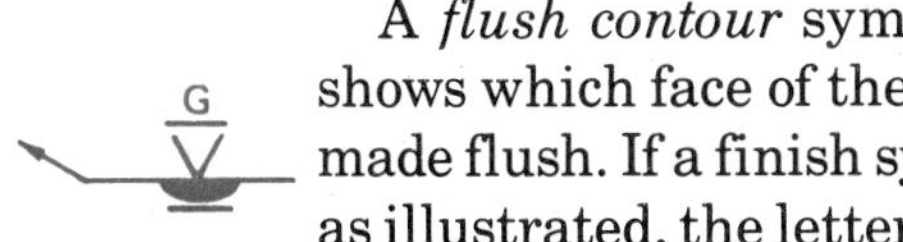

WELD DIMENSIONS FOR FILLET AND GROOVE WELDS

Dimensions such as length of weld leg, radius or size of weld, root opening, root face, depth of penetration, and length of weld, are generally included with symbols on a welding drawing. Some dimensions appear before the symbol. When both sides are to be welded to the same size, only one side is dimensioned. Both sides are dimensioned when different size welds are required.

The *weld length* on a fillet weld is placed after the symbol. For example, 6 50 denotes a 6 mm fillet weld which is 50 mm long. When center-to-center spacing is required for an intermittent weld, the *pitch* dimension follows the weld length of the *increments*. For instance, 6 50-100 indicates a pitch of 100 mm. If the intermittent fillet welds are *staggered*, so too are the weld symbols. For example, 6 50-100 / 6 40-90 .

The *root opening* dimension of a square, **V**-, bevel, **U**-, and flare groove is placed within the symbol. A groove angle (where applicable) appears below the groove opening size within the groove symbol $\frac{1}{8}$ 40°

No angle dimensioning is used if a general drawing note indicates that

ALL BEVEL GROOVE WELDS ARE FORMED TO A XX° ANGLE

The *radius*, *height*, and *thickness* of edge flange welds appear to the left of the weld symbol ($\frac{1}{8} + \frac{5}{32}$ $\frac{3}{32}$).

A weld *reinforcement height* is represented on a drawing by the symbol and height.

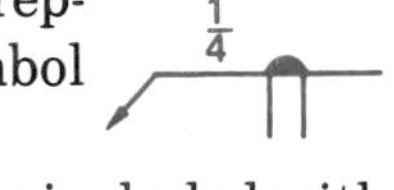

The radius of a flare **V** groove is included with the symbol ($\frac{3}{8}$).

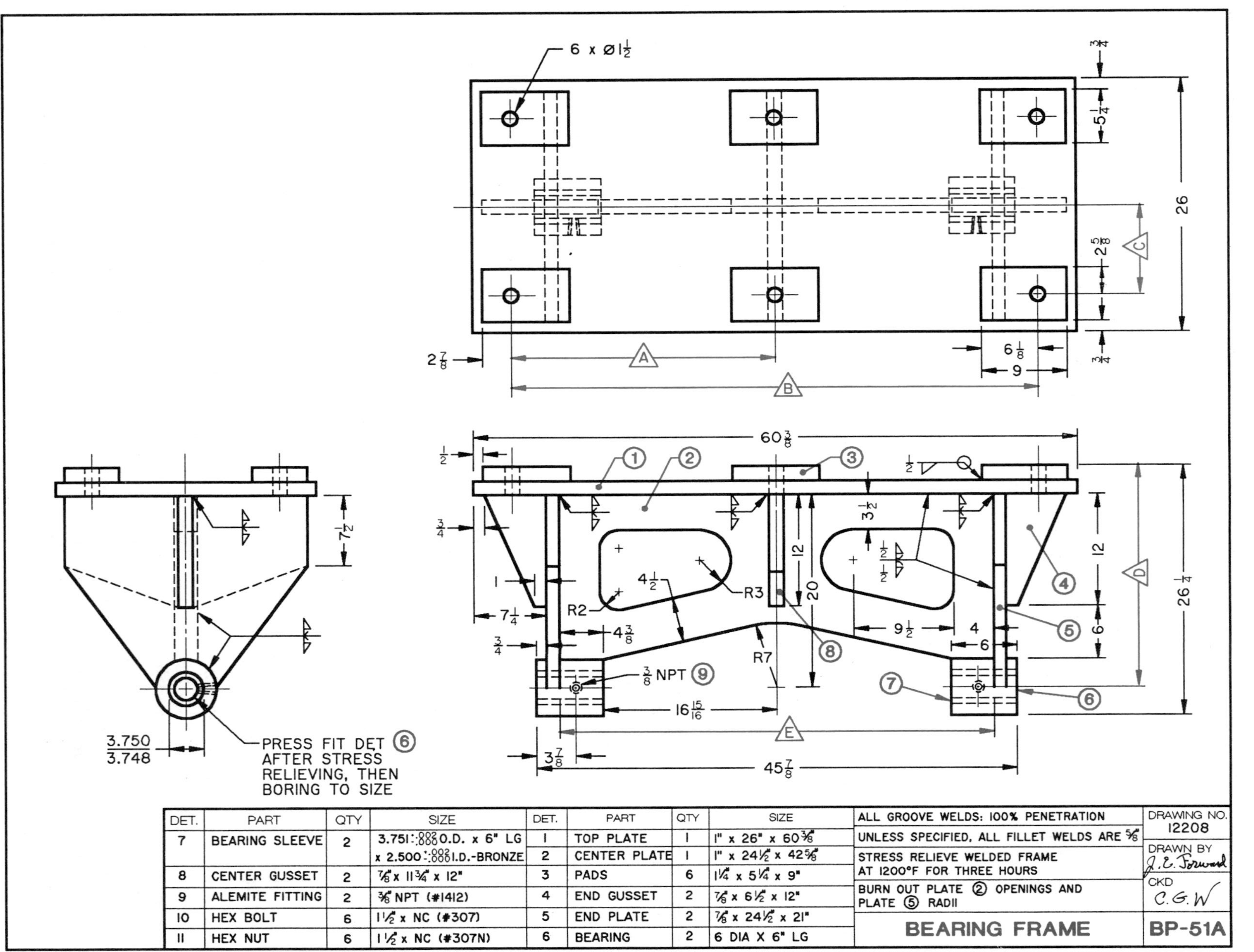
6 x Ø1 1/2
PRESS FIT DET ⑥ AFTER STRESS RELIEVING, THEN BORING TO SIZE
DET. | PART | QTY | SIZE
7 | BEARING SLEEVE | 2 | 3.751 O.D. x 6" LG x 2.500 I.D.-BRONZE
8 | CENTER GUSSET | 2 | 7/8 x 11 3/4 x 12"
9 | ALEMITE FITTING | 2 | 3/8 NPT (#1412)
10 | HEX BOLT | 6 | 1 1/2 x NC (#307)
11 | HEX NUT | 6 | 1 1/2 x NC (#307N)
1 | TOP PLATE | 1 | 1" x 26" x 60 3/8"
2 | CENTER PLATE | 1 | 1" x 24 1/2" x 42 5/8"
3 | PADS | 6 | 1 1/4 x 5 1/4 x 9"
4 | END GUSSET | 2 | 7/8 x 6 1/2 x 12"
5 | END PLATE | 2 | 7/8 x 24 1/2 x 21"
6 | BEARING | 2 | 6 DIA X 6" LG
ALL GROOVE WELDS: 100% PENETRATION
UNLESS SPECIFIED, ALL FILLET WELDS ARE 3/8
STRESS RELIEVE WELDED FRAME AT 1200°F FOR THREE HOURS
BURN OUT PLATE ② OPENINGS AND PLATE ⑤ RADII
BEARING FRAME
DRAWING NO. 12208
DRAWN BY J.E. Forward
CKD C.G.W
BP-51A

ASSIGNMENT A—UNIT 51

BEARING FRAME (BP-51A)

Student's Name ______________________

1. a. Identify a specific welding process for each of the following categories of welding:
   (1) Arc welding
   (2) Gas welding
   (3) Resistance welding
   b. Indicate the code letter used in the tail of the symbol for each process.
2. Identify the following supplementary symbols:

   a. b. c.

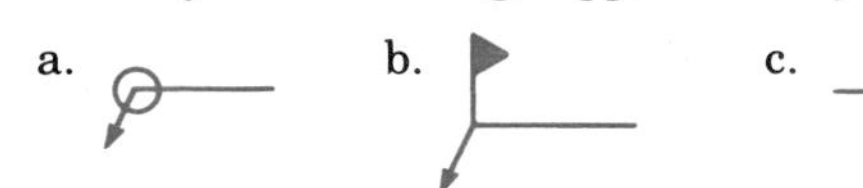

3. Identify the ten numbered elements of the welding symbol:

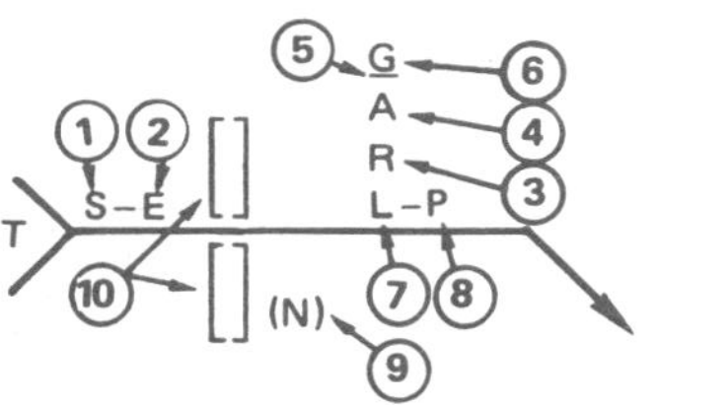

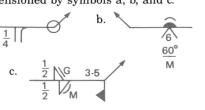

4. Give the welding specifications as represented and dimensioned by symbols a, b, and c.

   a. b. c.

5. List the sequence of welds specified by the accompanying welding symbol.
6. Determine the nominal length, width, and height of the Bearing Frame.
7. Compute center distances A, B, C, D, and E.
8. a. Sketch the welding symbol for joining the bearings to the end plates and the center plate.
   b. Explain the significance of the welding symbol elements.
   c. Give the general specification for all fillet welds.
9. a. Indicate the required percent penetration of all groove welds.
   b. State the condition for stress-relieving the weldment before machining the sleeve diameters.

1.

Welding Category	Welding Process (a)	Code Letter (b)
(1) Arc	______	______
(2) Gas	______	______
(3) Resistance	______	______

2. a. ______ c. ______
   b. ______

3. (1) ______ (6) ______
   (2) ______ (7) ______
   (3) ______ (8) ______
   (4) ______ (9) ______
   (5) ______ (10) ______

4. a. ______
   b. ______
   c. ______

5. (1) ______
   (2) ______
   (3) ______

6. Length ______ Width ______
   Height ______

7. A = ______ D = ______
   B = ______ E = ______
   C = ______

8. a.
   b. ______
   c. ______

9. a. ______
   b. ______

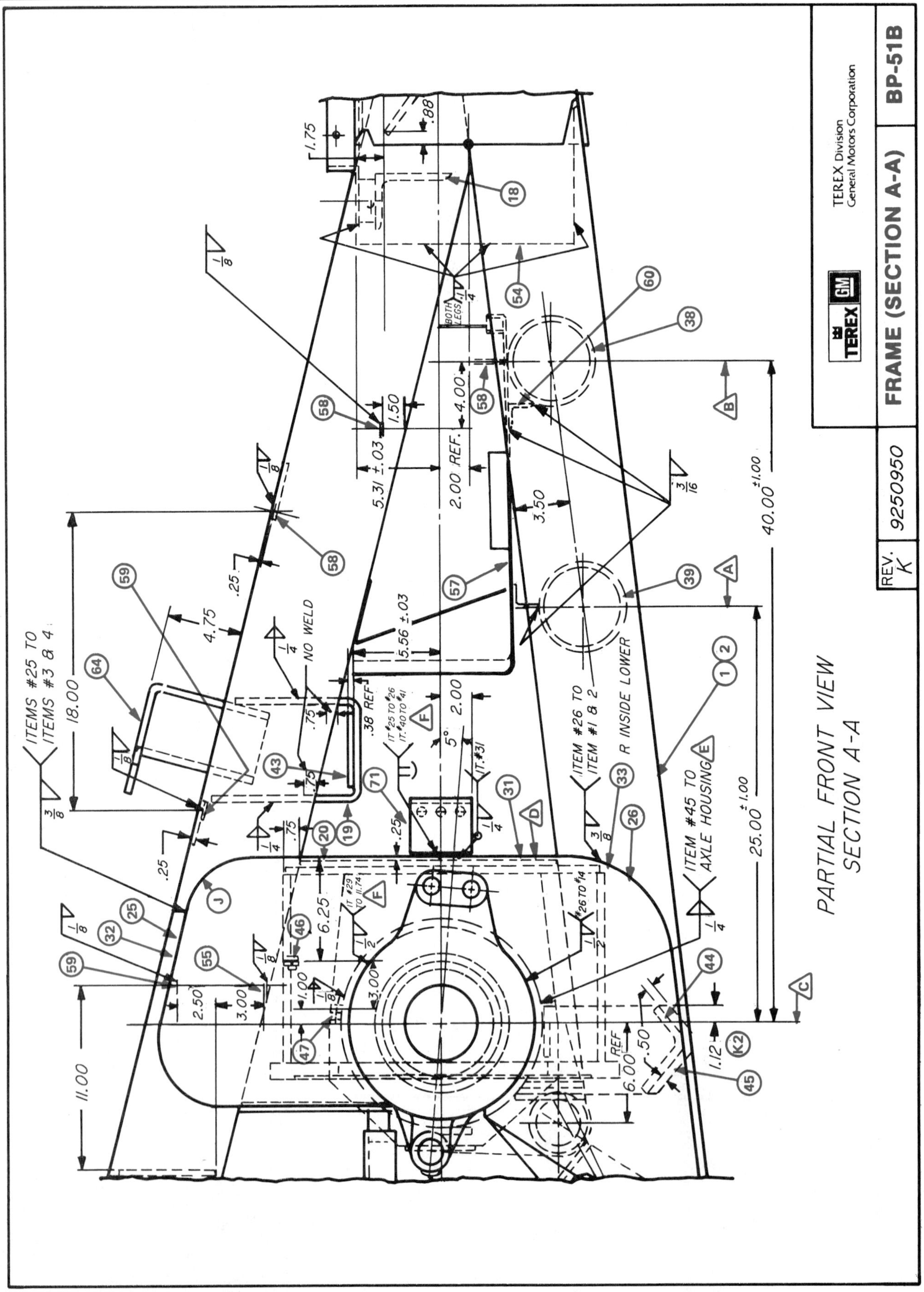

PARTIAL FRONT VIEW
SECTION A-A

TEREX Division
General Motors Corporation

FRAME (SECTION A-A)

BP-51B

REV. K

9250950

Courtesy of General Motors Corporation, Terex Division
(Modified Industrial Drawing)

ASSIGNMENT B—UNIT 51

FRAME (SECTION A-A) (BP-51B)

Student's Name ______________________

1. Sketch the symbol for the following welds.
   a. Plug or slot
   b. Scarf for a brazed joint
   c. Bevel groove
   d. Flange corner groove

2. a. Identify the symbols used in dimensioning items (1) through (4).

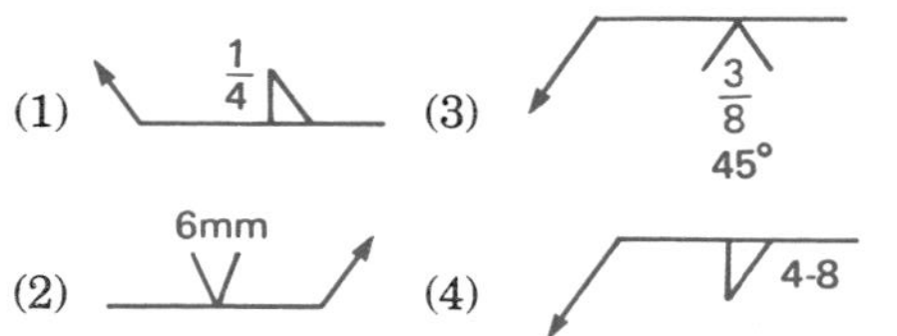

   b. Give the weld designation and size dimensions for each weld.

3. Determine the specifications, process, or other special information contained in the tail or added to the reference line for each of the following.

4. Give the welding specifications for each operation according to the dimensioning indicated on the multiple welding symbol.

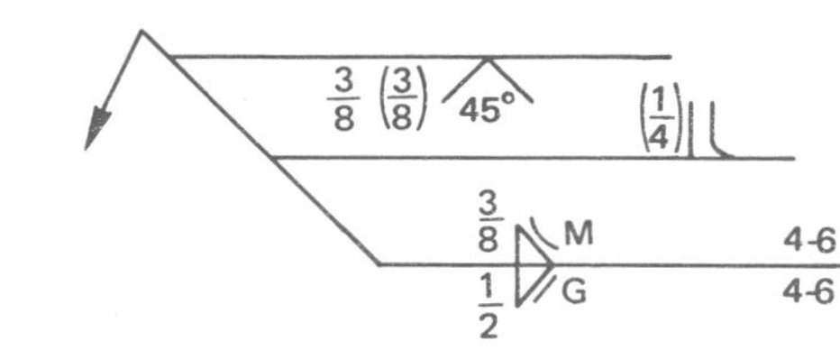

5. Determine the nominal, maximum, and minimum dimensions of the following features.
   a. Center line Ⓐ to center line Ⓑ
   b. Height from DET ⑤⑤ to ⑤⑨ using a tolerance of $\pm 0.06''$.
   c. Length between item ②⓪ and ④⑥, using a tolerance of $\pm 0.03''$.

6. Interpret the following welding specifications.

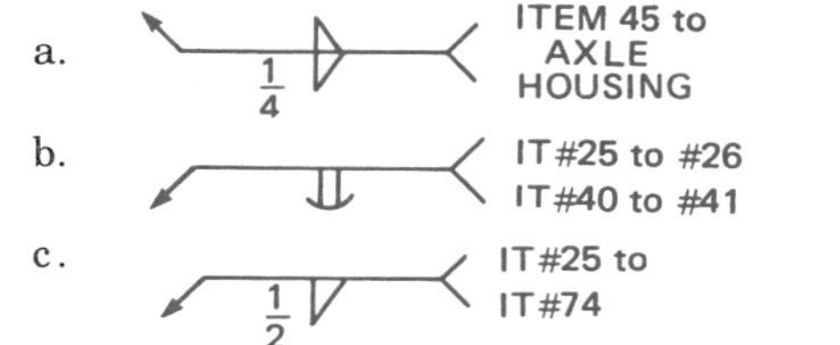

7. a. Draw the welding symbols for welds (1) and (2).
   (1) Brackets ⑲ and ④③
   (2) Items ②⑤ to items ③ and ④
   b. Give the dimensions and/or specifications for each weld.

1.

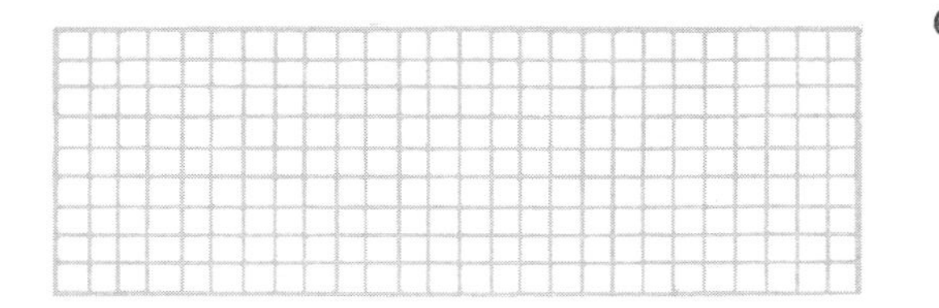

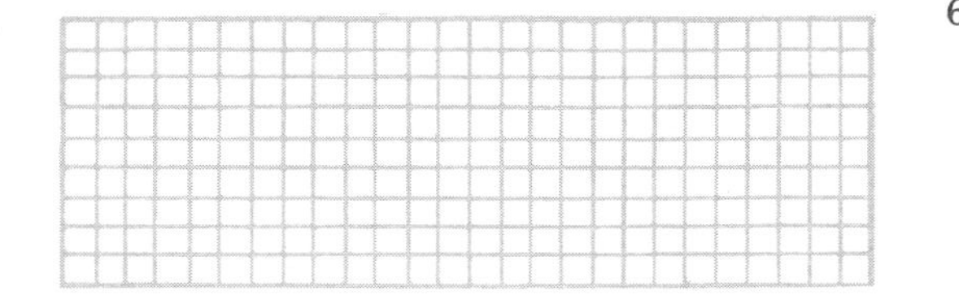

2.

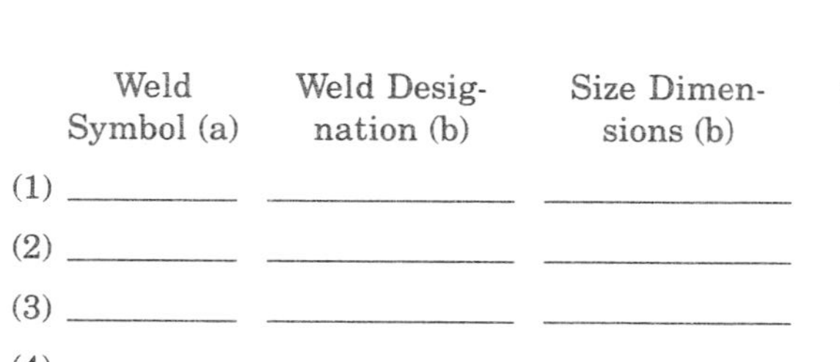

	Weld Symbol (a)	Weld Designation (b)	Size Dimensions (b)
(1)			
(2)			
(3)			
(4)			

3. a. ______
   b. ______
   c. ______
   d. ______

4.

Operation Sequence	Specifications
1	
2	
3	

5.

	Nominal	Maximum	Minimum
a.			
b.			
c.			

6. a. ______
   b. ______
   c. ______

7.

	Welding Drawing Symbol (a)	Weld Specifications (b)
(1)		
(2)		

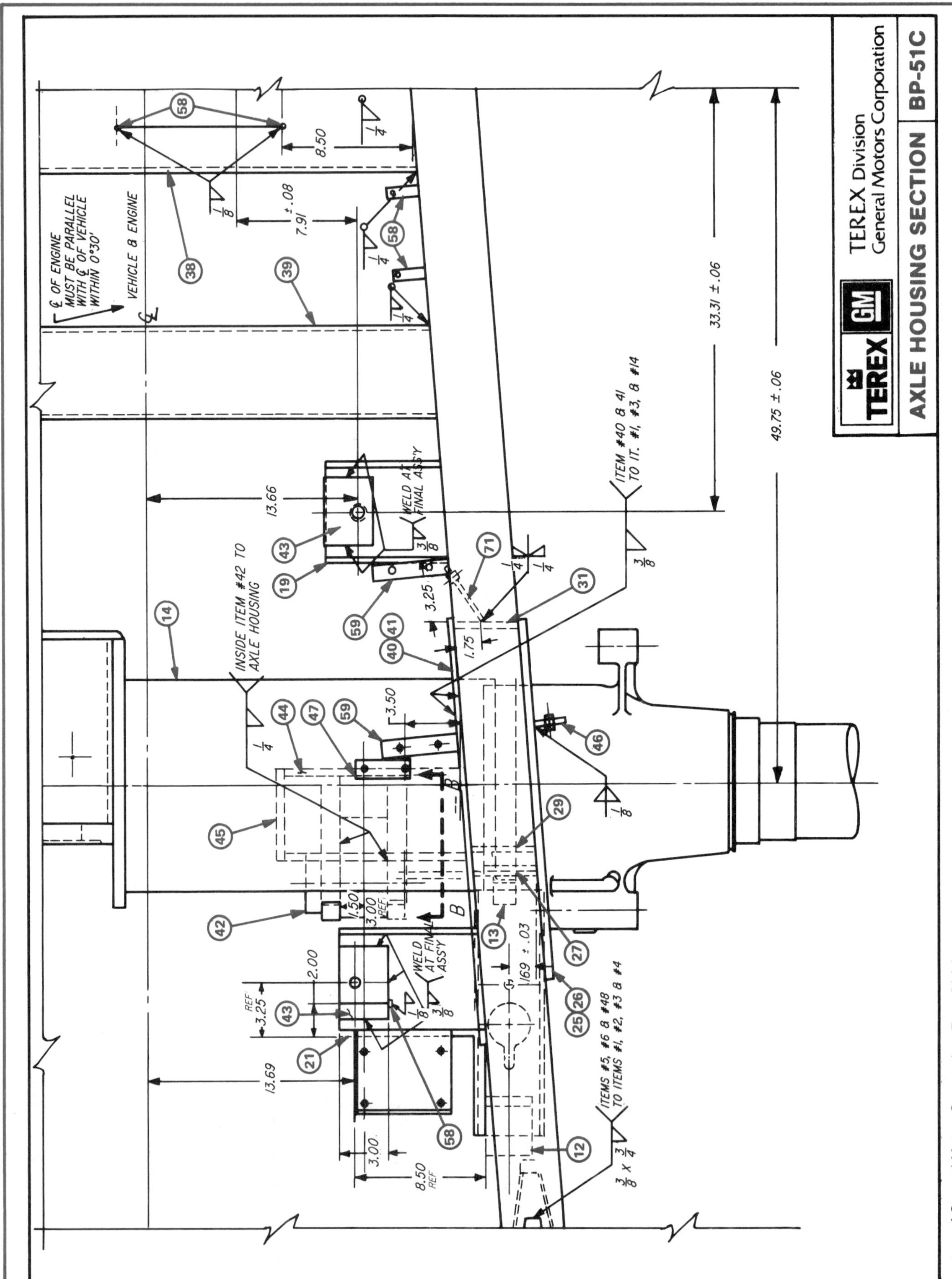

Courtesy of General Motors Corporation, Terex Division (Modified Industrial Drawing)

AXLE HOUSING SECTION (BP-51C)

Student's Name ______________________

1. Give the letter designations for each of the following cutting or welding processes:
   a. Oxyacetylene welding
   b. Gas tungsten-arc welding
   c. Resistance spot welding
   d. Metal arc cutting
   e. Plasma arc cutting

2. Sketch the symbol for welds a, b, c, and d, referenced to the arrow side.
   a. Spot or projection
   b. Flange edge
   c. V-groove
   d. Flare-V groove

3. State the tolerance for the parallelism of the engine and the vehicle center lines.

4. Determine the dimension between the center line of the vehicle and the threaded hole in (43).

5. Give the distance between the vertical center line of the holes in (59) and the right end of (31).

6. Establish the measurement from the underside of (40) to the location of the weld between parts (31) and (71).

7. Determine the maximum width of (43). Use a tolerance of ±0.03″.

8. Identify the nominal, maximum, and minimum dimensions between the center line of the vehicle and the center line of the hole in (43). Use a tolerance of ±0.06″.

9. Find the reference dimension from the center line of the hole in (43) to the point where part (21) touches the chassis (angular line).

10. a. Sketch the symbol used for the welding of detail (43).
    b. Interpret the meaning of the symbol and the dimension.
    c. Determine the number of similar parts (detail (43)) to be welded.

11. Give the size and type of weld used for welding the following details:
    a. (40) and (41) to items (1), (3), and (14)
    b. (46) to (26)
    c. (58) to (38)
    d. Inner surfaces of item (42) to the axle housing
    e. (5), (6), and (48) to items (1), (2), (3), and (4).

12. Give the welding specifications as represented and dimensioned in metric (mm) by symbols a, b, and c.

a.

b.

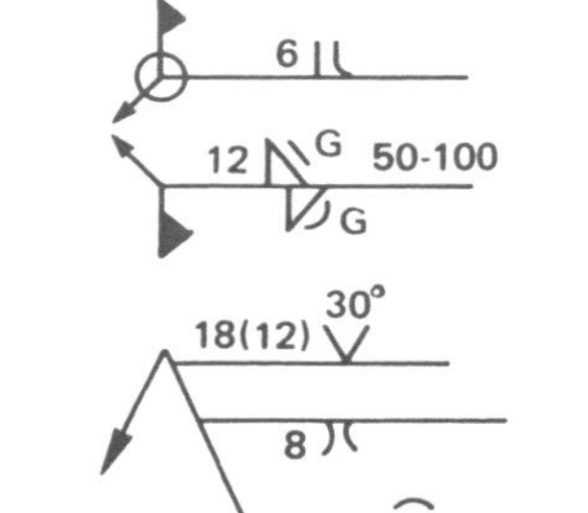

c.

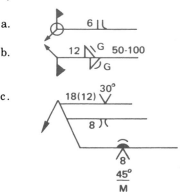

1. a. ______ d. ______
   b. ______ e. ______
   c. ______

2. a. c.
   b. d.

3. ______ 6. ______

4. ______ 7. ______

5. ______

8. Nominal ______ Max. ______ Min. ______

9. ______

10. a.

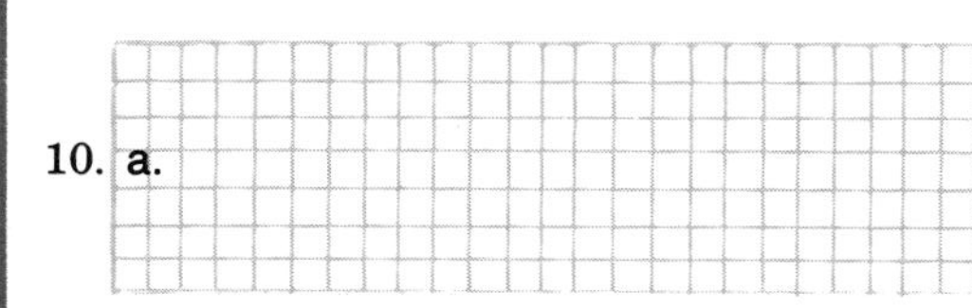

b. ______

c. ______

11. a. ______
    b. ______
    c. ______
    d. ______
    e. ______

12. a. ______

b. ______

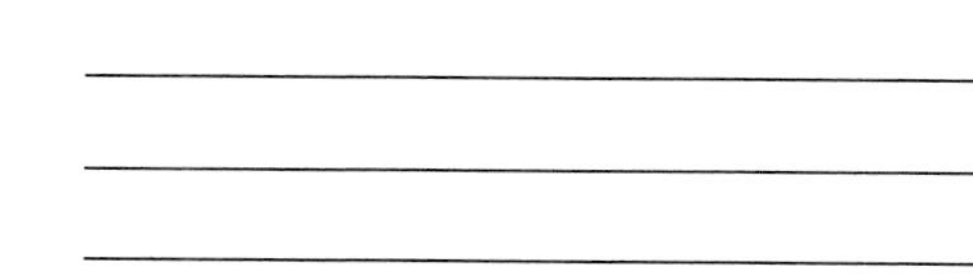

c. ______

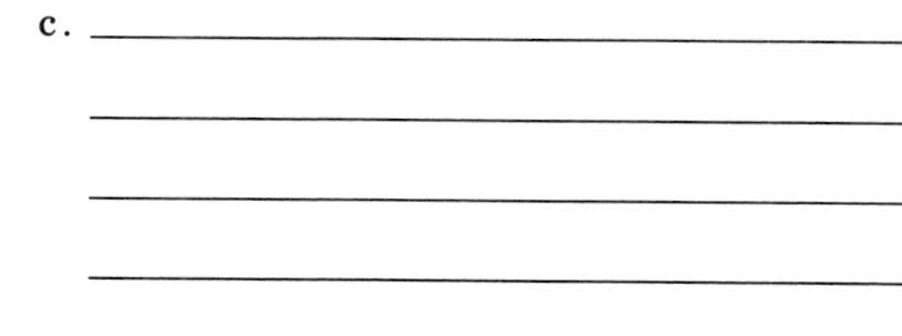

UNIT 52
Casting and Forging Processes and Detail Drawings

Casting is a practical industrial method of producing irregularly shaped parts. If such parts were to be machined from plate or other standard sizes of stock, the cost would be prohibitive. Casting processes may be used for small parts weighing a fractional part of an ounce to huge machine frames and beds weighing tons.

Complicated parts like an engine block are designed with hollowed internal sections and walls which otherwise are not machinable to shape.

FERROUS AND NONFERROUS METAL CASTINGS

Castings are produced by pouring a molten metal into a cavity of a particular size, shape, and design. Castings may be produced by using ferrous or nonferrous metals. Iron castings of *gray* cast iron, *white* cast iron, *malleable* cast iron, and *nodular (ductile)* cast iron are commonly produced by *sand mold casting.*

Steel Castings

Steel castings are especially adapted for machines and parts which must withstand heavy shocks and loads. Marine, rail, and mining equipment, turbine wheels, forging presses, and bridge components are examples of steel casting applications. Parts of such nonferrous materials as bronze, brass, and aluminum are also cast.

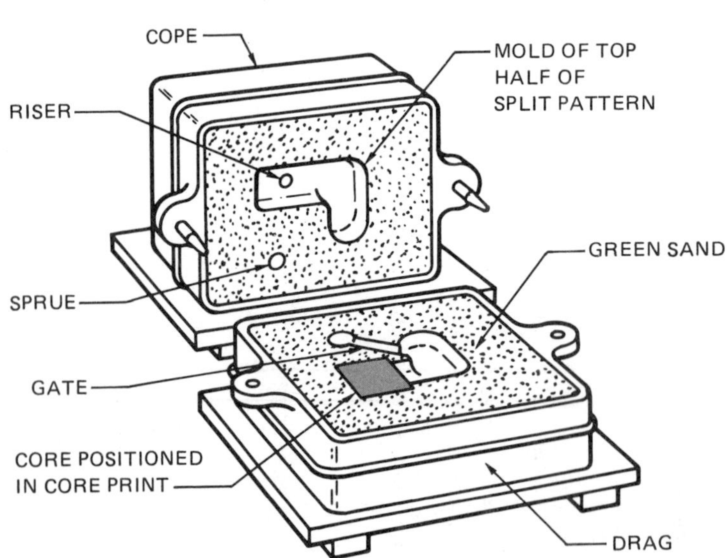

FIGURE 52-1 FLASK OPENED EXPOSING INTERNAL FEATURES OF A SAND MOLD WITH BAKED CORE SET IN POSITION

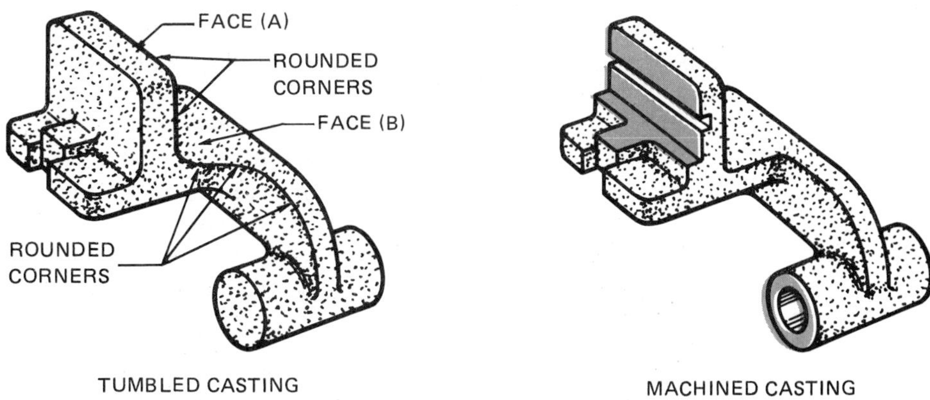

FIGURE 52-2 EXAMPLE OF ROUGH CASTING AND MACHINED CASTING

DESIGN FEATURES OF CASTINGS

In sand mold casting, a durable permanent pattern of wood or metal is required to form the cavity. Provision is made in the sand to pour the molten metal into a *sprue* (feeder opening). The metal then flows through *gates* and *runners* into the cavity. *Risers* above the cavity provide for the storage of surplus metal. As a casting cools and solidifies, additional liquid metal from the risers and gates is furnished to keep the casting free of shrinkage areas. Figure 52-1 shows the two parts (*cope* and *drag*) of a *flask* and internal features of a sand mold. The part to be cast and then machined is illustrated in figure 52-2.

Since all metals shrink a definite amount upon cooling, a *shrinkage allowance* is made. An *oversize pattern* is required. In addition, in order for the pattern to be removed from the mold, provision must be made for *draft*. Draft refers to the slight tapering of the sides.

Finally, an allowance is made to add metal to each surface which is to be machined. This *machining allowance* permits cuts to be taken below the outer irregular surface scale.

Dimensions and design features on *patternmaker's* (or *pattern shop*) drawings include information about and provide for shrinkage, draft, and machining allowances.

Cores for Cavities and Holes

Holes, cavities, and other recesses or cut-away areas within a casting are produced by using a *core*. A core is made from a mixture of sand and a binding agent. These materials are formed in a *core mold* to conform to the required internal shape and size. When heated, a hard, solid core is formed. Cores are designed to fit into support areas, called *core prints*, in the sand (figure 52-3).

After a casting is poured and the metal is cooled, the mold is broken and the casting is removed. The sprue, gates, risers, other excess metal, and the burned-out core material are removed. Castings are generally tumbled with small *shot* to get

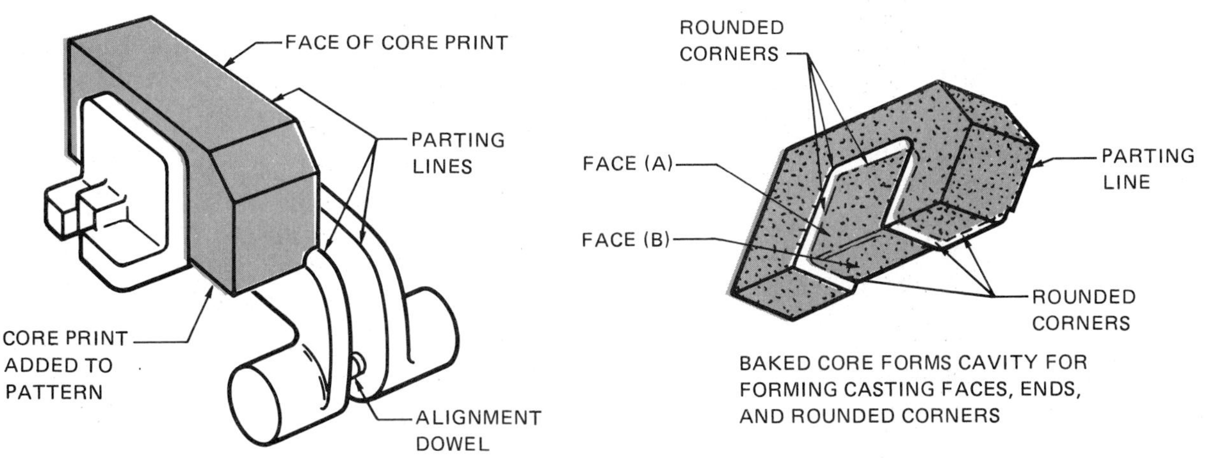

FIGURE 52-3 APPLICATION OF CORE AND CORE PRINT IN SAND MOLDING

rid of sharp edges and sand that tends to adhere to the metal casting. Rough surface irregularities are often ground away after tumbling.

DIE CASTING

Die casting is a production metal-forming process. Molten metal is forced into a permanent metal mold (die) cavity. The cast metal is then cooled quickly and removed. Some of the advantages of die casting follow:

- Economy in mass-producing interchangeable parts, including complicated forms which are difficult and costly to produce by other casting methods.
- Dimensional accuracy and quality of surface finish often eliminate additional machine finishing processes.
- Bushings, studs, other fasteners, and harder or different metals than the cast metal may be inserted in the dies. The inserts are then cast as an integral part of a die casting.
- Rapid chilling produces a fine grain metal with a quality surface finish and added strength.
- Engraved and ornamental surfaces are accurately reproduced without the loss of intricate details or sharpness.

DIE CAST METALS

There are six *base metals* used in die castings. The metals include zinc, tin, lead, aluminum, copper, and magnesium. Alloys of these base metals produce such desired qualities as minimum shrinkage, fluidity, and added strength.

Die-casting die designs provide allowance for the flow of metal, the escape of displaced air, shrinkage, draft for the removal of each part, and cores. *Ejector pins* are used to mechanically force (eject) the part when the die is opened.

MACHINE FORGING PROCESSES

There are six basic machine processes. These processes deal with: *upsetting, bending, punching, cutting, swaging,* and *welding.*

PHYSICAL PROPERTIES OF PRODUCTION FORGINGS

Forging processes are performed as a planned sequence of steps as a selected material moves through different stages in which the basic manufacturing processes are performed. As a result of these material forming and cutting processes, the forged products have the following advantages.

- A specially-shaped solid part may be manufactured by using a series of dies on one forging press.
- The forged parts may be held to comparatively uniform standards of dimensional accuracy and form geometry.
- Forging produces a fine grain structure, high resistance to shock, and other improved physical characteristics of the material.
- The internal structure of forged parts is uniform.
- *Upsetting.* In this process, a material is formed to a desired shape by applying force (to flow the metal) along the horizontal dimensions. As the length is decreased, the width is increased and the part is formed to a desired shape.
- *Bending,* as applied to forging, means that a heated flat stock is bent (by using a forming die) to produce a desired contour.
- *Punching.* In this forging process, a forging die penetrates the workpiece to produce a round or other contoured-shape hole to a specific depth (or through hole).
- *Cutting.* Surplus material is removed or a portion of a forging is severed. A forging die is used for the shearing or cutting process.
- *Swaging.* This process relates to the forming of a metal part by reducing the thickness of the material within forming dies or concave forming tools.
- *Welding.* As a process in forging, parts are joined together (welded) under extreme force. The area of the weld has the same physical characteristics as the material before welding.

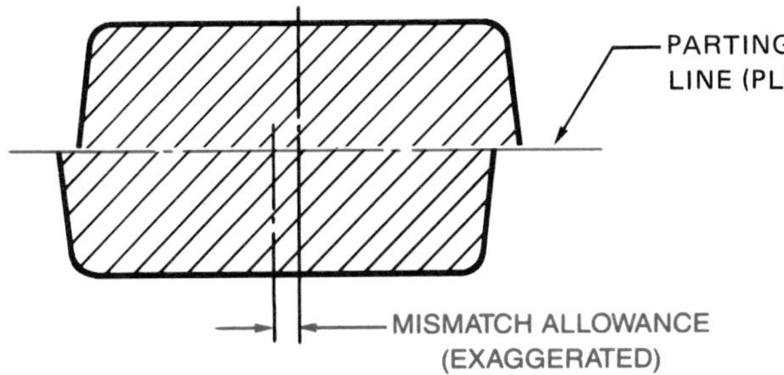

FIGURE 52-4 ALLOWANCE REQUIRED FOR DIE MISMATCH

FORGING DETAIL DRAWINGS

Parts may be *forged* by heating and flowing the metal under intense force. Forging dies are used to form the part within the die cavity. Forging detail drawings are similar to casting detail drawings. Provision is made for (1) draft, (2) metal shrinkage as the heated metal cools, (3) wear allowance on the dies, (4) machining allowance, and (5) *mismatch allowance.* Mismatch (figure 52-4) relates to surface machining errors which are produced by the misalignment of the die sections as they close to form the forging.

CHARACTERISTICS OF FORGING DIES

Sectional views are widely used with forging detail drawings. While external corners are forged rounded, they are projected and drawn as though they were sharp. Also, dimensions and features are generally referenced from a *parting line.* The parting line is the line around the forging where the parts of a die meet. The parting line is represented by the center line symbol, labeled **PL**. Figure 52-5 illustrates the shape and size of a machined forging, die cavity, allowances, and standard ANSI terms.

Other ANSI tables are used for tolerances of thickness, shrinkage, die wear, mismatch, and standard machining allowances. Notes on detail drawings of forgings generally include dimensions for draft angles, radii of corners and fillets, heat treating, and hardness testing specifications. It is common practice to use either a single drawing to show both the forging and machining details and dimensions, or a forging drawing and a machining drawing.

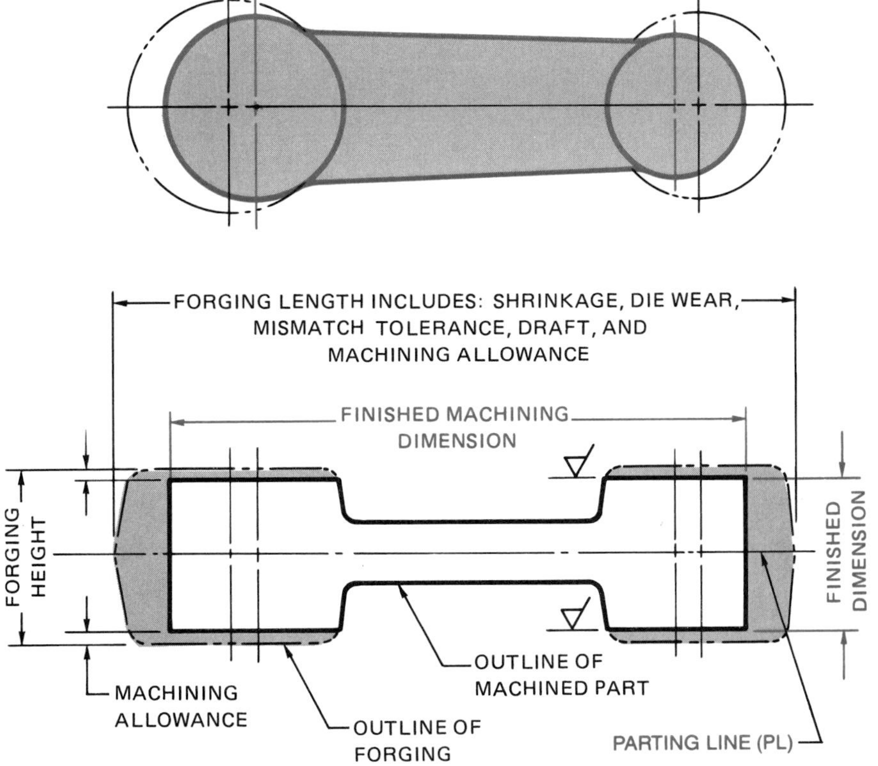

FIGURE 52-5 FORGING TERMS AND DESIGN FEATURES (ANSI)

NOTE: REFER TO APPENDIX B-8 FOR THE LARGER SCALE DRAWING TO USE WITH THIS ASSIGNMENT.

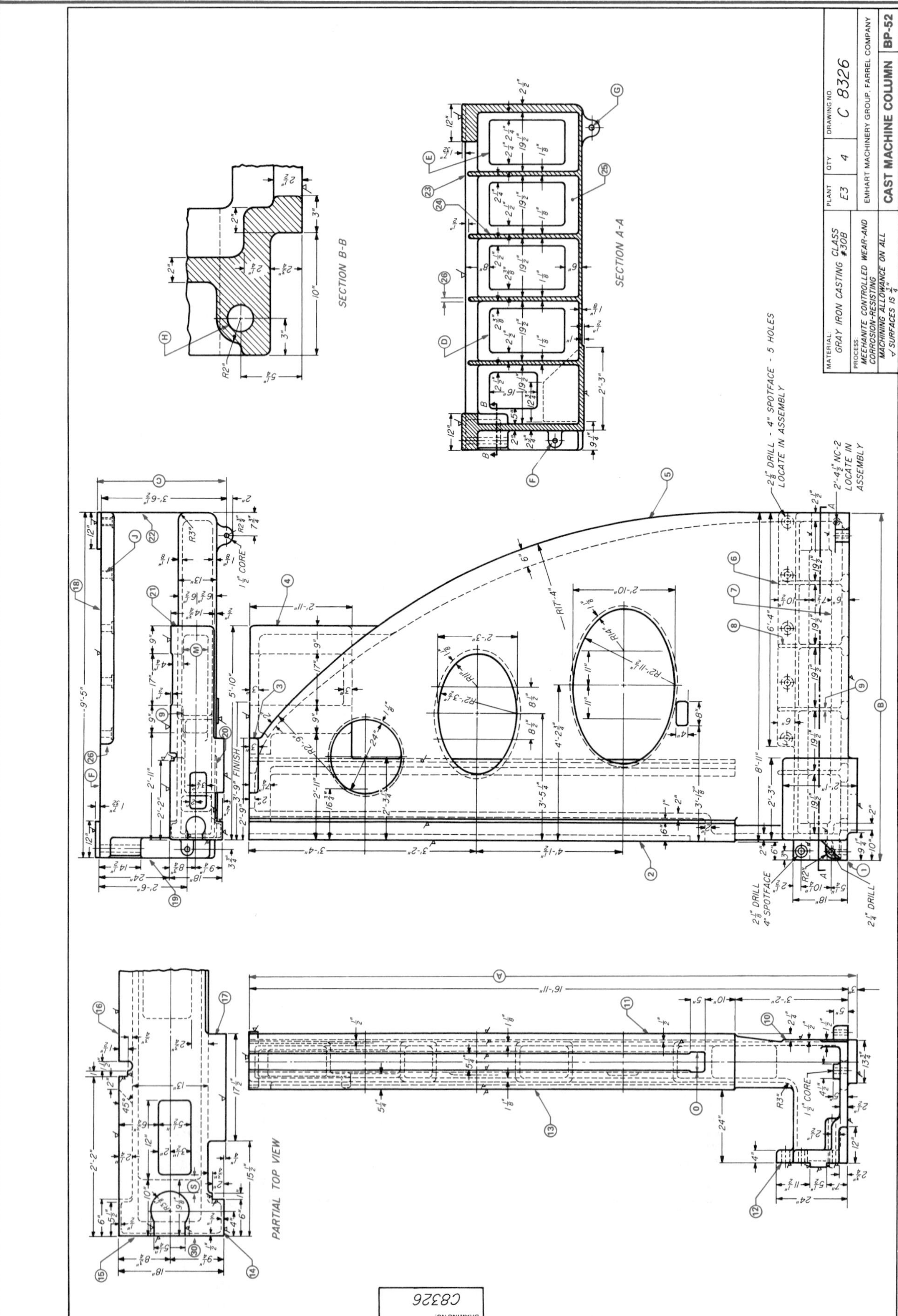

Courtesy of Emhart Machinery Group, Farrel Company (Modified Industrial Drawing)

ASSIGNMENT—UNIT 52

CAST MACHINE COLUMN (BP-52)

1. Specify the (a) material and (b) the process of producing the **CAST MACHINE COLUMN**
2. Calculate the nominal overall dimensions of the *rough casting* (including machining allowances) for features (A), (B), and (C).
3. a. Give the outside radius of the Column.
   b. Indicate the web thickness of the casting at (1) the center, (2) top, and (3) base of the curved back.
4. Indicate the areas and features for which set cores were used to produce hollow areas, openings, or holes.
5. Identify the surfaces or lines in the top view or partial top view which are represented by features (1) through (5).
6. Identify the features in **Section A–A** which correspond to items (6) through (9) in the front view.
7. Identify surfaces (10) through (13) in one of the top views.
8. Refer to **Section A–A**.
   a. Identify the design feature which strengthens the side walls.
   b. Give the thickness (depth) and width for the rectangular openings (D) and (E).
   c. Determine the core diameter of holes (F) and (G).
9. Refer to **Section B–B**.
   a. Identify where this section originates.
   b. Give the diameter and process for machining hole (H).
10. a. Identify the symbol which is used on the machined surfaces.
    b. State what the material allowance is for all finished surfaces.
11. Compute the nominal dimensions between the following features:
    a. (1) and (2)
    b. (11) and (13)
    c. (11) and (12)
12. a. Give the specifications for holes (J).
    b. Indicate the size and number of tapped holes.
13. State two conditions in forging which differ from sand mold casting.
14. List four physical properties or characteristics of forgings which must be planned for when designing a forged part.

Student's Name ____________________

1. a. ____________________
   b. ____________________
2. (A) = __________
   (B) = __________ (only one machined surface)
   (C) = __________
3. a. __________
   b. (1) ________ (2) ________ (3) ________
4. *Examples.*
   a. ____________________
   b. ____________________
   c. ____________________
   d. ____________________
5. (1) = __________ (4) = __________
   (2) = __________ (5) = __________
   (3) = __________
6. (6) = __________ (8) = __________
   (7) = __________ (9) = __________
7. (10) = __________ (12) = __________
   (11) = __________ (13) = __________
8. a. ____________________
   b. Thickness Width
   (D) = __________ x __________
   (E) = __________ x __________
   c. (F) and (G) = __________
9. a. ____________________
   b. (H) = ____________________
10. a. __________
    b. ____________________
11. a. __________ c. __________
    b. __________
12. a. ____________________
    ____
    b. ____________________
13. a. ____________________
    ____
    b. ____________________
    ____
14. a. ____________________
    b. ____________________
    c. ____________________
    d. ____________________

UNIT 53
Plastics: Materials, Fabrication, and Design Features

The widespread use of plastics is attributable to such advantages over other materials as the following:

- Adaptability to mass production at a comparatively low cost.
- Resistance to corrosion, solvents, water, etc.
- Quiet, smooth movement with limited, if any, lubrication required.
- Qualities of an electrical and thermal insulator.
- Light mass and great range of translucent and opaque colors.
- High quality of surface finish.
- Availability in solid form as powder, granules, sheets, tubes, and castings, or in liquid forms as adhesives.

CLASSIFICATION OF PLASTICS

Plastics are nonmetallic materials. Each specific base material is modified to form a *family of compositions*. While the products in each family have common properties, each composition has additional individual properties.

Thermoplastics and Thermosetting Plastics

Plastics are roughly classified as thermoplastic or thermosetting. The term *thermoplastic* refers to plastics which soften, liquify, or flow when heat is applied. The plastics set or solidify as heat is removed. Thermoplastics may be reheated, reformed, and reused. Common plastics in this group include acrylics, cellulosics, nylon, polyethylene, urethanes, and vinyls.

Thermosetting plastics are permanently changed chemically as heat is applied or when a catalyst or reactant is used. Thermosetting plastics become hard, may not be fused, and are not softened if heat is applied. Families of thermosetting plastics are identified as alkyds, amino (melamine and urea), casein, phenolic, epoxy, polyesters (fiberglass), and silicones.

Designers use tables of chemical and physical properties of plastics, commercially available forms and sizes, methods of forming, and recommended applications.

METHODS OF FORMING PLASTICS

Plastic parts are generally formed by molding processes which require heat and pressure. The basic molding processes include compression, injection, extruding, transfer, blow, and vacuum forming.

Injection Molding

Injection molding involves feeding the plastic material into a heating chamber where it softens and becomes fluid. The fluid plastic is then fed through a rotating, reciprocating screw to the nozzle of the mold as pictured in figure 53-1. The

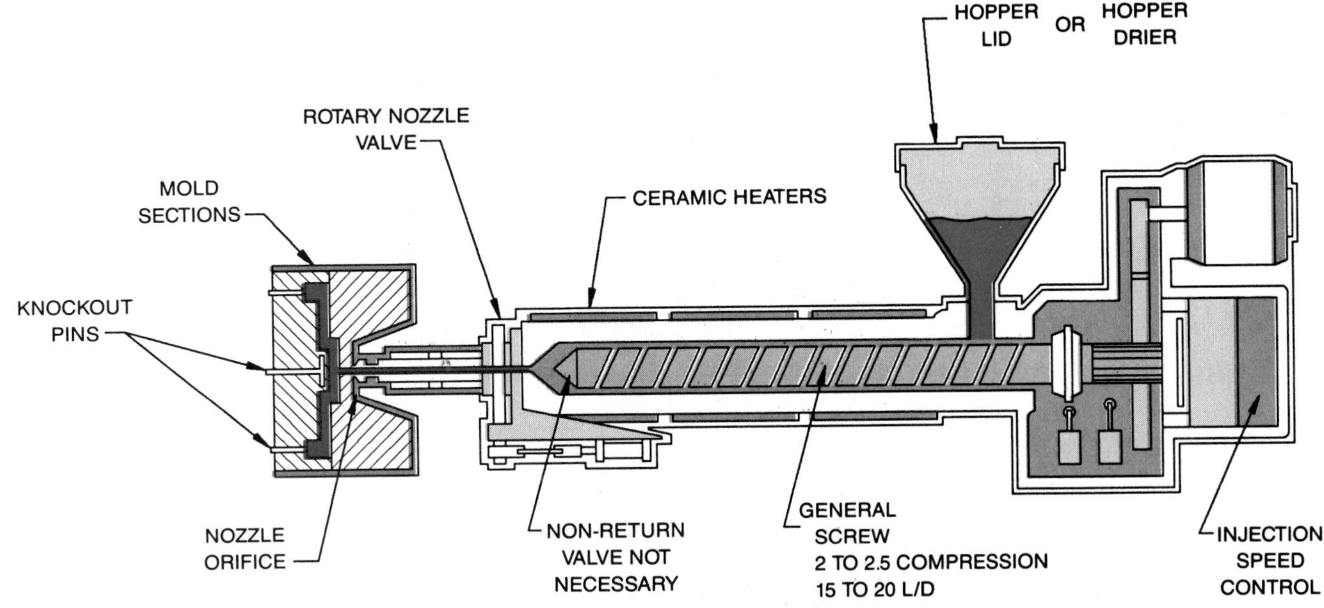

FIGURE 53-1 INJECTION MOLDING PROCESS: A ROTATING, RECIPROCATING SCREW-TYPE MOLDING MACHINE (COURTESY OF UNION CARBIDE CORPORATION)

plastic is forced at high pressure into the cool, closed mold. When the part and mold are cooled after injection, the mold is opened. The finished plastic part is ejected.

Extruding

The *extruding process* is used to produce sheets, rods, tubes, and other forms which are uniform and regular in cross-section. Dies of the required size and form are used. The plastic material is heated to a flow temperature and forced under pressure through the forming dies to produce a particular form.

Compression Molding

Compression molding requires a mold consisting of a heated cavity and plunger (figure 53-2). The plastic compound is inserted in the mold cavity. The part is formed as the plunger is forced against the heated plastic. The heat and pressure cause the plastic to liquify, flow, and form in the cavity and around the punch to produce the required form.

A number of parts are produced in what is called a multiple-cavity mold. A split-cavity mold is designed so that internal parts of the mold are movable and permit the removal of molded parts which are designed with indentations or other special forms.

Transfer Molding

The material in *transfer molding* is heated to a point of *plasticity*. Once in the mold, the plastic is hydraulically forced by a transfer plunger through a sprue in the mold cavity. The required part is produced as the plastic flows and fills the mold cavity.

Casting Method

The *casting method* differs from the molding processes in that no pressure is used. The plastic material is heated until fluid, poured into a mold, cured, and then removed.

PLASTIC COATINGS AND LAMINATING

Plastics are also *calendared*. This means the processing of thermoplastics into film as fine as 0.001″ (0.2 mm). The film is then applied as a coating to metals, wood, ceramics, and other materials.

Plastic laminating is used to strongly bond a plastic to such reinforcing materials as wood, glass fibers, and cloth. High heat and high pressure are required with thermosetting plastics for laminating. Reinforced plastics, which perform similar functions of binding the reinforcing material, require limited pressure.

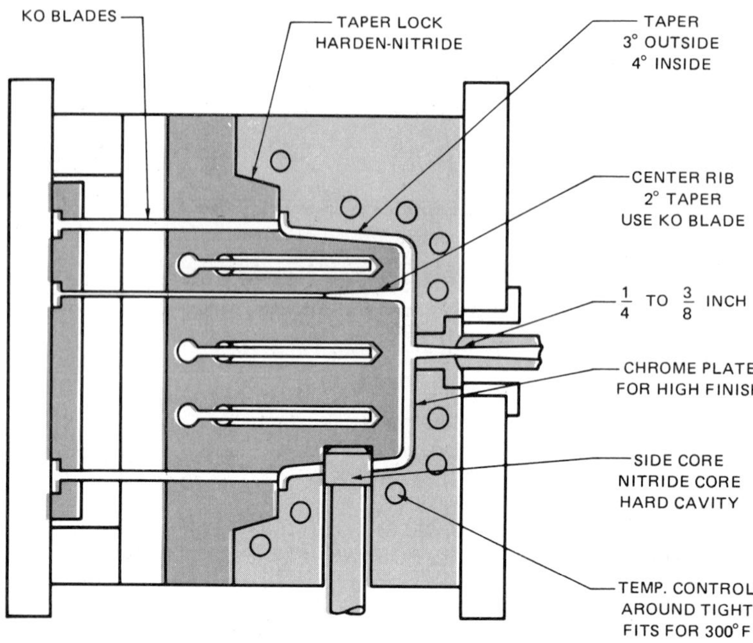

FIGURE 53-2 COMPRESSION MOLDING (COURTESY OF UNION CARBIDE CORPORATION)

DESIGN CHARACTERISTICS OF PLASTIC PARTS

Consideration is given when designing plastic parts to physical and chemical properties which affect the structure, accuracy of form and size, strength, and the actual functioning of a part.

Shrinkage

Standard allowances are made to compensate for differences in size and form between the dimensions of a mold and the finished plastic part. Attention is directed to the effect of shrinkage on warpage, residual stress, and moldability.

Also, since there is heat transfer between the plastic and the mold, sections of the part which vary in thickness may solidify at different rates. The factor of *solidification* requires that the mass of material in each section of a part be uniform.

Draft, Gates, and Flash Line

Each of these items must be considered by the designer. *Draft* means the internal and/or external tapering of the sides to permit the removal of the plastic form from the mold. *Gates* are used to uniformly distribute the flow of the plastic into a mold (figure 53-3). The *flash line* (often referred to as the *parting line*) represents the junction between the parts of a mold where a fine fin may be produced (figure 53-4).

Allowances for Fillets, Radii, Ribs, and Bosses

Fillets and *radii* permit the liquid plastic to flow easily and smoothly around corners and bends. Plastic parts are *ribbed* to provide added strength and increase rigidity without increasing wall thicknesses. *Raised bosses* are often used to increase the strength of the part, particularly in areas surrounding an insert.

Inserts

Metal inserts are widely used with plastics and serve functions similar to those used in die castings. Inserts are designed with knurls, grooves,

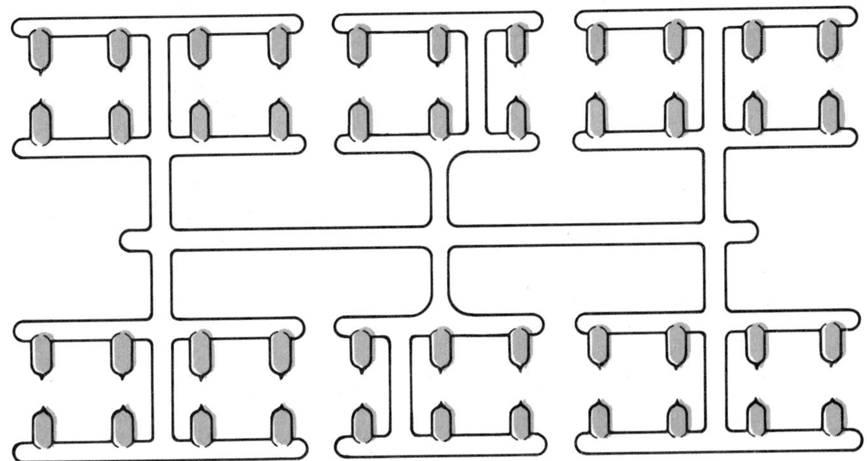

FIGURE 53-3 DESIGN AND POSITION OF GATES IN MULTIPLE MOLDING OF A PLASTIC PART (COURTESY OF UNION CARBIDE CORPORATION)

and other forms so they may be anchored securely to withstand torque and tension forces. Inserts are secured in a plastic part by a *press fit*, a *shrink fit*, or by *plastic shrinkage*.

ASSEMBLING PLASTIC PARTS

One plastic part may be assembled and secured to another part by applying heat and pressure. *Heat-forming* and *heat-sealing* processes eliminate the need for *adhesive bonding*, *riveting*, and other *mechanical fasteners*.

Mechanical fasteners are available for applications requiring assembly and disassembly. Serrated spring washers, speed clips and nuts, and spring-type metal hinges are common types of fasteners.

Bonding: Adhesive and Ultrasonic

Adhesive bonding provides a practical method of joining plastics or a plastic with a dissimilar material. The adhesives are usually applied as a liquid. When hardened, any tensile failure normally does not occur at the adhesive-bonded surface area.

Ultrasonic bonding refers to the fast process of welding by ultrasonic vibration as contrasted with adhesive bonding. *Ultrasonic staking* is used when one of the parts in an assembly is metal. A projecting plastic stud extends through a hole as shown in figure 53-5. The *horn* (or punch) above the stud operates at a high amplitude over a small contact area. This action causes the plastic stud to soften and flow, following the design of the horn.

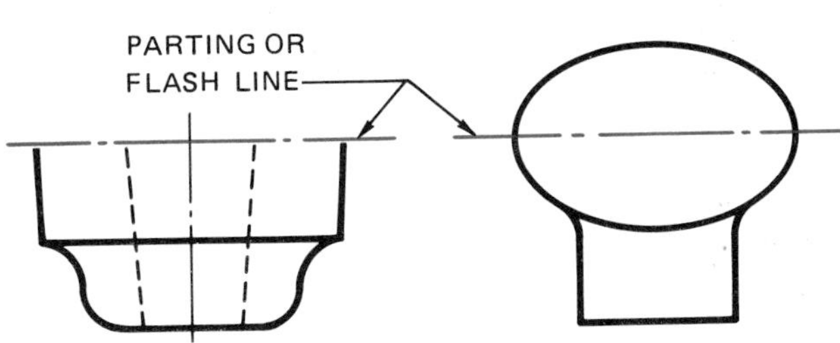

FIGURE 53-4 FLASH LINE ON MOLDED PLASTIC PARTS

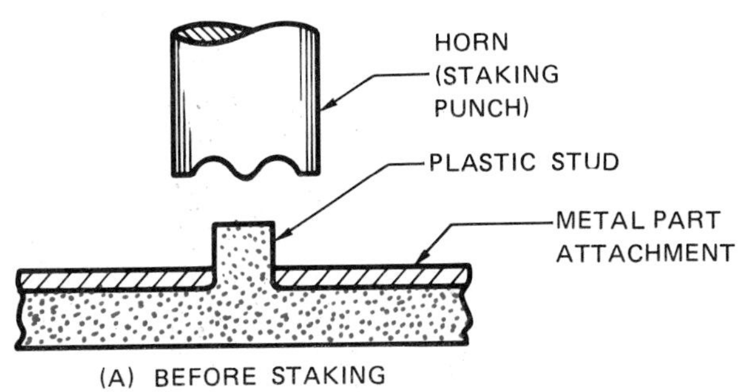

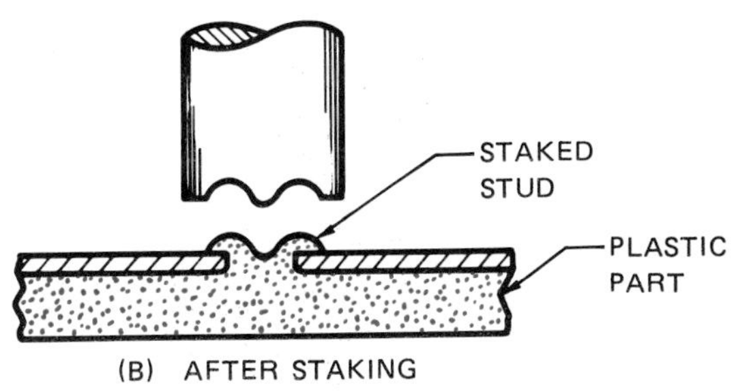

FIGURE 53-5 ULTRASONIC STAKING OF METAL AND PLASTIC PARTS OF AN ASSEMBLY

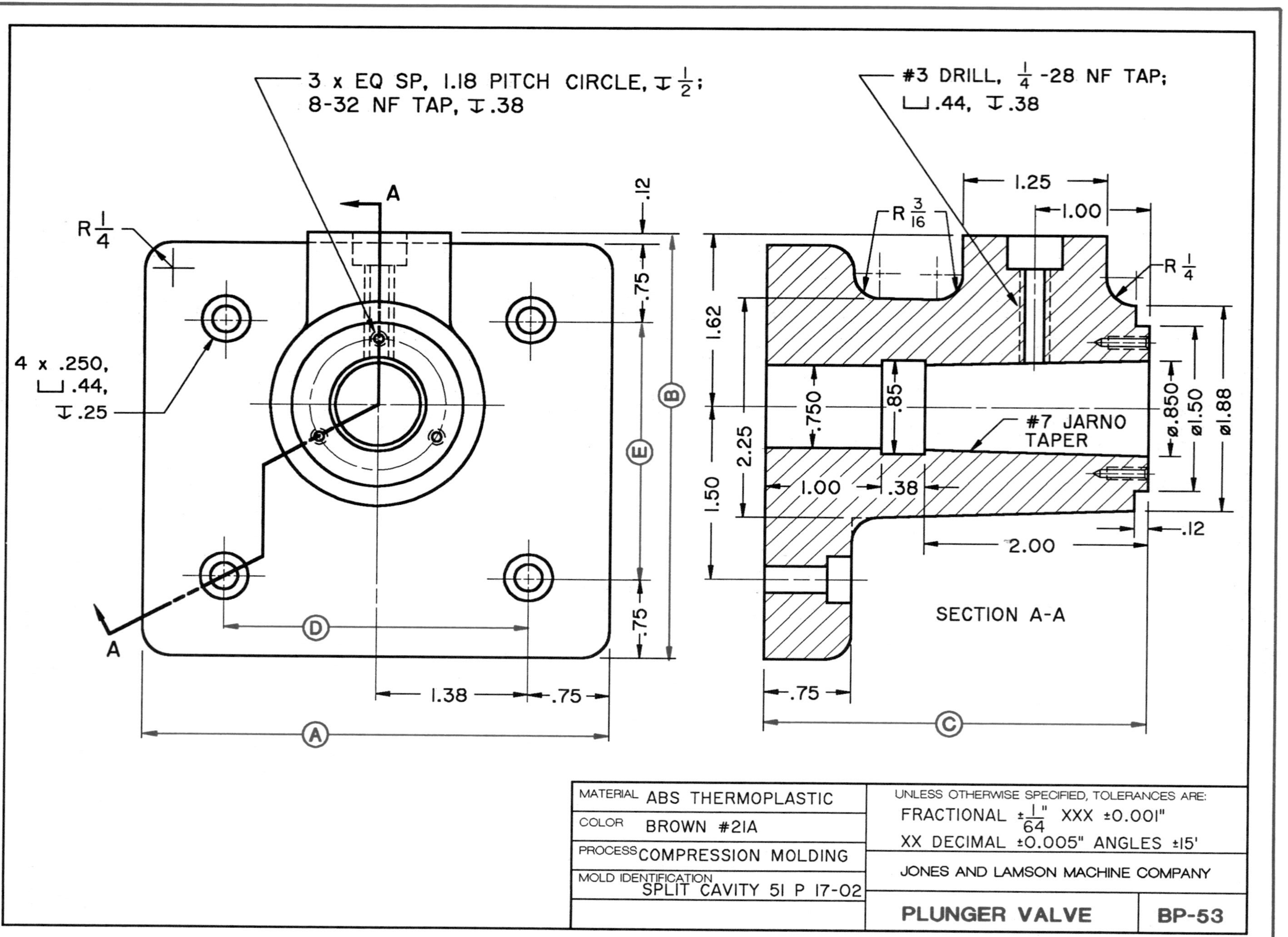
3 x EQ SP, 1.18 PITCH CIRCLE, ↧ 1/2;
8-32 NF TAP, ↧ .38
#3 DRILL, 1/4 -28 NF TAP;
⌴ .44, ↧ .38
R 1/4
4 x .250,
⌴ .44,
↧ .25
A
.12
.75
.75
1.38
R 3/16
1.25
1.00
R 1/4
1.62
2.25
1.50
.750
.85
#7 JARNO TAPER
ø.850
ø1.50
ø1.88
1.00
.38
.12
2.00
SECTION A-A
.75
MATERIAL ABS THERMOPLASTIC
COLOR BROWN #21A
PROCESS COMPRESSION MOLDING
MOLD IDENTIFICATION SPLIT CAVITY 51 P 17-02
UNLESS OTHERWISE SPECIFIED, TOLERANCES ARE:
FRACTIONAL ±1/64" XXX ±0.001"
XX DECIMAL ±0.005" ANGLES ±15'
JONES AND LAMSON MACHINE COMPANY
PLUNGER VALVE
BP-53

ASSIGNMENT—UNIT 53

PLUNGER VALVE (BP-53)

Student's Name ______________________

1. State why a front and section view are used instead of other conventional views.
2. Give the maximum dimensions for the width Ⓐ, height Ⓑ, and depth Ⓒ of the Plunger Valve.
3. Determine the upper- and lower-limit dimensions for center distances Ⓓ and Ⓔ.
4. Give the following maximum and minimum diameters:
   a. The machine-reamed hole.
   b. The large end of the taper hole.
5. Identify (a) the taper design, (b) taper number, and (c) taper per foot.
6. Determine the upper and lower limit dimensions for (a) the pitch circle and (b) the included angle between the three threaded holes.
7. Cite three advantages for using plastics over metal for such a part as the **Plunger Valve**.
8. Cite one difference between a thermoplastic and a thermosetting plastic.
9. State four general methods of forming plastics.
10. Identify one purpose for using metal inserts in plastic-formed parts.
11. Explain briefly the functions served by the corner radii and fillets.
12. a. Name (1) the material and (2) the color of the **Plunger Valve**.
    b. State which plastic molding process is used to produce the **Plunger Valve**.
    c. Identify (1) the type and (2) the identifying number of the mold.

1. ______________________
____
____

2. Maximum Dimension
a. Width Ⓐ ______________
b. Height Ⓑ ______________
c. Depth Ⓒ ______________

3.

	Upper Limit	Lower Limit
Ⓓ =	________	________
Ⓔ =	________	________

4. a. Max. ________ Min. ________
b. Max. ________ Min. ________

5. a. ________ b. ________
c. ______________________

6.

	Upper Limit	Lower Limit
a. Pitch Circle (PC) =	________	________
b. Included Angle =	________	________

7. a. ______________________
____
b. ______________________
____
c. ______________________
____

8. ______________________
____
____
____

9. a. ______________________
b. ______________________
c. ______________________
d. ______________________

10. ______________________
____
____

11. ______________________
____
____

12. a. (1) ______________________
(2) ______________________
b. ______________________
c. (1) ______________________
(2) ______________________

UNIT 54
Electrical Circuits, Symbols, and Drawings

Electrical and electronic drawings provide designers, engineers, and other technical workers with full information about the design, connections, installation, control, and maintenance of electrical or electronic devices.

In the construction trades, electrical drawings show the power source and the location and function of lines, switches, receptacles, and other electrical equipment. The electronics technician uses wiring diagrams and schematic diagrams to identify, trace, and maintain electronic circuits and electronic devices which require the use of electricity.

GRAPHIC SYMBOLS AND CIRCUIT COMPONENTS

Electrical and electronic drawings are simplified by using standardized graphic symbols. The ANSI symbols appear on a drawing in a size which is proportional to other equipment sizes and remains the same size throughout the circuit. Electronic/electrical drawings are generally not drawn to scale. The important requirement is legibility.

Templates are available with symbols grouped for specific requirements like general construction installations, power stations, electrical equipment, industrial electronics, and electronic communications components and circuitry. Figure 54-1 shows a simple inexpensive template for drawing standardized electrical and electronic ANSI symbols.

TYPES OF ELECTRICAL DIAGRAMS

The simplest types of standard ANSI electrical drawings include single-line, schematic, and wiring diagrams. The *single-line diagram* (figure 54-2) uses single lines and graphic symbols to present information about a circuit. A *schematic diagram* shows the electrical connections and the functions of each part. Graphic symbols are used. Figure

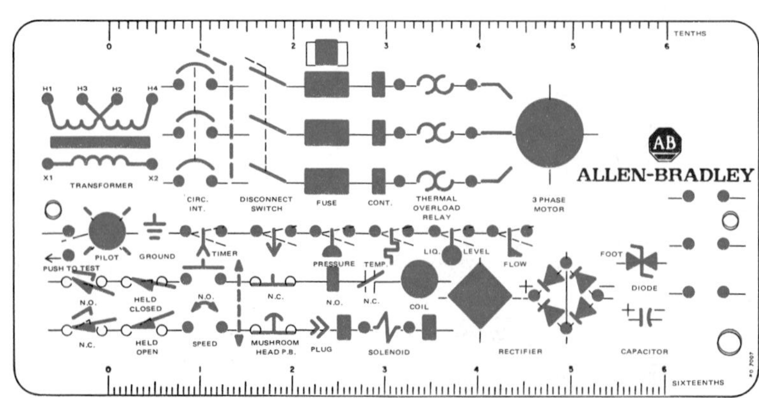

FIGURE 54-1 TEMPLATE USED IN PREPARING ELECTRICAL-ELECTRONIC CIRCUIT DRAWINGS (COURTESY OF THE ALLEN-BRADLEY COMPANY)

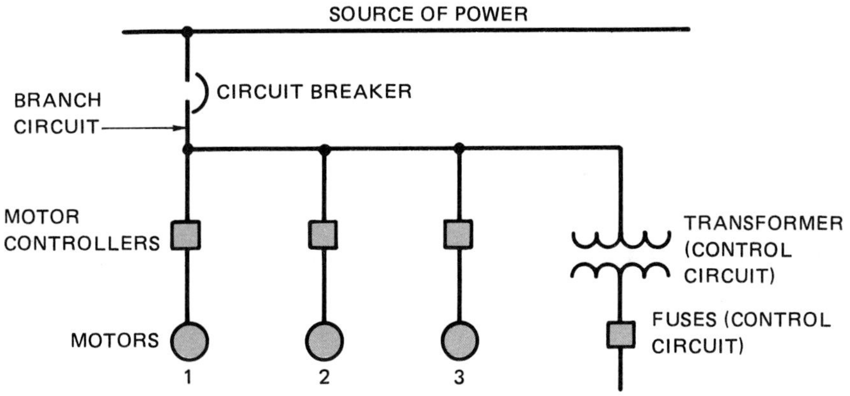

SOURCE OF POWER
BRANCH CIRCUIT
CIRCUIT BREAKER
MOTOR CONTROLLERS
TRANSFORMER (CONTROL CIRCUIT)
FUSES (CONTROL CIRCUIT)
MOTORS
1
2
3

FIGURE 54-2 SINGLE-LINE DIAGRAM OF A MULTIPLE MOTOR CIRCUIT PROTECTION ARRANGEMENT

54-3 is an example of a schematic drawing of a motor control circuit. Each major component (in this case) is named and represented by a symbol. The two vertical lines represent the power source. No attempt is made to show the actual positions of the devices in the circuit. The lines between the symbols represent the actual wires which connect the electrical components.

A *wiring diagram* includes the graphic symbols arranged in the same physical position which each component occupies in the equipment. The lines between the symbols represent the actual wires which connect the electrical components. Figure 54-4 shows the relationship of a schematic diagram (figure 54-3) and the wiring diagram of a step-down transformer in a motor starter control circuit. Wiring and schematic diagrams generally include a parts list with descriptive information which complements that provided by the drawing.

BLOCK DIAGRAM FORM

Block diagrams are used to divide electrical and electronic circuits into two or more squares, rectangles, or blocks. Each *block* shows the working relationships of the components at each stage. Arrowheads on drawings at the terminal ends of the lines show the direction of the signal path from input to output.

Heavy weight lines are used to represent the components in the *power portion* of the electrical circuit. Light weight lines display the components in the *control portion*. Figure 54-4 provides an example of the power and control portions of a block diagram of an industrial electronics circuit.

Sometimes, pictures or technical illustrations are used instead of blocks. Connections inside and outside the components are shown in wiring or *connection diagrams*.

COLOR CODES

Color codes are applied in electrical and electronics work to identify wire leads, to show where

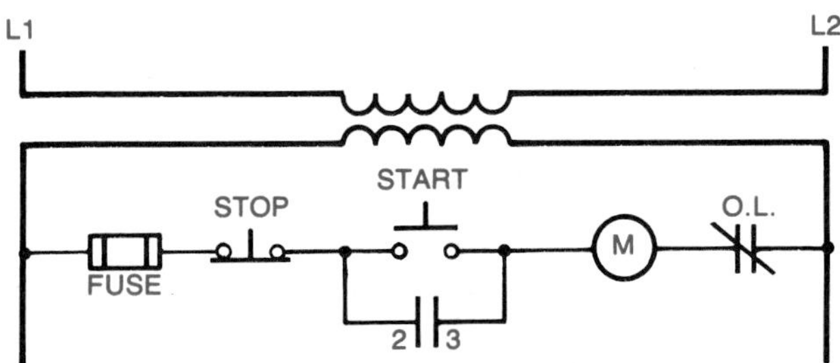

FIGURE 54-3 SCHEMATIC (LINE) DIAGRAM OF A CONTROL CIRCUIT USING A STEP-DOWN TRANSFORMER (COURTESY OF THE ALLEN-BRADLEY COMPANY)

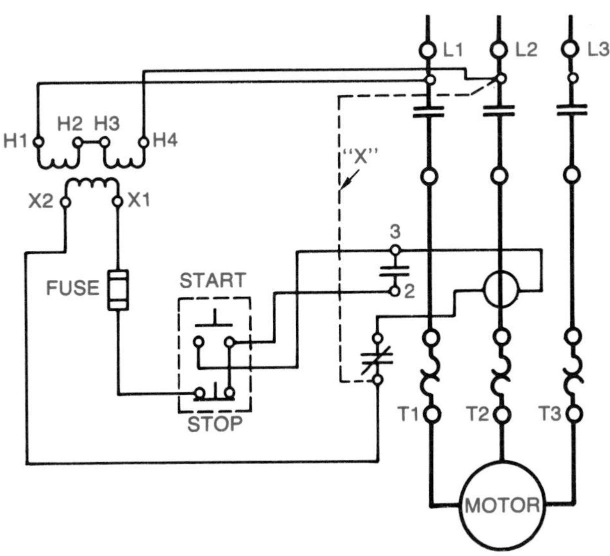

FIGURE 54-4 WIRING DIAGRAM PREPARED FROM A SCHEMATIC OF A SPECIFIC MOTOR CONTROL CIRCUIT (COURTESY OF THE ALLEN-BRADLEY COMPANY)

the wires are connected, and to indicate particular characteristics of components. Numbers are often assigned to each specific wire in a circuit. A color code may be included in a chart similar to figure 54-5.

COLOR	ABBREVIATION	NUMERAL*
Black	BLK	0
Brown	BRN	1
Red	RED	2
Orange	ORN	3
Yellow	YEL	4
Green	GRN	5
Blue	BLU	6
Violet	VIO	7
Gray	GRA	8
White	WHT	9

**Note.* The color code numerical sequence is often changed by manufacturers. Also, additional shades of colors are used.

FIGURE 54-5 STANDARD EIA COLOR CODE

ELECTRICAL DRAWING SYMBOLS

The uniform interpretation of electrical drawings requires the use of standardized symbols. Some of the most commonly used symbols are displayed in figure 54-6. More complete tables are identified in architectural, electrical, industrial electronics, and communications handbooks. Manufacturers' tables and technical literature also provide complementary information to the simplified specifications which appear on drawings. For example, complete information is provided for different types of switches, conductors, coils, resistors, and so on.

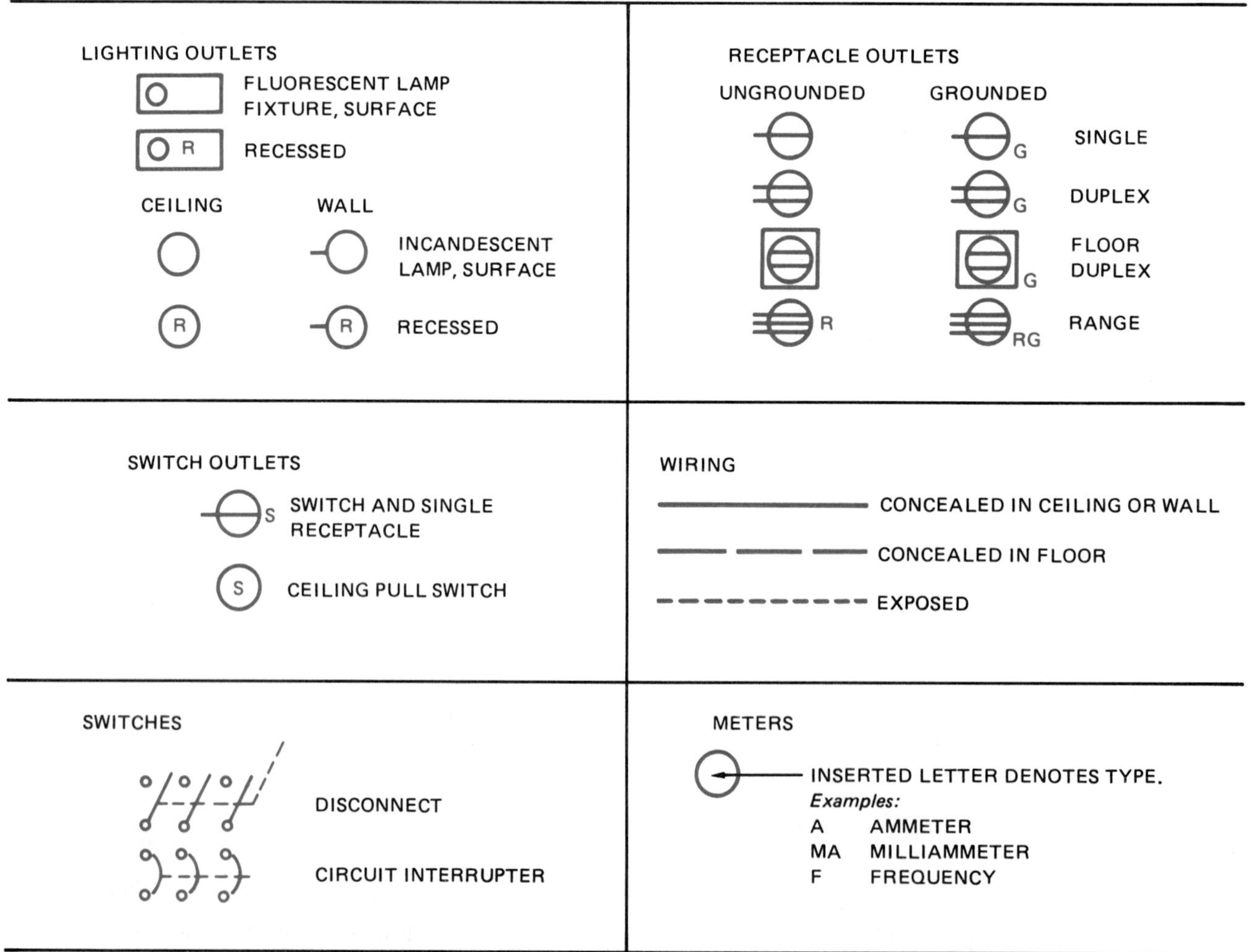

FIGURE 54-6 SELECTED EXAMPLES OF SYMBOLS COMMONLY USED FOR ELECTRICAL CIRCUIT DIAGRAMS

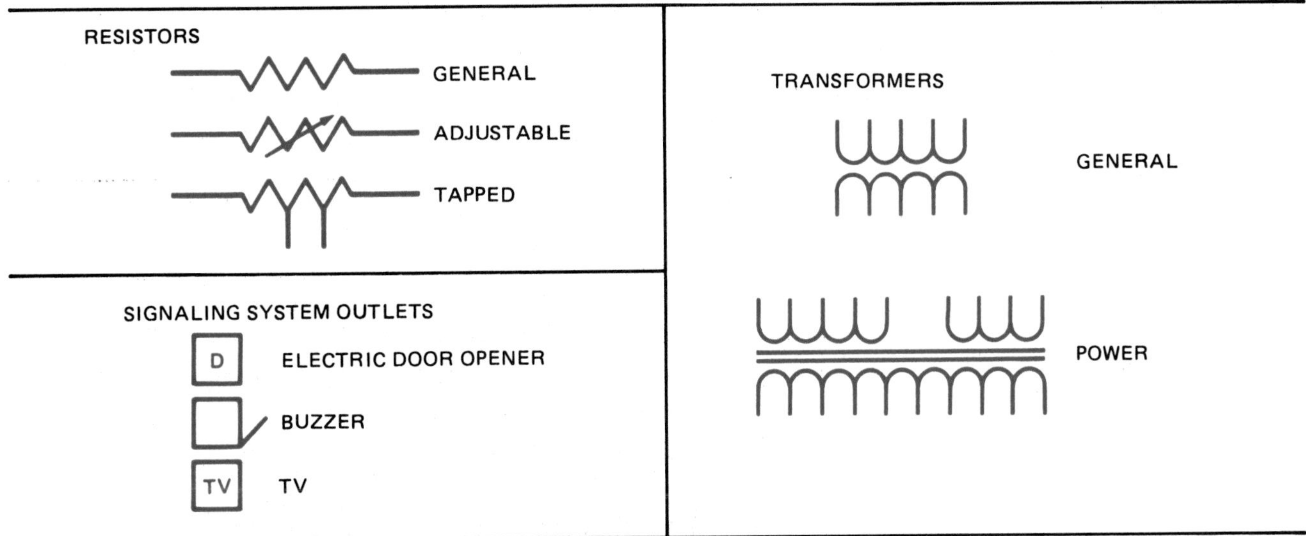

FIGURE 54-6 SELECTED EXAMPLES OF SYMBOLS COMMONLY USED FOR ELECTRICAL CIRCUIT DIAGRAMS

POINT-TO-POINT PICTORIAL PRODUCTION DRAWINGS

The *point-to-point pictorial production* (working) *drawing* is one of the easiest drawings to read. It is widely used to show each component in its place, the required wiring connections, and the completed unit as the worker sees it. Figure 54-7 illustrates the use of this type of drawing. Note that each component is identified by a three-digit number. Some of the parts are identified by a leader and number. Specifications for the numbered items are found in the parts list. The numerals **1**, **2**, **3**, and **4** represent terminals on the overload protector **(28)**

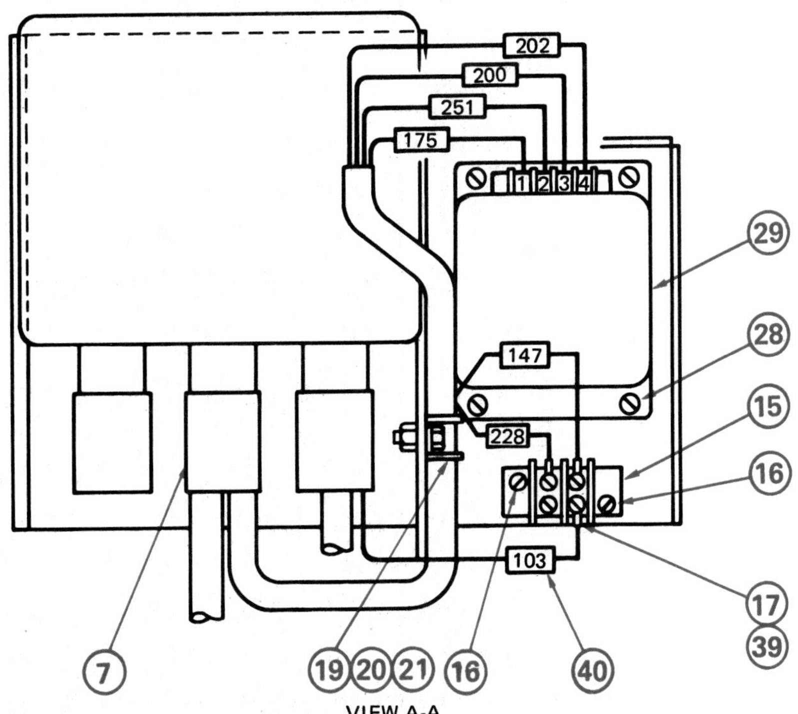

FIGURE 54-7 MODIFIED POINT-TO-POINT PICTORIAL PRODUCTION DRAWING (COURTESY OF TEREX DIVISION, GENERAL MOTORS CORPORATION)

NOTE: REFER TO APPENDIX B-9 FOR THE LARGER SCALE DRAWING TO USE WITH THIS ASSIGNMENT.

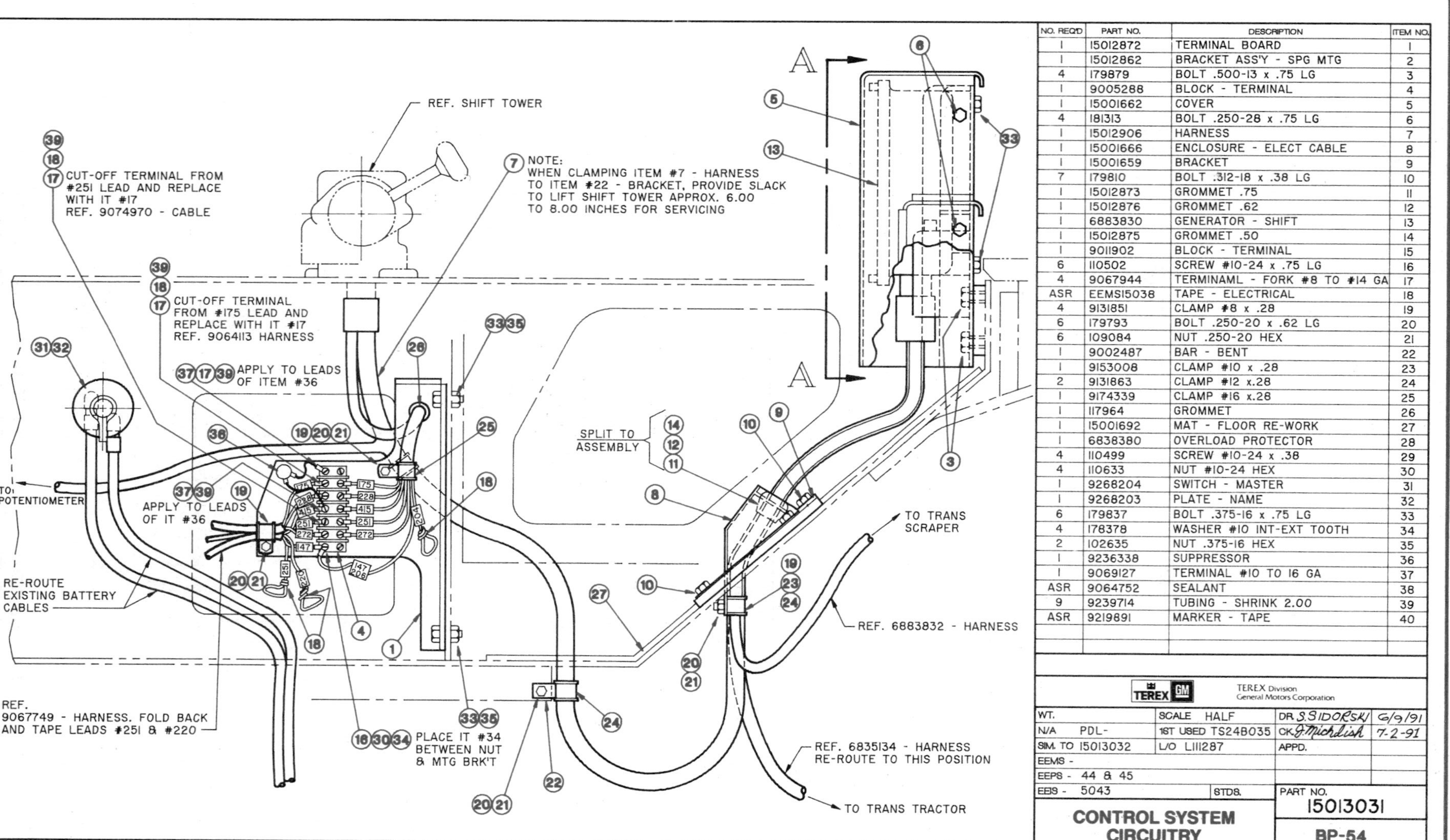

NO. REQD | PART NO. | DESCRIPTION | ITEM NO.
1 | 15012872 | TERMINAL BOARD | 1
1 | 15012862 | BRACKET ASS'Y - SPG MTG | 2
4 | 179879 | BOLT .500-13 x .75 LG | 3
1 | 9005288 | BLOCK - TERMINAL | 4
1 | 15001662 | COVER | 5
4 | 181313 | BOLT .250-28 x .75 LG | 6
1 | 15012906 | HARNESS | 7
1 | 15001666 | ENCLOSURE - ELECT CABLE | 8
1 | 15001659 | BRACKET | 9
7 | 179810 | BOLT .312-18 x .38 LG | 10
1 | 15012873 | GROMMET .75 | 11
1 | 15012876 | GROMMET .62 | 12
1 | 6883830 | GENERATOR - SHIFT | 13
1 | 15012875 | GROMMET .50 | 14
1 | 9011902 | BLOCK - TERMINAL | 15
6 | 110502 | SCREW #10-24 x .75 LG | 16
4 | 9067944 | TERMINAML - FORK #8 TO #14 GA | 17
ASR | EEMS15038 | TAPE - ELECTRICAL | 18
4 | 9131851 | CLAMP #8 x .28 | 19
6 | 179793 | BOLT .250-20 x .62 LG | 20
6 | 109084 | NUT .250-20 HEX | 21
1 | 9002487 | BAR - BENT | 22
1 | 9153008 | CLAMP #10 x .28 | 23
2 | 9131863 | CLAMP #12 x.28 | 24
1 | 9174339 | CLAMP #16 x.28 | 25
1 | 117964 | GROMMET | 26
1 | 15001692 | MAT - FLOOR RE-WORK | 27
1 | 6838380 | OVERLOAD PROTECTOR | 28
4 | 110499 | SCREW #10-24 x .38 | 29
4 | 110633 | NUT #10-24 HEX | 30
1 | 9268204 | SWITCH - MASTER | 31
1 | 9268203 | PLATE - NAME | 32
6 | 179837 | BOLT .375-16 x .75 LG | 33
4 | 178378 | WASHER #10 INT-EXT TOOTH | 34
2 | 102635 | NUT .375-16 HEX | 35
1 | 9236338 | SUPPRESSOR | 36
1 | 9069127 | TERMINAL #10 TO 16 GA | 37
ASR | 9064752 | SEALANT | 38
9 | 9239714 | TUBING - SHRINK 2.00 | 39
ASR | 9219891 | MARKER - TAPE | 40
TEREX GM
TEREX Division General Motors Corporation
WT. N/A
PDL-
SIM. TO 15013032
EEMS -
EEPS - 44 & 45
EES - 5043
SCALE HALF
1ST USED TS24B035
L/O L111287
DR. S.SIDORSKI 6/9/91
CK. 7-2-91
APPD.
STDS.
PART NO. 15013031
CONTROL SYSTEM CIRCUITRY
BP-54
REF. SHIFT TOWER
NOTE: WHEN CLAMPING ITEM #7 - HARNESS TO ITEM #22 - BRACKET, PROVIDE SLACK TO LIFT SHIFT TOWER APPROX. 6.00 TO 8.00 INCHES FOR SERVICING
SPLIT TO ASSEMBLY
TO TRANS SCRAPER
REF. 6883832 - HARNESS
REF. 6835134 - HARNESS RE-ROUTE TO THIS POSITION
TO TRANS TRACTOR
PLACE IT #34 BETWEEN NUT & MTG BRK'T
CUT-OFF TERMINAL FROM #251 LEAD AND REPLACE WITH IT #17 REF. 9074970 - CABLE
CUT-OFF TERMINAL FROM #175 LEAD AND REPLACE WITH IT #17 REF. 9064113 HARNESS
APPLY TO LEADS OF ITEM #36
APPLY TO LEADS OF IT #36
TO POTENTIOMETER
RE-ROUTE EXISTING BATTERY CABLES
REF. 9067749 - HARNESS. FOLD BACK AND TAPE LEADS #251 & #220
A
A

Courtesy of General Motors Corporation, Terex Division (Modified Industrial Drawing)

ASSIGNMENT—UNIT 54

CONTROL SYSTEM CIRCUITRY (BP-54)

1. State two functions of standardized graphic symbols for electrical/electronic drawings.
2. a. Identify two types of standard ANSI electrical drawings.
   b. Give the function of each type of drawing.
3. a. State two purposes for color coding in electrical/electronics circuits.
   b. Give the ANSI abbreviation and number for the following color wires:
   (1) Black (2) Orange (3) White
4. List two advantages of printed circuit boards over the conventional drawing of a circuit.
5. Refer to a table of electrical/electronics symbols. Sketch the symbol (or use a template) for each of the following items:
   a. Fuse
   b. Ground
   c. Heater
   d. Circuit breaker (general)
   e. Antenna
   f. Milliammeter
   g. Transformer
   h. Motor (general)
   i. Wall junction box
   j. Grounded range outlet
6. Identify two unique characteristics of the combination drawing for the Control System Circuitry.
7. Explain briefly what purpose is served by the phantom lines on the drawing.
8. Give (a) the identification number, (b) part number, and (c) required quantity of the following items:
   a. Terminal board
   b. Harness
   c. Electrical cable enclosure
9. Identify (a) the part name, (b) number, and (c) quantity for items (28), (31), and (36).
10. Give the item number and parts list information for securing the electrical cable enclosure.
11. Tell how the ends of wires [204], [220], and [251] are treated.
12. State what the worker is to do in relation to each of the following items:
    a. **REF 9067749–HARNESS**
    b. **REF 9074970–CABLE**
    c. **REF 9064113–HARNESS**
13. Determine the required slack to leave to the lift shift tower for harness (7) for servicing.
14. Identify the parts to which the following REF items lead:
    a. **REF 6883832–HARNESS**
    b. **REF 6835134–HARNESS**

Student's Name __________

1. (1) __________
   (2) __________

2.

	Drawing Type (a)	Function (b)
(1)		
(2)		

3. a. (1) __________
   (2) __________
   b. (1) _____ _____
   (2) _____ _____
   (3) _____ _____

4. a. __________
   b. __________

5. a. _____ f. _____
   b. _____ g. _____
   c. _____ h. _____
   d. _____ i. _____
   e. _____ j. _____

6. a. __________
   b. __________

7. __________

8.

	Item	Part #	Qty
a.			
b.			
c.			

9.

	Part Name	Part #	Qty
(28)			
(31)			
(36)			

10. __________

11. __________

12. a. __________
    b. __________
    c. __________

13. __________

14. a. _____ b. _____

UNIT 55
Electronic Components, Circuits, and Schematic Drawings

Electronics, as a branch of electricity, deals with the behavior and applications of semiconductor devices and the circuits in which they are used. Generally, electronics is associated with low voltage, amperage, and signal paths.

Schematic, pictorial, highway (harness), and cable diagrams (with slight modifications) are commonly used for both electrical and electronic drawings. Electrical drawings usually relate to the transfer of electricity and to high-voltage applications.

Attention is directed in this unit to electronic diagrams, block diagrams, conventional and CADD schematic and connection (wiring) diagrams, integrated and logic circuit schematics, semiconductor and other electronic symbols, printed board circuitry, and digital electronic circuits.

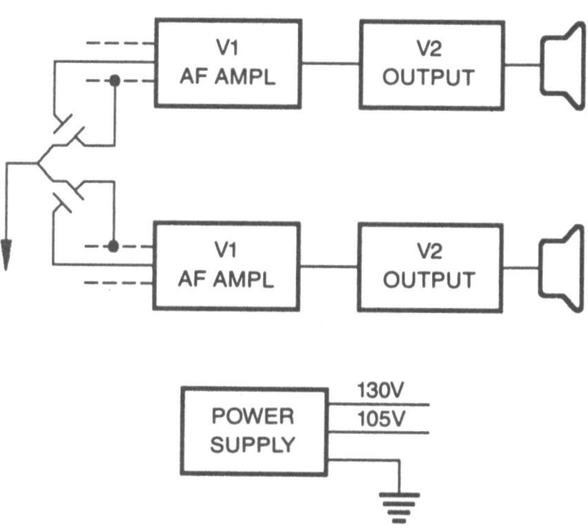

FIGURE 55-1 BLOCK DIAGRAM PREPARED FROM A SCHEMATIC DRAWING

ELECTRONIC DIAGRAMS

Block Diagrams

A block diagram contains a series of block symbols. Each symbol identifies a different component in a piece of equipment, an electronic product, or a complete system. Details are omitted so that the relationship between the different components and the major function of the system may be quickly interpreted. The block also includes basic standard graphic symbols that represent electronic components, power sources, switching devices, etc. A simple block diagram is illustrated in figure 55-1.

Schematic Diagrams

A *schematic diagram* provides technical information about basic electronic circuit connections. Schematic diagrams are often drawn to show various *stages (functions)*. For instance, one stage of a schematic diagram may show the components and circuitry for *amplification*. The diagram may continue with a second stage relating to *output* and a third stage dealing with power supply.

A letter or letters that are encircled with a number are known as *reference designations.* Schematic diagrams include values and other supplier information that relate to each component. Table 55-1 identifies reference designations of a few major electronic components.

TABLE 55-1 REFERENCE DESIGNATIONS FOR SELECTED ELECTRONIC COMPONENTS

COMPONENT	REFERENCE DESIGNATION	COMPONENT	REFERENCE DESIGNATION
Semiconductor	CR	Inductor	L
Resistor	R	Transistor	Q
Capacitor	C	Transformer	T
Diode	D	Switch	S

ELECTRONIC COMPONENT SYMBOLS

Symbols are used to represent components and connections on electronic schematic drawings. A brief description follows of the functions of some common semiconductors that are included on electronic drawings. Examples of basic symbols that represent semiconductors and other electronic devices on *elementary (schematic) drawings* are displayed graphically in figure 55-2. A more complete Table of Electronic Symbols is provided in the Appendix (Table A-11).

- **Semiconductor** (**CR**). This device is designed to control the *degree of resistance* in an electronic circuit. There are a number of different types of semiconductors; for example, diodes, transistors, integrated circuits, silicon controlled rectifiers, DIAC's, and TRIAC's. Semiconductor devices allow for the free passage of current under certain conditions and blocking the current under other conditions.
- **Capacitor Components** (**C**). A capacitor in an electronic circuit serves to *oppose a change* in voltage and to store electronic energy.
- **Inductors** (**L**). As a conductor, an inductor is designed to *concentrate a magnetic field.* The windings of an inductor have *inductance properties* in an electronic circuit. Inductance properties are needed to oppose a change in current flow or for the storage of electron energy in such equipment as a transformer.
- **Resistors** (**R**). Resistor devices *reduce the flow of current* to protect a circuit. Resistors are available with fixed values or as variable resistors that permit adjusting the resistance.

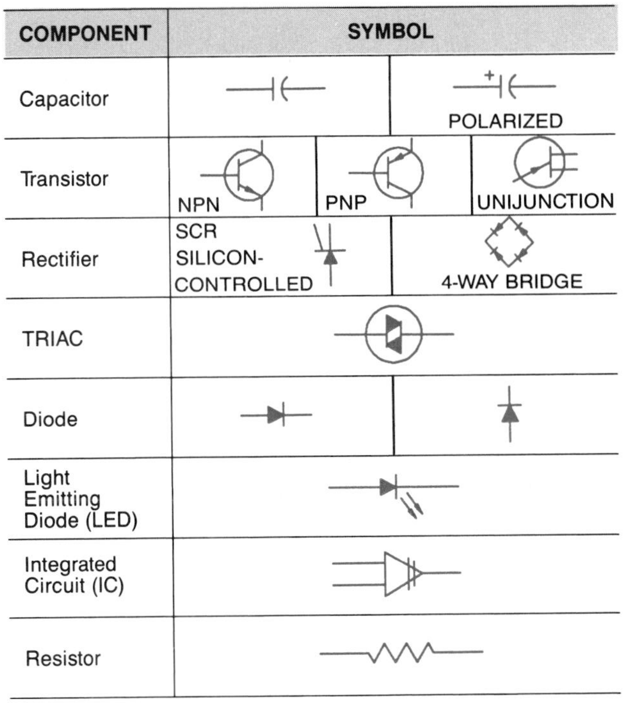

COMPONENT	SYMBOL	
Capacitor		POLARIZED
Transistor	NPN	PNP / UNIJUNCTION
Rectifier	SCR SILICON-CONTROLLED	4-WAY BRIDGE
TRIAC		
Diode		
Light Emitting Diode (LED)		
Integrated Circuit (IC)		
Resistor		

FIGURE 55-2 SAMPLE ELECTRONIC COMPONENT SYMBOLS

- **Diodes** (**D**). A diode is a semiconductor that permits *current to flow only in one direction.* Diodes rectify a current and are applied in switching processes. Two common types of diodes include power diodes and switching diodes. Power diodes and rectifiers change AC to pulsating DC in power supplies. Signal diodes are designed for light-duty signal applications (related to detection, modulation, and curve changing).
- **Silicon Controlled Rectifiers** (**SCR**). Another name for SCR's is *thyristors.* Thyristors are semiconductors that are designed for *speed controls,* applications for *controlling AC and DC power* in both AC and DC circuits, *regulating* a power load, and *phase control.*

A silicon controlled rectifier controls current in one direction by a positive gate signal. There are three terminals on an SCR; anode, cathode, and gate.

- A **TRIAC** (three element AC switch) is a bidirectional triode thyristor. A TRIAC is employed to *control current in either direction* by either a positive or a negative gate signal.
- A **DIAC** (two-element AC switch) is used in combination with a TRIAC, or other

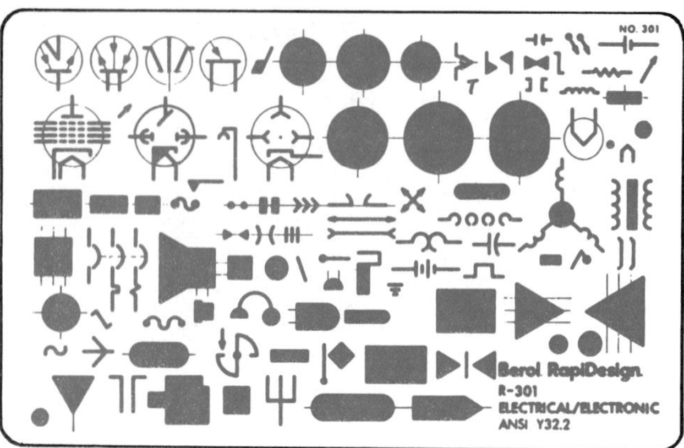

FIGURE 55-3 STANDARD DRAFTING TEMPLATE

SCR, to *trigger* these devices, using either positive or negative pulses.

- **Transistors** (Q). A *transistor* is a device for controlling electron flow for purposes of *amplifying* power or voltage, *oscillation*, and *switching*. The control function is accomplished by varying the amount of voltage applied to the three elements of a transistor.

GENERATING ELECTRONIC SCHEMATIC DIAGRAMS

The two prime methods of generating electronic schematic diagrams include the use of a drafting template and computer aided design and drafting (CADD). The drafting template (figure 55-3) contains recommended ANSI Y 32.2 symbols. Generally, a layout sketch is prepared on a graph sheet. The sketch translates design and engineering information into a schematic diagram.

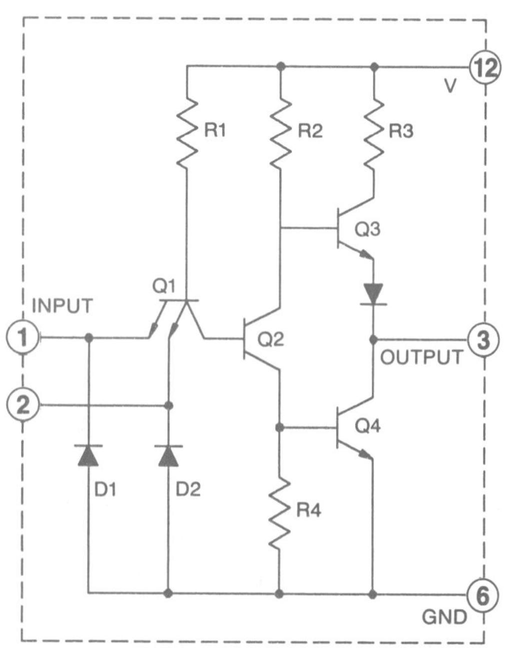

FIGURE 55-4 SCHEMATIC SHOWING INTERNAL CIRCUITRY OF AN INTEGRATED CIRCUIT

Labels and circuit notations are then added. Starting at the left of a schematic, a letter symbol of a component is followed by a number that gives the sequence of a part in a circuit. Components are numbered in sequence according to *level (layer)* or by the grouping of components according to *series*.

Numbering as applied to layers means that at each level represented on a drawing numbers appear in sequence, starting at the left. *Series designations* include the part symbol and a sequential number in the series. For example, three capacitors in a **200** series are identified on a schematic diagram as **C201**, **C202**, and **C203**.

CADD INTEGRATED ELECTRONIC CIRCUIT (IC) SCHEMATICS

An *integrated electronic circuit* (IC) refers to a microminiature component that is manufactured into a compacted, inseparable unit called a *chip*. Integrated electronic circuits are represented on schematic drawings with two exceptions: (1) transistors are not circled and (2) a dash outline box may be used to represent the unit containing the integrated circuit (IC). An integrated circuit schematic drawing is illustrated by figure 55-4.

Integrated Circuit Symbols and Logic Circuits

Logic circuits are computer-oriented circuits. A schematic for a logic circuit consists of an electronic schematic diagram and a *flow diagram*. Symbols are used in the flow diagram to identify the *self-contained components* for a specific function. In other words, a logic schematic diagram shows the logical sequence of events that cut across an electrical or electronic system.

The integrated circuit symbol (figure 55-5) represents the integrated circuit shown in figure 55-4. The IC symbol, in this case, is known as a *gate*. The gate permits an electronic system to operate when the required input conditions are fulfilled. The numbers that appear in figure 55-4

FIGURE 55-5 IC SYMBOL REPRESENTING AN INTEGRATED CIRCUIT SCHEMATIC

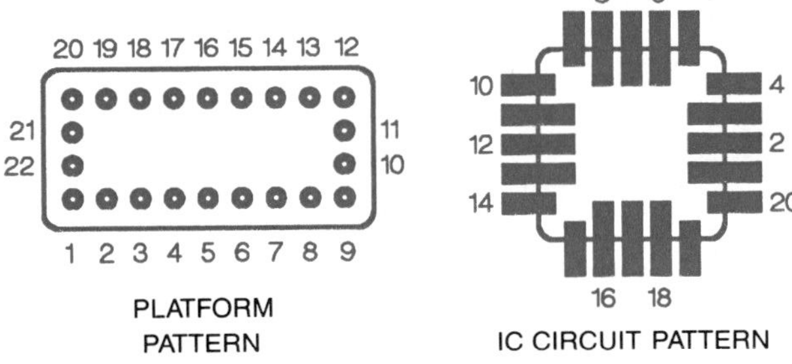

FIGURE 55-6 EXAMPLES OF MOUNTING PADS

with input and output lines *(pins)* identify the connections for the integrated circuit.

PRINTED CIRCUIT BOARD TECHNOLOGY

The term *printed circuit* refers to the circuitry (interconnections) between electronic components (devices). The circuits are printed on paper, glass, and plastic products. Electronic circuits may be printed on one or both sides of a circuit board. More than one circuit board may be required for complex electronic equipment.

Printed circuit boards include *pads (lands)* that provide termination locations for attaching electronic devices. Pads may be located individually or to conform to a particular pattern depending on the connection requirements of the electronic devices. Two examples of mounting patterns are illustrated in figure 55-6.

Conductor Traces (Lines)

Conductor trace (lines) are used to connect pads to complete the circuit pattern. Commercially available art work provides different *configurations* for using conductor traces in designing printed circuits (figure 55-7). Full information is given on circuit art work for size reduction, datums, referencing marks, identification, etc.

Photoprinting and Etching Circuit Boards

A final printed circuit board is produced from a printed circuit layout by a photoprinting process. In this process, a copper conductor or other material is deposited to form a printed circuit. The board is then lacquer coated to prevent corrosion.

The printed circuit board may also be produced by the *etched process*. A photographically reduced printed circuit layout negative is used when exposing a copper-covered base that is coated with a light-sensitive emulsion. The printed circuit is then etched, rinsed, and given a protective coating.

PICTORIAL ELECTRONIC DIAGRAMS

Pictorial assembly drawings are widely used in the manufacture of electronic components and products and for marketing purposes. Pictorial drawings provide a simplified technique for displaying general features and physical arrangements of components within an electronic product.

Pictorial assembly drawings may be of many different types. Examples include: a line drawn picture of a part; an exploded technical illustration; a composite pictorial and wiring diagram, and other combinations of drawings that represent objects as they appear.

DIGITAL ELECTRONIC CIRCUITS

A *digital electronic circuit* (sometimes referred to as a *binary circuit*) permits a two-state switching operation such as *on* or *off* and *high* or *low*. Signals that represent characters or numbers are used in digital electronics. Digital electronics is based on pulse-type signals as compared with *analog electronic equipment* and processes that require uniformly changing signals (like sine waves).

Digital equipment incorporates two basic logic circuits. These relate to *decision-making* and *memory* circuits. Data is stored in binary circuits whereas in decision-making circuits dependence is placed on maintaining binary inputs and the characteristics of the logic circuit to produce outputs.

LOGIC DIAGRAMS

This type of electronic drawing uses graphic symbols for logic elements and relationships to represent a device or a system. Logic circuits are employed as *gates or switching circuits* in computer and control systems. Logic symbols appear on a drawing to graphically represent a logic circuit. Diodes, transistors, and other electronic devices are included in the design of logic circuits.

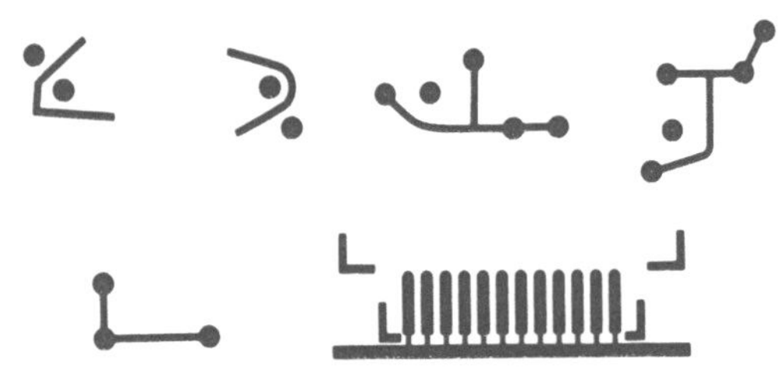

FIGURE 55-7 CONDUCTOR TRACES FOR PRINTED CIRCUITS

NOTE: REFER TO APPENDIX B-10 FOR THE LARGER SCALE DRAWING TO USE WITH THIS ASSIGNMENT.

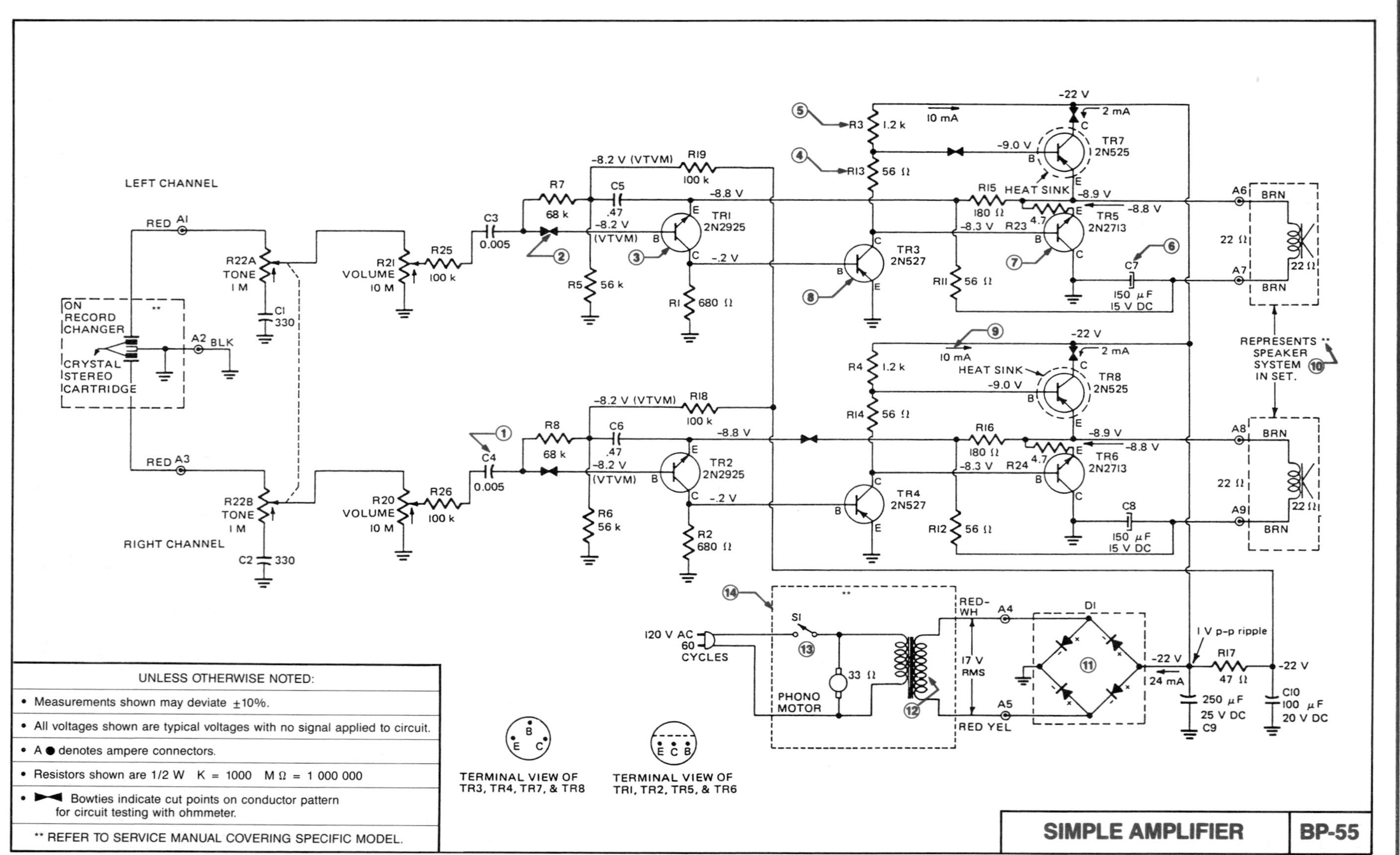

Adapted Drawing (Courtesy of General Electric)

ASSIGNMENT—UNIT 55

SIMPLE AMPLIFIER (BP-55)

1. Name the type of electronics drawing used to represent the **SIMPLE AMPLIFIER (BP-55)**.
2. State the prime function of a block diagram.
3. a. Identify the letter symbols that represent the following components in an electronic diagram.
   (1) Silicon-controlled rectifier
   (2) Transistor (NPN type)
   (3) Integrated circuit
   b. Provide a sketch of each symbol.
   c. Describe briefly the function of each component.
4. Tell (a) how a logic circuit is represented and (b) what a logic schematic shows.
5. Identify the functions of (a) lands (pads) and (b) conductor traces on a printed circuit board.
6. State the difference between digital electronic and analog electronic equipment signals.

Refer to the **SIMPLE AMPLIFIER** diagram (**BP-55**).

7. List the input current specifications.
8. Provide the name and specifications or values of the following components or features.
   ① ② ③ ④ ⑤ ⑥ ⑦ ⑧ ⑨
9. Give the letter designation and the total number of ampere connectors.
10. Identify the following electronic components.
   ⑪ ⑫ ⑬
11. State what the following details represent on the drawing.
   ⑩ Double bullet symbol (••).
   ⑭ Dotted block outline.
12. Give the allowable tolerance for all measurements.

Student's Name ____________________

1. ____________________
2. ____________________
____
3. a. (1) ________ (2) ________ (3) ________

b.(1) | (2) | (3)

c. (1) ____________________
(2) ____________________
____
(3) ____________________
____
4. a. ____________________
____
b. ____________________
____
5. a. ____________________
b. ____________________
6. ____________________
____
7. ____________________
8. ① ____________________
② ____________________
③ ____________________
④ ____________________
⑤ ____________________
⑥ ____________________
⑦ ____________________
⑧ ____________________
⑨ ____________________
9. ____________________
10. ⑪ ____________________
⑫ ____________________
⑬ ____________________
11. ⑩ ____________________
⑭ ____________________
12. ____________________

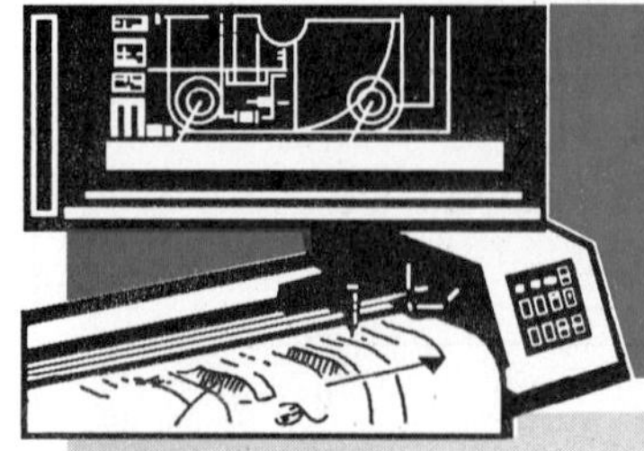

HIGH TECHNOLOGY applications

HT-16 CADD/CNC/CAM-NONTRADITIONAL MACHINING: ELECTRICAL DISCHARGE MACHINING (EDM)

Solid electrode and wire-cut models of electrical discharge machines are identified as *nontraditional machine tools.* These pieces of equipment and accessories are compatible with NC and CNC and may be interlocked into a flexible manufacturing system (FMS).

NC AND CNC WIRE-CUT ELECTRICAL DISCHARGE MACHINES (EDM)

Wire-cut electrical discharge machines (EDM) include: (1) a *NC* and/or *CNC* unit to perform such tasks as: program editing, wire/workpiece positioning, table movement, and feed rate setting; (2) an *EDM power supply* to provide controlled spark energy; and (3) a *dielectric fluid system* for maintaining flow rate and fluid temperature, removing chips, and controlling the purity of the dielectric fluid.

Metal cutting takes place by *erosion.* Erosion occurs by action of the EDM sparking energy that produces a spark between an electrode wire and the surface of a workpiece. The intense heat causes erosion to take place on both the electrode wire and the workpiece. The greater amount of erosion on the workpiece permits the wire to advance continuously along a controlled path in the workpiece.

The sparking area is flooded by a dielectric fluid that serves to do three things: (1) flow (flush) the fine metal erosion (chip) particles away; (2) maintain a constant electrode and workpiece temperature; and (3) surround the sparking area with a clean electrical conductor.

SOLID ELECTRODE MODEL EDM

Solid-electrode EDM models are designed for contour (sculpturing) machining; forming cavities; producing helixes for gears, worm shafts, and internal threads; and for sharp corner (edge) machining.

A preformed solid electrode is used to machine such forms as those illustrated at (A) through (E). The electrodes are made of materials that are good conductors of electricity. Examples include: graphite, copper, tungsten copper, and brass, to name a few.

Commonly used dielectric fluids include deionized water, low-viscosity petroleum oils, ethylene glycol, and other solutions. Metal removal rates and quality of surface finish depend on the electrode material, amperage, arcing frequency, and the nature of the machining process.

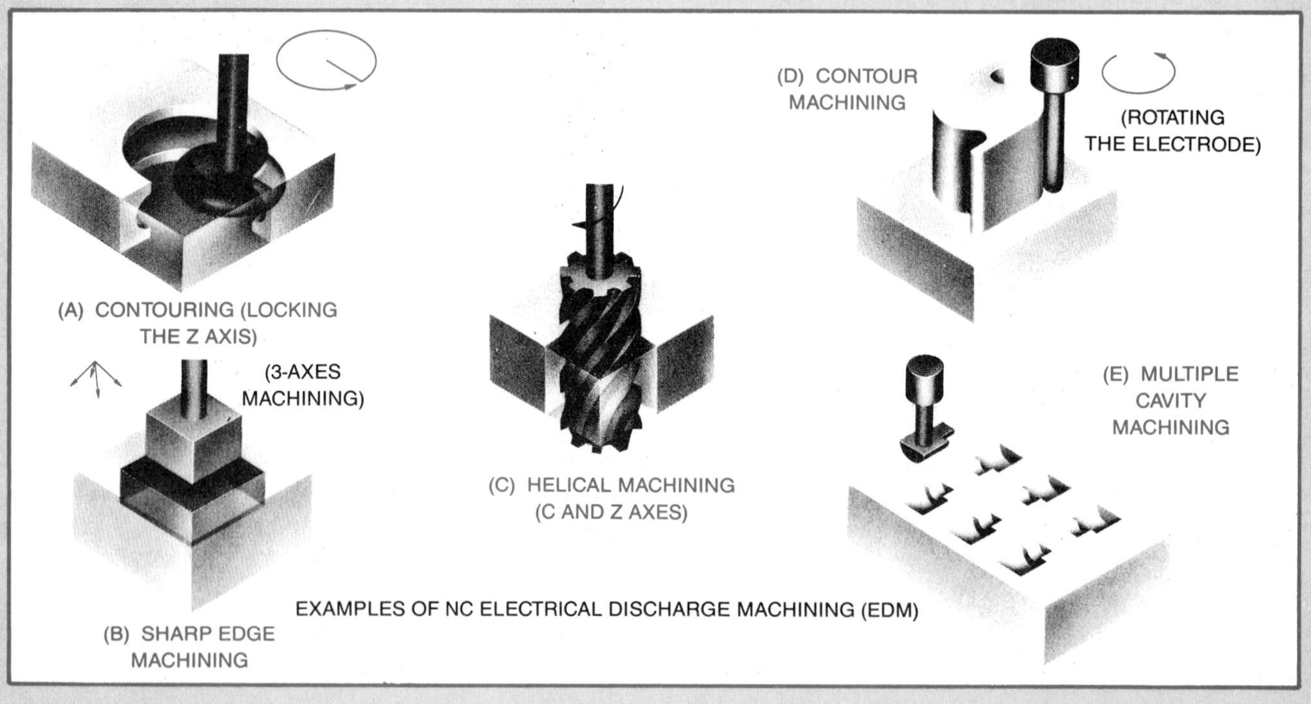

EXAMPLES OF NC ELECTRICAL DISCHARGE MACHINING (EDM)

Courtesy of SODICK INC.

UNIT 56
Aeronautical and Aerospace Industries Drawings

Full details about the design and features of aeronautical and aerospace parts, with dimensions and specifications for construction, assembly, and testing, are provided on conventional drawings for a particular industry or product. For example, standard drafting room practices are followed in representing and dimensioning parts which are die cast, mold cast, forged, stamped, machined, or produced by other manufacturing methods.

Aerospace products also require the use of specialized types of drawings such as mechanical, electrical, electronics, instrumentation, and hydraulics. Many different techniques of representation and kinds of drawings are used. Single- and multiple-view orthographic layout drawings; regular and cut-away sections; phantom outlines and exploded views; schematic and installation sketches; detail, working, and assembly drawings; and other types of drawings and drafting room practices are used. In addition, the aerospace industry uses standardized terminology, representation, and dimensioning techniques which are unique. These are applied to aircraft and other aerospace vehicles. One of the general three-view aerospace drawings is displayed in figure 56-1.

VIEWING OF PARTS AND ASSEMBLIES

External views on aircraft drawings are generally drawn from one of three positions. These are also shown in figure 56-1.

PRODUCTION DRAWING IDENTIFICATION

Seven-Digit Part-Numbering System

One of the systems used for identifying design and production drawings contains seven digits. A chart, developed by the company, shows the classifications for *model number* of the aircraft *(digits* 1 *and* 2*), major assembly group (digit* 3*),* and *subdivision* or *minor assembly group (digit* 4*).* Usually, *digits* 5, 6, *and* 7 distinguish the *type number* and the *drawing number.* Figure 56-2 provides an example of how a layout drawing is identified.

Identifying Parts of an Assembly

A seven-digit part or assembly identification number may be followed by the use of a dash and numerals. Each dash and numeral is circled and with a leader is used to identify each separate part (figure 56-2).

AIRCRAFT LAYOUT DRAWINGS

Aircraft layout drawings are, generally, the last design development drawings which include all the parts within a particular mechanism or unit. Layout drawings relate all design, production, and maintenance problems to the overall operation, construction, adjustment, assembly, and installation of the mechanism or unit.

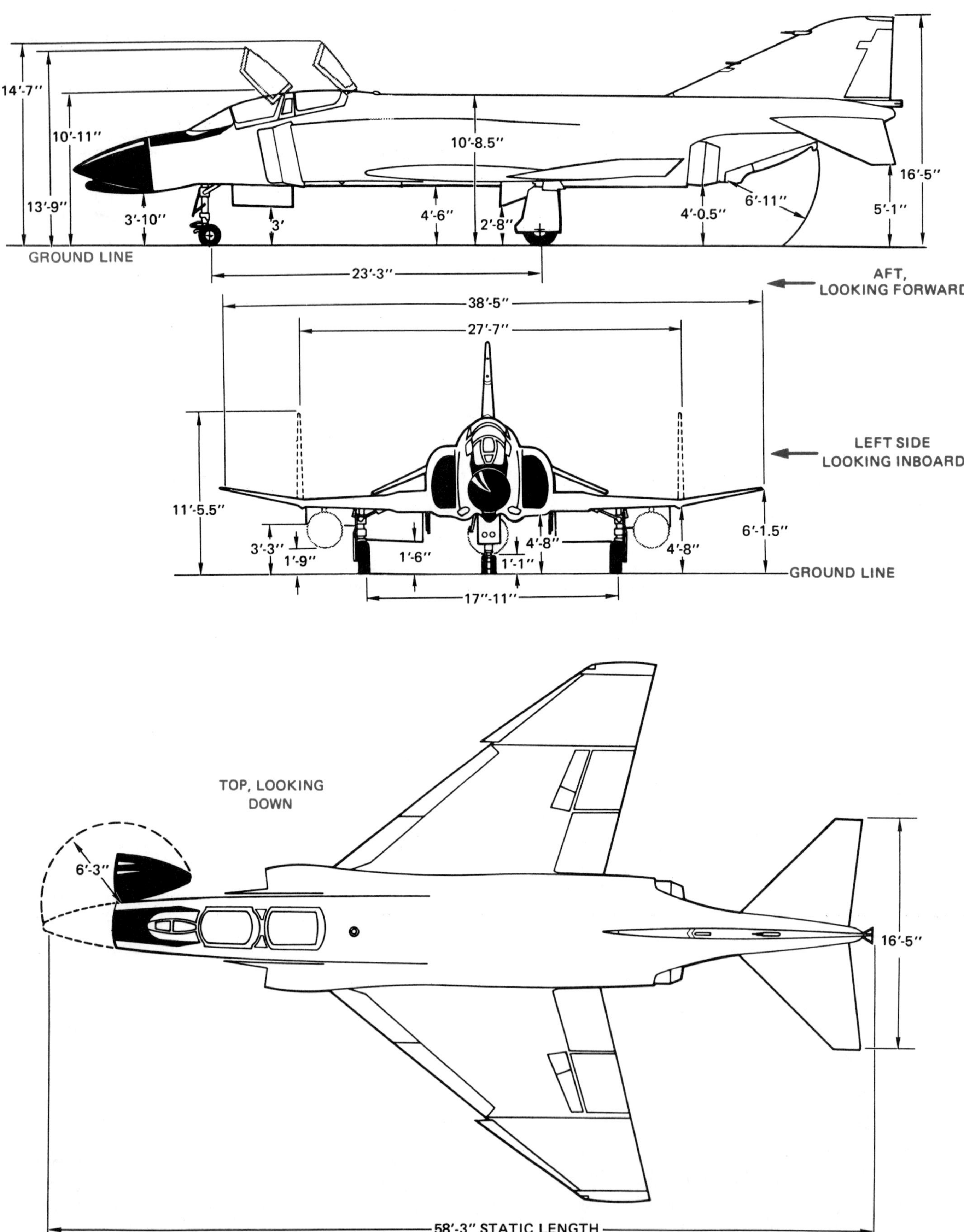

FIGURE 56-1 GENERAL PURPOSE THREE-VIEW AEROSPACE DRAWING SHOWING POSITIONS FROM WHICH EXTERNAL FEATURES ARE REPRESENTED (COURTESY OF MCDONNELL DOUGLAS CORPORATION)

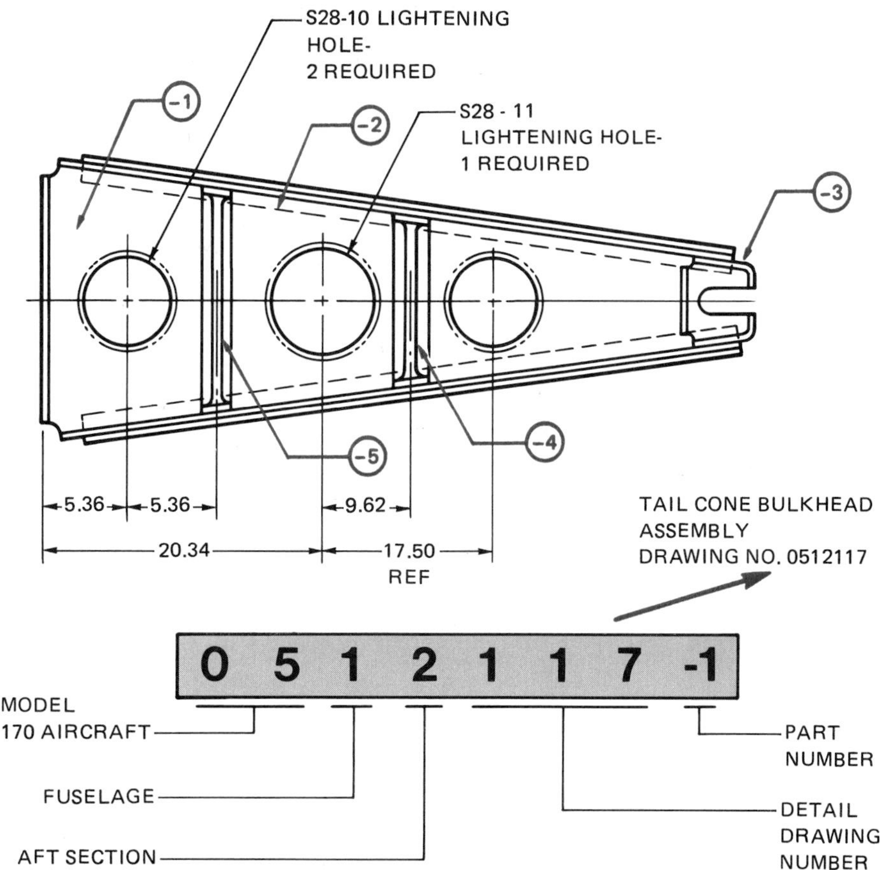

FIGURE 56-2 IDENTIFICATION OF LAYOUT DRAWING FOR A BULKHEAD ASSEMBLY AND PARTS

LOFTING DRAWINGS

Lofting means that the shapes, sizes, and contours of an aerospace part, structure, or mechanism are laid out full size. Parts ranging from simple, flat (plane) surfaces to complex bends and contours are prepared by lofting.

Ordinate tables of standard station dimensions and other lofting specifications have been standardized by industry advisory committees and the National Aeronautics and Space Administration (NASA).

Reference Lines

This means the loftsperson uses the basic dimensions developed by engineering and establishes *reference lines* from which the actual contours will be developed. Reference lines are identified as ① *water lines*, ② *buttock lines*, and ③ *station lines* (Figure 56-3).

Water lines (as shown at A) are edge (side) views of horizontal planes which cut through the aircraft or aerospace vehicle. Water lines are viewed from the side. Water lines are numbered sequentially from point zero (**WL 0**). Lines on planes above point zero are positive (**WL 1**, **WL 2**, etc.); below, negative (**WL-1**, **WL-2**, etc.). Generally, each water line designation represents the number of inches or the dimension from **WL 0**. For example, **WL 6** indicates **6″** above the water line.

Buttock Lines (B) are parallel lines as seen by looking down on the aircraft. The buttock lines originate at **BUTT 0**. Buttock lines to the right of the zero reference (looking forward) are identified as **BL 1R**, **BL 2R**, etc. Buttock lines **BL 1L**, **BL 2L**, etc. are drawn to the left of the zero reference. Again, the number represents inches or the dimension (position) from **BUTT 0**.

Station lines represent edge views of vertical planes (figure 56-3C). The planes (section lines) are perpendicular to water and buttock lines. In the illustration, the number indicates the distance in inches or the dimension of the station line from the forward zero reference point.

(C) SAMPLE STATION LINES
(B) BUTTOCK LINES
BL24R
BL18R
BL12R
BL6R
℄
BL6L
BL12L
BL18L
BL24L
0
25
50
83
116
162
182
206
WL+30
WL+24
WL+18
WL+12
WL+6
WL0
WL−6
WL−12
WL−18
(A) WATER LINES

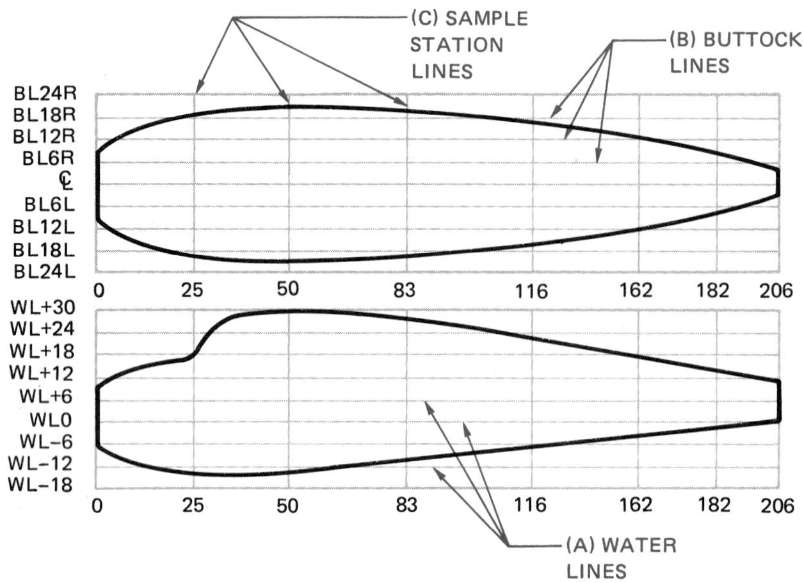

FIGURE 56-3 APPLICATION OF WATER, BUTTOCK, AND STATION LINES ON AN ENGINEERING LAYOUT

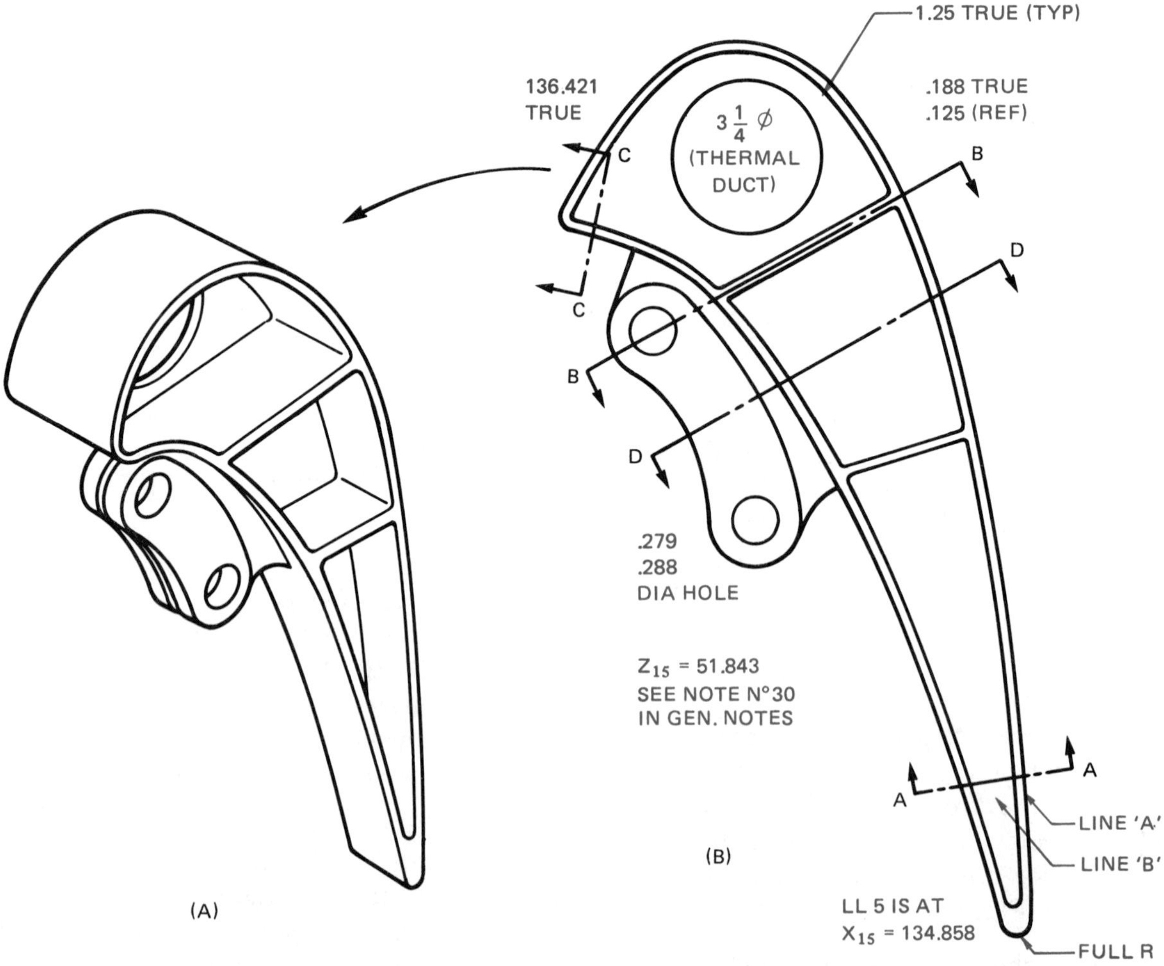

1.25 TRUE (TYP)
136.421 TRUE
$3\frac{1}{4}$ ⌀ (THERMAL DUCT)
.188 TRUE
.125 (REF)
C
C
B
B
D
D
.279
.288
DIA HOLE
Z_{15} = 51.843
SEE NOTE N°30
IN GEN. NOTES
A
A
LINE 'A'
LINE 'B'
LL 5 IS AT
X_{15} = 134.858
FULL R
(A)
(B)

FIGURE 56-4 UNDIMENSIONED AEROSPACE MACHINED PART (COURTESY OF MCDONNELL DOUGLAS CORPORATION)

METHODS OF SECURING PARTS

Common fasteners and methods of fastening are modified to meet particular requirements. Due to material thickness and special conditions governing the use of external parts, exacting specifications must be met for riveting, welding, bolting, securing, and bonding together sheet metal and other parts.

The aerospace industry utilizes standard tables of data and specifications such as rivet or weld type, design features for flush forming rivets and joints, numbers of rivets and welds, center distances, and other important information.

UNDIMENSIONED DRAWINGS

Full-size *undimensioned drawings* are used to check the fit of certain flat-pattern development and complicated-form machine parts. Figure 56-4 provides an example of a machined part (A) and the undimensioned drawing (B). It is apparent that if the drawing were fully dimensioned, due to the continuously changing profile a number of measurements would be unclear (and, possibly, wrong). Since parts produced from undimensioned drawings are machined on numerical control machines, positioning and machining information may be fed through a machine control unit. The machined part then conforms to the full-size, accurately drawn, undimensioned drawing.

PICTORIAL DRAWINGS GENERATED BY CADD/CAE/CAM SYSTEMS

Pictorial drawings are used to clearly illustrate the position of components and parts within different sections and subsystems in aeronautical and aerospace drawings. Figure 56-5 provides a good illustration of a pictorial drawing that shows in phantom outline the locations and relative relationship of principal components within a specific airplane.

This particular drawing was produced (in enlarged size) on a pencil plotter from the output of CADD/CAE/CAM systems from drawings related to major components within different sections of the aircraft. Generally, a phantom pictorial drawing contains leaders identifying the major components, section lines, and other specifications. This information is referenced to separate section, systems, components, and parts drawings. Each worker interprets the appropriate drawing for essential data, materials specifications, manufacturing processes information, quality controls, assembly procedures, and the like.

Courtesy of JAKA, INC.

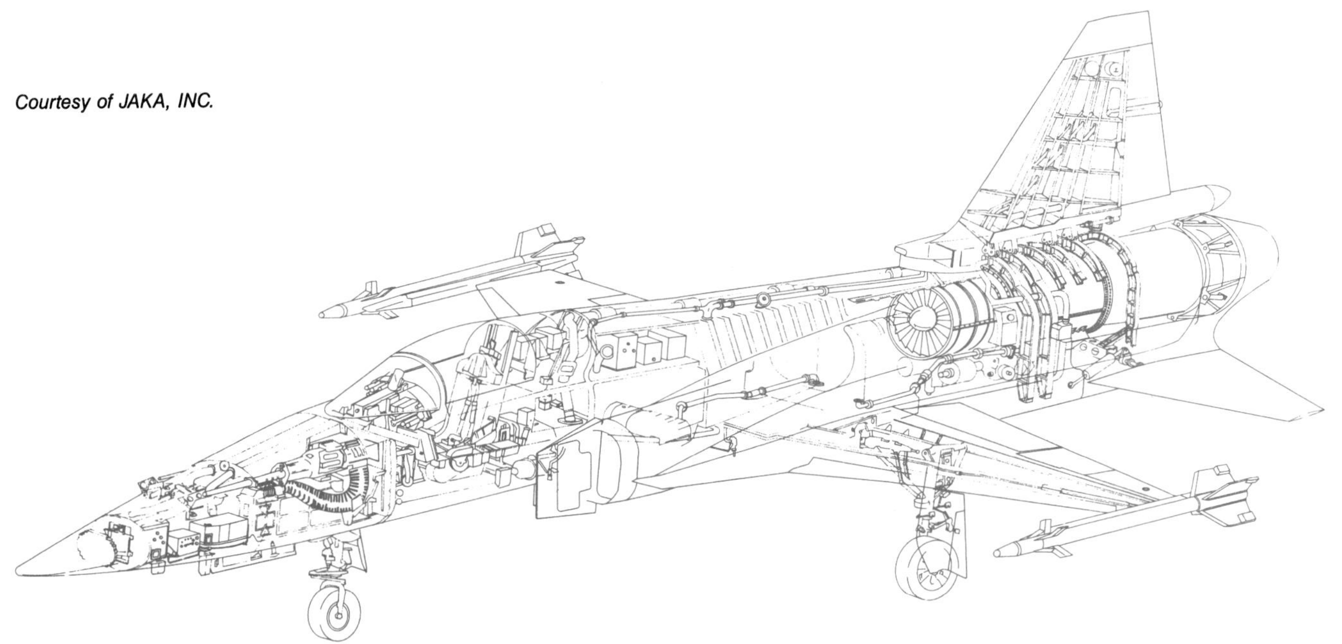

FIGURE 56-5 PICTORIAL DRAWING GENERATED BY CADD/CAE/CAM SYSTEMS AND REPRODUCED ON A PENCIL PLOTTER

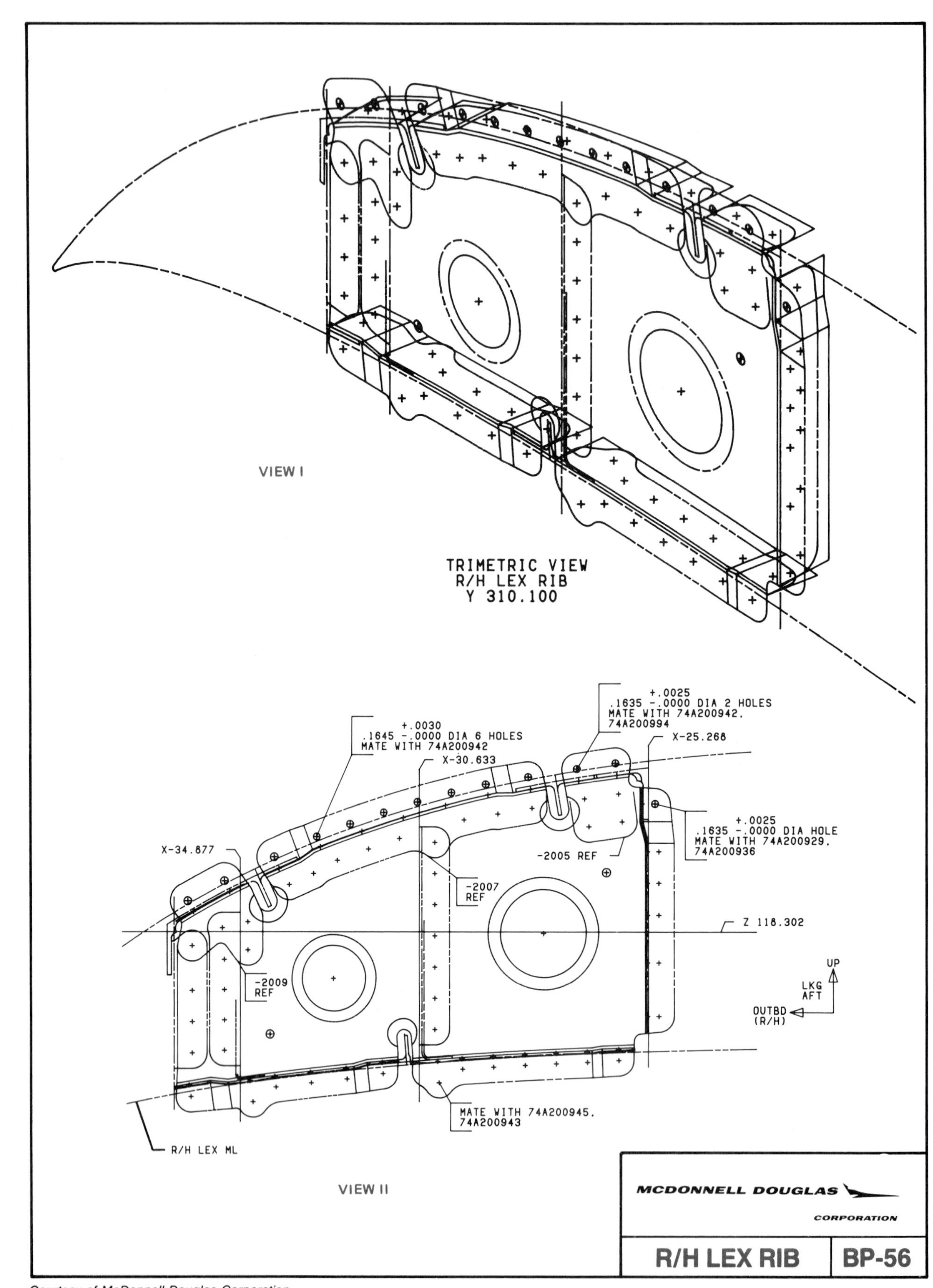

Courtesy of McDonnell Douglas Corporation

ASSIGNMENT—UNIT 56

R/H LEX RIB (BP-56)

Student's Name ______________________

1. Identify the type of drawing used in **VIEW I** to represent the rib subassembly.
2. Name the type of drawing which represents detail features, references, and dimensions in the second view.
3. Indicate the two directions from which the subassembly is viewed.
4. Establish all **X**, **Y**, and **Z** values.
5. State briefly what lofting means.
6. Tell why complicated profiles of aerospace parts are often undimensioned on drawings.
7. Identify the reference points.
8. Give the specifications of the holes which mate with **74A200942**.
9. Determine the upper and lower limit diameters and the number of holes which mate with **74A200942** and **74A200994**.
10. Identify the mates for the single $.1635 \begin{smallmatrix} +.0025 \\ -.0000 \end{smallmatrix}$ hole.
11. State what views are generally used in aerospace drawings to represent external features.
12. List what each item in a seven-digit identification code represents on an aircraft part or subassembly drawing.
13. State the difference between (a) water lines and (b) station lines as used in lofting.
14. Explain, briefly, what use is made of reference lines in lofting.

1. ______________________
2. ______________________
3. a. ______________________
   b. ______________________
4. **X** ________ ________ ________
   Y ________
   Z ________
5. ______________________
6. ______________________
7. **REF** ________ ________ ________
8. ______________________
9. Upper Limit ________ Lower Limit ________ # of Holes ________
10. ______________________
11. ______________________
12.

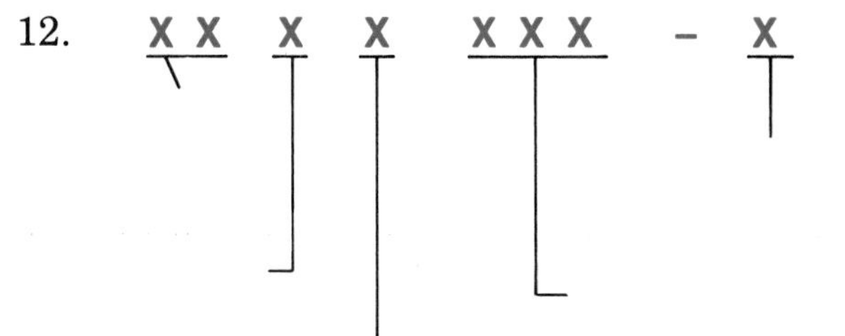

13. a. ______________________
    b. ______________________
14. ______________________

UNIT 57
Sheet Metal Development and Precision Products Drawings

A great many parts in the construction, automotive, aerospace, and other major industries are produced from sheet metals, sheet plastics, wood, fiberboard, and similar materials which can be cut and formed. Ducts, pipes, and other sheet metal objects consist of sides and surfaces which are square, cylindrical, prism- or cone-shaped, or a combination. The surfaces may also intersect at a specified angle.

Each form may be produced by first laying out each section or area in full size as a single form on a flat, plane surface. This technique of layout is known as *surface development*. The part as produced may be called a *template*, a *stretchout*, or a *pattern*.

The layout is then cut to the required size and contour. The pattern is next formed to shape by folding, bending, rolling, wiring, or beading. In custom sheet metal work, these operations are performed by machines which are named according to the process performed. Examples include folding machines, wiring machines, form-rolling machines, and beading machines. Cone and cylindrical contours are often spun to shape.

Surfaces which have complicated curvatures are produced by drop hammers, hydraulic presses, stretching presses, and other die-stamping or forming processes. In the aerospace industry, contoured parts are generally formed with dies which match contours on loft drawings.

PATTERN DEVELOPMENT PROCESSES

A flat-form pattern in sheet metal work may be developed by *parallel line development*, *radial line development*, or *triangulation*. Sheet metal (or nonmetallic) parts which connect pipes or openings of different sizes and shapes are known as *transition pieces*. Two common transition pieces are shown in figure 57-1.

Parallel Line Development

Parallel line development means that the pattern is drawn full size using parallel lines for the width, height, and depth features of an object. If the object is cylindrical and it is cut by a plane at an angle *(truncated)*, the pattern is developed by dividing the circle of the cylinder into an equal number of parts *(elements)*. The height of the truncated cylinder at each element is transferred to the stretchout. Figure 57-2 shows the technique of developing a pattern for a truncated cylinder.

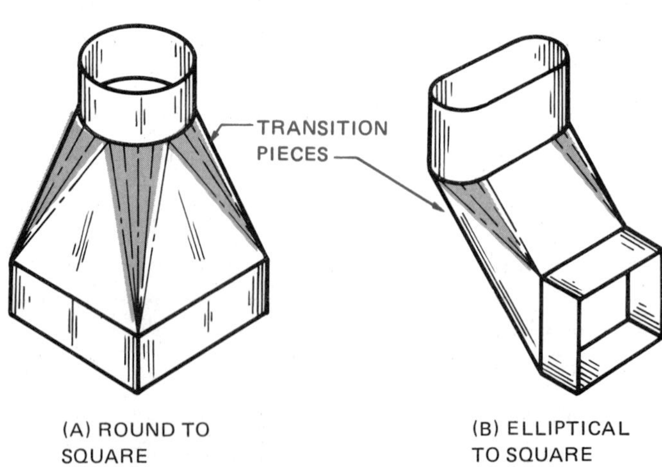

FIGURE 57-1 EXAMPLES OF COMMON DUCT FORMS CONNECTED BY TRANSITION PIECES

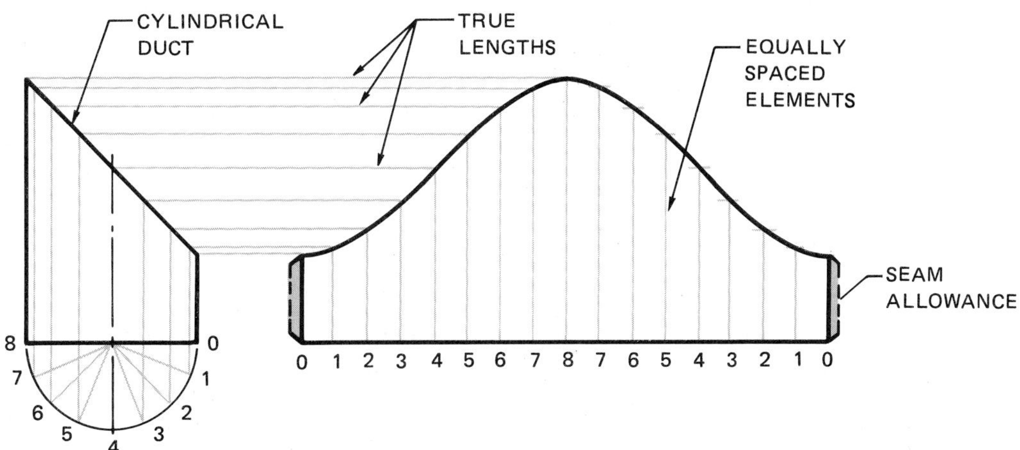

FIGURE 57-2 PARALLEL LINE DEVELOPMENT OF A PATTERN FOR A TRUNCATED CYLINDRICAL DUCT

Using the same method, it is possible to develop the pattern or template for two, three, or more pieces of the same diameter (like an elbow) which are to fit together.

Radial Line Development

This technique is used to form the pattern of full or truncated cone or pyramid-shaped objects. *Radial line developments* require line layouts of elements which are equally spaced around the cylindrical part. The true length of each line is projected radially as shown in figure 57-3.

Triangulation

Triangulation is a convenient method of producing a layout of an object whose elements are

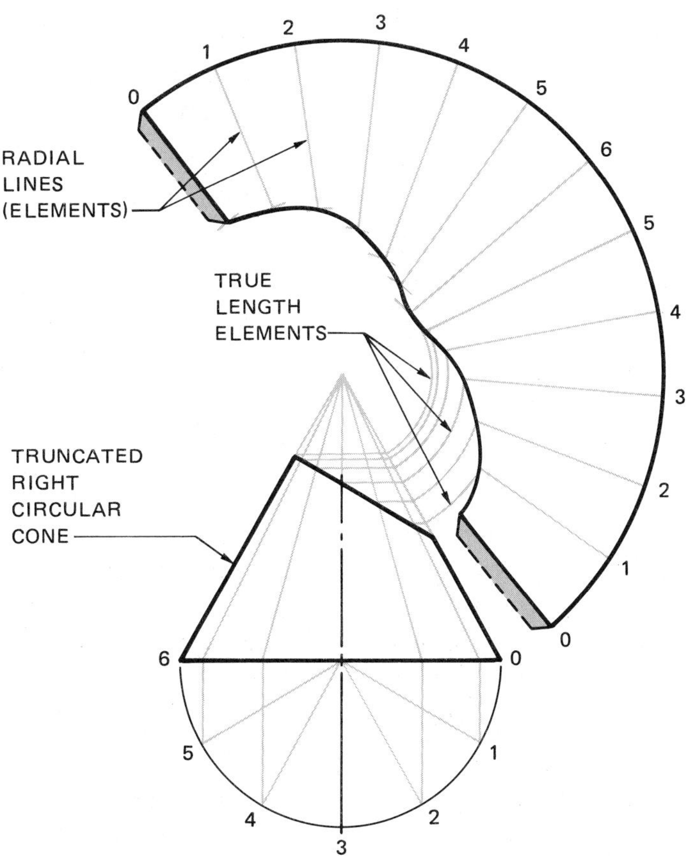

FIGURE 57-3 RADIAL LINE DEVELOPMENT OF A CONE-SHAPED DUCT

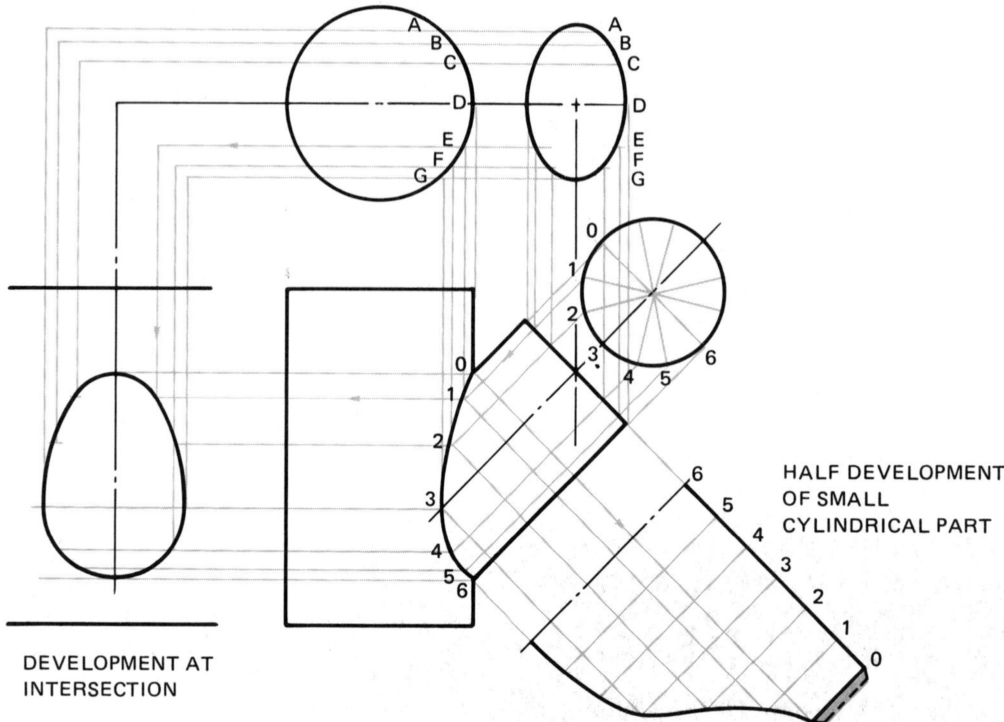

FIGURE 57-4 DEVELOPMENTS OF TWO CYLINDRICAL PARTS WHICH INTERSECT AT AN ANGLE

not shown in *true length*. Each surface of the object is divided into triangles. The true length of each element on a surface is established and transferred in successive order within each triangular area. A line is then drawn connecting the true lengths to form the pattern. Transition pieces, as illustrated earlier in figure 57-1, are often laid out on a flat surface by triangulation. Each complete layout is a combination of the different forms which make up the object.

Intersections

The point or place where the several parts of a sheet metal object join is known as the *line of intersection* or *cutting plane line*. The line may be straight, as in the case of two equal square ducts which intersect. In other instances, like a square duct being cut by a cylindrical duct, the line of intersection or the hole where the separate ducts join is represented by curved lines on both pieces.

Intersections are developed by transferring true length heights from successive elements onto a plane surface layout for each part. The principle used in developing intersections is illustrated in figure 57-4. While the example shows two different size cylinders intersecting at an angle, the same technique may be used with intersecting prisms, cylinders, cones, and other forms.

Major components in a heating and cooling air conditioning system and the sheet metal duct work for a typical installation are shown in figure 57-5.

BEND, SEAM, AND LAP ALLOWANCES

Allowances must be provided on layouts for *bends*, *seams*, and *laps*. Each different sheet metal or sheet material, each thickness, and the form and size of each bend calls for a specific bend allowance. Charts provide standardized information about allowances as used in such industries as automotive, aerospace, sheet metal products, and construction.

Bend allowances for straight, right-angle, and open or closed bends appear on drawings as a dimension. Drawings of precision sheet metal parts usually show the *bend line* for the starting

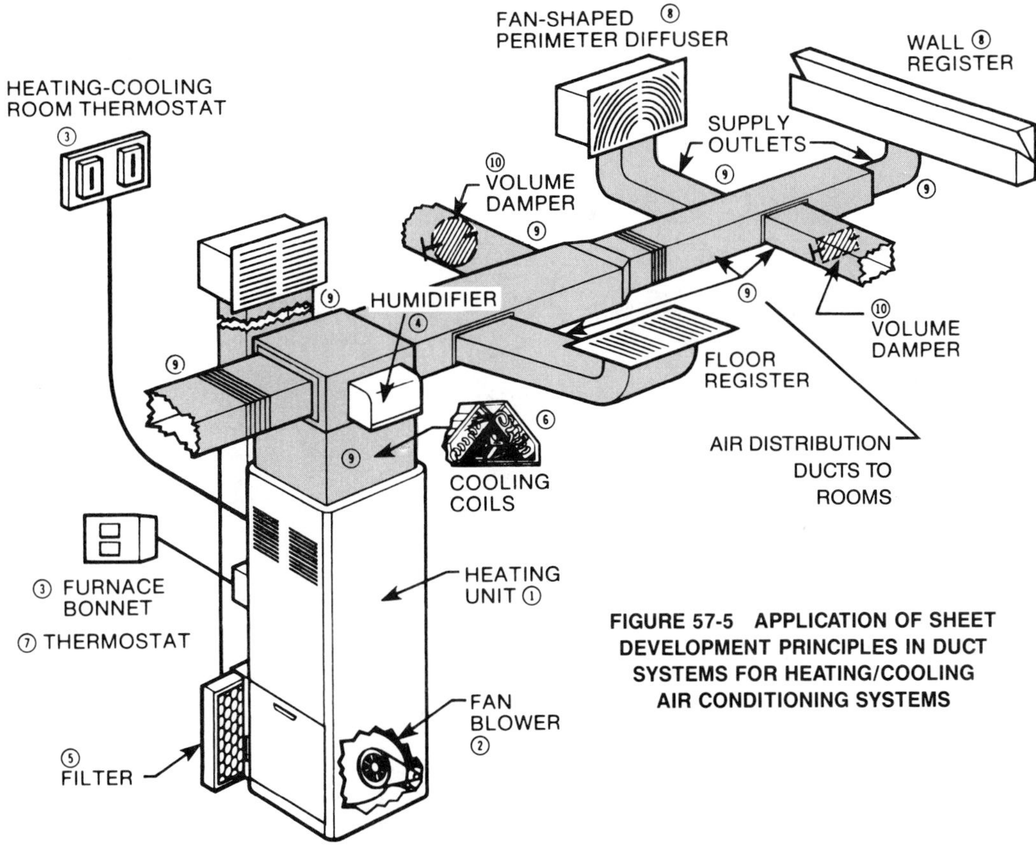

FIGURE 57-5 APPLICATION OF SHEET DEVELOPMENT PRINCIPLES IN DUCT SYSTEMS FOR HEATING/COOLING AIR CONDITIONING SYSTEMS

of the bend and a dimension which gives the allowance for the bend.

Drawing developments for sheet metal ducts and pipes make provision for *wiring*, *seaming*, *hemming*, and other processes for joining the ends together and locking them in position (figure 57-6).

PRECISION SHEET METAL PARTS PRODUCTION

The increasing number of applications of thin-gage metals for parts within instruments, gages, electronic, and other precision equipment requires layouts and parts which are produced to precise machine shop standards. Such sheet metal parts are often blanked, perforated, formed, and trimmed using punches and dies.

In other instances, parts may be manually or computer programmed and produced on numerical control machines. Ordinate drawings are commonly used for producing precision sheet metal parts. Ordinate dimensions are referenced to zero datums to provide position and size dimensions for production of the part on NC or CNC equipment.

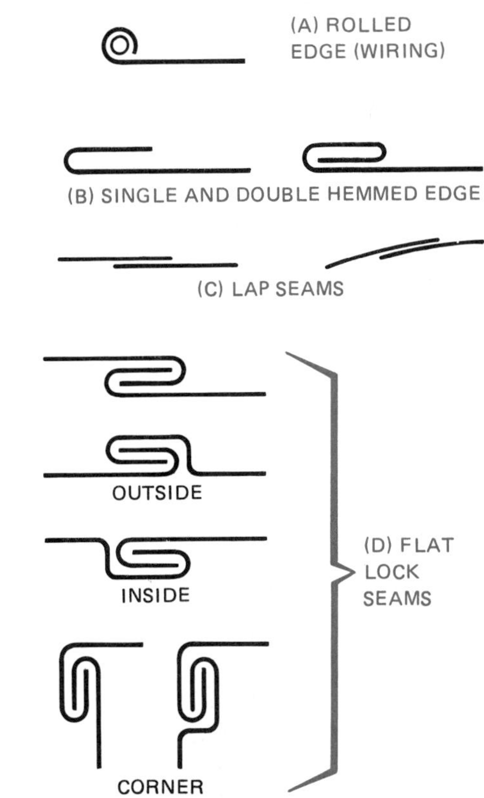

FIGURE 57-6 WIRING, SEAMING, AND HEMMING EDGES OF SHEET METAL PARTS

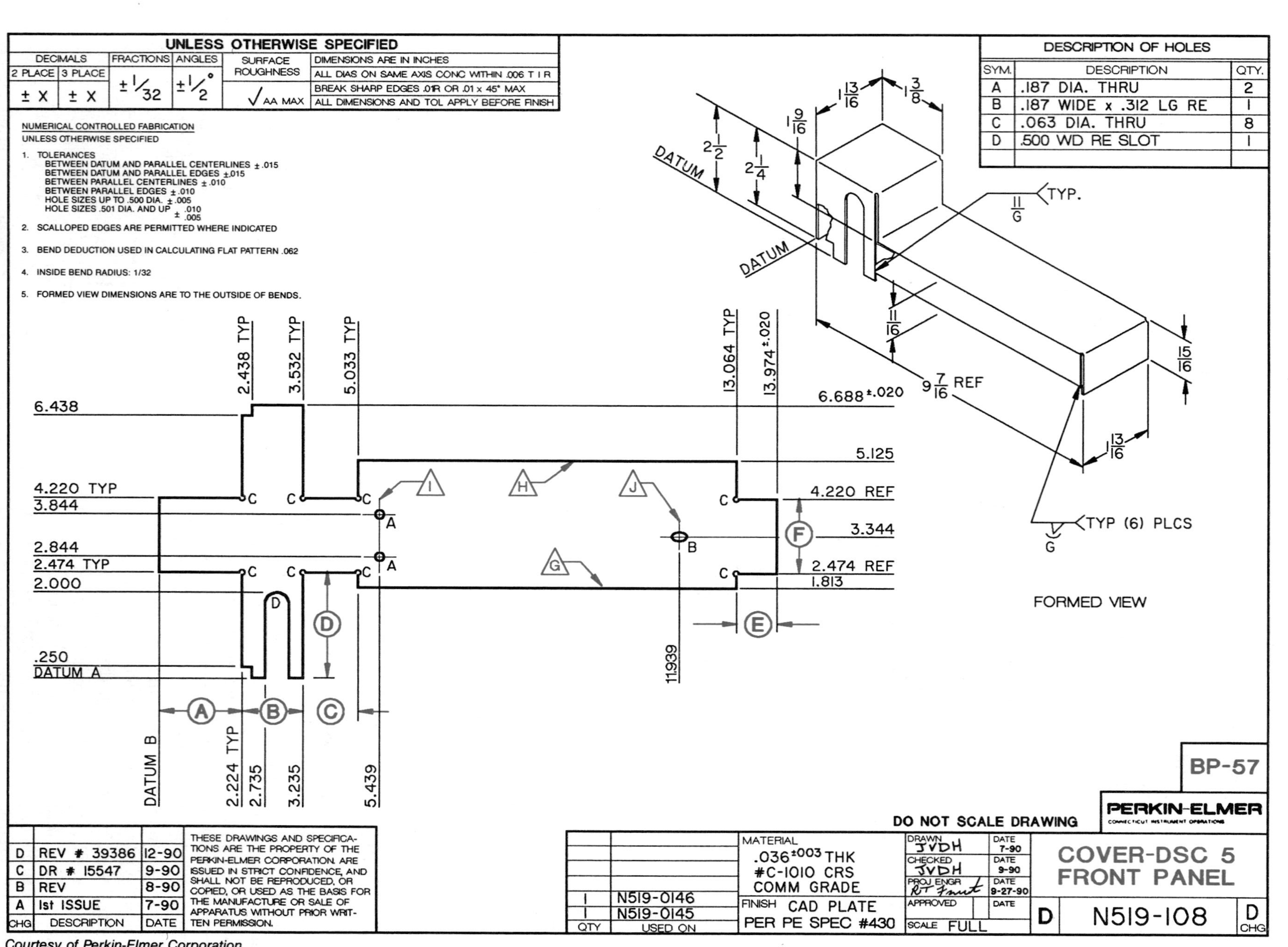

Courtesy of Perkin-Elmer Corporation
(Modified Industrial Drawing)

ASSIGNMENT—UNIT 57

COVER – DSC 5 FRONT PANEL (BP-57)

Student's Name ______________________

1. State the type of dimensioning which is used for the (a) pictorial (formed view) sketch and (b) the layout drawing.
2. Give the specifications for (a) the material used for the **Panel Cover** and (b) the required finish.
3. Identify the concentricity tolerance (unless otherwise specified) for all diameters on the same axis.
4. Specify the kind of fabrication equipment which is to be used in producing the part.
5. Provide the following information (unless otherwise specified):
   a. Inside bend radius
   b. Bend allowance for laying out the flat pattern
   c. Location of the dimensions in the formed view
6. Give the following general tolerances:
   a. Hole diameters up to 0.500″
   b. Between parallel center lines
   c. Between a datum and parallel center lines
   d. Between parallel edges
7. Indicate the dimensions to which sharp edges are to be broken.
8. a. Show how the two different types of welds are represented.
   b. Interpret the meaning of each specification.
9. Establish the maximum and minimum dimensions for each of the following features:
   a. Elongated hole **B**
   b. Slot **D**
   c. Holes **A**
   d. Corner radius **C**
   e. Length and width of the flat pattern from the datums.
10. Identify the **REFERENCE** dimensions for the flat pattern and the formed view.
11. Establish the following nominal dimensions: (A) through (F) (including the bend allowances) of the finish-formed part. Give the dimensions in fractional inch values.
12. Compute the maximum and minimum dimensions between the following features:
    a. Surfaces △G and △H
    b. Datum **B** and center line △I
    c. Center lines △I and △J

1. a. ______________________
   b. ______________________
2. a. Material ______________________
   ____
   b. Finish ______________________
   ____
3. ______________________
4. ______________________
5. a. __________ b. __________
   c. ______________________
6. a. __________ b. __________
   c. __________ d. __________
7. ______________________

	(1)	(2)
8. a.		
b.		

	Maximum	Minimum
9. a. Width (DIA)	______	______
Length	______	______
b. Width	______	______
c.	______	______
d.	______	______
e. Length	______	______
Width	______	______

10. a. Flat Pattern ______________________
    b. Formed View ______________________

11. (A) __________ (D) __________
    (B) __________ (E) __________
    (C) __________ (F) __________

	Maximum	Minimum
12. a.	______	______
b.	______	______
c.	______	______

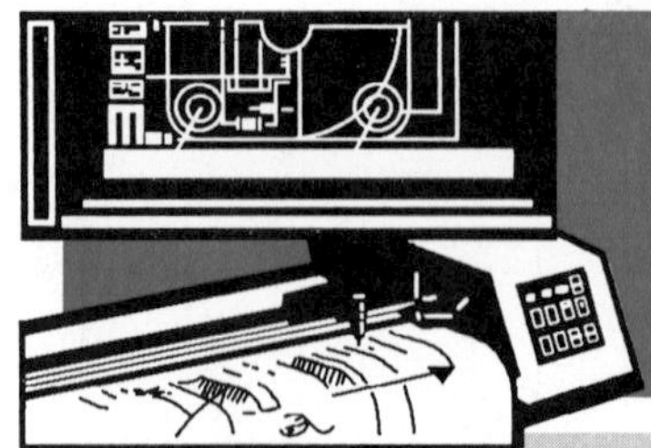

HIGH TECHNOLOGY applications

HT-17 CADD/CNC/CAM: SUPERABRASIVE MACHINING

Increased demands for heavier-duty, higher-speed, multi-axis, more precise machine tools, machining centers, and factory integrated manufacturing systems, are paralleled with revolutionary developments in new superabrasive cutting materials.

Superabrasives have the capacity to grind at speeds in excess of 25,000 surface feet per minute (sfpm) and to machine at rates above 10,000 sfpm. Importantly, it is possible even at these high cutting speeds to maintain quality controls for surface texture and to meet precise dimensional and geometric tolerancing requirements. The two superabrasives that are described briefly include cubic boron nitride (CBN) and synthetic polycrystalline diamonds (PCD).

CUBIC BORON NITRIDE (CBN) SUPERABRASIVES

Cubic boron nitride (CBN) crystals may be formed into wheels for grinding, honing, lapping, and polishing and as CBN inserts for cutting tools. CBN superabrasives are efficient for machining hardened steels, difficult-to-machine tough superalloys, as well as composites and laminates. Although machining is done at high cutting speeds, CBN inserts are not affected chemically nor are they subject to oxidation or any reduction in hardness.

High removal rates (without metallurgical damage to a workpiece) are possible when grinding due to the fact that the sharp, long-life cutting edges of CBN crystals *microfracture* when dull, exposing new, sharp, cutting edges. Limited CBN wheel wear means that fewer machining adjustments are needed, there is less machine *down time*, and the process is cost effective.

POLYCRYSTALLINE DIAMOND (PCD) CUTTING TOOLS

This classification of superabrasive cutting tools is produced by forming a mixture of fine powder diamond tungsten carbide which is subjected to high temperature (3,000° F) and extremely high pressure (one million psi). The sintered polycrystalline diamond crystals are then bonded to a carbide substrate.

While PCD cutting tools have excellent tool wear life and shock resistance (when compared with carbides), polycrystalline cutting tools react chemically at temperatures above 1200° F with metals that contain carbon. PCD cutting tools outperform carbide cutting tools for machining brass and bronze alloys, nonabrasive plastics, presintered ceramic products, and other non-carbon materials.

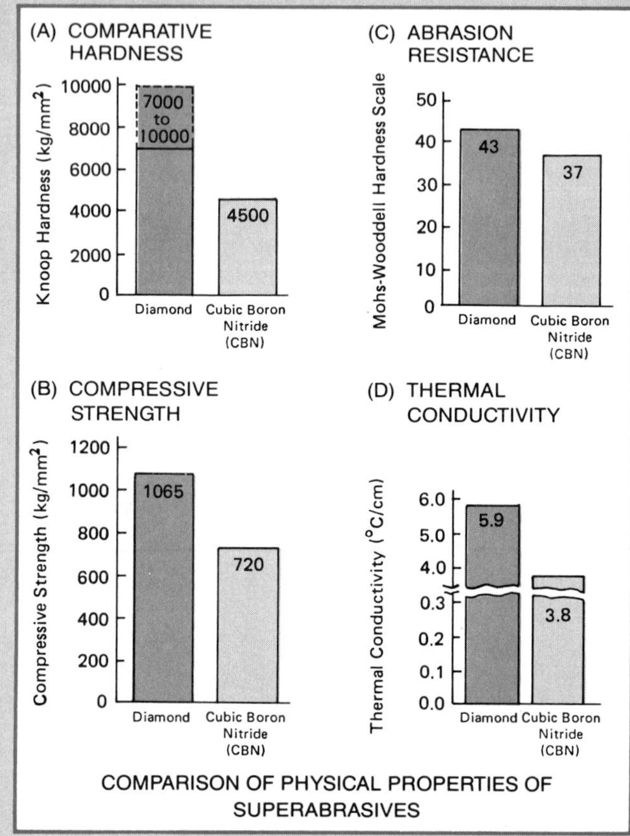

COMPARISON OF PHYSICAL PROPERTIES OF SUPERABRASIVES

CBN SUPERABRASIVES AND DIAMOND PROPERTIES

The ability of CBN and PCD superabrasives to machine the hardest known materials depends on four main physical properties: (1) hardness, (2) compressive strength, (3) abrasion resistance, and (4) thermal conductivity. These properties are visually compared in the separate charts (A, B, C, and D).

Other factors affecting performance include: whether a wet or dry grinding mode is used; the concentration of superabrasives (weight in carats per cubic inch or cubic centimeter); and the bonding system. Micron PCD and CBN superabrasive powders are used for heavy-duty lapping, honing, and polishing operations.

Drawings adapted from ADVANCED MACHINING TOOL TECHNOLOGY

Courtesy of C. THOMAS OLIVO ASSOCIATES

UNIT 58
Piping Drawings

The piping trades and related occupations depend on conventional as well as specialized drawings for information about design, construction, installation, and maintenance features of systems through which gases, liquids, and solids flow. Parts and different units within a system which are hydraulically, mechanically, pneumatically, electrically, or electronically controlled depend on drawings which follow special drafting room practices for that occupation.

Piping drawings are used in structural and construction work and where piping is an integral part of a system involving other occupations. Standard symbols are used for pipes, fittings, valves, and other general parts. In addition to these symbols, special representation techniques are employed for diagramming, dimensioning, placing, and identifying the functions of all parts within a system.

INTERLOCKING SYMBOLS OF OTHER OCCUPATIONS

Where piping systems interlock with heating, air conditioning, ventilating, heat power, and other installations and apparatus, still other graphic symbols are required. Figure 58-1 provides a few examples of graphic symbols used in four different occupations. Complete graphic symbol tables are given in trade handbooks and architectural standards publications. These, and other ANSI standards, are referred to in order to establish mechanical and physical requirements for parts made of different materials, components within piping installations, and standardized symbols.

LINE METHODS OF REPRESENTATION

Single-line and *double-line drawings* are common for piping installations. A *double- (two-) line*

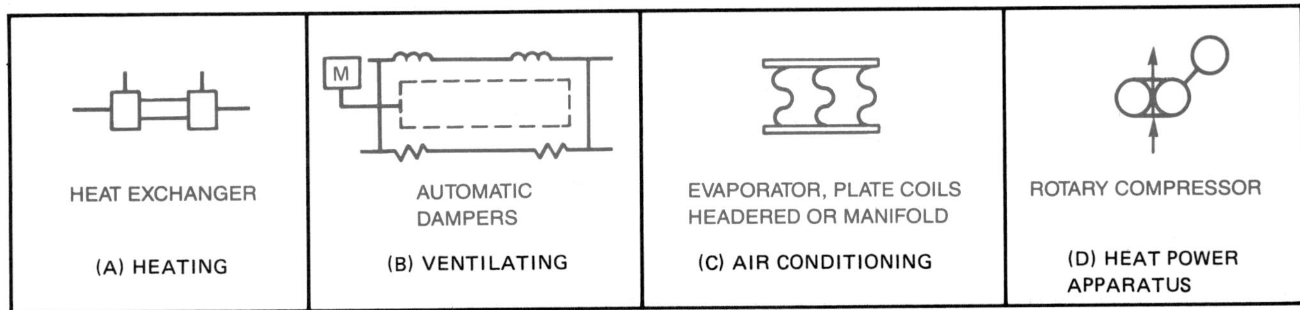

FIGURE 58-1 SAMPLE GRAPHIC SYMBOLS USED IN RELATED OCCUPATIONS

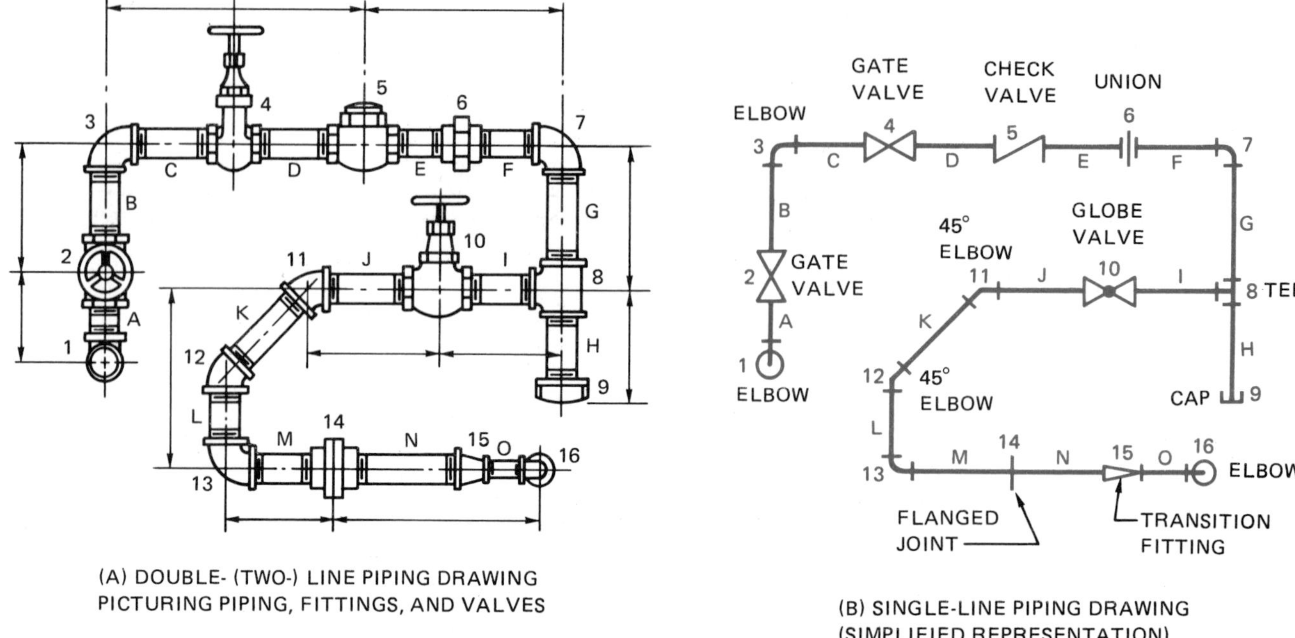

(A) DOUBLE- (TWO-) LINE PIPING DRAWING PICTURING PIPING, FITTINGS, AND VALVES

(B) SINGLE-LINE PIPING DRAWING (SIMPLIFIED REPRESENTATION)

FIGURE 58-2 COMPARISON OF A SINGLE-LINE AND DOUBLE-LINE PIPING DRAWING

piping drawing furnishes a picture of the piping, valves, and fittings in a system (figure 58-2A). This type of drawing may be simplified by using graphic symbols and identifying the piping, fittings, and valves, using a single-line piping drawing (figure 58-2B).

PLAN, ELEVATION, AND OTHER VIEWS USED ON PIPING DRAWINGS

Piping drawings are generally prepared as either orthographic or pictorial drawings. The principal views used for orthographic projection drawings include *plan*, *elevation*, and *bottom views*.

The terms *in elevation*, *turned up*, and *turned down* are used in representing fittings and valves. *In elevation* means the symbol is viewed from the front position (figure 58-3A) as in other orthographic projection drawings. When the piping system is viewed from the left so the flow is away from the viewer, the elbow and tee are seen as represented at (B).

If the flow through the elbow is viewed from the right toward the viewer, one connection is represented as a circle; the other, as a straight line (figure 58-3C). The dot represents the end of the pipe.

The *plan view* at (D) shows the graphic symbol of an elbow and a tee as viewed from the top. *Pictorial views* of an elbow and tee are illustrated in figure 58-3E.

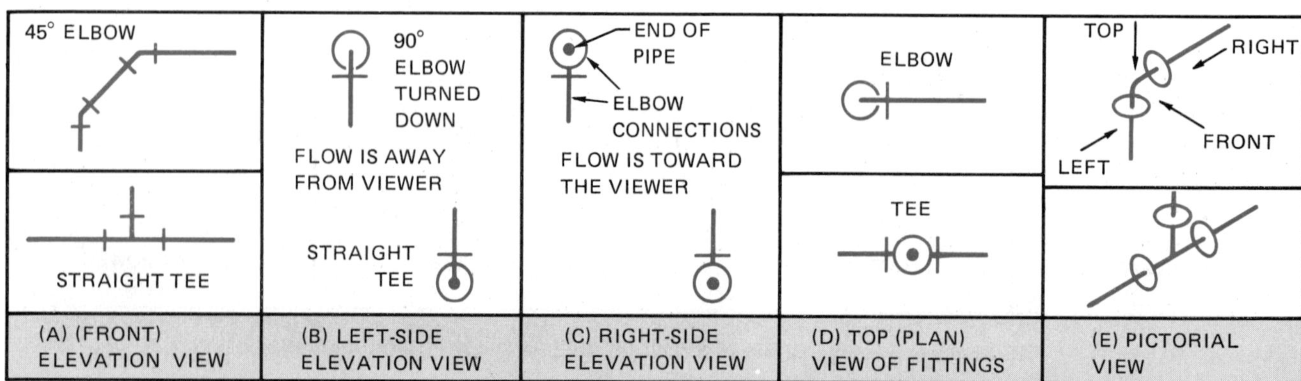

FIGURE 58-3 REPRESENTATION OF SAMPLE FITTINGS AS APPLIED TO CONVENTIONAL VIEWS ON PIPING DRAWINGS

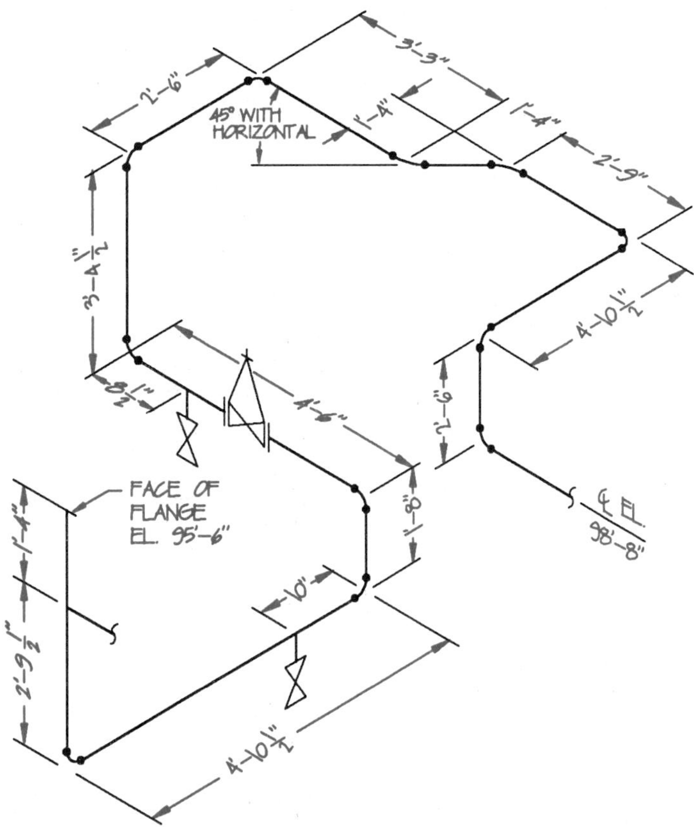

FIGURE 58-4 DIMENSIONING APPLIED TO A PICTORIAL (ISOMETRIC) PIPING DRAWING

STANDARD VIEWS AND DIMENSIONING

Piping drawings usually include two views. A *plan view* (top view) and one *elevation view* usually provide full information. The elevation views are identified as *front elevation*, *right-* or *left-side elevation*, or *rear elevation*. These correspond with the placement of conventional views.

Importantly, each view is dimensioned with certain principal measurements. The *front elevation* provides width and height dimensions; *right elevation*, length and height. The *top plan view* contains dimensions of length and width. The *bottom view* (when used) is not a plan view. Rather, the view shows the structure (building) and installation looking up and the dimensions of length and width.

Center lines are particularly important in piping drawings. Pipes, valves, and other parts in a system are generally dimensioned in relation to center lines.

FULL DIMENSIONING ON PICTORIAL (ISOMETRIC) DRAWINGS

Pictorial (isometric) single-line piping drawings are comparatively easy to draw and interpret. Pictorial drawings may be fully dimensioned for width, length, and height. Figure 58-4 shows how dimensioning practices are applied on a section of piping.

GRAPHIC SYMBOLS FOR PIPE FITTINGS AND VALVES

Like other industries, specific graphic symbols are used to simplify the representation of fittings, valves, and other parts and components. Full tables are provided in handbooks with complete lists of graphic symbols. A limited sampling is provided in figure 58-5. The symbols in tables are grouped into such categories as cross, elbow, joint, reducer, tee, angle, other types of valves, and so forth. Major types within each category are then identified. Figure 58-6 shows the application of graphic symbols on an isometric piping drawing.

Changed markings for each graphic symbol tell how the parts are secured together. Note in the same symbols in figure 58-5 that there are five columns. Each column represents a different fastening method. Attention is directed to the use of (||) to denote *flanged*, (|) *screwed*, (( ) Bell and Spigot, (•) welded, and (○) soldered parts.

PIPING ITEM	METHOD OF FASTENING				
	FLANGED	THREADED	BELL & SPIGOT	WELDED	SOLDERED
3. CROSS 3.1 REDUCING	2 6 6 4	2 6 6 4	2 6 6 4	2 6 6 4	2 6 6 4
22. GLOBE VALVE 22.1					
6.2 EXPANSION					

FIGURE 58-5 STANDARD GRAPHIC SYMBOLS FOR SAMPLE PIPING ITEMS USING GENERAL FASTENING DESIGNS

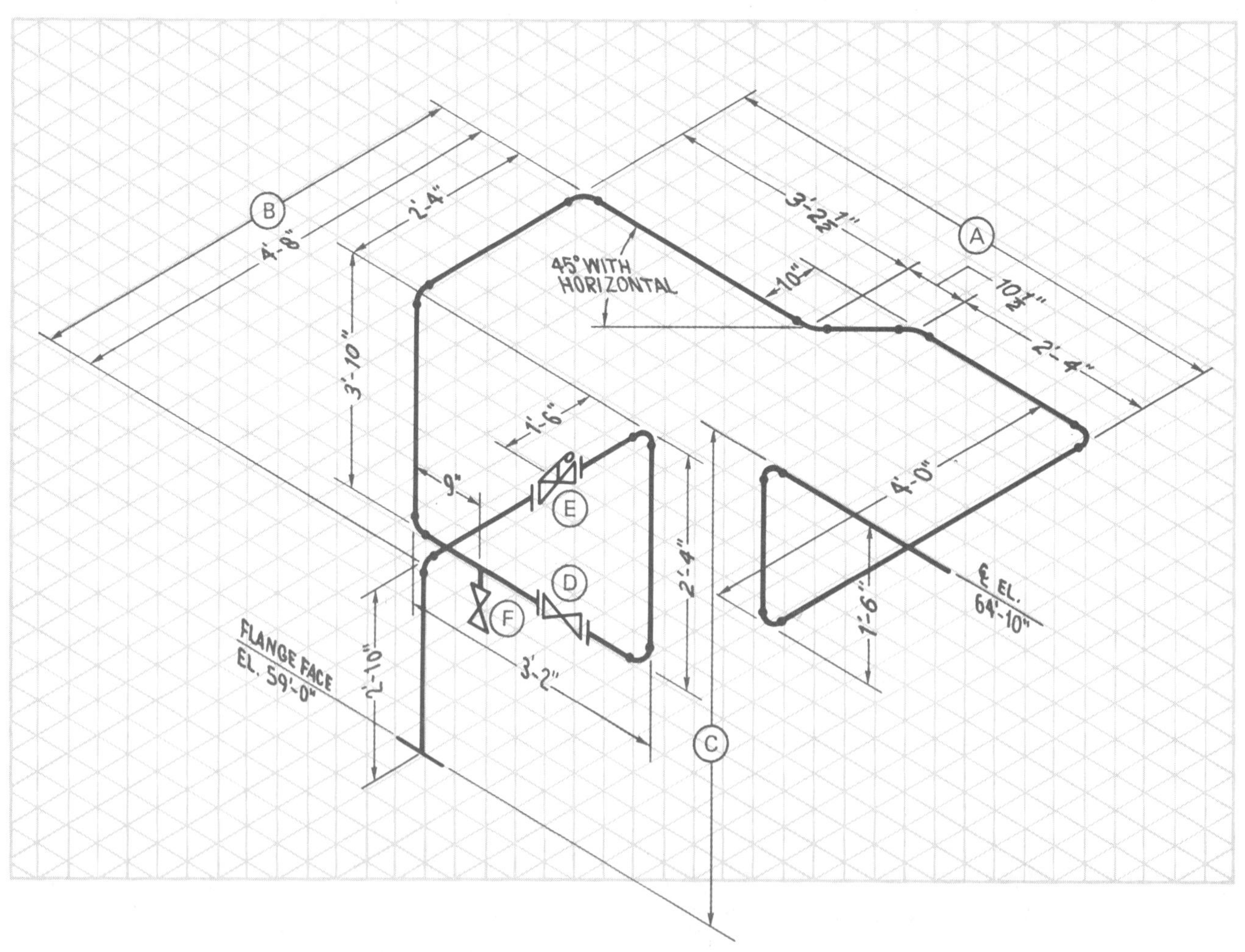

FIGURE 58-6 ISOMETRIC VIEW OF A SECTION OF PIPING AND VALVES

FLOW AND SPOOL SHEETS

Single-line schematic piping drawings may also be planned for use as a *flow sheet.* A flow sheet shows the direction of flow of the liquid or gas and the sequence and function of each part in the system. For example, a flow sheet is referred to for the locations (not the actual measurements) for valves, vessels, or instruments and to identify the types used in the system. A flow sheet is sometimes called a P.&I.D. for *Piping and Instrument Drawing.*

Spool Sheets

The use of pictorial drawings is often extended to include fabrication and installation. Small sections or areas of a piping drawing are also called *spools* of a system. *Spool sheets* are dimensioned pictorial drawings with each item numbered or lettered. A material list is generally included on the drawing. The required quantity, specification, and part identification are given in the material list.

GRAPHIC SYMBOLS FOR PIPING

Piping drawings often require the use of lines and letter codes which differ from conventional orthographic drawings. The shape and weight of the line and letters (where required) clearly designate the nature and function of the piping.

For example, a ——-—-——-—-—— line indicates a *hot water line* in a plumbing installation. In air conditioning, a ——-——H——-—— line identifies a *humidification line.* The symbol ——OO——OO—— on a heating system drawing indicates a *feedwater pump discharge line.* The symbol ——S————S——, when used on a drawing of a sprinkler system, denotes the *branch* and *head.*

Graphic symbols, as stated earlier, are standardized for piping and related industries. Complete tables are found in trade handbooks, manufacturers' technical literature, and among architectural and engineering standards.

DRAFTING TEMPLATES FOR ISOMETRIC DIAGRAMS

Piping diagrams may be prepared by CAD or with the aid of a *symbol template (guide).* Templates are available for drawing residential, commercial, and industrial piping installations where regular plan views or isometric diagrams are required. Figure 58-7 illustrates a small portion of a larger, complete isometric schematic boiler piping diagram. The drawing was prepared by using an isometric piping template.

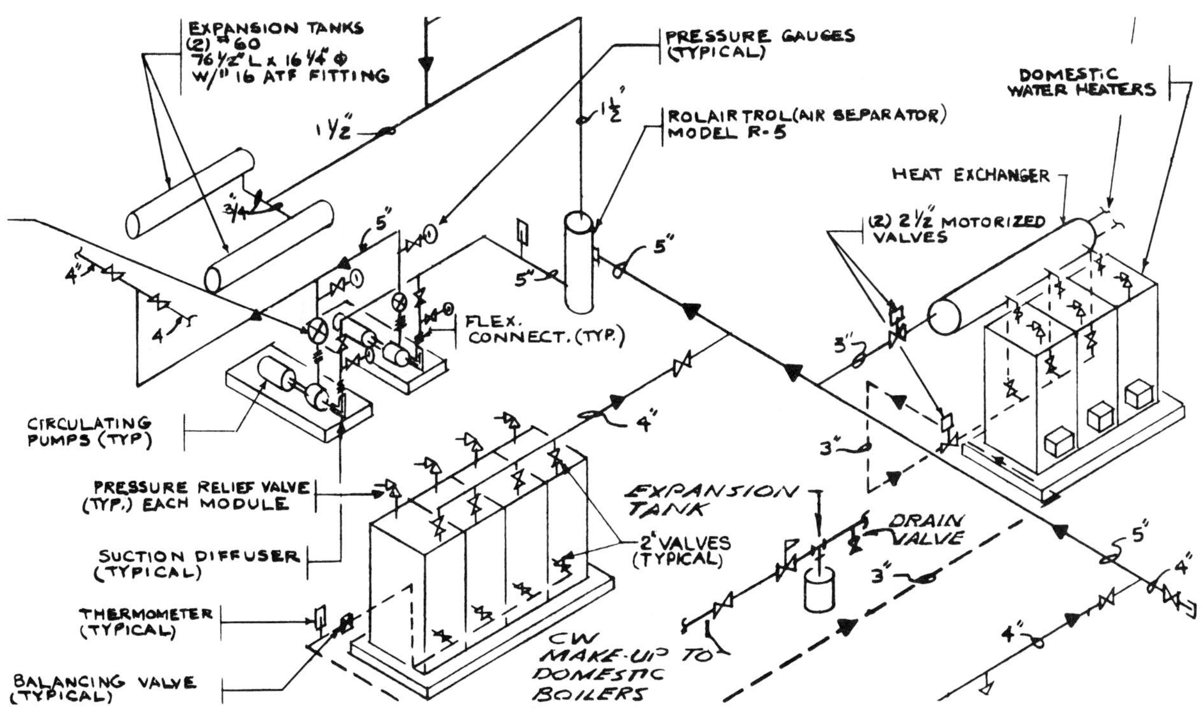

FIGURE 58-7 PARTIAL ISOMETRIC SCHEMATIC PIPING DIAGRAM PREPARED WITH AN ISOMETRIC PIPING TEMPLATE

NOTE: REFER TO APPENDIX B-11 FOR THE LARGER SCALE DRAWING TO USE WITH THIS ASSIGNMENT.

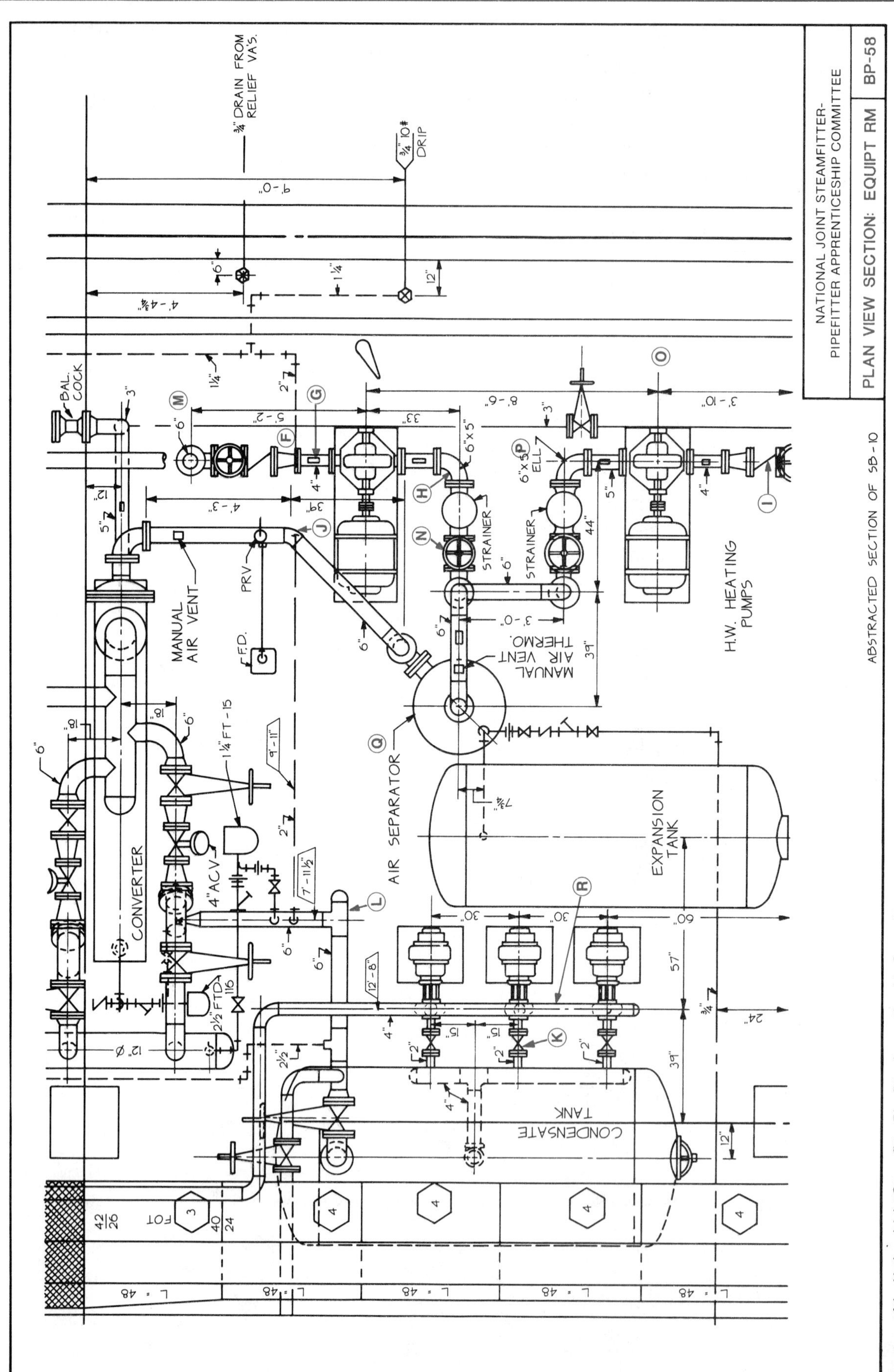

Courtesy of the National Joint Steamfitter-Pipefitter Apprenticeship Committee (Modified Industrial Drawing)

PLAN VIEW SECTION: EQUIPMENT ROOM (BP-58)

1. Make a simple one-line drawing of the graphic symbol used on piping drawings for each of the following pipe fittings and valves. Use a table of symbols, if required.
   a. Threaded cap
   b. Welded 45° elbow
   c. Threaded concentric reducer
   d. Soldered straight tee
   e. Threaded angle check valve
   f. Flanged gate hose valve

2. Refer to the isometric view of the section of piping and valves (figure 58-6).
   a. Identify (1) the different types of elbows and (b) the number of each type.
   b. Calculate the overall center-to-center piping dimensions for Ⓐ, Ⓑ, and Ⓒ.
   c. Determine the letter on the drawing which identifies the following valves:
      (1) Threaded gate hose valve
      (2) Welded gate valve
      (3) Welded quick opening valve

3. Determine the elevation to (a) the center line of the connecting pipe (—⟆) and (b) the face of the flange.

4. Refer to the Plan View Section of the Equipment Room. Identify fittings Ⓕ through Ⓛ.

5. Name two pieces of equipment that are used in the hot water system.

6. Calculate the following center-to-center dimensions:
   a. Pipe Ⓜ and valve Ⓝ
   b. Valve Ⓝ and the hot water heating pump Ⓞ
   c. Ell Ⓟ and the air separator Ⓠ
   d. The expansion tank and the condensate tank
   e. Piping Ⓡ and the condensate tank

7. State two differences between this Plan View Section and a Spool Sheet Drawing.

ASSIGNMENT—UNIT 58

Student's Name ____________________

1.

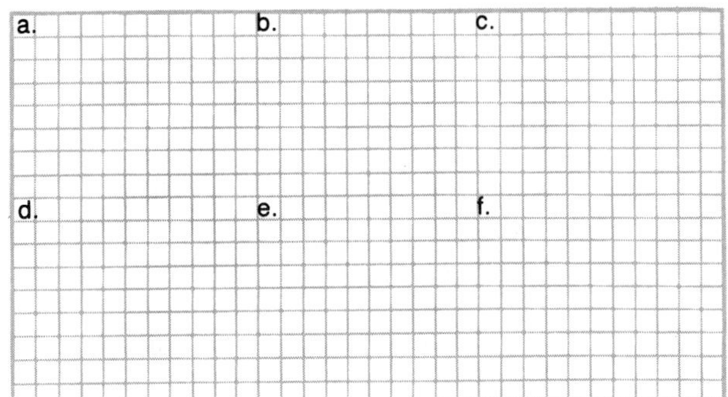

2. (1) Types (2) Number

a. ____________ ____________

____________ ____________

b. Ⓐ = ____________ Ⓑ = ____________

Ⓒ = ____________

c. (1) ________ (2) ________ (3) ________

3. a. ____________ b. ____________

4. Ⓕ ____________

Ⓖ ____________

Ⓗ ____________

Ⓘ ____________

Ⓙ ____________

Ⓚ ____________

Ⓛ ____________

5. a. ____________

b. ____________

6. a. ____________

b. ____________

c. ____________

d. ____________

e. ____________

7. a. ____________

b. ____________

UNIT 59
Instrumentation and Industrial Control Drawings

Instrumentation drawings and *diagrams* relate to mechanical, electrical, electronic, fluid power, and other mechanisms, components, and equipment. Instrumentation units and industrial controls are designed to actuate, convert, direct, and control different forms of power to produce desired motions and movements.

A good example of instrumentation is demonstrated in a machine control unit (MCU) for a numerically controlled machine. Input to the MCU, as described in an earlier unit, is provided by electronic instructions which control servomotors. These motors drive longitudinal, transverse, and vertical (**X**, **Y**, and **Z** axes) lead screws. Additional signals control speeds and feeds for the spindle and other machine units.

The instrument drawing for the MCU is a composite of electrical-electronic circuit diagrams and drawings of mechanical, pneumatic, hydraulic, and other mechanisms. Each drawing requires the use of symbols and terms which apply specifically to each occupation. For instance, electrical/electronic symbols and lines are used to identify the parts, locations, and functions which are electrically/electronically controlled.

A *fluid power installation* requires the manual, mechanical, electronic, and pressure control of motors, cylinders, valves, heaters, coolers, and indicators; and temperature, flow, and other controllers. Particular lines are used to describe fluid power applications. A main line on a fluid power diagram appears as a solid line (————). The symbol (——)(——) represents a fluid power line with fixed restriction. The closed arrow symbol (——▶——) shows the flow direction of a hydraulic line; the open arrow (——▷——), a pneumatic line. Like the symbols used in other industries, those found on drawings for liquid or gas fluid power instruments or components are also standardized.

COMPLEMENTARY INFORMATION FOUND ON INSTRUMENTATION DRAWINGS

Parts and units on instrument drawings are usually numbered or lettered. The identification of each item on the drawing (except for duplicate items which bear the same identity) is keyed to a *component list*. This list corresponds to a parts list or bill of materials.

SEQUENCE OF OPERATIONS

The various functions of different parts of instrumentation equipment are described in the *order of occurrence*. Each operational phase is numbered or coded. A brief but complete description of the functions of each part and/or the sequence of operations is then provided on the drawing or in a companion technical manual or data sheet.

FLUIDIC CONTROL SYSTEMS

Fluidics utilizes a low pressure system as a means of sensing and controlling fluid power circuits. Fluid movement is used to produce a pressure signal to operate a fluidic control circuit.

Again, another set of terms is used. A fluidic component is called a *gate*. An *output signal* means that another fluidic device is operated by an *output port* which has sufficient pressure. As a final example, an *output device* (which actuates pneumatic and hydraulic components to perform the actual work) is one that is controlled by a fluidic device. Each new term and device requires a graphic symbol, an industrially standardized definition, and conventions for representation on a drawing. These are provided in handbooks and on technical data sheets.

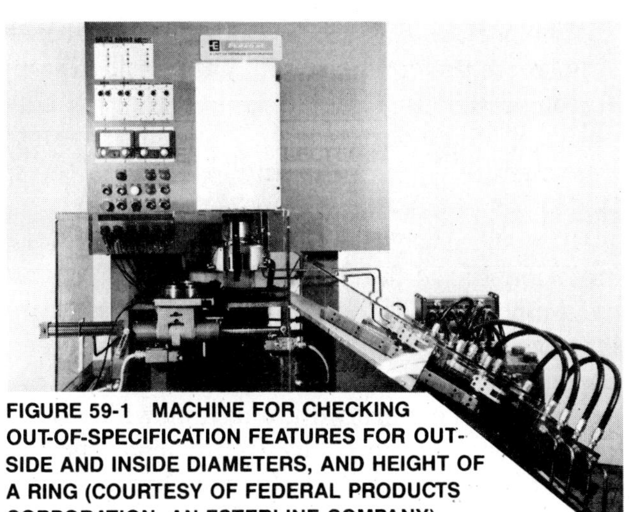

FIGURE 59-1 MACHINE FOR CHECKING OUT-OF-SPECIFICATION FEATURES FOR OUTSIDE AND INSIDE DIAMETERS, AND HEIGHT OF A RING (COURTESY OF FEDERAL PRODUCTS CORPORATION, AN ESTERLINE COMPANY)

TYPES OF INSTRUMENTATION AND CONTROL DRAWINGS

A number of drafting room practices are used to represent instrumentation and control devices, circuits, and complete machines. Four different types of instrumentation drawings are illustrated for the design, construction, and operation of a particular machine for gaging steel rings. Figure 59-1 pictures the machine for checking out-of-specification features for the outside and inside diameters and the height of a manufactured ring. Each ring as it comes from a conveyor is automatically fed into the gage, cycled through the gage (machine) automatically, and disposed out exit chutes into one of five disposal categories. The categories depend on the part features which are out of specification.

The instrumentation drawings are identified according to function as: *power distribution schematic* (figure 59-2), *logic schematic* (figure 59-3), *signal schematic* (figure 59-4), and hydraulic dia-

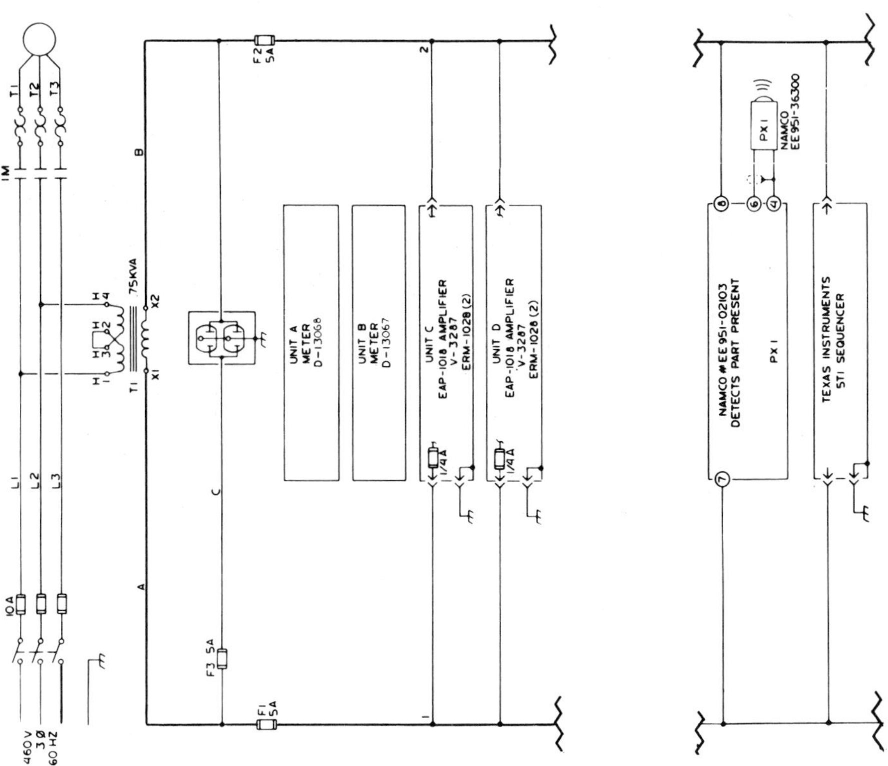

FIGURE 59-2 PARTIAL POWER DISTRIBUTION SCHEMATIC OF RING GAGING MACHINE (COURTESY OF FEDERAL PRODUCTS CORPORATION, AN ESTERLINE COMPANY)

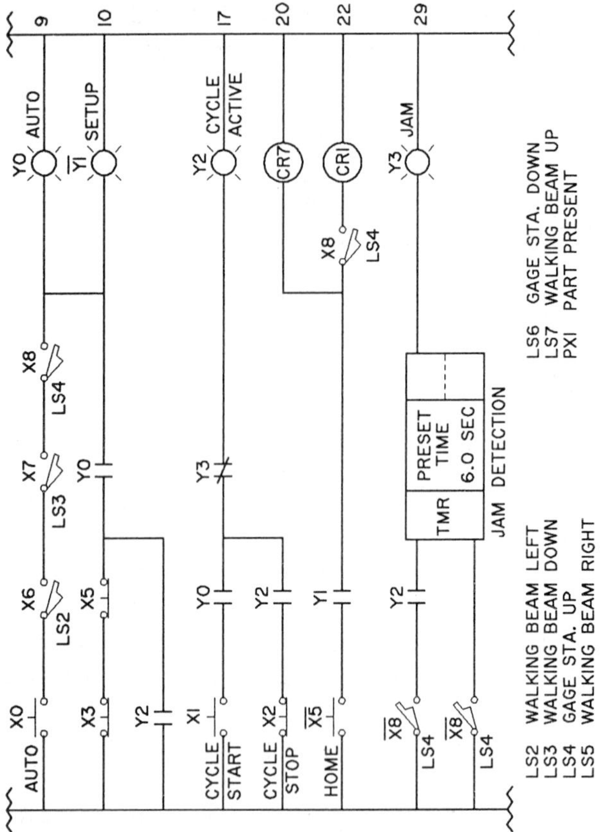

FIGURE 59-3 PARTIAL LOGIC SCHEMATIC OF RING GAGING MACHINE (COURTESY OF FEDERAL PRODUCTS CORPORATION, AN ESTERLINE COMPANY)

gram (figure 59-5). Greatly reduced line drawings show portions of the machine. Cut-away diagrams which show internal features, pictorial diagrams, combination schematics, conventional mechanical drawings, and sketches may be used as part of a series of drawings for each instrument.

The *power distribution schematic* (figure 59-2) provides the electric power circuitry beginning with a three-phase (3 ∅), **460** volts, **60**-hertz input. After reduction through transformers, the power is distributed among successive units, instruments, pumps, solenoids, etc.

The partial *logic schematic* (figure 59-3) drawing traces the circuitry from automatic to setup, through various gage controls for inside diameter salvage of the part ($\overline{\text{X12}}$); outside diameter salvage ($\overline{\text{X13}}$); to the final opening of the inside diameter salvage trap (Y15).

The *signal schematic* (figure 59-4) shows the relationship of the different parts, components, and circuits for the five disposal categories, depending on the out-of-specification features of each part.

The complete hydraulic diagram, which is only partially represented by figure 59-5, includes symbols and connections for each hydraulic control component in the completed machine.

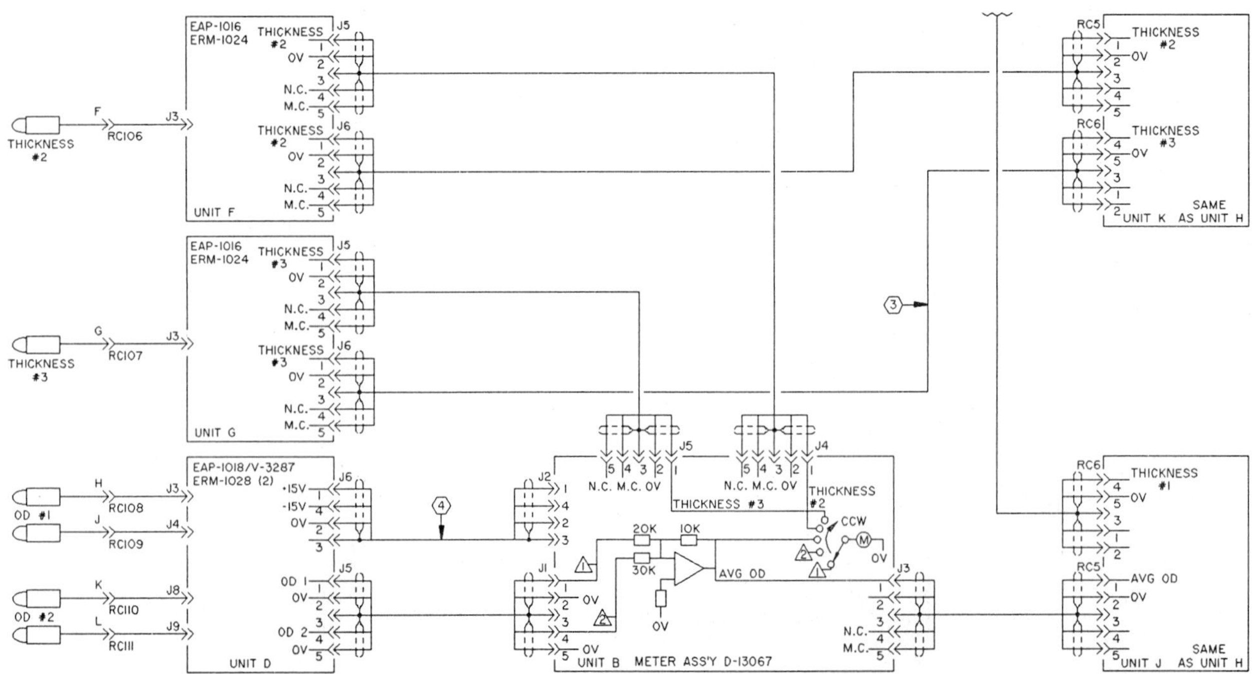

FIGURE 59-4 EXAMPLE OF SIGNAL SCHEMATIC (COURTESY OF FEDERAL PRODUCTS CORPORATION, AN ESTERLINE COMPANY)

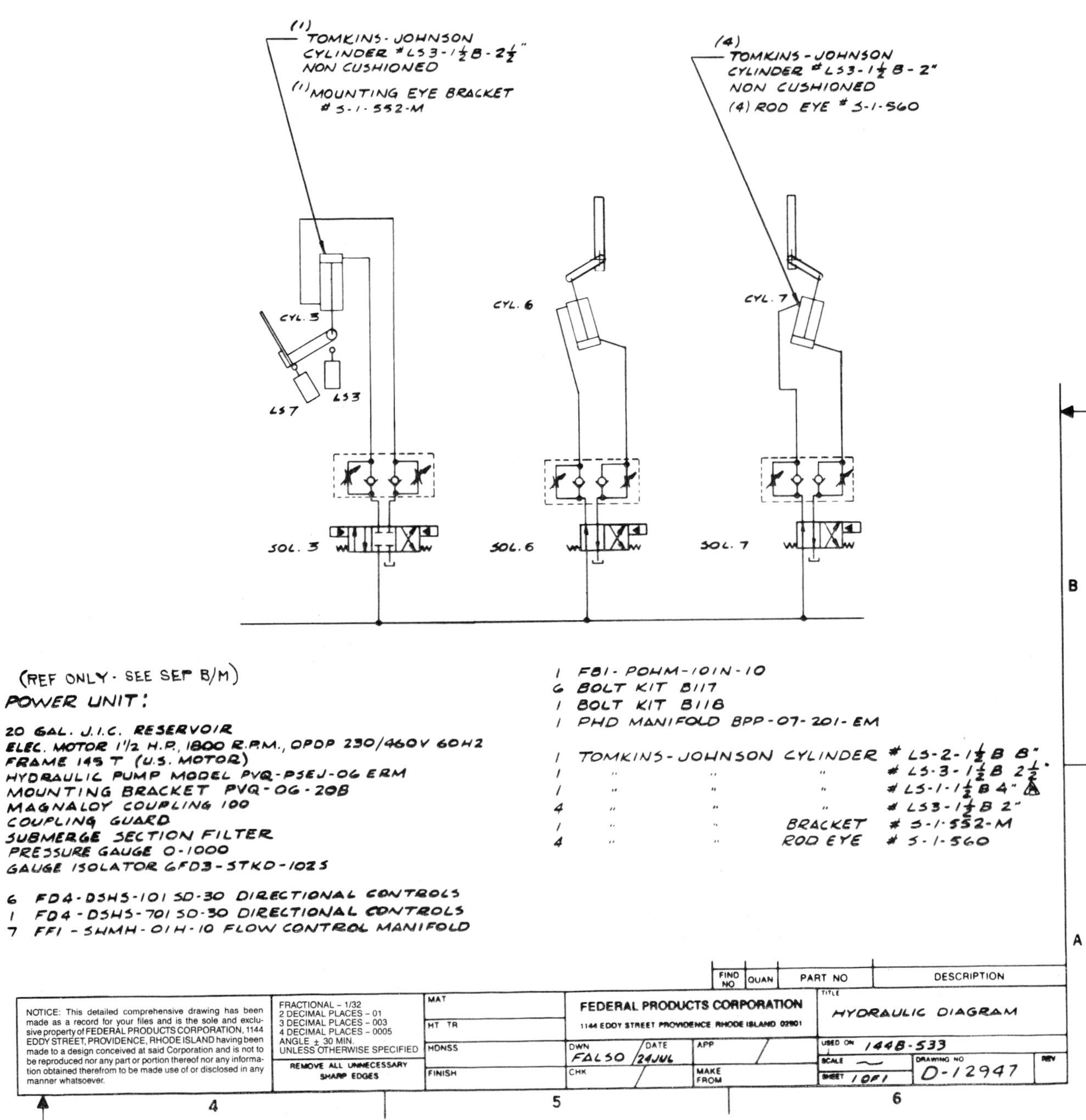

FIGURE 59-5 (PARTIAL) HYDRAULIC DIAGRAM (COURTESY OF FEDERAL PRODUCTS CORPORATION, AN ESTERLINE COMPANY)

READING AN INSTRUMENTATION AND CONTROL DRAWING

There are two basic processes involved in reading an instrumentation drawing:

- Identifying each part, component, and line represented on the drawing. This means recognition of each graphic symbol and the accompanying specifications. Descriptions of the parts may be provided on the component list.
- Tracing the sequence of each operation. This requires that each circuit be followed according to descriptive explanatory information provided on the drawing or in complementary technical literature. The operation of each phase is described in terms of how action is initiated and the resulting action.

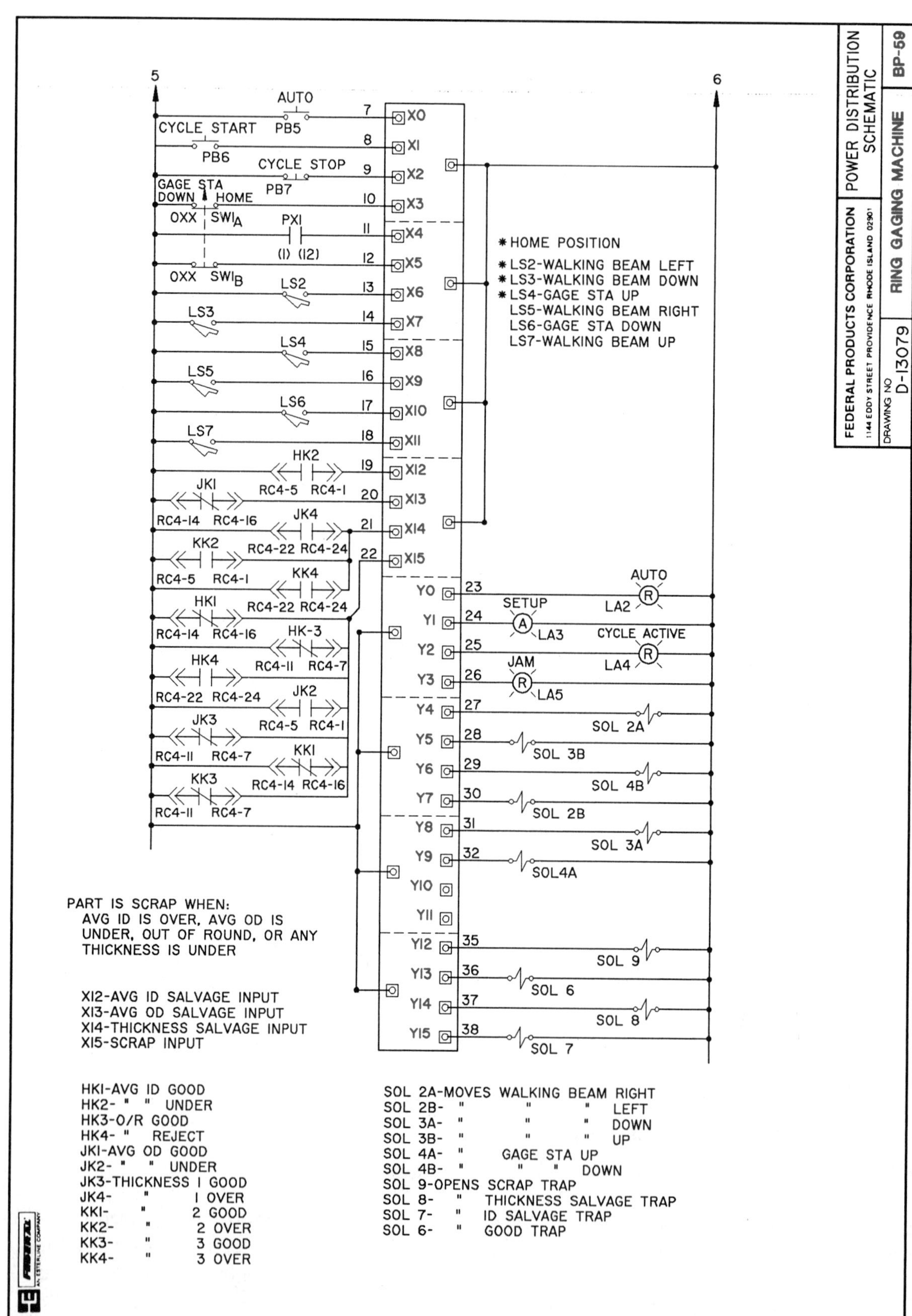

Courtesy of Federal Products Corporation, an Esterline Company (Modified Industrial Drawing)

ASSIGNMENT—UNIT 59

ID, OD, AND THICKNESS GAGING MACHINE (BP-59)

Student's Name ______________________

1. a. Describe briefly what the designation *instrumentation drawings* means.
   b. List five different types of instrumentation drawings.

2. a. Classify the type of instrument drawing which is represented by BP-59.
   b. Describe briefly what function is served by the drawing.

3. Tell what functions are served by conventional drawings on instrumentation equipment drawings.

4. Identify the design features of the ring part which the instrument (machine) checks and scraps.

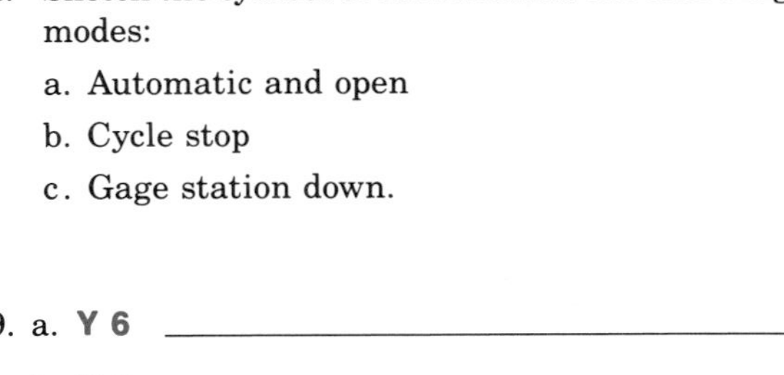

5. Determine the power circuits which relate to salvage input for (a) average inside diameter, (b) average outside diameter, (c) thickness, and (d) scrap input.

6. a. Identify what dimension or geometric characteristic is gaged at each of the following stations:
   (1) **HK 1** (2) **HK 2** (3) **HK 3** (4) **JK 2** (5) **KK 3**
   b. State the quality of each measurement.

7. Designate the **HK** and/or **JK** station power distribution units at which the part is scrapped or rejected.

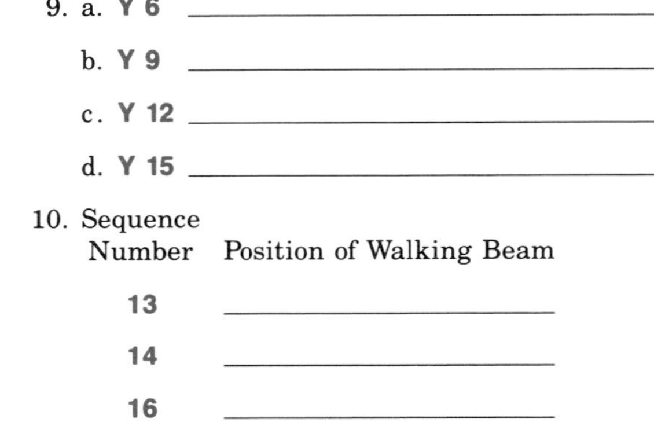

8. Indicate the functions of solenoids **2A**, **2B**, **3A**, and **3B**.

9. State the function of the circuit at the following positions:
   a. **Y 6** b. **Y 9** c. **Y 12** d. **Y 15**

10. Determine the position of the walking beam at sequence **13**, **14**, **16**, and **18**.

11. Sketch the symbol of the switch in the following modes:
    a. Automatic and open
    b. Cycle stop
    c. Gage station down.

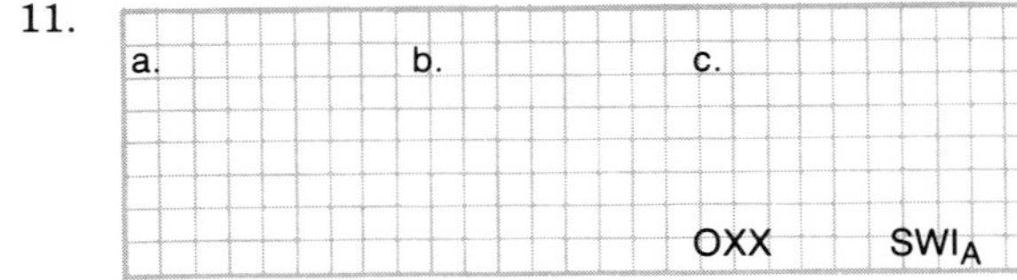

1. a. ______________________
   b. Types of Drawings
   (1) ______________________
   (2) ______________________
   (3) ______________________
   (4) ______________________
   (5) ______________________

2. a. ______________________
   b. ______________________

3. ______________________

4. a. ______________________
   b. ______________________
   c. ______________________
   d. ______________________

5. a. ________ b. ________ c. ________ d. ________

6.

	Gage Features (a)	Quality of Measurement (b)
(1)		
(2)		
(3)		
(4)		
(5)		

7. ____________ ____________

8. 2A ______________________
   2B ______________________
   3A ______________________
   3B ______________________

9. a. **Y 6** ______________________
   b. **Y 9** ______________________
   c. **Y 12** ______________________
   d. **Y 15** ______________________

10.

Sequence Number	Position of Walking Beam
13	
14	
16	
18	

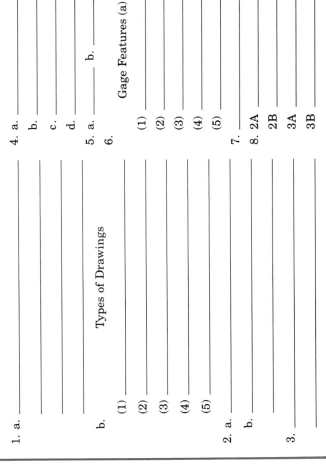

11. a. b. c.

OXX SWI$_A$

UNIT 60
Fluid Power Technology and Diagrams

The science of hydraulics (liquids) and pneumatics (gases), with applications, is referred to as *fluid power* or *fluidics*. The contents of this unit complement the technology presented in relation to instrumentation control drawings.

The reading of drawings (representative of fluid power components, circuits, and systems) requires an understanding of basic rules governing the use of fluid power graphic symbols and definitions of common terms; working with simple formulas; converting values from one system of measurement units to another; and interpreting standard drawings and fluid power diagrams.

BASIC FLUID POWER DEFINITIONS

The National Fluid Power Association (NFPA) and the American National Standards Institute (ANSI) have produced a complete glossary of fluid power terms, abbreviations, and symbols. A selected number of terms and definitions (adopted as ANSI standards) follow.

- *Fluid Power* The controlled transmission of energy produced through the use of a pressurized fluid.
- *Cycle* A complete operation or process that consists of a progressive sequence of steps that begin and end at a fixed neutral position.
- *Fluid Power System* An enclosed circuit for the transmission and control of a pressurized fluid (liquid or gas).
- *Hydraulic Power Unit* Components of a fluid power system that are used for fluid storage and conditioning and fluid delivery (under controlled pressure and flow) to the pump discharge port. Hydraulic power units include applicable maximum pressure and sensing devices. Circuitry components are generally not part of the power unit.
- *Pressure, Back* The measured pressure on the return side of the fluid power system.
- *Pressure, Differential* The difference in pressure between any two points in a fluid power system or component.
- *Pressure, System* The pressure required to overcome resistance in a fluid power system, inclusive of system losses and useful work.
- *Pump, Fixed Displacement* Converting mechanical energy into fluid energy by causing a liquid to flow against a pressure at a constant displacement rate per hour.
- *Pump, Variable* A pump that permits varying the displacement rate per cycle.

EXAMPLES OF FLUID POWER FORMULAS

The reading of standard view drawings, pictorial sketches, graphic diagrams, and other visual descriptions of components and systems requires skill in applying physical science principles and in making computations that involve the use of formulas. Four selected formulas are given to illustrate some of the computations that apply to fluid power systems.

- TORQUE (T)
  T = HP × 5252 ÷ RPM; Where **T** values are in foot-pounds
- HORSEPOWER (HP)
  HP = T × RPM ÷ 5252
- FLUID POWER (HYDRAULIC) HORSEPOWER (HP)
  HP = PSI × GPM ÷ 1714; Where system gauge pressure (**PSI**) is given in pounds per square inch and **GPM** = flow in gallons per minute
- FLUID POWER (HEAT EQUIVALENT)
  BTU per Hour = PSI × GPM × 1 1/2

CALCULATING EQUIVALENT SI, METRIC, AND U.S. CUSTOMARY MEASUREMENT VALUES

Reference tables are available for interchanging fluid power measurement values for torque, linear, volume (cubic), area (square), force, velocity, fluid and mechanical pressure, power, and others, from one system to an equivalent in another system.

Table 60-1 provides information for torque conversion measurements, as an example. The Newton-Meters in the left column relate to *International Standard (SI)* torque measurements. The second column (Kilopond-Meters) deals with *Metric* units. The third and fourth columns relate to *U.S. Customary* units of measurement in foot-pounds and inch pounds.

Interchanging Torque Values

- Select the column in a torque values table for the given torque measurement unit.
- Locate the line with the factor (1).
- Follow horizontally on the same line to the column that contains conversion factors for the required measurement unit.
- Multiply the factor (on the horizontal line) by the given torque measurement value.
- State the measurement unit in the answer.

Example Determine the equivalent metric value (kilo-ponds) of 250 foot-pounds of torque that is produced in a fluid power system.

Solution Table 60-1 shows that one foot pound of torque is equal to 1.382×10^{-1} kilopond-meters.

Since the exponent 10^{-1} tells that the decimal point is moved one place to the left, the 1.382×10^{-1} equals .1382.

Substituting values,

kilopond-meters $= 250 \times .1382$
$= 34.55$ Answer.

Note The kilopond-meters answer may be checked by multiplying it by the U.S. Customary unit factor of 7.233.

$34.55 \times 7.233 = 250$ foot-pounds Ans.

GRAPHIC SYMBOLS: FLUID POWER DRAWINGS/DIAGRAMS (ANSI Y 32.10)

ANSI Standard Y 32.10 provides the graphic symbols that were prepared in collaboration with the National Fluid Power Association (NFPA) and as adopted by the American National Standards Institute. The graphic symbol standards fall into

TABLE 60-1 INTERCHANGE OF TORQUE SI, METRIC, AND U.S. CUSTOMARY UNITS

SI METRIC	METRIC	U.S. CUSTOMARY	
Newton-Meters	Kilopond-Meters	Foot-Pounds	Inch-Pounds
1	1.020×10^{-1}	7.376×10^{-1}	8.851
9.807	1	7.233	86.80
1.356	1.382×10^{-1}	1	12
1.130×10^{-1}	1.152×10^{-2}	8.333×10^{-2}	1

TABLE 60-2 CATEGORIES OF ANSI FLUID POWER GRAPHIC SYMBOLS

CATEGORY	SYMBOL GROUPING	GENERAL APPLICATIONS
1	Introduction	Basic Statements, Scope, and Purposes
2	Symbol Rules	Rules, Guidelines, Drafting Requirements
3	Fluid Conductors	Energy or Fluid Transmission Among Components
4	Energy and Fluid Storage	Pressurized and Nonpressurized Containers for Liquids and Gases
5	Fluid Conditioners	Devices Controlling the Physical Characteristics of Fluids
6	Linear Devices	Actuators, Cylinders, Positioners, and Intensifiers
7	Controls	Devices for Regulating Component or Systems Functions
8	Rotary Devices	Pumps, Compressors, Oscillators, and Motors
9	Instruments and Accessories	Indicating, Sensing, and Recording Devices
10	Valves	Flow Control, Direction Control, and Pressure Control
11	Composite Symbols	Miscellaneous Air and Hydraulic Components

eleven different categories (from 1 through 11). Introductory information is provided in the first category.

Table 60-2 lists the different groupings of fluid power graphic symbols (categories 1 through 11). General applications are given in the third column. The graphic symbols within each category are listed sequentially, like 3.1, 3.2, etc. In many cases, there is a further subdivision for particular applications (like 3.1.1, 3.1.2, or 3.1.1.1, 3.1.2.1, etc.).

RULES RELATING TO THE USE OF GRAPHIC SYMBOLS FOR FLUID POWER DIAGRAMS

A partial list follows of ANSI rules governing the use of symbols for fluid power diagrams and the interpretation of the symbols in reading drawings.

- Graphic symbols relate to functions of components, flow paths, connections, and conditions that occur during transition from one flow path arrangement to another.
  - Symbols are not used to indicate construction features nor values for pressure, rate flow, and other component settings.
- Generally, the meaning of a graphic symbol is unchanged when it is rotated or reversed.
- Line weights (thickness) are drawn approximately equal.
  - A *solid line* represents a main line, conductor, outline, or a shaft.
  - A *dash line* is used as a pilot line for control.
  - A *dotted line* represents an exhaust line or a drain line.
  - A *center line* shows an enclosure outline.
- Lines that intersect at 90°, or any other angle, appear on graphic diagrams as follows.

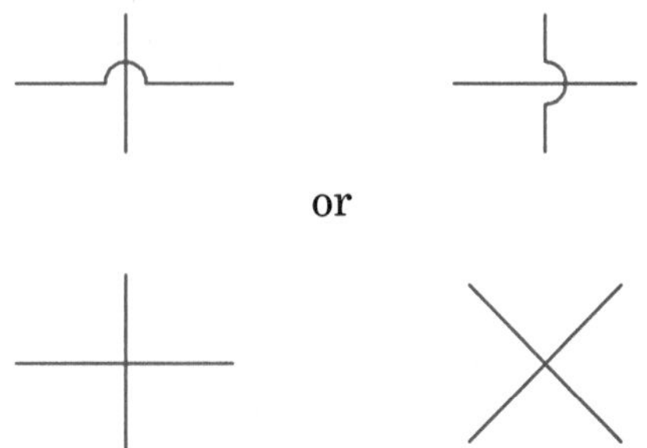

- Joining lines are represented by the following symbols.

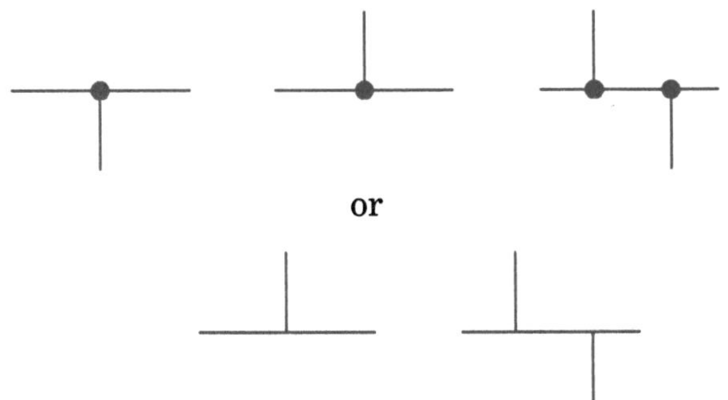

- Any suitable size symbol may be used on a graphic diagram for circles, triangles, arrows, squares, and rectangles. Relative sizes represent main components in comparison with other secondary or auxiliary components.

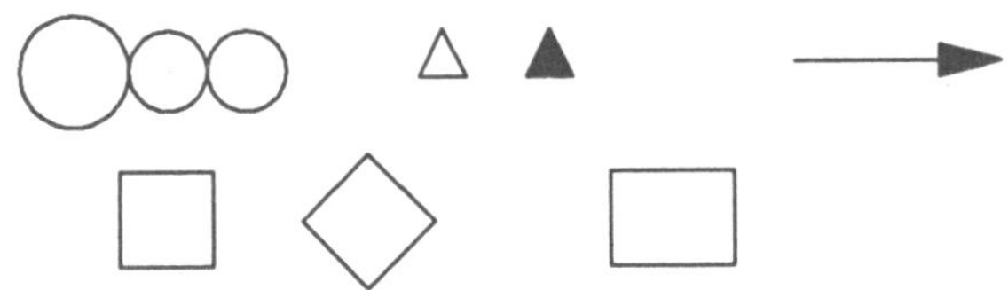

- Unless multiple fluid power diagrams are provided, each drawn symbol represents the normal, at-rest, or neutral condition of the component.

- Letter combinations that appear on drawings as a part of a graphic symbol are not necessarily abbreviations.

- A component that may be varied or adjusted is represented by an arrow that cuts through the symbol at about a 45° angle.

- A pressure-compensated component is represented by an arrow that is drawn within the symbol, parallel to the short side.

- Temperature cause or effect is represented by a line that terminates in a large dot.

- The direction of shaft rotation is indicated by an arrow. The arrow is assumed to be on the near side of the shaft.

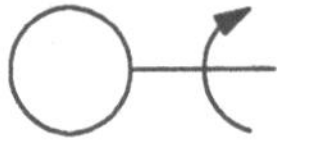 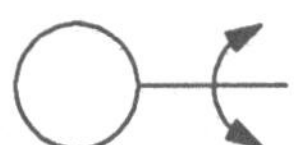

- A component enclosure is drawn around a symbol or a group of symbols that represent an assembly of components within a total fluid power system.

- ANSI standards for representation include the use of a *complete symbol* within a component or a *simplified symbol.*

COLOR CODING FLUID POWER CIRCUIT DRAWINGS

The reading of fluid power circuit drawings is simplified by using color coding to identify the systems and flow conditions within major components. One or more of five basic colors (and no-color white) are used to color code basic hydraulic circuit drawings.

- Red — Operating Pressure or System Pressure
- Orange — Intermediate Pressure Inside Flow Control
- Yellow — Reduced Pressure or Reduced Flow
- Green — Intake or Drain
- Blue — Return Line Flow
- White — Inactive Fluid

REPRESENTATIVE FLUID POWER DIAGRAMS

Fluid power systems are generally represented by one of four different types of drawings.

- A *graphic diagram* (figure 60-1A) uses simple line graphic shapes (symbols) with interconnecting lines to represent the function of each component in the circuit. This type diagram is functional for designing and troubleshooting fluid power circuits.
- A *pictorial fluid power diagram* (figure 60-1B) is generally used to show the piping system in relation to each component. Each component is represented pictorially by single or double lines. The pictorial diagram does not show functions or methods of operation.
- A *cutaway fluid power diagram* (figure 60-1C), as in conventional drafting practice, shows internal details of working parts, component functions, and flow paths. Double lines are used.
- A *combination fluid diagram* (figure 60-1D) is produced by using interconnecting lines with graphic, pictorial, and cutaway symbols. This type of drawing presents information about the functions of components, essential piping, and flow paths.

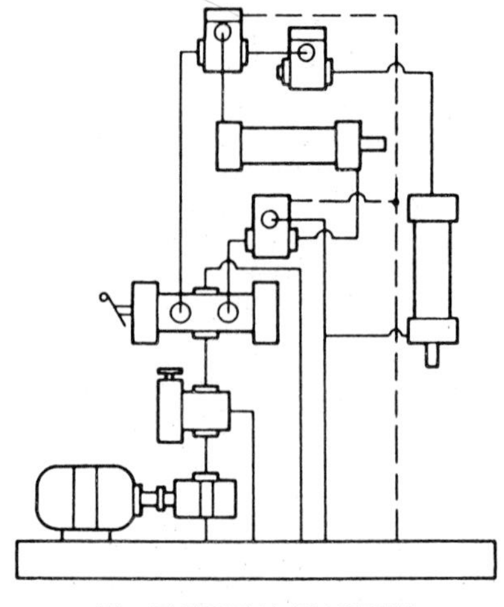

(B) PICTORIAL DIAGRAM

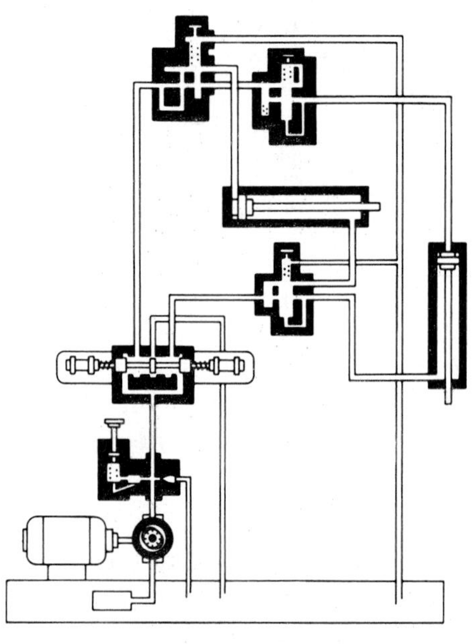

(C) CUTAWAY DIAGRAM

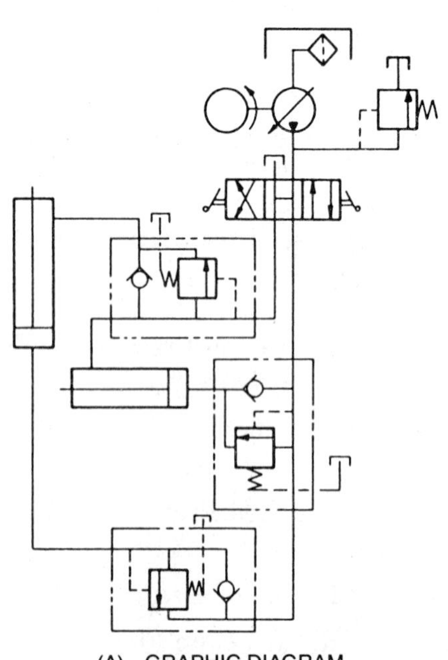

(A) GRAPHIC DIAGRAM

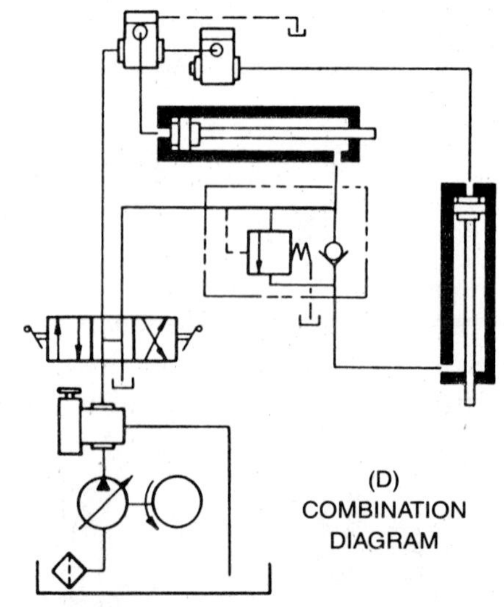

(D) COMBINATION DIAGRAM

FIGURE 60-1 BASIC TYPES OF FLUID POWER DIAGRAMS (COURTESY OF PARKER HANNIFIN CORPORATION)

SELECTED GRAPHIC SYMBOLS FOR FLUID POWER DIAGRAMS

The complete ANSI standard Y 32.10 is available in professional and craft handbooks and other technical publications. Selected examples of fluid power graphic symbols are displayed in table 60-3.

TABLE 60-3 SELECTED EXAMPLES OF GRAPHIC SYMBOLS FOR FLUID POWER DIAGRAMS

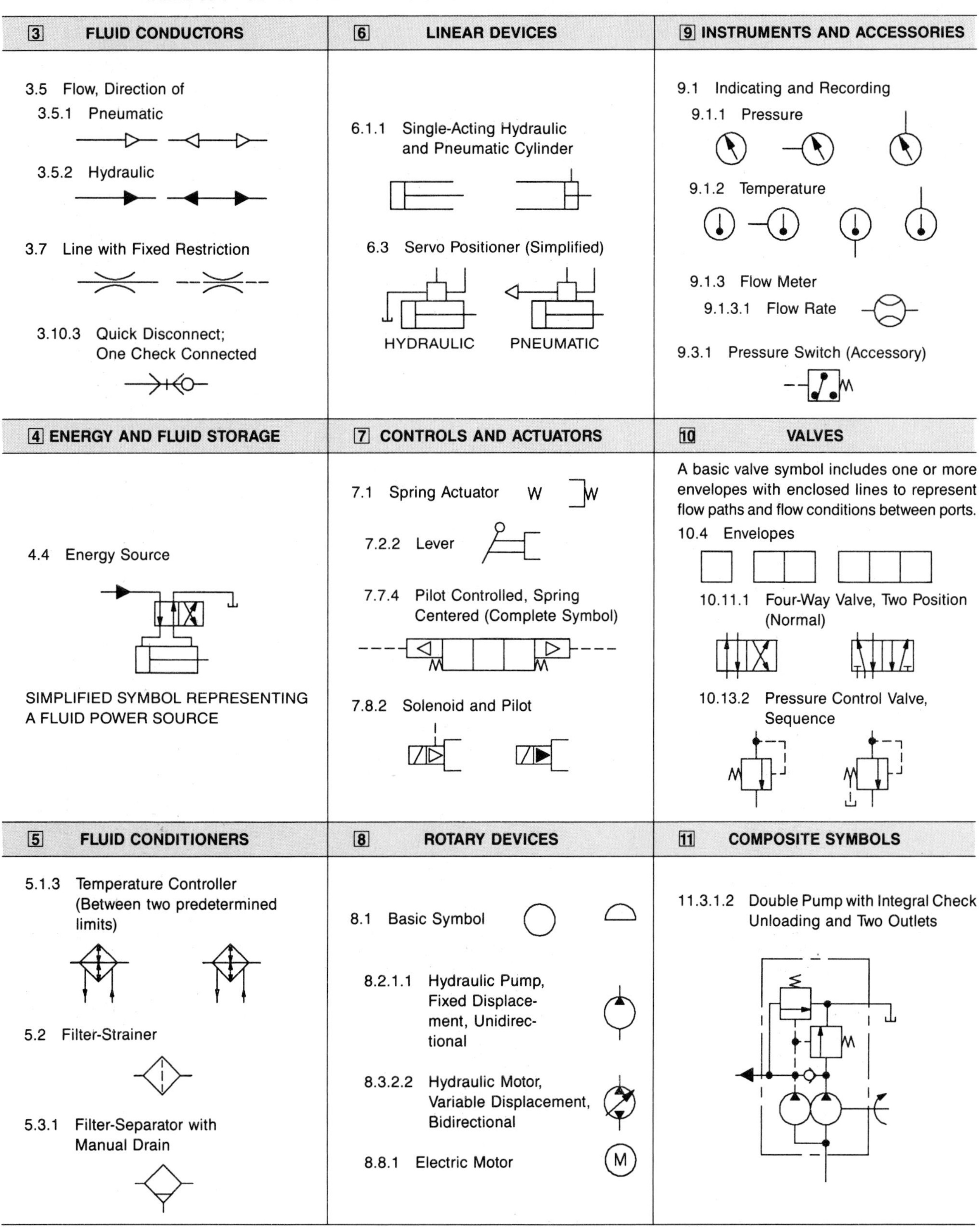

3 FLUID CONDUCTORS	6 LINEAR DEVICES	9 INSTRUMENTS AND ACCESSORIES
3.5 Flow, Direction of 3.5.1 Pneumatic 3.5.2 Hydraulic 3.7 Line with Fixed Restriction 3.10.3 Quick Disconnect; One Check Connected	6.1.1 Single-Acting Hydraulic and Pneumatic Cylinder 6.3 Servo Positioner (Simplified) HYDRAULIC PNEUMATIC	9.1 Indicating and Recording 9.1.1 Pressure 9.1.2 Temperature 9.1.3 Flow Meter 9.1.3.1 Flow Rate 9.3.1 Pressure Switch (Accessory)
4 ENERGY AND FLUID STORAGE	**7 CONTROLS AND ACTUATORS**	**10 VALVES**
4.4 Energy Source SIMPLIFIED SYMBOL REPRESENTING A FLUID POWER SOURCE	7.1 Spring Actuator W 7.2.2 Lever 7.7.4 Pilot Controlled, Spring Centered (Complete Symbol) 7.8.2 Solenoid and Pilot	A basic valve symbol includes one or more envelopes with enclosed lines to represent flow paths and flow conditions between ports. 10.4 Envelopes 10.11.1 Four-Way Valve, Two Position (Normal) 10.13.2 Pressure Control Valve, Sequence
5 FLUID CONDITIONERS	**8 ROTARY DEVICES**	**11 COMPOSITE SYMBOLS**
5.1.3 Temperature Controller (Between two predetermined limits) 5.2 Filter-Strainer 5.3.1 Filter-Separator with Manual Drain	8.1 Basic Symbol 8.2.1.1 Hydraulic Pump, Fixed Displacement, Unidirectional 8.3.2.2 Hydraulic Motor, Variable Displacement, Bidirectional 8.8.1 Electric Motor (M)	11.3.1.2 Double Pump with Integral Check Unloading and Two Outlets

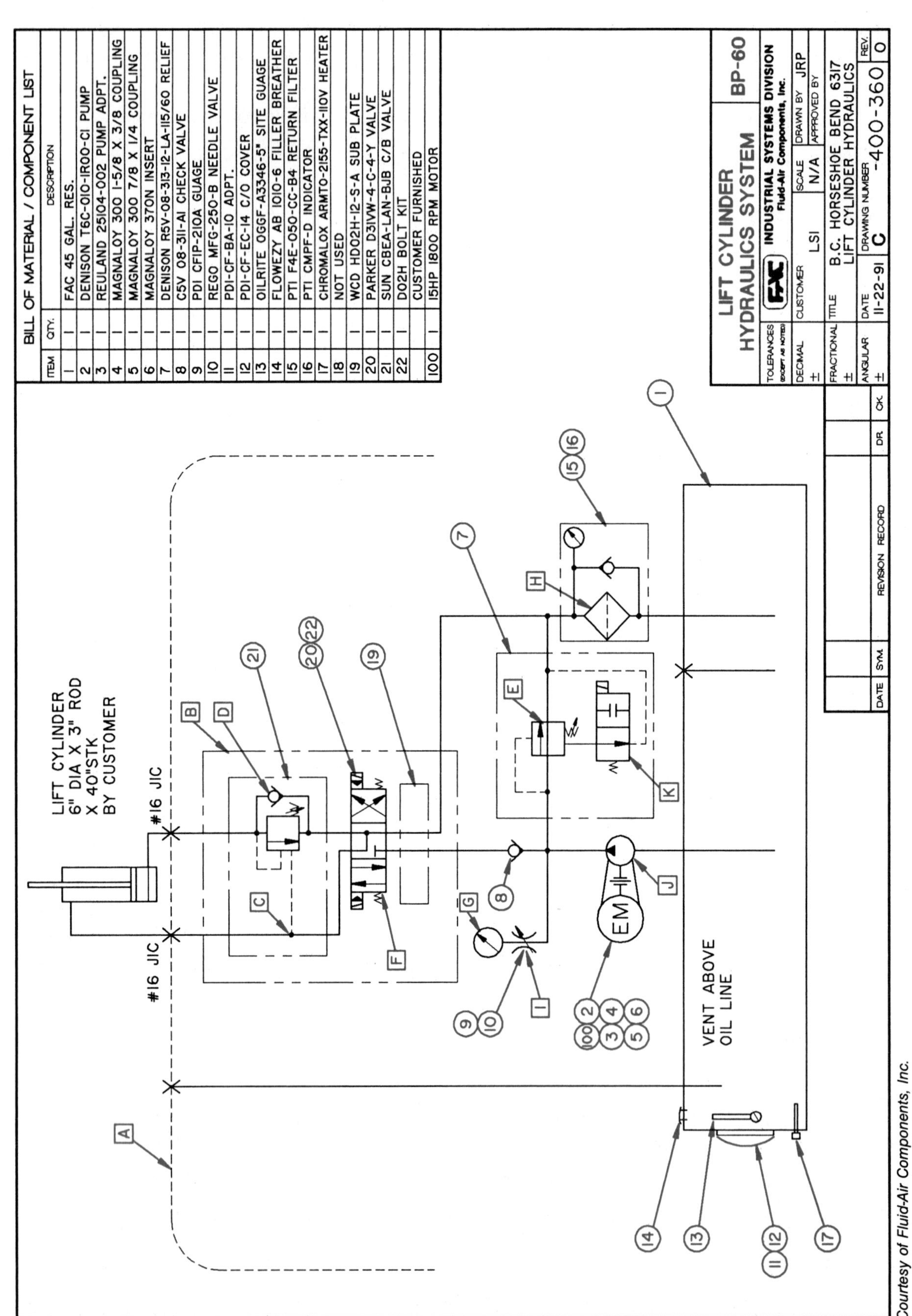

Courtesy of Fluid-Air Components, Inc.

ASSIGNMENT—UNIT 60

LIFT CYLINDER HYDRAULICS SYSTEM (BP-60)

Student's Name ______________________

1. Describe fluid power terms (a) and (b).
   a. Fluid power system
   b. System pressure
2. a. Compute the torque generated by a 15 horsepower motor rotating at 1800 RPM. Round the answer to the nearest foot-pound.

$$T = \frac{HP \times 5252}{RPM}$$

   b. Convert the answer (a) to its equivalent SI value, rounded to the nearest whole unit.
   c. Convert the SI answer (b) to its equivalent metric value. Label the unit.
3. State two basic rules for using graphic symbols on fluid power diagrams.
4. Tell what each of the following lines represents on a fluid power diagram.
   a. Dotted line b. Center line
5. State how a component that may be adjusted or varied is represented.
6. Explain briefly how main and secondary or auxiliary components are represented.
7. Give the color that is used in color-coded diagrams for conditions/systems (a), (b), and (c).
   a. Intermediate pressure inside flow control
   b. Return line flow
   c. Inactive fluid
8. a. Name four types of fluid power diagrams.
   b. Cite a drafting room application of two different diagram types.
9. Give the (1) category number and (2) description and (3) the function of the components, conditions, or functions represented by symbols (a), (b), and (c).

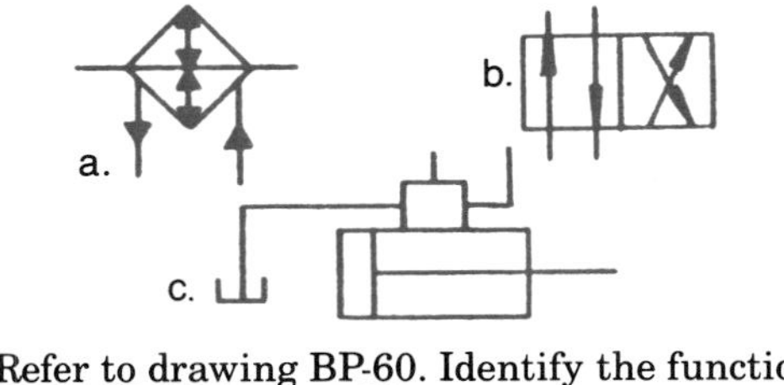

10. Refer to drawing BP-60. Identify the function of lines (A), (B), and (C).
11. State what each symbol ((D) through (K)) represents. Refer to ANSI Y 32.10, if needed.
12. Name the following parts, instruments, and components. Also, give the manufacturer's identification and specifications.
   a. (7) d. (15) g. (100)
   b. (9) e. (16)
   c. (10) f. (21)

1. a. ______________________
   b. ______________________
2. a. Torque ______________
   b. SI equivalent ______________
   c. Metric equivalent ______________
3. a. ______________________
   b. ______________________
4. a. ______________________
   b. ______________________
5. ______________________
6. ______________________
7. a. ________ b. ________ c. ________
8. a. (1) ________ (3) ________
   (2) ________ (4) ________
   b. (1) ______________________
   (2) ______________________
9.

#	Description	Function
a.		
b.		
c.		

10. a. (A) ______________________
   (B) ______________________
   (C) ______________________
11. (D) ________ (H) ________
   (E) ________ (I) ________
   (F) ________ (J) ________
   (G) ________ (K) ________
12. (7) ______________________
   (9) ______________________
   (10) ______________________
   (15) ______________________
   (16) ______________________
   (21) ______________________
   (100) ______________________

UNIT 61
Architectural Drawings

ARCHITECTURAL DRAWINGS

The word *architecture* is usually associated with buildings, structures, interior systems designs; mechanical and civil engineering; and construction planning in relation to the environment.

The term *buildings* relates to residential and light and heavy-duty commercial structures. Regardless of the nature of the structure, the same basic principles of drawing, standards, and drafting room techniques are applied to architectural drawings.

WORKING DRAWINGS

Working drawings represent the end product of research and design studies. Working drawings provide a graphic picture to different contractors, engineers, and others in terms of building sizes, quantities, location, and relationship of all parts of a building, dimensions, and other important information.

Each construction project requires a *set of working drawings.* A set may include any combination of the following types of drawings, depending on the size and complexity of each building.

- Location Plans (Plot, Landscape, Survey)
- Floor Plans
- Foundation Plans
- Elevation Plans
- Detail Section Drawings

Major Drawing Supplements

A complete set of building plans includes drawings of the various *systems* that are a part of the whole structure. For residential or light commercial construction (where installations for light, heat, power, etc. are simple), sufficient information may be provided on standard plan, elevation, and section drawings.

Buildings that have complex installations to be accommodated within the physical structure require drawing supplements. Four examples follow as illustrative of systems drawings that are incorporated in a working set of building plans.

- Electrical system drawings are used to identify circuits, power distribution panels, and the special requirements for equipment installation.
- Electronic systems, as they impact on building design features, require separate drawings.
- Plumbing and piping systems require drawings and schematics to represent systems for transporting liquids, gases, and other solids to particular locations in a building.
- Heating, ventilating, air conditioning (HVAC) and disposal systems require still other drawings.

Additional information is provided on plan and system working drawings. Included are *specifications, schedules, codes, inspection* and *testing procedures,* and other data. A brief description follows for basic types of building plans, systems, and complementary information generally provided in a set of working drawings.

Location Site Plans

Location site plans include plot, landscape, and

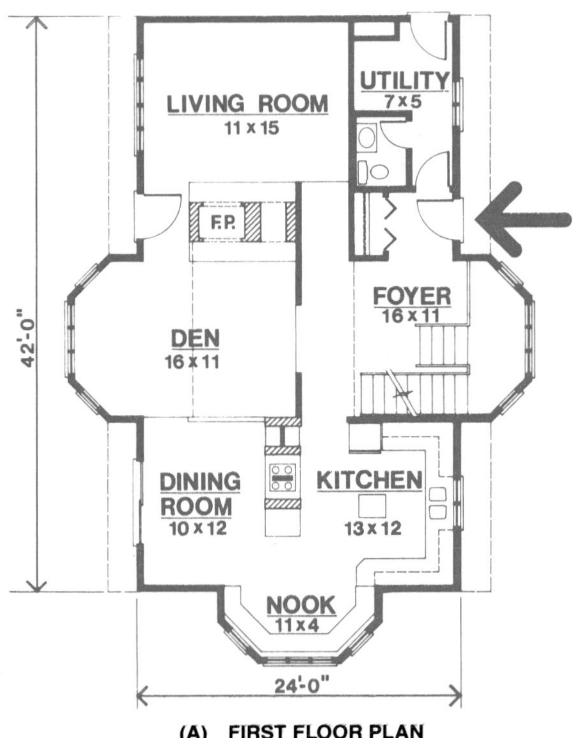

(A) FIRST FLOOR PLAN

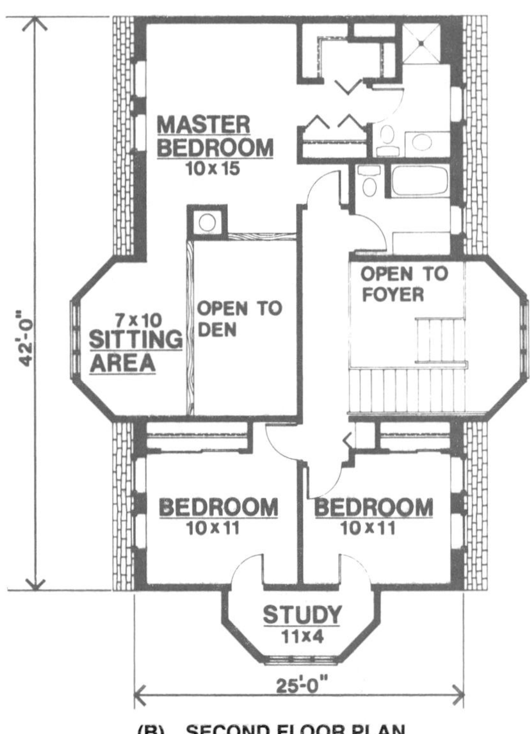

(B) SECOND FLOOR PLAN

FIGURE 61-1 FIRST AND SECOND FLOOR RESIDENTIAL PLAN DRAWINGS (COURTESY OF LANGER, DION, AND MORSE ASSOCIATES)

survey plans. Overall, these plans provide information about the characteristics of the property.

Plot Plan. A *plot plan* relates to the compass orientation of the property, overall lot and building dimensions, and external walkways, driveways, etc.

Landscape Plan. The *landscape plan* shows land contours in relation to building position for purposes of site appearance, gardening, and landscaping designs.

Survey Plan. The *survey plan* is a drawing that is prepared by a surveyor and/or civil engineer. This plan serves as a legal document that accurately provides dimensions, contour information, and features of a building lot and surrounding property.

Standard *symbols* are widely used on plot, landscape, and survey drawings to graphically represent and describe features of the site and surrounding areas.

BASIC TYPES OF ARCHITECTURAL DRAWINGS

Architectural design details on building structures are represented by four basic types of drawings: (1) Plan, (2) Elevation, (3) Perspective, and (4) Cross-Section Detail.

Foundation Plans

Foundation Plans, as the name suggests, deal with the foundations of a building, levels, and to uniformly distributed support structures. Information is provided about excavating, waterproofing, and design features of the building foundations.

Some foundation plans often serve as a *basement plan.* Typical forms and sizes for walls, beams, and columns, and their locations and openings and construction materials are identified on basement plans. Information is also provided about the various power, light, heating and cooling, and other systems that may be located in the basement, or controlled from that area.

Basement plans are complemented by other drawings of building and structural components that are *above* the basement level.

Floor Plans

A *floor plan* shows graphically the outline, partitions, and other general design features of a structure as it would look when viewed from above on a horizontal cutting plane that is at a level around four feet above the floor line. Figure 61-1 illustrates *first floor* and *second floor plans.* Floor plans show all exterior and interior walls, doors, windows, stair location, built-in cabinets,

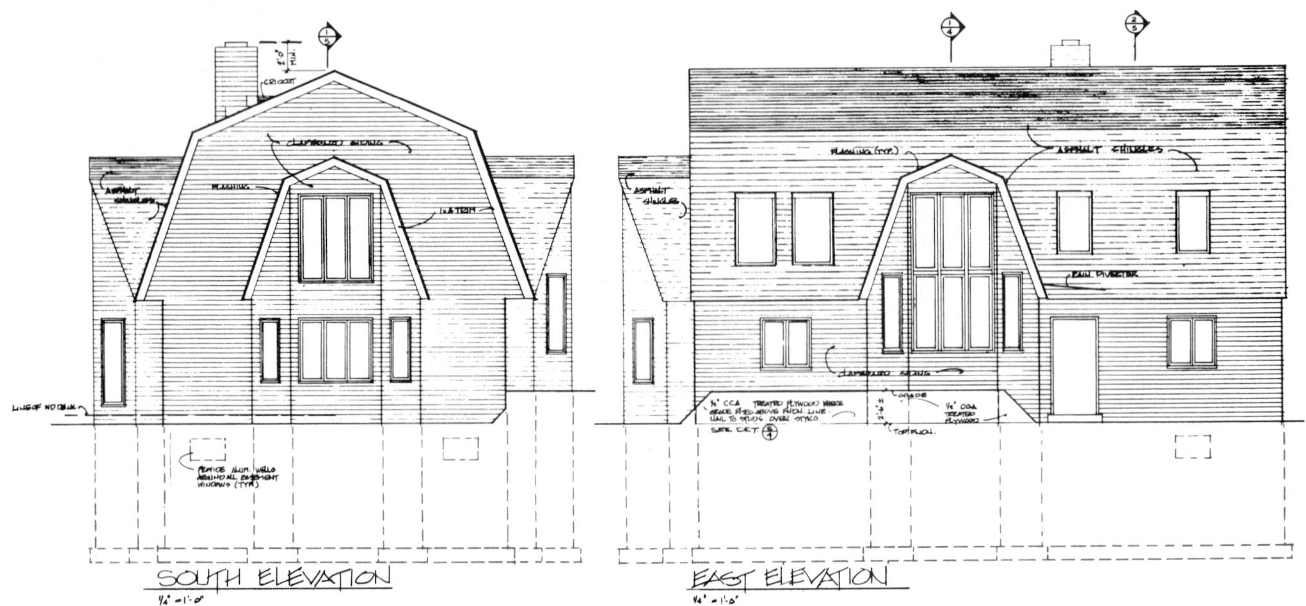

FIGURE 61-2 ELEVATION DRAWINGS OF A TWO FLOOR RESIDENCE (COURTESY OF LANGER, DION, AND MORSE ASSOCIATES)

appliances, bathroom and other fixtures, fireplace, and the arrangement of all rooms.

Working floor plans include location, size, and shape dimensions, specifications, and other essential details that are needed in construction. Since a relatively small scale (such as: 1/8″ or 1/4″, or 3/4″ equals 1′) is used for architectural drawings, only major dimensions are shown on a floor plan.

Dimension and other design features that are too small to show clearly on a floor plan are drawn in enlarged detail, usually as a cross-section. Each *level* of a building requires a detailed floor plan.

Elevation Drawings (Plans)

An *elevation drawing* defines the structural form and architectural style. Windows, doors, and other openings and building features are represented as in a regular view. Each view is identified according to direction as a *north elevation, south elevation, east elevation, or west elevation.* The south and east elevations for a residence are illustrated in figure 61-2.

Elevation plans are drawn as in standard orthographic views. Vertical dimensions may be placed on elevation views to show height above a ground datum line; distances from the floor line to ceilings; and other detail dimensions as for doors, windows, ridges, chimney, etc. Dimensions of features below the ground line are represented by *dotted* lines (figure 61-2).

Perspective (Presentation) Drawings

Perspective drawings (sometimes referred to as *presentation drawings*) provide a pictorial image of a building designer's ideas as to how the features of the completed structure will look. Perspective (presentation) drawings are used primarily to provide a property owner with a graphic illlustration that may be used for building zoning, environmental impact, and other study purposes.

Perspective drawings are often prepared as freehand sketches. While such drawings are generally not drawn to scale, there is an attempt to show all features in proportion. There are increasing numbers of perspective drawings being generated as CADD drawings.

Common types of presentation (perspective) drawings include *renderings* for plot plans, elevations, floor plans, and sections. *Rendering* refers to a drafting room technique of applying artistic style to a perspective drawing (figure 61-3). A drawing is rendered to provide a more realistic picture of a building, internal structural components, and plantings.

Renderings include the addition of landscape features for exterior view drawings and the relationship of inside features. Rendering is accomplished by shading to highlight particular design features. Renderings may be produced in black or white or in color. Drawings are rendered by different line weights using black pencil or ink, or with colored pencils or markers, or by painting

FIGURE 61-3 PERSPECTIVE (PICTORIAL) ARCHITECTURAL DRAWING OF AN HISTORICAL PRESERVATION BUILDING (COURTESY OF LANGER, DION, AND MORSE ASSOCIATES)

with pastel colors. Usually, it is the print and not the original drawing that is highlighted.

Presentation (Perspective) Floor and Plot Plans

A *rendered floor plan* is another example of a presentation drawing. This type of floor plan shows relationships among the rooms; window, door, and other openings; and external areas and plantings. Room sizes and overall dimensions are given. Plot plans may be rendered to show roads, waterways, plantings, and other design features of the structure in relation to a job site.

Cross-Detail Section Drawings

A *cross-detail section drawing* is a section view that cuts across a number of different structural members at a particular location. Figure 61-4 provides an example of a typical section of a wall.

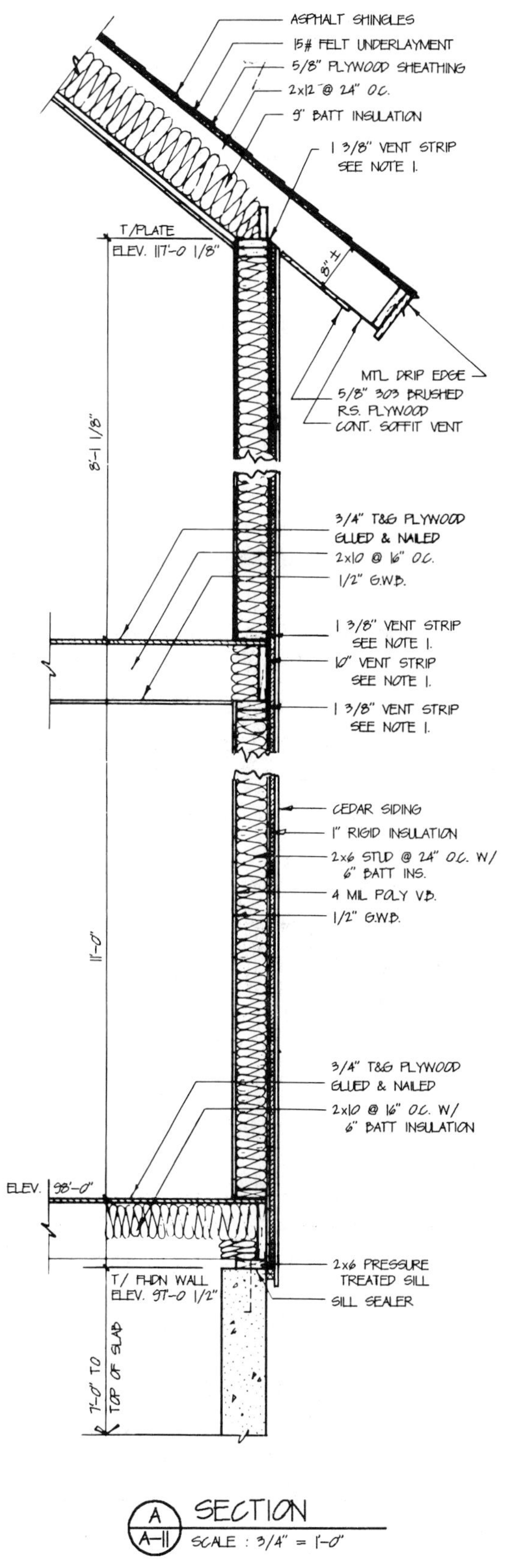

FIGURE 61-4 TYPICAL CROSS-DETAIL SECTION DRAWING OF A WALL (COURTESY OF LANGER, DION, AND MORSE ASSOCIATES)

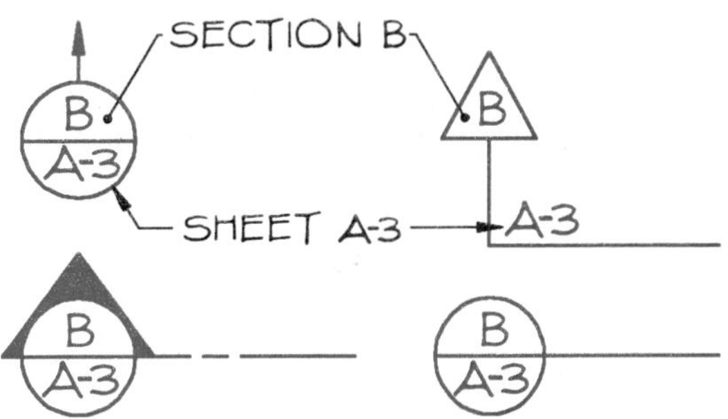

FIGURE 61-5 SAMPLE REFERENCE SYMBOLS

The section shows the location and the vertical relationship of the different parts and features from the floor to the roofing. Working detail drawings may also be rendered to emphasize structural and other design features and space utilization.

Section Reference Symbols

Reference symbols are found on section working drawings. The symbols provide a cross-reference for easily and quickly identifying a specific detail that may require an additional sketch or drawing in order to give adequate dimensional, form, and full construction information.

While each architect uses an individual style, each section reference symbol serves the function of identifying the section to which it referenced on a drawing and the number of the drawing sheet. Four simple examples are provided in figure 61-5.

SCALES FOR ARCHITECTURAL DRAWINGS

Customary Inch and SI Metric Scales

The most frequently used customary inch architectural scales are: 1/4" = 1'-0" or 3/4" = 1'-0". Civil engineering scales for site plans are: 1" = 10', 1" = 50', and 1" = 300'. On drawings, where dimensions are given in the metric system of measurement, the following scales are used:
1:5 for construction details; 1:50 for floor plans, elevations, and foundations; and 1:500 for plot plans.

Generally, *soft conversion* (using the rounded-off 25 mm = 1″ conversion factor) is used to change a measurement from one system to the other. As an example, a 16" center-to-center stud spacing (using soft conversion) equals 400 mm in metric.

With the wide diversity of scales that are used on construction drawings and prints, it is important to check the title block for the correct scale for each part. The stated dimension is used. A drawing should not be scaled to obtain a measurement.

SPECIFICATIONS, SCHEDULES, AND KEYS

Specifications are written standards that relate to all materials and all construction conditions a building contractor must fulfill. Specifications are designed to provide uniformity of understanding and as a check into the meeting of the requirements of building codes; safety, sanitation, and ecology requirements; and other building construction standards. Specifications provide the legal foundations against which claims may be made.

Specifications include lists of building details, construction materials; product sizes, grades, features, color, style, etc., and other design characteristics. Specifications are important to quality control.

A *schedule* generally provides a listing of materials, hardware, fixtures, supplies, etc. Sizes and quantities are noted, along with type or model and finish materials.

A *key symbol* is placed on drawings to identify doors and windows. Each *door* on a floor plan is given a key number that appears within a circle or other form of enclosure (figure 61-6 at A). *Windows* are identified by a capital letter that is usually enclosed in a geometric form as shown at (B), figure 61-6.

Modifications of keys are illustrated at (C) and (D), figure 61-6. A circle is divided and the letter F (for floor) and W (for window) is placed above

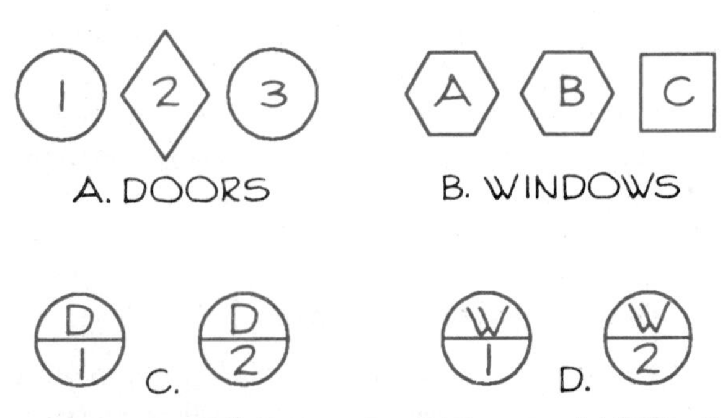

FIGURE 61-6 SAMPLE ENCLOSED SYMBOLS

the dividing line. The consecutive number of a door or window is placed below the dividing line.

BUILDING CODES

The three most widely used building codes include the *Uniform Building Code* (UBC), *Basic National Building Code* (BNBC), and the *Standard Building Code* (SBC). Further modifications are provided by the Department of Housing and Urban Development (HUD) and the Federal Housing Authority (FHA). While the building codes have variations, each provides requirements related to safety, a healthful environment, exit facilities, sanitation, and other conditions that must be met.

DIMENSIONING AND TERMINOLOGY

An unbroken dimension is preferred. Dimensions appear above the line. Dimension lines usually terminate in arrowheads, dots (1′-6″), or dashes (1′-6″).

When *customary inch modular units* are used, building materials and products are standardized around the ANSI *four-inch module* (meaning multiples of 4″). The comparable SI metric module is *100 mm*.

Standard terms are used to designate main parts, major structures, or design features in houses and industrial and commercial construction work. For example, houses are *framed*. Framing starts with the *sill*, *header*, and *floor plate* on the foundation wall. *Studding*, *girders*, *joists*, *braces*, and *sheathing* are common terms with special industrial meanings.

Figure 61-7 shows the symbols for some common building materials and design features. Complete tables are provided in architectural handbooks and technical literature prepared by building, construction, and related industries.

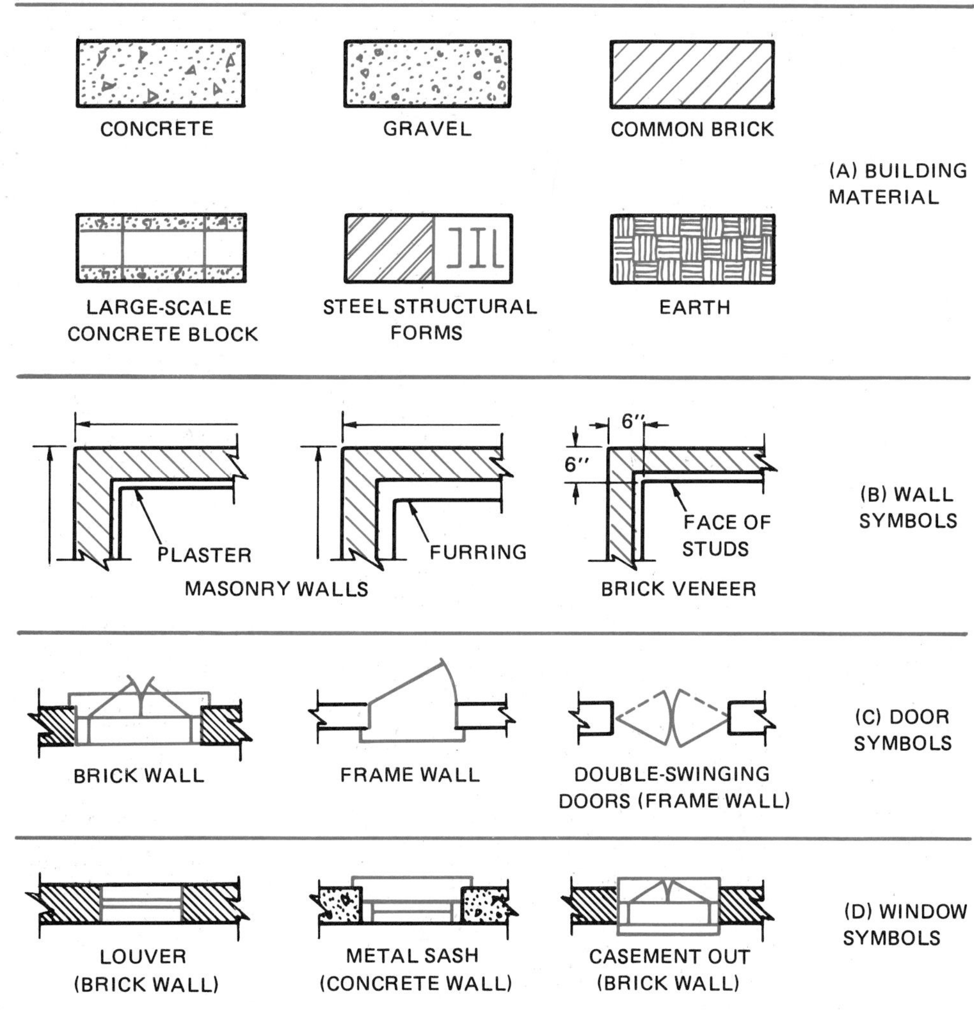

FIGURE 61-7 SYMBOLS OF SELECTED COMMON BUILDING MATERIALS AND ACCESSORIES

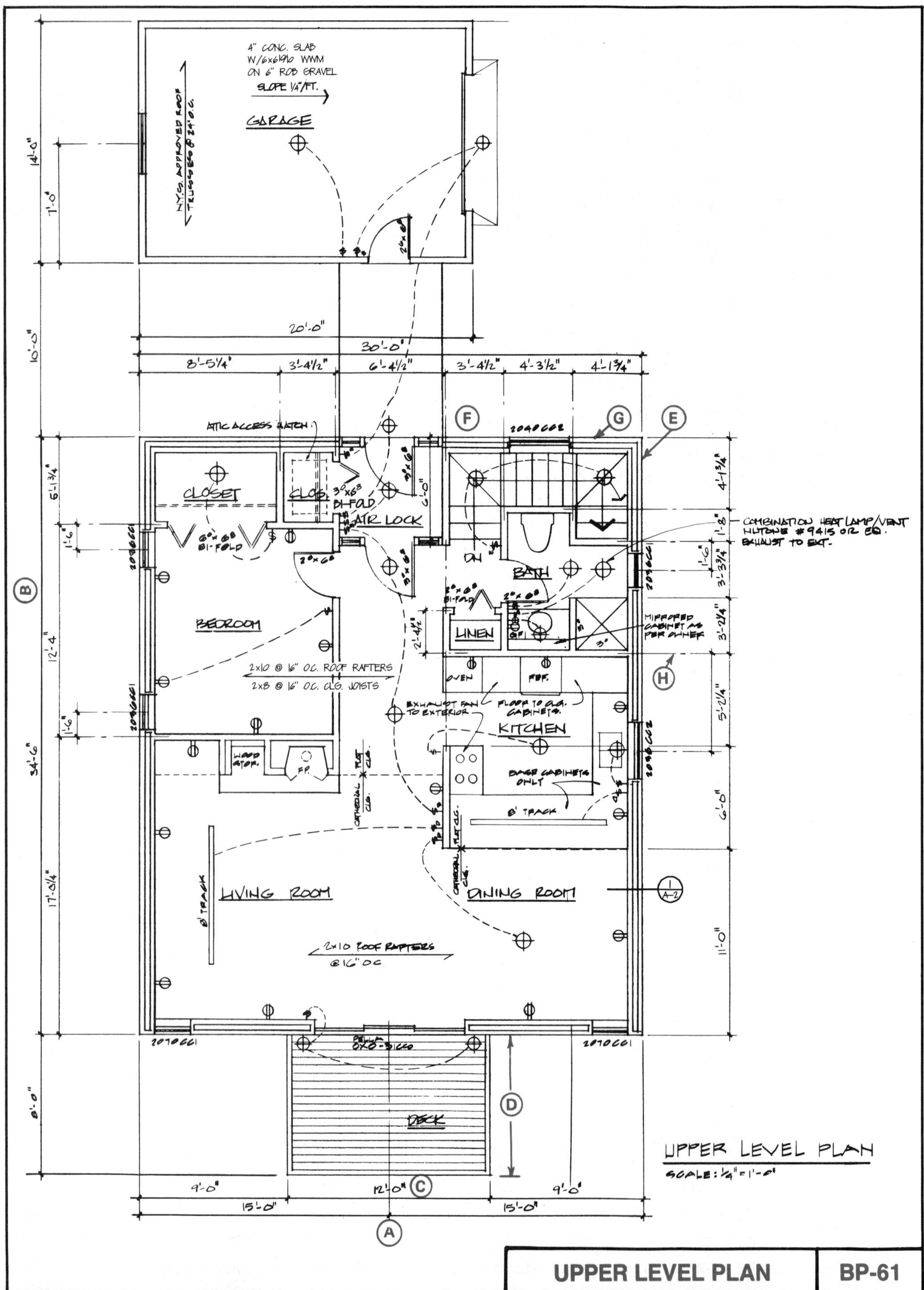

Courtesy of Langer, Dion, and Morse Associates

ASSIGNMENT—UNIT 61

UPPER LEVEL PLAN (BP-61)

Student's Name ______________________

1. a. List four basic types of drawings used in architecture.
   b. Identify the type of drawing illustrated by BP-61.
2. Sketch the drawing symbols which represent the following construction features:
   a. External windows
   b. Bi-fold closet doors
   c. Ceiling lighting fixtures
3. Explain briefly what the term 4″ (or 100 mm) modular standard means.
4. Calculate the following dimensions:
   a. Overall width (A) and depth (B) from the front of the house to the rear of the garage.
   b. Overall width (C) and depth (D) of the deck.
   c. Width of the bathroom from outside wall (E) to center line (F).
   d. Depth of the bathroom from outside wall (G) to center line (H).
5. Give the size and on-center distances of the roof rafters for the house portion.
6. Identify the different sizes or specifications of the following features:
   a. Bi-fold closet doors
   b. Hinged doors between rooms
   c. Windows
7. State the specifications for the garage floor.
8. Sketch the switch symbols for controlling the dining room, living room, and two hall lights from the outside kitchen wall.
9. Translate the symbol ——● (1/A-2) .
10. a. State two functions that are served by a rendered perspective (presentation) drawing.
    b. List two techniques of rendering.
11. Convert the following drawing dimensions. Use soft conversion.
    a. (2″), (5″), (1′ = 2″) to equivalent metric mm values.
    b. (75 mm), (162 mm), (800 mm) to equivalent customary inch or feet/inch values.
12. a. List three building construction items that are covered by specifications on drawings.
    b. Describe briefly two functions that building specifications serve.

1. a. (1) ______
   (2) ______
   (3) ______
   (4) ______
   b. ______
2. a. b. c.
3. ______
4. a. (A) = ______ (B) = ______
   b. (C) = ______ (D) = ______
   c. ______ d. ______
5. ______
6. a. ______ ______ ______
   b. ______ ______
   c. ______ ______
7. ______
8. ______
9. ______
10. a. (1) ______
    (2) ______
    b. (1) ______
    (2) ______
11. a. ______ ______ ______
    b. ______ ______ ______
12. a. (1) ______
    (2) ______
    (3) ______
    b. (1) ______
    (2) ______

UNIT 62
Structural Drawings

STRUCTURAL DRAWINGS

Structural drawings relate to the design, fabrication, and erection of primarily steel members. These members are used for framing buildings, bridges, dams, communication towers, and other structures.

While steel is widely used in construction work, there are a number of applications where structural clay products, concrete, laminated wood, and aluminum are functional and desirable. Figure 62-1 is a pictorial drawing showing structural details for a section of a building, using several different building materials.

The three common types of structural drawings are steel framing, erection, and shop drawings. Steel *framing drawings* are prepared by engineers to show the location, sizes, and numbers of beams, columns, trusses, and other structural members. The *erection plan* is an assembly drawing on which each part is numbered for position within the framework. The *shop drawing* details the features and provides dimensioning.

Shop drawings provide information about the fabrication of each structural member, design features, details about connectors, and bills of materials of all items required during fabrication

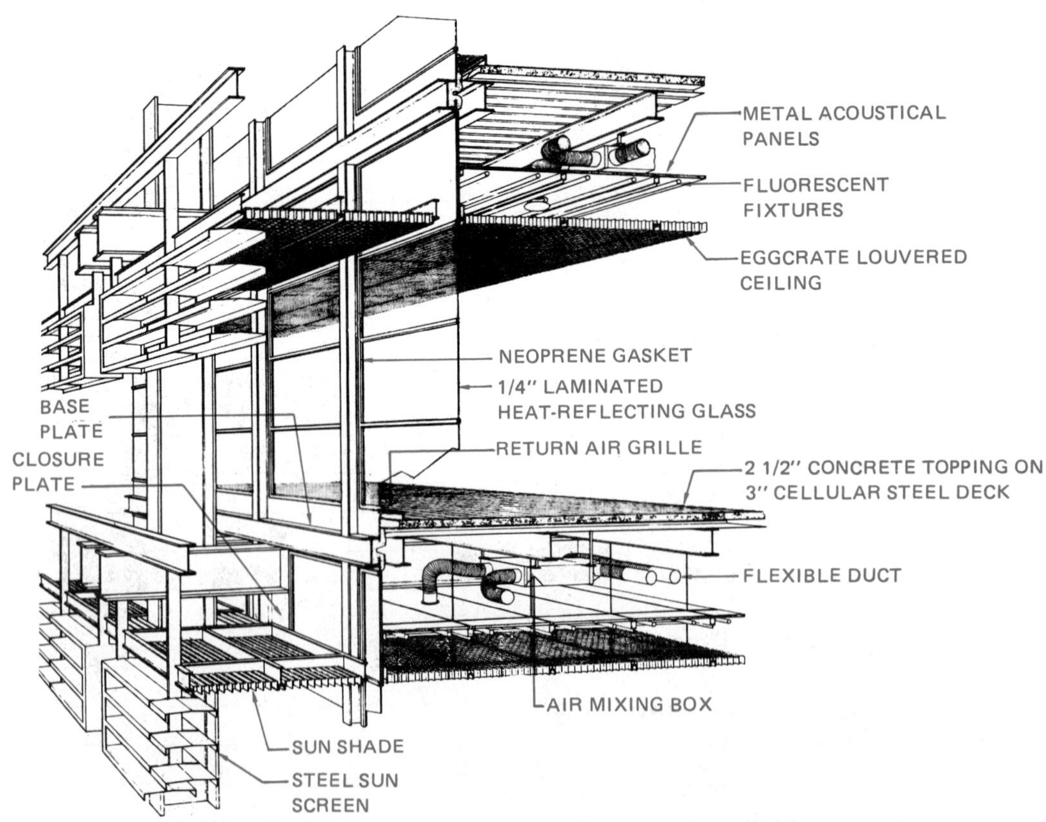

FIGURE 62-1 PICTORIAL DRAWING: SECTION SHOWING STRUCTURAL DETAILS (COURTESY OF KEVIN ROCHE JOHN DINKELOO AND ASSOCIATES)

TABLE 62-1 STANDARD STRUCTURAL STEEL SHAPES, SYMBOLS, AND NEW AND OLD DESIGNATIONS

STRUCTURAL STEEL		DESIGNATION	
SHAPE	SYMBOL	NEW	OLD
I	W	W 24 × 76	24 WF 76
I	S	S 24 × 120	24 I 120
I	M	M 8 × 22.5	8 M 22.5
[	C*	C 12 × 32.4	12 32.4
I	HP	HP 14 × 86	14 BP 86
L	L*	L 9 × 6 × 3/4	∠9 × 6 × 3/4
Structural Tees Cut from			
T	WT	WT 12 × 42	ST 12 WF 42
⊥	ST	ST 12 × 60	ST 12 I 60
Plates, Bars, and Pipes			
—	PL	PL 3/4 × 24	PL 24 × 3/4
■	Bar □	Bar 1 ⌽	Bar 1 ⌽
O	Pipe	Pipe 6 Std.	Pipe 6 Std.
Structural Tubing			
▭	TS	TS 8 × 6 × .500	Tube 8 × 6 × .500

C* Amer. Std. Channel L* Unequal Leg Angle

and the erection of the structure. Table 62-1 identifies standard structural shapes, symbols, and new and old techniques of specifying structural members on drawings.

STRUCTURAL CONNECTORS

Prefabricated steel and precast concrete structural members require *structural connectors* during on-site erection. Structural members are connected by one or both basic methods: bolting or welding. The third method (riveting) is almost obsolete. Standard steel bolts and extra length threaded steel rods are used for steel and precast concrete structures. High-strength steel bolts permit greater tightening capacity, replacing hot-driven rivets.

Split Ring and Shear Plate Connectors

Split ring and shear plate connectors serve with fasteners for heavy timber and laminated wood members. A *split ring connector* is a circular metal band. The connector fits into a circular groove that is machined in each member.

A *shear plate connector* is a cup-shaped circular plate that fits into a machined groove in one member. Split ring and shear plate connectors serve aligning functions and permit easy assembly and disassembly. Bolts with washers are used as fasteners in connector applications.

Welded Connections

The four common welds that are used for connecting structural members are: *fillet weld*, *U-groove*, *back weld*, and *plug* or *slot weld*. Full specifications for welded connections are provided on drawings, using basic welding standards and procedures. Welding drawing reference lines, symbols, abbreviations, positioning information, and dimensions are given on drawings, including welding electrodes.

BEAM FRAMING PLANS FOR STRUCTURAL STEEL FRAMING

Engineering drawings are used primarily for design, marketing, and to provide technical data and specifications for erection purposes. Engineering drawings include framing plans, specifications about structural steel products, and sections of all construction members in a building.

Beam framing plans show each beam, joist, and girder. A structural connection that requires a connector is coded with a letter. The letter is referenced to similar connection details. Drawings with two or more connectors that share the same detail may be designated by letter, followed by the abbreviation Sim. Structural connections that are opposite (right or left) hand or exactly opposite so they share a common connection detail on a drawing are marked OPP HD

SECTION DRAWINGS FOR STRUCTURAL STEEL STRUCTURES

Full sections, partial sections, and offset sections show relationships in a structure and provide height and other dimensional information. While structural sections are drawn on separate sheets, they are identified on framing plans by section-cutting symbols (figure 62-2).

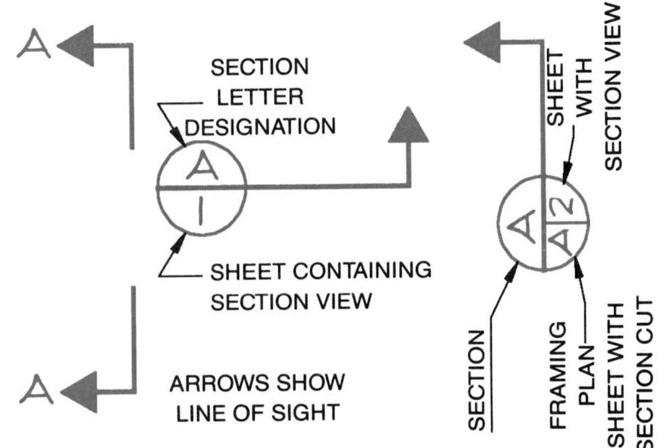

FIGURE 62-2 SAMPLE SECTION CUT DESIGNATION AND SYMBOLS

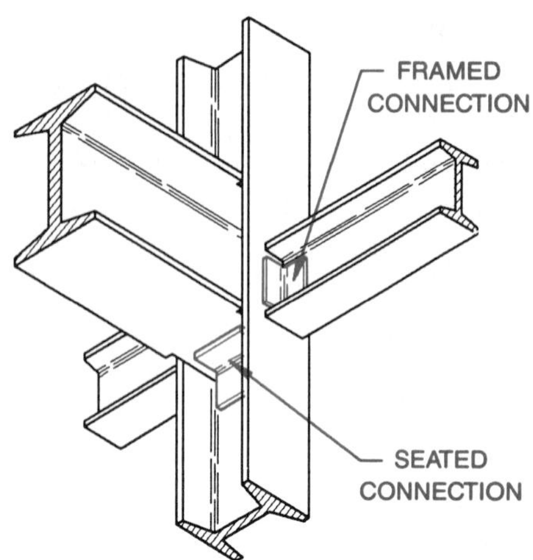

FIGURE 62-3 FRAMED AND SEATED BEAM TO COLUMN CONNECTIONS

Connection Details

Structural steel connections are used to connect: beams to beams, beams to columns, column baseplates to a foundation, and column splices. Drawings of structural beams that are fabricated in a shop provide complete dimensions and other specifications for each individual beam.

By contrast, a field installation/assembly drawing provides all features that bear a relationship to the location on the supporting members. Generally, beam to column connections are either *framed* or *seated*. Examples of these two types are shown in figure 62-3.

Standard *connection angles* (angle plates) are bolted or welded to the web of a beam in the fabricating plant and secured to the supporting member on the job.

Bolt Hole Dimensioning

The longitudinal location of bolt holes that are uniformly spaced along a center line is referred to as *pitch* or *bolt pitch*. The vertical center-to-center dimension for staggered holes is called *gage*. Tables of standardized dimensions for pitch and gage are available in customary inch and metric measurement units.

REINFORCED AND PRESTRESSED CONCRETE FORMS AND DRAWINGS

When concrete is used, it may be *reinforced* or *prestressed*. In *reinforced concrete members*, structural steel reinforcing bars are positioned to support required loads as shown in figure 62-4. The reinforcing bars are positioned and held in place during casting by using *bar supports, bent stirrups*, or metal *wraparound ties*.

Structural, *prestressed concrete members* are cast with steel reinforcing bars (stretched under tension) placed in predetermined locations.

Structural precast concrete installations require engineering/architectural framing plans and shop drawings. A set of working drawings includes column, beam, wall, floor, and roofing plans. Framing plans include *mark numbers, product schedules*, and a *master legend.*

Mark Numbers. Precast concrete structure members are identified by a number (*mark*). The same mark number is used for members that have identical dimensional sizes, shape, and fabricating features. A range of numbers like: 1 to 99 for floor, wall, and roof members; 100 to 199 for columns; and 200 to 299 for beams, provide for the identification of structural members within a series.

If members have the same geometry features, but differ slightly, a letter following the mark number (like 200 A, 200 B, etc.) indicates a modification.

Product Schedules on a framing plan provide information that is similar to that contained on a regular Bill of Materials; product identification, dimensions, quantity, and other information.

General Notes contain complementary information that broadly applies across a set of framing plans. These are additional to notes that apply to specific items and relate to a particular drawing.

Master Legend. This legend generally appears on a cover sheet for a set of plans. The *legend* identifies and explains the symbols and abbrevia-

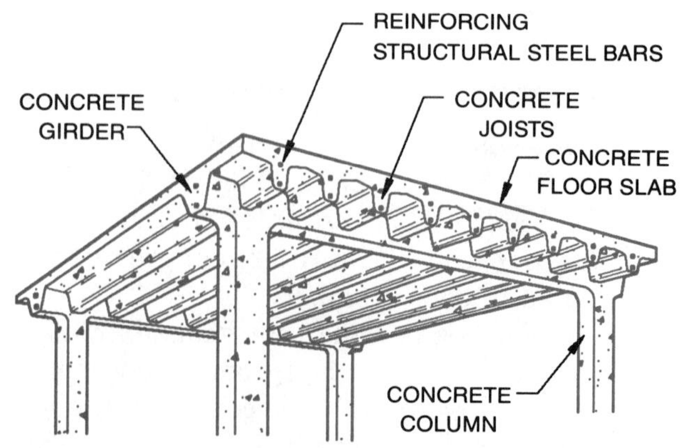

FIGURE 62-4 REINFORCED CONCRETE STRUCTURAL FRAME MEMBERS

tions that appear on the drawings. While there are standard precast concrete structure symbols, abbreviations, etc., variations among companies make it necessary to use a legend.

One important legend relates to *blockouts.* A blockout refers to a particular shape and size opening in a precast structure. Openings are provided for doors, windows, conduit channels, etc. A legend identifies each different blockout by letter and by width, depth, and length dimensions.

PRECAST CONCRETE SECTIONS

Internal structural relationships that are not represented on framing plans are clarified by using full, partial, or offset sections. *Full sections* are longitudinal sections that cut through the total length of a building; *full cross sections,* across the total width. Full sections that are cut on framing plans are cross referenced by symbols. A symbol incorporates information on *section designation* (letter), *section number,* and *section cut* on the framing plan drawing, and the *direction of viewing* (sight).

Partial sections provide additional information to clarify fabrication, construction, and other details in a specific area of a precast concrete structure.

Offset sections show the interior details of a structure as viewed when a cutting plane line is offset. An offset section is represented as a section drawing just as though a straight line plane cut through the structure from the point of origin.

STRUCTURAL CLAY PRODUCTS

Structural clay members (such as brick walls) depend on overlapping and interlocking brick patterns and the strength of the mortar joints.

Reinforced masonry is required when compression and other structural strength requirements exceed those of brick and the mortar joints. Masonry is *reinforced* by embedding wire screen or steel rods in the mortar. Concrete masonry and brick are also used to enclose structural steel components to meet building codes for fireproofing.

Marble, granite, limestone, and other stone products are used for ornamental facings in *curtain-wall* building construction.

STRUCTURAL WOOD SYSTEMS

Structural members include *wood laminated* posts, girders, joists, and arches. A laminated wood structure that is subject to tension, compression, and torque forces has the same engineering predictability as steel structures.

Laminated Wood Structures and Construction

A laminated structural member consists of several pieces (layers) of kiln-dried lumber that are pressed together under tremendous force. The individual pieces are permanently sealed into a solid structural member by applying a high-strength, waterproof glue under high pressure. A structural member may then be fabricated into a desired shape.

Unlike structural steel drawings that are generated by a structural engineer or architect, design features and specifications of laminated wood arches, beams, rigid frames, and other structural members are prepared by designers and fabricators. Structural architectural drawings, therefore, relate to space allocation for each laminated wood component.

Figure 62-5 provides an example of structural details of a continuous wood *three-hinged laminated arch.* The manufacturer establishes and publishes tables of sizes, properties, and characteristics. Drawings generated by a structural designer and architect provide on-site details and dimensioned information about wall heights, spaces, slope, bearings, and other specifications.

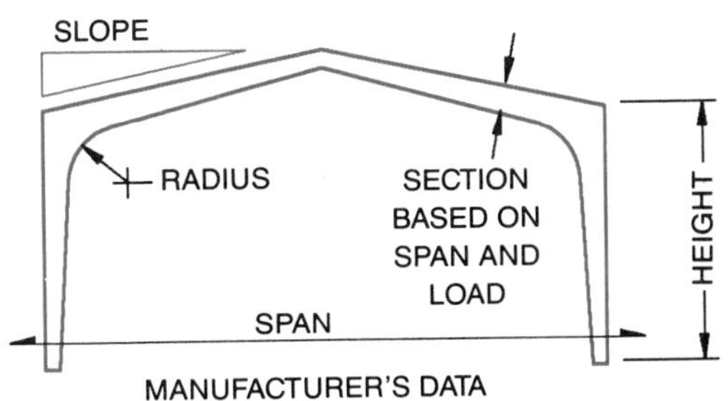

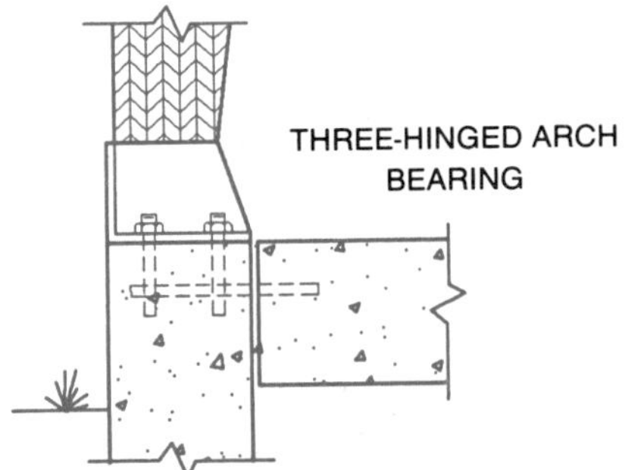

FIGURE 62-5 STRUCTURAL DETAILS OF LAMINATED THREE HINGED ARCH

NOTE: REFER TO APPENDIX B-12 FOR THE LARGER SCALE DRAWING TO USE WITH THIS ASSIGNMENT.

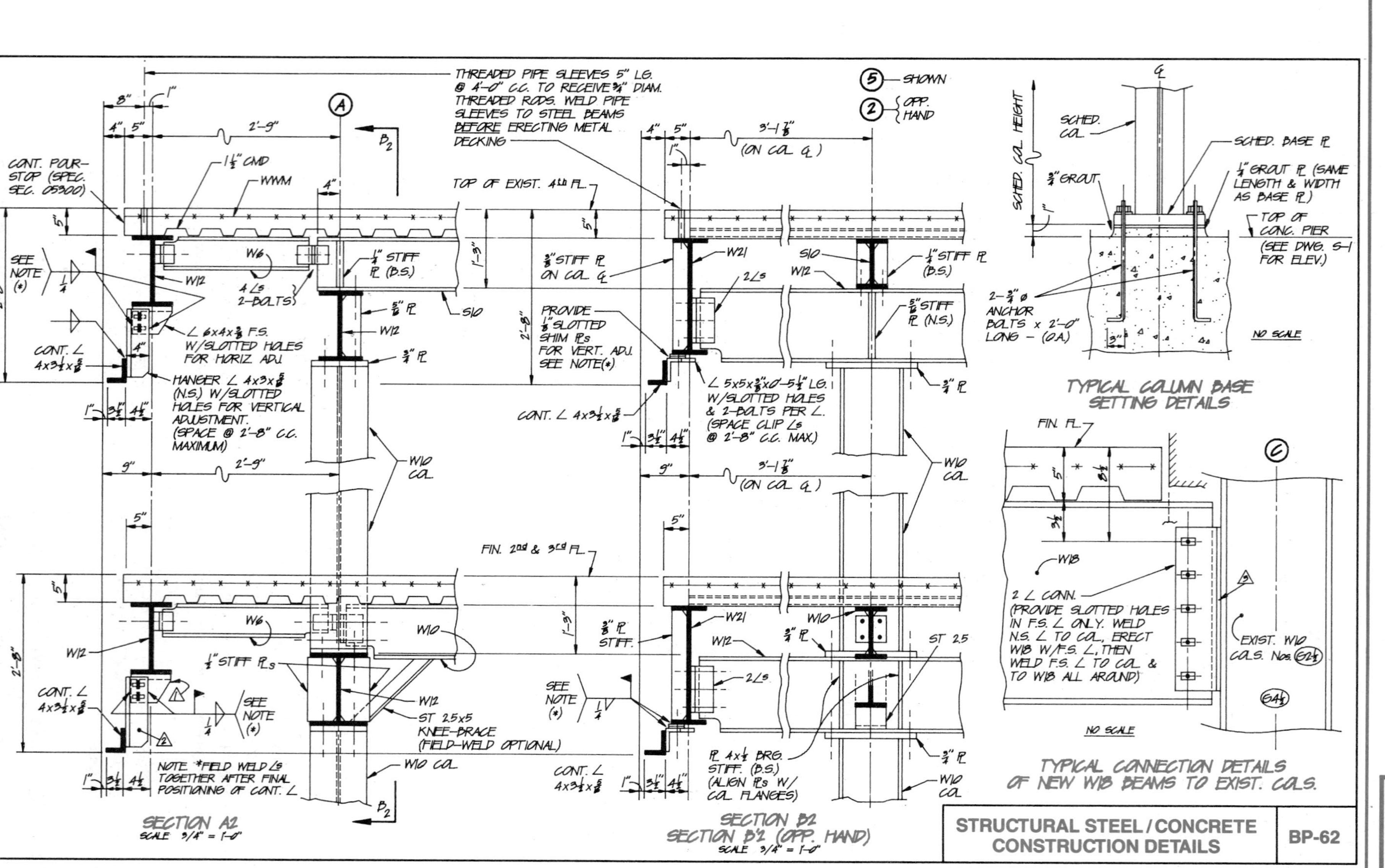

Adapted from Structural Framing Plans. Courtesy of Dodge, Chamberlain, and Luzine Associates, Architects

STRUCTURAL STEEL/CONCRETE CONSTRUCTION DETAILS (BP-62)

1. Sketch the shape of the following steel members.
   a. American Standard Channel C.
   b. Structural T shape.
   c. Unequal leg L.
2. a. State two functions of connectors that are used in heavy timber construction.
   b. Name two connectors for such construction.
3. Identify three kinds of information provided on structural concrete framing plans.
4. a. Name structural parts Ⓘ, Ⓙ, and Ⓚ.

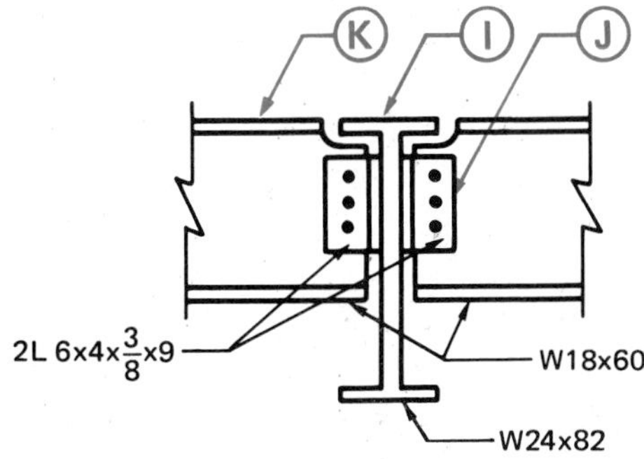

   b. Give the dimensional sizes or specifications of each structural member.
5. a. Sketch the shape of the reinforced concrete floors.
   b. Add sizes and/or specifications.
6. Illustrate how *Section B2* is identified.
7. Refer to *Section A2* (up to the second floor).
   a. List the different *marked* angles, beams, columns, and plates.
   b. Give the size of the welded L ⚠1.
   c. Interpret the information provided by the welding symbol and specifications.
   d. Give the size and specifications of hanger ⚠2.
8. Refer to *Section B2*.
   a. Give the size and shape of each beam at each floor level that supports the *W12* cross beam.
   b. Determine the thickness of the stiffener plates for the *S10* and *W12* beams and the *W10* columns.
   c. Give the center-to-center distance between the *W21* beam and the *W10* column.
9. Refer to the *Column Base Details*. State the dimensions and quantity of the column base anchor bolts.
10. Refer to the *Connection Details* drawing Ⓒ.
    a. Identify member ⚠3.
    b. Provide full information for securing new beams to existing columns *G2½* and *G4½*.

ASSIGNMENT—UNIT 62

Student's Name ______________________

1. a. b. c.

2. a. (1) ______________________
   (2) ______________________
   b. (1) ____________ (2) ____________
3. (1) ______________________
   (2) ______________________
   (3) ______________________
4.

	Part Name (a)	Sizes and/or Specifications (b)
Ⓘ	____________	____________
Ⓙ	____________	____________
Ⓚ	____________	____________

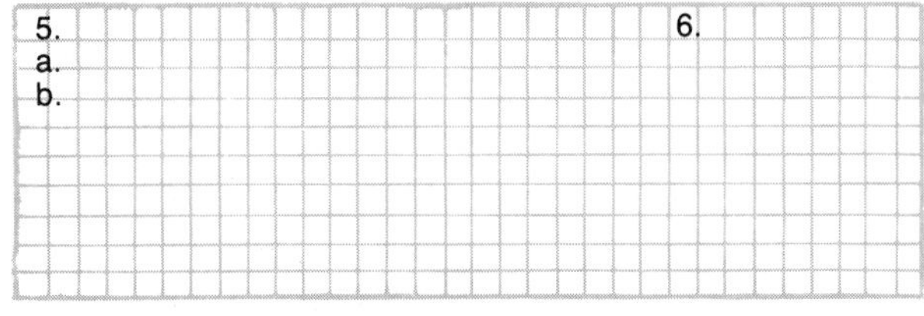

7. a. (1) ____________ (4) ____________
   (2) ____________ (5) ____________
   (3) ____________ (6) ____________
   b. ______________________
   c. ______________________
   d. ______________________
8. a. ______________________
   b. ______________________
   c. ______________________
9. ______________________
10. a. ______________________
    b. ______________________

PART 2

Technical Illustration: Sketching

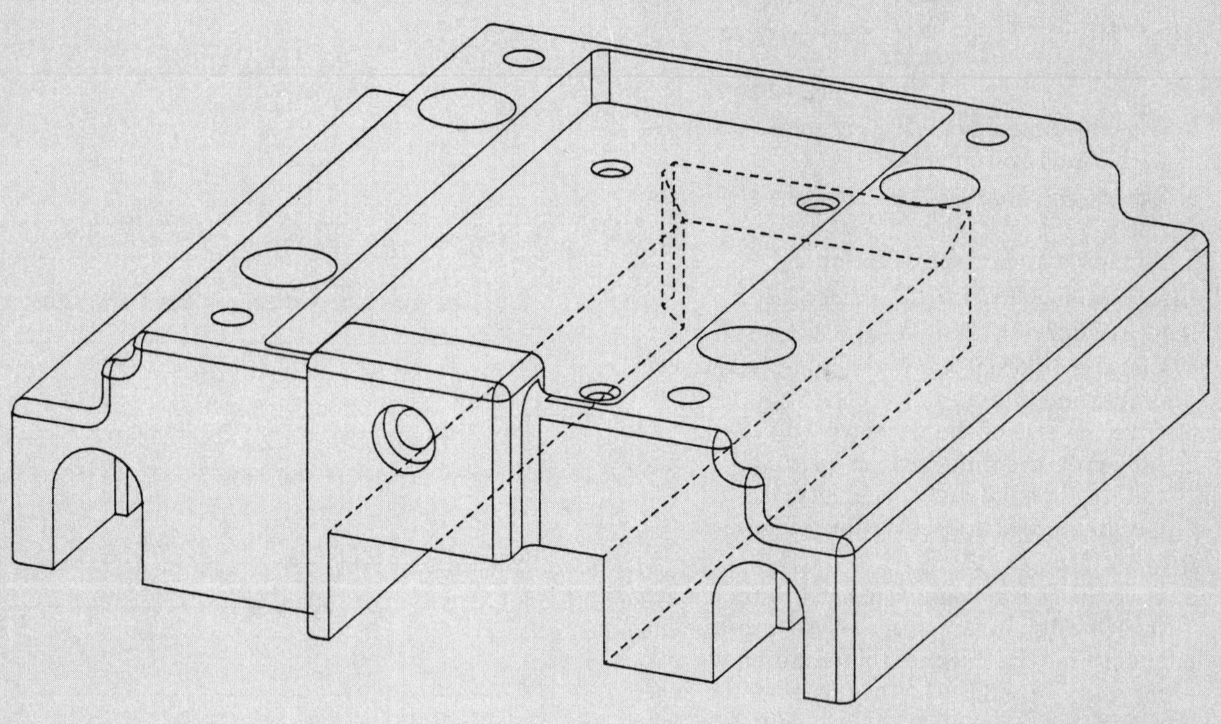

SECTION 15
Basic Lettering Techniques

UNIT 63
Techniques of Freehand Lettering

Engineering drawings and technical sketches contain dimensions, notes, data, specifications, and other written instructions. This kind of information is communicated among engineers, designers, craftspersons and technicians by *lettering*.

Thus, skill in lettering is important as each worker makes complementary sketches and is called on to dimension and letter notes and other descriptive details. Over the years, individual styles and sizes of letters and numerals have been developed for application within different industries and professions. However, the style and form which was adopted and standardized by ANSI is covered in this unit and the next one.

Reference is made to the approved *single-stroke vertical* and *inclined styles of alphabets* for letters and numerals using a *Gothic form*. Each style has both *uppercase (capital) letters and numerals*. The selection of style is usually a matter of individual preference. The exceptions are in such cases as military and aerospace drawings involving government contracts which often require conformity within prescribed standards, and structural and topographical (land area) drawings where lowercase letters are preferred. Mechanical, electrical, and manufacturing industries generally use only uppercase letters. Architectural drawings use uppercase and lowercase letters.

DEVELOPING LETTERING SKILLS

Two prime factors are involved in developing an ability to letter freehand:

- A knowledge of the required proportion, weight, and form of individual letters and numerals and the sequence of the strokes.
- An understanding of the spacing of letters, words, and lines which is required to produce accurate dimensions and other legible information.

FORMING UPPERCASE SINGLE-STROKE GOTHIC LETTERS AND NUMERALS

Vertical letters are considered to be more readable than slant (inclined) letters. To acquire skill in lettering, it is necessary to follow certain strokes for forming each letter and numeral. The shapes of both vertical and inclined Gothic capital (uppercase) letters and numerals are shown in figures 63-1 and 63-2. The fine lines with arrows which follow the form of each letter and numeral give the direction of the strokes. The small numbers indicate the sequence and the number of strokes. The same strokes are used to produce vertical and inclined single-stroke Gothic letters. In general, the same direction of lines and number of strokes are used for lettering with the right or left hand.

Lettering is done with a single stroke to keep the width and density of each line uniform. A soft lead pencil is best for freehand lettering because it is easy to guide and produces dark, visible lines. Inclined letters are drawn at a slope between 60° to 70° with the horizontal. Beginners are encouraged to use horizontal and inclined guide lines

FIGURE 63-1 FORMING FREEHAND SINGLE-STROKE VERTICAL GOTHIC CAPITAL LETTERS AND NUMERALS

to correctly form letters and numerals and space words and lines.

PROPORTIONS OF LETTERS AND NUMERALS

Letters and numerals have a certain relationship (proportion) between height and width for readability and balance. In order to understand this relationship, the letters and numerals in figure 63-1 and figure 63-2 are enclosed in a six-unit square. Notice that almost all letters and numerals are five or six units wide (except the letter W) and all are six units high.

SPACING OF LETTERS, WORDS, AND LINES

Readability also depends on the spacing between letters, words, and lines. The background area around letters and numerals *must appear to be equal.* This is the case in figure 63-3A. Actually, the space between each letter cannot really be

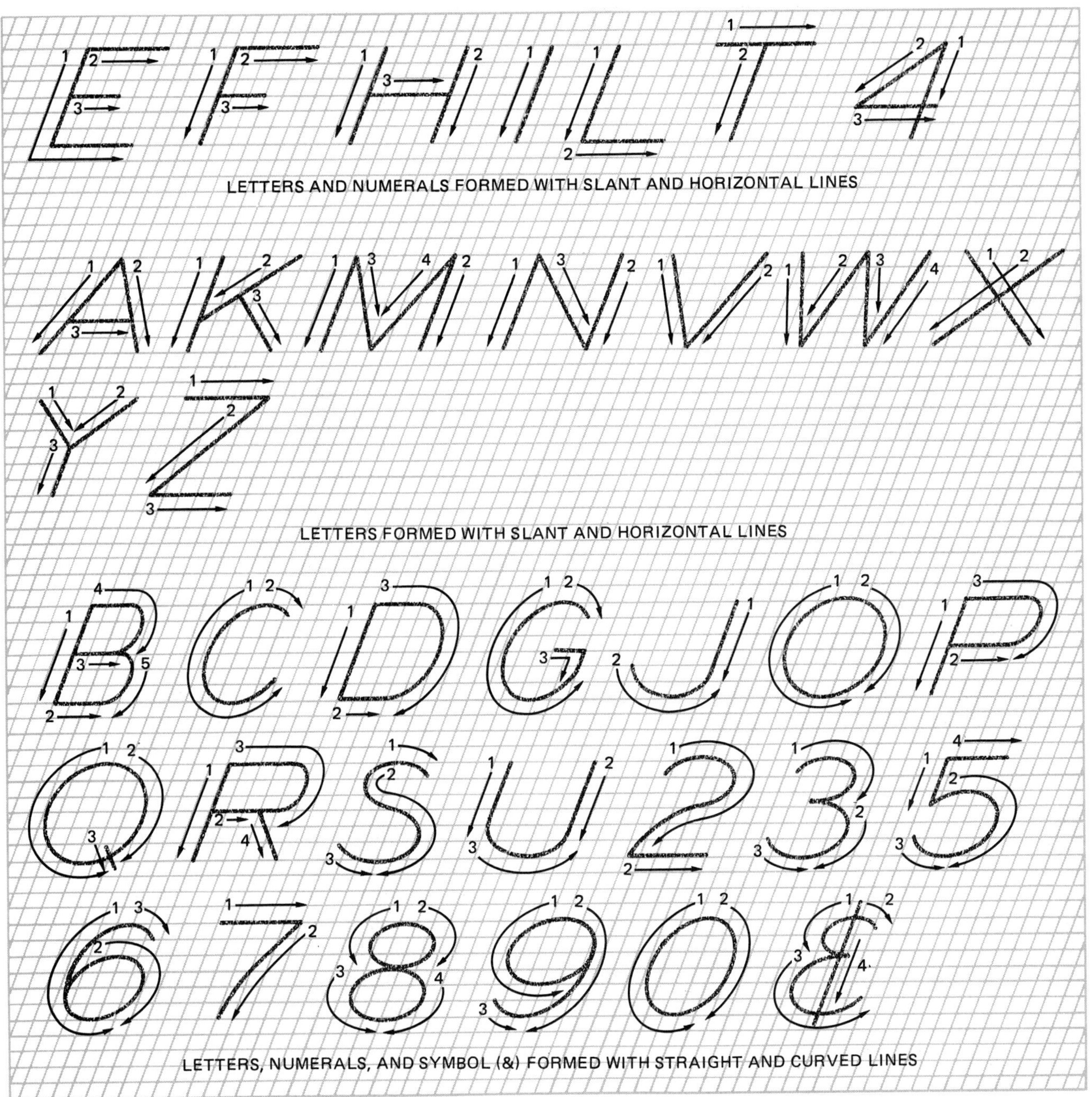

FIGURE 63-2 FORMING UPPERCASE SLANT LETTERS AND NUMERALS

equal due to the straight, curved, and angular lines which are used to form each different letter and numeral. Developing the ability to space correctly involves practice in equalizing the space by eye.

Spacing between words is generally equal to the capital letter (O). Lines are easiest to read when the vertical space is no less than one-half the height of the letters and no more than the full height of the letters. The effects of poor word and line spacing is illustrated in figure 63-3B.

LETTERING FRACTIONS

The lettering of fractions is particularly important. Numerals must be well formed so that there is limited possibility of error in reading a dimension or other numerical value.

Fractions are formed freehand as shown in figure 63-4. Fractions are lettered with a horizontal bar. The numerals in the numerator and the denominator are about three-fourths the height of the whole number *(integer)*. The whole number

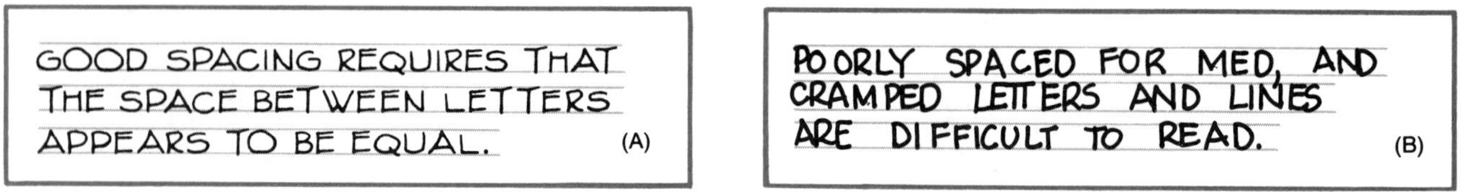

GOOD SPACING REQUIRES THAT
THE SPACE BETWEEN LETTERS
APPEARS TO BE EQUAL. (A)

POORLY SPACED FOR MED, AND
CRAMPED LETTERS AND LINES
ARE DIFFICULT TO READ. (B)

FIGURE 63-3 EFFECTS OF SPACING BETWEEN LETTERS, WORDS, AND LINES

is the same height as other capital letters to which the dimension relates.

STROKES IN LETTERING

Lettering is another form of freehand sketching and drawing. Lettering skills require practice and continuous review of how letters and numerals are formed, proportion, spacing, and uniform density and widths of all letters/numerals. Good lettering requires optical balancing. This means that allowances are made in forming letters so they do not appear to be top-heavy.

Six basic steps are used in forming letters. These were incorporated earlier in the formation of letters and numerals and are summarized in figure 63-5. Note that there are three strokes for vertical lettering and three for slant lettering. The arrows in each instance show the direction of the strokes.

LETTERING DEVICES

There are two common types of freehand lettering devices.

- *Lettering Triangle.* This triangle is used for drawing light vertical and horizontal guide lines for straight and slant letters, whole numbers, and fractions. A lettering triangle has a series of holes that are equally-spaced, usually 1/16″ (1.6 mm) apart.

  The spacing permits light guide lines to be drawn; the number depending on whether just capitals and lower-case letters, and/or whole numbers, and/or fractions are to be included on a drawing. The guide lines are important in producing quality freehand lettering.
- *Lettering Guides.* Guides are commonly used to produce drawings that require more accurate, standardized, mechanically drawn letters. Lettering guides (devices) generally consist of three separate components.
  - A set of varying sizes and lettering styles that are impressed in a *template*; referred to as a *lettering guide.*
  - A series of different diameter (for varying line widths) ink *lettering pens.*
  - A *lettering guide scriber* with an adjustable arm. A solid guide pin in one leg of the adjustable, movable arm fits into and is moved in each required letter or numeral groove on the lettering guide. Freehand, mechanical letters are produced by the inking pen (that is mounted on the second arm of the scriber) as the scriber is moved to form each letter. The path followed on the template by the pin is reproduced by the inking pen on the drawing. The scriber may be adjusted to form vertical or slant letters.

Lettering on many engineering drawings is mechanically created on lettering machines which produce a *tape.* The lettering on the tape is then applied on drawings. Also, lettering may be generated through CADD systems directly on a drawing or as press-on lettering.

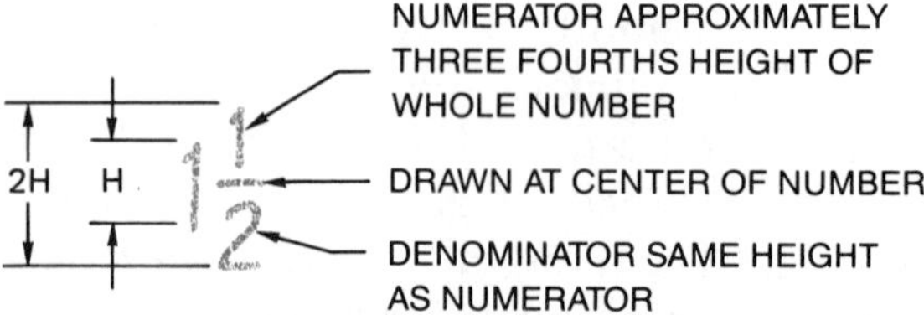

FIGURE 63-4 SIZE OF FRACTION PARTS

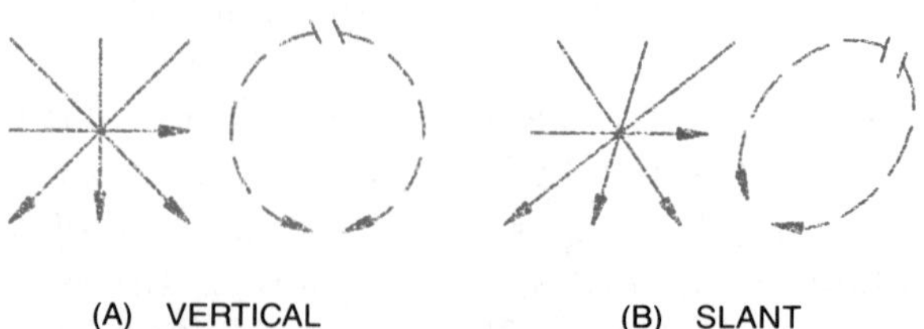

FIGURE 63-5 FREEHAND STROKES

DET.	#REQ.	NAME	MAT'L	DESCRIPTION
18	1	SHIM	STEEL	
17	4	DOWEL	STEEL	$\frac{1}{4}$ DIA × $\frac{5}{8}$ LONG
16	4	SOCKET HEAD CAP SCREW	STEEL	#10-24 × $\frac{5}{8}$ LONG
15	1	HEAD PRESS FIT BUSHING	STEEL	ID .1540 $^{+.0001}_{-.0004}$ × OD $\frac{5}{16}$ × $\frac{5}{8}$ LONG
14	1	CLAMP	STEEL	
13	2	DOWEL	STEEL	$\frac{3}{16}$ DIA × 1 LONG
12	2	SOCKET HEAD CAP SCREW	STEEL	$\frac{1}{4}$-20 × 1 LONG
11	1	DOWEL	STEEL	$\frac{1}{16}$ DIA × $\frac{3}{4}$ LONG
10	2	POST	STEEL	
9	1	STUD	STEEL	
8	1	DOWEL	STEEL	$\frac{1}{8}$ DIA × $\frac{1}{2}$ LONG
7	4	SOCKET HEAD CAP SCREW	STEEL	#10-24 × $\frac{1}{4}$ LONG
6	1	HEAD PRESS FIT BUSHING	STEEL	ID .3100 $^{+.0001}_{-.0005}$ × OD $\frac{9}{16}$ × $\frac{3}{4}$ LONG
5	1	HAND KNOB FIT BUSHING	CI	KNOB, $\frac{5}{16}$-18 TAP
4	1	DOWEL	STEEL	
3	1	BUSHING PLATE	STEEL	
2	1	PLATE	STEEL	
1	1	BASE	STEEL	

MATERIAL: SEE DETAILS

NAME: **DRILL JIG (.316 or .154 HOLES)**

HELI-COIL
Heli-Coil Products, Div. of Mite Corp., Danbury, Conn.

BP-63

Courtesy of Heli-Coil Products, division of Mite Corporation

ASSIGNMENT—UNIT 63

Student's Name ______________________

(1) VERTICAL LETTERING

(2) INCLINED (SLANT) LETTERING

DET.	#REQ.	NAME	MAT'L	DESCRIPTION

FREEHAND LETTERING ASSIGNMENT

1. USE VERTICAL LETTERS AND NUMERALS TO LETTER THE SPECIFICATIONS OF DETAILS 1 THROUGH 9.
2. USE INCLINED (SLANT) LETTERS AND NUMERALS TO LETTER THE SPECIFICATIONS OF DETAILS 10 THROUGH 18.

UNIT 64
Freehand Lettering: Lowercase Gothic and Architectural

FORMING LOWERCASE SINGLE-STROKE GOTHIC LETTERS AND NUMERALS

Lowercase letters are often used for lettering minor notes on drawings. The bodies of single-stroke vertical lowercase letters are made two-thirds the height of capital letters. The *ascenders* and *descenders* (tails) of the letters extend from the body until the height is equal to the height of an uppercase letter.

The form of lowercase vertical letters is based upon a circle, arc, and/or a straight line. The shape and the sequence of strokes for forming each vertical lowercase letter are indicated in figure 64-1.

Lettering Inclined Lowercase Letters

When forming inclined lowercase single-stroke letters and numerals, the shape, proportion, and the form strokes are the same as for vertical lowercase letters. However, inclined letters are formed by combinations of straight lines and ellipses. The 60° to 70° slope angle of the letters is the same as for capital letters.

ARCHITECTURAL STYLE LETTERING

Architects and others in related fields often develop an individual style of lettering after mastering the standard Gothic style. Frequently,

FIGURE 64-1 FORMING LOWERCASE GOTHIC LETTERS AND NUMERALS

drawings and sketches utilize a combination of Gothic and architectural lettering. Although an architect may develop a characteristic style, there is a certain similarity between styles because of basic standards.

Architectural lettering must carry out the objectives of all technical lettering. Letters, numerals, and words must be clear, attractive, easy to read and must be quickly and accurately formed. A unique style of lettering may be created by changing the design and proportion and by expanding or condensing the letters and numerals.

Variations of Lettering Styles

Three widely used styles of architectural, structural, and other construction drawing alphabets of uppercase and lowercase letters and numerals are illustrated in figure 64-2. Lettering on drawings may be straight or inclined and in condensed or extended form. While different architects, designers, and other construction workers follow general principles and guidelines for lettering, drawings reflect individualized styles.

STYLE	CHARACTERISTICS OF SAMPLE LETTERS AND NUMERALS
CONVENTIONAL STYLE	ABCDEFGHIJKLMNOPQRS TUVWXYZ 1234567890 CIRCULAR VERTICAL LETTERING
	ABCDEFGHIJKLMOPQRST UVWXYZ 1234567890 abcdefghijklmnopqrstuvwxyz CIRCULAR INCLINED LETTERING
	A B C D E F G H I J K L M N O P Q R S T U V W X Y Z 1 2 3 4 5 6 7 8 9 0 CONDENSED LETTERING
TRIANGLE STYLE	ABCDEFGHIJKLMNOPQRS TUVWXYZ 1234567890 TRIANGLE USED FOR VERTICAL STROKES
ROMAN STYLE	ABCDEFGHIJKLMNOPQRS TUVWXYZ 1234567890 OPEN AND PARTLY SOLID LINES

FIGURE 64-2 EXAMPLES OF COMMON LETTERING STYLES

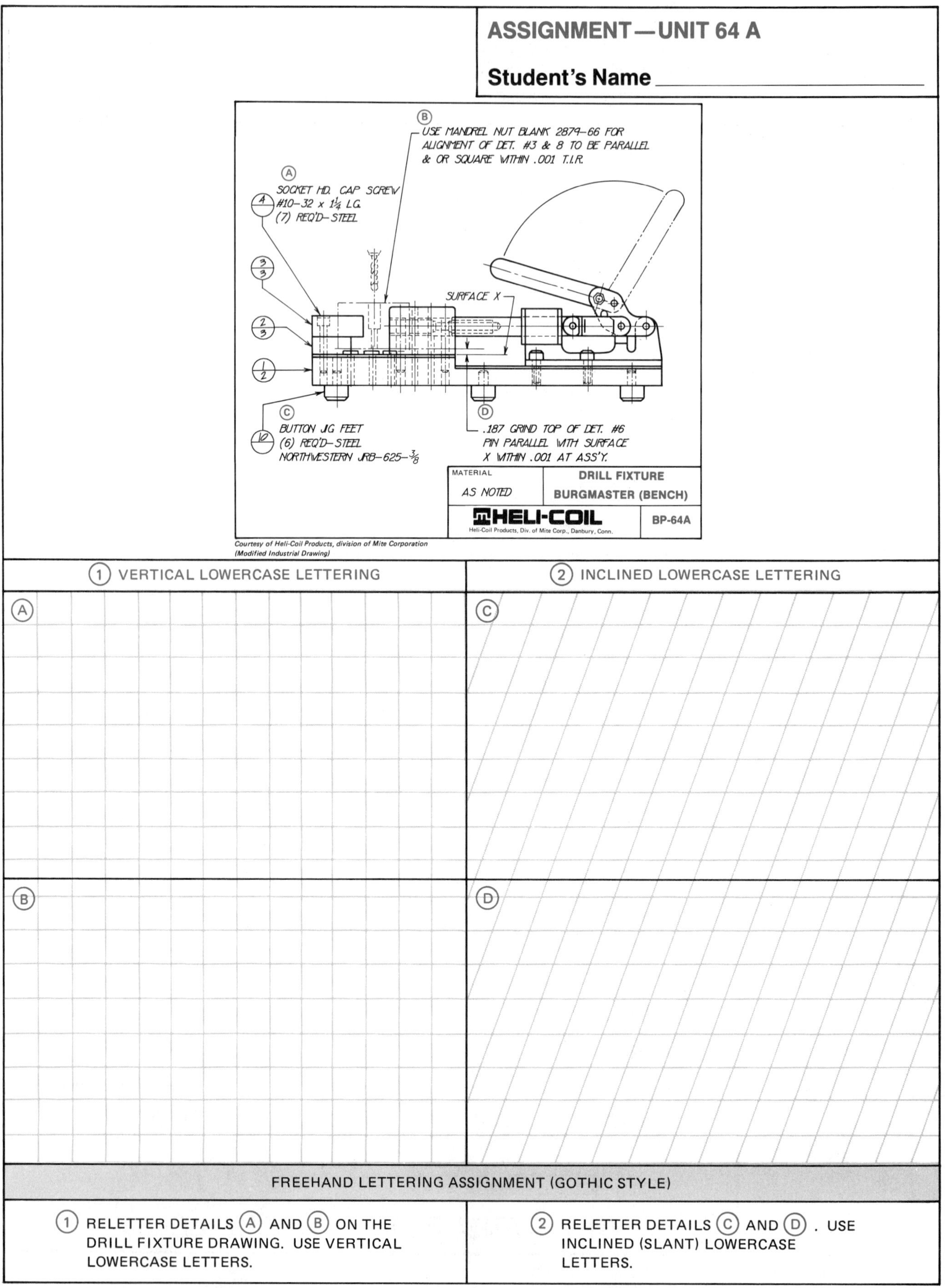
ASSIGNMENT—UNIT 64 A
Student's Name
(B) USE MANDREL NUT BLANK 2879-66 FOR ALIGNMENT OF DET. #3 & 8 TO BE PARALLEL & OR SQUARE WITHIN .001 T.I.R.
(A) SOCKET HD. CAP SCREW #10-32 x 1¼ LG. (7) REQ'D-STEEL
SURFACE X
(C) BUTTON JIG FEET (6) REQ'D-STEEL NORTHWESTERN JRB-625-3/8
(D) .187 GRIND TOP OF DET. #6 PIN PARALLEL WITH SURFACE X WITHIN .001 AT ASS'Y.
MATERIAL
AS NOTED
DRILL FIXTURE
BURGMASTER (BENCH)
HELI-COIL
Heli-Coil Products, Div. of Mite Corp., Danbury, Conn.
BP-64A
Courtesy of Heli-Coil Products, division of Mite Corporation (Modified Industrial Drawing)
(1) VERTICAL LOWERCASE LETTERING
(2) INCLINED LOWERCASE LETTERING
(A)
(C)
(B)
(D)
FREEHAND LETTERING ASSIGNMENT (GOTHIC STYLE)
(1) RELETTER DETAILS (A) AND (B) ON THE DRILL FIXTURE DRAWING. USE VERTICAL LOWERCASE LETTERS.
(2) RELETTER DETAILS (C) AND (D). USE INCLINED (SLANT) LOWERCASE LETTERS.

FINISH NOTES :	ASSIGNMENT—UNIT 64 B Student's Name ______

2. BASE MOULDING SHALL BE 1×5 'P' SELECT PINE WITH IROQUOIS #8535 PANEL MOLDING UNLESS NOTED OTHERWISE. DOOR CASINGS SHALL BE 1×3 'P' SELECT PINE WITH IROQUOIS #8535 PANEL MOLDING UNLESS NOTED OTHERWISE.

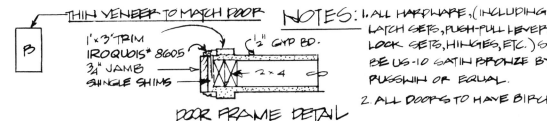

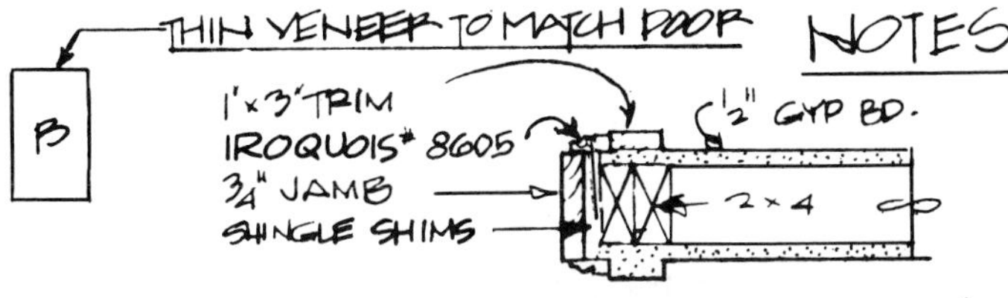

DOOR FRAME DETAIL

NOTES: 1. ALL HARDWARE, (INCLUDING LATCH SETS, PUSH-PULL LEVERS, LOCK SETS, HINGES, ETC.) SHALL BE US-10 SATIN BRONZE BY STANLEY RUSSWIN OR EQUAL.

2. ALL DOORS TO HAVE BIRCH VENEER.

OFFICE RENOVATIONS & ALTERATIONS	F. A. EVANS architect troy, n.y.	BP-64B

①

②

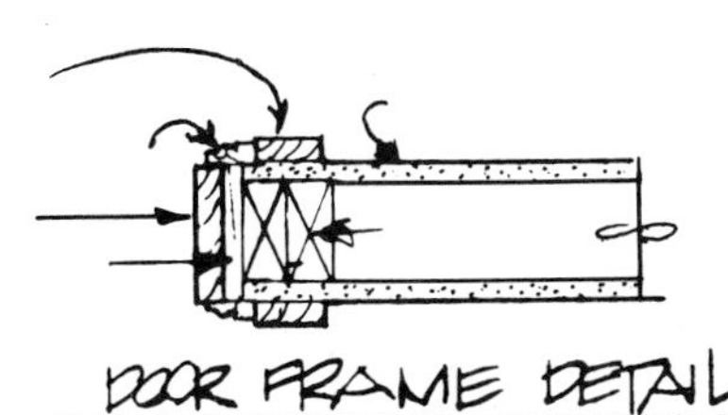

DOOR FRAME DETAIL

FREEHAND LETTERING ASSIGNMENT (ARCHITECTURAL LETTERING)

① RELETTER THE SECOND FINISH NOTE USING INCLINED LOWERCASE ARCHITECTURAL STYLE LETTERING.

② ADD THE DIMENSIONS AND NOTES TO THE DOOR FRAME DETAIL DRAWING. USE CAPITAL LETTERING, ARCHITECTURAL STYLE (VERTICAL).

SECTION 16 Fundamentals of Industrial Sketching

UNIT 65 Sketching Straight Lines

Freehand drawing and sketching provide a functional, practical method of visually representing parts, machines, mechanisms, circuits, and systems. Freehand drawings and sketches are widely used in manufacturing, construction, assembling, servicing, marketing, and other occupations. Sketches are made and used by scientists, engineers, product designers, industrial and business technicians, craftspersons, marketing specialists, and other workers.

Technical sketching is not dependent upon artistic ability. Rather, sketching depends on the ability to read industrial blueprints; to understand sketching techniques for drawing different lines, surfaces, and sections; and to apply sketching skills in making orthographic, isometric, and perspective sketches.

TYPES OF SKETCHES AND SCALES

Technical sketches generally conform to one of four standard types of projection as illustrated in figure 65-1. Sketching, as covered in this Section, relates to orthographic, multi-view projection (figure 65-1A).

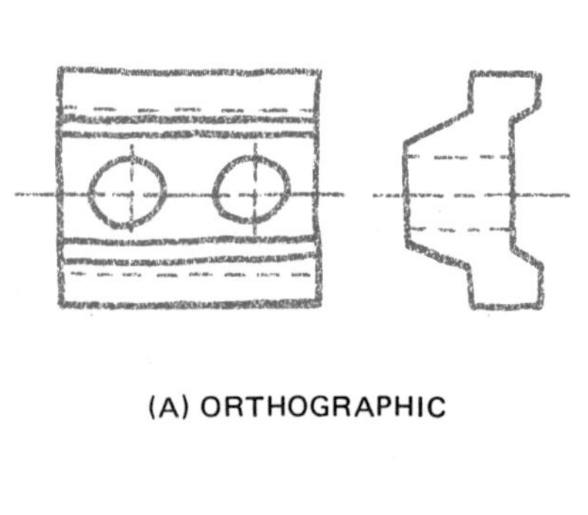

(A) ORTHOGRAPHIC

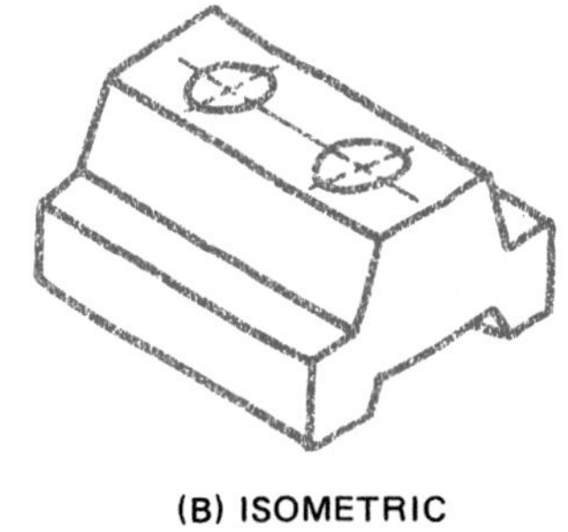

(B) ISOMETRIC

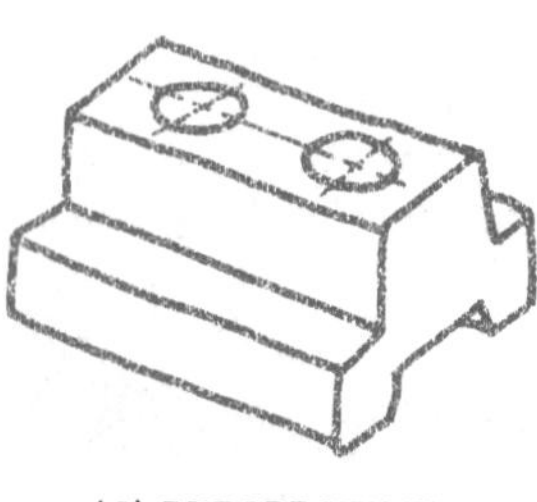

(C) PERSPECTIVE

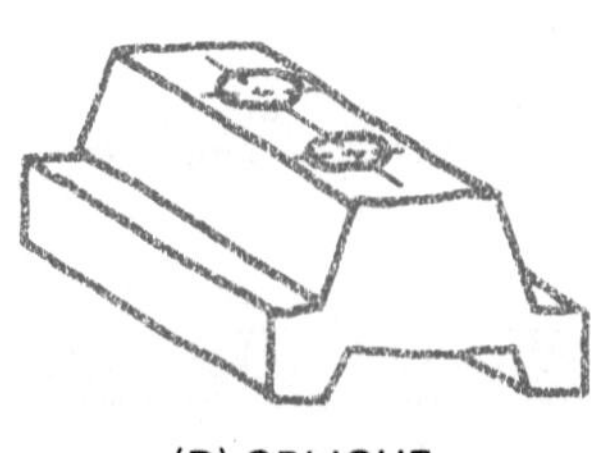

(D) OBLIQUE

FIGURE 65-1 STANDARD TYPES OF PROJECTION USED IN SKETCHING

Sketches are rarely drawn to any scale. Objects are sketched so the features are shown in correct proportion as estimated by eye. Grid spacings are used as a dimensional guide. The size of the sketch depends on the complexity of the object. Frequently, small objects or complicated features are sketched oversize to clearly show the details.

SKETCHING TOOLS AND MATERIALS

Sketching requires a pencil and an eraser. A common HB or #2 medium-soft lead is preferred to a hard lead. The soft lead may be used to make both light lines for blocking and, later, dark lines for clearly defining the object. A sharp, cone-shaped point is used for making fine lines. The point is rounded *(blunted)* for wide lines. Corrections and changes are made by erasing with a pencil eraser which will not smudge the copy or abrade the paper. With an adjustable lead pencil, a .5 mm HB (or harder) lead may be used for drawing fine lines; a .7 mm lead, wide lines.

Coordinate Grid Sheets

The making of a sketch is simplified by using sheets of paper which are ruled vertically and horizontally with fine lines. The sheets are called *plain* and *coordinate grid paper.* The fine grid lines are printed in color which may be nonreproducible. Commercial grid papers are available with accurate line spacings of 2, 4, 8, 10, and 16 squares per inch and in metric equivalents. The fine coordinate lines provide guide lines for lettering, for making notes, dimensioning, and for proportionally representing each feature of an object. Grid papers with angular lines are available for making isometric, oblique, and perspective sketches.

SKETCHING TECHNIQUES

Line Thickness and Darkness

Light, thin lines are used for laying out an object. Light lines can be easily erased or altered. Light guide lines are often left on a drawing. After the features and part are correctly laid out, all permanent lines are then darkened and drawn to the correct thickness following the same standards as used in mechanical or architectural drawing.

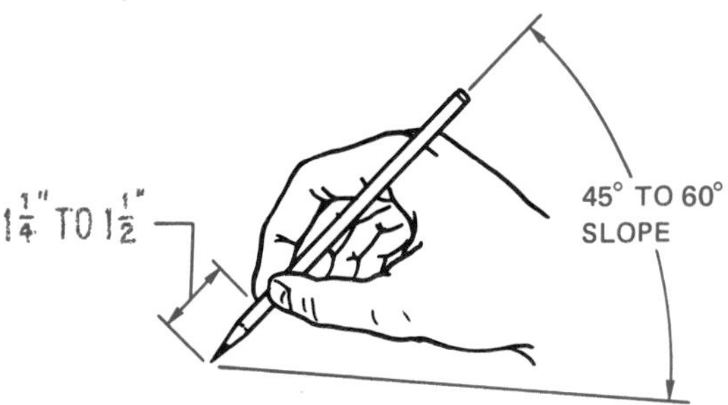

FIGURE 65-2 GENERAL POSITION TO HOLD A PENCIL FOR SKETCHING

Sketching Straight Lines

In sketching, the pencil is usually held between 1 1/4" to 1 1/2" from the point to permit easy movement and turning so the point does not become blunt. The pencil is held at an angle of 45° to 60° for drawing straight lines as shown in figure 65-2. For making circles, the angle is steepened to 30°. The place and position for holding the pencil should be as natural as possible to permit free arm movement. Finding this correct position requires some experimentation, especially for persons who draw with the left hand.

Sketching Horizontal Lines

The three methods frequently used to sketch horizontal lines include (A) *single stroke,* (B) *short (overlapping) strokes,* and (C) the *long stroke.* All three methods require marking off two points which represent the start and finish position of a surface or edge as shown in figure 65-3. *Short* horizontal lines (A) are drawn between points A and B by placing the pencil point on the first dot. Keeping an eye on dot B and with the hand placed on the drawing surface, a line is produced by a short movement (stroke) of the fingers guiding the pencil between points A and B.

Long lines (figure 65-3B) may be produced by using a free arm movement. With eyes focused on point B the pencil is drawn in one stroke from point A to B. The third method, as illustrated in figure 65-3C, uses short, overlapping strokes.

Fine lines are recommended for beginners. Then, when the sketch is completed, the lines are darkened. Attention is called to controlling the pencil point by resharpening or rotating it during sketching.

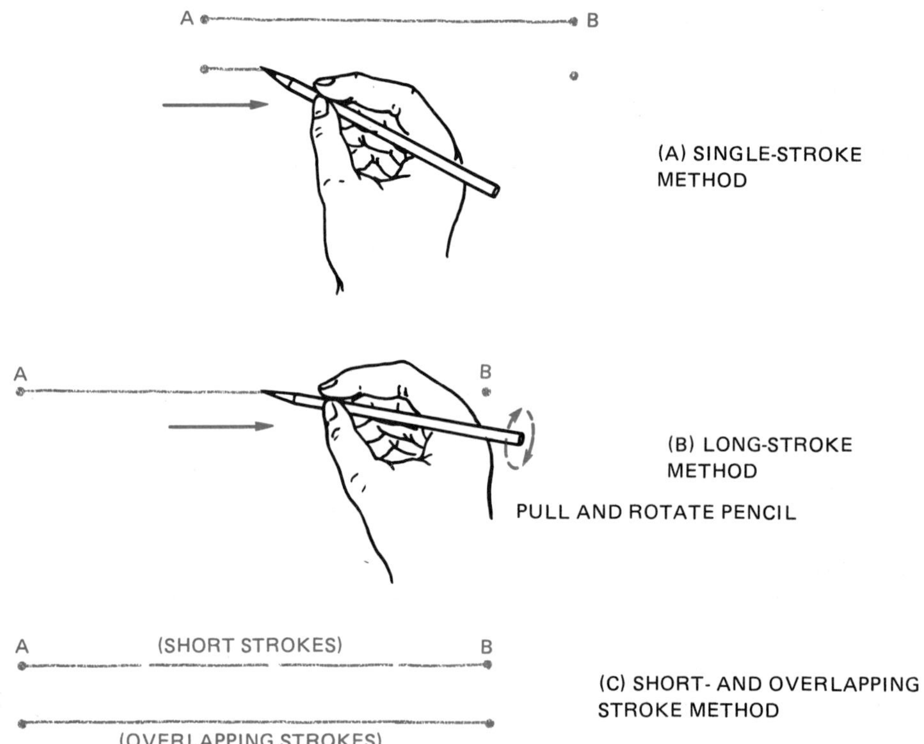

FIGURE 65-3 TECHNIQUES USED TO SKETCH STRAIGHT LINES

Sketching Vertical and Inclined Lines

Vertical and inclined lines may be sketched by using the same three methods. Once again, the line is started by establishing the end points A and B as shown in figure 65-4. The pencil point is placed on point A and moved toward point B. If it is easier to sketch lines horizontally with more freedom and accuracy than at an angle, turn the paper (if practical) so the vertical or slant lines assume and are drawn from a horizontal position.

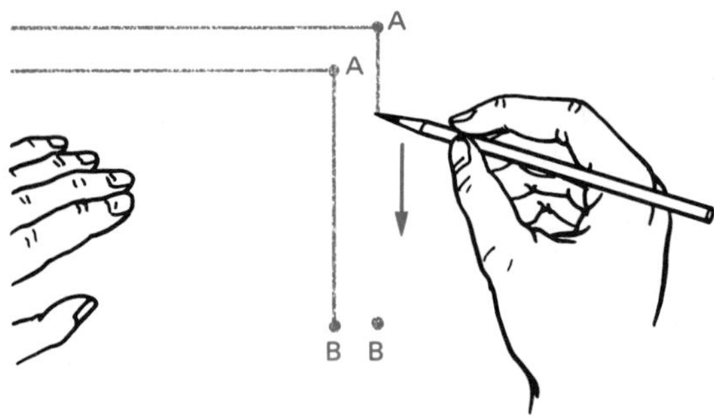

(A) SKETCHING VERTICAL LINES

Sketching Horizontal, Vertical, and Inclined Lines with the Left Hand

Left-handed persons sometimes find the sketching of lines and letters difficult especially if the motion is toward the body. This movement partially hides the work and it is harder to join strokes or to draw lines of uniform thickness. As with lettering, the left-handed person must experiment with the control of the pencil and the method until a natural sketching technique is developed.

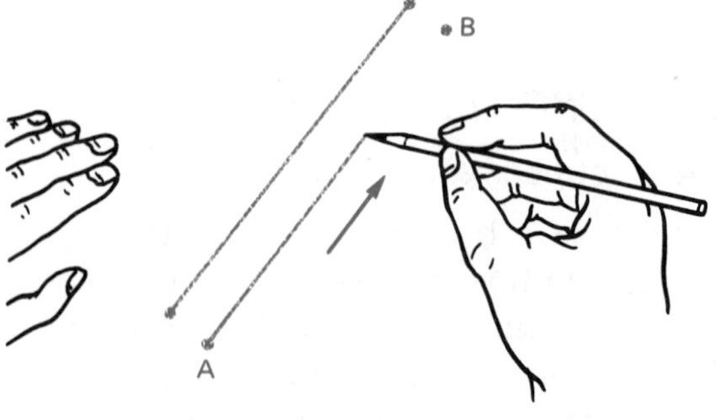

(B) SKETCHING INCLINED (SLANT) LINES

FIGURE 65-4 SKETCHING VERTICAL AND INCLINED LINES

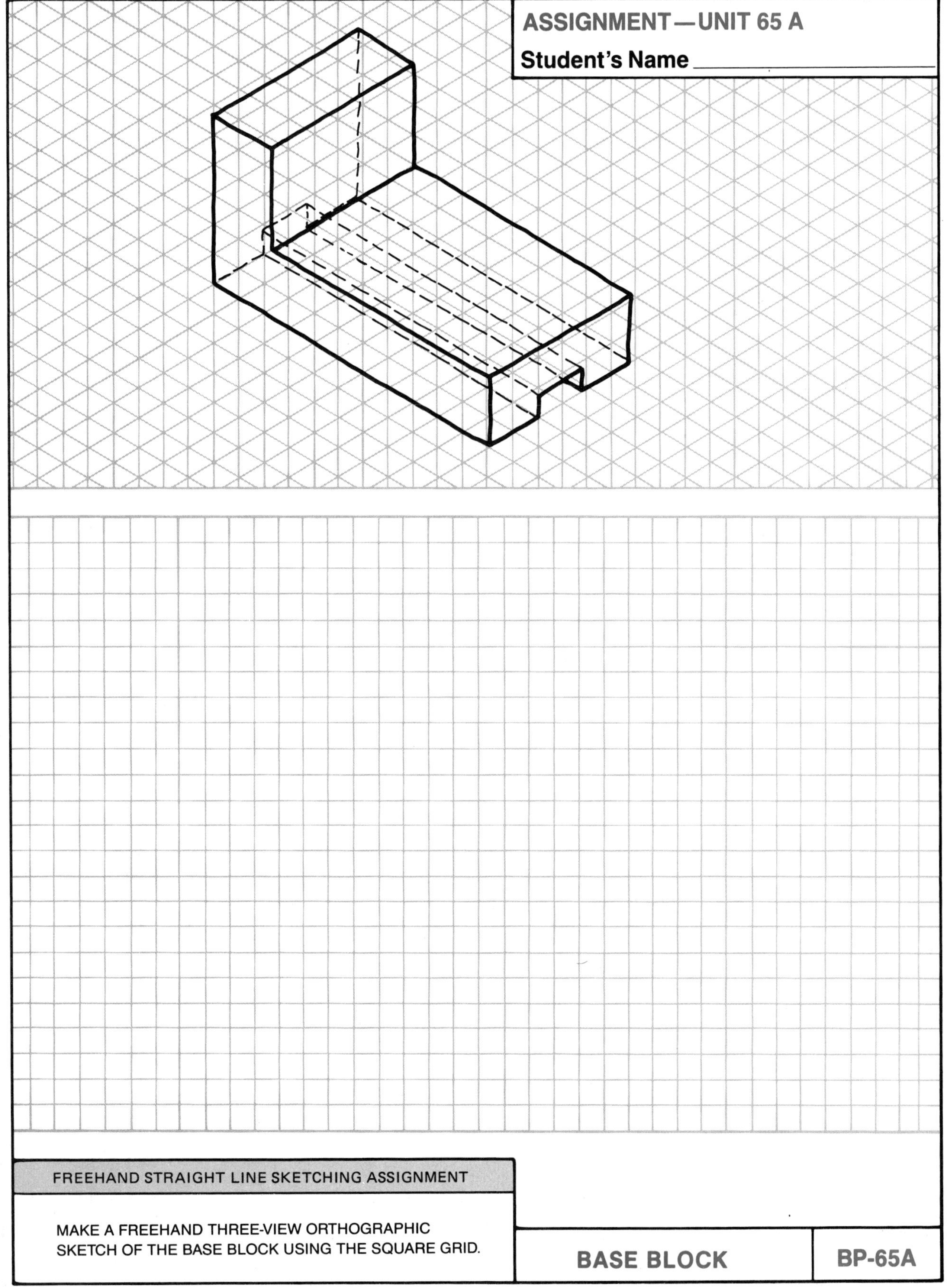

ASSIGNMENT—UNIT 65 A
Student's Name
FREEHAND STRAIGHT LINE SKETCHING ASSIGNMENT
MAKE A FREEHAND THREE-VIEW ORTHOGRAPHIC SKETCH OF THE BASE BLOCK USING THE SQUARE GRID.
BASE BLOCK
BP-65A

ASSIGNMENT — UNIT 65 B

Student's Name ____________

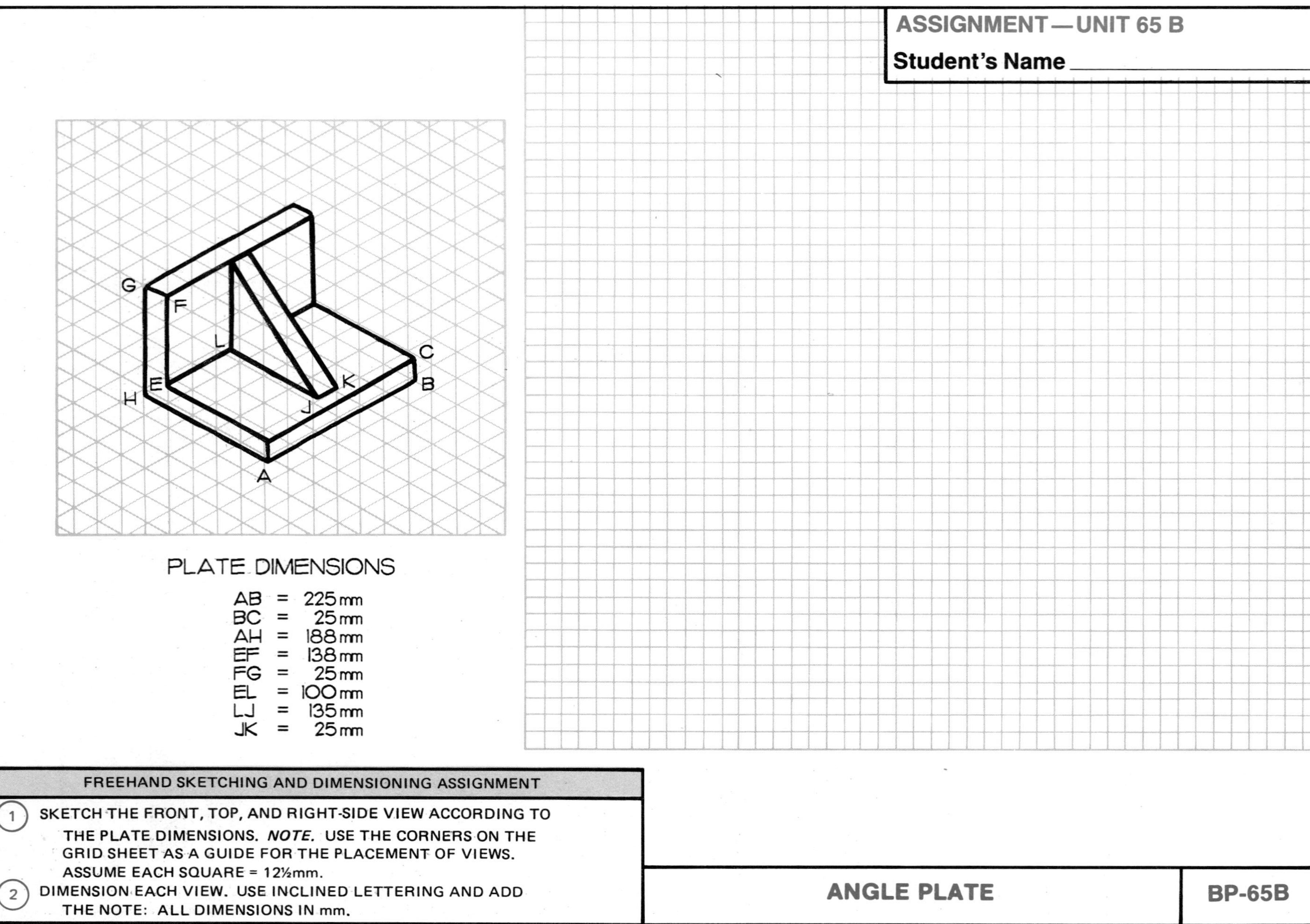

PLATE DIMENSIONS

AB	=	225 mm
BC	=	25 mm
AH	=	188 mm
EF	=	138 mm
FG	=	25 mm
EL	=	100 mm
LJ	=	135 mm
JK	=	25 mm

FREEHAND SKETCHING AND DIMENSIONING ASSIGNMENT

1. SKETCH THE FRONT, TOP, AND RIGHT-SIDE VIEW ACCORDING TO THE PLATE DIMENSIONS. *NOTE.* USE THE CORNERS ON THE GRID SHEET AS A GUIDE FOR THE PLACEMENT OF VIEWS. ASSUME EACH SQUARE = 12½mm.
2. DIMENSION EACH VIEW. USE INCLINED LETTERING AND ADD THE NOTE: ALL DIMENSIONS IN mm.

ANGLE PLATE

BP-65B

UNIT 66
Sketching Curved Lines

The first step in planning a sketch is establishing the basic geometric forms which represent each feature of the part to be sketched. For example, the spindle pulley bracket subassembly drawing (figure 66-1) identifies the geometric form of each part. In this instance, the separate parts are either cylindrical, square, rectangular, or pyramidal in shape.

Further examination of the features reveals that every detail may be sketched by using either straight or curved lines or a combination of these two lines. Mastering the ability to accurately draw these two types of lines in combination makes it possible to sketch objects which are plane or of solid geometric shapes.

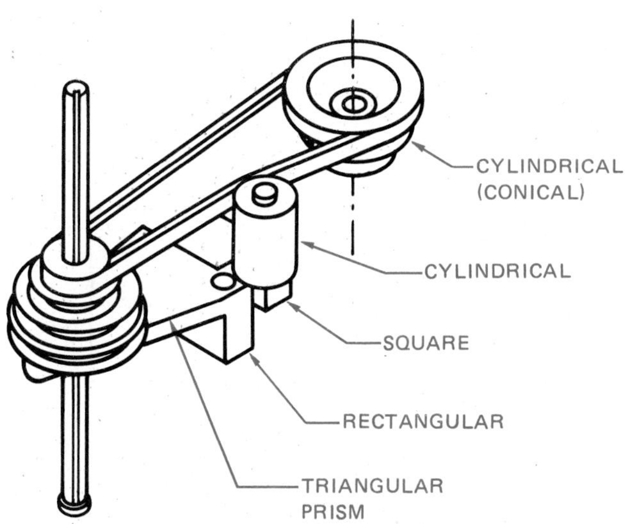

FIGURE 66-1 COMBINATION OF GEOMETRIC SHAPES IN A SUBASSEMBLY OF SPINDLE PULLEY BRACKET

SKETCHING CURVED LINES

Sketching Full Circles

The shapes of curved lines include the *circle*, an *arc*, an *ellipse*, and any combination of these lines. The sketch of an object takes on shape when curved and straight lines are combined. Curved lines are sketched by using certain guide lines or reference points. Circles are drawn in relation to center lines, other curved lines, and the edges of square, angular, and other surfaces.

One common technique of sketching full circles that are larger in diameter than approximately 5/8″ (16 mm) is to first sketch a square. This technique is illustrated in figure 66-2 and explained in the steps which follow:

STEP 1 → Lay out two center lines which cross and are at right angles to each other.

STEP 2 → Mark off the radius of the required circle on the leg of each center line.

STEP 3 → Sketch the vertical and horizontal sides of the square.

STEP 4 → Draw diagonal lines between either the center lines or to the corners of the square.

STEP 5 → Locate the center of each triangle or the radius along each diagonal.

STEP 6 → Sketch a light line to draw a quarter of the circle.
NOTE: Rest the side of the hand on the paper. Extend the fingers to bring the point of the pencil to a

STEP 3 FORMING A SQUARE	STEPS 4 AND 5 DRAWING DIAGONALS AND REFERENCE POINTS	STEP 6 SKETCHING ARC	STEP 7 SKETCHING CIRCLE	STEP 8 DARKENING CIRCLE

FIGURE 66-2 STEPS IN DRAWING A CIRCLE

reference point on a center line. Move the fingers as a compass (figure 66-3). The pencil is then guided in a circular movement through the midpoint (second reference dot) to the next center line.

STEP 7 ➤ Continue to draw the second, third, and fourth quarters of the circle. Depending on the diameter, it is often easier to sketch large circles by turning the drawing sheet a quarter turn for each arc.

NOTE: Small circles of less than 5/8″ (16 mm) diameter are generally sketched by just marking off the diameter on a center line. The circle is then drawn by using a full-circle or two part-circle strokes.

STEP 8 ➤ Darken the circle outline. Erase guide lines if necessary and where practical.

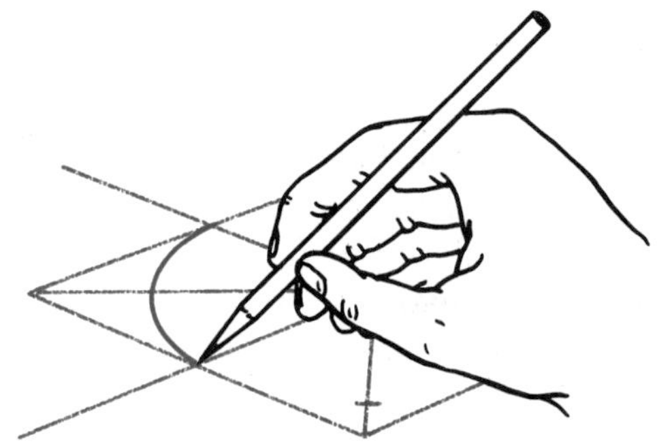

FIGURE 66-3 GUIDING THE POINT TO FORM A CIRCLE

Sketching Ellipses

A *regular ellipse* is a curved line which changes constantly between its width and height. One method of sketching an ellipse is to start with a rectangular form. The size of the rectangle is determined by the width and height of the ellipse. The steps required to sketch an ellipse are illustrated in figure 66-4. The steps are similar to

STEP 1 DRAW RECTANGLE	STEP 2 DRAW DIAGONALS	STEP 3 LOCATE REFERENCE POINTS	STEP 4 SKETCH ELLIPSE

FIGURE 66-4 STEPS IN FORMING A REGULAR ELLIPSE

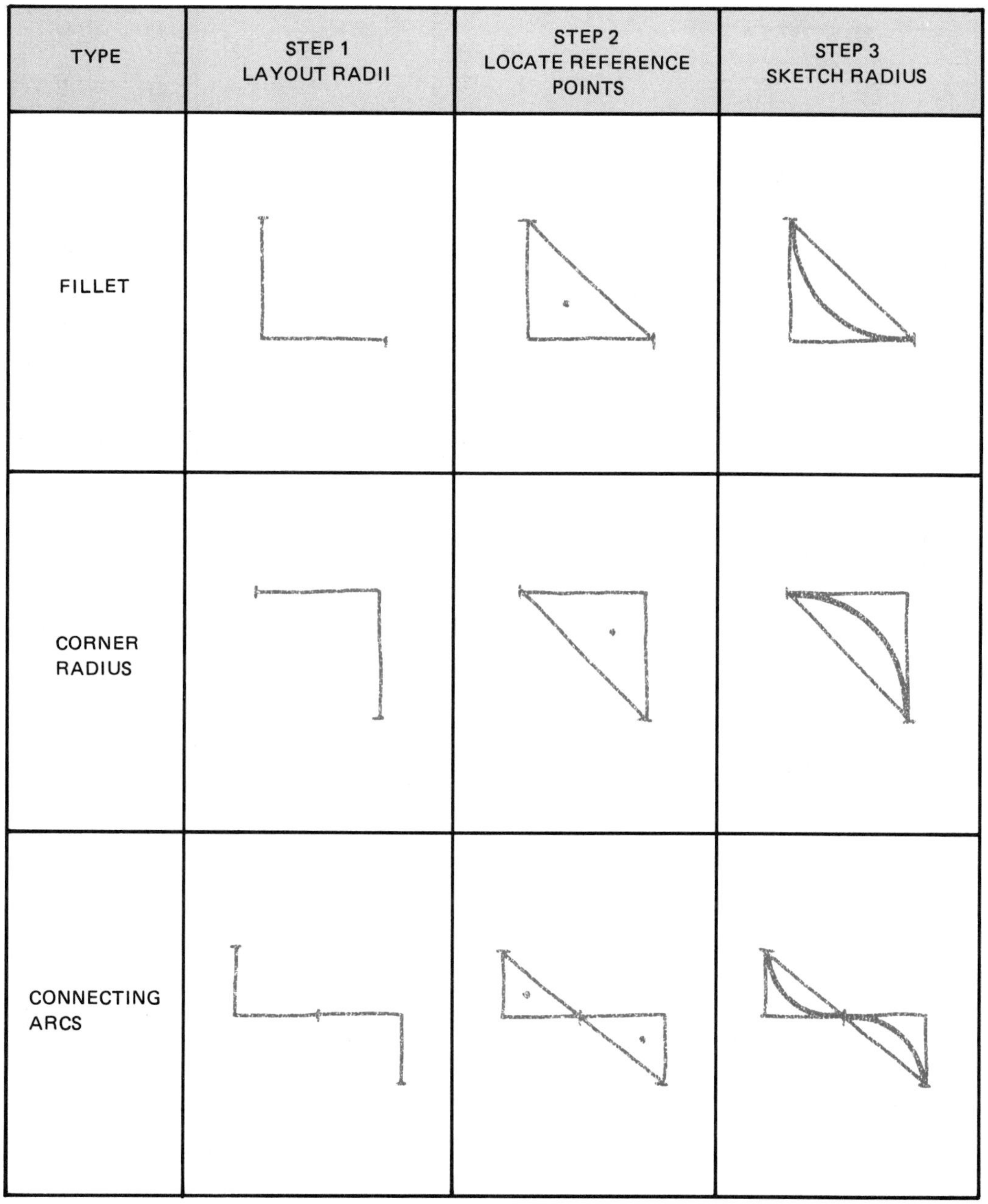

TYPE	STEP 1 LAYOUT RADII	STEP 2 LOCATE REFERENCE POINTS	STEP 3 SKETCH RADIUS
FILLET			
CORNER RADIUS			
CONNECTING ARCS			

FIGURE 66-5 TECHNIQUES OF SKETCHING ARCS

those used for a circle except that additional diagonals and reference points are required. A light curved line is drawn through each reference point to form the ellipse.

Sketching Rounded Corners, Fillets, and Connecting Arcs

The sharp corners or noncutting edges are usually eliminated by "rounding the corner." By contrast, intersecting corners are designed with a radius (*fillet*) to increase the strength of a part or to permit easy withdrawal from a mold. In still other cases, radii are tangent. Regardless of the corner radius, fillet, or connecting arcs, the procedure for sketching is almost identical. The steps in each instance are shown graphically in figure 66-5.

DRAWING NO.
W5-1S TO W5-28S

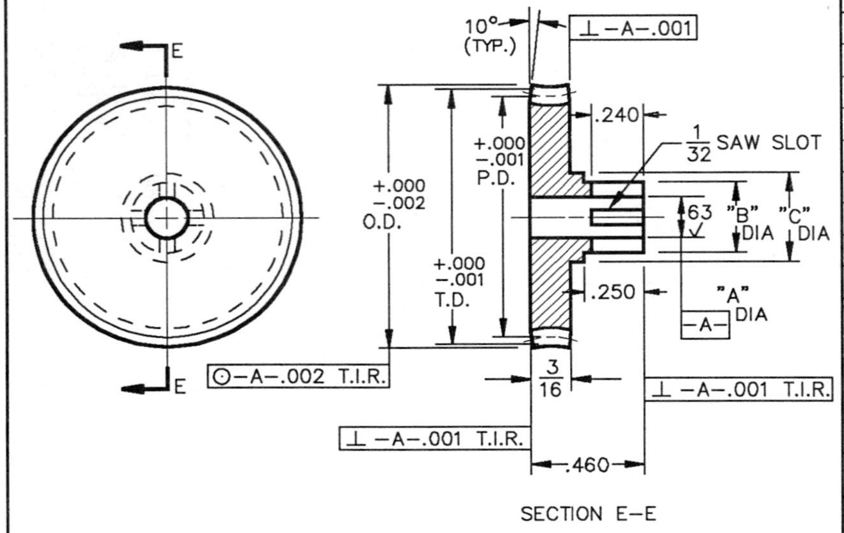

WORM WHEEL DATA (MATES WITH WII)							
PART NO	NO. OF TEETH	PITCH DIA	THROAT DIA	O.D.	"A" DIA	"B" DIA	"C" DIA
W5-1S	30	.6250	.6624	.6832	+.0005 −.0000 .1248 DIA	+.000 −.003 .1875 DIA	±.005 .250 DIA
−2S	40	.8333	.8707	.8915			
−3S	50	1.0417	1.0791	1.0999			
−4S	60	1.2500	1.2874	1.3082			
−5S	70	1.4583	1.4957	1.5165			
−6S	80	1.6667	1.7041	1.7249			
−7S	90	1.8750	1.9124	1.9332			
−8S	100	2.0833	2.1208	2.1415			
W5-9S	120	2.5000	2.5374	2.5582			
W5-10S	30	.6250	.6624	.6832	+.0005 −.0000 .1873 DIA	+.000 −.003 .2500 DIA	±.005 .312 DIA
−11S	40	.8333	.8707	.8919			
−12S	50	1.0417	1.0791	1.0999			
−13S	60	1.2500	1.2874	1.3082			
−14S	70	1.4583	1.4957	1.5165			
−15S	80	1.6667	1.7041	1.7249			
−16S	90	1.8750	1.9124	1.9332			
−17S	100	2.0833	2.1208	2.1415			
W5-18S	120	2.5000	2.5374	2.5582			

10° (TYP.)
⊥ −A−.001
.240
$\frac{1}{32}$ SAW SLOT
+.000 −.001 P.D.
+.000 −.002 O.D.
+.000 −.001 T.D.
63 √
"B" DIA
"C" DIA
.250
"A" DIA
−A−
⊙−A−.002 T.I.R.
$\frac{3}{16}$
⊥ −A−.001 T.I.R.
⊥ −A−.001 T.I.R.
.460
E
E

SECTION E−E

WORM WHEEL DATA	
PITCH (DIAM)	48
THREAD	FOUR (R.H.)
LEAD OF WORM	.2618
PRESSURE ∡	25°
LEAD ANGLE	14°−2'
WHOLE DEPTH	.0407
AGMA	PREC 1
TESTING PRESSURE	20 OZ.
TOOTH FORM	INVOLUTE

ITEM NO.	SYMBOL	NO REQ'D	DESCRIPTION	PART OR CATALOG NO	SPEC NO OR MFG

LIST OF MATERIAL

DO NOT SCALE THIS DRAWING
REMOVE ALL BURRS AND SHARP EDGES UNLESS OTHERWISE SPEC

SURFACE FINISH 125 √ UNLESS OTHERWISE SPECIFIED — WEIGHT — BP-66

TOLERANCES UNLESS SPECIFIED — FRACT DIM: ± 1/64 — DECIMAL DIM: ±.005 — ANGULAR DIM: ∽ — SCALE: NONE

DRAWN BY: 9-17-91 A.E.H. — APPROVED BY
CHECKED BY 9-19-91 A.J.C. — ENG'R
CLASS NO — JOB NO
HEAT TREAT ∽
FINISH ∽
MAT'L BRONZE AS PER QQ-B-637, COMP. 1

STERLING INSTRUMENT
DIVISION OF DESIGNATRONICS, INC.
New Hyde Park, New York 11040

TITLE WORM WHEEL (CLAMP TYPE)
48 PITCH R.H. 3/16 FACE

NEXT ASSEMBLY — DRAWING NUMBER B. W5-1S TO W5-28S — REV F

REVISION	DATE	APP'D	CHANGE
F	12/4	GUY	1321-2314 O.D. CONC. .002 WAS. .0005
E	1/11	NU	ECO 1321-2119

Courtesy of Sterling Instrument, Division of Designatronics, Inc.
(Modified Industrial Drawing)

ASSIGNMENT—UNIT 66

Student's Name ____________________

FREEHAND SKETCHING ASSIGNMENT ON CIRCLES AND ARCS

1. MAKE A FREEHAND SKETCH OF THE FRONT VIEW AND A HALF-SECTION SIDE VIEW OF WORM WHEEL W5-11S, ACCORDING TO THE DATA PROVIDED. *NOTE.* MAKE THE SKETCHES APPROXIMATELY TWO TIMES THE ACTUAL SIZE.
2. DIMENSION THE VIEWS, INCLUDING GEOMETRIC TOLERANCING AND REF. SURFACE [—A—].
3. ADD A NOTE SPECIFYING (a) THE MATERIAL AND (b) THE UNSPECIFIED DIMENSIONAL AND SURFACE FINISH TOLERANCES.

SECTION 17
Industrial Pictorial Sketching

UNIT 67
Orthographic Sketching

Orthographic sketching simply means that one or more regular views or sections of an object are sketched freehand instead of being drawn mechanically with instruments. An orthographic sketch is generally drawn because it is the simplest of the four basic types of sketches to make.

Some objects are drawn in one view; others require two or more views to accurately describe the shape and to dimension the features properly. A multiview sketch need not be drawn to any scale; however, each feature must be in proportion. Sketches are prepared using the same kinds of lines, line weights, and dimensioning as for conventional mechanical drawings. Interior details are often represented by sketching one or more sections freehand.

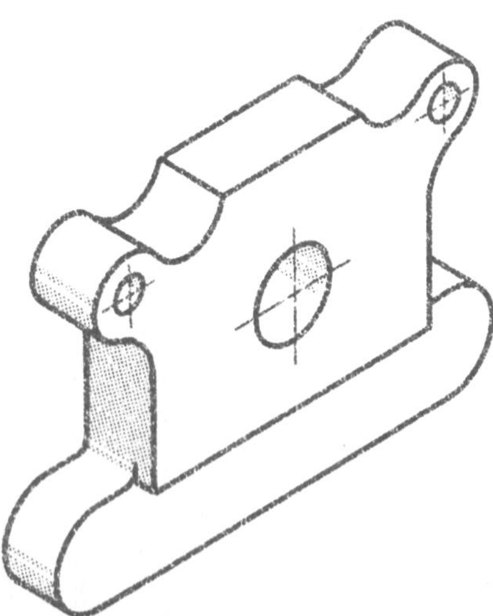

FIGURE 67-1 EXAMPLE OF A MULTIPLE-FORM BASE PLATE

ORTHOGRAPHIC SKETCHING USING COMBINATIONS OF LINES

Like all other drawings, orthographic sketches are made by using straight lines, curved lines, and combinations of these lines. Using the irregular-shaped Base Plate which is illustrated in figure 67-1 as an example, the basic steps required to produce an orthographic sketch follow. Some of the major steps are shown in figure 67-2.

STEP 1 → Determine the number of views required to completely describe the Base Plate.

STEP 2 → Outline the major geometric forms in proportion.

STEP 3 → Layout the main center lines and object lines.

STEP 4 → Draw the guide lines and mark off the reference points for all of the circular features.

STEP 5 → Sketch the required circles, arcs, and curves. Then sketch the flat portion of the Base Plate. *Use fine layout lines.*

STEP 6 → Sketch extension lines, dimension lines, and leaders.

STEP 7 → Add dimensions, notes, and other essential information.

STEP 8 → Check the representation of each detail and dimension for technical accuracy.

STEP 9 → Go back over the sketch and sharpen or widen (darken) each line according to use.

STEP 10 → Erase extra lines or unnecessary markings on the sketch.

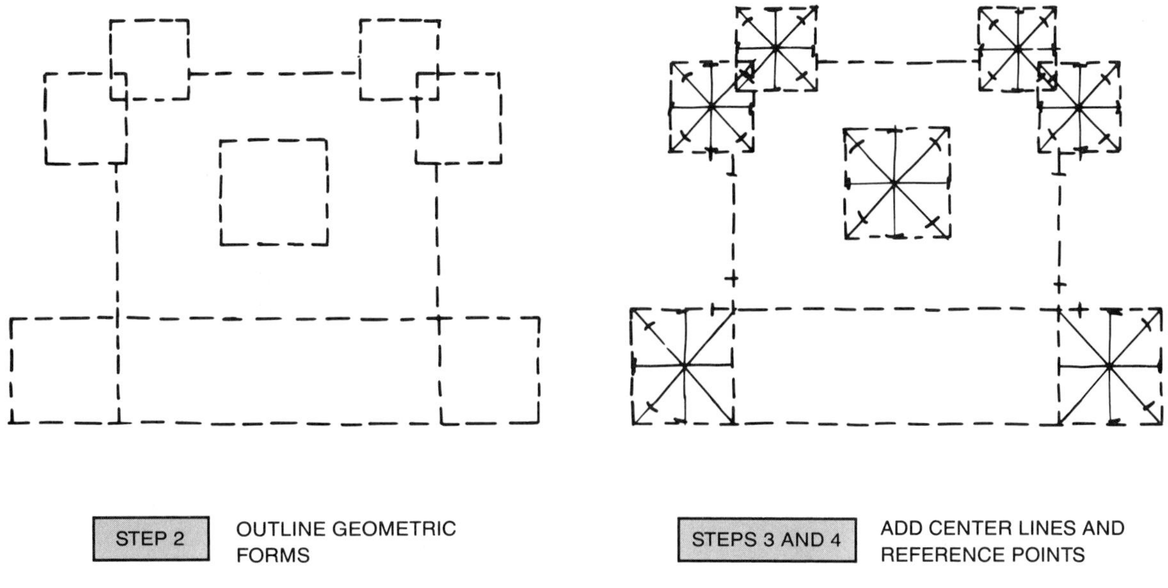

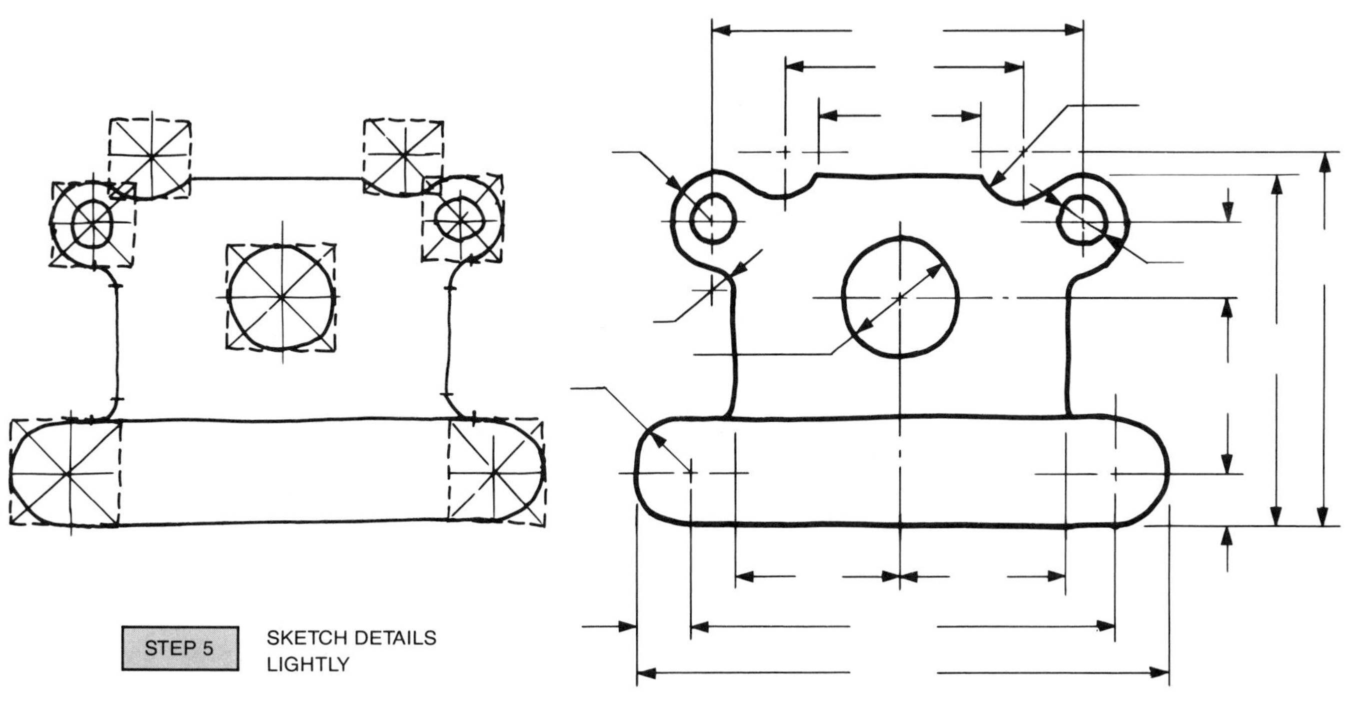

STEPS 6 AND 7 ADD DIMENSIONS AND NOTES AND DARKEN THE FEATURES

FIGURE 67-2 STEPS IN SKETCHING A MULTIPLE-FORM (IRREGULAR SHAPE) OBJECT

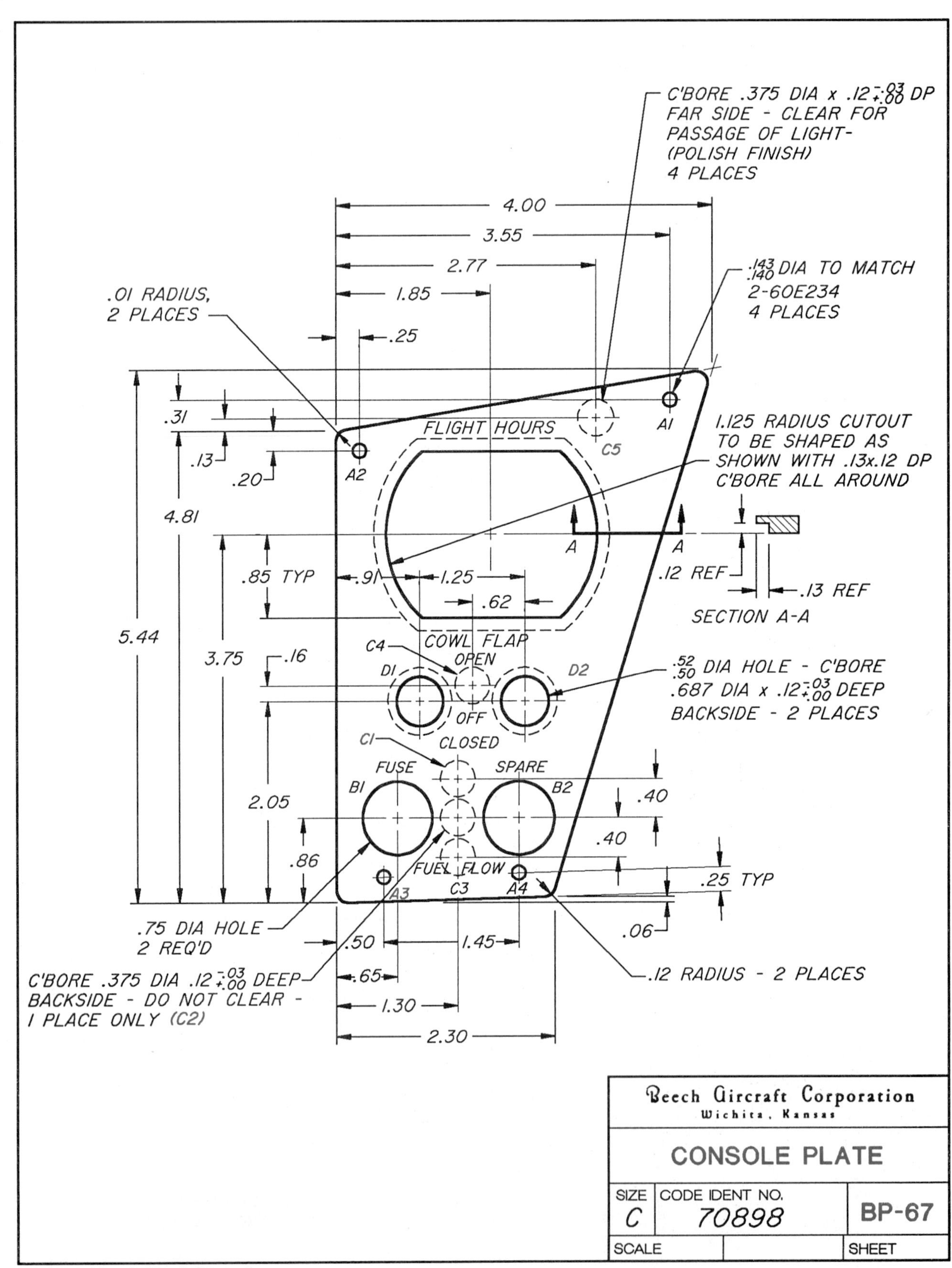

Courtesy of Beech Aircraft Corporation
(Modified Industrial Drawing)

ASSIGNMENT—UNIT 67

Student's Name ____________________

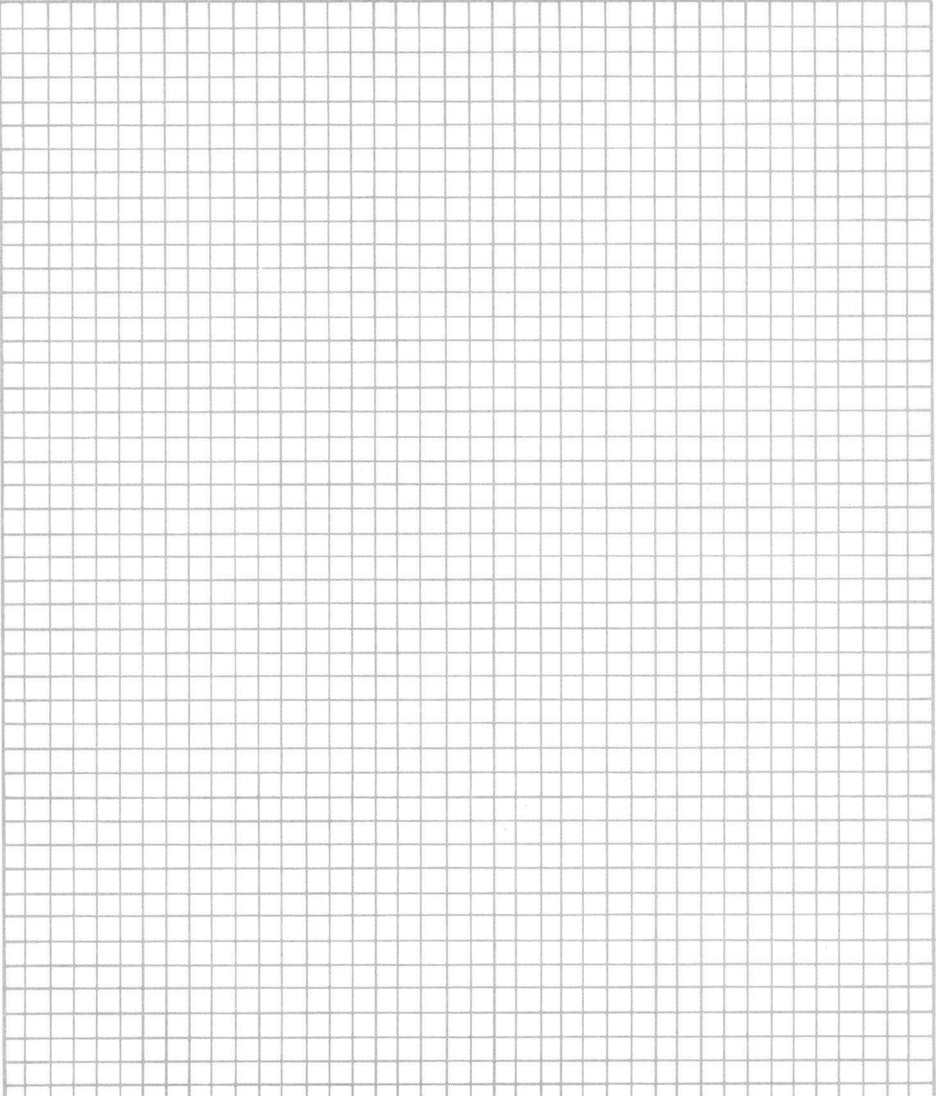

(3)

HOLE ID	LOCATION		DIMENSIONS
	X	Y	
A3			
C2			
D2			
C5			

FREEHAND SKETCHING AND COORDINATE DIMENSIONING
(1) SKETCH THE *BACK VIEW* OF THE CONSOLE PLATE *FULL SIZE*.
(2) IDENTIFY AND LABEL EACH HOLE.
(3) DETERMINE THE POSITIONAL DIMENSIONS FOR HOLES A3, C2, D2, AND C5. LETTER THIS INFORMATION IN THE CHART USING SLANT LETTERING.

HIGH TECHNOLOGY applications

HT-18 COMPUTER-AIDED SUBSYSTEMS INTERFACED IN FMS

One of the most productive computer-directed, random-order, highly-automated, flexible manufacturing systems (FMS) that has been in successful operation over many years is graphically illustrated below. This system interfaces CADD/CAM/CIM; NC/CNC/DNC; and other major central unit processing capability.

FLEXIBLE MANUFACTURING SYSTEM (FMS) SUBSYSTEMS

The flexible manufacturing system, as modeled, includes eight machining centers. Each center is equipped with a 90-station cutter and other tool storage matrix (with capacity to automatically replace worn or damaged cutters). Additional equipment includes: two coordinate measuring machines, a parts cleaner, and an automatic removal system.

Palletized equipment, guided-cart and robotic on- and off-loading systems that respond to computer commands provide automated materials and tool handling.

Interactive computer graphics and database management, analysis, and manipulation of data in memory permit the continuous processing of engineering information. As a result, computer commands are generated that permit simultaneous modification of product design and manufacturing processes in FMS.

ADAPTIVE CONTROL SUBSYSTEMS

The flexible manufacturing system incorporates part family group (cellular manufacturing), the use of automated sensing probes, automated tool setting equipment, in-process and post-processor gaging, and other *adaptive control subsystems.* This last named subsystem senses torque in relation to tooling, material, and machining condition overloads, as well as dimensional inaccuracies.

An *adaptive control unit* automatically adjusts feed and speed rates to compensate for changing machining variations under actual operating conditions. A final note: all on-floor and off-floor subsystems are interfaced in flexible manufacturing systems (FMS).

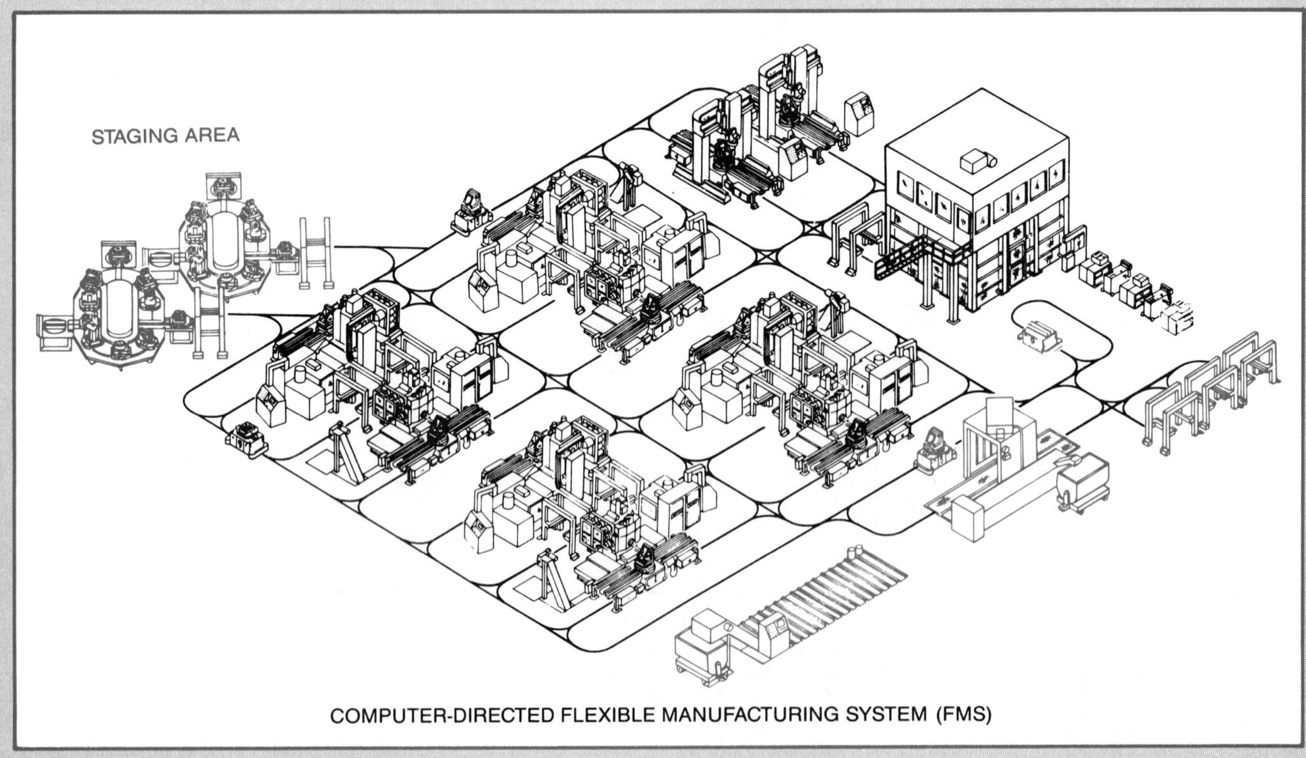

COMPUTER-DIRECTED FLEXIBLE MANUFACTURING SYSTEM (FMS)

Courtesy of CINCINNATI MILACRON INC.

UNIT 68
Oblique Sketching

Oblique sketches are prepared by a pictorial projection method in which one face or plane of projection (front view) is at a right angle to the observer. Features represented on the front view appear in true size and shape as shown by the cabinet drawing (figure 68-1).

Oblique sketches are commonly used in cabinet making, woodworking and construction, and other occupations when it is important to show details and dimensions on the frontal plane. Objects having primarily cylindrical or curved surfaces on one side are sketched by this method because these features appear as circles and arcs rather than ellipses.

CAVALIER, CABINET, AND GENERAL OBLIQUE SKETCHES

The three common types of oblique sketches which were introduced earlier under pictorial drawings are *cavalier*, *cabinet*, and *general*. Regardless of type, all oblique sketches require three axes of projection. The horizontal and vertical axes (representing features and dimensions of width ((length)) and height, respectively) are at right angles and appear on the *frontal plane*. Features along the third (receding) axis are drawn at angles of 30° to 45° for *cavalier oblique drawings*, 30° to 60° for *cabinet oblique drawings*, and at any angle for *general oblique sketches*.

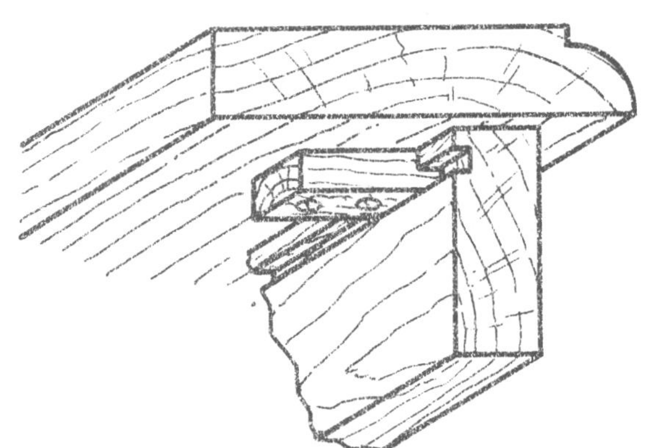

FIGURE 68-1 CABINET OBLIQUE SKETCH SHOWS FEATURES ON THE FRONTAL PLANE IN TRUE SHAPE AND SIZE

Right and Left Oblique Projection

The receding axis for an oblique projection drawing may be at an angle to the right (figure 68-2A) or to the left (figure 68-2B) of the front face.

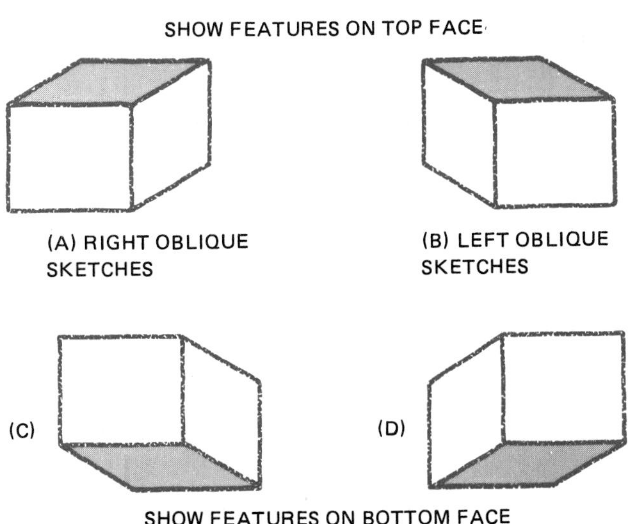

FIGURE 68-2 RIGHT AND LEFT OBLIQUE PROJECTION SKETCHING

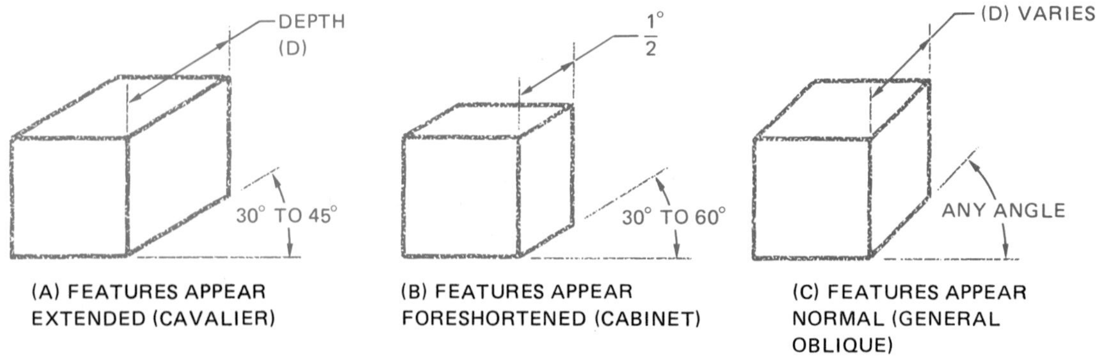

FIGURE 68-3 EFFECT ON FEATURES WITHIN PLANES ON RECEDING AXES

The receding axis may also be reversed as illustrated in figures 68-2C and 2D to show features which appear from the bottom side.

FORESHORTENING RECEDING AXIS FEATURES

Although depth dimensions are accurately laid out along the third axis, the object in *cavalier oblique sketches* appears distorted, as pictured in figure 68-3A. This condition is sometimes corrected by drawing features along the receding axis to approximately one-half scale (figure 68-3B). This is a characteristic of *cabinet oblique sketches. General oblique* sketches permit varying the angle of the third axis and the amount of foreshortening. The cube represented by the general oblique drawing in figure 68-3 at (C) creates a more natural, less distorted appearance.

CIRCULAR FEATURES ON OBLIQUE SKETCHES

To repeat, circular features on the frontal plane are represented as full circles or arcs. A circle which is projected to the top, bottom, or side planes of an oblique drawing forms an *ellipse* within a geometric form called a *parallelogram*. The length of one side of the parallelogram represents the *major* (reference) *diameter*. The height or width represents the *minor diameter* of the ellipse. The reference diameter of the circle is measured along the horizontal center line of the parallelogram in the top or bottom view. The minor diameter is represented by the foreshortened dimension along the center line, which is parallel to the receding axis line.

CHARACTERISTICS OF AN ELLIPSE

A circle may be sketched as an ellipse on the foreshortened faces of an oblique drawing, or on any other isometric or perspective drawing by following six basic steps. The forming of an ellipse on the side face of an oblique drawing is illustrated and detailed in figure 68-4. It should be noted that the same steps may also be applied to the sketching of an ellipse in the top, bottom, or other side view.

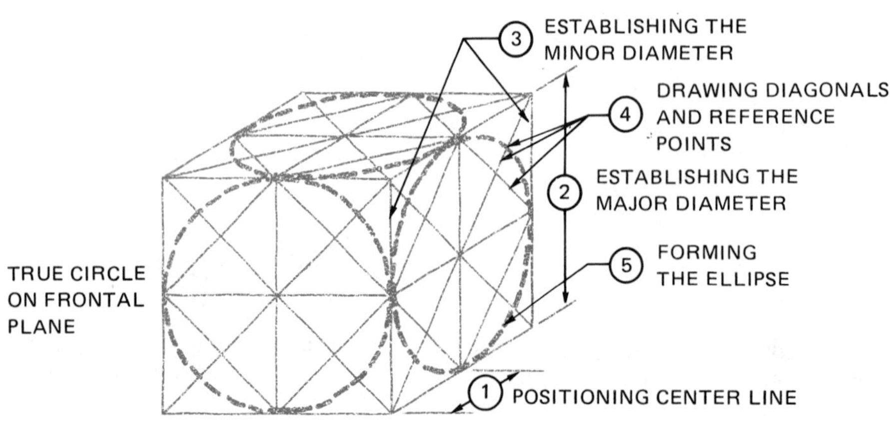

FIGURE 68-4 CONSTRUCTION OF AN ELLIPSE IN OBLIQUE PROJECTION SKETCHING

Forming an Ellipse on any Receding Plane

STEP 1 → Refer to the front face. Determine the location of the vertical center line of the circle in relation to the foreshortened plane.

STEP 2 → Lay out or project the center line of the major (true) circle diameter along the vertical center line.

STEP 3 → Determine the foreshortened width of the circle. Sketch parallel lines to form the parallelogram.

STEP 4 → Draw diagonals from each corner and between the center lines to form triangles. Locate the reference points at the centers of the corner triangles.

STEP 5 → Sketch the ellipse lightly so the changing shape passes through each reference point.

STEP 6 → Darken the outline of the ellipse.

DIMENSIONING OBLIQUE SKETCHES

The dimensioning of width and height features on the frontal plane of an oblique sketch is the same as for dimensioning these same features on a mechanical drawing. Other dimensions for features on the top or side receding faces are lettered as *aligned* (figure 68-5A) or *unidirectional dimensions* (figure 68-5B). These two methods of dimensioning are approved by ANSI. Oblique dimension lines are always drawn parallel to the basic axes.

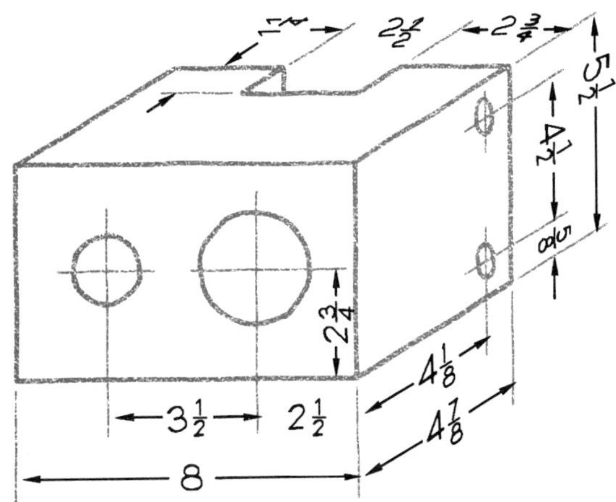

(A) ALIGNED DIMENSIONING

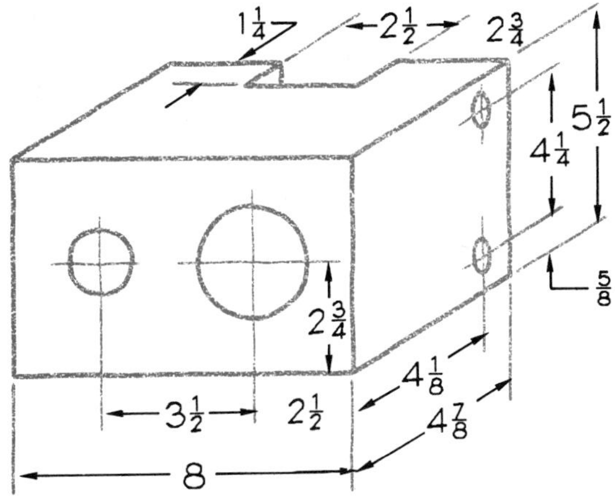

(B) UNIDIRECTIONAL DIMENSIONING

FIGURE 68-5 ALIGNED AND UNIDIRECTIONAL METHODS OF DIMENSIONING AN OBLIQUE SKETCH

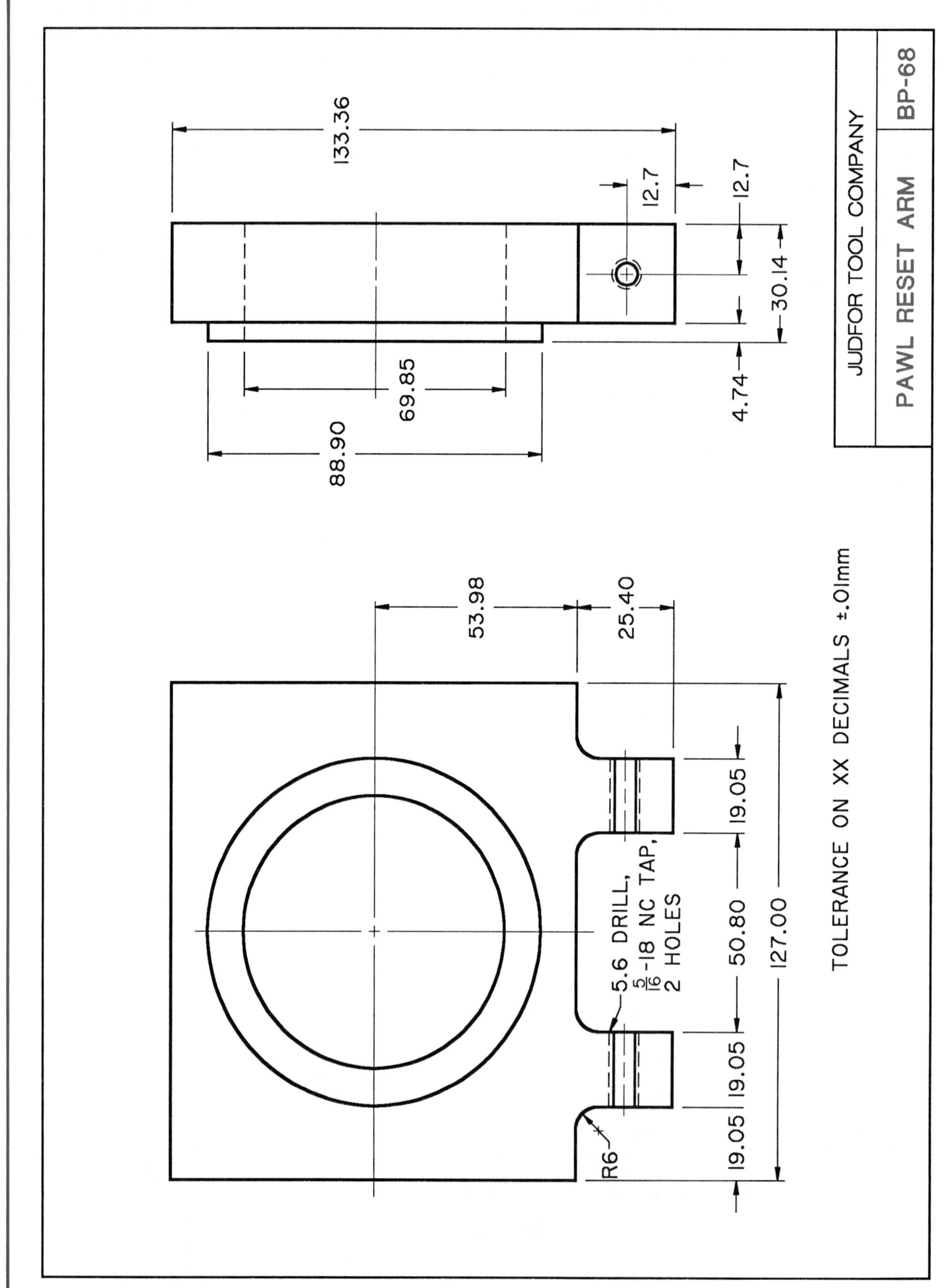
JUDFOR TOOL COMPANY
PAWL RESET ARM
BP-68
133.36
12.7
12.7
30.14
4.74
69.85
88.90
53.98
25.40
19.05
50.80
127.00
19.05
19.05
R6
5.6 DRILL,
$\frac{5}{16}$-18 NC TAP,
2 HOLES
TOLERANCE ON XX DECIMALS ±.01mm

ASSIGNMENT—UNIT 68

Student's Name ____________

FREEHAND OBLIQUE PROJECTION SKETCHING ASSIGNMENT

1. MAKE A FREEHAND SKETCH OF THE PAWL RESET ARM USING ANY ONE OF THE OBLIQUE PROJECTION METHODS. *NOTE*. DRAW TO A ONE-HALF SIZE SCALE.
2. ADD *NOMINAL, ALIGNED DIMENSIONS* AND NOTES.

UNIT 69
Isometric Sketching

The isometric drawing or sketch is a widely used method of projection for pictorial drawings. Isometric sketches are generally prepared from features and other information found on an orthographic drawing or as visualized by the designer or maker of the object.

POSITIONS OF MAJOR AXES

Isometric drawings are prepared from the location of the *three major axes* which are used to establish the features of the object. The four positions which the object may take are shown in figure 69-1. Notice that one axis is drawn either horizontally or vertically. The other two axes are 120° apart and are drawn in a direction which makes the object appear to be viewed from the bottom, top, left side, or right side (figure 69-1, positions A, B, C, and D).

REPRESENTATION OF FEATURES

All features which are represented on an orthographic projection drawing appear on an isometric sketch. As can be seen in figure 69-2, each of the lines, points, and planes which are shown in the orthographic views are represented on the isometric sketch. Isometric sheets with accurately ruled grids are used to simplify the making of a sketch.

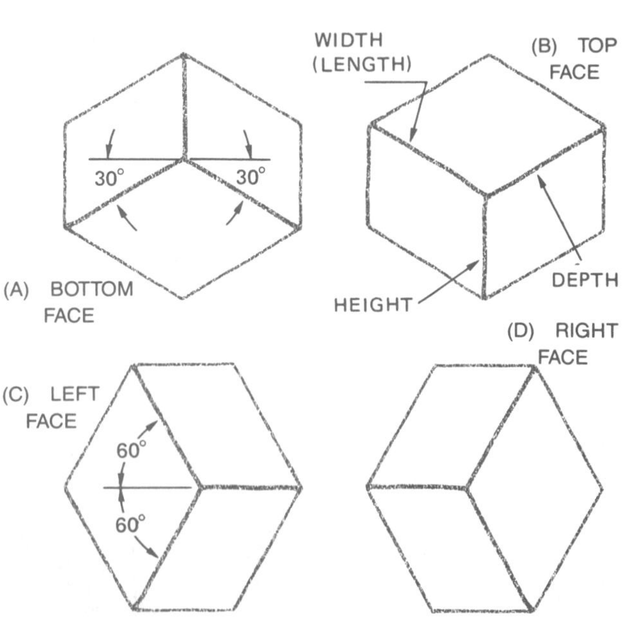

FIGURE 69-1 POSITIONS OF ISOMETRIC AXES FOR EMPHASIZING FEATURES ON DIFFERENT FACES OF AN OBJECT

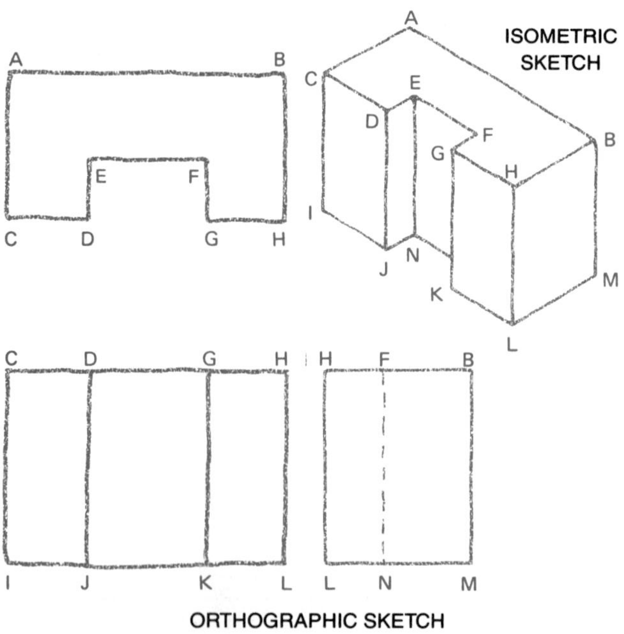

FIGURE 69-2 REPRESENTATION OF FEATURES ON BOTH ORTHOGRAPHIC AND ISOMETRIC SKETCHES

ISOMETRIC SECTIONS AND ASSEMBLY SKETCHES

Internal details are often represented on an isometric sketch by a *section view*. The area in which the features are located is identified by the conventional section line, arrowhead to indicate direction, and letter identification. The features are shown pictorially as they appear on the cutting plane(s). The same standards and techniques of representation are used as for the drawing of the part mechanically.

Mechanisms and objects consisting of more than one part often require a pictorial assembly drawing. Simple straight lines and offset lines are used to show the sequence and details of each part of a complete assembly. The parts may be sketched using oblique, isometric, or perspective projection. Each part is numbered sequentially and further identified in the parts list.

STEPS IN MAKING AN ISOMETRIC PROJECTION SKETCH

An isometric sketch is usually started by positioning an object so that one corner seems to rest on either the horizontal or vertical axis. The steps which follow provide a guide for making an isometric sketch. Figure 69-3 illustrates the procedure.

STEP 1 → Decide which of the four location positions of the isometric axes will produce the clearest view of the object. Lay out the axes in this location (figure 69-3A).

STEP 2 → Lay out the overall sizes: length, width, and depth, on the isometric axes.

STEP 3 → Draw a frame which will enclose the object. Make the isometric lines parallel to the isometric axes.

STEP 4 → Lay out other features by measuring parallel to the axis lines. Circular features are formed according to the description and steps which were identified for oblique sketching.

STEP 5 → Dimension the object according to the same rules which were applied to oblique sketches (figure 69-3B).

STEP 6 → Darken all lines. Erase guide lines.

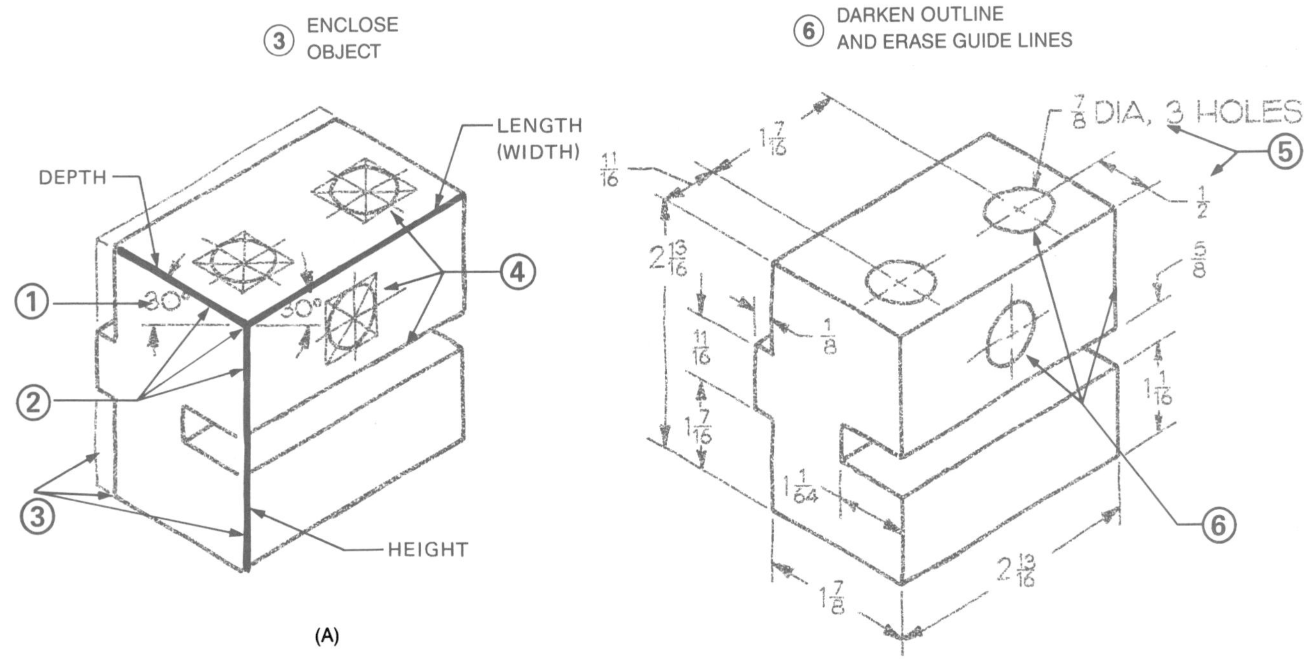

FIGURE 69-3 MAJOR STEPS IN PREPARING AN ISOMETRIC PROJECTION SKETCH

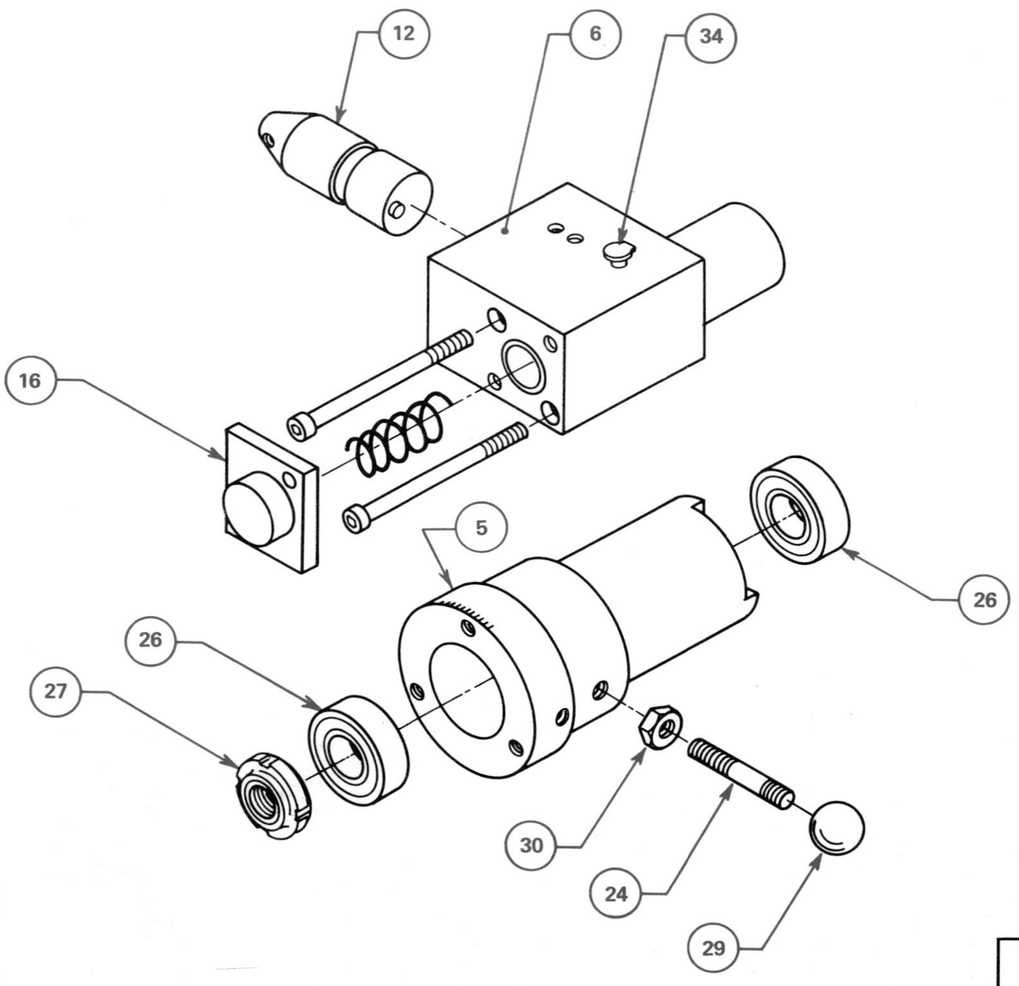

ITEM	QTY	NAME	PART NO.
34	1	OIL CUP	#501
33	1	SPRING	LC-067H-5
32	1	JAM NUT	$\frac{9''}{16}$-18 N.F.
31	1	WOODRUFF KEY	#404
30	1	JAM NUT	$\frac{5''}{16}$ - 18 N.C.
29	2	BALL KNOB	70143
28	2	THUMB SCREW	70872
27	1	BALL BRG. LOCKNUT	BL-N-04
26	2	BALL BEARINGS	773L04
25	1	HANDLE	3A0063
24	1	HANDLE	3A0062
23	2	THUMB SCREW	3A0052
22	1	ROLL PIN	R.P.-316-214
21	1	CAM LOCK HANDLE KNOB	3A0051
20	1	LOWER CAM LOCK PIN	3A0050
19	1	UPPER CAM LOCK PIN	3A0049
18	1	CAM LOCK SCREW	3A0048
17	1	BACKLASH ADG. SCREW	3A0044
16	1	PLUNGER HOUSING CAP	3A0043
15	2	DOWEL PIN	D.P.-14-1
14	1	VERNIER SCALE PLATE	3B0053
13	2	TABLE CLAMP	3B0052
12	1	PLUNGER CAM	3B0051
11	1	PLUNGER	3B0050
10	1	ADJUSTING PLATE	3C0043
9	1	GRADUATED COLLOR	3C0042
8	1	HAND WHEEL	3C0044
7	1	NOTCH PLATE	3C0041
6	1	PLUNGER HOUSING	3C0030
5	1	WORM SHAFT HOUSING	3C0029
4	1	WORM SHAFT	3C0039
3	1	WORM WHEEL	3C0040
2	1	TABLE TOP	3D0009
1	1	BASE HOUSING	3E0000

VISE & TOOL DIVISION
UNIVERSAL VISE & TOOL CO. PARMA, MICHIGAN

DWG NO. 9D0025	BP-69
PARTS DRAWING 9 INCH DIAMETER	
COMBINATION TABLE	

Courtesy of Universal Vise & Tool Company

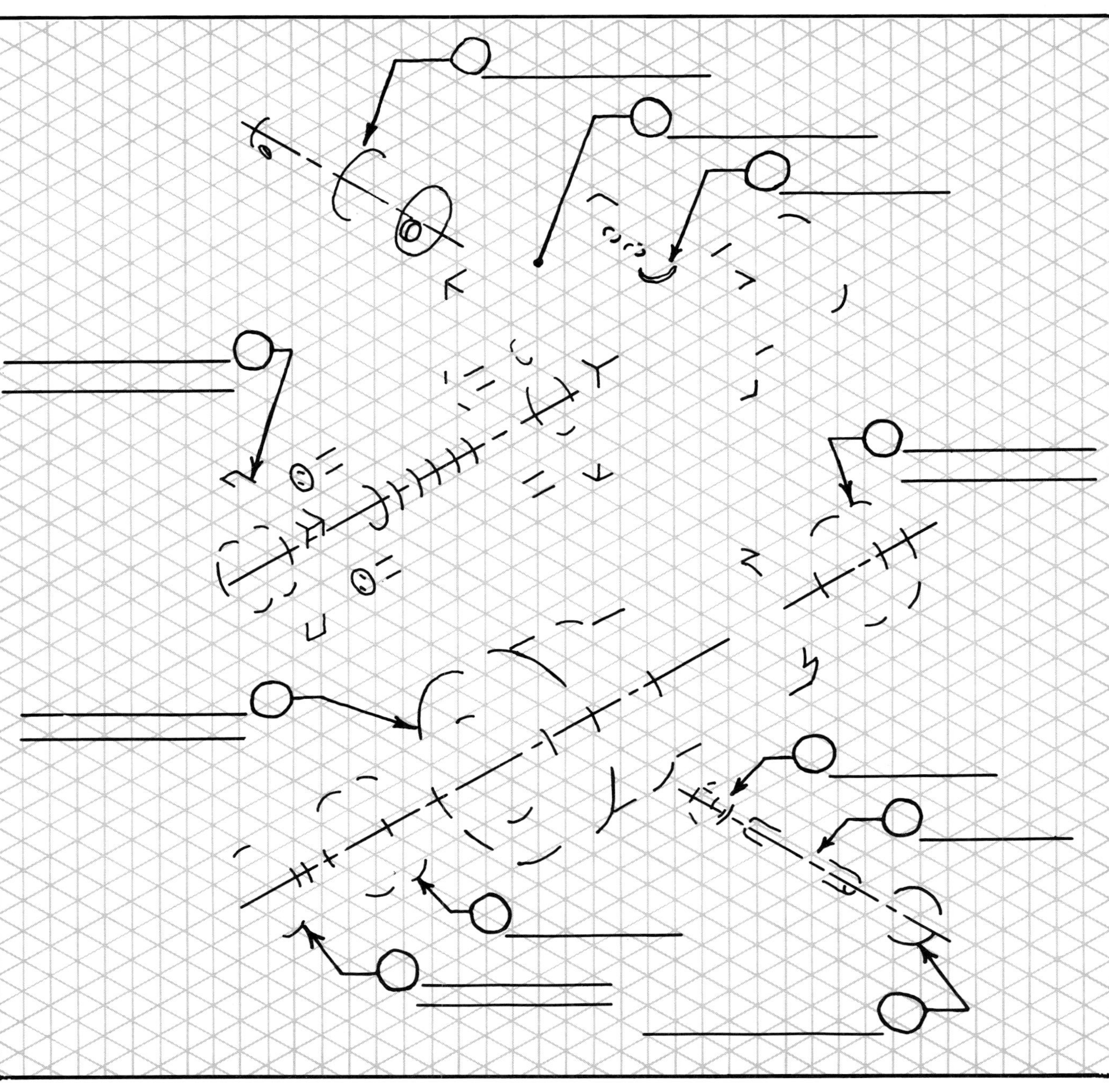

ASSIGNMENT—UNIT 69

Student's Name ______________________

FREEHAND ISOMETRIC ASSEMBLY SKETCHING ASSIGNMENT

1. REFER TO THE ISOMETRIC DRAWING BP-69. SKETCH THE FEATURES FREEHAND TO COMPLETE EACH PART OF THE ASSEMBLY.
2. IDENTIFY EACH PART BY NUMBER AND NAME. USE INCLINED LETTERING.

UNIT 70
Perspective Sketching

The *perspective sketch* has the advantage over other projection methods in that objects are represented as they are seen. The lines and forms on the drawing converge toward one, two, or three vanishing points. The perspective sketch is widely used in architecture, construction drawings, mechanical and electrical design, and other applications where an object is to be drawn with the least distortion of image.

However, there are some limitations of perspective drawings. Since only three surfaces are shown, an object may not be completely described. Also, dimensioning a complex part or construction details is difficult.

CHARACTERISTICS OF PERSPECTIVE DRAWINGS

As described and illustrated in Section 10 on pictorial drawings, the object lines in perspective drawing may be *parallel (single-point)*, *angular* (as in *two-point perspective*), or *oblique (three-point perspective)*. Perspective drawings may display *interior* or *exterior features* or *constructions*.

A perspective sketch requires the use of certain basic elements, such as *picture plane, horizon line, ground line, vanishing points*, and *station points*.

One of the first considerations in producing a perspective sketch is to determine the position from which the object is to be viewed. Figure 70-1 provides an example of how the position *above, on,* and *below the horizon line* changes the shape and faces which are produced.

PRODUCING A TWO-POINT PERSPECTIVE SKETCH

A *two-point (angular) perspective sketch* requires two vanishing points (VP). These are generally located on a horizontal line, called a *horizon line* (HL). The steps required to produce a perspective sketch of a rectangular block with a groove follow. The description is complemented by the line drawings in figure 70-2.

STEP 1 → Determine the scale to be used with the grid paper. Draw a horizontal line to represent the *horizon line* (figure 70-2A).

STEP 2 → Draw a vertical line to represent one edge of the part at the approximate center of the sketch (figure 70-2B).

STEP 3 → Locate the right and left vanishing points from the major edge (figure 70-2C).

STEP 4 → Lay out the true height of the edge of the object (figure 70-2D).

STEP 5 → Draw light lines from the ends of the vertical line to each vanishing point (figure 70-2E).

STEP 6 → Lay out the width and depth features. Draw parallel vertical lines (figure 70-2F).

STEP 7 → Draw light lines for the width and depth features, using the reference points and the vanishing points (figure 70-2G).

STEP 8 → Dimension the features. Darken the object lines. Erase the guide lines (figure 70-2H).

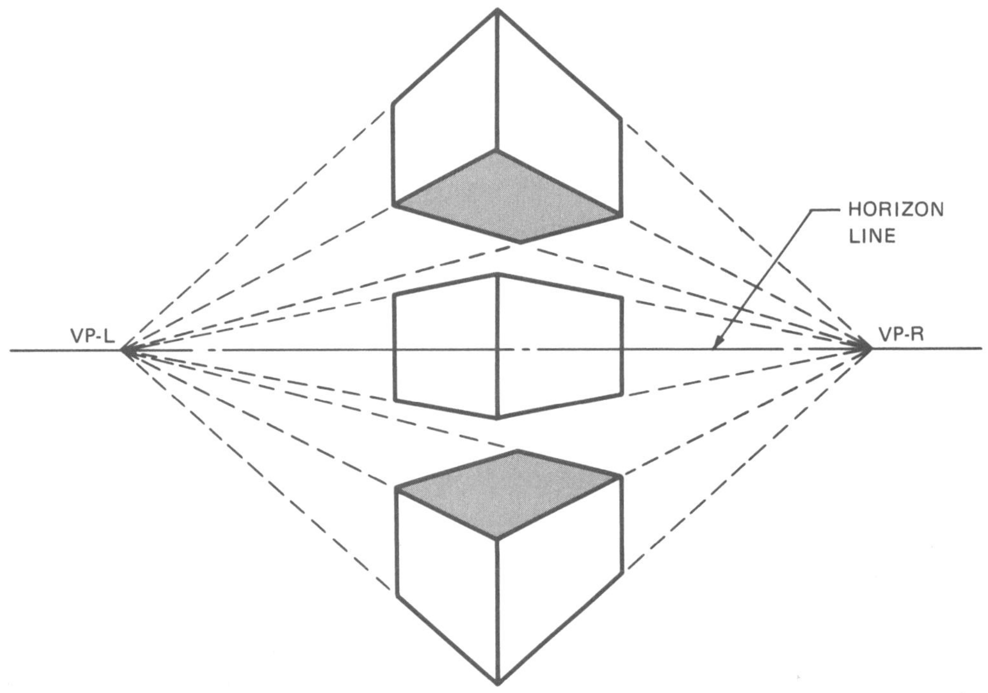

FIGURE 70-1 SHAPE OF FACES DEPENDS ON THE POSITION OF THE OBJECT TO THE HORIZON LINE

Cylindrical or curved profile features which are to be represented on a two- or three-point perspective drawing are drawn as ellipses and curved lines. These forms are produced by following steps which are similar to those described for oblique and isometric sketches.

SKETCHING IN CORRECT PROPORTIONS

Freehand sketching requires judgment about how the sizes of different features are to be correctly proportioned. Usually, an approximation by eye is adequate. If a grid paper is used, each

(A) LAY OUT HORIZON LINE	(B) DRAW VERTICAL EDGE	(C) LOCATE VANISHING POINTS	(D) MARK TRUE HEIGHT OF EDGE
(E) DRAW SIGHT LINES	(F) MARK OFF PART FEATURES	(G) DRAW FEATURES LIGHTLY	(H) DIMENSION AND DARKEN LINES

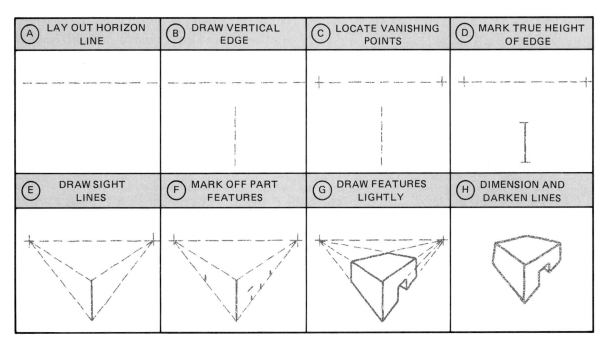

FIGURE 70-2 STEPS FOR PRODUCING A TWO-POINT (ANGULAR) PERSPECTIVE SKETCH

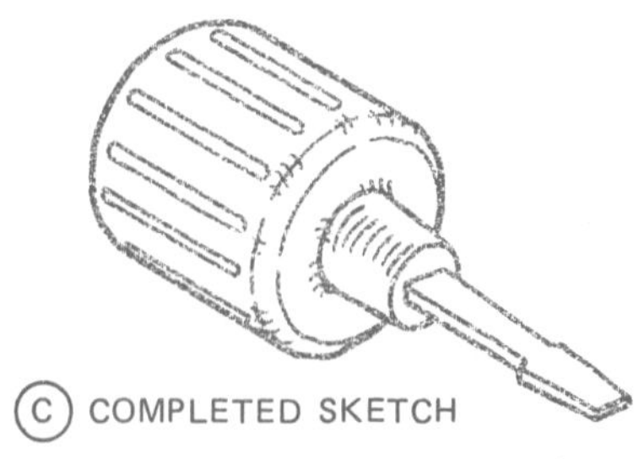

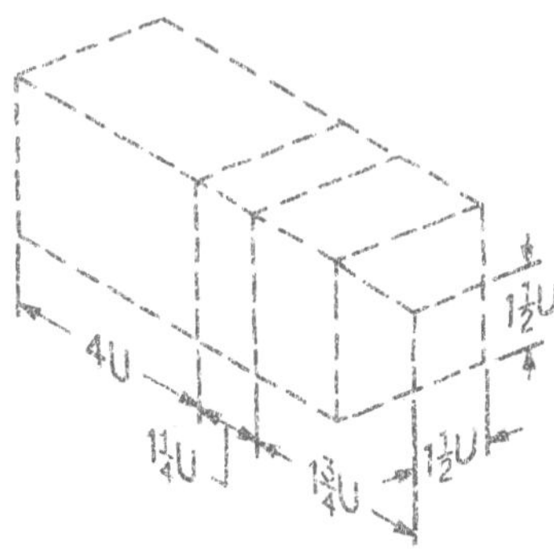

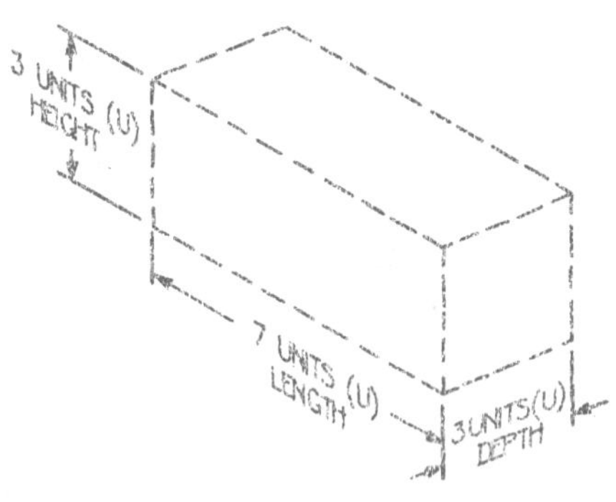

FIGURE 70-3 STEPS IN SKETCHING A STUB NOSE SCREWDRIVER IN CORRECT PROPORTION

block generally represents a definite measurement. Still more precise sketches may be made by the actual use of measuring instruments.

One simple technique of sketching a part in correct proportion is illustrated in figure 70-3. In this case, the handle diameter is equivalent to three units for height and depth and four units for length (width). The metal ferrule is proportionally 1 1/4 units wide, and the blade is 1 3/4 units long.

The three major steps in drawing this short blade screwdriver in proportion include:

- Blocking out the major sizes.
- Laying out the units for each detail.
- Roughing out and then completing the sketch.

SIMPLE SHADING OF CIRCULAR FEATURES

Drawings are *shaded* to bring out details about the shape of a part and to improve the appearance of a drawing. *Shading* is usually done by varying the spacing of straight and/or curved lines or by darkening the features in certain areas of a drawing. Typical examples of shading by these methods are pictured in figure 70-4 on a series of cutting tools.

Another technique of shading corners, fillets, and rounds is illustrated in figure 70-5. Guide (reference) lines are used as the boundary within which the curves for shading are contained. Note that the spacing between each shade line varies and the curved lines tend to straighten toward the center lines and the corners.

Two other methods of shading fillets and rounds are shown in figure 70-6. The simplest of the two

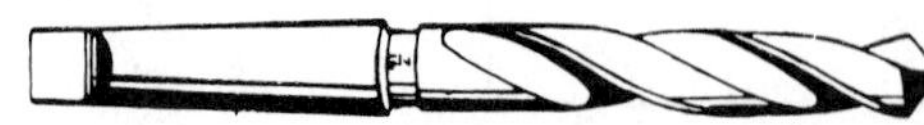

FIGURE 70-4 SHADING TECHNIQUES APPLIED TO REPRESENTATIVE CUTTING TOOLS

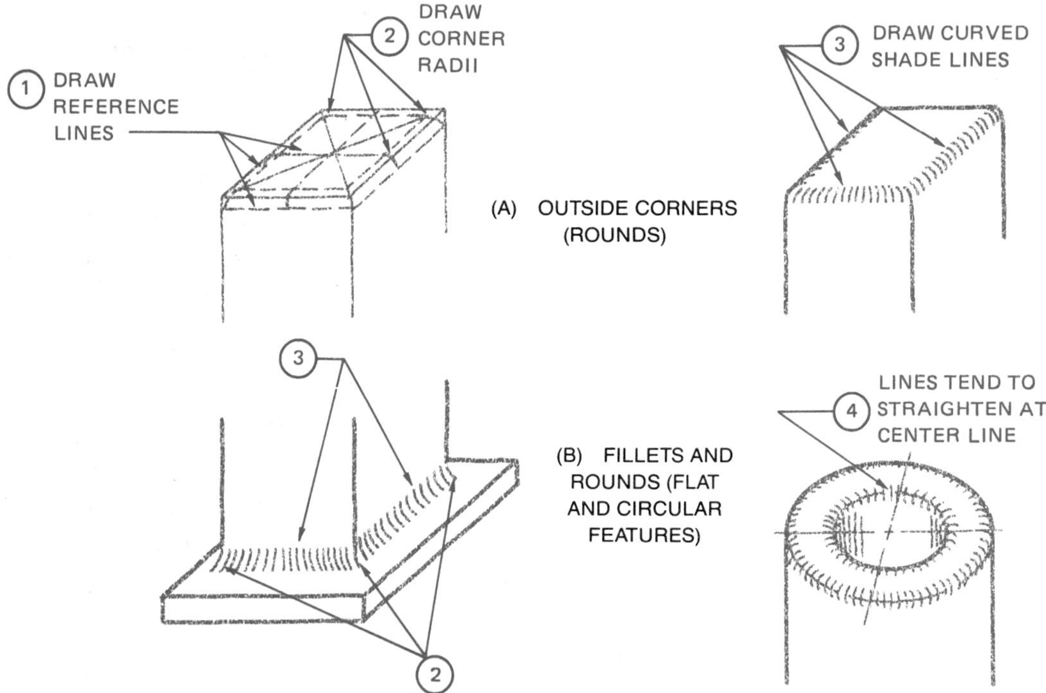

FIGURE 70-5 SHADING INSIDE AND OUTSIDE CURVED SURFACES

methods is to use a single broken line at the axis of the radius to denote a rounded surface (figure 70-6A). At B, three lines are drawn to depict the fillets and rounded edges. Two of the lines represent the starting or tangent points of a radius; the third line, the center line.

STIPPLE AND PENCIL-TONE SHADING

Stipple shading is another technique used on pictorial drawings to provide a shape description of a part or to show the relationship of many parts or components within an assembly unit.

Stipple shading consists of a series of dots. Varying degrees of darkness or lightness are created by increasing or decreasing the number of dots within a particular area of a part. The greater the number of dots the darker the area is shaded.

Pencil-tone or *smudge shading* is still another technique of highlighting specific features of a part. A pencil or charcoal stick is rubbed over the area to be shaded to create a smudged effect on vellum or paper.

ADHESIVE SHADING FILMS

A wide variety of shading patterns may be produced on drawings with commercially available *adhesive shading films.* The films permit a more uniform, consistent, standard shade pattern to be duplicated on a drawing.

As the name implies, shading films are made with the shade pattern on one side of a film; an adhesive on the other side. In practice, the shading film is cut to the approximate size and peeled off from the sheet. The shade film is then placed over the drawing and cut to conform to the required shape. The excess film is pulled away from the drawing. The film remaining on the shaded area is pressed securely to provide a strong bond to adhere to the drawing.

Adhesive shading films come in a variety of grid patterns and textures. The patterns vary from a number of dot sizes and densities to solid and dotted lines and squares to special shapes.

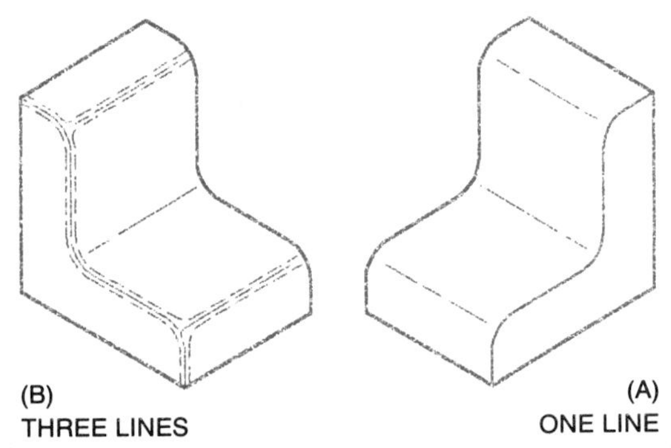

FIGURE 70-6 SHADING FILLETS AND ROUNDED EDGES

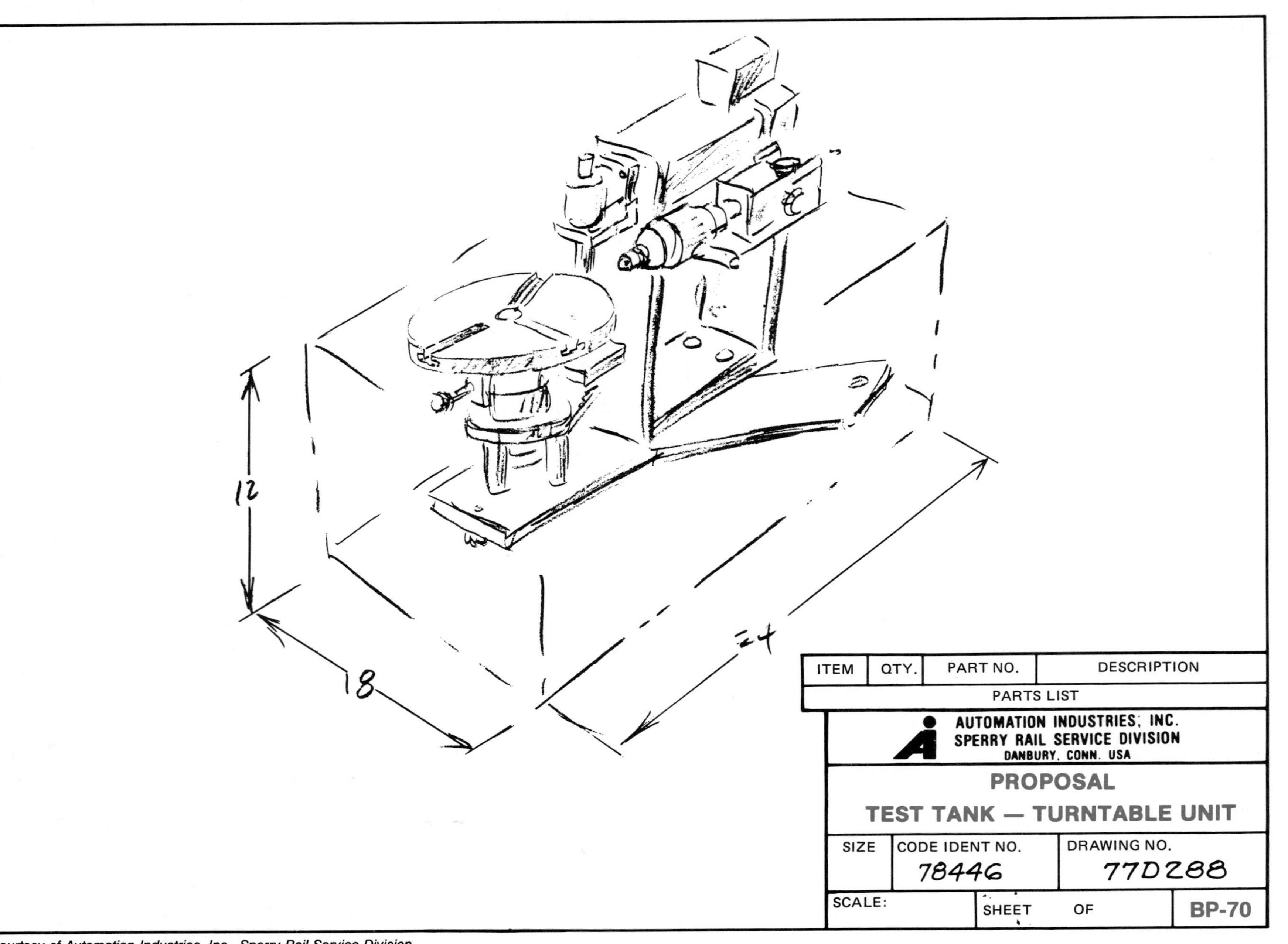

Courtesy of Automation Industries, Inc., Sperry Rail Service Division

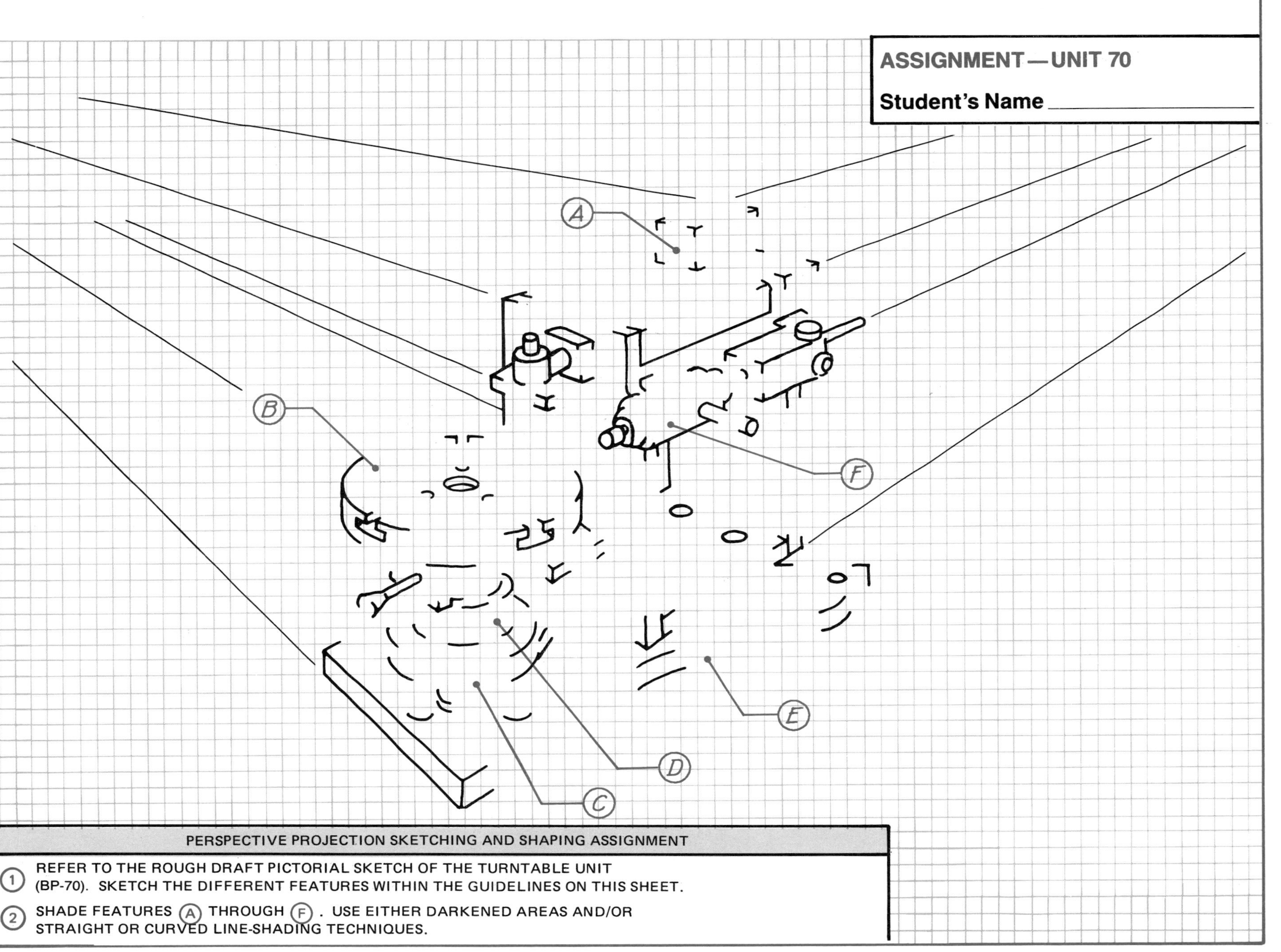
ASSIGNMENT—UNIT 70
Student's Name
A
B
C
D
E
F
PERSPECTIVE PROJECTION SKETCHING AND SHAPING ASSIGNMENT
(1) REFER TO THE ROUGH DRAFT PICTORIAL SKETCH OF THE TURNTABLE UNIT (BP-70). SKETCH THE DIFFERENT FEATURES WITHIN THE GUIDELINES ON THIS SHEET.
(2) SHADE FEATURES (A) THROUGH (F). USE EITHER DARKENED AREAS AND/OR STRAIGHT OR CURVED LINE-SHADING TECHNIQUES.

APPENDIX

Glossary of Technical Terms

Absolute System. A numerically controlled machining system. Measurement of all coordinates (point-to-point) from a fixed point of origin or zero point.

Accumulator. A container for storing a liquid under pressure as a power source.

Accuracy. A number of measurements which conform to a specific standard.

Addendum. Radial distance from the top of a gear tooth to the pitch circle.

Aligned Dimensioning. A method of dimensioning a drawing. All dimensions are placed to read from the bottom or right side of the sheet.

Allowance. A prescribed, intentional difference in the dimensions of mating parts. *Positive allowance* represents the minimum clearance between two mating parts. *Negative allowance* represents the maximum interference between parts.

Alphabet of Lines. A universally accepted standard set of lines used alone or in combination in making a technical drawing.

American System of Manufacturing. An economically sound, effective system of mass production of interchangeable parts.

Anneal. Application of heat to cause the internal structure of a metal to change. Heating a metal to remove internal stresses and to soften the metal.

ANSI (American National Standards Institute). A major professionally recognized body in the United States which establishes industrial standards. Represents the United States internationally in standards setting.

Anti-Friction Bearing Manufacturers Association Standards (AFBMA). Standards used in the production of interchangeable ball and roller bearings.

Assembly Drawing. A drawing containing all the parts of a structure or machine when it is assembled. Each part is drawn in the relative position in which it functions.

ASTM (American Society for Testing Materials). Largest professional body devoted to establishing standards for materials, products, systems, and services; for sampling and testing; and for studying effects.

Auxiliary View. The true projection of a surface which does not lie in either a frontal, horizontal, or profile plane. A *primary auxiliary view* lies in a plane which is parallel to the inclined surface from which features are projected in true size and shape.

Axis. An imaginary line which relates to particular features or around which they rotate.

Axonometric Drawing. A classification of pictorial drawing which includes isometric, dimetric, and trimetric projections. One-plane projection in which the drawing shows the object inclined at an angle to the frontal plane.

Base-Line Dimensioning. A system of dimensioning the features of a part from a common set of datums.

Basic Dimension. A theoretically exact dimension which is used to describe the size, shape, or location of a feature.

Basic Size. The exact theoretical size from which all limiting allowances and tolerances are made.

Bearings. A support member which permits movement between mating surfaces in three basic conditions: support of a rotating shaft or journal; support of axial loads on rotating members; and guidance of moving parts along a straight path. The two major classifications of bearings are *sliding contact* and *rolling contact.*

Bend Allowance. The amount of material required to form a sheet metal part to a specific radius.

Bevel Gears. Conical-shaped gears used to transmit power and motion between two shafts which are at an angle to each other. Two basic forms of bevel gears are *straight* and *hypoid tooth.*

Bilateral Tolerance. An equal or unequal tolerance permitted above or below a basic size.

Blanking. A production stamping operation requiring a punch and die set to stamp parts from flat sheets or rolls of material.

Boring. A machining process of forming a hole to a specified dimension by using a boring tool.

Boss. A raised area of a casting or forging for the purpose of increasing the thickness of the part.

Broaching. Removing material by pulling a preformed, stepped cutter through the workpiece to produce a desired shape.

Building Code. Legal guidelines for designing, fabricating, constructing, and maintaining structural, mechanical, plumbing, electrical, and other areas of a structure to ensure public protection.

Bushing. A lining which serves as a bearing between rotating parts. In other applications, particularly named bushings, like a *drill bushing,* may be used to guide a drill or other cutting tool.

Butt Joint. A square-cut joint formed at the junction of two members, either end-to-end or edge-to-edge.

Cam. A machine element which transmits rotating, oscillating, or reciprocal motion through another element known as a *follower.* The motion may be full cycle or interrupted for part of a cycle.

CAP (Computer-Assisted Program). A computer system using symbols and precise meaning English words and with computational capability to generate input for a machine control unit (MCU). A program which provides for such functions as automatic, multiple axis, fixed machining cycles, linear and circular coordinate data, and others.

Carburizing. Changing the properties of a low carbon steel by heating the part at a temperature where the outer case may absorb carbon from a carbonaceous material.

Casting. Forming an object by pouring molten metal, heated plastic, or other fluid resin into a preformed mold and cooling.

Chain Dimensioning. Successive dimensions which extend from one feature to another. Tolerances accumulate unless a note indicates, "Tolerances are not cumulative."

Chamfer. A bevel or angle to which an external edge or corner is machined.

Circuit. A closed path through which electrical current flows from and back to the point of generation (generator).

Circuit Board. A panel or plate or other backing on which components are mounted and interconnected to produce a functional unit.

Circuit Diagram. A line drawing which uses graphic symbols to show the complete path of flow in a hydraulic, pneumatic, fluidic, or electronic system.

Circular Pitch. The length of the arc between the center of one gear tooth to the center of the next tooth, measured along the pitch circle.

Coil Spring. A spring steel (or other spring material) wire wound into a continuous spiral. Identified by function as *compression, tension,* or *torsion.*

Cold Molding. The process of shaping a plastic composition at room temperature and curing it by subsequent baking.

Color Coding (Fluid Power). Use of different colors on diagrams to denote piping, flow paths, and other functions.

Combination Diagram (Fluid Power). Uses interconnecting lines with graphic, pictorial, and cutaway symbols. Drawing that displays information about functions of components, essential piping, and flow paths.

Component Drawing (Electronics). A drawing which shows the top side of a circuit board with the location of each component. Used in assembling parts of a circuit layout.

Composite Drawing. A single drawing which provides full information for multiple production processes, such as forging, casting, and machining an object. The outline of the finished part is usually represented by solid lines. The original forging or casting is represented by phantom lines.

Compression Molding. The application of heat and pressure to a molding compound that is placed in an open mold cavity, closing the mold, applying heat and pressure, and cooling.

Computer. An electronic device capable of making logical decisions according to programmed control input.

Computer-Aided Design (CAD). The interlocking of functions, products, and systems of individual computers (software and hardware) into hierarchical order applications. Control of design, processes, and products by a hierarchical order of computer input.

Computer-Aided (Complete) Manufacturing (CAM). Application of NC or CAD to all facets of manufacturing inclusive of and ranging from product/process design, production, inspection and assembly, marketing, etc., to plant management.

Computer Compensation Feature. A compensation feature of an NC machine which permits command adjustments for variations of workpieces, cutter diameter, and tool length.

Concentric. Part features having a common center.

Conversion Tables. Cross-referenced equivalent mathematical values in two or more systems of measurement. Tables used to convert from U.S. customary units to SI and metric values, and vice versa.

Coordinate Dimensioning. Rectangular datum dimensioning in which all dimensions are referenced from two or three mutually perpendicular datum planes. Dimensions are represented by using regular extension and dimension lines and arrowheads.

Counterbored (C'BORED) Hole. A hole that has been cylindrically enlarged and recessed to a specific depth. A recessed hole with a flat machined surface, usually for seating a bolt or a nut.

Countersunk Hole (CSK). A hole which is machined with a cone-shaped opening to a specified outside diameter and depth. A conical recess which provides an angular seat for flat head screws and other cone-shaped objects.

Cutaway Diagram (Fluid Power). Drawing that shows the internal details of working parts, component functions, and flow paths.

Cutting Plane Line. A thick, broken line representing an imaginary plane cutting through an object. The resulting view is a *section view*.

Cycle. Complete operation or process consisting of a progressive sequence of steps that begin and end at a fixed neutral position.

Cylinder (Fluid Power). A mechanical device consisting of a cylinder bore with end cap enclosures and a movable element. A device for converting a fluid power (force) into a mechanical linear force.

Datum. An exact reference point from which a line, plane, or feature dimension is taken or geometric relationship is established.

Decimal-Inch System of Dimensioning. A system in which dimensions are given with subdivisions of the inch in multiples of ten; like, 0.1″, 0.01″, and 0.001″.

Detail Drawing. A drawing which includes all views needed to accurately describe a part and its features. Dimensions, notes, materials, heat treatment, quantity, etc. are included.

Diametral Pitch. The number of gear teeth for each inch of pitch diameter.

Die Casting. A production method of casting metal parts by flowing a low melting point die cast metal into a preformed mold.

Die-Cutting and Forming Processes. Production of parts by blanking, piercing, lancing, bending, forming, cutting off, shaving, trimming, and drawing.

Digital. The use of signals produced by one current or voltage value to represent characters or numbers.

Digital Computer. An electronic device for performing high-speed mathematical operations and data storage by representing and processing information in the form of digit combinations in the binary system.

Dimension. Size and location measurements which appear on a drawing.

Dimensional Measurements. Measurements relating to plane surfaces, multiple surface objects, straight and curved lines, areas, angles, and volumes.

Displacement Diagram. A visual representation of cam motion. A diagram which provides information about the kind of motion and the distance a cam causes the follower to rise and fall during the working cycle. *Displacement* is the distance a cam moves a follower.

Dowel Pin. A precision-ground pin which accurately fits into a hole to ensure accurate location and to secure parts in a fixed relationship.

Draft. An angle or taper on a pattern, casting, or die. An angle which permits easy removal of a part from a mold or a stamping die.

Drawing Callout. A note on a drawing which provides a dimension, specification, or other information about a feature or process.

Dual Dimensioning. A technique of giving all dimensions on a drawing (using soft conversion dimensions) in both customary inch and SI metric units of measurement.

Eccentric, Eccentricity. A surface deviation from a center. Off center.

Electronic Rectifier. A rectifier consisting of tubes or semiconductor devices for purposes of changing alternating current to direct current.

Electronics. A branch of electricity dealing with characteristics and applications of tubes and semiconductor devices and circuitry.

Elevation. The height of one point in relation to another specific point. Exterior views on a drawing of a structure.

Elevation Drawing. A view, generally used in architecture, to define the architectural style and structure in a frontal or profile plane.

Engineering Work Order. Instructions which give permanent change information from an original drawing. A drawing change may be called a *revision*, *change notice*, or *alteration*. Information recorded in the *change block* area of the title block.

Exploded Drawing. A pictorial drawing showing each part of an assembled unit in its location in the order of assembly.

Extrusion. The process of continuous compacting and forcing of a plastic material through an orifice or dies of a particular form.

Fabrication. Layout and assembly of parts of a structure away from a job site.

Fastener. A mechanical device for securely holding two or more bodies in a fixed position.

Feature. The geometric form of a portion of a part which identifies particular characteristics at that location.

Feature Control Symbols. A simplified form of a geometric characteristic(s) combined with tolerance information which appears on blueprints in an enclosed rectangular box.

Fillet. A concave area between two reference surfaces.

First-Angle Projection. The SI (ISO) metric standard of projection from quadrant I. Each view is drawn as projected *through* the object.

First-Angle Projection Symbol. The standard SI symbol ( ) indicating that the drawing is made by the metric first angle projection method as opposed to third angle projection.

Fit. The relationship between two mating parts produced by interference or clearance.

Fixture. A device used to accurately position and to hold a piece part for subsequent machining processes.

Flange. An edge, rim, or collar, usually at a right angle to the body of the part.

Flash. Excess plastic that forms at the parting line and generally requires removal to produce the finished part.

Flow Sheet (Piping). Single-line, schematic piping drawings. Shows the direction of flow of a liquid or gas and the sequence and function of each part in the system.

Fluidics (Fluid Power). The technology and application of scientific principles relating to hydraulics (liquids) and pneumatics (gases); pressure and flow.

Fluid Power. Controlled transmission of energy produced through the use of a pressurized fluid.

Fluid Power System. An enclosed circuit for the transmission and control of a pressurized fluid (liquid or gas).

Forging. The forming of heated metals by hammering between forging dies (or rolling).

Form Tolerancing. The permitted variation of a part feature from the exact form identified on the drawing.

Foundation. Footing of a building that supports structures and loads and consists of walls, footings, and piers.

Fusion Welding. The joining at a particular location of metals which are brought up to fusion temperatures.

Gaging. A process of measuring manufactured products to assure specified uniformity of size and other geometric features, as required.

Geometric Control Symbols. ANSI geometric characteristic symbols grouped according to *form tolerance*, *location tolerance*, and *runout tolerance*.

Geometric Tolerancing. Specified tolerances which apply to shape, form, or position of specific features of a part.

Glue-Laminated Timber (Glu-Lam). Multiple layers of lumber that are glued together to form a structural member.

Graphic Diagram (Fluid Power). A drafting technique using simple line graphic shapes (symbols) with interconnecting lines to represent the function of each component in a circuit. Functional for designing and troubleshooting fluid power circuits.

Hard Conversion. Features of an object designed to conform to standard sizes in the other system of measurement (SI metric or customary inch). Objects designed by hard conversion are *not interchangeable* between two measurement systems.

Hardness Testing. Measurement of the degree of hardness of heat treated and other materials.

Harmonic Curve Motion. Acceleration and deceleration at a constantly changing speed following the motion of the driver (usually a cam).

Helical Gear. A gear used to transmit motion and/or load between two shafts. Generally designed for use on parallel shafts at 45° or 90° angles. The teeth are designed to gradually overlap to permit continuous engagement action.

Honing. A machine process for precision finishing a hole or other surface by moving an abrasive material, hone, and the workpiece together.

Hydraulic Power Unit. Components of a fluid power system that facilitates fluid storage and conditioning and fluid delivery under controlled pressure and flow.

I-Beam. An American Standard steel beam with a wide flange and an I-shape cross section.

Incremental System. A numerically controlled machining system which references each movement to the preceding point. The system is also known as the *continuous path* or *contouring* method of NC machining.

Injection Molding. The process of forcing a heat-softened plastic to flow from a cylinder into a cool mold cavity to produce a desired part.

Interference (Plastics). Planned negative allowance on mating parts to ensure a shrink or press fit.

Involute. A spiral curve with specific geometric characteristics which are generated by a point on a chord as it is unwound from a circle or a regular-sided part.

Isometric Drawing. A pictorial drawing of an object with all three axes making equal (120°) angles with the picture plane. Measurements on each axis are made to the same scale.

Jig. A work-holding and tool-positioning and guiding device.

Key. A fastener used between two parts to prevent them from turning and to permit assembly and disassembly.

Keyslot. A slot machined in a shaft to receive a Woodruff (semi-circular type) key.

Keyway. A groove cut into the hub area of a mating part to receive that portion of a key which extends beyond the shaft.

Knurled Surface. A diamond or straight-line pattern which is impressed by hardened, formed steel rolls into the outer surface of a round part.

Laminated Plastics. Bonding layers of sheet plastics that are impregnated with a resin and curing the laminated product by applying heat and/or pressure to produce a dense, tough, strong plastic.

Lapping. A final surface-finishing process produced by the movement of a lap, lubricant, and abrasive on or in the part to be finished.

Lay. The direction of the dominant surface pattern of the cutting or forming tool in the manufacture of a part.

Limits. The maximum and minimum or extreme permissible sizes of acceptance of a part.

Location Dimension. A dimension which identifies the position of a surface, line, or feature in relation to other features.

Logical Diagram (Digital-Computer). A visual (schematic) diagram that represents the interconnection between gates of a logic circuit.

Machine Actuators. Devices that transmit movement from a driver to a movable part. Gears and cams are examples of actuators which create three common motions: *constant velocity, parabolic,* and *harmonic.*

Maximum Material Condition (MMC). A condition where a part feature contains the maximum amount of material; for example, minimum hole diameter and maximum diameter of the mating shaft.

Measured Point. The specific point at which each measurement ends.

Measurement Conversion (Fluid Power). A process for changing torque, linear, volume, force, velocity, and other measurements from one measurement system to another. Interchanging values among SI, metric, and U.S. customary units of measurement.

Metrication. The international movement toward the universal adoption of the SI metric system.

Milling. A material removal process requiring a rotating cutting tool. Machining process performed on the milling machine.

Motor (Fluid Power). A device for delivering torque and rotary motion to convert fluid power to mechanical power.

Multiview Projection. The projection of two or more views of an object upon the picture plane in orthographic projection.

Nominal Size. A size which represents a particular dimension or unit of length without specific limits of accuracy.

Nominal Surface. A surface which is theoretically geometrically perfect. Surface texture deviations are measured from the nominal surface.

Normalizing. A process in which metals are heated and then cooled under controlled conditions to produce a uniform grain structure free from the strains produced by machining, hammering, drawing, or other processes.

Notes. A supplemental message found on a drawing which provides technical information. *General notes* relate to the entire drawing. *Local notes* give processing information relating to a specific area of the object.

Numerical Control System Functions. Basic NC system functions include controlling machine and tool positions, speeds and feeds, shutdown, and recycling; establishing operation sequences; monitoring tool performance, and producing readouts.

Oblique Drawing. A pictorial drawing of an object which has one of its principal faces parallel to the plane of projection. Features on this face are seen and projected in true size and shape. The features on other planes are drawn oblique to the plane of projection at a selected angle.

Offset Section. A type of section view used to show features which are not aligned. The cutting plane is shifted at one or more places to cut through a desired detail.

Ordinate Dimensioning. An arrowless rectangular datum dimensioning system. Dimensions originate from two or three mutually related datums or zero coordinates.

Orthographic Projection. A standardized method of representing an object with details by projecting straight lines perpendicular from the object to two or more planes. Third-angle orthographic drawings are standard in the United States and Canada.

Parabolic Curve Motion. The acceleration and deceleration of a driven object (like a cam follower) at a constant speed following the curve motion of the driver (cam).

Part Programming. Producing complete input information to machine a part by numerical control or to develop precision art work to produce a printed circuit board.

Perspective Drawing. A pictorial drawing on which the features of an object appear to converge in the distance toward one, two, or three vanishing points.

Phantom Section. A hidden section used to represent interior and exterior features of an object in one view. Using thin, evenly spaced broken lines for a sectional view which is superimposed on a regular view.

Pictorial Drawing. A one-view drawing showing two or more faces of an object. An easily understood drawing which provides a quick visual image of an object. The three major types of pictorial drawing are *axonometric, oblique,* and *perspective.*

Pictorial Drawing (Fluid Power). Representing each component of a piping system by using single or double lines. Operating functions are not shown.

Pilot. The end of a cutting tool which is designed to fit into a hole to guide the cutting tool portion to machine a particular feature.

Pitch. The linear distance from a point on one thread to a corresponding point on the next thread. The slope (angle) of a surface or roof.

Plane of Projection. An imaginary plane placed between the observer and the object. Assumed to be perpendicular to the line of sight.

Plan View. The top view of an object as generally used for architectural drawings.

Plastic (Noun). A rigid or flexible solid of synthetic or semisynthetic substances that can be formed into desired shapes by applying heat and pressure; retaining the shapes after cooling.

Positional Tolerancing. The extent to which a feature is permitted to vary from the true or exact position indicated on the drawing.

Postprocessor Statements. Preparation of a sequence of words which in a CNC setup control the machine tool and cutting tool positions, machining setups, and machining processes.

Precision. A stated quality of required accuracy or mechanical exactness.

Prefabricated/Precast. Structural components that are laid out and assembled off the job site and transported for erection on-site.

Pressure Differential. A pressure difference between two points of a component or system.

Prestressed. Applying compression forces to a concrete component during casting for purposes of resisting deflection.

Print. A copy of an original technical drawing. Generally called a *blueprint* regardless of the type or reproduction process.

Printed Circuit. A layout of conductors on an insulated board by use of pressure-sensitive preforms, by photo-etching, or other processes.

Printed Circuit Template. A circuit drawing used to produce printed-circuit boards photographically.

Process Specification. Complete information about the exact procedures, materials, equipment, etc. required to perform a single or series of operations.

Profile. The shape or contour of a surface. The profile is viewed on a plane which is perpendicular to a machined surface.

Program (Computer). Step-by-step instructions to a computer for solving a problem using input information from manual operation or as contained in computer memory.

Progressive Die. A punch and die set by means of which two or more power press operations are performed for each stroke.

Projecting. Extending from one reference point to another.

Proportion. Size and form relationships between features and parts.

Pump (Fixed Displacement). An energy conversion device in which the displacement per cycle is constant.

Pump (Fluid Power). A device for converting mechanical energy into physical energy by causing a liquid to flow against a pressure.

Quenching. The rapid cooling, generally of heat treated metal parts, by immersion in a liquid bath.

Reaming. A machining process for finishing a hole to close dimensional and geometric tolerancing standards.

Rectangular (Cartesian) Coordinate System. A four-quadrant (X, Y, and Z axis) system with vertical and horizontal planes. A system of representing + and − values to control NC machine tools and processes.

Rectifier (rect). A tube, electromechanical device, or semiconductor used in changing ac to dc.

Reference Dimension (REF). Dimension given on a drawing to provide general information. Not used in the manufacture of a part.

Reference Point. The specific point or plane at which each measurement of a dimension begins.

Regardless of Feature Size (RFS). A condition of conforming to the tolerance of position or form regardless of the feature size within its tolerance.

Reinforced Concrete. Concrete with structural steel reinforcement to resist tension forces.

Resistance Welding. A metal-joining process which depends on the resistance of a metal to the flow of electricity to produce the heat for fusing the metal parts.

Rib (Plastics). A design feature (reinforcing member) to strengthen a fabricated or molded part.

Roughness. The fine surface irregularities produced by cutting tools, feed marks, or other production processes.

Scale. The outer rough surface produced by forging or sand casting. Also, the ratio of a drawing size to the actual size of the object.

Schematic. A diagram of a pneumatic, fluidic, electronic, or other circuit showing connections and flow patterns and the identification of major components.

Section (Building Construction). A drawing technique showing a structure or component as if it had been cut through to expose interior construction features.

Sectional View. A view produced by passing an imaginary cutting plane into an object at a selected location, removing the cut-away section, and exposing the internal details.

Serrations. A surface having a series of notches, teeth, or other machined indentations.

Shrinkage Allowance. Compensation for the amount a metal shrinks when changing from a liquid to solid state.

Shrink Fit (Plastics). A planned dimensional allowance between two parts. A heating (expansion) and cooling (shrinking) process of joining a plastic and an insert.

SI Metric System. The internationally accepted decimal system of measurements built upon seven basic units of measurement. The SI metric system has been adopted by the ISO as the basis for its drafting standards.

Soft Conversion. Direct conversion of a measurement in one system to its equivalent in the other system. Provides equivalent measurements in two systems without regard to the standardization of parts in both systems.

Specifications. Stated requirements (standards) that must be satisfied by a product, material, process, structure, component, or system.

Spline. A series of uniform, parallel surfaces (teeth) which are machined in a shaft or a hub or other mating part. A machine element which permits transmitting heavy loads between mating parts without slippage, sliding a member along a shaft without changing position, and positioning parts accurately.

Spotfacing. Producing a flat bearing surface over a small area of a part for purposes of seating a bolt or shoulder of a shaft.

Stretchout. A flat pattern development which, when formed, produces a sheet metal product of a particular size and shape.

Subfloor. A base flooring surface layer for a finished floor which is laid on the floor joist.

Surface Texture. The designation of the characteristics and conditions of a surface. Deviations from a nominal profile which forms the pattern of a surface. Deviations produced by *roughness, waviness, lay*, and *flaws*.

Symbol. A letter, numeral, character, or geometric design which is accepted as a standard to represent a feature, operation, part, component, etc.

Symmetrical (SYM). A condition where features on both sides of a center line are of equal form, dimensional size, and surface quality.

Tabular Dimensioning. An ordinate dimensioning system in which all dimensions are taken from datums. Each dimension on a drawing is identified by a reference letter with a corresponding numerical value found in a table or chart.

Taper. The uniform change in size along the length of a cylindrical part, as in the case of a taper shank; or on a flat surface like a wedge-shaped key.

Tapping. Forming a uniform internal thread using a hand or machine tap. Threading according to ANSI, Unified, and SI metric standards for thread form and size.

Template. An accurately formed pattern which is used as a guide or checking gage.

Thermoforming. Forming thermoplastic sheet by heating and pulling the sheet down onto a mold surface.

Thermoplastic. A polymer capable of being repeatedly softened and hardened by heating and cooling.

Thread Class Number. A number appearing on a drawing with a thread note to indicate the required degree of accuracy between mating features of an external and internal thread.

Thread Form. The shape or profile of a thread. The most common thread form is based on a 60° included angle. American National, Unified, and SI metric threads use this angular form.

Title Block. An area on a drawing containing details and specifications complementing the information given on the drawing. An information block usually located at the bottom right corner of a drawing.

Tolerance. The total permissible variation in the size of a part. The dimensional range within which a feature or part will perform a required function. The difference between the upper and lower limits of a dimension.

Tolerances. Maximum and minimum dimensions for machining.

Transfer Molding. A molding process consisting of softening a plastic by heat and pressure in a transfer chamber and forcing it to flow under high pressure through sprues, runners, and gates in a closed mold; and cooling.

Transformer. A device for transferring electrical energy from one circuit to another by electromagnetic induction.

Transistor. A semiconductor device (generally three-terminal) used for purposes of amplification, oscillation, and switching action.

True Position. The theoretically exact reference location of a feature.

Truncated. The form of an object having part of the top (vertex) cut through by a plane.

Typical (TYP). A dimension or feature which applies to the locations where the dimension or feature is identical, unless otherwise noted.

Undimensioned Drawings. A technique used to produce parts and components which have a continuously changing profile where the use of conventional dimensioning is impractical. Undimensioned parts (particularly in the aerospace industry) are often NC machined by controlling input information to the machine control unit (MCU).

Unidirectional Dimensions. Dimensions placed to read in one direction.

Unidirectional Tolerance. A single-direction tolerance which is specified as either above (+) or below (−) a basic size.

Viscosity. A measure of the internal friction caused as one layer of fluid moves in relation to another layer.

Welding Symbol. A standard designation consisting of a reference line, arrow, and tail. Other symbols, code letters, dimensions, and notes are added.

Working Drawing. A multiview orthographic drawing which provides details for the design, manufacture, construction, or assembly of a part, machine, or structure.

Worm Gearing. Two gears used as a set. A *worm* with teeth similar to those of an Acme screw thread serves as a driver. The worm is mated to a *worm gear*. Worm gearing is used to transmit power efficiently and to obtain a large reduction in velocity between two nonintersecting shafts.

Standard Abbreviations Used on Drawings

Actual	ACTL
Addendum	ADD
Adjust	ADJ
Advance	ADV
Alignment	ALIGN
Allowance	ALLOW
Alteration	ALTRN
Alternating Current	AC
Aluminum	AL
American National Standards Institute	ANSI
Amplifier	AMPL
Anneal	ANL
Anodize	ANDZ
Approved	APVD
Approximate	APPROX
Architectural	ARCH
Arrange	ARR
As Required	AR
Assemble	ASSEM
ASTM Building Code	BC
Automatic	AUTO
Automatic Frequency Control	AFC
Auxiliary	AUX
Average	AVG
Base Line	BL
Battery	BTRY
Bearing	BRG
Bend Radius	BR
Bevel	BEV
Bill of Material	B/M
Bolt Circle	BC
Bracket	BRKT
Brass	BRS
Brinnell Hardness Number	BHN
Bronze	BRZ
Brown & Sharpe (Gages)	B&S
Bushing	BSHG
Cabinet	CAB
Canadian Standards Association	CSA
Capacitor	CAP
Carburize	CARB
Case Harden	CH
Cast Concrete	C.CONC
Casting	CSTG
Cast Iron	CI
Cathode Ray Tube	CRT
Center Line	℄
Center to Center	C TO C
Centigrade (or Celsius)	°C
Centimeter	cm
Centrifugal	CNTFGL
Chamfer	CHAM
Check Valve	CV
Circuit	CIR
Circuit Breaker	CB
Circular	CIRC
Circumference	CRCMF
Clearance	CL
Clockwise	CW
Cold-drawn Steel	CDS
Cold-rolled Steel	CRS
Column	COL
Computer-Aided Drawing	CAD
Computer-Aided Engineering	CAE
Computer-Aided Manufacturing	CAM
Concentric	CNCTRC
Condition	CONDTN
Construction Standards Institute	CSI
Contour	CTR
Counterbore	CBORE
Counterclockwise	CCW
Countersink	CSK
Coupling	CPLG
Cross Section	S
Cube	$(^3)$
Cylinder	CYL
Datum	DAT ISO [A] ANSI [-A-]
Decimal	DEC
Degree (angle)	DEG (°)
Detail	DET
Deviation	DEV
Diagonal (or Diagram)	DIAG
Diameter	(ϕ) or DIA
Diameter Bolt Circle	DBC
Diametral Pitch	DP
Dimension	DIM
Direct Current	DC

Double-pole, Double-throw	DPDT
Dowel	DWL
Draft	DFT
Drawing	DWG
Drawing Change Notice	DCN
Drill	DR
Drop Forge	DF
Eccentric	ECC
Effective	EFF
Electrical	ELECT
Elevation	ELEV
Enclosure	ENCL
Engineering	ENGRG
Engineering Change Order	ECO
Equal	EQL
Equivalent	EQUIV
Estimate	EST
Fabricate	FAB
Fillet	FIL
Finish	FNSH
Finish All Over	FAO
Flange	FLG
Flat Head	FLH
Flexible	FLEX
Floor Plan	FLPL
Fluid	FL
Forging	FORG
Foundation Plan	FDN
Gage	GA
Galvanized	GALV
Gasket	GSKT
Generator	GEN
Girder	G
Glued-Laminated Timber	GLU-LAM
Grind (or Ground)	GRD
Harden	HDN
Head	HD
Heat Treat	HT TR
Hexagon	HEX
High Carbon Steel	HCS
High Frequency	HF
High Speed	HS
High Speed Steel	HSS
Horizontal	HOR or HORIZ
Hot-rolled Steel	HRS
Housing	HSG
Hydraulic	HYDR
Hydrostatic	HYDRST
Identification	IDENT
Impregnate	IMPRG
Inclined	INCLN
Independent	INDEP
Indicator	IND
Information	INFO
Inside Diameter	ID
Installation	INSTL
Integrated Circuit or Gate	IC
International Organization for Standardization	ISO
Interrupt	INTRPT
Joggle	JOG
Junction	JCT
Keyway	KWY
Laminate	LAM
Left Hand	LH
Letter	LTR
Light Emitting Diode	LED
Limit Switch	LIM SW
Linear	LIN
Liquid	LIQ
Liquid Crystal Display	LCD
Low Carbon	LC
Low Frequency	LF
Low Voltage	LV
Lubricate	LUB
Lumber	LBR
Machine Control Unit (Computerized)	MCU
Machine (ing or ed)	√ or MACH
Magnaflux	M
Maintenance	MAINT
Major	MAJ
Malleable Iron	MI
Manufacturing	MFG
Master Switch	MSW
Material	MATL
Maximum	MAX
Maximum Material Condition	MMC
Medium	MDM
Meter	M or m
Millimeter	mm
Minimum	MIN
Minute (Angle)	(′)
Miscellaneous	MISC
Modification	MOD
Mold Line	ML
Motor	MOT
Mounting	MTG
Multiple	MULT
Nominal	NOM
Normalize	NORM
Not to Scale	NTS
N-Type Semiconductor	NPN
Number	NO
Obsolete	OBS
Opposite	OPP
Oscillator	OSC
Outside Diameter	OD
Overall	OA
Package	PKG
Parallel	//
Parting Line (castings)	PL
Pattern	PATT
Perpendicular	⊥
Piece	PC

Pilot .... PLT
Pitch .... P
Pitch Circle .... PC
Pitch Diameter .... PD
Plan View .... PV
Plot Plan .... PL
Plumbing .... PLMB
Pounds per Square Inch .... PSI
Power Amplifier .... PA
Power Supply .... PWR SPLY
Prefabricated .... PRE FAB
Preferred .... PFD
Pressure .... PRESS
Printed Circuit .... PC
P-Type Semiconductor .... PNP

Quality .... QUAL
Quantity .... QTY

Radius .... RAD or R
Ream .... RM
Receptacle .... RCPT
Reference (or Reference Dimension) .... REF
Regardless of Feature Size .... RFS
Reinforced .... REINF
Release .... RLSE
Required .... REQD
Revision .... REV
Revolutions per Minute .... RPM
Right Hand .... RH
Rivet .... RVT
Rockwell Hardness .... RH

Schedule .... SCH
Schematic .... SCHEM
Screw Threads
American National Coarse .... NC
American National Extra Fine .... NEF
American National Fine .... NF
American National Pitch Series ..... N
American Standard Straight (Dryseal) .... NPSF
American Standard Straight Pipe ..... NPSC
American Standard Taper (Dryseal) .... NPTF
American Standard Taper Pipe .... NTP
Unified Screw Thread Coarse .... UNC
Unified Screw Thread Extra Fine ...... UNEF
Unified Screw Thread Fine .... UNF
Unified Screw Thread; (8) Pitch 8UN
Second (Arc) .... (″)
Section .... SECT
Sequence .... SEQ
Serial .... SER
Serrate .... SERR
Sheathing .... SHTHG
Single-pole, Double-throw .... SPDT
Slotted .... SLTD
Society of Automotive Engineers .... SAE
Special .... SPCL
Specification .... SPEC
Spotface .... SFACE (old) or SF

Spring .... SPR
Square .... SQ
Stainless Steel .... SST
Standard .... STD
Steel .... STL
Stock .... STK
Structure .... STR
Surface .... SUR
Switch .... SW
Symmetrical .... SYMM
System .... SYS

Tabulate .... TAB
Tangent .... TAN
Tapping .... TPG
Technical Manual .... TM
Tee .... ⊤
Teeth .... T
Temper .... TEM
Tensile Strength .... TS
Thick .... THK
Thread .... THD
Through .... THRU
Tolerance .... TOL
Tool Steel .... TS
Torque .... TRQ
Total Indicator Reading .... TIR
Transformer .... XFMR
Transistor .... XSTR
Transmitter .... XMTR
Triode (AC) .... TRIAC
True Involute Form .... TIF
True Profile .... TP
Tubing .... ◎ or TUB
Typical .... TYP

Ultra-high Frequency .... UHF
Unfinished .... UNFIN
Unit .... U
Universal .... UNIV
Unless Otherwise Specified .... UOS

Valve V
Variable .... VAR
Vernier .... VERN
Vertical .... VERT
Very-high Frequency .... VHF
Vibrate .... VIB
Vitreous .... VIT
Volt V

Watt .... W
Watthour .... Wh
Wattmeter .... WM
Weight .... WT
Welded Wire Fabric .... WWF
Wide .... W
Width .... WD
Wire Wound .... WW
Wrought Iron .... WI

Yard .... YD
Yield Point (PSI) .... YP
Yield Strength (PSI) .... YS

Standards, Codes, Symbols: Reference Sources

The interchangeability of parts, components, structures, and systems depends upon the uniform application of standards, codes, and symbols. These are used in every facet of design and drafting, layout, production, fabrication, manufacturing, inspection and testing, installation, and maintenance.

A selected number of professional organizations and bodies who are involved in preparing standards, codes, and symbols and overseeing their universal adoption and use follow. The list begins with the American National Standards Institute (ANSI) and the American Society of Mechanical Engineers (ASME). Some of the most widely used standards are given. Where applicable, the last date of a revised standard and/or the most recent data that a standard was *reaffirmed* (R) is provided.

- American National Standards Institute (ANSI)
  1430 Broadway New York, NY 10018
- American Society of Mechanical Engineers (ASME)
  22 Law Drive Fairfield, NJ 07007-2300

Drafting and Drawing Standards and Terminology

Y 14.2M–1979 (R 1987) Line Conventions and Lettering

Y 14.3–1975 (R 1987)
Multi- and Sectional View Drawings

Y 14.4M–1989
Pictorial Drawings

Y 14.8M–1989
Castings and Forgings

Y 14.15–1966 (R 1988)
Electronics and Electrical Diagrams

Y 14.17–1966 (R 1987)
Fluid Power Diagrams

Y 14.36–1978 (R 1987)
Surface Texture Symbols

Graphics Symbols Standards

Y 32.2–1975 and 315 A–1986 Supplement Symbols for Electronic and Electrical Diagrams

Y 32.2.4–1949 (R 1984)
Symbols for Heating, Ventilating, and Air Conditioning

Y 32.4–1977 (R 1987)
Graphic Symbols: Diagrams in Architecture and Building Construction for Plumbing Fixtures

Y 32.10–1967 (R 1987)
Graphic Symbols for Fluid Power Diagrams

Fasteners: Metric

B 18.2.3.2M–1979 (R 1989)
Metric Formed Hexagon Screws

B 18.6.7M–1985
Metric Machine Screws

B 18.12–1962 (R 1981)
Glossary of Terms for Mechanical Fasteners

Fasteners: Customary Units

B 18.3–1986
Socket Cap, Shoulder, and Set Screws

B 18.6.3–1972 (R 1983)
Machine Screws and Machine Screw Nuts

B 18.8.2–1978 (R 1989)
Pins: Taper, Dowel, Straight, Grooved

B 18.13–1987
Screw and Washer Assemblies-Seams

Standards for Keys and Keyseats

B 17.1–1967 (R 1989)
Keys and Keyseats

B 17.2–1967 (R 1978)
Woodruff Keys and Keyseats

Measurement: Dimensional Metrology

B 89.1.9–1984 (R 1989)
Length Measurement: Precision Inch Gage Blocks

B 89.1.10M–1987
Linear Measurement: Dial Indicators

Piping and Plumbing Standards

A 13.1–1981 (R 1985)
Identification of Piping Systems

A 112.18.1M–1989
Plumbing Fixture Fittings

Screw Thread Standards: Dimensional Data

B 1.3M–1986
Screw Thread Gaging Systems for Dimensional Acceptability: Inch and Metric Screw Threads

B 1.7M–1984
Nomenclature, Definitions, and Letter Symbols for Screw Threads

B 1.3M–1983 (R 1989)
Metric Screw Threads-M Profile

B 1.20.1–1983
Pipe Threads, General Purpose (Inch)

Machine Tools, Cutting Tools, Hand Tools

B 5.10–1981 (R 1987)
Machine Tapers: Self Holding and Steep Tapers

B 5.18–1972 - 1972 (R 1984)
Spindle Noses and Tool Shanks: Milling Machines

B 5.25–1978 (R 1986)
Punch and Die Sets

B 5.43–1977 (R 1988)
Modular Machine Tool Standards

B 94.2–1983 (R 1988)
Reamers

B 94.9–1987
Taps: Cut and Ground Threads

B 107.9–1978 (R 1987)
Wrenches: Box, Open, End, Flare Nut (Metric)

ADDITIONAL STANDARDS RESOURCES

AIA — American Institute of Architects
1735 New York Avenue NW
Washington, DC 20006

AISI — American Iron and Steel Institute
1133 15th Street NW
Washington, DC 20005

AITC — American Institute for Timber Construction
11818 SE Mill Plain Blvd
Vancouver, WA 98684

ASHRAE — American Society of Heating, Refrigeration, and Air Conditioning Engineers
1791 Tullie Circle, NE
Atlanta, GA 30329-2305

ASTM — American Society for Testing and Materials
1916 Race Street
Philadelphia, PA 19103

AWS — American Welding Society
550 NW Lejoune Road
Miami, FL 33135

DOD — Department of Defense, Office of Engineering and Research
Washington, DC 20301

ISO — International Organization for Standardization
1 rue de Varembe, CH 1211
Geneve 20 Switzerland/Suisse

NASA — National Aeronautical and Space Administration
400 Maryland Avenue SW
Washington, DC 20546

NBS — National Bureau of Standards, Standards Information Service (SIS)
Building 225, Room B 162
Washington, DC 20234

NCWM — National Conference on Weights and Measures
c/o National Bureau of Standards
Washington, DC 20234

NFPA — National Fluid Power Association
3333 North Mayfair Road
Milwaukee, WI 53222

SPE — Society of Plastic Engineers, Inc.
14 Fairfield Drive
Brookfield Center, CT 06805

Selected Graphic Symbols, Measurement Conversion, and Other Reference Tables

TABLE A-1 CONVERSION OF METRIC TO INCH-STANDARD UNITS OF MEASURE

mm Value	Inch (decimal) Equivalent	mm Value	Inch (decimal) Equivalent	mm Value	Inch (decimal) Equivalent	mm Value	Inch (decimal) Equivalent
.01	.00039	.34	.01339	.67	.02638	1	.03937
.02	.00079	.35	.01378	.68	.02677	2	.07874
.03	.00118	.36	.01417	.69	.02717	3	.11811
.04	.00157	.37	.01457	.70	.02756	4	.15748
.05	.00197	.38	.01496	.71	.02795	5	.19685
.06	.00236	.39	.01535	.72	.02835	6	.23622
.07	.00276	.40	.01575	.73	.02874	7	.27559
.08	.00315	.41	.01614	.74	.02913	8	.31496
.09	.00354	.42	.01654	.75	.02953	9	.35433
.10	.00394	.43	.01693	.76	.02992	10	.39370
.11	.00433	.44	.01732	.77	.03032	11	.43307
.12	.00472	.45	.01772	.78	.03071	12	.47244
.13	.00512	.46	.01811	.79	.03110	13	.51181
.14	.00551	.47	.01850	.80	.03150	14	.55118
.15	.00591	.48	.01890	.81	.03189	15	.59055
.16	.00630	.49	.01929	.82	.03228	16	.62992
.17	.00669	.50	.01969	.83	.03268	17	.66929
.18	.00709	.51	.02008	.84	.03307	18	.70866
.19	.00748	.52	.02047	.85	.03346	19	.74803
.20	.00787	.53	.02087	.86	.03386	20	.78740
.21	.00827	.54	.02126	.87	.03425	21	.82677
.22	.00866	.55	.02165	.88	.03465	22	.86614
.23	.00906	.56	.02205	.89	.03504	23	.90551
.24	.00945	.57	.02244	.90	.03543	24	.94488
.25	.00984	.58	.02283	.91	.03583	25	.98425
.26	.01024	.59	.02323	.92	.03622	26	1.02362
.27	.01063	.60	.02362	.93	.03661	27	1.06299
.28	.01102	.61	.02402	.94	.03701	28	1.10236
.29	.01142	.62	.02441	.95	.03740	29	1.14173
.30	.01181	.63	.02480	.96	.03780	30	1.18110
.31	.01220	.64	.02520	.97	.03819		
.32	.01260	.65	.02559	.98	.03858		
.33	.01299	.66	.02598	.99	.03898		

TABLE A-2 CONVERSION OF FRACTIONAL INCH VALUES TO METRIC UNITS OF MEASURE

Fractional Inch	mm Equivalent	Fractional Inch	mm Equivalent	Fractional Inch	mm Equivalent	Fractional Inch	mm Equivalent
1/64	0.397	17/64	6.747	33/64	13.097	49/64	19.447
1/32	0.794	9/32	7.144	17/32	13.494	25/32	19.844
3/64	1.191	19/64	7.541	35/64	13.890	51/64	20.240
1/16	1.587	5/16	7.937	9/16	14.287	13/16	20.637
5/64	1.984	21/64	8.334	37/64	14.684	53/64	21.034
3/32	2.381	11/32	8.731	19/32	15.081	27/32	21.431
7/64	2.778	23/64	9.128	39/64	15.478	55/64	21.828
1/8	3.175	3/8	9.525	5/8	15.875	7/8	22.225
9/64	3.572	25/64	9.922	41/64	16.272	57/64	22.622
5/32	3.969	13/32	10.319	21/32	16.669	29/32	23.019
11/64	4.366	27/64	10.716	43/64	17.065	59/64	23.415
3/16	4.762	7/16	11.113	11/16	17.462	15/16	23.812
13/64	5.159	29/64	11.509	45/64	17.859	61/64	24.209
7/32	5.556	15/32	11.906	23/32	18.256	31/32	24.606
15/64	5.953	31/64	12.303	47/64	18.653	63/64	25.003
1/4	6.350	1/2	12.700	3/4	19.050	1	25.400

TABLE A-3 EQUIVALENT METRIC AND UNIFIED STANDARD UNITS OF MEASURE

Metric Unit of Measure		Equivalent Unified Standard Unit of Measure
1 millimeter	mm	0.03937079″
1 centimeter	cm	0.3937079″
1 decimeter	dm	3.937079″
1 meter	m	39.37079″ 3.2808992′ 1.09361 yds.
1 decameter	dkm	32.808992′
1 kilometer	km	0.6213824 mi.
1 square cm	cm^2	0.155 sq. in.
1 cubic cm	cm^3	0.061 cu. in.
1 liter	1	61.023 cu. in.
1 kilogram	kg	2.2046 lbs.

English Unified Unit of Measure	Metric Equivalent Unit of Measure
1 inch	25.4mm or 2.54cm
1 foot	304.8mm or 0.3048m
1 yard	91.14cm or 0.9114m
1 mile	1.609km
1 square inch	6.452 sq. cm
1 cubic inch	16.393 cu. cm
1 cubic foot	28.3171
1 gallon	3.7851
1 pound	0.4536kg

TABLE A-4 GENERAL NC TAP SIZES AND RECOMMENDED TAP DRILLS *(NC STANDARD THREADS)*

SIZE (OUTSIDE DIAMETER INCH)	THREADS PER INCH	TAP DRILL SIZE* (75% THREAD DEPTH)
1/4	20	#7
5/16	18	"F"
3/8	16	5/16
7/16	14	"U"
1/2	13	27/64
9/16	12	31/64
5/8	11	17/32
11/16	11NS	19/32
3/4	10	21/32
13/16	10NS	23/32
7/8	9	49/64
15/16	9NS	53/64
1	8	7/8
1 1/8	7	63/64
1 1/4	7	1 7/64
1 3/8	6	1 13/64
1 1/2	6	1 11/32
1 5/8	5 1/2NS	1 29/64
1 3/4	5	1 35/64
1 7/8	5NS	1 11/16
2	4 1/2	1 25/32

*Nearest commercial drill size to produce a 75% thread depth.

TABLE A-5 GENERAL NF TAP SIZES AND RECOMMENDED TAP DRILLS *(NF STANDARD THREADS)*

SIZE (OUTSIDE DIAMETER INCH)	THREADS PER INCH	TAP DRILL SIZE* (75% THREAD DEPTH)
1/4	28	#3
5/16	24	"I"
3/8	24	"Q"
7/16	20	"W"
1/2	20	29/64
9/16	18	33/64
5/8	18	37/64
11/16	11NS	19/32
3/4	16	11/16
13/16	10NS	23/32
7/8	14	13/16
15/16	9NS	53/64
1	12	59/64
1	14NS	15/16
1 1/8	12	1 3/64
1 1/4	12	1 11/64
1 3/8	12	1 19/64
1 1/2	12	1 27/64

*Nearest commercial drill size to produce a 75% thread depth.

TABLE A-6 GENERAL METRIC STANDARD TAP SIZES AND RECOMMENDED TAP DRILLS *(METRIC STANDARD THREADS)*

METRIC THREAD SIZE (NOMINAL OUTSIDE DIAMETER AND PITCH) (MM)	RECOMMENDED TAP DRILL SIZE			
	METRIC SERIES DRILLS		INCH SERIES DRILLS	
	SIZE (MM)	EQUIVALENT (INCH)	NOMINAL SIZE	DIAMETER (INCH)
M1.6×0.35	1.25	.0492		
M2×0.4	1.60	.0630	#52	.0635
M2.5×0.45	2.05	.0807	#45	.0820
M3×0.5	2.50	.0984	#39	.0995
M3.5×0.6	2.90	.1142	#32	.1160
M4×0.7	3.30	.1299	#30	.1285
M5×0.8	4.20	.1654	#19	.1660
M6×1	5.00	.1968	#8	.1990
M8×1.25	6.80	.2677	"H"	.2660
M10×1.5	8.50	.3346	"Q"	.3320
M12×1.75	10.25	.4035	13/32	.4062
M14×2	12.00	.4724	15/32	.4688
M16×2	14.00	.5512	35/64	.5469
M20×2.5	17.50	.6890	11/16	.6875
M24×3	21.00	.8268	53/64	.8281
M30×3.5	26.50	1.0433	1 3/64	1.0469
M36×4	32.00	1.2598	1 1/4	1.2500

TABLE A-7 DECIMAL EQUIVALENTS OF FRACTIONAL, WIRE GAGE, LETTER, AND METRIC SIZE DRILLS 0.0059″ (0.15mm) TO 1.000″ (25 mm)

Decimal	Inch	Wire	mm.
.0059		97	.15
.0063		96	.16
.0067		95	.17
.0071		94	.18
.0075		93	.19
.0079		92	.20
.0083		91	.21
.0087		90	.22
.0091		89	.23
.0095		88	.24
.0098			.25
.0100		87	
.0102			.26
.0105		86	
.0106			.27
.0110		85	.28
.0114			.29
.0115		84	
.0118			.30
.0120		83	
.0122			.31
.0125		82	
.0126			.32
.0130		81	.33
.0134			.34
.0135		80	
.0138			.35
.0145		79	
.0156	1/64		
.0158			.4
.0160		78	
.0177			.45
.0180		77	
.0197			.5
.0200		76	
.0210		75	

Decimal	Inch	Wire	mm.
.0532			1.35
.0550		54	
.0551			1.4
.0571			1.45
.0591			1.5
.0595		53	
.0610			1.55
.0625	1/16		
.0630			1.6
.0635		52	
.0650			1.65
.0669			1.7
.0670		51	
.0689			1.75
.0700		50	
.0709			1.8
.0728			1.85
.0730		49	
.0748			1.9
.0760		48	
.0768			1.95
.0781	5/64		
.0785		47	
.0787			2.
.0807			2.05
.0810		46	
.0820		45	
.0827			2.1
.0847			2.15
.0860		44	
.0866			2.2
.0886			2.25
.0890		43	
.0906			2.3
.0925			2.35
.0935		42	

Decimal	Inch	Wire and Letter	mm.
.1495		25	
.1496			3.8
.1520		24	
.1535			3.9
.1540		23	
.1562	5/32		
.1570		22	
.1575			4.
.1590		21	
.1610		20	
.1614			4.1
.1654			4.2
.1660		19	
.1673			4.25
.1693			4.3
.1695		18	
.1719	11/64		
.1730		17	
.1732			4.4
.1770		16	
.1772			4.5
.1800		15	
.1811			4.6
.1820		14	
.1850		13	4.7
.1870			4.75
.1875	3/16		
.1890		12	4.8
.1910		11	
.1929			4.9
.1935		10	
.1960		9	
.1969			5.
.1990		8	
.2008			5.1
.2010		7	
.2031	13/64		

Decimal	Inch	Letter	mm.
.2717			6.9
.2720		I	
.2756			7.
.2770		J	
.2795			7.1
.2810		K	
.2812	9/32		
.2835			7.2
.2854			7.25
.2874			7.3
.2900		L	
.2913			7.4
.2950		M	
.2953			7.5
.2969	19/64		
.2992			7.6
.3020		N	
.3032			7.7
.3051			7.75
.3071			7.8
.3110			7.9
.3125	5/16		
.3150			8.
.3160		O	
.3189			8.1
.3228			8.2
.3230		P	
.3248			8.25
.3268			8.3
.3281	21/64		
.3307			8.4
.3320		Q	
.3347			8.5
.3386			8.6
.3390		R	
.3425			8.7
.3438	11/32		

Decimal	Inch	mm.
.4331		11.
.4375	7/16	
.4528		11.5
.4531	29/64	
.4688	15/32	
.4724		12.
.4844	31/64	
.4921		12.5
.5000	1/2	
.5118		13.
.5156	33/64	
.5312	17/32	
.5315		13.5
.5469	35/64	
.5512		14.
.5625	9/16	
.5709		14.5
.5781	37/64	
.5906		15.
.5938	19/32	
.6094	39/64	
.6102		15.5
.6250	5/8	
.6299		16.
.6406	41/64	
.6496		16.5
.6562	21/32	
.6693		17.
.6719	43/64	
.6875	11/16	
.6890		17.5
.7031	45/64	
.7087		18.
.7188	23/32	
.7283		18.5
.7344	47/64	
.7480		19.

.0217			.55
.0225		74	
.0236			.6
.0240		73	
.0250		72	
.0256			.65
.0260		71	
.0276			.7
.0280		70	
.0292		69	
.0295			.75
.0310		68	
.0312	1/32		
.0315			.8
.0320		67	
.0330		66	
.0335			.85
.0350		65	
.0354			.9
.0360		64	
.0370		63	
.0374			.95
.0380		62	
.0390		61	
.0394			1.
.0400		60	
.0410		59	
.0413			1.05
.0420		58	
.0430		57	
.0433			1.1
.0453			1.15
.0465		56	
.0469	3/64		
.0472			1.2
.0492			1.25
.0512			1.3
.0520		55	

.0938	3/32		
.0945			2.4
.0960		41	
.0965			2.45
.0980		40	
.0984			2.5
.0995		39	
.1015		38	
.1024			2.6
.1040		37	
.1063			2.7
.1065		36	
.1083			2.75
.1094	7/64		
.1100		35	
.1102			2.8
.1110		34	
.1130		33	
.1142			2.9
.1160		32	
.1181			3.
.1200		31	
.1221			3.1
.1250	1/8		
.1260			3.2
.1280			3.25
.1285		30	
.1299			3.3
.1339			3.4
.1360		29	
.1378			3.5
.1405		28	
.1406	9/64		
.1417			3.6
.1440		27	
.1457			3.7
.1470		26	
.1476			3.75

.2040		6	
.2047			5.2
.2055		5	
.2067			5.25
.2087			5.3
.2090		4	
.2126			5.4
.2130		3	
.2165			5.5
.2188	7/32		
.2205			5.6
.2210		2	
.2244			5.7
.2264			5.75
.2280		1	
.2284			5.8
.2323			5.9
.2340		A	
.2344	15/64		
.2362			6.
.2380		B	
.2402			6.1
.2420		C	
.2441			6.2
.2460		D	
.2461			6.25
.2480			6.3
.2500	1/4	E	
.2520			6.4
.2559			6.5
.2570		F	
.2598			6.6
.2610		G	
.2638			6.7
.2656	17/64		
.2658			6.75
.2660		H	
.2677			6.8

.3445			8.75
.3465			8.8
.3480		S	
.3504			8.9
.3543			9.
.3580		T	
.3583			9.1
.3594	23/64		
.3622			9.2
.3642			9.25
.3661			9.3
.3680		U	
.3701			9.4
.3740			9.5
.3750	3/8		
.3770		V	
.3780			9.6
.3819			9.7
.3839			9.75
.3858			9.8
.3860		W	
.3898			9.9
.3906	25/64		
.3937			10.
.3970		X	
.4040		Y	
.4062	13/32		
.4130		Z	
.4134			10.5
.4219	27/64		

.7500	3/4	
.7656	49/64	
.7677		19.5
.7812	25/32	
.7874		20.
.7969	51/64	
.8071		20.5
.8125	13/16	
.8268		21.
.8281	53/64	
.8438	27/32	
.8465		21.5
.8594	55/64	
.8661		22.
.8750	7/8	
.8858		22.5
.8906	57/64	
.9055		23.
.9062	29/32	
.9219	59/64	
.9252		23.5
.9375	15/16	
.9449		24.
.9531	61/64	
.9646		24.5
.9688	31/32	
.9843		25.
.9844	63/64	
1.0000	1	

TABLE A-8 ANSI SPUR GEAR RULES AND FORMULAS FOR REQUIRED FEATURES OF 20° AND 25° FULL-DEPTH INVOLUTE TOOTH FORMS

Required Feature	Rule	Formula*
Diametral pitch (P_d)	Divide the number of teeth by the pitch diameter	$P_d = \frac{N}{D}$
	Add 2 to the number of teeth and divide by the outside diameter	$P_d = \frac{N + 2}{O}$
	Divide 3.1416 by the circular pitch	$P_d = \frac{3.1416}{P_c}$
Number of teeth (N)	Multiply the diametral pitch by the pitch diameter	$N = P_d \times D$
	Multiply the diametral pitch by outside diameter and then subtract 2	$N = (P_d \times O) - 2$
Pitch diameter (D)	Divide the number of teeth by the diametral pitch	$D = \frac{N}{P_d}$
	Multiply the addendum by 2 and subtract the product from the outside diameter	$D = O - 2A$
Outside diameter (O)	Add 2 to the number of teeth and divide by the diametral pitch	$O = \frac{N + 2}{P_d}$
	Add 2 to the number of teeth and divide by the quotient of the number of teeth divided by the pitch diameter	$O = \frac{N + 2}{N/D}$
Circular pitch (P_c)	Divide 3.1416 by the diametral pitch	$P_c = \frac{3.1416}{P_d}$
	Divide the pitch diameter by the product of 0.3183 times the number of teeth	$P_c = \frac{D}{0.3183 \times N}$
Addendum (A)	Divide 1.000 by the diametral pitch	$A = \frac{1.000}{P_d}$
Working depth (W^1)	Divide 2.000 by the diametral pitch	$W^1 = \frac{2.000}{P_d}$
Clearance (S)	Divide 0.250 by the diametral pitch	$S = \frac{0.250}{P_d}$
Whole depth of tooth (W^2)	Divide 2.250 by the diametral pitch	$W^2 = \frac{2.250}{P_d}$
Thickness of tooth (T)	Divide 1.5708 by the diametral pitch	$T = \frac{1.5708}{P_d}$
Center distance (C)	Add the pitch diameters and divide the sum by 2	$C = \frac{D^1 + D^2}{2}$
	Divide one-half the sum of the number of teeth in both gears by the diametral pitch	$C = \frac{1/2\,(N^1 + N^2)}{P_d}$

*Dimensions are in customary inch units.

TABLE A-9 MICROINCH (μ in.) AND MICROMETER (μ m) RANGES OF SURFACE ROUGHNESS FOR SELECTED MANUFACTURING PROCESSES

Manufacturing Process	Roughness Height in Microinches and Micrometers (μm)*														
	4000 (101.60)	3000 (76.20)	2000 (50.80)	1000 (25.40)	500 (12.70)	250 (6.35)	125 (3.18)	63 (1.60)	32 (0.81)	16 (0.41)	8 (0.20)	4 (0.10)	2 (0.05)	1 (0.03)	0.5 (0.01) μin. / μm
Flame cutting															
Snagging															
Sawing															
Planing, shaping															
Drilling															
Electrical discharge machining															
Milling (chemical)															
Milling (rough)															
Broaching															
Reaming															
Boring, turning (finish)															
Turning (rough)															
Barrel finishing															
Electrolytic grinding															
Burnishing (roller)															
Grinding (commercial)															
Grinding (finish)															
Honing															
Polishing															
Lapping															
Superfinishing															
Sand casting															
Hot rolling															
Forging															
Mold casting (permanent)															
Extruding															
Cold rolling (drawing)															
Die casting															

Code ■ General manufacturing (average) surface finish range

▒ Higher or lower range produced by using special processes

*Values rounded to nearest second place μm decimal

Representative Graphic Symbols for Drawings

TABLE A-10 COMMON GRAPHIC SYMBOLS USED ON BUILDING CONSTRUCTION DRAWINGS

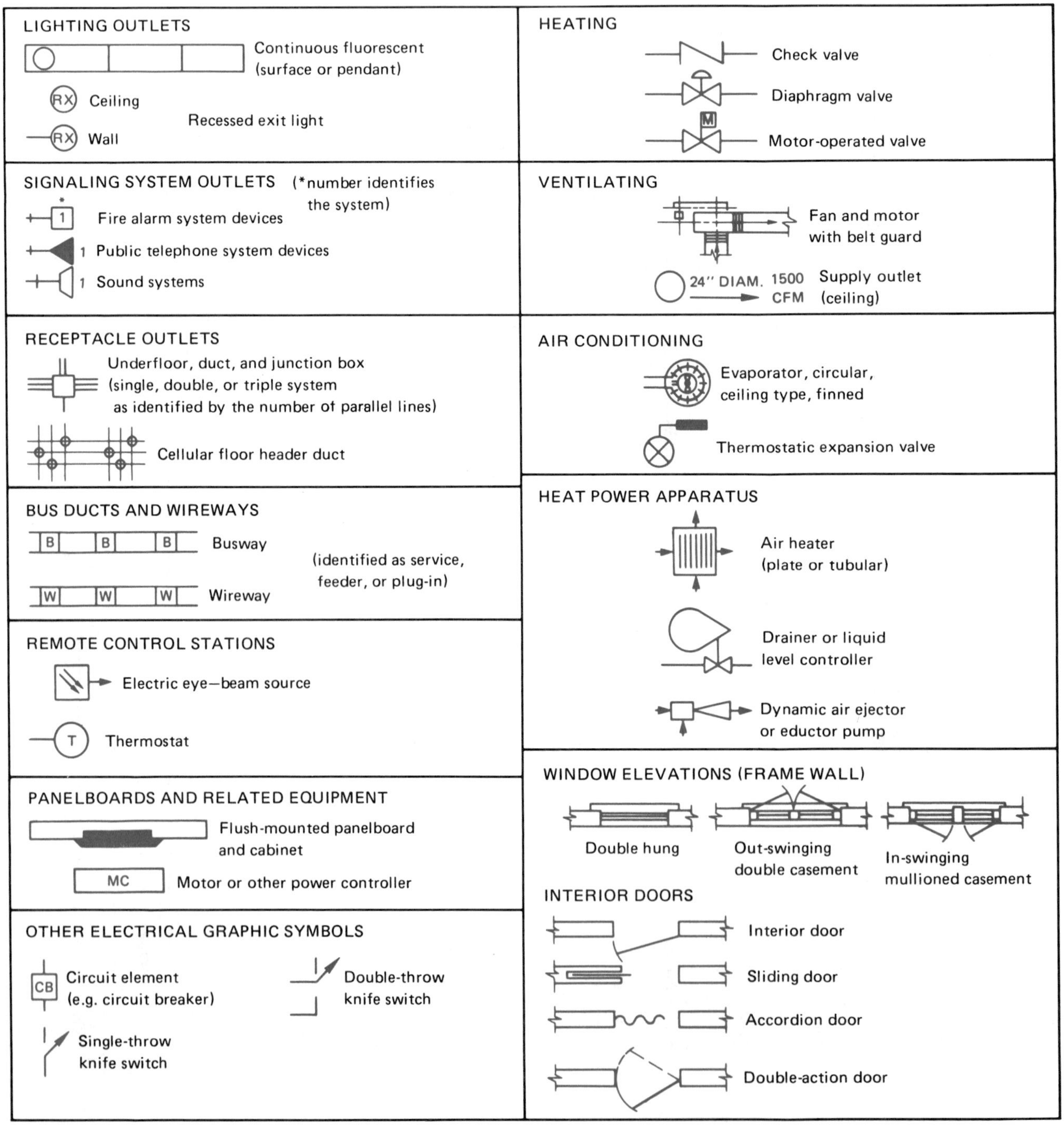

LIGHTING OUTLETS

Continuous fluorescent (surface or pendant)

RX Ceiling

RX Wall

Recessed exit light

SIGNALING SYSTEM OUTLETS (*number identifies the system)

1 Fire alarm system devices

1 Public telephone system devices

1 Sound systems

RECEPTACLE OUTLETS

Underfloor, duct, and junction box (single, double, or triple system as identified by the number of parallel lines)

Cellular floor header duct

BUS DUCTS AND WIREWAYS

B B B Busway

W W W Wireway

(identified as service, feeder, or plug-in)

REMOTE CONTROL STATIONS

Electric eye—beam source

T Thermostat

PANELBOARDS AND RELATED EQUIPMENT

Flush-mounted panelboard and cabinet

MC Motor or other power controller

OTHER ELECTRICAL GRAPHIC SYMBOLS

CB Circuit element (e.g. circuit breaker)

Double-throw knife switch

Single-throw knife switch

HEATING

Check valve

Diaphragm valve

M Motor-operated valve

VENTILATING

Fan and motor with belt guard

24" DIAM. 1500 CFM Supply outlet (ceiling)

AIR CONDITIONING

Evaporator, circular, ceiling type, finned

Thermostatic expansion valve

HEAT POWER APPARATUS

Air heater (plate or tubular)

Drainer or liquid level controller

Dynamic air ejector or eductor pump

WINDOW ELEVATIONS (FRAME WALL)

Double hung

Out-swinging double casement

In-swinging mullioned casement

INTERIOR DOORS

Interior door

Sliding door

Accordion door

Double-action door

TABLE A-11 GENERALLY USED ELECTRONIC SCHEMATIC SYMBOLS

ANTENNA	
Balanced, Dipole	
General	or or
Loop, Shielded	
Battery	+ −
CAPACITOR	
Feedthrough	or
Ganged, Variable	
Split-Stator, Variable	
CATHODE	
Directly Heated	
Indirectly Heated	or
Cold	
CABLE	
Coaxial	or
Coaxial: Grounded Shield	or
DIODE	
General	
Light-Emitting (LED)	
Pin	or
Tunnel	or

HEADPHONE	
Double	
Single	
Stereo	L R
INDUCTOR	
Air-Core	
Iron-Core	
Tapped	
Variable	or
JACK	
Coaxial or Phono	
Phone, Two-Conductor	
Phone, Two-Conductor Interrupting	
Phone, Three Conductor	
Microammeter	μA
Milliammeter	mA
OUTLET	
Nonpolarized	
Polarized	
Utility, 117 V, Nonpolarized	
Utility, 234 V	

TABLE A-11 (Continued) GENERALLY USED ELECTRONIC SCHEMATIC SYMBOLS

Photocell (Tube)	
PLUG	
Nonpolarized	
Polarized	
Phone-Two-Conductor	
Phone-Three-Conductor	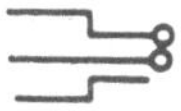
Utility, 117 V	
Utility, 234 V	
Potentiometer, Variable Resistor or Rheostat	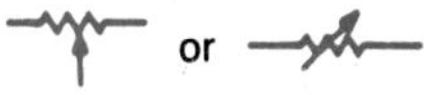 or
RECTIFIER	
Semiconductor	
Silicon-Controlled	
Tube Type	
RELAY	
DPDT	
DPST	
SPDT	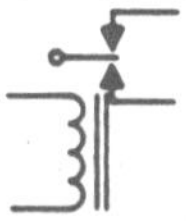
SPST	
RESISTOR	
Fixed	
Preset	
Tapped	

TRANSFORMER	
Air-Core, Adjustable	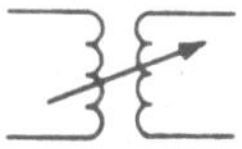
Iron-Core	
Tapped-Primary	
Tapped-Secondary	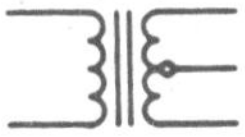
TRANSISTOR	
Bipolar, NPN	
Bipolar, PNP	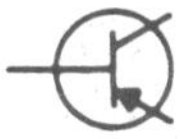
Junction Field-Effect or JFET	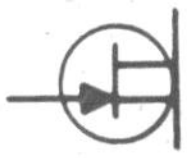
Metal-Oxide, Dual Gate	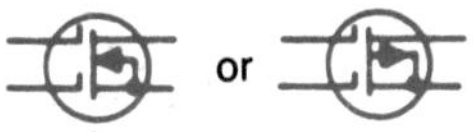 or
Photosensitive	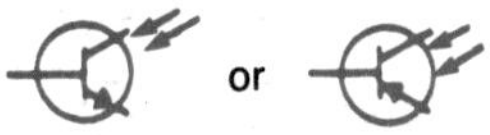 or
Unijunction	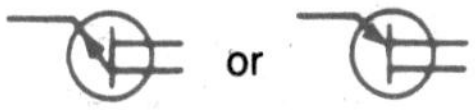 or
TUBE	
Diode	
Triode	
Voltmeter	 V
Wattmeter	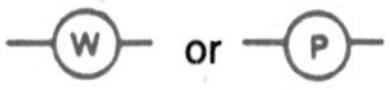 W or P
WAVEGUIDE	
Flexible	
Twisted	

Index

Note: Underlined and bold entries are references to non-text material.